Edited by
Krishna C. Majumdar and
Shital K. Chattopadhyay

**Heterocycles in Natural
Product Synthesis**

Related Titles

Joule, J. A., Mills, K.

Heterocyclic Chemistry

640 pages
2010
Softcover
ISBN: 978-1-4051-3300-5

Royer, J. (ed.)

Asymmetric Synthesis of Nitrogen Heterocycles

425 pages with 19 figures and 1 tables
2009
Hardcover
ISBN: 978-3-527-32036-3

Dubois, P., Coulembier, O., Raquez, J.-M. (eds.)

Handbook of Ring-Opening Polymerization

425 pages approx.
2009
Hardcover
ISBN: 978-3-527-31953-4

Padwa, A., Pearson, W. H. (eds.)

Synthetic Applications of 1,3-Dipolar Cycloaddition Chemistry Toward Heterocycles and Natural Products

2008
E-Book
ISBN: 978-0-470-24827-0

Rosowsky, A.

The Chemistry of Heterocyclic Compounds, Volume 26, Seven-Membered Heterocyclic Compounds Containing Oxygen and Sulfur

950 pages
2007
Hardcover
ISBN: 978-0-471-38210-2

Mosby, W. L.

The Chemistry of Heterocyclic Compounds, Part 1, Volume 15, Heterocyclic Systems with Bridgehead Nitrogen Atoms

758 pages
2007
Hardcover
ISBN: 978-0-470-38049-9

Weissberger, A., Taylor, E. C.

The Chemistry of Heterocyclic Compounds, Volume 30, Special Topics in Heterocyclic Chemistry

602 pages
2007
Hardcover
ISBN: 978-0-471-67253-1

*Edited by Krishna C. Majumdar
and Shital K. Chattopadhyay*

Heterocycles in Natural Product Synthesis

WILEY-VCH Verlag GmbH & Co. KGaA

The Editors

Prof. Krishna C. Majumdar
University of Kalyani
Department of Chemistry
Kalyani, W.B. 741235
Indien

Prof. Shital K. Chattopadhyay
University of Kalyani
Department of Chemistry
Kalyani, W.B. 741235
Indien

All books published by **Wiley-VCH** are carefully produced. Nevertheless, authors, editors, and publisher do not warrant the information contained in these books, including this book, to be free of errors. Readers are advised to keep in mind that statements, data, illustrations, procedural details or other items may inadvertently be inaccurate.

Library of Congress Card No.: applied for

British Library Cataloguing-in-Publication Data
A catalogue record for this book is available from the British Library.

Bibliographic information published by the Deutsche Nationalbibliothek
The Deutsche Nationalbibliothek lists this publication in the Deutsche Nationalbibliografie; detailed bibliographic data are available on the Internet at <http://dnb.d-nb.de>.

© 2011 Wiley-VCH Verlag & Co. KGaA, Boschstr. 12, 69469 Weinheim, Germany

All rights reserved (including those of translation into other languages). No part of this book may be reproduced in any form – by photoprinting, microfilm, or any other means – nor transmitted or translated into a machine language without written permission from the publishers. Registered names, trademarks, etc. used in this book, even when not specifically marked as such, are not to be considered unprotected by law.

Cover Design Adam-Design, Weinheim
Typesetting Toppan Best-set Premedia Limited, Hong Kong
Printing and Binding Fabulous Printers Pte Ltd

Printed in Singapore
Printed on acid-free paper

ISBN: 978-3-527-32706-5
ePDF ISBN: 978-3-527-63490-3
ePub ISBN: 978-3-527-63489-7
oBook ISBN: 978-3-527-63488-0

Contents

Preface *XV*
List of Contributors *XVII*

Part One   Strained Heterocycles in the Synthesis of Natural Products *1*

1   Aziridines in Natural Product Synthesis *3*
Candice Botuha, Fabrice Chemla, Franck Ferreira and Alejandro Pérez-Luna
1.1   Introduction   *3*
1.2   Synthesis of Natural Products Containing Aziridine Units   *3*
1.2.1   Synthesis of Aziridine-2,3-Dicarboxylic Acid   *3*
1.2.2   Synthesis of (Z)-Dysidazirine   *5*
1.2.3   Syntheses of Mitomycins   *5*
1.2.4   Syntheses of FR-900482 and FR-66979   *8*
1.3   Synthesis of Natural Products Involving the Transformation of an Aziridine Moiety   *10*
1.3.1   Nucleophilic Ring-Opening of Aziridines for Natural Product Synthesis   *10*
1.3.1.1   Carbon-Centered Nucleophiles   *11*
1.3.1.2   Nitrogen-Centered Nucleophiles   *15*
1.3.1.3   Oxygen-Centered Nucleophiles   *18*
1.3.1.4   Halogen Nucleophiles   *24*
1.3.1.5   Reductions   *25*
1.3.2   Cycloaddition Reactions and Rearrangements   *25*
1.3.2.1   Aziridines in [3 + 2] Cycloadditions   *26*
1.3.2.2   Aziridines in [2,3]-Wittig Rearrangements   *27*
1.3.2.3   Aziridines in Iodide-Mediated Rearrangements   *27*
1.3.2.4   Aziridines in Miscellaneous Rearrangements   *28*
1.3.3   Synthesis of Natural Products Involving the Transformation of an Aziridinium Moiety   *31*

1.4	Conclusion	32
	References	33

2 Azetidine and Its Derivatives 41
Hidemi Yoda, Masaki Takahashi and Tetsuya Sengoku

2.1	Introduction	41
2.2	Structural Description of Azetidines	41
2.3	Synthetic Methodologies for the Formation of Azetidine Rings	43
2.4	Synthesis of Mugineic Acids	44
2.5	Synthesis of Penaresidins	46
2.6	Structural Description of Azetidin-2-ones	50
2.7	Synthetic Methodologies for the Formation of Azetidin-2-ones	50
2.8	Synthesis of Penicillin	52
2.9	Synthesis of Cephalosporin	54
2.10	Conclusion	56
	Acknowledgment	56
	References	57

3 Epoxides and Oxetanes 63
Biswanath Das and Kongara Damodar

3.1	Introduction	63
3.2	Epoxides in Natural Product Synthesis	63
3.2.1	Synthesis of Natural Products Possessing an Epoxide Moiety	68
3.2.1.1	Synthesis of (−)-Posticlure	68
3.2.1.2	Synthesis of Natural Polyethers	68
3.2.1.3	Synthesis of (+)-11,12-Epoxysarcophytol A	68
3.2.1.4	Synthesis of (−)-Scyphostatine	68
3.2.1.5	Synthesis of Arenastatin A	70
3.2.1.6	Synthesis of (+)-Ambuic Acid	70
3.2.1.7	Total Synthesis of Epocarbazolin A	71
3.2.1.8	Synthesis of Multiplolide A	71
3.2.2	Synthesis of Natural Products Involving the Transformation of the Epoxide Moiety	71
3.2.2.1	Synthesis of Dodoneine	71
3.2.2.2	Synthesis of (−)-Pericosin B	73
3.2.2.3	Synthesis of (−)-Peucedanol	73
3.2.2.4	Synthesis of (+)-Bourgeanic Acid	74
3.2.2.5	Synthesis of (6S)-5,6-Dihydro-6-([2R]-2-Hydroxy-6-Phenylhexyl)-2H-Pyran-2-one	74
3.2.2.6	Synthesis of Verbalactone	74
3.2.2.7	Synthesis of (2S, 3S)-(+)-Aziridine-2,3-Dicarboxylic Acid	75
3.2.2.8	Synthesis of D-*erythro*-Sphingosine	75
3.2.2.9	Synthesis of (+)-L-733,060	76
3.2.2.10	Synthesis of (+)-Chelidonine	76
3.2.2.11	Synthesis of (−)-Pironetin	76

3.2.2.12	Synthesis of (−)-Codonopsinine	77
3.2.2.13	Synthesis of Sesamin and Dihydrosesamin	77
3.2.2.14	Synthesis of (9S, 12R, 13S)-Pinellic Acid	78
3.2.2.15	Synthesis of (Z)-Nonenolide	78
3.2.2.16	Synthesis of (−)-Cubebol	79
3.2.2.17	Synthesis of (+)-Schweinfurthins B and E	79
3.2.2.18	Synthesis of (−)-Cleistenolide	80
3.2.2.19	Synthesis of Decarestricine J	80
3.2.2.20	Synthesis of (−)-Gloeosporone	82
3.2.2.21	Synthesis of (S)-Dihydrokavain	82
3.2.2.22	Synthesis of (−)-Phorocantholide-J	83
3.3	Oxetane in Natural Product Synthesis	85
3.3.1	Synthesis of Natural Products Possessing an Oxetane Moiety	85
3.3.1.1	Synthesis of Epi-oxetin	85
3.3.1.2	Synthesis of Dioxatricyclic Segment of Dictyoxetane	86
3.3.1.3	Synthesis of (−)-Merrilactone A	86
3.3.1.4	Total Synthesis of (+)-(Z)-Laureatin	87
3.3.1.5	Synthesis of Taxol	87
3.3.2	Synthesis of Natural Products Involving Transformation of the Oxetane Moiety	88
3.3.2.1	Synthesis of Erogorgiaene	88
3.3.2.2	Synthesis of trans-Whiskey Lactone	88
3.3.2.3	Synthesis of (±)-Sarracenin	89
3.4	Conclusion	89
	Acknowledgment	90
	References	90

Part Two Common Ring Heterocycles in Natural Product Synthesis 97

4	**Furan and Its Derivatives**	99
	Alicia Boto and Laura Alvarez	
4.1	Introduction	99
4.2	Natural Products Containing the Furan Ring	100
4.2.1	Occurrence of Furan Rings in Natural Products	100
4.2.2	Synthesis of Furans in Natural Products	104
4.3	Furan Derivatives as Reagents in the Synthesis of Natural Products	106
4.3.1	Metallation	107
4.3.2	Reduction and Oxidation	111
4.3.3	Furan Derivatives as Electrophiles and Nucleophiles	118
4.3.4	Furan in Cycloadditions	124
4.3.4.1	[2 + 1], [2 + 2] and [3 + 2] Cycloadditions	124
4.3.4.2	Diels–Alder ([4 + 2] Cycloadditions)	127

4.3.4.3	[4 + 3], [6 + 4], [8 + 2] and [5 + 2] Cycloadditions 133
4.3.5	Furan in Other Reactions 136
4.3.6	Other Uses of Furan in Synthesis 138
4.4	Summary 139
	References 140
5	**Pyran and Its Derivatives** 153
	Hideto Miyabe, Okiko Miyata and Takeaki Naito
5.1	Introduction 153
5.2	Application of Pyran Moieties in the Synthesis of Natural Products 158
5.2.1	2,6-Disubstituted Pyran Natural Products 158
5.2.2	2,6-Cyclic Pyran Compounds 161
5.2.3	Complex Pyran Natural Products 165
5.2.4	Fused Pyran Compounds with Aromatic Rings 168
5.2.5	Fused Pyran Compounds with Aliphatic Rings 171
5.3	Conclusion 176
	References 176
6	**Pyrrole and Its Derivatives** 187
	Dipakranjan Mal, Brateen Shome and Bidyut Kumar Dinda
6.1	Introduction 187
6.2	Synthesis of Pyrrole Natural Products 193
6.2.1	Monopyrrolic Natural Products 193
6.2.2	Dipyrrolic Natural Products 203
6.2.3	Tripyrrolic Natural Products: Prodigiosins 205
6.3	Synthesis of Non-pyrrole Natural Products from Pyrrole Derivatives 209
6.4	Conclusion 214
	Acknowledgments 214
	References 215
7	**Indoles and Indolizidines** 221
	Sarah M. Bronner, G.-Yoon J. Im and Neil K. Garg
7.1	Introduction 221
7.2	Applications of Indoles and Indolizidines in the Synthesis of Natural Products 222
7.2.1	Indoles and Oxindoles 222
7.2.1.1	Total Synthesis of Actinophyllic Acid (Overman) 222
7.2.1.2	Total Synthesis of Dragmacidin F (Stoltz) 226
7.2.1.3	Total Synthesis of Penitrem D (Smith) 230
7.2.1.4	Total Synthesis of Welwitindolinone A Isonitrile (Baran, Wood) 232
7.2.2	Indolines 237
7.2.2.1	Total Synthesis of 11,11'-Dideoxyverticillin A (Movassaghi) 237

7.2.2.2	Total Synthesis of Minfiensine (Overman, Qin, MacMillan)	240
7.2.2.3	Total Synthesis of Norfluorocurarine (Vanderwal)	245
7.2.2.4	Total Synthesis of Psychotrimine (Baran)	247
7.2.3	Indolizidines	249
7.2.3.1	Total Synthesis of Myrmicarins 215A, 215B and 217 (Movassaghi)	249
7.2.3.2	Total Synthesis of Serratezomine A (Johnston)	252
7.3	Conclusion	254
	Acknowledgment	254
	References	254

8 Pyridine and Its Derivatives 267
Paula Kiuru and Jari Yli-Kauhaluoma

8.1	Introduction	267
8.2	Application of the Pyridine Moiety in the Synthesis of Natural Products	268
8.2.1	Pyridines	268
8.2.1.1	Synthesis of Noranabasamine Enantiomers	268
8.2.1.2	Synthesis of Quaterpyridine Nemertelline	268
8.2.1.3	Synthesis of Caerulomycin C	276
8.2.1.4	Synthesis of the Spongidine Isomer	277
8.2.2	2-Alkylpyridines	278
8.2.2.1	Synthesis of Montipyridine	278
8.2.2.2	Synthesis of Piericidin A1	278
8.2.3	3-Alkylpyridine, 3-Alkylpyridinium and 3-Alkyltetrahydropyridine Compounds	281
8.2.3.1	Synthesis of Xestamines	281
8.2.3.2	Synthesis of Pyrinadine A	282
8.2.3.3	Synthesis of Pyrinodemin A	282
8.2.3.4	Synthesis of Haliclamine A	284
8.2.4	Piperidines	285
8.2.4.1	Synthesis of Coniine and Pipecoline	285
8.2.4.2	Synthesis of Stenusine	286
8.2.5	Pyridones	287
8.2.5.1	Synthesis of (±)-Cytisine	287
8.2.5.2	Synthesis of Iromycin A	288
8.3	Conclusion	289
	Acknowledgment	289
	References	290

9 Quinolines and Isoquinolines 299
Antonio Garrido Montalban

9.1	Introduction	299
9.2	Application of Quinolines and Isoquinolines in the Synthesis of Natural Products	300

9.2.1	Quinoline-Containing Natural Products	308
9.2.1.1	Quinine	308
9.2.1.2	Sandramycin	311
9.2.1.3	Lavendamycin	313
9.2.2	Isoquinoline-Containing Natural Products	317
9.2.2.1	Morphine	317
9.2.2.2	Emetine	320
9.2.2.3	Protoberberines	326
9.2.2.4	Nitidine	331
9.3	Conclusion	332
	References	332

10 Carbazoles and Acridines 341
Konstanze K. Gruner and Hans-Joachim Knölker

10.1	Introduction to Carbazoles	341
10.2	Total Synthesis of Carbazole Alkaloids	341
10.2.1	Palladium-Catalyzed Synthesis of Carbazoles	350
10.2.1.1	Total Synthesis of Pityriazole	350
10.2.1.2	Total Synthesis of Euchrestifoline and Girinimbine	351
10.2.2	Iron-Mediated Synthesis of Carbazoles	353
10.2.2.1	Total Syntheses of the Antiostatins	353
10.2.2.2	Total Synthesis of R-(–)-Neocarazostatin B and Carquinostatin A	355
10.2.3	Total Syntheses of Ellipticine and Staurosporinone	356
10.2.3.1	Synthesis of Ellipticine	356
10.2.3.2	Synthesis of Staurosporinone	357
10.3	Introduction to Acridines	358
10.4	Synthesis of Acridines and Acridones	361
10.4.1	Total Synthesis of Acronycine	361
10.4.2	Synthesis of Amsacrine	362
10.4.3	Total Syntheses of Amphimedine	362
	References	364

11 Thiophene and Other Sulfur Heterocycles 377
Krishna C. Majumdar and Shovan Mondal

11.1	Introduction	377
11.2	Synthesis of Natural Products Containing Thiophene	378
11.2.1	Synthesis of Natural Products from Thiophene-Based Substrates	378
11.2.2	Synthesis of Natural Products by Construction of the Thiophene Nucleus	386
11.3	Synthesis of Natural Products Containing Other Sulfur Heterocycles	393
11.4	Conclusion	395
	Acknowledgments	396
	References	397

12	**Oxazole and Its Derivatives** *403*	
	David W. Knight	
12.1	Introduction *403*	
12.2	Mono-Oxazoles *404*	
12.2.1	Pimprinin *404*	
12.2.2	Texamine and Relatives *405*	
12.2.3	Synthesis of Sulfomycin Fragments *406*	
12.2.4	Ajudazol A and B *408*	
12.2.5	Rhizoxin *409*	
12.2.6	The Calyculins *410*	
12.2.7	Leucascandrolide A, B and Neopeltolide *413*	
12.2.8	Chivosazole *416*	
12.2.9	Madumycin II *416*	
12.2.10	14,15-Anhydropristinamycin II$_B$ *418*	
12.2.11	Griseoviridin *418*	
12.2.12	Thiangazole *419*	
12.3	Unconnected Bis- and Tris-Oxazoles *420*	
12.3.1	Disorazole C$_1$ *420*	
12.3.2	Phorboxazoles *421*	
12.3.3	Leucamide A *424*	
12.3.4	Promothiocin A *425*	
12.3.5	Berninamycin A *425*	
12.4	Cyclic Polyheterocyclic Metabolites Containing Single Oxazole Residues *426*	
12.4.1	Dendroamide A *426*	
12.4.2	Nostocyclamide *427*	
12.4.3	Bistratamides *428*	
12.4.4	Tenuecyclamides A–D *428*	
12.4.5	Dolastatin I *429*	
12.5	Conjugated Bis-Oxazoles *429*	
12.5.1	(−)-Hennoxazole A *429*	
12.5.2	Muscoride A *432*	
12.5.3	Diazonamide A *433*	
12.5.4	Bengazole A *437*	
12.5.5	Siphonazole *439*	
12.6	Tris- and Poly-Oxazoles *440*	
12.6.1	Ulapualide A *440*	
12.6.2	(R)-Telomestatin *443*	
12.6.3	IB-01211 *445*	
12.6.4	YM-216391 *445*	
	References *446*	
13	**Thiazoline and Thiazole and Their Derivatives** *459*	
	Zhengshuang Xu and Tao Ye	
13.1	Introduction *459*	

13.2	General Methods for the Synthesis of Thiazoline and Thiazole Derivatives *460*	
13.2.1	Methods for the Preparation of Thiazolines *460*	
13.2.1.1	Using Vicinal Amino Thiols as Starting Materials *461*	
13.2.1.2	From Vicinal Amino Alcohol *468*	
13.2.1.3	Miscellaneous *473*	
13.2.2	Methods for Preparation of Thiazoles *474*	
13.2.2.1	Dehydrogenation of Thiazolines or Thiazolidines *474*	
13.2.2.2	The Hantzsch Method and Its Modifications *479*	
13.2.2.3	Alkylation of Thiazole or Thiazole Derivatives *483*	
13.2.2.4	Miscellaneous *483*	
13.3	Thiazole and Thiazoline-Containing Natural Products *485*	
13.3.1	Thiazoline and Thiazole Embedded in Polyketides *485*	
13.3.2	Thiazoline and Thiazole Embedded in Peptides *491*	
13.4	Conclusions *494*	
	References *494*	

14 Pyrimidine and Imidazole *507*
Vipan Kumar and Mohinder P. Mahajan

14.1 General Introduction *507*
14.2 Pyrimidine-Based Natural Products *507*
14.2.1 Introduction *507*
14.2.2 Synthesis of Pyrimidine-Based Natural Products *508*
14.3 Imidazole-Based Natural Products *518*
14.3.1 Introduction *518*
14.3.2 Synthesis of Imidazole-Based Natural Products *520*
14.4 Conclusion *527*
Acknowledgment *528*
References *529*

Part Three Natural Products Containing Medium and Large Ring-Sized Heterocyclic Systems *535*

15 Oxepines and Azepines *537*
Darren L. Riley and Willem A.L. van Otterlo

15.1 Introduction *537*
15.2 Synthesis of the Heterocyclic Core of Selected Natural Products Containing Oxepines *538*
15.3 Synthesis of the Heterocyclic Core of Selected Natural Products Containing Azepines *549*
15.4 Synthesis of the Heterocyclic Core of Selected Natural Products Containing Oxazapines *559*
15.5 Conclusion *561*

Acknowledgments 562
References 562

16 Bioactive Macrocyclic Natural Products 569
Siti Mariam Mohd Nor, Zhengshuang Xu and Tao Ye
16.1 General 569
16.2 Natural Products Containing Azoles 569
16.2.1 Apratoxin A 569
16.2.2 Halipeptins A and D 572
16.2.3 Largazole 573
16.2.4 Bistratamide H and Didmolamide A 575
16.2.5 IB-01211 576
16.2.6 (R)-Telomestatin 578
16.3 Pyridine- and Piperidine-Containing Natural Products 581
16.3.1 Micrococcin P1 581
16.3.2 GE2270s 584
16.4 Indole- and Imidazole-Containing Natural Products 587
16.4.1 Celogentin C 587
16.4.2 Complestatin (Chloropeptin II) 589
16.5 Pyran- and Furan-Containing Natural Products 590
16.5.1 Phorboxazole B 590
16.5.2 Sorangicin A 592
16.5.3 Kendomycin 594
16.5.4 Bryostatin 16 597
16.5.5 IKD-8344 598
16.5.6 Deoxypukalide 599
16.5.7 Norhalichondrin B 601
16.6 Piperazic Acid-Containing Natural Products 605
16.6.1 Piperazimycin A 605
16.6.2 Azinothricin and Kettapeptin 607
16.7 Mixed Heterocyclic Systems 608
16.7.1 (−)-Nakadomarin A 608
16.8 Conclusions 610
References 611

Index 621

Preface

Heterocycles are significant because of their biological activity and their applications in diverse fields, reflected by the fact that heterocycles constitute more than half of known organic compounds. Heterocyclic moieties are ubiquitous in important classes of naturally occuring compounds, for example, alkaloids, vitamins, hormones, and antibiotics. They are also widely prevalent in pharmaceuticals, herbicides, dyes and many other application-oriented materials.

Many natural products possess heterocyclic moieties, and many of them are important in terms of their biological activities and application in different fields; for example, "taxol" isolated from the yew tree is an anticancer drug.

To the best of our knowledge, there is currently no comprehensive publication which deals with systematic survey of literature on the heterocycles for the synthesis of natural products in the form of a ready reference. The present book is an attempt to bridge this gap. Emphasis has been given to the current literature while including the earlier literature as references. Different heterocycles are classified according to their ring size under different sections and subsections. Sources, synthetic aspects, relevant biological activities and important physical properties and applications of the natural products are described. Unfortunately not all the important heterocycles, such as coumarins and flavones, could be included due to the restricted size of the book. However, available literature in these areas has already been adequately reviewed. This may be the subject of a future edition on the basis of the feedback of this edition received from the users.

We are grateful to our contributors for their valuable contributions which have made this project successful. We are also grateful to the members of our research groups for their untiring assistance during the course of this project. Finally we do hope that this will be a valuable secondary source book for the concerned readers.

K.C. Majumdar
S.K. Chattopadhyay

List of Contributors

Laura Alvarez
Universidad Autónoma del Estado de Morelos
Centro de Investigaciones Químicas
Avenida Universidad 1001
Col. Chamilpa 62209
Cuernavaca, Morelos
Mexico

Alicia Boto
Instituto de Productos Naturales y Agrobiología CSIC
Avda. Astrofísico Fco. Sánchez, 3
38206 La Laguna, Tenerife
Spain

Candice Botuha
IPCM–UPMC Univ Paris 6
Case Courrier 183
4 place Jussieu
75252 Paris Cedex 05
France

Sarah M. Bronner
University of California
Department of Chemistry and Biochemistry
607 Charles Young Drive East, Box 951569
Los Angeles, CA 90095-1569
USA

Fabrice Chemla
IPCM–UPMC Univ Paris 6
Case Courrier 183
4 place Jussieu
75252 Paris Cedex 05
France

Kongara Damodar
Indian Institute of Chemical Technology
Organic Chemistry Division-I
Uppal Road
Hyderabad 500007
India

Biswanath Das
Indian Institute of Chemical Technology
Organic Chemistry Division-I
Uppal Road
Hyderabad 500007
India

Bidyut Kumar Dinda
Indian Institute of Technology
Department of Chemistry
Kharagpur W.B. 721302
India

Franck Ferreira
IPCM–UPMC Univ Paris 6
Case Courrier 183
4 place Jussieu
75252 Paris Cedex 05
France

Neil K. Garg
University of California
Department of Chemistry and Biochemistry
607 Charles Young Drive East, Box 951569
Los Angeles, CA 90095-1569
USA

Antonio Garrido Montalban
Arena Pharmaceuticals, Inc.
6166 Nancy Ridge Drive
San Diego, CA 92121
USA

Konstanze K. Gruner
Technische Universität Dresden
Department Chemie
Bergstr. 66
01069 Dresden
Germany

G.-Yoon J. Im
University of California
Department of Chemistry and Biochemistry
607 Charles Young Drive East, Box 951569
Los Angeles, CA 90095-1569
USA

Paula Kiuru
University of Helsinki
Faculty of Pharmacy, Division of Pharmaceutical Chemistry
PO Box 56 (Viikinkaari 5 E)
FI-00014 Helsinki
Finland

David W. Knight
Cardiff University
School of Chemistry, Main College
Park Place
Cardiff CF10 3AT
UK

Hans-Joachim Knölker
Technische Universität Dresden
Department Chemie
Bergstr. 66
01069 Dresden
Germany

Vipan Kumar
Guru Nanak Dev University
Department of Applied Chemistry
Amritsar 143005
India

Mohinder P. Mahajan
Guru Nanak Dev University
Department of Applied Chemistry
Amritsar 143005
India

Krishna C. Majumdar
University of Kalyani
Department of Chemistry
Kalyani, W.B. 741235
India

Dipakranjan Mal
Indian Institute of Technology
Department of Chemistry
Kharagpur W.B. 721302
India

Hideto Miyabe
Hyogo University of Health Sciences
School of Pharmacy
Minatojima
Kobe 650-8530
Japan

Okiko Miyata
Kobe Pharmaceutical University
4-19-1 Motoyamakita, Higashinada
Kobe 658-8558
Japan

Siti Mariam Mohd Nor
Universiti Putra Malaysia
Department of Chemistry
Faculty of Science
43400 UPM Serdang
Selangor
Malaysia

Shovan Mondal
University of Kalyani
Department of Chemistry
Kalyani, W.B. 741235
India

Takeaki Naito
Kobe Pharmaceutical University
4-19-1 Motoyamakita, Higashinada
Kobe 658-8558
Japan

Willem A.L. van Otterlo
University of the Witwatersrand
Molecular Sciences Institute
School of Chemistry
PO Wits
2050 Johannesburg
South Africa

Alejandro Pérez-Luna
IPCM – UPMC Univ Paris 6
Case Courrier 183
4 place Jussieu
75252 Paris Cedex 05
France

Darren L. Riley
University of the Witwatersrand
Molecular Sciences Institute
School of Chemistry
PO Wits
2050 Johannesburg
South Africa

Tetsuya Sengoku
Shizuoka University
Department of Materials Science,
Faculty of Engineering
3-5-1 Johoku
Naka-ku
Hamamatsu
Shizuoka 432-8561
Japan

Brateen Shome
Indian Institute of Technology
Department of Chemistry
Kharagpur W.B. 721302
India

Masaki Takahashi
Shizuoka University
Department of Materials Science,
Faculty of Engineering
3-5-1 Johoku
Naka-ku
Hamamatsu
Shizuoka 432-8561
Japan

Zhengshuang Xu
Peking University Shenzhen Graduate School
School of Chemical Biology and Biotechnology
University Town of Shenzhen
Xili, Nanshan District
Shenzhen 518055
China
and
The Hong Kong Polytechnic University
Department of Applied Biology and Chemical Technology
Hung Hom, Kowloon, Hong Kong
China

Tao Ye
Peking University Shenzhen Graduate School
School of Chemical Biology and Biotechnology
University Town of Shenzhen
Xili, Nanshan District
Shenzhen 518055
China
and
The Hong Kong Polytechnic University
Department of Applied Biology and Chemical Technology
Hung Hom, Kowloon, Hong Kong
China

Jari Yli-Kauhaluoma
University of Helsinki
Faculty of Pharmacy, Division of Pharmaceutical Chemistry
PO Box 56 (Viikinkaari 5 E)
FI-00014 Helsinki
Finland

Hidemi Yoda
Shizuoka University
Department of Materials Science, Faculty of Engineering
3-5-1 Johoku
Naka-ku
Hamamatsu
Shizuoka 432-8561
Japan

Part One
Strained Heterocycles in the Synthesis of Natural Products

1
Aziridines in Natural Product Synthesis
Candice Botuha, Fabrice Chemla, Franck Ferreira and Alejandro Pérez-Luna

1.1
Introduction

The aziridinyl ring has attracted considerable attention over the last 20 years. A number of reviews dedicated to the synthesis of aziridines has appeared, dealing mainly with the preparation of this small heterocycle [1] and its reactions [2]. Very few have focused on natural product synthesis using aziridines.

This review is devoted to the occurrence of the aziridine moiety in the total synthesis of natural products (Table 1.1). Aziridines can be present in the natural product itself, or can serve as intermediates in the course of the natural product synthesis. Only the synthesis of truly natural products will be discussed here, and the synthesis of analogs or non-natural products will not be developed [3]. This review covers the literature from 1986, and follows the previous review [4] focused on aziridines in natural product synthesis.

1.2
Synthesis of Natural Products Containing Aziridine Units

The ability of aziridines to undergo highly stereo- and regioselective ring-opening reactions has been exploited in nature. A number of natural products possessing an aziridine ring have been shown to possess potent biological activity, which is closely associated with the reactivity of the strained heterocycle.

1.2.1
Synthesis of Aziridine-2,3-Dicarboxylic Acid

Among natural products containing aziridines, C_2-symmetric aziridine-2,3-dicarboxylic acid **4**, a metabolite of *Streptomyces* MD 398-A1 [5a] was prepared in enantiopure form in few steps from L-(+)-diethyl tartrate [5b]. An optimized synthetic pathway to aziridine-2,3-dicarboxylate precursor **3** of naturally-occurring aziridine **4** was published later, removing the epimerization tendency

Table 1.1 Aziridine-based natural products.

Serial No.	Trivial name	Structure	Source	Isolation [Ref]	Biological activity	Synthesis [Ref]
1	Aziridine-2,3-dicarboxylic acid		*Streptomyces* MD 398-A1	[5a]	Antibacterial activity	[5b, 6]
2	(Z)-Dysidazirine		*Dysidea fragilis*	[7a]	Cytotoxicity, antifungal activity	[7b, 8]
3	FR-900482		*Streptomyces sandaensis*	[9a]	Antibacterial, anticancer activity	[10–15]
4	FR-66979		*Streptomyces sandaensis*	[9b]	Antibacterial, anticancer activity	[15, 16]
5	Mitomycin A		*Streptomyces cuespitasus*	[17b]	Antibacterial, anticancer activity	[18, 19]
6	Mitomycin C		*Streptomyces cuespitasus*	[17a]	Antibacterial, anticancer activity	[19, 20]
7	Mitomycin K		*Streptomyces verticillatus*	[17c]	Antibacterial, anticancer activity	[21, 22]

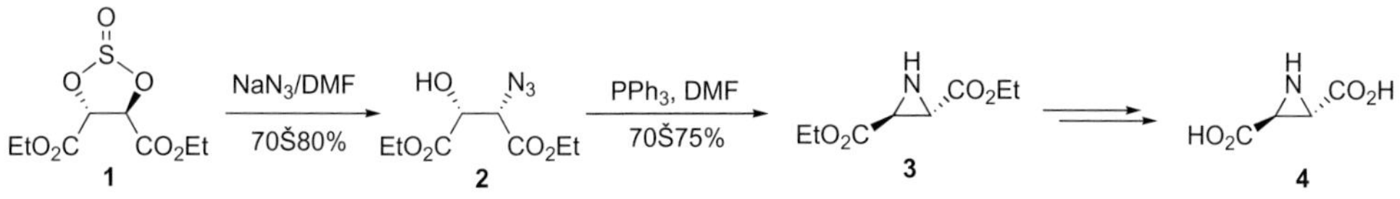

Scheme 1.1 Aziridine-2,3-carboxylic acid.

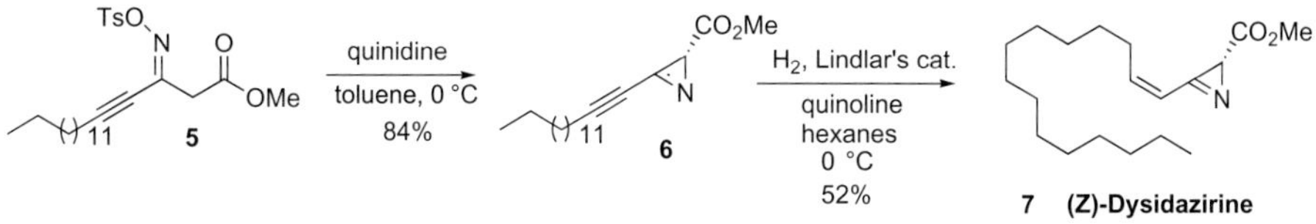

Scheme 1.2 (Z)-Dysidazirine.

of synthetic intermediates by using *anti*-3-azido-2-hydroxy-succinate **2** (Scheme 1.1) [6].

1.2.2
Synthesis of (Z)-Dysidazirine

While the first enantioselective synthesis of antifungal active (Z)-dysidazirine **7** isolated from the marine sponge *Dysidea fragilis* [7a] was achieved by a Darzens-type synthesis of *cis*-N-sulfinylaziridine carboxylic acid [7b], a very recent synthesis includes the transformation of tosylated imine **5** into azirine carboxylate **6** as a key step (Scheme 1.2) [8].

1.2.3
Syntheses of Mitomycins

In the field of total synthesis of natural products containing an aziridine ring, most efforts of synthetic organic chemists have been dedicated to mitomycins, a class of very potent antibacterial and anticancer compounds isolated [17] from extracts of genus *Streptomyces*, a filamentous gram-positive soil bacterium. The most abundant mitomycins in nature are represented in Scheme 1.3. Mitomycin C [17a], besides mitomycin A [17b] and K [17c], has become the most effective drug of the series against non-small-cell lung carcinoma and other soft and solid tumors [23]. Despite significant medicinal features, syntheses of this class of compounds (furthermore in racemic form) have been reported only four times over the past 30 years [24].

Since the first synthesis of a mitomycin by Kishi [18], only few organic chemists have succeeded to achieve a total synthesis of a mitomycin since serious difficulties occurred during the construction of both reactive quinone and aziridine rings with the elimination of methanol from the 9a position (Scheme 1.3). Mitomycins A and

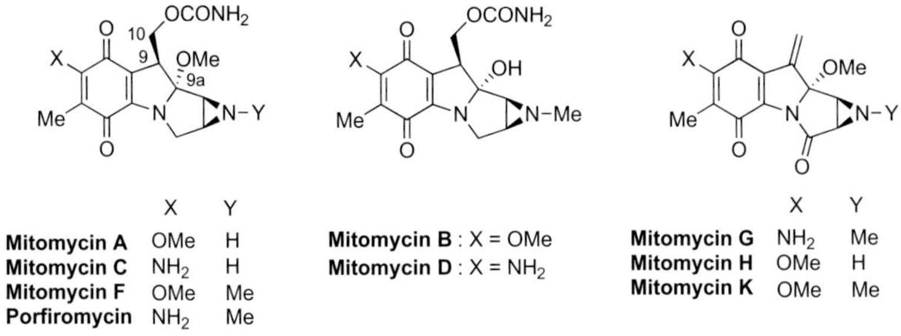

1 Aziridines in Natural Product Synthesis

Scheme 1.3 General mitomycins.

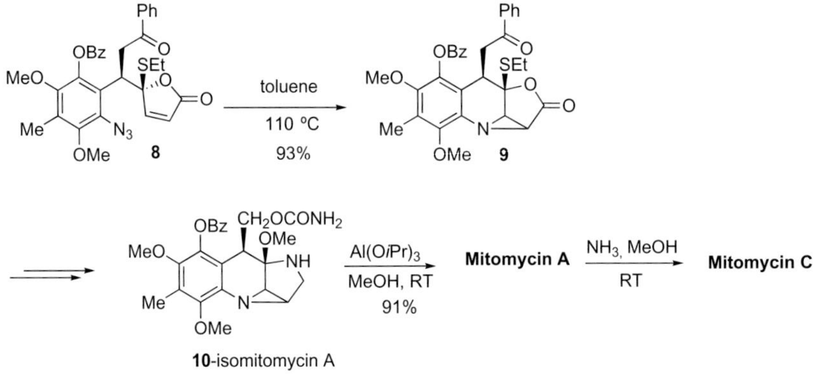

	X	Y
Mitomycin A	OMe	H
Mitomycin C	NH_2	H
Mitomycin F	OMe	Me
Porfiromycin	NH_2	Me

Mitomycin B : X = OMe
Mitomycin D : X = NH_2

	X	Y
Mitomycin G	NH_2	Me
Mitomycin H	OMe	H
Mitomycin K	OMe	Me

Scheme 1.4 Mitomycin A and C.

C were successfully synthesized by Fukuyama and co-workers in 18 steps from a readily available chalcone [19]. An intramolecular azide-olefin cycloaddition on **8** gave exclusively tetracyclic aziridine **9**. Isomitomycin intermediate **10** was then obtained in few steps providing mitomycin A by a subsequent reaction with Al(i-OPr)$_3$. A final ammonolysis step gave mitomycin C (Scheme 1.4).

Subsequent improvement for the total synthesis of mitomycin C was reported later by the same authors using highly reactive bridgehead iminium species in the key steps [20]. The required C-9a methoxy group was introduced under mild acidic conditions to give **13** in 60% yield via highly strained iminium ion **12** which was obtained from compound **11** through acidic treatment. Transformation of the aromatic ring of **13** using hydrogenolysis followed by oxidation with DDQ afforded the desired quinone ring of isomitomycin A in 77% yield. Isomitomycin A was then converted to (±)-mitomycin C via isomitomycin C **15** in 85% yield by treatment with NH$_3$ in methanol (Scheme 1.5).

Danishefsky and colleagues have designed a short total synthesis of the densely functionalized mitomycin K from parent mitomycins A and C by elimination of the carbamate at position 10 [21]. Introduction of N-methyl aziridine from an olefin was achieved in only 3 steps by 1,3-dipolar cycloaddition. Reaction of meth-

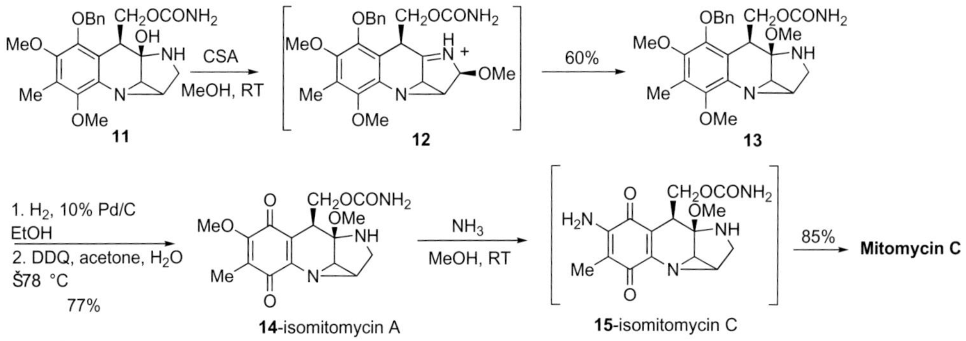

Scheme 1.5 Mitomycin C.

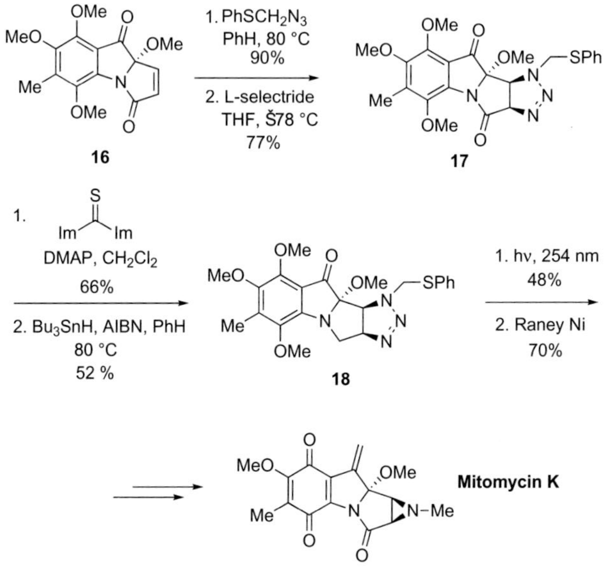

Scheme 1.6 Mitomycin K.

ylthiophenyl azide with imide **16** provided triazoline **17** which was then transformed to **18** in two steps. N-methylaziridine **19**, an advanced intermediate to mitomycin K, was obtained by irradiation at 254 nm (Scheme 1.6).

A facile method was set up for the transformation of azidomitosenes **20** into mitomycins (introduction of the C9a methoxy group) by using an oxidation reaction of the C9-9a double bond with $MoO_5 \cdot$ hexamethylphosphoramide (HMPA). Specifically, oxidation of **20** afforded in 46% **21** from which the fused N-methylaziridine ring could be constructed. **22** led to the mitomycin K in two steps. (Scheme 1.7) [22].

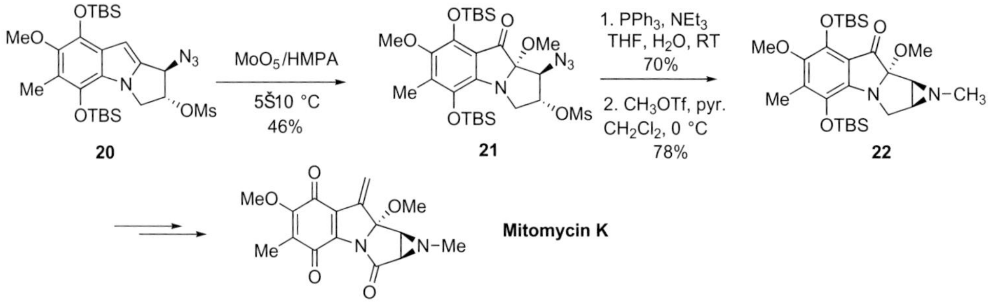

Scheme 1.7 Mitomycin K.

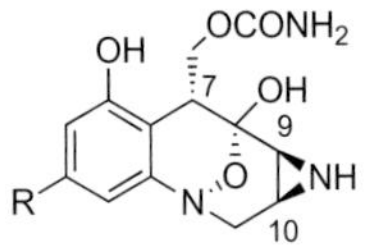

Scheme 1.8 FR-66979 and FR-900482.

FR-66979, R = CH$_2$OH
FR-900482, R = CHO

1.2.4
Syntheses of FR-900482 and FR-66979

Antitumor antibiotic natural products FR-900482 and FR-66979 (Scheme 1.8) isolated from *Streptomyces sandaensis* [9] are structurally related to mitomycin C and possess similar biological activities. They have proven to be less toxic than mitomycins in clinical cancer chemotherapeutics. In addition of the biomedical potential, the uncommon structure and the synthetic problem related to the construction of both aziridine ring and the hemiacetal functionality of FR-900482 and FR-66979 have attracted the attention of a number of synthetic chemists. Although several approaches have been explored to construct these highly functionalized structures, only six total syntheses have been accomplished and two formal syntheses [10, 25] have been reported to date.

The first total synthesis of (±)-FR-900482 was realized in 41 steps from readily available *N*-benzylamine [11]. In this strategy, the authors introduced the aziridine ring at the end of the synthesis to prevent any lability of the three-membered ring under acidic condition. Azide **24** was prepared by ring opening of epoxide **23** with NaN$_3$ followed by the transformation of the resulting alcohol to a mesylate. Upon oxidative treatment of **24** to exchange the PMB-protecting group on the aromatic ring for a dimethyl acetal, reduction of azide **25** by PPh$_3$ in the presence of a base furnished aziridine **26** which was then advanced to the target compound (Scheme 1.9). An enantioselective total synthesis of (+)-FR-900482 was reported later on by the same group with a slight modification of the initial route [12].

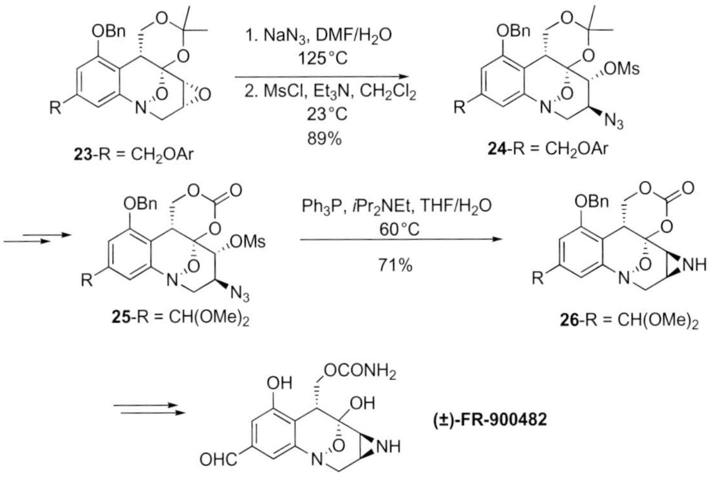

1.2 Synthesis of Natural Products Containing Aziridine Units

Scheme 1.9 (±)-FR-900482.

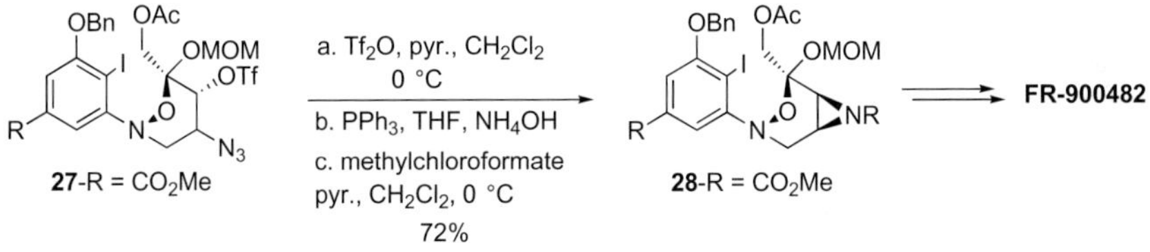

Scheme 1.10 (±)-FR-900482.

Danishefsky reported a total synthesis of racemic FR-900482 [13] in which the installation of the 9,10-aziridine ring was carried out following the precepts founded by Kishi and colleagues in their synthesis of mitomycins [26]. Compound **27** was treated in sequence with Tf$_2$O and pyridine, PPh$_3$, NH$_4$OH and finally methylchloroformate to furnish aziridine **28** in 72% overall yield (Scheme 1.10).

Naturally occurring FR-66979 has been synthesized in racemic form with a method comprising an original homo-Brook-mediated aziridine fragmentation of compound **30** providing the eight-membered ring benzazocenol **31**. The aziridine ring and the end of the total synthesis [16] were achieved subsequently from epoxide **32** following a described procedure (Scheme 1.11) [11].

The first enantioselective synthesis of the enantiomer of FR-900482 was developed by Terashima [27] starting from FK-973 [10], the more stable semisynthetic triacetyl derivatives of FR-900482. In the same time, a second total synthesis of both enantiomers was accomplished in a convergent manner in a 57 steps sequence using classical methods for the introduction of the aziridine ring [14].

Shorter enantioselective total syntheses of (+)-FR-900482 and (+)-FR-66979 were reported by Williams and co-workers and described the installation of the labile

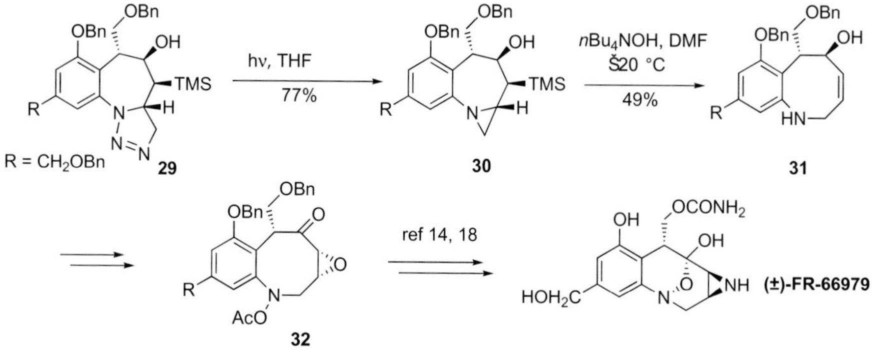

Scheme 1.11 (±)-FR-66979.

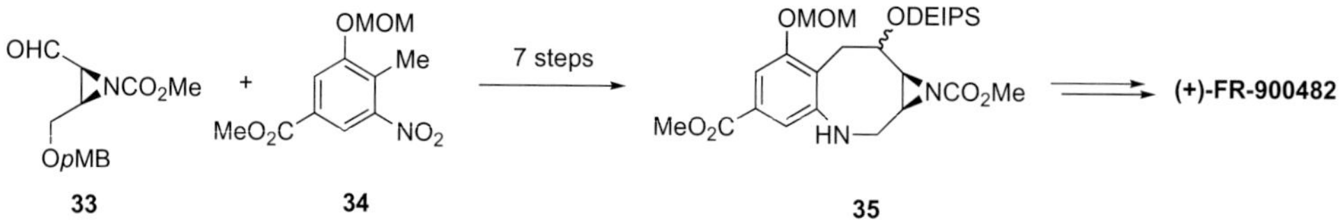

Scheme 1.12 (+)-FR-900482.

aziridine ring in the very beginning of the synthesis. In this purpose, aziridine **35** was prepared in 7 steps from compounds **33** and **34** (Scheme 1.12) [15].

1.3
Synthesis of Natural Products Involving the Transformation of an Aziridine Moiety

1.3.1
Nucleophilic Ring-Opening of Aziridines for Natural Product Synthesis

The usefulness of aziridines in organic synthesis is greatly related to their facility to undergo nucleophilic ring opening to relieve ring strain. In the case of "unactivated" alkyl-substituted or unsubstituted aziridines, acid catalysis is usually required. Conversely, it is well known that aziridines bearing nitrogen electron-withdrawing substituents such as carbonyl, sulfonyl, sulfinyl, phosphoryl, and phosphinyl are activated towards nucleophilic ring opening as the developing negative charge on nitrogen is stabilized. A wide array of nucleophiles can be used and in most cases, the regio- and stereoselectivity of the ring opening is predictable, a matter of paramount relevance when it comes to design a synthetic plan. While selectivity issues are highly dependent on the substrate and the reaction conditions, steric congestion is often at the origin of regioselectivity (nucleophilic attack occurring at the less congested terminus) and stereoselectivity is frequently related to a ring opening resulting from an *anti* attack (stereospecific S_N2). A tremendous body of work has been disclosed in the field and the subject has been comprehensively reviewed. The use of nucleophilic ring opening of aziridines in the context of natural product synthesis, illustrated hereafter through representative examples, provides a good indication of the reaction's synthetic potential.

1.3.1.1 Carbon-Centered Nucleophiles

Nucleophilic ring opening of aziridines with carbon nucleophiles is mainly achieved using organometallic reagents. In spite of an early report in the 1970s [28], this type of reaction only started to become customary in natural product synthesis following seminal reports in the late 1980s concerning the use of organocuprate or Grignard reagents to open *N*-alkyl aziridines in the presence of BF_3 Et_2O [29] or to open *N*-tosyl aziridines (with no need of Lewis acidic activation) [30].

Ring opening of aziridines with organocopper reagents has proved useful for a number of syntheses. In a first example, as part of their studies on the synthesis of carbapenem antibiotics using aziridine ring openings [2a], Tanner and coworkers developed an enantioselective entry to naturally occurring (+)-PS-5 (Scheme 1.13) [31]. Reaction of $LiEt_2Cu$ with chiral *trans*-2,3-aziridino alcohol **36** afforded in 70% yield sulfonamido alcohol **37** which proved suitable for β-lactam ring construction and was advanced to a known intermediate of (+)-PS-5. As a result of the complexation of the reagent to the free C-1 hydroxyl group, nucleophilic attack occurred highly regioselectively at the C-2 carbon, in a behavior analogous to that of related epoxy alcohols. Ring opening took place with inversion of configuration and thus provided the requisite stereochemistry for the stereogenic centers of the final product. Interestingly, the related ring opening of the analogous 3-benzyloxyethyl substituted aziridino alcohol using excess Me_3Al took place at the C-3 carbon [32], a transformation that proved useful for the preparation of non-natural 1β-methylcarbapenem antibiotics [33].

In addition to cuprates, addition of Grignards has also been often used, most of the time under copper catalysis. For instance, Harrity and co-workers disclosed a formal synthesis of (−)-dihydropinidine wherein the tetrahydropyridine ring was constructed through a two-step [3 + 3] annelation sequence involving the reaction between alkylmagnesium reagents such as **38** bearing a pendant 1,3-dioxolane moiety and terminal *N*-sulfonyl aziridines (Scheme 1.14) [34]. Specifically, **38** was reacted with aziridine **39** in the presence of 20 mol % CuBr to afford intermediate **40**, which, upon acidic hydrolysis without intermediate purification, led to piperidine **41** in 78% overall yield.

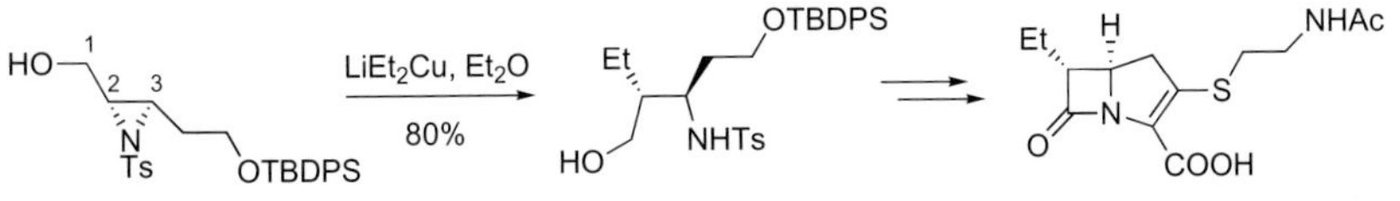

Scheme 1.13 (+)-PS-5.

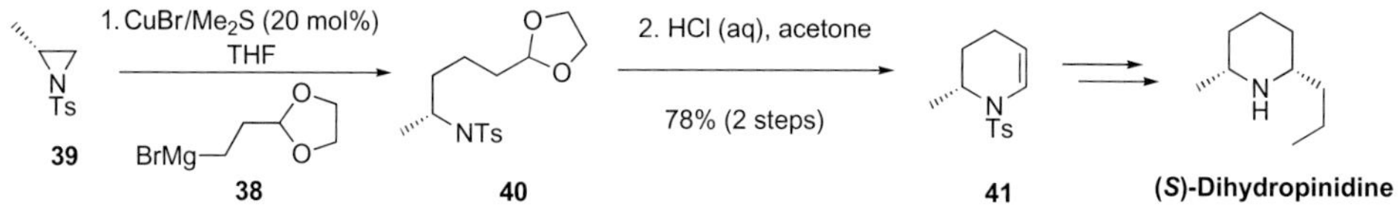

Scheme 1.14 (−)-Dihydropinidine.

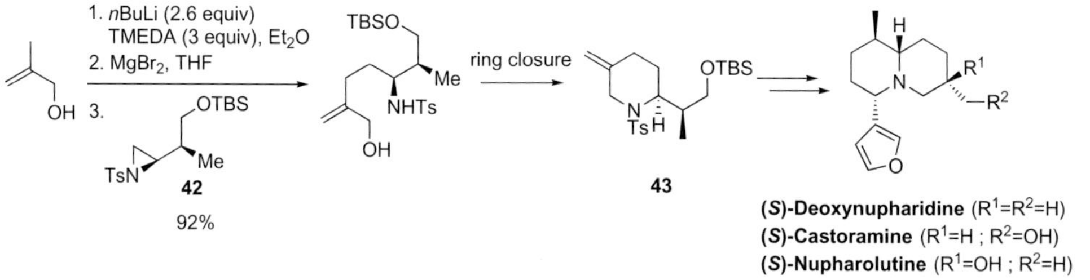

Scheme 1.15 Nuphar alkaloids.

In a related [3 + 3] annelation strategy, the allylmagnesium reagent obtained following deprotonation/transmetallation of methallyl alcohol was found to react with terminal N-sulfonyl aziridines, but this time with no need of copper salt addition (Scheme 1.15). The resulting 1,5-aminoalcohol could undergo ring closure to piperidines using palladium catalysis or Mitsunobu conditions, thus affording 2-alkylpiperidines with an exo methylene moiety at C-5, in what represents an improved approach to the related [3 + 3] cycloaddition reaction between aziridines and palladium-trimethylenemethane (Pd-TMM) species (see Section 1.3.2.4). In its application to synthesis, nucleophilic ring opening of enantiopure aziridine 42 and subsequent cyclization led to piperidine 43 that could be advanced to *nuphar* alkaloids (−)-deoxynupharidine, (−)-castoramine and (−)-nupharolutine [35, 36]. The same approach has also been used for the construction of the spiropiperidine motif in the formal synthesis of (±)-perhydrohistrionicotoxin [37].

Copper-bromide promoted coupling of arylmagnesium derivatives with terminal enantiopure N-butoxycarbonyl (N-Boc) aziridines is pivotal in the design of a very recent approach to renieramycin alkaloids based on the use of aziridines derived from (S)-serine as linchpins to bring together both tetrahydroisoquinoline parts of the target molecules (Scheme 1.16) [38]. First, homologation of densely substituted arylmagnesium 44 was effected by nucleophilic ring opening of enantiopure terminal N-Boc aziridine 45. Intermediate 46 was then further elaborated through a Pictet–Spengler reaction involving aziridine 47. The right and left part of the molecules were connected through a second copper -promoted coupling between 48 and arylgrignard 49 which afforded 50, reaction of which with benzyloxyacetaldehyde led to a bistetrahydroisoquinoline common intermediate to (−)-renieramycins M and G and (−)-jorumycin, three alkaloids with potent antitumor antibiotic activities.

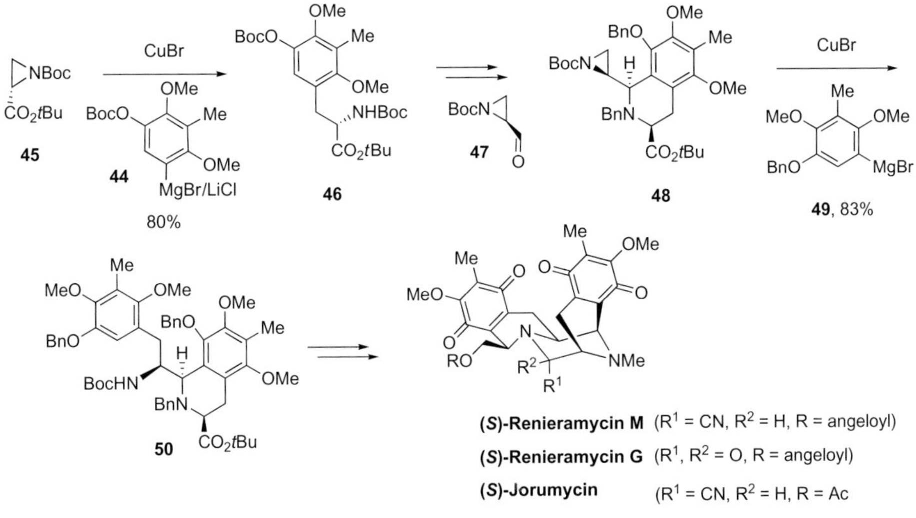

1.3 Synthesis of Natural Products Involving the Transformation of an Aziridine Moiety

Scheme 1.16 Renieramycins.

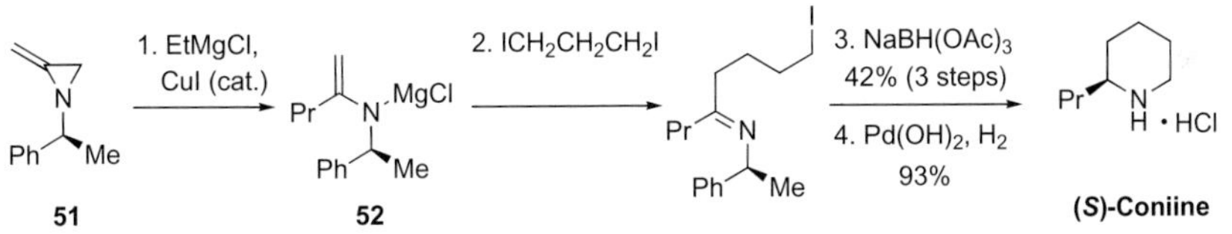

(S)-Renieramycin M (R^1 = CN, R^2 = H, R = angeloyl)
(S)-Renieramycin G (R^1, R^2 = O, R = angeloyl)
(S)-Jorumycin (R^1 = CN, R^2 = H, R = Ac)

Scheme 1.17 (S)-Coniine.

Interestingly, copper-catalyzed nucleophilic ring opening of N-alkyl 2-methyleneaziridines by Grignard reagents does not require Lewis-acidic activation of the heterocycle if carried out at room temperature (Scheme 1.17). It results in the highly regioselectivity formation of a metalloenamine that can then be alkylated in the presence of electrophiles. Building on this chemistry, a concise synthesis of the hemlock alkaloid (S)-coniine was achieved from enantiopure methyleneaziridine **51** using a multicomponent strategy involving: ring opening, alkylation of the resulting metalloenamine **52** with a 1,3-difunctionalized electrophile, and diastereoselective reduction followed by *in situ* cyclization [39].

In addition to organocopper and organomagnesium reagents, ring opening of aziridines by lithium carbanions has also been used for synthetic purposes. The reaction between α-(phenylsulfonyl) lithio anions and N-phosphinyl or N-sulfonyl aziridines plays a key role in the strategy developed by Craig and co-workers for the synthesis of pyrrolidine-containing alkaloids (+)-monomorine I [40], (+)-preussin [41] and (±)-lepadiformine [42], built around the base-mediated 5-endo-trig intramolecular addition of amides onto vinylic sulfones. The case of (±)-lepadiformine is illustrative (Scheme 1.18). Ring opening at the least-substituted carbon of N-SES aziridine **53** by PhSO$_2$CH$_2$Li led to 3-amino sulfone **54**. *In situ* generation

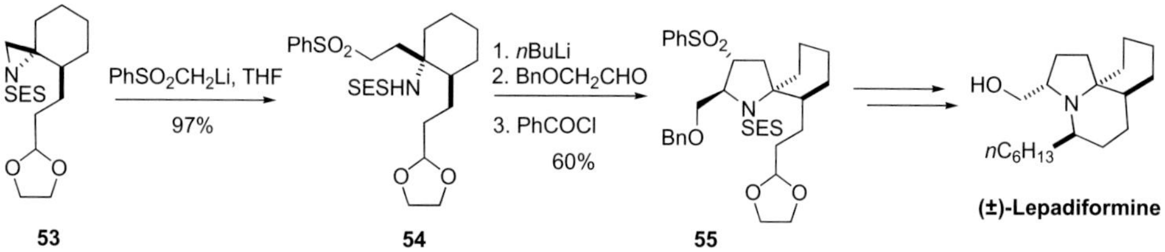

Scheme 1.18 (±)-Lepadiformine.

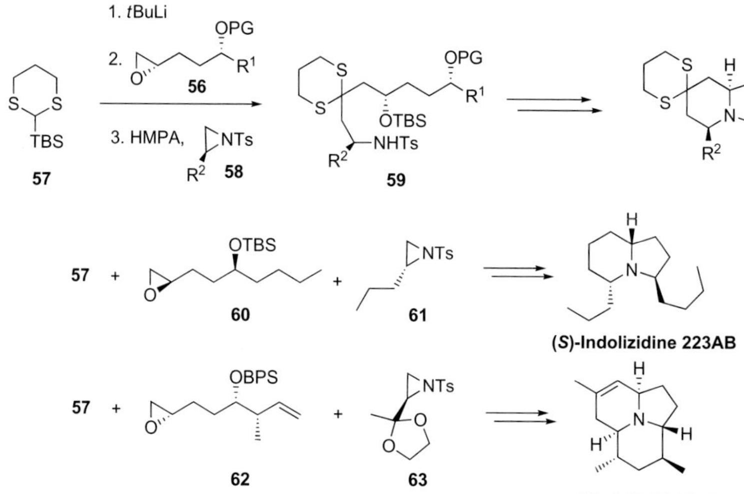

Scheme 1.19 (−)-Indolizidine 223AB and alkaloid (−)-205B.

of the requisite vinylic sulfone from this precursor enabled pyrrolidine formation following 5-*endo*-trig cyclization and afforded **55** that was advanced to the target compound.

Smith III and Kim have designed a very elegant strategy for the construction of indolizidine alkaloids using a three-component linchpin coupling involving a lithiated silyl dithiane, an epoxide and a *N*-tosyl aziridine (Scheme 1.19) [43, 44]. Nucleophilic ring opening of 5-alkoxy epoxides **56** by the lithio anion **57** of 1,3-dithiane resulted in the formation of a lithium alcoholate intermediate that underwent a solvent-controlled Brook rearrangement upon addition of HMPA. The lithiated dithiane derivative thus produced, could then be used to effect the nucleophilic ring opening of *N*-tosyl aziridines **58**, to afford, in fine, stereodefined 1,5-amino alcohols **59** that are suitable for indolizidine construction. This convergent approach has already proved successful for the preparation of (−)-indolizidine 223AB from enantiopure epoxide **60** and aziridine **61**, and for the preparation of alkaloid (−)-205B from enantiopure epoxide **62** and aziridine **63**.

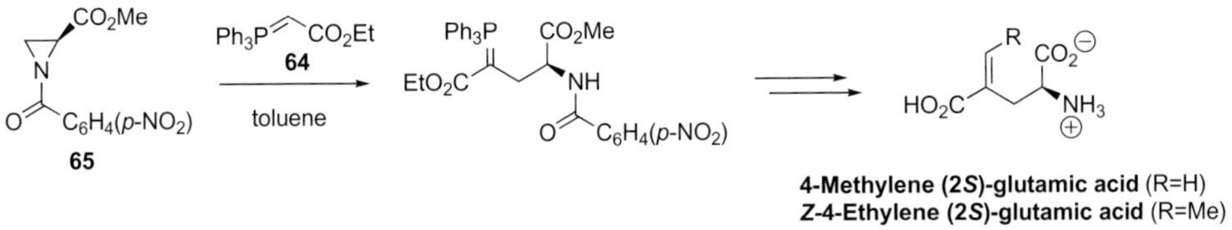

1.3 Synthesis of Natural Products Involving the Transformation of an Aziridine Moiety | 15

Scheme 1.20 Alkylideneglutamic acids.

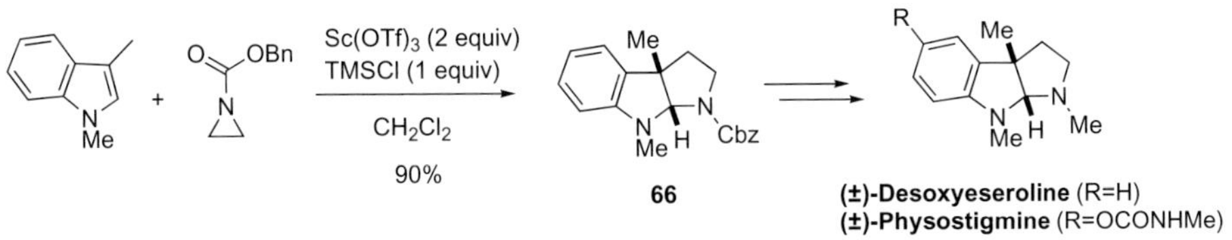

Scheme 1.21 (±)-Physostigmine.

Ring opening of aziridines with carbon nucleophiles other than carbanions, though less common, has also proved useful for the synthesis of natural products. In an early example, Baldwin and co-workers disclosed an access to γ-alkylidene glutamates based on the reaction of carbonyl stabilized Wittig reagent **64** with enantiopure N-acyl and N-sulfonyl aziridine-2-carboxylates derived from serine (Scheme 1.20) [45]. For instance, the enantiopure ylide resulting from ring opening of aziridine **65** at the less-substituted carbon was further modified to provide 4-methylene (2S)-glutamic acid and Z-4-ethylene (2S)-glutamic acid.

Nakagawa and co-workers disclosed a concise approach to Calabar alkaloids based on the nucleophilic ring opening of N-benzyloxy-aziridine by 1,3-dimethylindole and subsequent intramolecular interception by nitrogen of the resulting indolenium (Scheme 1.21) [46]. In the optimized conditions involving activation of the aziridine by Sc(OTf)$_3$ in the presence of trimethylsilyl chloride (TMSCl), adduct **66** was obtained in 90% yield. This intermediate could be advanced to (±)-desoxyeseroline, a known intermediate of naturally occurring physostigmine.

Potassium cyanide was recently used efficiently for the ring opening of monosubstituted allylglycine derived N-sulfonyl aziridine **67** (used as mixture of diastereoisomers) (Scheme 1.22) [47]. The resulting β-aminonitrile moiety of adduct **68** could then be used for the construction of the 2-amino-tetrahydropyrimidine ring required to complete the synthesis of tetrahydrolathyrine.

1.3.1.2 Nitrogen-Centered Nucleophiles

Ring opening of aziridines with nitrogen nucleophiles, mainly amines and azides, has attracted considerable attention from organic chemists as a result of the increasing interest in diamine compounds for synthetic and pharmaceutical purposes. Quite surprisingly however, its use for the synthesis of natural products has so far been limited. As probable reasons, one might put forward both a limited

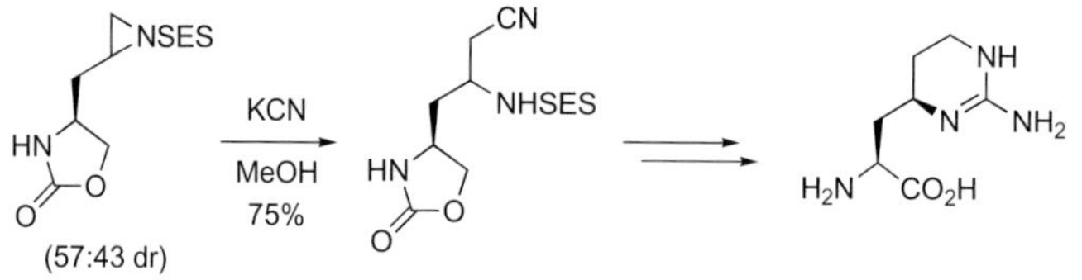

Scheme 1.22 Tetrahydrolathyrine.

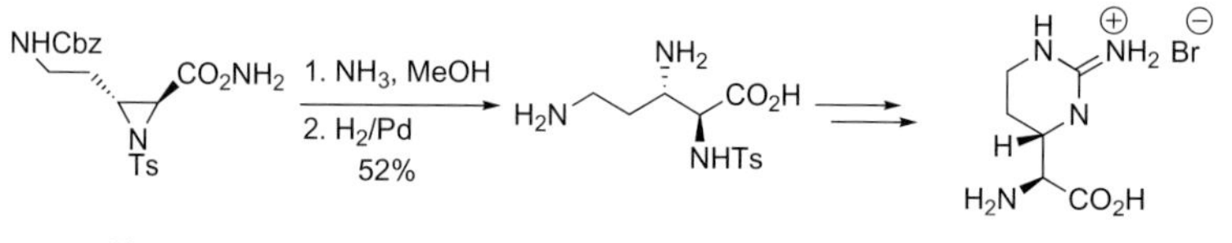

Scheme 1.23 L-*epi*-Capreomycidin.

number of targets bearing the diamino moiety and the fact that amines are generally not inert to mainstream Lewis or Brønsted acidic aziridine ring opening promoters. Nonetheless, specially in the field of amino acid synthesis using aziridine-2-carboxylates as substrates, some elegant applications have been disclosed.

Amines are sufficiently nucleophilic to attack aziridines without the presence of a promoter. Shiba and co-workers took advantage of it to perform the ring opening of enantiopure *trans*-N-tosyl aziridine-2-carboxylate **69** with ammonia (Scheme 1.23) [48]. The attack took place regioselectively at C-3 in an S_N2 manner and thus afforded diamine **70** in 52% yield which was then converted into the hydrobromide salt of naturally occurring L-epicapreomycidin.

The reaction between amino acid carboxylates and N-tosyl aziridine-2-carboxylates has been used for the preparation of peptidyl derivatives (Scheme 1.24). In spite of moderate levels of regioselectivity, ring opening in methanol of aziridines **71** and **72** by nucleophilic attack of L-histidine at C-3 in the presence of 1M sodium hydroxide, was used for the preparation of FR900490 [49], feldamycin [50] (from **71**) and melanostatin [50] (from **72**).

In another related example, sodium pipecolinate was found to react highly efficiently with monosubstituted enantiopure N-tosyl aziridine **73** (Scheme 1.25). The corresponding piperidine adduct was obtained in 88% yield and was used to prepare verruculotoxin [51].

Other nucleophiles that have been used for the opening of aziridines without promoter are imidazole and 1,2,4-oxadiazolidine-3,5-dione (Scheme 1.26). Their reaction with (*S*)-serine derived enantiopure aziridine **74**, which has also proved useful in the context of ring opening with carbon nucleophiles (see Section 1.3.1.1), was carried out by Baldwin and co-workers to synthesize (*S*)-β-pyrazolylalanine and (*S*)-quiscalic acid [52].

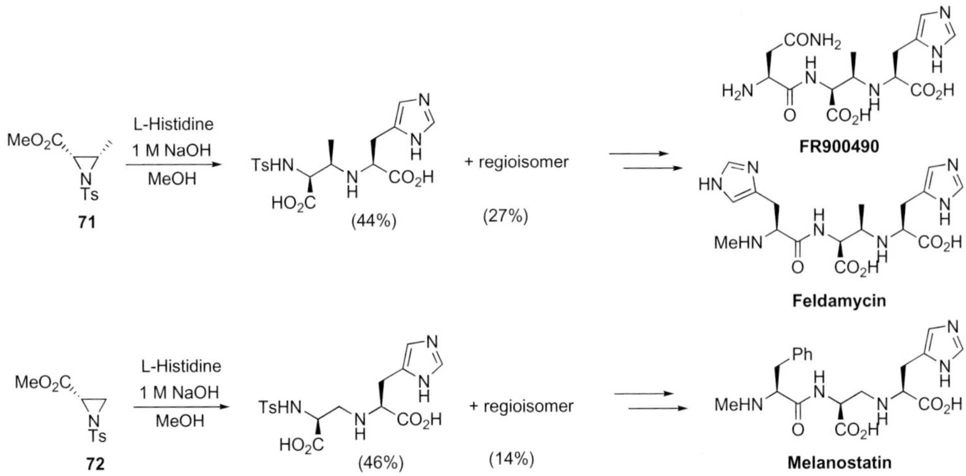

Scheme 1.24 FR-900490, feldamycin and melanostatin.

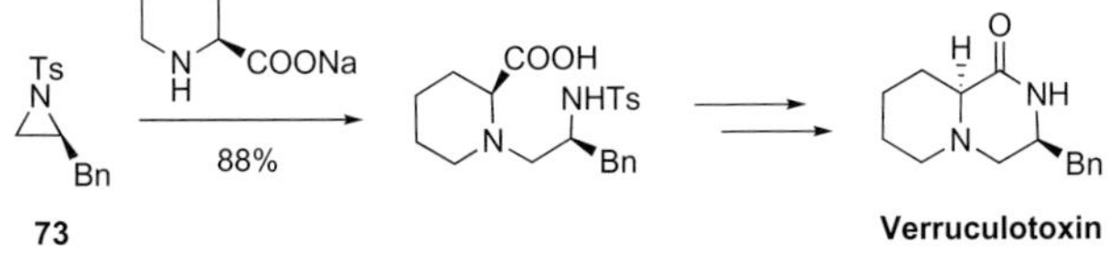

Scheme 1.25 Verruculotoxin.

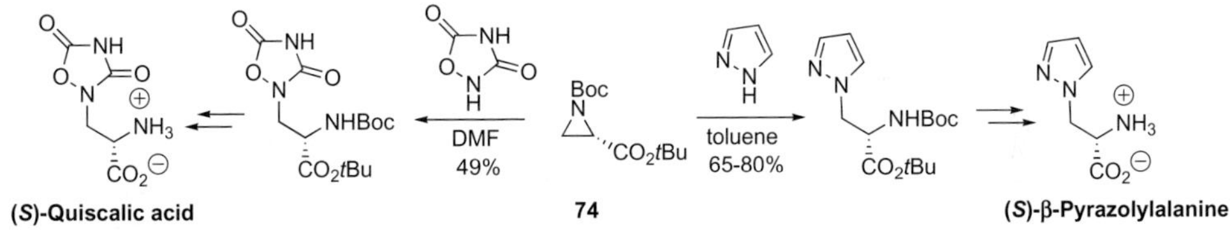

Scheme 1.26 (S)-β-Pyrazolylalanine.

From another perspective, as part of their studies on palladium-catalyzed dynamic kinetic asymmetric transformations, Trost and co-workers have developed a strategy towards pyrrolopiperazinones based on an annulation reaction between 5-bromopyrrole-2-carboxylate esters and vinyl aziridines (for a related cycloaddition reaction between vinyl aziridines and isocyanates see Section 1.3.2.4) (Scheme 1.27) [53]. In this transformation, following an asymmetric allylic alkylation step wherein the nitrogen of the pyrrole behaves as a good nucleophile for the regioselective ring opening of the aziridine, the pendant ester group serves as electrophile for lactam formation. Specifically, enantioselective palladium-catalyzed annulation

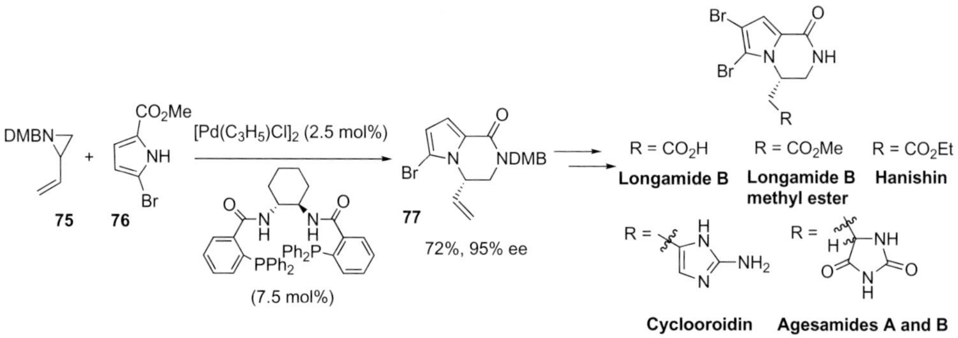

Scheme 1.27 Longamides and agesamides.

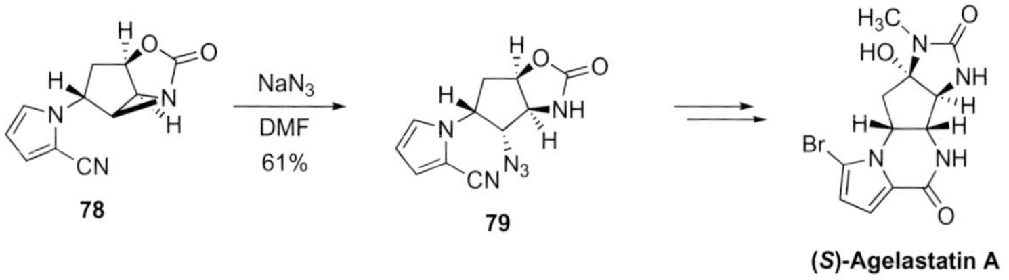

Scheme 1.28 (−)-Agelastatin A.

of racemic **75** and **76** provided pyrrolopiperazinone **77** in 72% yield and 95% enantiomeric excess. **77** was then found to be a suitable intermediate for the synthesis of pyrrole alkaloids longamide B (and thus longamide B methyl ester, hanishin and cyclooroidin) and agesamides A and B.

Finally, as illustrated in the synthesis of (−)-agelastatin A disclosed by Yoshimitsu, Ino and Tanaka, azides are also efficient nucleophiles to perform the ring opening of aziridines (Scheme 1.28) [54]. In the present example, unlike basic nitrogen nucleophiles such as ammonia and benzylamine that led to exclusive addition to the oxazolidinone core of **78**, sodium azide attacked the aziridine ring selectively to produce the azidated product **79** following cleavage of the weak outer bond of the tricyclic system.

1.3.1.3 Oxygen-Centered Nucleophiles

Nucleophilic ring opening of aziridines with oxygen nucleophiles provides direct access to amino alcohol units that are ubiquitous in nature. In general, aziridines, that show a lower reactivity towards this type of nucleophiles than structurally related epoxides, require activation of the nitrogen atom to undergo ring opening either by attachment of an electron-withdrawing group or by Brønsted or Lewis acids.

Ring opening using water as nucleophile leads directly to hydroxy amine derivatives and was promoted by strong organic Brønsted acids such as *p*-toluenesulfonic

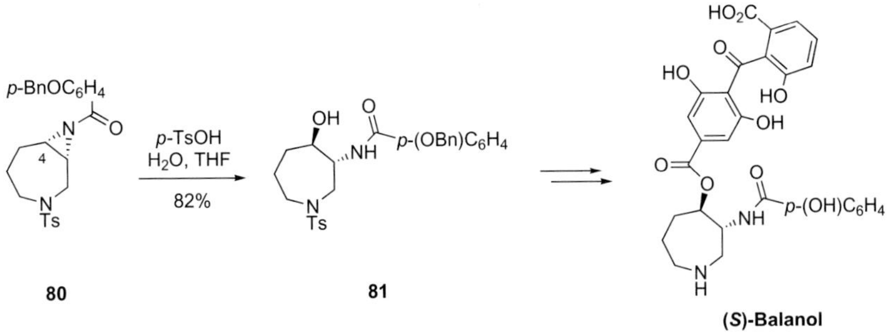

1.3 Synthesis of Natural Products Involving the Transformation of an Aziridine Moiety

Scheme 1.29 (−)-Balanol.

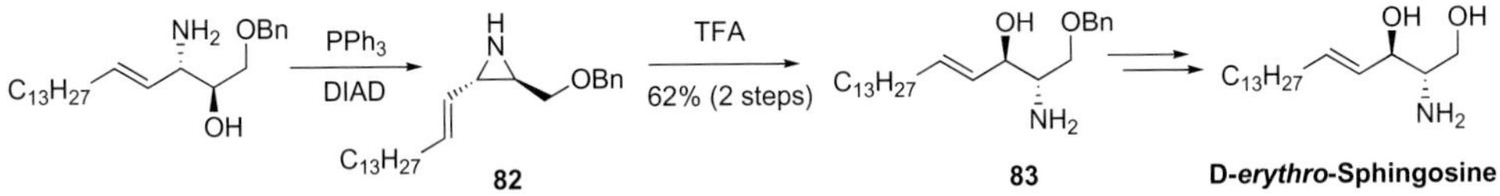

Scheme 1.30 D-erythro-Sphingosine.

acid (TsOH) and trifluoroacetic acid (TFA). Tanner and co-workers used this strategy in their synthesis of (−)-balanol (Scheme 1.29) [55–57], wherein treatment of enantiomerically pure N-acyl aziridine **80** with water in the presence of TsOH resulted in the near-exclusive ring opening at C-4 and led in 71% yield to amino alcohol **81** that was advanced to the natural product. The remarkable regioselectivity of the opening seemed to be related to both conformational and electronic effects of the fused bicyclic aziridine as it proved general for other nucleophiles and for ring opening of the analogous epoxide.

Olofsson and Somfai used a TFA-mediated ring opening of a vinyl aziridine by water to synthesize D-erythro-sphingosine in enantiomerically pure form (Scheme 1.30) [58]. Specifically, crude unprotected aziridine **82** prepared from the parent vinylic 1,2-amino alcohol by Mitsunobu ring closure afforded regio- and stereoselectively *anti* amino alcohol **83** in 62% yield.

In the same vein, Trost and Dong performed the regio- and stereoselective hydrolytic ring opening of enantiopure fused bicyclic tosyl aziridine **84** on route to the non-natural (+) enantiomer of agelastatin A (Scheme 1.31) [59, 60][1]. Of the several conditions examined (CAN or BF$_3$ Et$_2$O in aqueous acetonitrile, variable amounts of TFA in aqueous acetone or dioxane at different temperatures), only treatment with TFA in dioxane/water under microwave heating (150 °C) gave high (84%) reproducible results. Interestingly, as the synthetic plan required oxidation

---

1) Intermediate **84** is prepared from achiral starting materials using enatioselective catalytic reactions with chiral ligands available in both enantiomeric forms.

Synthesis of the natural (−) enantiomer of agelastatin should thus mirror the sequence used for the (+) enantiomer.

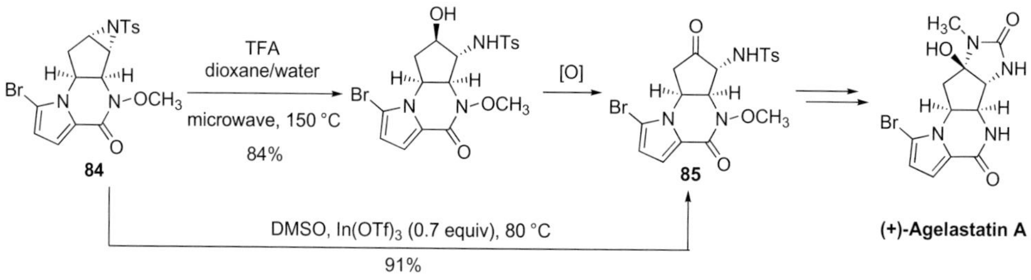

20 | *1 Aziridines in Natural Product Synthesis*

Scheme 1.31 (+)-Agelastatin A.

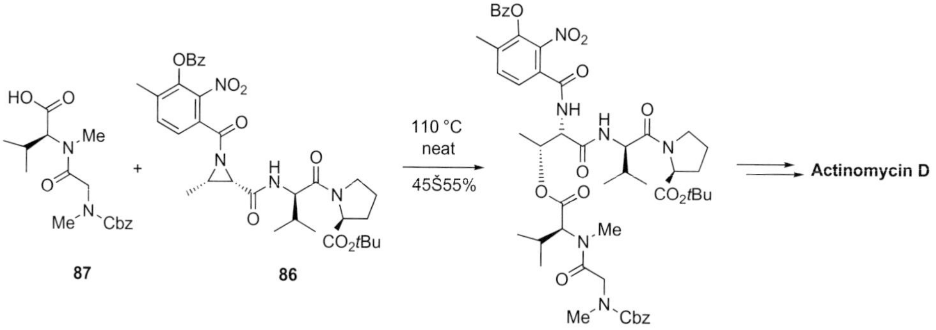

Scheme 1.32 Actinomycin D.

of the hydroxy amine to the corresponding α-amino ketone, direct oxidative ring opening was also developed. Reaction of **84** in DMSO in the presence of 0.7 equiv In(OTf)$_3$ led directly to **85** in 91% yield through a mechanism believed to involve attack of DMSO to the In(III) activated aziridine.

Nucleophilic ring opening of aziridines using carboxylate anions is rather common in the context of total synthesis. Very popular is the opening of aziridines containing 2-carboxylate or 2-carboxamide functionalities as it generally occurs regioselectively at C-3 and affords α-amino acid derivatives. Carboxylic acids themselves can operate this reaction in neutral conditions. For instance, Okawa and co-workers disclosed a total synthesis of cyclic peptide actinomycin D wherein ester bond formation between two dipeptide fragments was achieved in 45–55% yield through nucleophilic ring opening of 2-carboxamide-*N*-acyl aziridine **86** by the acid moiety of dipeptide **87** (Scheme 1.32) [61].

Alternatively, Cardillo and co-workers operated the ring opening of a similar aziridine in milder conditions using acetic anhydride in the presence of pyridine (Scheme 1.33) [62].

A nice illustration of the synthetic complementarity between opening by carboxylate and hydrolytic ring opening of *N*-benzyl aziridine-2-carboxylates can be found in the work of Ishikawa and co-workers (Scheme 1.34) [63]. Treatment of enantiopure *cis*-aziridine **88** with TsOH in the presence of water led in excellent

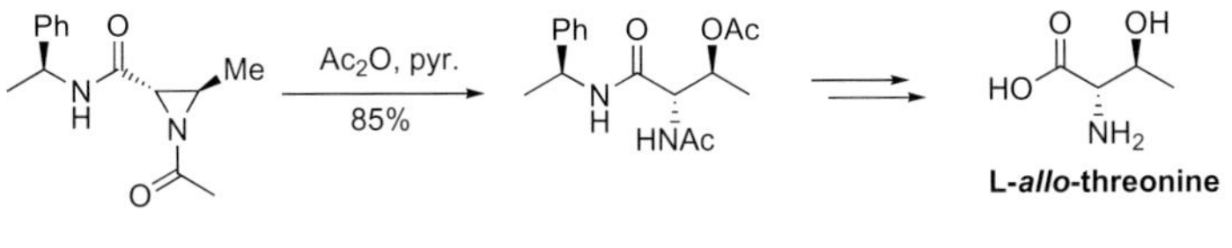

Scheme 1.33 L-*allo*-Threonine.

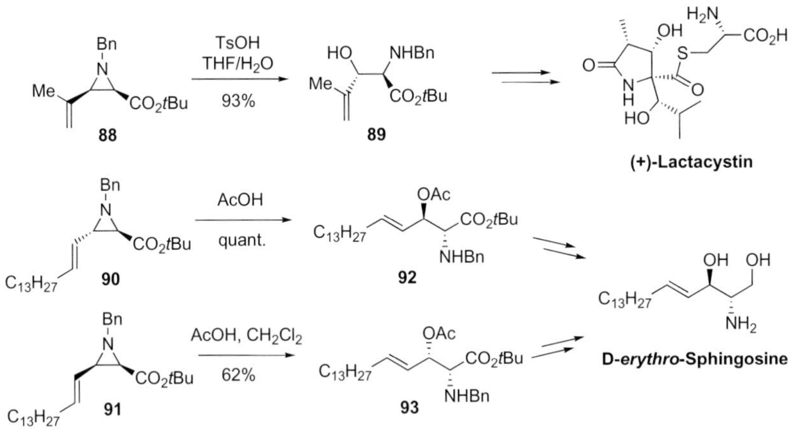

Scheme 1.34 (+)-Lactacystin and D-*erythro*-sphingosine.

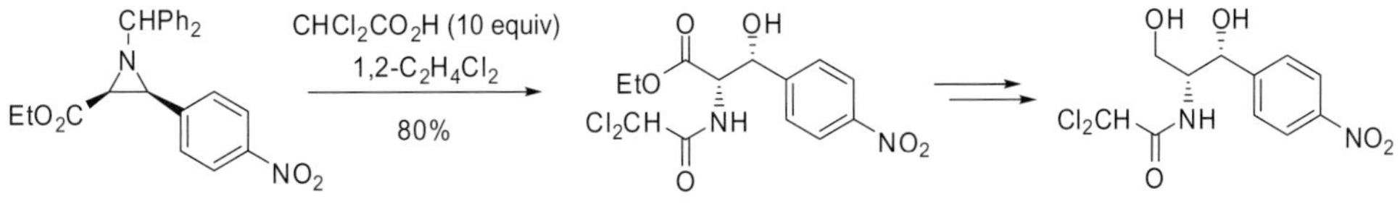

Scheme 1.35 (−)-Chloramphenicol.

regio- and stereoselectivity to hydroxy amine **89** that was then used to prepare a known intermediate of (+)-lactacystin. However, when a similar approach was projected to access D-*erythro*-sphingosine, ring opening under analogous conditions of *cis*- or *trans*-aziridines **90** and **91** occurred with lower regioselectivity, a result that also contrasts with the opening of **82** in Somfai's approach (see above Scheme 1.30). Conversely, reaction with acetic acid was completely regio- and stereoselective and afforded acetoxy amines **92** and **93** that were used to reach the desired target.

Interestingly, Loncaric and Wulff evidenced that in the case of *N*-benzhydryl aziridine-2-carboxylate **94**, treatment with excess carboxylic acid resulted not only in ring opening at the benzylic position, but also in the subsequent *in situ* deprotection and acetyl transfer (Scheme 1.35). This sequence was elegantly exploited to prepare (−)-chloramphenicol enantioselectively [64].

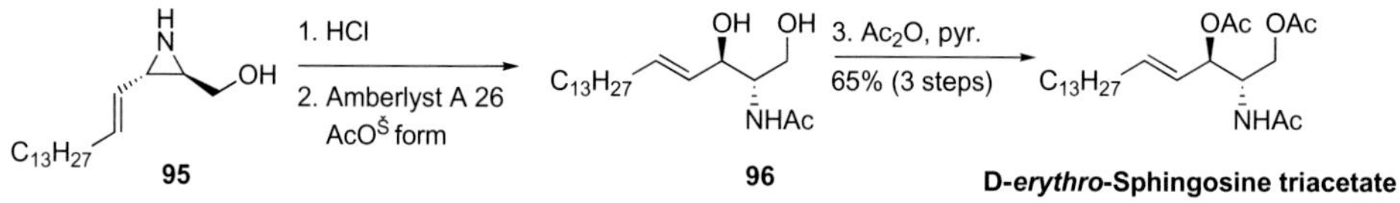

Scheme 1.36 D-*erythro*-Sphingosine triacetate.

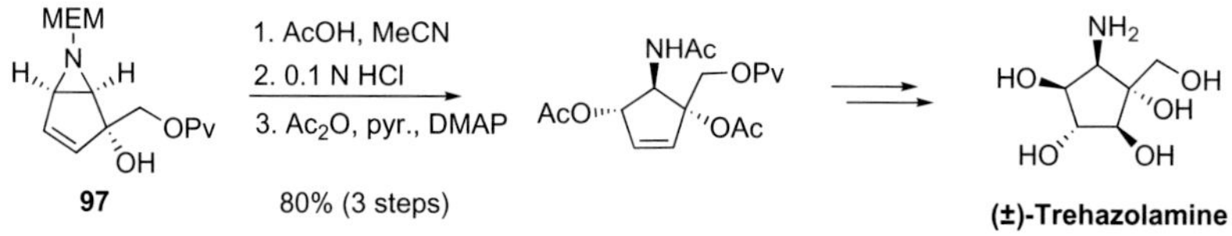

Scheme 1.37 (±)-Trehazolamine.

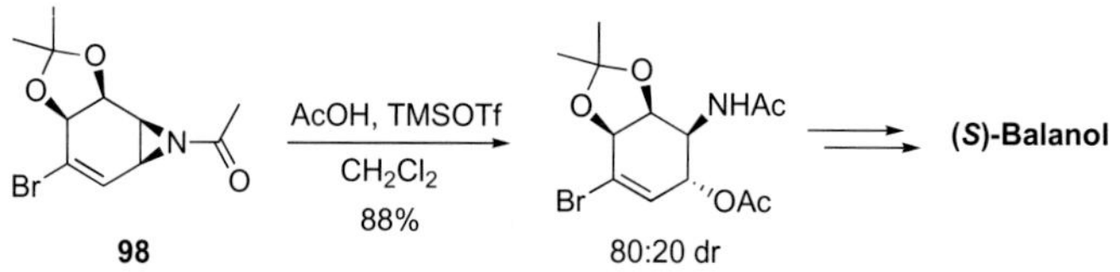

Scheme 1.38 (−)-Balanol.

A similar mechanism might be operating in the ring opening of unprotected aziridine **95** to afford *erythro*-sphingosine N-acetate **96** following treatment in refluxing benzene first with HCl and then with amberlyst A 26 in the acetate form (Scheme 1.36) [65].

Worthy of note, in addition to the above mentioned examples starting form aziridine-2-carboxylates, some examples of nucleophilic ring opening by acetates of fused bicyclic vinyl aziridines have been disclosed. In an approach to potent trehalase inhibitor (±)-trehazoline, Mariano and co-workers used acetic acid to achieve the ring opening of bicyclic N-MEM aziridine **97** obtained by photocyclization of 1-MEM-3-pivaloylmethylpyridinium perchlorate (Scheme 1.37) [66]. The use of a chiral inductor for the aziridination instead of the MEM group provided an enantiodivergent formal synthesis of the natural (+) and unnatural (−) enantiomers.

Similarly, in order to determine the optical rotation of a well established intermediate to (−)-balanol with absolute optical purity, Hudlicky and Sullivan carried out the ring opening of bicyclic N-acetyl vinyl aziridine **98** with acetic acid in the presence of trimethylsilyl triflate (TMSOTf) (Scheme 1.38) [67]. While nucleophilic attack of the acetate occurred unsurprisingly at the allylic position, only moderate levels of diastereoselectivity were achieved.

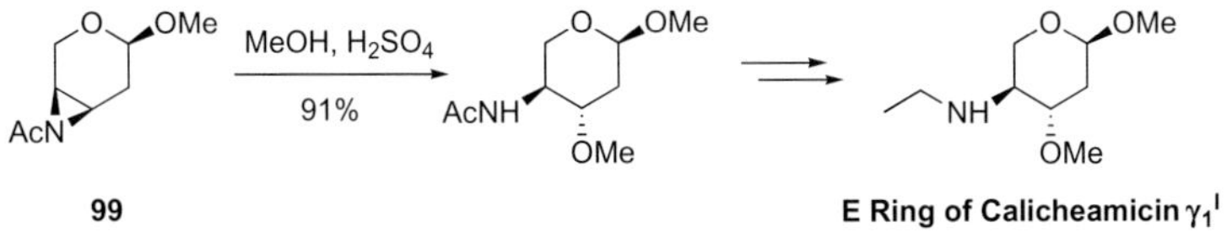

Scheme 1.39 Calicheamicin.

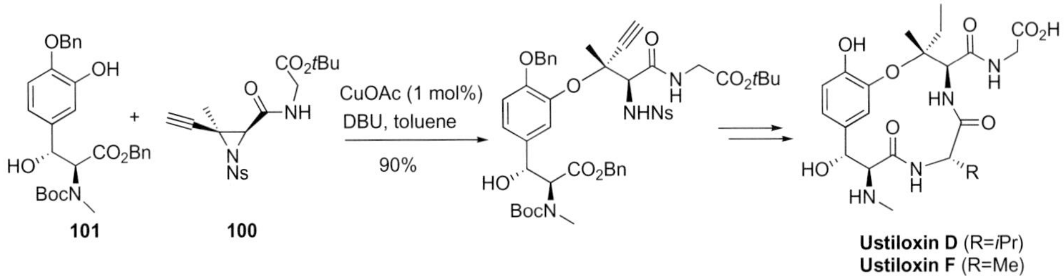

Scheme 1.40 Ustiloxins.

The use of nucleophilic ring opening of aziridines with alcohols for the synthesis of natural products has been rather limited so far. As previously, activation by Brønsted or Lewis acids is generally required. In an example of the first type, opening of N-acyl aziridine **99** by methanol promoted by sulfuric acid was found to be highly regio- and stereoselective, presumably as a result of conformational constraints of the bicyclic structure (Scheme 1.39) [68]. The resulting amino ether was used for the synthesis of E ring monosaccharide unit of calicheamicin γ_1^I. In an example of the second type, Lewis-acidic activation with $BF_3 \cdot Et_2O$ was used to promote the addition of allyl alcohol to a 2,3-aziridino-γ-lactone in the synthesis of the non-natural enantiomer of polyoxamic acid [69].

From a different angle, Jouillé and co-workers have developed a highly convergent route to ustiloxins through the copper-catalyzed ethynyl aziridine ring opening by phenol derivatives (Scheme 1.40) [70, 71]. In a representative application to the synthesis of cyclopeptide ustiloxins D and F, 2-carboxamide-aziridine **100** was coupled regio- and stereoselectively with β-hydroxy tyrosine derivative **101** in 90% yield.

Remarkably, clean regio- and stereoselective nucleophilic ring opening of enantiopure 2-carboxylate-N-nosyl aziridine **102** by methanol, useful to synthesize β-methoxytyrosine, could be carried out with no activation, probably due to the fact that the electron-rich phenol moiety weakens the C–N benzylic bond (Scheme 1.41) [72]. As a matter of fact, in the presence of TFA or copper salts, the diastereoselectivity of the reaction was moderate, as ring opening seemed to be occurring both via S_N1 and S_N2 mechanisms.

In a last example involving this time nucleophilic ring opening in basic conditions, aziridino alcohol **103** led to the regio- and stereoselective formation of the known intermediate of bestatin **104**, upon treatment with formaldehyde in the

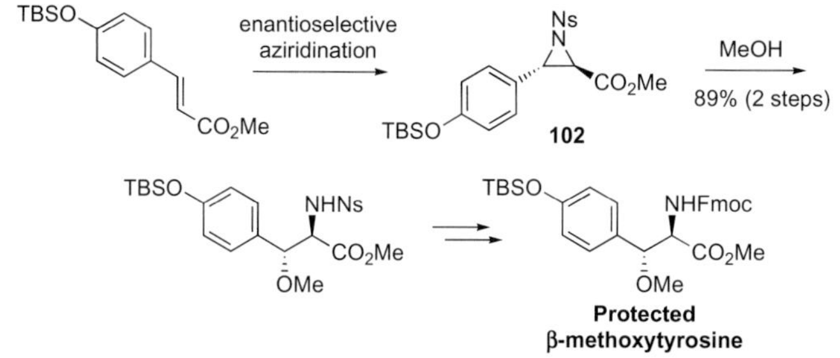

Scheme 1.41 β-Methoxytyrosine.

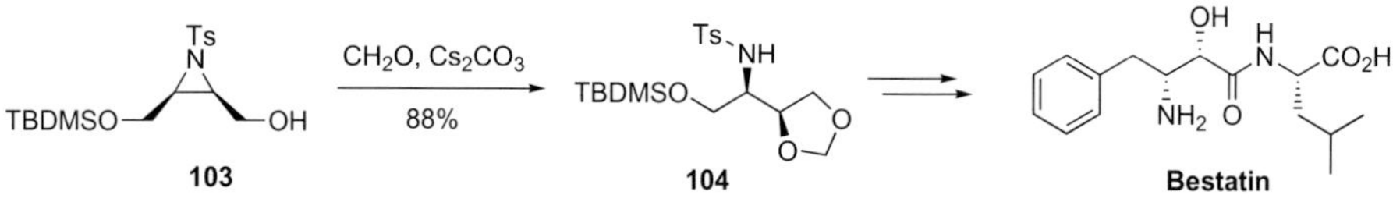

Scheme 1.42 Bestatin.

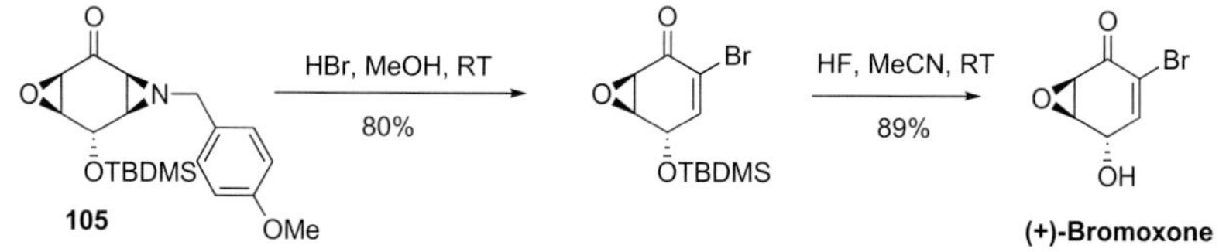

Scheme 1.43 (+)-Bromoxone.

presence of cesium carbonate via internal attack of the hydroxide anion of the hemiacetal formed *in situ* (Scheme 1.42) [73].

1.3.1.4 Halogen Nucleophiles

Ring opening of aziridines by halogen nucleophiles has been used only very recently for the purpose of natural product synthesis. Hydrogen bromide has been extensively used in the ring opening of aziridines to generate vicinal bromo amines. In the context of the synthesis of (+)-bromoxone, Maycock and co-workers described the use of 0.1 M HBr in MeOH to open selectively without any activation on the aziridine nitrogen N-PMB-aziridine **105** in the presence of an epoxide. The strategy was successfully applied to install the vinyl bromide moiety of the naturally occurring (+)-bromoxone (Scheme 1.43) [74].

Ring opening of aziridine **106** with tetraethylammonium chloride and TFA afforded with inversion and a total regio- and stereoselectivity the desired tetrahydroquinoline core of the virantmycin (Scheme 1.44) [75].

In the total synthesis of chlorodysinosin A, Hanessian and co-workers used cerium (III) chloride, an inexpensive and non-toxic inorganic salt to promote selec-

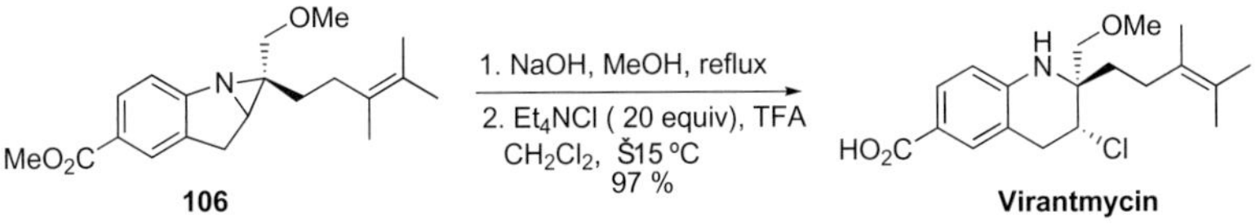

1.3 Synthesis of Natural Products Involving the Transformation of an Aziridine Moiety

Scheme 1.44 Virantmycin.

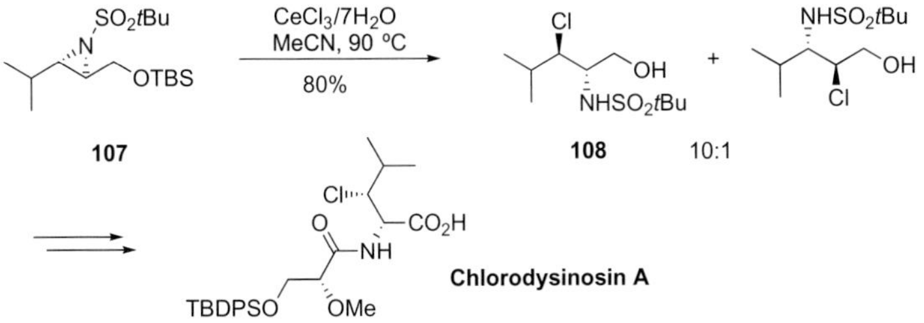

Scheme 1.45 Chlorodysinosin A.

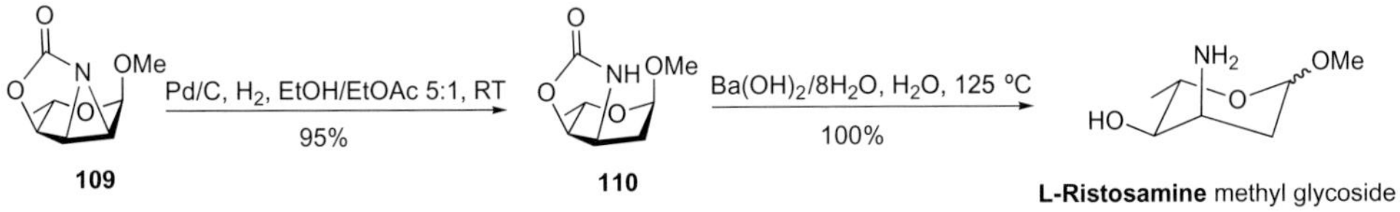

Scheme 1.46 L-Ristosamine.

tive ring opening of *N*-Bus aziridine **107** to form the corresponding 3-chloro sulfonamide **108** with an excellent regioselectivity (Scheme 1.45) [76].

1.3.1.5 Reductions

Reductive ring opening of aziridines can be highly regioselective by using catalytic hydrogenation. A very selective C–N bond cleavage was achieved on aziridine **109** using palladium on charcoal to produce carbamate protected L-ristosamine derivative **110**, a useful synthon which after carbamate deprotection afforded L-ristosamine methyl glycoside, a 2,3,6-trideoxy-3-amino-hexopyranose member of 2,6-dideoxy sugars. A similar series of transformation successfully afforded L-daunosamine methyl glycoside (Scheme 1.46) [77].

1.3.2 Cycloaddition Reactions and Rearrangements

Aziridines have been used as partners in various cycloaddition reactions in order to obtain cyclic adducts. Intramolecular [3 + 2] cycloadditions of aziridines with olefinic moieties have been reported, as well as the [2,3]-Wittig rearrangement of vinylaziridines.

1.3.2.1 Aziridines in [3 + 2] Cycloadditions

Aziridinyl esters are known to be precursors of azomethine ylides [78] and to react with olefinic moieties in a [3 + 2] cycloaddition reaction. This key step was used by Takano and co-workers in the total synthesis of acromelic acid A (Scheme 1.47) [79]. Aziridine **111** reacted under thermal conditions to form the corresponding pyrrolidinyl cycloadduct stereoselectively, an advance intermediate to acromelic acid A.

The same key step has also been used [80] by the same authors in a total synthesis of (−)-kainic acid, as depicted in Scheme 1.48. This key step has also been used in a synthesis of the unnatural N-demethyl mesembrine [81].

Related to these [3 + 2] cycloaddition reactions, the thermal rearrangement of vinylaziridines such as **112** bearing an electron-withdrawing group (prepared through triazene pyrolysis) into pyrrolines was used by Hudlicky and co-workers in a formal synthesis of various pyrrolizidinyl heterocycles such as (±)-supinidine, (±)-isoretronecanol and (±)-trachelantamidine (Scheme 1.49) [82].

The same key step was applied by the same authors to the preparation of bicyclic compound **113**, which is an advanced intermediate in the synthesis of (±)-retronecine and (±)-heliotridine, as well as (±)-platynecine, (±)-hastanecine, (±)-turnefor- cidine and (±)-dihydroxyheliotridane (Scheme 1.50) [83].

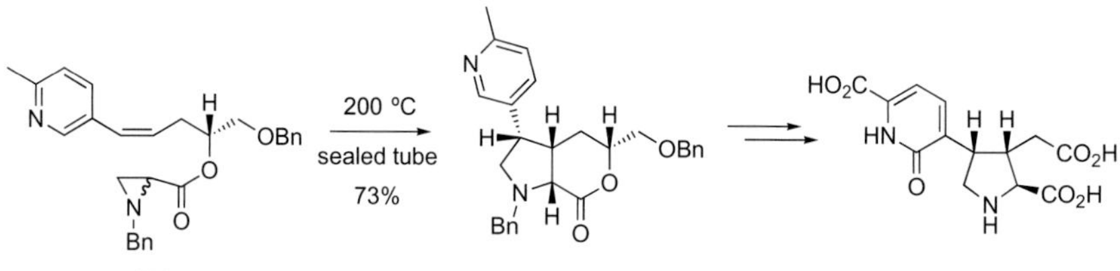

Scheme 1.47 Acromelic acid A.

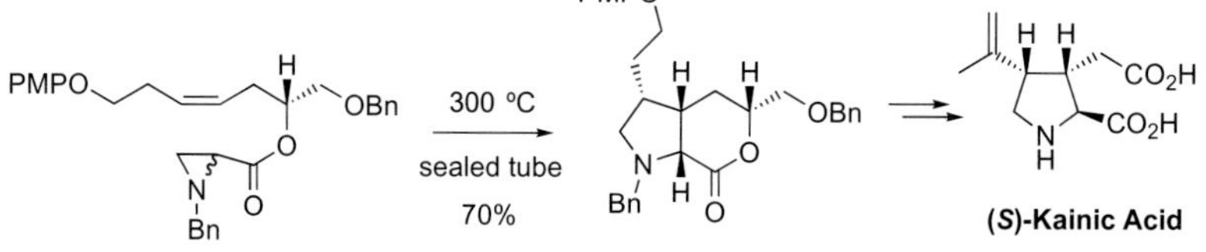

Scheme 1.48 (−)-Kainic acid.

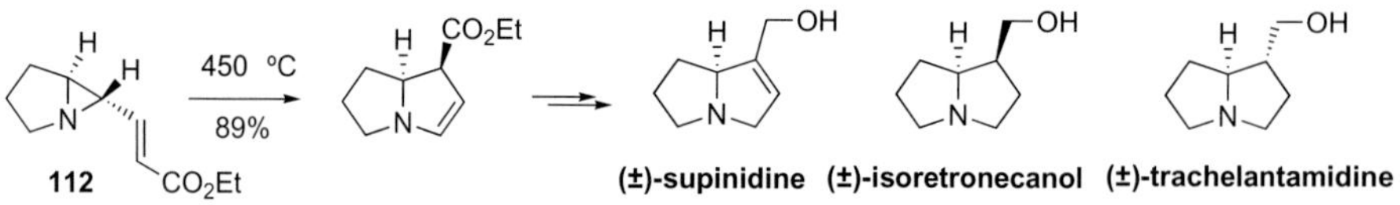

Scheme 1.49 (±)-Supinidine, (±)-isoretronecanol and (±)-trachelantamidine.

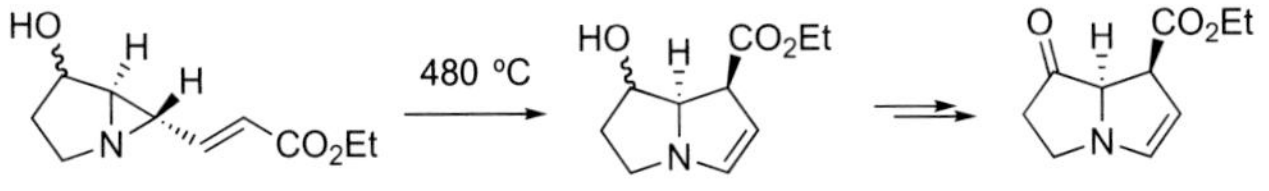

Scheme 1.50 Retronecine.

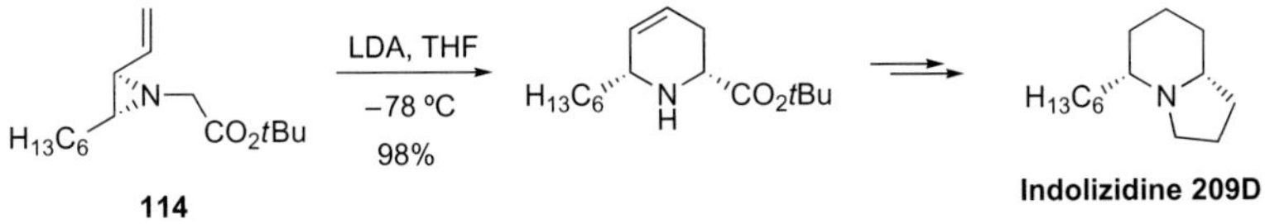

Scheme 1.51 Indolizidine 209D.

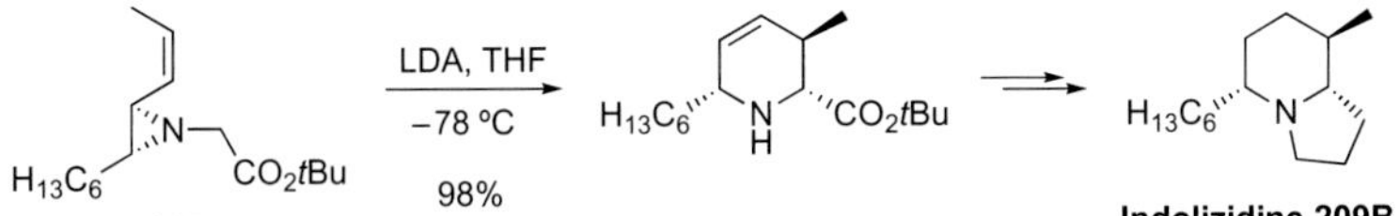

Scheme 1.52 Indolizidine 209B.

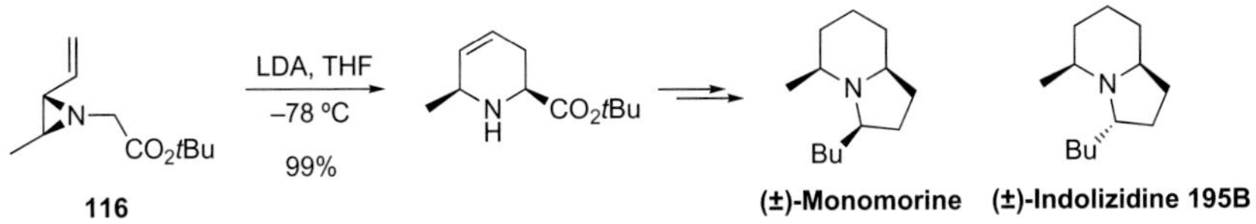

Scheme 1.53 (±)-Monomorine and (±)-indolizidine 195B.

1.3.2.2 Aziridines in [2,3]-Wittig Rearrangements

The utility of the [2,3]-Wittig rearrangement of vinyl aziridines [84] in natural product synthesis was largely demonstrated by Somfai and co-workers. For example, an enantioselective total synthesis of indolizidine 209D was achieved using the [2,3]-Wittig rearrangement of the enolate derived from aziridine **114** (Scheme 1.51) as a key step [85, 86].

The same key step was applied to the total synthesis of the indolizidine 209B. In this case, the [2,3]-Wittig rearrangement of the enolate derived from the substituted vinylaziridine **115** was used (Scheme 1.52) [86].

The total synthesis of racemic monomorine and indolizidine 195B have also been achieved using the [2,3]-Wittig rearrangement of vinyl aziridine **116** by the same authors (Scheme 1.53) [87].

1.3.2.3 Aziridines in Iodide-Mediated Rearrangements

Another way to realize the rearrangement of vinyl aziridines into pyrrolines is the S_N2' ring opening of vinyl aziridines with iodide ion followed by ring closure. This

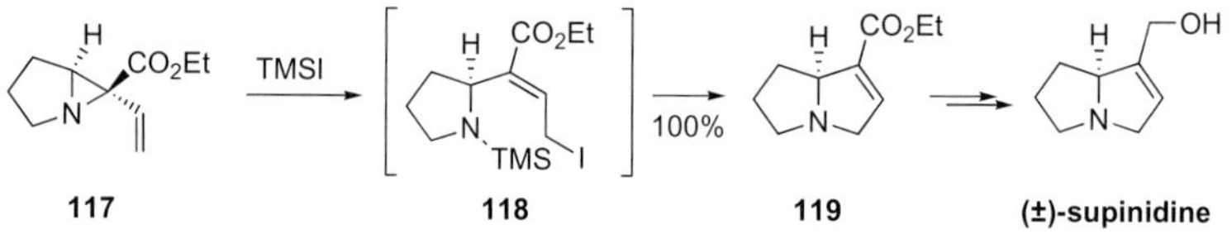

Scheme 1.54 (±)-Supinidine.

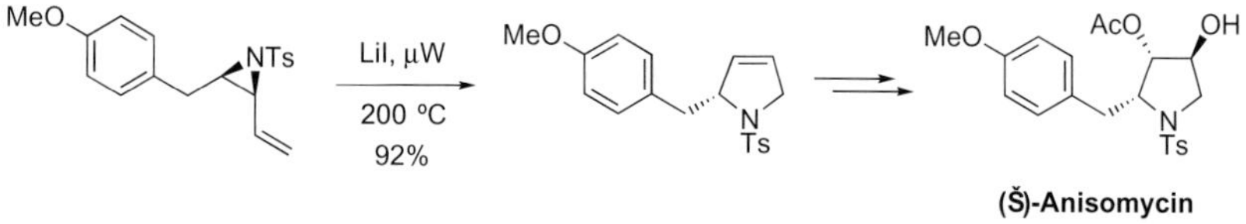

Scheme 1.55 (−)-Anisomycin.

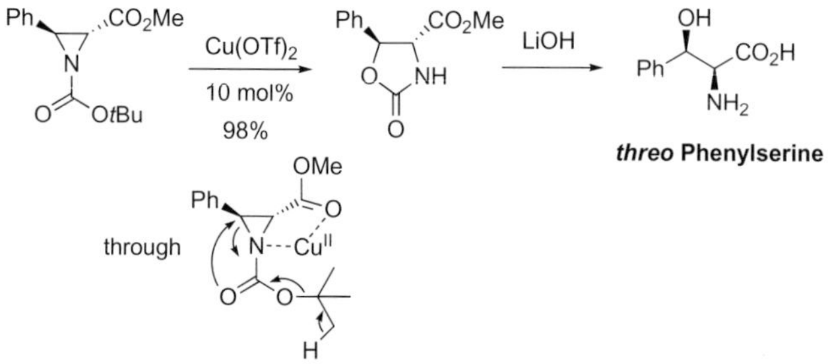

Scheme 1.56 *threo*-Phenylserine.

methodology was early recognized by Hudlicky and co-workers as an efficient way to achieve the synthesis of pyrrolizidinyl alkaloids (Scheme 1.54) [82, 88]. The S_N2' ring opening of aziridine **117** afforded the iodo compound **118** which gave the bicyclic compound **119** upon ring closure.

A similar reaction was later applied by Somfai in a formal synthesis of (−)-anisomycin (Scheme 1.55) [89].

1.3.2.4 Aziridines in Miscellaneous Rearrangements

The Lewis acid-mediated rearrangement of *N*-acyl aziridines into oxazolidin-2-ones was used in a straightforward synthesis of *threo* phenylserine (Scheme 1.56) [90]. Interestingly this rearrangement was shown to occur with retention of the configuration. By contrast, the same rearrangement, when applied to *N*-acyl aziridines, leads to the oxazoline formation (and not the oxazolidinone). This was applied to the total synthesis of (non-natural) Bn-protected (*L*)-*threo* sphingosine [58].

The stereoselective Lewis acid-catalyzed rearrangement of aziridinyl alcohols **120** and **121** into β-amino aldehydes was applied in a total synthesis of (±)-crinane and (±)-mesembrine (Scheme 1.57) [91].

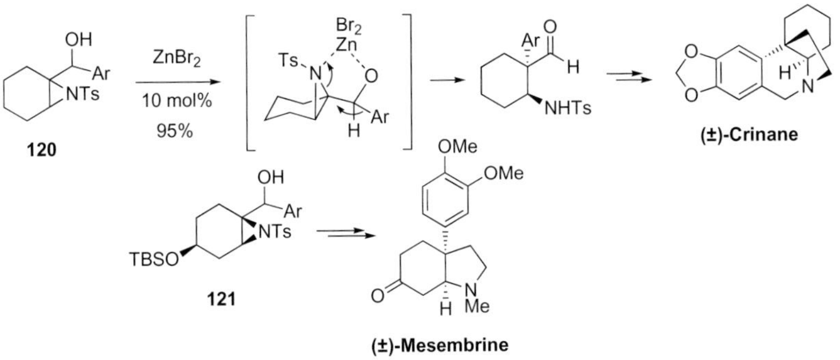

1.3 Synthesis of Natural Products Involving the Transformation of an Aziridine Moiety | 29

Scheme 1.57 (±)-Crinane and (±)-mesembrine.

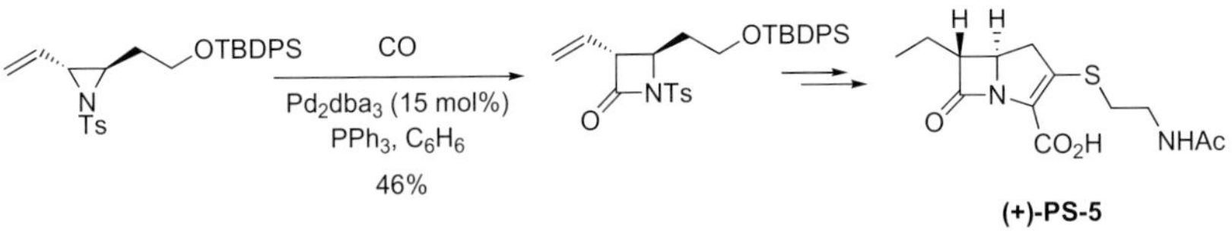

Scheme 1.58 (+)-PS-5.

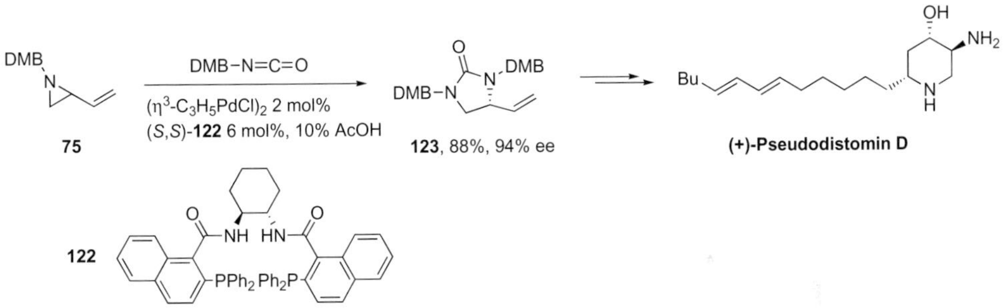

Scheme 1.59 (+)-Pseudodistomin D.

Vinylaziridines have been reported to undergo carbon monoxide insertion under palladium (0) catalysis. This method allowed the formation of β-lactams and was applied by Tanner and Somfai for the total synthesis of the carbapenem (+)-PS-5 (Scheme 1.58) [92].

The insertion of isocyanates can also been achieved under palladium (0) catalysis, leading to imidazolidin-2-ones. This strategy was used by Trost and co-workers in the total synthesis of (+)-pseudodistomin D (Scheme 1.59) [93]. The insertion occurred enantioselectively in the presence of chiral ligand **122** to afford enantioenriched imidazolidin-2-one **123**.

(−)-Pseudoconhydrin was prepared by Harrity and co-workers through a [3 + 3] cycloaddition strategy involving the aziridine **124** and a palladium-trimethylenemethane complex (Scheme 1.60) [94].

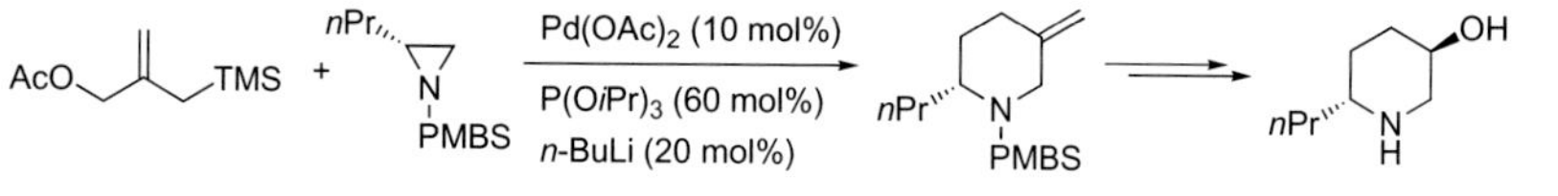

124

(S)-Pseudoconhydrin

Scheme 1.60 (−)-Pseudoconhydrin.

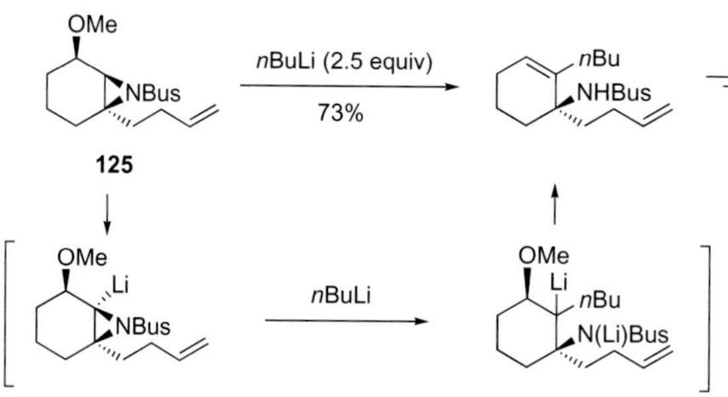

125

(±)-Perhydrohistrionicotoxin

Scheme 1.61 (±)-Perhydrohistrionicotoxin.

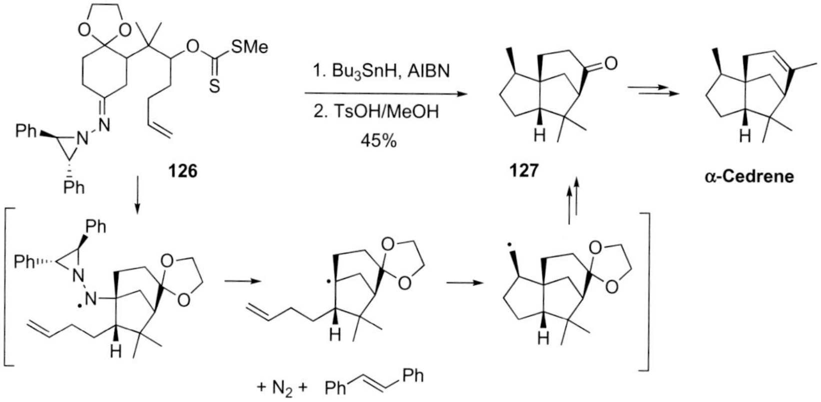

126

127

α-Cedrene

Scheme 1.62 α-Cedrene.

A formal synthesis of (±)-perhydrohistrionicotoxin has been reported through an interesting carbenoid rearrangement of a lithiated aziridine prepared by deprotonation of **125**, insertion of the resulting carbenoid into BuLi and subsequent β-elimination (Scheme 1.61) [95].

An elegant synthesis of α-cedrene was reported using the ability of aziridinyl hydrazones to behave as alkyl radical acceptors/precursors. Hydrazone **126** underwent a radical cascade upon reduction of the xanthate moiety to afford after hydrolysis tricyclic compound **127** (Scheme 1.62) [96].

1.3.3
Synthesis of Natural Products Involving the Transformation of an Aziridinium Moiety

The chemistry of aziridiniums has been reviewed recently [97]. In this section only the examples where a true aziridinium ion is involved will be reviewed, since the activation of aziridines under Brønsted conditions has been developed in Section 1.3.1. Aziridiniums are prepared mainly through intramolecular substitution of a leaving group by nitrogen atom engaged in an aziridinyl moiety. Their reactivity is mainly related to ring enlargement reactions.

Biomimetic syntheses of (−)-vincadifformine and (−)-tabersonine have been reported by Kuehne and co-workers through the ring opening reactions of the aziridinium derived from the pentacyclic chloro compound **128** (Scheme 1.63) [98].

The *in situ* mesylation of amino alcohol **129** followed by the reaction with dibutylcuprate was reported by Tanner and Somfai to occur with retention of the configuration. This was explained by a mechanism involving the intramolecular displacement of mesylate by nitrogen. Ring opening of the resulting aziridinium ion **130** with dibutylcuprate afforded an advanced intermediate in the total synthesis of depentylperhydrohistrionicotoxin derivatives (Scheme 1.64) [99].

Aziridinium ion **132** formed through iodoamination of **131** followed by nucleophilic anchimeric assistance by the vicinal tertiary amine was intramolecularly opened by a proximal ester moiety to afford (+)-croomine in an elegant single step (Scheme 1.65) [100].

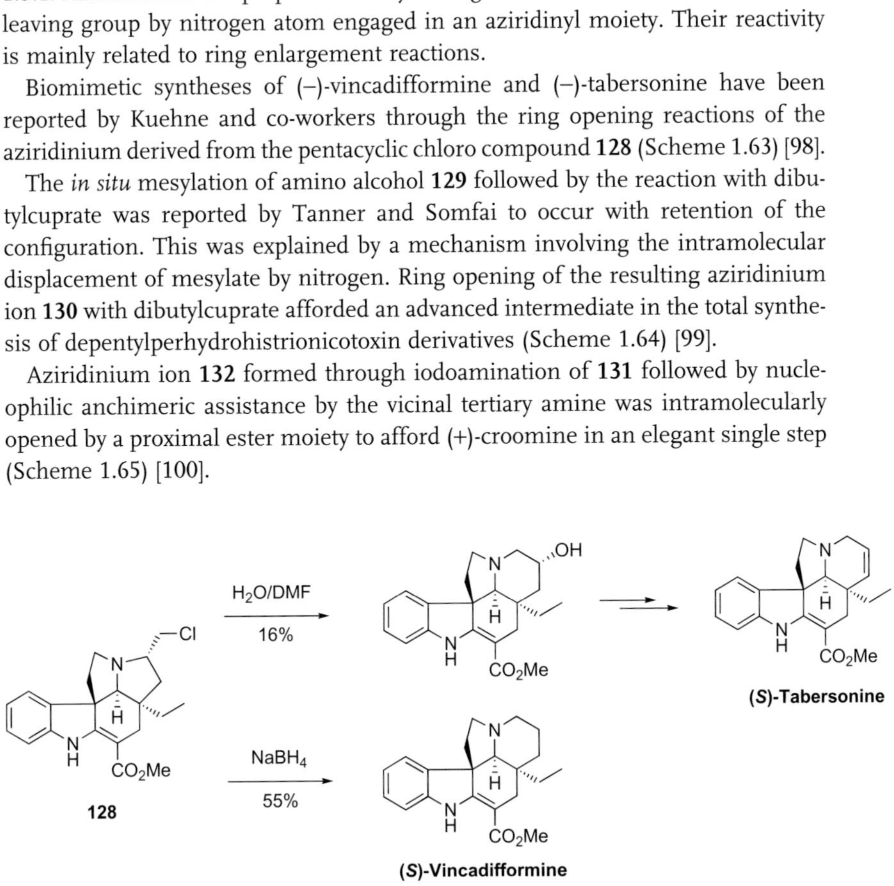

Scheme 1.63 (−)-Vincadifformine and (−)-tabersonine.

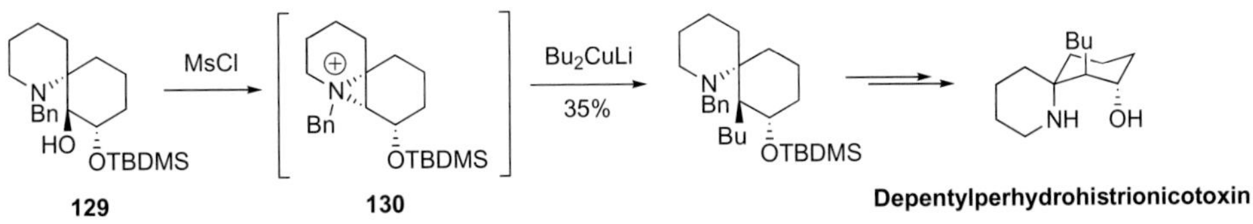

Scheme 1.64 Depentylperhydrohistrionicotoxin.

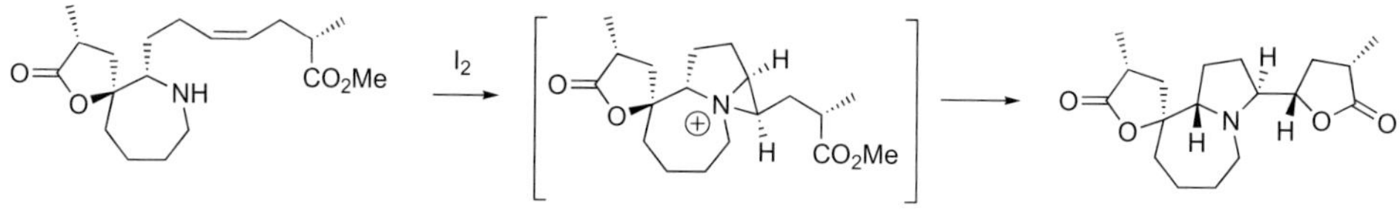

32 | *1 Aziridines in Natural Product Synthesis*

Scheme 1.65 (+)-Croomine.

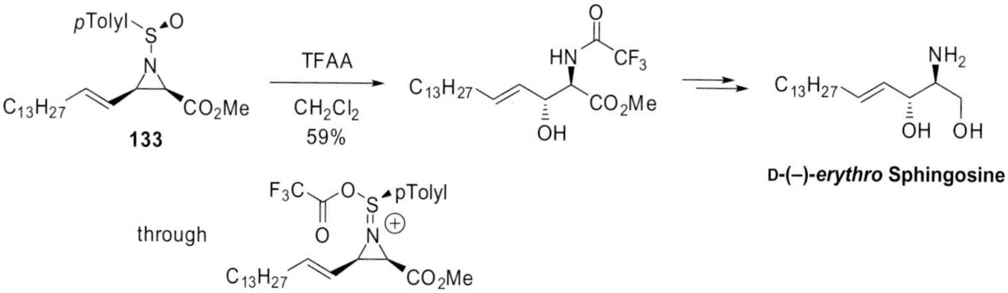

Scheme 1.66 D-(−)-erythro-Sphingosine.

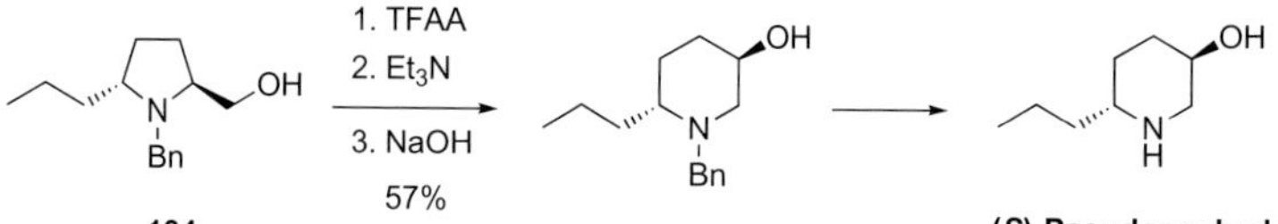

Scheme 1.67 (S)-Pseudoconhydrin.

The formation of an aminoalcohol unit through the Pummerer-like rearrangement of *N*-sulfinylaziridine **133** was used in a total synthesis of D-(−)-*erythro*-sphingosine by Davis and co-workers (Scheme 1.66) [101].

The stereoselective rearrangement of pyrrolidine-2-methanols into 3-hydroxypiperidines has been extensively studied by Cossy and co-workers [97], and has found several efficient applications in total synthesis. The substituted prolinol **134** has served in the total synthesis of (−)-pseudoconhydrin (Scheme 1.67) [102].

The synthesis of the piperidinyl core of (−)-velbanamine [103], as well as the synthesis or formal synthesis of various drugs such as (−)-zamifenacin [104], (−)-paroxetine [105] and reboxetine [106] have also been reported using the same type of methodology.

1.4
Conclusion

The aziridines have been named "the epoxides' ugly cousins" [2], and in fact the reactions of aziridines have received little interest compared with those of epoxides. However, the use of aziridines in the total synthesis of natural products has attracted increasing interest over the past 15 years. Besides the considerable amount of work devoted to aziridine-containing natural products (mitomycins and azinomycins) as well as to related synthetic targets, the aziridine ring has been used as an efficient precursor for the synthesis of 1,2-amino alcohols and other polyfunctionalized structures. Undoubtedly this synthetic potential is directly related to the ability to prepare stereo- and enantioselectively the aziridinyl core and will increase in the future.

References

1 (a) Tanner, D. (1994) Chiral aziridines–their synthesis and use in stereoselective transformations. *Angew. Chem. Int. Ed. Engl.*, **33** (6), 599–619; (b) Atkinson, R.S. (1999) 3-Acetoxyaminoquinazolinones (QNHOAc) as aziridinating agents: ring-opening of N-(Q)-substituted aziridines. *Tetrahedron*, **55** (6), 1519–1559; (c) Chemla, F. and Ferreira, F. (2002) Alkynyl-oxiranes and aziridines: synthesis and ring opening reactions with carbon nucleophiles. *Curr. Org. Chem.*, **6** (6), 539–570; (d) Ciufolini, M.A. (2005) Synthetic studies on heterocyclic natural products. *Farmaco*, **60** (8), 627–641; (e) Padwa, A. and Murphree, S.S. (2006) Epoxides and aziridines–a mini review. *Arkivoc*, (iii), 6–33; (f) Pellissier, H. (2010) Recent developments in asymmetric aziridination. *Tetrahedron*, **66** (8), 1509–1555.

2 (a) Tanner, D. (1993) Stereocontrolled synthesis via chiral aziridines. *Pure Appl. Chem.*, **65** (6), 1319–1328; (b) Stamm, H. (1999) Nucleophilic ring opening of aziridines. *J. Prakt. Chem.*, **341** (4), 319–331; (c) McCoull, W. and Davis, F.A. (2000) Recent synthetic applications of chiral aziridines. *Synthesis*, (10), 1347–1365; (d) Sweeney, J.B. (2002) Aziridines: epoxides' ugly cousins? *Chem. Soc. Rev.*, **31** (5), 247–258; (e) Hu, X.E. (2004) Nucleophilic ring opening of aziridines. *Tetrahedron*, **60** (12), 2701–2743; (f) Pineschi, M. (2006) Asymmetric ring-opening of epoxides and aziridines with carbon nucleophiles.

Eur. J. Org. Chem., (22), 4979–4988; (g) Lu, P. (2010) Recent developments in regioselective ring opening of aziridines. *Tetrahedron*, **66** (14), 2549–2560.

3 (a) Dahanukar, V.H. and Zavialov, I.A. (2002) Aziridines and aziridinium ions in the practical synthesis of pharmaceutical intermediates–a perspective. *Curr. Opin. Drug Discov. Devel.*, **5** (6), 918–927; (b) Ismail, F.M.D., Levisky, D.O. and Dembitsky, V.M. (2009) Aziridine alkaloids as potential therapeutic agents. *Eur. J. Med. Chem.*, **44** (9), 3373–3387.

4 Kametani, T. and Honda, T. (1986) Application of aziridines to the synthesis of natural products. *Adv. Heterocycl. Chem.*, **39**, 181–236.

5 (a) Naganawa, H., Usui, N., Takita, T., Hamada, M. and Umezawa, H. (1975) (S)-2, 3-Dicarboxy-aziridine, a new metabolite from a *steptomyces*. *J. Antibiot.*, **28** (10), 828–829; (b) Legters, J., Thijs, L. and Zwanenburg, B. (1991) Synthesis of naturally occurring (2S,3S)-(+)-aziridine-2,3-dicarboxylic acid. *Tetrahedron*, **47** (28), 5287–5294.

6 Breuning, A., Vicik, R. and Schirmeister, T. (2003) An improved synthesis of aziridine-2,3-dicarboxylates via azido alcohols–epimerization studies. *Tetrahedron Asymmetry*, **14** (21), 3301–3312.

7 (a) Molinski, T.F. and Ireland, C.M. (1988) Dysidazirine, a cytotoxic azacyclopropene from the marine sponge *Dysidea fragilis*. *J. Org. Chem.*, **53** (9), 2103–2105; (b) Davis, F.A., Reddy, G.V. and Liu, H. (1995)

Asymmetric synthesis of 2H-azirines: first enantioselective synthesis of the cytotoxic antibiotic (R)-(−)-dysidazirine. *J. Am. Chem. Soc.*, **117** (12), 3651–3652.

8 Skepper, C.K., Dalisay, D.S. and Molinski, T.F. (2008) Synthesis and antifungal activity of (−)-(Z)-dysidazirine. *Org. Lett.*, **10** (22), 5269–5271.

9 (a) Iwami, M., Kiyoto, S., Terano, H., Kohsaka, M., Aoki, H. and Imanaka, H. (1987) A new antitumor antibiotic, FR-900482. *J. Antibiot.*, **40** (5), 589–593; (b) Terano, H., Takase, S., Hosoda, J. and Kohsaka, M. (1989) A new antitumor antibiotic, FR-66979. *J. Antibiot.*, **42** (1), 145–148.

10 Paleo, M.R., Aurrecoechea, N., Jung, K.-Y. and Rapoport, H. (2003) Formal enantiospecific synthesis of (+)-FR900482. *J. Org. Chem.*, **68** (1), 130–138.

11 Fukuyama, T., Xu, L. and Goto, S. (1992) Total synthesis of (±)-FR-900482. *J. Am. Chem. Soc.*, **114** (1), 383–385.

12 Suzuki, M., Kambe, M., Tokuyama, H. and Fukuyama, T. (2002) Facile construction of N-hydroxybenzazocine: enantioselective total synthesis of (+)-FR900482. *Angew. Chem. Int. Ed.*, **41** (24), 4686–4688.

13 Schkeryantz, J.M. and Danishefsky, S.J. (1995) Total synthesis of (±)-FR-900482. *J. Am. Chem. Soc.*, **117** (16), 4722–4723.

14 Katoh, T., Nagata, Y., Yoshino, T., Nakatani, S. and Terashima, S. (1997) Total synthesis of an enantiomeric pair of FR900482. 3. Completion of the synthesis by assembling the two segments. *Tetrahedron*, **53** (30), 10253–10270.

15 Judd, T.C. and Williams, R.M. (2002) Concise enantioselective synthesis of (+)-FR66979 and (+)-FR900482: dimethyldioxirane-mediated construction of the hydroxylamine hemiketal. *Angew. Chem. Int. Ed.*, **41** (24), 4683–4685.

16 Ducray, R. and Ciufolini, M.A. (2002) Total synthesis of (±)-FR66979. *Angew. Chem. Int. Ed.*, **41** (24), 4688–4691.

17 (a) Wakaki, S., Marumo, H., Tomioka, K., Shimizu, G., Kato, E., Kamada, H., Kudo, S. and Fujimoto, T. (1958) *Antibiot. Chemother.*, **8**, 228–232; (b) Hata, T., Sano, R., Sugawara, Y., Matsumae, A., Kanamori, K., Shima, T. and Hoshi, T. (1956) Mitomycin, a new antibiotics from *Streptomyces*. *J. Antibiot.*, **9** (4), 141–146; (c) Urakawa, C., Tsuchiya, H. and Nakano, K.-I. (1981) New mitomycin, 10-decarbamoyloxy-9-dehydromitomycin B from *Streptomyces caespitosus*. *J. Antibiot.*, **34** (2), 243–244; (d) Osborn, H.M.I. and Sweeney, J. (1997) The asymmetric synthesis of aziridines. *Tetrahedron Asymmetry*, **8** (11), 1693–1715.

18 Kishi, Y. (1979) The total synthesis of mitomycins. *J. Nat. Prod.*, **42** (6), 549–568.

19 Fukuyama, T. and Yang, L. (1987) Total synthesis of (±)-mitomycins via isomitomycin A. *J. Am. Chem. Soc.*, **109** (25), 7881–7882.

20 Fukuyama, T. and Yang, L. (1989) Practical total synthesis of (±)-mitomycin C. *J. Am. Chem. Soc.*, **111** (21), 8303–8304.

21 (a) Benbow, J.W., Schulte, G.K. and Danishefsky, S.J. (1992) The total synthesis of (±)-mitomycin K. *Angew. Chem. Int. Ed. Engl.*, **31** (7), 915–917; (b) Benbow, J.W., McClure, K.F. and Danishesky, S.J. (1993) Intramolecular cycloaddition reactions of dienyl nitroso compounds: application to the synthesis of mitomycin K. *J. Am. Chem. Soc.*, **115** (26), 12305–12314.

22 Wang, Z. and Jimenez, L.S. (1996) A total synthesis of (±)-mitomycin K. Oxidation of the mitosene C9-9a double bond by (hexamethylphosphoramido) oxodiperoxomolybdenum (VI) (MoO$_5$ HMPA). *Tetrahedron Lett.*, **37** (34), 6049–6052.

23 Bradner, W.T. (2001) Mitomycin C: a clinical update. *Antibiot. Chemother.*, **27** (1), 35–50.

24 Andrez, J.C. (2009) Mitomycins syntheses: a recent update. *Beilstein J. Org. Chem.*, **5**, N°33.

25 Fellows, I.M., Kaelin, D.E., Jr. and Martin, S.F. (2000) Application of ring-closing metathesis to the formal total synthesis of (+)-FR900482. *J. Am. Chem. Soc.*, **122** (44), 10781–10787.

26 (a) Fukuyama, T., Nakatsubo, F., Cocuzza, A. and Kishi, Y. (1977)

Synthetic studies toward mitomycins. III. Total syntheses of mitomycins A and C. *Tetrahedron Lett.*, **18** (49), 4295–4298; (b) Nakatsubo, F., Fukuyama, T., Cocuzza, A.J. and Kishi, Y. (1977) Synthetic studies toward mitomycins. 2. Total synthesis of dl-porfiromycin. *J. Am. Chem. Soc.*, **99** (24), 8115–8116.

27 (a) Katoh, T., Itoh, E., Yoshino, T. and Terashima, S. (1997) Total synthesis of an enantiomeric pair of FR900482. 1. Synthetic and end-game strategies. *Tetrahedron*, **53** (30), 10229–10238; (b) Yoshino, T., Nagata, Y., Itoh, E., Hashimoto, M., Katoh, T. and Terashima, S. (1997) Total synthesis of an enantiomeric pair of FR900482. 2. Syntheses of the aromatic and the optically active aliphatic segments. *Tetrahedron*, **53** (30), 10239–10252.

28 Hassner, A. and Kascheres, A. (1970) Competitive attack of nucleophiles at ring carbon vs carbonyl. Reactions of aziridinecarbamates. *Tetrahedron Lett.*, **11** (53), 4623–4626.

29 Eis, M.J. and Ganem, B. (1985) BF_3-etherate promoted alkylation of aziridines with organocopper reagents: a new synthesis of amines. *Tetrahedron Lett.*, **26** (9), 1153–1156.

30 Baldwin, J.E., Adlington, R.M., O'Neil, I.A., Schofield, C., Spivey, A.C. and Sweeney, J.B. (1989) The ring-opening of aziridine-2-carboxylate esters with organometallic reagents. *J. Chem. Soc. Chem. Commun.*, (23), 1852–1854.

31 Tanner, D. and Somfai, P. (1988) From aziridines to carbapenems via a novel β-lactam ring closure: an enantioselective synthesis of (+)-PS-5. *Tetrahedron*, **44** (2), 619–624.

32 Tanner, D., He, H.M. and Somfai, P. (1992) Regioselective nucleophilic ring opening of 2,3-aziridino alcohols. *Tetrahedron*, **48** (29), 6069–6078.

33 Tanner, D. and He, H.M. (1992) Enantioselective routes toward 1β-methylcarbapenems from chiral aziridines. *Tetrahedron*, **48** (29), 6079–6086.

34 Pattenden, L.C., Wybrow, R.A.J., Smith, S.A. and Harrity, J.P.A. (2006) A [3 + 3] annelation approach to tetrahydropyridines. *Org. Lett.*, **8** (14), 3089–3091.

35 Moran, W.J., Goodenough, K.M., Raubo, P. and Harrity, J.P.A. (2003) A concise asymmetric route to nuphar alkaloids. A formal synthesis of (−)-deoxynupharidine. *Org. Lett.*, **5** (19), 3427–3429.

36 Goodenough, K.M., Moran, W.J., Raubo, P. and Harrity, J.P.A. (2005) Development of a flexible approach to nuphar alkaloids via two enantiospecific piperidine-forming reactions. *J. Org. Chem.*, **70** (1), 207–213.

37 Provoost, O.Y., Hedley, S.J., Hazelwood, A.J. and Harrity, J.P.A. (2006) Formal synthesis of (±)-perhydrohistrionicotoxin via a stepwise [3 + 3] annelation strategy. *Tetrahedron Lett.*, **47** (3), 331–333.

38 Wu, Y.-C. and Zhu, J. (2009) Asymmetric total syntheses of (−)-renieramycin M and G and (−)-jorumycin using aziridine as a lynchpin. *Org. Lett.*, **11** (23), 5558–5561.

39 Hayes, J.F., Shipman, M. and Twin, H. (2001) Asymmetric synthesis of 2-substituted piperidines using a multi-component coupling reaction: rapid assembly of (S)-coniine from (S)-1-(1-phenylethyl)-2-methyleneaziridine. *Chem. Comm.*, (18), 1784–1785.

40 Berry, M.B., Craig, D., Jones, P.S. and Rowlands, G.J. (1997) The enantiospecific synthesis of (+)-monomorine I using a 5-endo-trig cyclization strategy. *Chem. Commun.*, (22), 2141–2142.

41 Caldwell, J.J., Craig, D. and East, S.P. (2001) A sulfone-mediated synthesis of (+)-preussin. *Synlett*, (10), 1602–1604.

42 Caldwell, J.J. and Craig, D. (2007) Sulfone-mediated total synthesis of (±)-lepadiformine. *Angew. Chem. Int. Ed.*, **46** (15), 2631–2634.

43 Smith, A.B., III and Kim, D.-S. (2005) Total synthesis of the neotropical poison-frog alkaloid (−)-205B. *Org. Lett.*, **7** (15), 3247–3250.

44 Smith, A.B., III and Kim, D.-S. (2006) A general, convergent strategy for the construction of indolizidine alkaloids: total syntheses of (−)-indolizidine 223AB

and alkaloid (−)-205B. *J. Org. Chem.*, **71** (7), 2547–2557.

45 Baldwin, J.E., Adlington, R.M. and Robinson, N.G. (1987) Nucleophilic ring-opening of aziridine-2-carboxylates with wittig reagents – an enantioefficient synthesis of unsaturated amino acids. *J. Chem. Soc. Chem. Commun.*, (3), 153–155.

46 Nakagawa, M. and Kawahara, M. (2000) A concise synthesis of physostigmine from skatole and activated aziridine via alkylative cyclization. *Org. Lett.*, **2** (7), 953–955.

47 Benohoud, M., Leman, L., Cardoso, S.H., Retailleau, P., Dauban, P., Thierry, J. and Dodd, R.H. (2009) Total synthesis and absolute configuration of the natural amino acid tetrahydrolathyrine. *J. Org. Chem.*, **74** (15), 5331–5336.

48 Teshima, T., Konishi, K. and Shiba, T. (1980) Synthesis of L-epicapreomycidine. *Bull. Chem. Soc. Jpn.*, **53** (2), 508–511.

49 Shigematsu, N., Setoi, H., Uchida, I., Shibata, T., Terano, H. and Hashimoto, M. (1988) Structure and synthesis of FR900490, a new immunomodulating peptide isolated from a fungus. *Tetrahedron Lett.*, **29** (40), 5147–5150.

50 Imae, K., Kamachi, H., Yamashita, H., Okita, T., Okuyama, S., Tsuno, T., Yamasaki, T., Sawada, Y., Ohbayashi, M., Naito, T. and Oki, T. (1991) Synthesis, stereochemistry and biological properties of the depigmenting agents, melanostatin, feldamycin and analogs. *J. Antibiot.*, **44** (1), 76–85.

51 Martens, J. and Scheunemann, M. (1991) EPC-synthese von verruculotoxin. *Tetrahedron Lett.*, **32** (11), 1417–1418.

52 Farthing, C.N., Baldwin, J.E., Russell, A.T., Schofield, C.J. and Spivey, A.C. (1996) Syntheses of (S)-β-pyrazolylalanine and (S)-quisqualic acid from a serine-derived aziridine. *Tetrahedron Lett.*, **37** (29), 5225–5226.

53 Trost, B.M. and Dong, G. (2007) Asymmetric annulation toward pyrrolopiperazinones: concise enantioselective syntheses of pyrrole alkaloid natural products. *Org. Lett.*, **9** (12), 2357–2359.

54 Yoshimitsu, T., Ino, T. and Tanaka, T. (2008) Total synthesis of (−)-agelastatin A. *Org. Lett.*, **10** (23), 5457–5460.

55 Tanner, D., Almario, A. and Högberg, T. (1995) Total synthesis of balanol, part 1. Enantioselective synthesis of the hexahydroazepine ring via chiral epoxides and aziridines. *Tetrahedron*, **51** (21), 6061–6070.

56 Tanner, D., Tedenborg, L., Almario, A., Pettersson, I., Csöregh, I., Kelly, N.M., Andersson, P.G. and Hogberg, T. (1997) Total synthesis of balanol, part 2. Completion of the synthesis and investigation of the structure and reactivity of two key heterocyclic intermediates. *Tetrahedron*, **53** (13), 4857–4868.

57 For a formal synthesis of (−)-balanol based on Tanner's synthesis, see: Unthank, M.G., Hussain, N. and Aggarwal, V.K. (2006) The use of vinyl sulfonium salts in the stereocontrolled asymmetric synthesis of epoxide- and aziridine-fused heterocycles: application to the synthesis of (−)-balanol. *Angew. Chem. Int. Ed.*, **45** (42), 7066–7069.

58 Olofsson, B. and Somfai, P. (2003) Divergent synthesis of D-erythro-sphingosine, L-threo-sphingosine, and their regioisomers. *J. Org. Chem.*, **68** (6), 2514–2517.

59 Trost, B.M. and Dong, G. (2006) New class of nucleophiles for palladium-catalyzed asymmetric allylic alkylation. Total synthesis of agelastatin A. *J. Am. Chem. Soc.*, **128** (18), 6054–6055.

60 Trost, B.M. and Dong, G.B. (2009) A stereodivergent strategy to both product enantiomers from the same enantiomer of a stereoinducing catalyst: agelastatin A. *Chem. Eur. J.*, **15** (28), 6910–6919.

61 Tanaka, T., Nakajima, K. and Okawa, K. (1980) Studies on 2-aziridinecarboxylic acid. IV. Total synthesis of actinomycin-D (C1) via ring-opening reaction of aziridine. *Bull. Chem. Soc. Jpn.*, **53** (5), 1352–1355.

62 Cardillo, G., Gentilucci, L., Tolomelli, A. and Tomasini, C. (1998) Formation of aziridine-2-amides through 5-halo-6-methylperhydropyrimidin-4-ones. A route to enantiopure L- and D-threonine

and *allo*-threonine. *J. Org. Chem.*, **63** (10), 3458–3462.
63 Disadee, W. and Ishikawa, T. (2005) Chirality transfer from guanidinium ylides to 3-alkenyl (or 3-alkynyl) aziridine-2-carboxylates and application to the syntheses of (2R,3S)-3-hydroxyleucinate and D-*erythro*-sphingosine. *J. Org. Chem.*, **70** (23), 9399–9406.
64 Loncaric, C. and Wulff, W.D. (2001) An efficient synthesis of (–)-chloramphenicol via asymmetric catalytic aziridination: a comparison of catalysts prepared from triphenylborate and various linear and vaulted biaryls. *Org. Lett.*, **3** (23), 3675–3678.
65 Cardillo, G., Orena, M., Sandri, S. and Tomasini, C. (1986) A novel, efficient synthesis of (±)-*erythro*-sphingosine. *Tetrahedron*, **42** (3), 917–922.
66 Feng, X., Duesler, E.N. and Mariano, P.S. (2005) Pyridinium salt photochemistry in a concise route for synthesis of the trehazolin aminocyclitol, trehazolamine. *J. Org. Chem.*, **70** (14), 5618–5623.
67 Sullivan, B. and Hudlicky, T. (2008) Chemoenzymatic formal synthesis of (–)-balanol. Provision of optical data for an often reported intermediate. *Tetrahedron Lett.*, **49** (35), 5211–5213.
68 Crotti, P., Di Bussolo, V., Favero, L., Macchia, F. and Pineschi, M. (1996) An efficient stereoselective synthesis of the amino sugar component (E ring) of calicheamicin γ_1I. *Tetrahedron Asymmetry*, **7** (3), 779–786.
69 Tarrade, A., Dauban, P. and Dodd, R.H. (2003) Enantiospecific total synthesis of (–)-polyoxamic acid using 2,3-aziridino-γ-lactone methodology. *J. Org. Chem.*, **68** (24), 9521–9524.
70 Li, P., Evans, C.D. and Joullié, M.M. (2005) A convergent total synthesis of ustiloxin D via an unprecedented copper-catalyzed ethynyl aziridine ring-opening by phenol derivatives. *Org. Lett.*, **7** (23), 5325–5327.
71 Li, P., Evans, C.D., Wu, Y., Cao, B., Hamel, E. and Jouillé, M.M. (2008) Evolution of the total syntheses of ustiloxin natural products and their analogs. *J. Am. Chem. Soc.*, **130** (7), 2351–2364.
72 Cranfill, D.C. and Lipton, M.A. (2007) Enantio- and diastereoselective synthesis of (R,R)-β-methoxytyrosine. *Org. Lett.*, **9** (18), 3511–3513.
73 Fuji, K., Kawabata, T., Kiryu, Y. and Sugiura, Y. (1996) Ring opening of optically active *cis*-disubstituted aziridino alcohols: an enantiodivergent synthesis of functionalized amino alcohol derivatives. *Heterocycles*, **42** (2), 701–722.
74 Barros, M.T., Matias, P.M., Maycock, C.D. and Ventura, M.R. (2003) Aziridines as a protecting and directing group. Stereoselective synthesis of (+)-bromoxone. *Org. Lett.*, **5** (23), 4321–4323.
75 (a) Morimoto, Y., Matsuda, F. and Shirahama, H. (1991) Total synthesis of (±)-virantmycin and determination of its stereochemistry. *Synlett*, (3), 202–203; (b) Morimoto, Y., Matsuda, F. and Shirahama, H. (1996) Synthetic studies on virantmycin. 1. Total synthesis of (±)-virantmycin and determination of its relative stereochemistry. *Tetrahedron*, **52** (32), 10609–10630.
76 (a) Hanessian, S., Del Valle, J.R., Xue, Y. and Blomberg, N. (2006) Total synthesis and structural confirmation of chlorodysinosin A. *J. Am. Chem. Soc.*, **128** (32), 10491–10495; (b) Hanessian, S., Del Valle, J.R., Xue, Y. and Blomberg, N. (2006) Total synthesis and structural confirmation of chlorodysinosin A. *J. Am. Chem. Soc.*, **128** (35), 11727–11728.
77 Mendlik, M.T., Tao, P., Hadad, C.M., Coleman, R.S. and Lowary, T.L. (2006) Synthesis of L-daunosamine and L-ristosamine glycosides via photoinduced aziridination. Conversion to thioglycosides for use in glycosylation reactions. *J. Org. Chem.*, **71** (21), 8059–8070.
78 (a) Dauban, P. and Malik, P.G. (2009) A masked 1,3-dipole revealed from aziridines. *Angew. Chem. Int. Ed.*, **48** (48), 9026–9029 and ref. cit.; (b) Najera, C. and Sansano, J.M. (2009) 1,3-Dipolar cycloadditions: applications to the synthesis of antiviral agents. *Org. Biomol. Chem.*, **7** (22), 4567–4581.

79 (a) Takano, S., Iwabuchi, Y. and Ogasawara, K. (1987) Concise enantioselective synthesis of acromelic acid A. *J. Am. Chem. Soc.*, **109** (18), 5523–5524; (b) Takano, S., Tomita, S., Iwabuchi, Y. and Ogasawara, K. (1989) A concise enantioselective synthesis of acromelic acid-B from (S)-O-benzylglycidol. *Heterocycles*, **29** (8), 1473–1476.

80 Takano, S., Iwabuchi, Y. and Ogasawara, K. (1988) A concise enantioselective route to (−)-kainic acid from (S)-2-(benzyloxymethyl) oxirane. *J. Chem. Soc. Chem. Commun.*, (17), 1204–1206.

81 Takano, S., Samizu, K. and Ogasawara, K. (1990) Enantiospecific construction of quaternary carbon center via intramolecular 1,3-dipolar cycloaddition. A new route to natural (−)-mesembrine from (S)-O-benzylglycidol. *Chem. Lett.*, (7), 1239–1242.

82 Hudlicky, T., Frazier, J.O., Seoane, G., Tiedje, M., Seoane, A., Kwart, L.D. and Beal, C. (1986) Topological selectivity in the intramolecular [4 + 1] pyrroline annulation. Formal total stereospecific synthesis of (±)-supinidine, (±)-isoretronecanol, and (±)-trachelanthamidine. *J. Am. Chem. Soc.*, **108** (13), 3755–3762.

83 Hudlicky, T., Seoane, G. and Lovelace, T.C. (1988) Intramolecular [4 + 1] pyrroline annulation via azide-diene cycloadditions. 2: formal stereoselective total syntheses of (±)-platynecine, (±)-hastanecine, (±)-turneforcidine, and (±)-dihydroxyheliotridane. *J. Org. Chem.*, **53** (9), 2094–2099.

84 Åhman, J. and Somfai, P. (1994) Aza-[2,3]-Wittig rearrangements of vinylaziridines. *J. Am. Chem. Soc.*, **116** (21), 9781–9782.

85 Åhman, J. and Somfai, P. (1996) A novel rearrangement of N-propargyl vinylaziridines. Mechanistic diversity in the aza-[2,3]-Wittig rearrangement. *Tetrahedron Lett.*, **37** (14), 2495–2498.

86 Åhman, J. and Somfai, P. (1995) Enantioselective total synthesis of (−)-indolizidines 209B and 209D via a highly efficient aza-[2,3]-Wittig rearrangement of vinylaziridines. *Tetrahedron*, **51** (35), 9747–9756.

87 Somfai, P., Jarevång, T., Lindström, U.M. and Svensson, A. (1997) Total synthesis of (±)-monomorine I and (±)-indolizidine 195B by an aza-[2,3]-Wittig rearrangement of a vinylaziridine. *Acta Chem. Scand.*, **51** (10), 1024–1029.

88 Hudlicky, T., Sinai-Zingde, G. and Seoane, G. (1987) Mild methodology for [4 + 1] pyrroline annulation. Second generation synthesis of pyrrolizidine alkaloids. *Synth. Commun.*, **17** (10), 1155–1163.

89 Hirner, S. and Somfai, P. (2005) Microwave-assisted rearrangement of vinylaziridines to 3-pyrrolines: formal synthesis of (−)-anisomycin. *Synlett*, (20), 3099–3102.

90 Tomasini, C. and Vecchione, A. (1999) Novel synthesis of 4-carboxymethyl 5-alkyl/aryl oxazolidine-2-ones by rearrangement of 2-carboxymethyl 3-alkyl/aryl N-tertbutoxy carbonyl aziridines. *Org. Lett.*, **1**, 2153–2156.

91 Song, Z.L., Wang, B.M., Tu, Y.Q., Fan, C.A. and Zhang, S.Y. (2003) A general efficient strategy for cis-3a-aryloctahydroindole alkaloids via stereocontrolled $ZnBr_2$-catalyzed rearrangement of 2,3-aziridino alcohols. *Org. Lett.*, **5** (13), 2319–2321.

92 Tanner, D. and Somfai, P. (1993) Palladium-catalyzed transformation of a chiral vinylaziridine to a β-lactam. A enantioselective route to the carbapenem (+)-PS-5. *Bioorg. Med. Chem. Lett.*, **3** (11), 2415–2418.

93 Trost, B.M. and Fandrick, D.R. (2005) DYKAT of vinyl aziridines: total synthesis of (+)-pseudodistomin D. *Org. Lett.*, **7** (5), 823–826.

94 Hedley, S.J., Moran, W.J., Prenzel, A.H.G.P., Price, D.A. and Harrity, J.P.A. (2001) Synthesis of functionalised piperidines through a [3 + 3] cycloaddition strategy. *Synlett*, (10), 1596–1598.

95 Coote, S.C., Moore, S.P., O'Brien, P., Whitwood, A.C. and Gilday, J. (2008) Organolithium-mediated conversion of β-alkoxy aziridines into allylic sulfonamides: effect of the N-sulfonyl group and a formal synthesis of (±)-perhydrohistrionicotoxin. *J. Org. Chem.*, **73** (19), 7852–7855.

96 Lee, H.-Y., Lee, S., Kim, D., Kim, B.K., Bahn, J.S. and Kim, S. (1998) Total synthesis of α-cedrene: a new strategy utilizing N-aziridinylimine radical chemistry. *Tetrahedron Lett.*, **39** (42), 7713–7716.

97 (a) Métro, T.-X., Duthion, B., Gomez Pardo, D. and Cossy, J. (2010) Rearrangement of β-aminoalcohol via aziridiniums: a review. *Chem. Soc. Rev.*, **39**, 89–102; (b) Cossy, J., Gomez Pardo, D., Dumas, C., Mirguet, O., Déchamps, T., Métro, T.-X., Burger, B., Roudeau, R., Appenzeller, J. and Cochi, A. (2009) Rearrangement of β-aminoalcohol and application to the synthesis of biologically active compounds. *Chirality*, **21** (9), 850–856; (c) Cossy, J. and Gomez Pardo, D.G. (2002) Synthesis of substituted piperidines via aziridinium intermediates: synthetic applications. *ChemTracts*, **15** (11), 579–605.

98 (a) Kuehne, M.E., Okuniewicz, F.J., Kirkemo, C.L. and Bohnert, J.C. (1982) Studies in biomimetic alkaloid syntheses. 8. Total syntheses of the C-14 epimeric hydroxyvincadifformines, tabersonine, a (hydroxymethy1)-D-norvincadifforminea and the C-20 epimeric pandolines. *J. Org. Chem.*, **47** (7), 1335–1343; (b) Kuehne, M.E. and Podhorez, D.E. (1985) Studies in biomimetic alkaloid syntheses. 12. Enantioselective total syntheses of (−)-and (+)-vincadifformine and of (−)-tabersonine. *J. Org. Chem.*, **50** (7), 924–929.

99 Tanner, D. and Somfai, P. (1986) New routes to perhydrohistrionicotoxin. *Tetrahedron*, **42** (20), 5657–5664.

100 Williams, D.R., Brown, D.L. and Benbow, J.W. (1989) Studies of *stemona* alkaloids. Total synthesis of (+)-croomine. *J. Am. Chem. Soc.*, **111** (5), 1923–1925.

101 Davis, F.A. and Reddy, G.V. (1996) Aziridine-2-carboxylic acid mediated asymmetric synthesis of D-*erythro* and L-*threo*-sphingosine from a common precursor. *Tetrahedron Lett.*, **37** (25), 4349–4352.

102 Cossy, J., Dumas, C. and Gomez Pardo, D. (1997) Synthesis of (−)-pseudoconhydrine through ring enlargement of a L-proline derivative. *Synlett*, (8), 905–907.

103 Cossy, J., Mirguet, O. and Gomez Pardo, D. (2001) Ring expansion: synthesis of the velbanamine piperidine core. *Synlett*, (10), 1575–1577.

104 Cossy, J., Dumas, C. and Gomez Pardo, D. (1997) A short and efficient synthesis of zamifenacin, a muscarinic M_3 receptor antagonist. *Bioorg. Med Chem. Lett.*, **7** (10), 1343–1344.

105 Cossy, J., Mirguet, O., Gomez Pardo, D. and Desmurs, J.-R. (2001) Ring expansion: formal total synthesis of (−)-paroxetine. *Tetrahedron Lett.*, **42** (33), 5705–5707.

106 Metro, T.-X., Gomez Pardo, D. and Cossy, J. (2008) Syntheses of (S,S)-reboxetine via a catalytic stereospecific rearrangement of β-amino alcohols. *J. Org. Chem.*, **73** (2), 707–710.

2
Azetidine and Its Derivatives

Hidemi Yoda, Masaki Takahashi and Tetsuya Sengoku

2.1
Introduction

Since the historic discovery by Fleming of antibiotic penicillin [1], whose structure was proposed by Crowfoot Hodgkin [2], azetidines, four-membered nitrogen-containing heterocycles, have received considerable attention because of their potential therapeutic values as antibiotics [3]. The structural outcomes of azetidines appear to fall configurationally into aza analogs of cyclobutanes that possess considerable ring strain attributed to the orthogonally twisted σ-bonds. Although the strained nature of the small-sized ring systems are responsible for antibacterial activities [4], the problems encountered with the four-membered ring synthesis have been posed in various ways by numerous natural product families over the past few decades, thus making it difficult to provide general and versatile methods for their preparations [5]. This chapter is devoted to azetidines and azetidin-2-ones, which are selected as representative classes of the four-membered nitrogen-containing heterocycles because these two structures can typically be found in important natural products and potential drugs for bacterial pathogens (Table 2.1).

2.2
Structural Description of Azetidines

From a synthetic point of view, an understanding of structural features in azetidine ring systems is of fundamental importance because synthetic applications of the structural characteristics provide accurate information on reliable ways of constructing strained bonds of the four-membered rings. It needs to be emphasized that all the sp^3 configurations of the ring components should give a certain degree of flexibility in the orthogonally twisted σ-bonds with the deformation angle of 19.5°, resulting in a preference for the azetidine rings to adopt puckered conformations, that may interconvert with a barrier to the ring inversion of 1.26 kcal mol^{-1} [6]. The thermodynamic stability of the relatively flexible ring structures allows synthetic access to numerous azetidine derivatives through a variety of possible

Heterocycles in Natural Product Synthesis, First Edition. Edited by Krishna C. Majumdar and Shital K. Chattopadhyay.
© 2011 Wiley-VCH Verlag GmbH & Co. KGaA. Published 2011 by Wiley-VCH Verlag GmbH & Co. KGaA.

Table 2.1 Azetidine-based natural products.

Serial No.	Trivial name	Structure	Source	Isolation [Ref]	Biological activity	Synthesis [Ref]
1	2′-Deoxymugineic acid		*Triticum aestivum* L. (wheat)	[15]	Iron chelation [15]	[17, 18]
2	Cephalosporin C		*Cephalosporium acremonium*	[40]	Antimicrobial [40]	[44]
3	Mugineic acid		*Hordium vulgare* L. (barley)	[14a]	Iron chelation [13]	[19]
4	Penaresidin A		*Penares sp.* (Okinawan sponge)	[20a]	Actomyosin ATPase activation [20a]	[21, 22]
5	Penaresidin B		*Penares sp.* (Okinawan sponge)	[20a]	Actomyosin ATPase activation [20a]	[21c, 23, 25]
6	Penicillins		*Penicillium notatum*	[1, 35]	Antibiotic [1, 35]	[38]

2.3 Synthetic Methodologies for the Formation of Azetidine Rings

strategies involving cyclization and cycloaddition methodologies. Therefore, practical synthesis of the azetidine-based natural products has in principle no restrictions in planning the four-membered ring construction at some stage, where various transformations can be conducted on the azetidine subunits to achieve the target-directed synthesis.

2.3 Synthetic Methodologies for the Formation of Azetidine Rings

The convenient and general methodology for the preparation of azetidine-containing compounds can be divided into two categories: intramolecular cyclization [7] and intermolecular [2 + 2] cycloaddition [8]. The former category typically employs intramolecular nucleophilic attacks of deprotonated amides onto reactive γ-carbon atoms bearing suitable leaving groups. In this regard, γ,δ-oxyranyl moieties can also be an excellent electrophile for this type of 1,4-cyclization, providing γ-hydroxymethyl-substituted azetidines in a stereospecific manner [9]. One particularly attractive strategy would involve palladium-catalyzed intramolecular C–N coupling of γ-allenyl sulfonamide **1** (Scheme 2.1) [10]. Presumably, this reaction proceeds by direct insertion of the aryl-palladium species **2** into the internal alkene of allenyl system to generate (π-allyl) palladium complexes **3** which result in stereocontrolled ring closure, giving rise to cis-fused γ-arylalkenyl azetidines **4**. The latter category consists of metal-catalyzed [2 + 2] cycloadditions of imines to enol ethers, which can be recognized as an analogous Staudinger reaction and a practical method for preparing α-oxy-substituted azetidines (Scheme 2.2) [11].

Mechanistically, it is proposed that this reaction proceeds via well-stabilized 1,4-zwitterions **5**, which might be expected to cyclize reversibly to give the thermodynamically most stable cis-isomers of 2,4-disubstituted azetidines **6**. Moreover, a few cases of enantioselective [2 + 2] cycloadditions have been reported for closely related systems employing allenylsilanes **7** instead of the enol ethers, where a

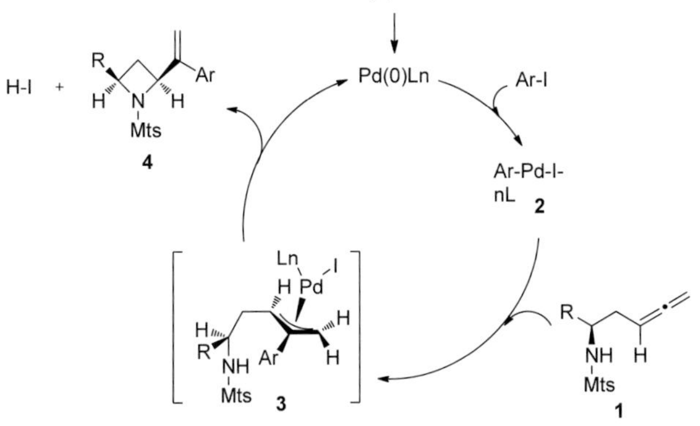

Scheme 2.1

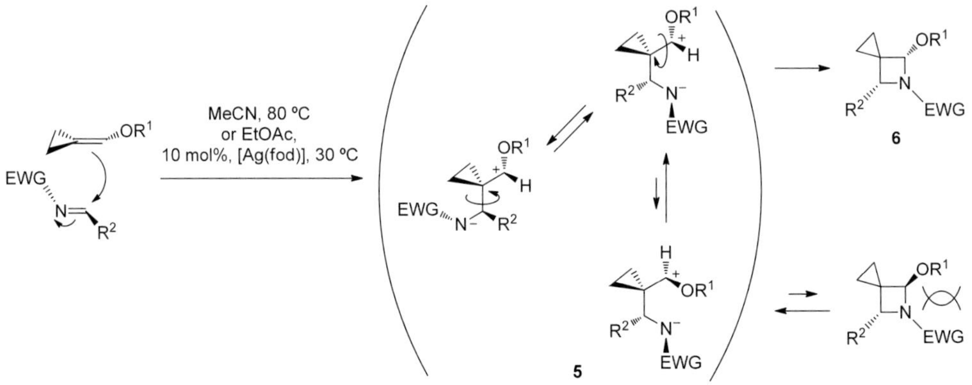

Scheme 2.2

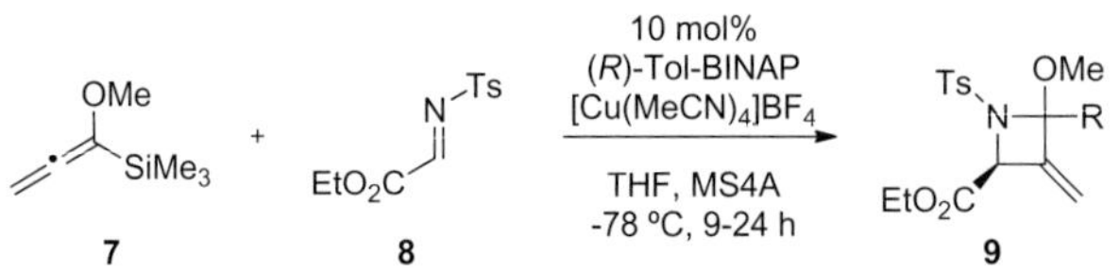

Scheme 2.3

chiral copper catalyst mediates efficient and enantioselective transformations of the achiral reactants **8** into chiral 3-methylene-substituted azetidines **9** (Scheme 2.3) [12].

Consequently, the elaboration of preparative methods outlined above can be applied to access a wide variety of azetidine analogs, which may provide synthetically useful approaches to any targeted molecules. With regard to the azetidine-based natural products, only a few examples that include mugineic acids and penaresidins are known at present. Accordingly, a major focus of contemporary azetidine research surrounds the development of synthetic strategies for these two natural product families.

2.4
Synthesis of Mugineic Acids

In higher plants, the biosynthetic processes of chlorophylls require uptake and transport of ion, whose deprivation results in plant disorder. Several amino acids including mugineic acids exhibit chelating properties for iron and other metals, which serve, therefore, as essential participants in the growth and maintenance of plant cells [13]. Mugineic acid and its 2'-deoxy derivative are individual members of the naturally occurring mugineic acid family and have been isolated from barley [14] and wheat [15], respectively. These compounds share common structural features highlighted for azetidine-2-carboxylic acid moiety except for the substitu-

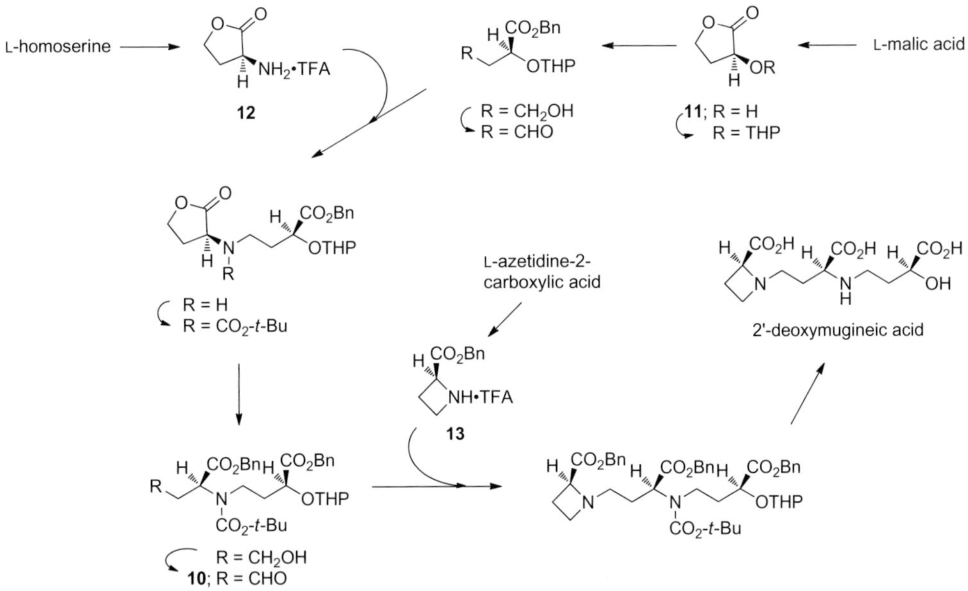

2.4 Synthesis of Mugineic Acids

Scheme 2.4

tion pattern of the N-alkyl chains. Since L-azetidine-2-carboxylic acid occurs naturally in members of the *Liliaceae* family and can be envisioned as a commercially available compound [16], elaboration of mugineic acids have been achieved by covalent installation of this chiral building unit in the linear-chain residues at intentional stages of modular synthetic approaches. Shown in Scheme 2.4 is a representative example of the synthesis of 2'-deoxy mugineic acid reported by Ohfune and co-workers [17].

In this work, the synthetic strategy for the preparation of the requisite linear-chain segment **10** involved the use of L-α-hydroxy γ-butyrolactone **11** and L-homoserine lactone trifluoroacetate **12** as the chiral sources for the required stereocenters, which are readily available from L-malic acid and L-homoserine, respectively. After preparation of the appropriate linear-chain subunit carrying an aldehyde moiety **10**, completion of the synthesis was achieved in three steps through reductive amination of the subunit with C-protected L-azetidine-2-carboxylic acid **13** and sodium cyanoborohydride followed by removal of the protecting groups. While several reports on the synthesis of mugineic acid congeners are available [18], the first synthesis of mugineic acid reported by Shioiri and co-workers represents another important example of the studies, which describe the development of alternative strategy for elaboration of this class of compounds [19].

Their synthetic approach to mugineic acid started with N-alkylation of C-protected L-azetidine-2-carboxylic acid **14** (Scheme 2.5). Crucial to the success of this route was stereochemical control in the creation of the chiral center at the C2' position, since the new stereo-center needs to be created. To satisfy this requirement, they employed the direct C-acylation protocol to generate 4-(alkoxycarbonyl)

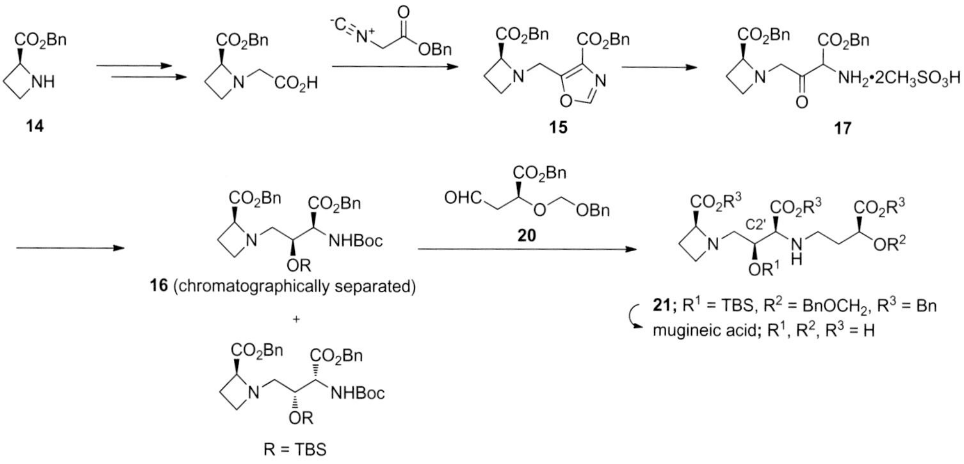

Scheme 2.5

oxazole intermediate **15**, which can serve as a latent *erythro*-β-hydroxy-α-amino acid. Based on this synthetic planning, elaboration of azetidine-containing *erythro*-β-hydroxy-α-amino acid **16** was achieved by stereoselective reduction of α-amino ketone **17**, that was obtained through treatment of **15** with methane sulfonic acid. Finally, **16** was coupled with the L-malic acid-derived chiral building block **20** to form the fully protected mugineic acid **21**, whose deprotection led to completion of the synthesis.

2.5
Synthesis of Penaresidins

Marine sponges have frequently afforded a wide variety of sphingosine-related compounds. In 1991, two types of sphingosine-derived alkaloids, penaresidin A and B, were isolated from Okinawan marine sponge *Penares* sp. by Kobayashi and co-workers [20]. Attractive features of these natural products are that they possess an azetidine ring structure as well as potent actomyosin ATPase-activating activity, as demonstrated by biological evaluation of their inseparable mixture. Mori and co-workers, who have devoted considerable efforts to study unresolved stereochemistry of the two penaresidins [21], pioneered the chemical approach to these compounds.

According to the synthetic works, their strategy for the construction of side-chain structures involves the use of L-isoleucine and L-leucine as the chiral sources for the required stereocenters of penaresidin A and B, respectively (Scheme 2.6). Thus, L-isoleucine was converted to the chiral epoxide **22** which was then coupled with lithium acetylide to form β-alkynyl alcohol **23**. The protected derivative of the resulting product **23** underwent triple bond isomerization through the acetylene zipper reaction to terminal alkynes **24**, which serve as the side-chain units. On the

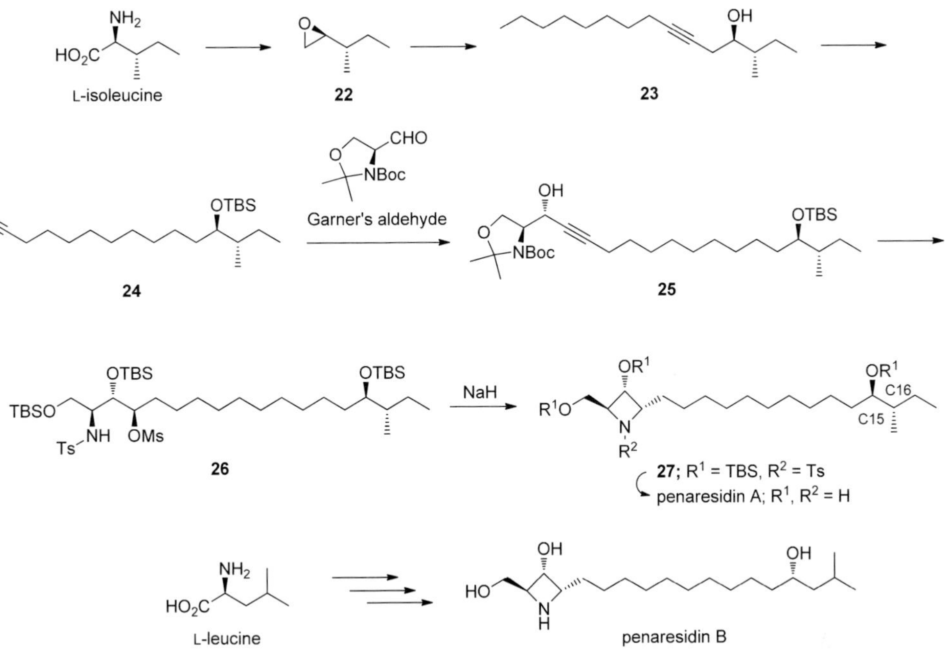

2.5 Synthesis of Penaresidins

Scheme 2.6

other hand, they used Garner's aldehyde to introduce new stereogenic centers that were to be incorporated into the azetidine-ring structures. The reaction of lithium acetylide derived from **24** with Garner's aldehyde afforded α-alkynyl alcohols **25**. After various transformations to obtain N-tosyl-β-amino mesylates **26**, the azetidine-ring formations were achieved by deprotonation of the amide NH groups with sodium hydride via the successive intramolecular cyclization processes to give fully protected penaresidin A and B **27**. As a consequence of their synthetic study, three stereoisomers of penaresidin A as well as two stereoisomers of penaresidin B could be accessed synthetically in fewer than 20 steps. Nevertheless, the spectrometric identification of these compounds failed to provide definitive support for the stereochemical assignment of natural penaresidins, whose absolute configurations remained unresolved. Among the more recent contributions to this subject [22], our demonstration of the first synthetic elaboration led to desired success in the stereochemical assignment of penaresidin B [23]. During the course of early investigations on the absolute structure of penaresidins, Mori and co-workers established the *syn*-relationship between the two substituents at C15 and C16 of the side-chain moiety of penaresidin A, while Kamikawa and co-workers determined 2*R*, 3*R*, 4*S*-configurations of the azetidine ring structure using a simplified analog of penaresidins [24]. With these stereochemical outcomes as a preliminary assumption, our synthetic study has been aimed to identify the absolute structure of penaresidin B through asymmetric transformations of chiral

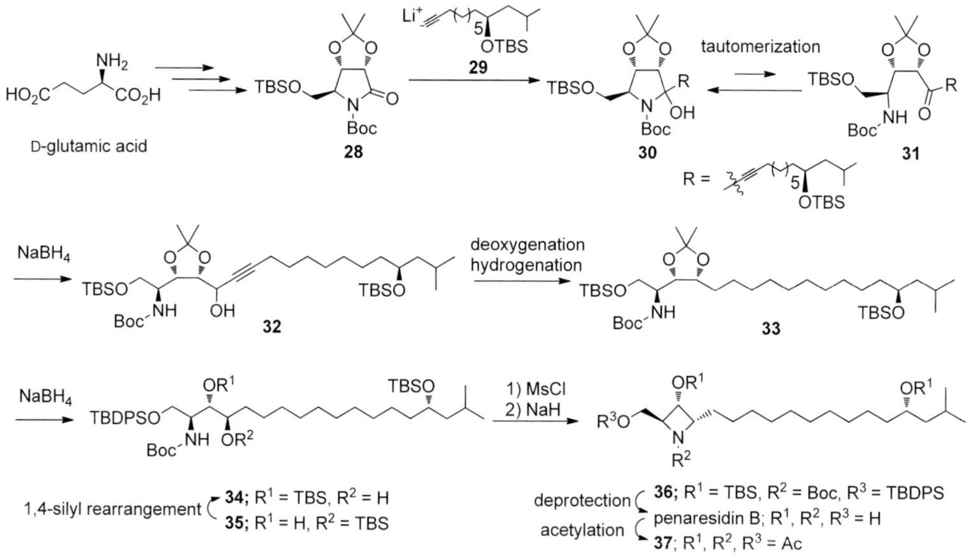

Scheme 2.7

compounds that have completely ordered structures with precisely controlled stereocenters (Scheme 2.7).

Thus, the synthesis of penaresidin B commenced with D-glutamic acid as the chiral source for the required stereocenter of the azetidine ring structure. This chiral building block can be transformed into N-t-butoxycarbonyl γ-lactam **28** through the established synthetic route. Because the carbonyl groups of N-carbonyl-substituted γ-lactams are potentially reactive toward nucleophilic partners, the reaction of this substrate could be carried out with lithium acetylide **29**, prepared according to the method of Mori and co-workers, to provide N-t-butoxycarbonyl α-hydroxy pyrrolidine **30**. In fact, this particular type of molecule has been shown to undergo facile tautomerization in part to produce the ring-opened ketoamides **31**, which can be reduced to the alcohol **32**. The later was converted into deoxygenated molecule **33** by free radical cleavage of thioimidazolide and successive hydrogenation of the alkyne moiety with tributyltin hydride. After deprotection and chemoselective protection of the oxygen-bearing functional groups of substrates, N-t-butoxycarbonyl α-silyloxy-β-hydroxylamine **34** was obtained through a 1,4-silyl rearrangement of the relevant regioisomer **35**. The hydroxyl group in **34** was then mesylated to promote the intramolecular cyclization upon deprotonation of the amide NH group with sodium hydride, giving rise to the fully protected penaresidin B **36**. Characterization data for the tetraacetate derivative of the synthetic penaresidin B **37** proved to be in good agreement with that reported in the literature, which has led to unambiguous demonstration of the accurate absolute stereochemistry of this natural product. Despite the noteworthy achievements, these synthetic strategies were nevertheless considered unsatisfactory because the multistep reactions resulted in low yield recovery of the

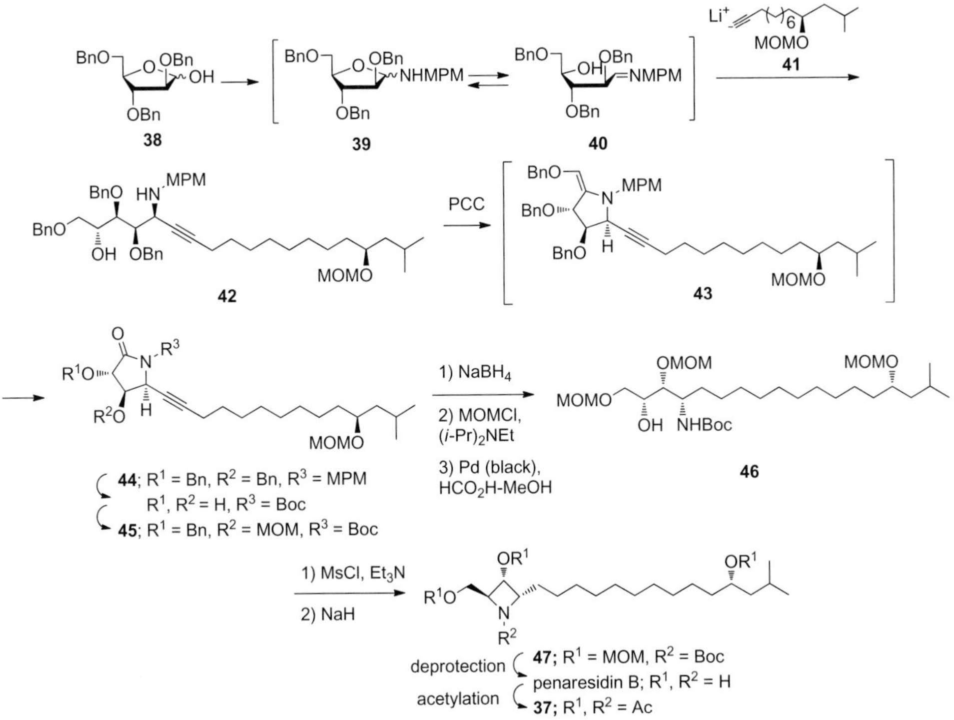

2.5 Synthesis of Penaresidins

Scheme 2.8

final product required. Thus, the rapid and economical generation of this fascinating molecule has remained a standing concern to synthetic chemists. To address this synthetic challenge, we decided to pursue the development of short and simplified routes to penaresidins as part of our continuing synthetic studies [25].

Scheme 2.8 depicts the revised synthesis of penaresidin B. Starting from tribenzyl-protected D-arabinose **38**, furanosylamine **39** was prepared according to an already published reaction protocol. Compound **39** exists in equilibrium with ring-opened γ-hydroxylimine **40** which underwent nucleophilic addition with the lithium acetylide **41** in a stereoselective fashion and to generate δ-hydroxy-α-alkynylamine **42** as a single stereoisomer. When the secondary hydroxyl group of this compound was converted into the carbonyl group upon treatment with pyridinium chlorochromate, *in situ* cyclization presumably took place to give the exocyclic enamine **43** which was oxidatively cleaved at the C=C double bond to alkynyl lactam **44**. At this point, it should be noted that the absolute configurations of the three stereogenic centers in this product are identical with those contained in the chiral skeletal framework of the azetidine ring. With this idea in mind, our efforts have focused on the generation of β-hydroxy *N-t*-butoxycarbonylamine **46**, as an appropriate precursor for the azetidine ring formation, through reductive ring opening of the alkynyl lactam **45**. Thus, the desired mesylate underwent *in situ* cyclization upon base treatment, allowing construction of the azetidine ring structure. Completion of the synthesis was achieved by conversion of the resulting product 47 to penaresidin B and its tetraacetate derivative 37 via removal of all the protecting groups and subsequent acetylation. As a consequence, this revised synthetic approach involving the use of 44 as an advanced template has proven to be a highly efficient pathway accessible to penaresidins without the need to separate the stereoisomers at every stage of the total synthesis, which proceeded in 12% overall yield from the starting material.

2.6
Structural Description of Azetidin-2-ones

Compared with the azetidine ring system, the structural description of azetidin-2-ones, commonly known as β-lactams, are better understood on the basis of the inherent difference in conformational constraints of four-membered heterocycles. Indeed, the general structure of azetidin-2-ones contains the amide group, whose geometry should be nearly planar because of the conjugation between the carbonyl π-system and the unpaired electrons of the nitrogen atom. This specific geometrical property of the azetidin-2-ones imposes conformational restrictions that limit the molecular mobility to lead to near co-planar alignment of the ring components. Thus, molecular rigidity due to the planar frameworks of the four-membered rings causes increased torsional strains which raise chemical reactivity of the carbonyl groups toward nucleophilic species. In fact, the potent antibacterial activity of the β-lactam-based agents, such as penicillins, can be attributed to enhanced susceptibility of the amide linkages to hydrolysis, whereas a type of monocyclic azetidin-2-ones is known to be rather thermodynamically stable, showing their resistance to hydrolysis [26]. The markedly different lability of the antibiotics from that shown by the simple azetidin-2-ones can be explained by weak conjugation of the C–N bond, which is fundamentally due to the out-of-plane twist of the unpaired electrons on the nitrogen atoms that adopt sp^3 configuration under the steric influence of ancillary rings fused to the β-lactam core [27]. Therefore, it is essential for the antibiotic β-lactams that the general structure involves the bicyclic system to have a high degree of potency, whose synthesis is problematic due to their lability.

2.7
Synthetic Methodologies for the Formation of Azetidin-2-ones

Since the discovery of penicillins, a considerable amount of effort has been directed towards synthesizing a wide range of functionalized azetidin-2-ones [28]. In this context, the Staudinger reaction between ketenes and imines provides convenient access to diversely functionalized azetidin-2-ones with high selectivity in the [2 + 2] cycloadditions [29]. Many examples of its potential applications in the synthesis of

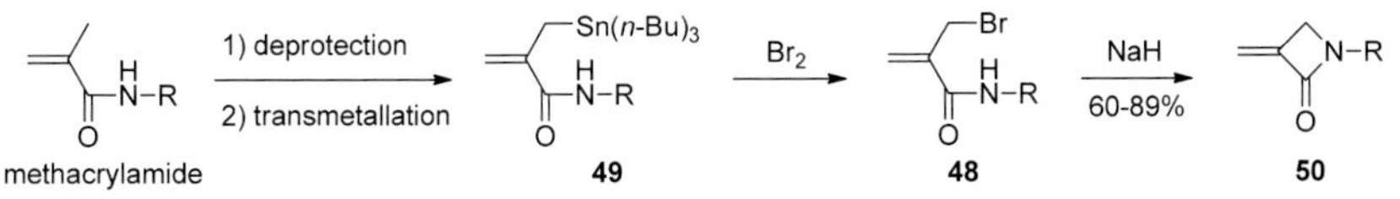

2.7 Synthetic Methodologies for the Formation of Azetidin-2-ones | 51

Scheme 2.9

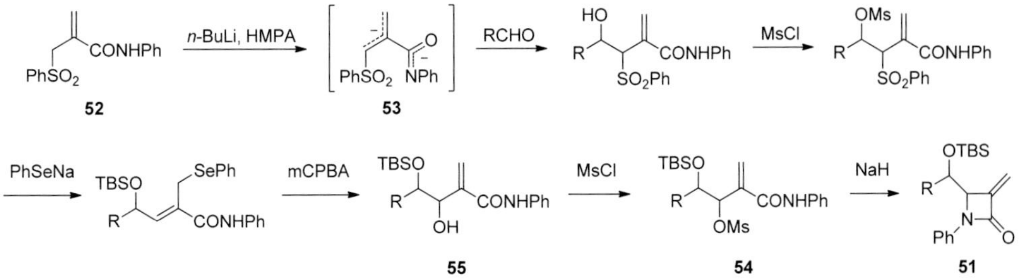

Scheme 2.10

a number of potent antibiotics have demonstrated that this type of reaction represents an important and versatile class of β-lactam-forming reactions [30]. Another important synthetic approach that may be advantageous for more general utility involves intramolecular 1,4-cyclization of deprotonated amide onto reactive carbon atom [31]. In this case, secondary amides bearing suitable electron-withdrawing substituents at the β-position would be appropriate precursors for the subsequent 1,4-cyclization reaction.

For instance, a 1,4-cyclization precursor, 2-(bromomethyl)-propenamides **48**, has been demonstrated to be accessible via bromination of 2-(stannylmethyl)-propenamides **49** which, in turn, can be prepared by deprotonation of methacrylamides and subsequent transmetallation (Scheme 2.9) [32]. Upon treatment with sodium hydride, these substances (e.g., **48**) undergo efficient formation of α-methylene azetidin-2-ones **50**, which serve as effective antibacterial agents due to their extremely strained ring structures imposed by in-plane bending deformation of the sp^2-hybridized olefinic α-carbons. In addition to the above example, a modular synthetic approach has been developed to access hydroxymethyl-functionalized α-methylene azetidin-2-ones **51** (Scheme 2.10) [33].

The main synthetic challenges associated with the target molecules are preparation of 2-(2-hydroxy-1-sulfonyl) propenamides **52**, in particular implementation of a practical strategy involving nucleophilic addition of β-sulfonyl-propenamide-derived dianion **53** with various aldehydes. After the addition products had been converted to 2-(1-methanesulfonyloxy) propenamides **54** through an addition-elimination sequence of sodium benzeneselenolate and successive mesylation of the resulting 2-(1-hydroxy)-propenamides **55**, intramolecular 1,4-cyclization has been accomplished, using sulfonates to generate **51**. Given the success in accessing the highly strained azetidin-2-one ring systems that allow a range of structural modification and functionalization, the synthetic methodology based on the

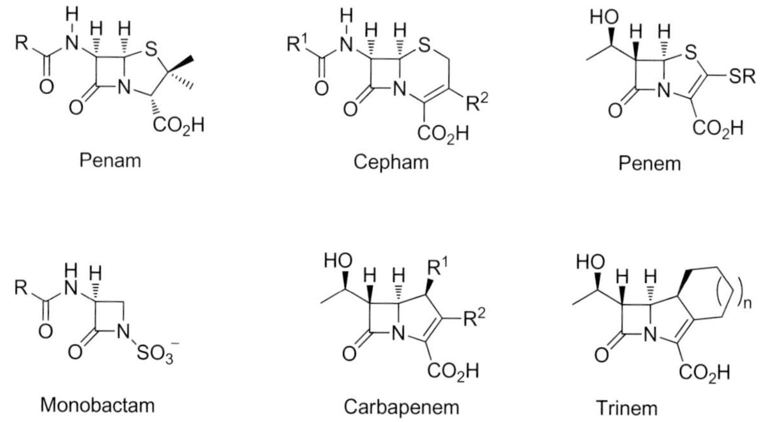

Figure 2.1 Representative structural classes of β-lactam antibiotics.

intramolecular 1,4-cyclization has been demonstrated to represent a potentially useful and generally applicable class of β-lactam-forming reactions.

In microbial drug discovery research, the azetidin-2-one heterocyclic core has been recognized as the key structural element for diverse analogs of prominent antimicrobial agents, which comprise natural and unnatural subclasses such as penams, cephems, monobactams, carbapenems and trinems (Figure 2.1) [30]. Despite the long history of considerable efforts to drug development and noteworthy achievements, β-lactam antibiotics synthesis remains a formidable challenge [34]. Among a myriad of the synthetic challenges of the azetidin-2-one-based molecules, total syntheses of penicillin and cephalosporin are selected as two representative examples of classic outstanding contributions to the fields of medicinal and synthetic organic chemistry.

2.8
Synthesis of Penicillin

Penicillin derived from *Penicillium chrysogenum* was discovered by Fleming in 1928 [1], and its remarkable *in vivo* antibiotic activity against a variety of pathogenic organisms was first demonstrated by Chain, Florey and co-workers [35]. As a matter of fact, a mixture of the structurally-related substances that contain azetidin-2-one fused to thiazolidine ring systems is referred to as the so-called penicillin, representing a member of the penam family. In industrial processes, the penicillin production has been based on semi-synthetic procedures [36], where penicillin G generated via fermentation of the *Penicillium* fungi is chemically converted by displacement of the phenylacetyl side-chain to various types of penicillins via the formation of 6-aminopenicillanic acid (Scheme 2.11) [37].

Historically, the chemical approach to penicillin production was developed in the 1950s. An enormous advancement of the field of synthetic organic chemistry

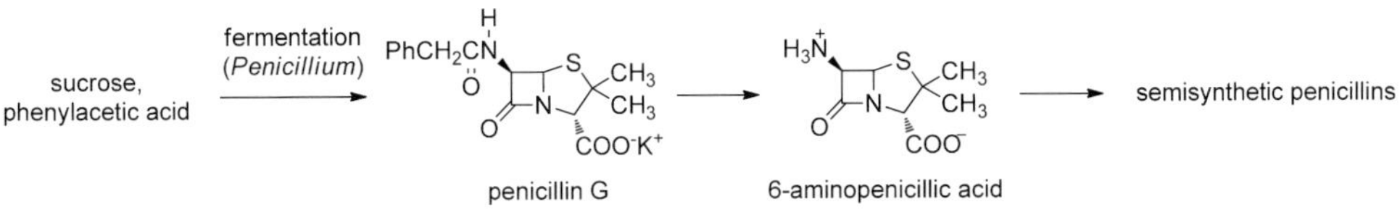

2.8 Synthesis of Penicillin

Scheme 2.11

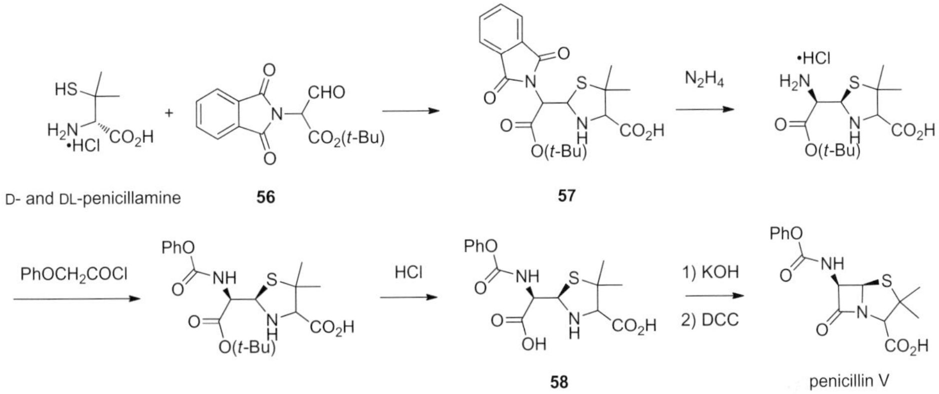

Scheme 2.12

prompted many researchers to achieve total synthesis of these natural products. The pioneering studies by Sheehan and co-workers paved the way for the total synthesis of penicillin V as well as to lay the foundations for the development of therapeutic tools which have been attracting intense interest for scientific research and industrial applications [38].

According to the studies established in 1957, their synthetic strategy involves the use of D-penicillamine, a chiral source for the required stereocenters (Scheme 2.12). The reaction of this material with t-butyl phthalimidomalonaldehydate **56** proceeds stereoselectively to form two stereoisomers of thiazolidines **57**. One of the constituents that appeared to have the desired absolute configurations was converted into penicilloic acid **58** through several transformations of functional groups attached to the phthalimide-derived nitrogen atom and subsequent acidic removal of t-butyl ester group. In the final step of the total synthesis, potassium salt of penicilloic acid underwent effective condensation upon treatment of the dilute solution with N,N′-dicyclohexylcarbodiimide to afford the target compound, potassium salt of penicillin V which proved to exhibit a potent microbiological activity equivalent to that of the natural product. It is interesting to note that they also synthesized L-antipode of penicillin V via optical resolution of the racemic product as well as demonstrated its microbiological inertness during the course of their target-directed synthesis. As a consequence, the above studies have served to extend the scope of synthetic organic chemistry, promoting the researches on biomedical and pharmaceutical applications directed toward development of improved antibiotic candidates for the antimicrobial therapy of drug-resistant strains.

2.9
Synthesis of Cephalosporin

In 1953, Newton and Abraham found that a species of *Cephalosporium acremonium* produced a new type of penicillin called cephalosporin N [39]. During the course of studying this antibacterial substance in 1955, they isolated a structurally-related natural product, cephalosporin C [40], whose structure was established through X-ray analysis by Crowfoot Hodgkin and co-workers in 1961 [41]. Interestingly, cephalosporin C has been shown to display a broad spectrum of antibiotic activity even against Gram-negative bacteria, albeit low potency of these primitive generations for therapeutic use. From this basis, the cephalosporin nucleus, which consists of azetidin-2-one fused to a sulfur-containing six-membered ring, had been modified to overcome this limitation and to gain more extended spectrum as well as greater Gram-negative antimicrobial activity, resulting in the development of analogous cephalosporins that constitute the cephem subclass. As has been noted by Morin and co-workers [42], productions of cephalosporins in the biological systems have been found to involve enzymatic transformation of penicillins, where the thiazolidine ring should be converted through oxidation, esterification, and dehydration to ring-expanded intermediates, allowing the access to this natural product (Scheme 2.13).

The large-scale industrial synthesis of the cephalosporin antibiotics can be achieved by a semi-synthetic procedure that follows the biomimetic route. In this process, penicillin G generated via the fermentation is chemically transformed to 7-aminodeacetoxycephalosporanic acid that undergoes amidic coupling with appropriate side-chains to produce various cephalosporins (Scheme 2.14) [36, 43]. With respect to the challenge for the chemical synthesis, Woodward and co-workers achieved the first total synthesis of cephalosporin C and its derivative in 1966 (Scheme 2.15) [44].

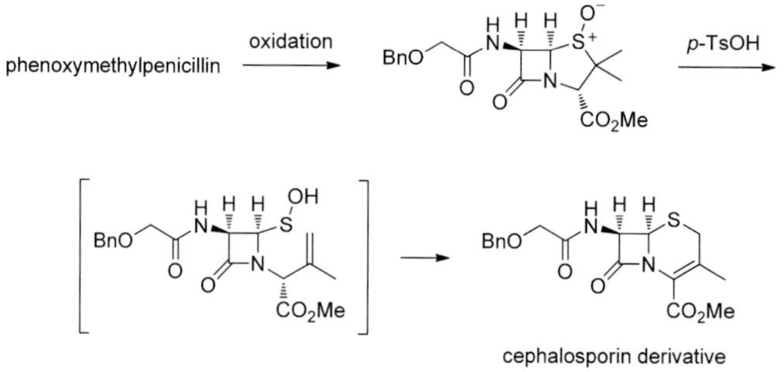

Scheme 2.13

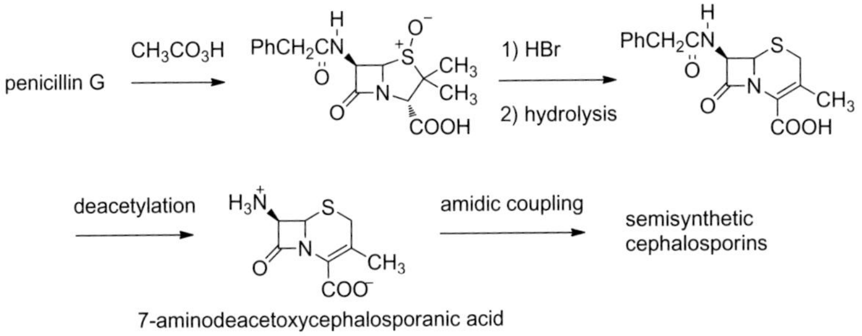

2.9 Synthesis of Cephalosporin

Scheme 2.14

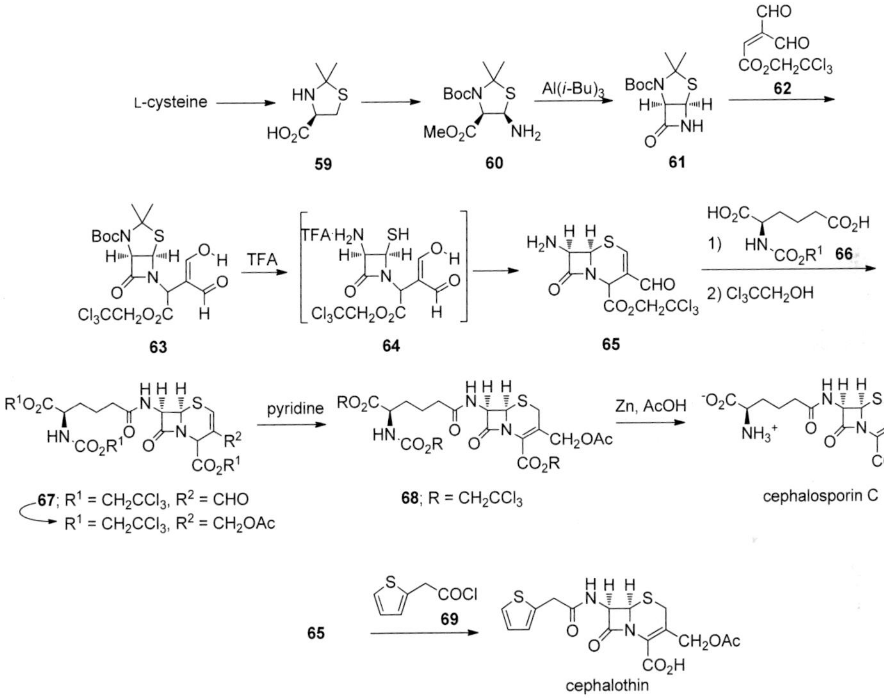

Scheme 2.15

Their approach to the cephalosporins started with the preparation of L-2,2-dimethylthiazolidine-4-carboxylic acid **59**, which was readily accessible according to their established procedure from L-cysteine, that served as a chiral source for the required stereocenters. Following structural modifications of the thiazolidine rings allowed access to *cis*-amino ester **60** which underwent intramolecular cyclization to afford thiazolidine-fused azetidin-2-one **61** upon treatment with triisobutylaluminum. The presence of a partially free electron pair on the less-conjugated amide nitrogen makes the azetidin-2-one reactive in nucleophilic processes, so the reaction of this compound with 2-methylenemalonaldehyde derivative **62** led to efficient production of the Michael adduct **63**. Since the thiazolidine ring system is labile under acidic conditions, this material decomposed to *cis*-amino thiol **64** through treatment with trifluoroacetic acid, which immediately formed bicyclic aminoaldehyde **65**. Condensation of this compound with N-protected D-α-aminoadipic acid **66** in the presence of N,N'-dicyclohexylcarbodiimide gave amidoaldehyde **67**, whose carbonyl group was subsequently converted by the reduction and acetylation processes to acetoxymethyl functionality. Furthermore, isomerization of the C=C double bond promoted by addition of pyridine occurred during equilibration between different structures, giving rise to the fully protected cephalosporin C **68**. Removal of all the protecting groups via reductive cleavage using zinc dust in 90% aqueous acetic acid provided cephalosporin C. The synthetic route illustrated above was also applied toward the total synthesis of cephalothin. Thus, this structurally related antibiotic agent could be approached by employing α-(thiophen-2-yl) acetyl chloride **69** instead of **66** for the amidation of **65** and by following the established synthetic route.

2.10
Conclusion

The studies described here exemplify the obvious importance of azetidine-based compounds, which have potential applications in the areas of both agrochemical and pharmaceutical developments. In particular, azetidin-2-ones represent a key stage in the ongoing evolution of drug discovery involving the synthesis and biological evaluation of a new class of antibiotics. Despite the great significance of the four-membered heterocyclic structures, development of a simple and efficient methodology for the construction of such compounds is substantially less advanced compared with medium-sized heterocycles. It should again be emphasized that the synthetic elaboration of the bicyclic β-lactams has more difficulty than that of the monocyclic azetidines and is subject to the inherent limitations that may arise due to the lability of the susceptible amide linkages. There have been numerous efforts to find potential applications for azetidines in organic synthesis, where the highly strained molecules can serve as synthetic intermediates for ring-opening [45] and ring-expansion [46] reactions, generating γ-amino butyric acids and larger ring heterocycles, respectively. Hopefully, new versatile methodologies giving many important applications in the construction of complex molecular systems will be developed in due course.

Acknowledgment

Fruitful conversations with Emeritus Professor Takebe, Shizuoka University, are gratefully acknowledged. We thank Emeritus Professor Mori, The University of Tokyo, and Emeritus Professor Tanaka, Wakayama University, for their useful advice and helpful comments on the preparative aspects of our synthetic works.

References

1 Fleming, A. (1929) The antibacterial action of cultures of a penicillium, with special reference to their use in the isolation of B influenzae. *Br. J. Exp. Path.*, **10**, 226–236.

2 (a) Abrahamsson, S., Crowfoot Hodgkin, D. and Maslen, E.N. (1963) The crystal structure of phenoxymethylpenicillin. *Biochem. J.*, **86**, 514–535; (b) Crowfoot Hodgkin, D. (1965) The X-ray analysis of complicated molecules. *Science*, **150**, 979–988.

3 (a) Nathwani, D. and Wood, M.J. (1993) Penicillins: a current review of their clinical pharmacology and therapeutic use. *Drugs*, **45**, 866–894; (b) Niccolai, D., Tarsi, L. and Thomas, R.J. (1997) The renewed challenge of antibacterial chemotherapy. *Chem. Commun.*, 2333–2342.

4 (a) Livermore, D.M. (1995) β-lactamases in laboratory and clinical resistance. *Clin. Microbiol. Rev.*, **8**, 557–584; (b) Ruddle, C.C. and Smyth, T.P. (2007) Exploring the chemistry of penicillin as a β-lactamase-dependent prodrug. *Org. Biomol. Chem.*, **5**, 160–168.

5 Brandi, A., Cicchi, S. and Cordero, F.M. (2008) Novel syntheses of azetidines and azetidinones. *Chem. Rev.*, **108**, 3988–4035.

6 (a) Carreira, L.A. and Lord, R.C. (1969) Far-infrared spectra of ring compounds. IV. Spectra of compounds with an unsymmetrical potential function for ring inversion. *J. Chem. Phys.*, **51**, 2735–2744; (b) Eliel, E.L. and Wilen, S.H. (1994) Configuration and conformation of cyclic molecules, in *Stereochemistry of Organic Compounds*, John Wiley & Sons, Inc., New York, USA, pp. 665–834.

7 (a) Puentes, C.O. and Kouznetsov, V. (2002) Recent advancements in the homoallylamine chemistry. *J. Heterocycl. Chem.*, **39**, 595–614; (b) Couty, F., Evano, G. and Rabasso, N. (2003) Synthesis of enantiopure azetidine 2-carboxylic acids and their incorporation into peptides. *Tetrahedron Asymmetry*, **14**, 2407–2412; (c) Breternitz, H.-J. and Schaumann, E. (1999) Ring-opening of *N*-tosylaziridines by heterosubstituted allyl anions. Application to the synthesis of azetidines and pyrrolidines. *J. Chem. Soc. Perkin Trans. 1*, 1927–1931; (d) Vargas, L.M., Rozo, W. and Kouznetsov, V. (2000) Synthesis of new spiro-*N*-heterocycles with cyclooctane fragment from *N*-(1-alkenylcyclooctyl)-*N*-aryl (benzyl) amines. *Heterocycles*, **53**, 785–796; (e) Robin, S. and Rousseau, G. (2000) Preparation of azetidines by 4-*endo trig* cyclizations of *N*-cinnamyl tosylamides. *Eur. J. Org. Chem.*, 3007–3011; (f) Wadsworth, D.H. (1973) Azetidine. *Org. Synth.*, **53**, 13–16.

8 (a) Uyehara, T., Yuuki, M., Masaki, H., Matsumoto, M., Ueno, M. and Sato, T. (1995) Lewis acid-promoted [2 + 2] azetidine annulation of *N*-acylaldimines with allyltriisopropylsilane. *Chem. Lett.*, 789–790; (b) Burtoloso, A.C.B. and Correia, C.R.D. (2006) A new entry to the synthesis of substituted azetidines: [2 + 2] cycloaddition reaction of four-membered endocyclic enamides to ketenes. *Tetrahedron Lett.*, **47**, 6377–6380; (c) Xu, X., Cheng, D., Li, J., Guo, H. and Yan, J. (2007) Copper-catalyzed highly efficient multicomponent reactions: synthesis of 2-(sulfonylimino)-4-(alkylimino) azetidine derivatives. *Org. Lett.*, **9**, 1585–1587.

9 (a) Moulines, J., Bats, J.-P., Hautefaye, P., Nuhrich, A. and Lamidey, A.-M. (1993) Substituent control in the synthesis of azetidines and pyrrolidines by *N*-tosyloxiraneethylamines base-mediated cyclization. *Tetrahedron Lett.*, **34**, 2315–2318; (b) Medjahdi, M., González-Gómez, J.C., Foubelo, F. and Yus, M. (2009) Stereoselective synthesis of azetidines and pyrrolidines from *N*-*tert*-butylsulfonyl (2-aminoalkyl) oxiranes. *J. Org. Chem.*, **74**, 7859–7865.

10 (a) Anzai, M., Toda, A., Ohno, H., Takemoto, Y., Fujii, N. and Ibuka, T. (1999) Palladium-catalyzed regio- and stereoselective synthesis of *N*-protected 2,4-dialkylated azacyclobutanes from amino allenes. *Tetrahedron Lett.*, **40**, 7393–7397; (b) Ohno, H., Anzai, M., Toda, A., Ohishi, S., Fujii, N., Tanaka, T.,

Takemoto, Y. and Ibuka, T. (2001) Stereoselective synthesis of 2-alkenylaziridines and 2-alkenylazetidines by palladium-catalyzed intramolecular amination of α- and β-amino allenes. *J. Org. Chem.*, **66**, 4904–4914; (c) Rutjes, F.P.J.T., Tjen, K.C.M.F., Wolf, L.B., Karstens, W.F.J., Schoemaker, H.E. and Hiemstra, H. (1999) Selective azetidine and tetrahydropyridine formation via Pd-catalyzed cyclizations of allene-substituted amines and amino acids. *Org. Lett.*, **1**, 717–720.

11 Nakamura, I., Nemoto, T., Yamamoto, Y. and de Meijere, A. (2006) Thermally induced and silver-salt-catalyzed [2 + 2] cycloadditions of imines to (alkoxymethylene) cyclopropanes. *Angew. Chem. Int. Ed.*, **45**, 5176–5179.

12 Akiyama, T., Daidouji, K. and Fuchibe, K. (2003) Cu(I)-catalyzed enantioselective [2 + 2] cycloaddition of 1-methoxyallenylsilane with α-imino ester: chiral synthesis of α,β-unsaturated acylsilanes. *Org. Lett.*, **5**, 3691–3693.

13 (a) Takemoto, T., Nomoto, K., Fushiya, S., Ouchi, R., Kusano, G., Hikino, H., Takagi, S., Matsuura, Y. and Kakudo, M. (1978) Structure of mugineic acid, a new amino acid possessing an iron. chelating activity from roots washings of water-cultured *Hordeum vulgare* L. *Proc. Jpn. Acad.*, **54** (B), 469–473; (b) Iwashita, T., Mino, Y., Naoki, H., Sugiura, Y. and Nomoto, K. (1983) High-resolution ^{1}H NMR analysis of solution structures and conformational properties of mugineic acid and its metal complexes. *Biochemistry*, **22**, 4842–4845; (c) Ripperger, H. and Schreiber, K. (1982) Nicotianamine and analogous amino acids, endogenous iron carriers in higher plants. *Heterocycles*, **17**, 447–461.

14 (a) Takagi, S. (1976) Naturally occurring iron-chelating compounds in oat- and rice-root washings. *Soil Sci. Plant Nutr.*, **22**, 423–433; (b) Murakami, T., Ise, K., Hayakawa, M., Kamei, S. and Takagi, S. (1989) Stabilities of metal complexes of mugineic acids and their specific affinities for Iron(III). *Chem. Lett.*, 2137–2140.

15 Nomoto, K., Yoshioka, H., Arima, M., Fushiya, S., Takagi, S. and Takemoto, T. (1981) Structure of 2′-deoxymugineic acid, a novel amino acid possessing an iron-chelating activity. *Chimia*, **35**, 249–250.

16 Fowden, L. (1956) Azetidine-2-carboxylic acid: a new cyclic imino acid occurring in plants. *Biochem. J.*, **64**, 323–332.

17 Ohfune, Y., Tomita, M. and Nomoto, K. (1981) Total synthesis of 2′-deoxymugineic acid, the metal chelator excreted from wheat root. *J. Am. Chem. Soc.*, **103**, 2409–2410.

18 (a) Fushiya, S., Sato, Y., Nakatsuyama, S., Kanuma, N. and Nozoe, S. (1981) Synthesis of avenic acid A and 2′-deoxymugineic acid, amino acids possessing an iron chelating activity. *Chem. Lett.*, 909–912; (b) Matsuura, F., Hamada, Y. and Shioiri, T. (1994) Total synthesis of 2′-deoxymugineic acid and nicotianamine. *Tetrahedron*, **50**, 9457–9470; (c) Shioiri, T., Irako, N., Sakakibara, S., Matsuura, F. and Hamada, Y. (1997) A new efficient synthesis of nicotianamine and 2′-deoxymugineic acid. *Heterocycles*, **44**, 519–530; (d) Klair, S.S., Mohan, H.R. and Kitahara, T. (1998) A novel synthetic approach towards phytosiderophores: expeditious synthesis of nicotianamine and 2′-deoxymugineic acid. *Tetrahedron Lett.*, **39**, 89–92; (e) Miyakoshi, K., Oshita, J. and Kitahara, T. (2001) Expeditious synthesis of nicotianamine and 2′-deoxymugineic acid. *Tetrahedron*, **57**, 3355–3360; (f) Singh, S., Crossley, G., Ghosal, S., Lefievre, Y. and Pennington, M.W. (2005) A short practical synthesis of 2′-deoxymugineic acid. *Tetrahedron Lett.*, **46**, 1419–1421; (g) Namba, K., Murata, Y., Horikawa, M., Iwashita, T. and Kusumoto, S. (2007) A practical synthesis of the phytosiderophore 2′-deoxymugineic acid: a key to the mechanistic study of iron acquisition by graminaceous plants. *Angew. Chem. Int. Ed.*, **46**, 7060–7063; (h) Jung, Y., Hoon, K. and Chang, M. (1999) A formal asymmetric synthesis of mugineic acid: an efficient synthetic route through chiral oxazolidinone. *Arch. Pharm. Res.*, **22**, 624–628.

19 (a) Hamada, Y. and Shioiri, T. (1986) Synthesis of mugineic acid through direct C-acylation using diphenyl phosphorazidate. *J. Org. Chem.*, **51**, 5489–5490; (b) Matsuura, F., Hamada, Y. and Shioiri, T. (1992) Efficient synthesis of mugineic acid, a typical phytosiderophore, utilizing the phenyl group as the carboxyl synthon. *Tetrahedron Lett.*, **33**, 7917–7920; (c) Matsuura, F., Hamada, Y. and Shioiri, T. (1993) Total synthesis of mugineic acid. efficient use of the phenyl group as the carboxyl synthon. *Tetrahedron*, **49**, 8211–8222.

20 (a) Kobayashi, J., Cheng, J.-F., Ishibashi, M., Wäichli, M.R., Yamamura, S. and Ohizumi, Y. (1991) Penaresidin A and B, two novel azetidine alkaloids with potent actomyosin ATPase-activating activity from the Okinawan marine sponge *Penares* sp. *J. Chem. Soc., Perkin Trans. 1*, 1135–1137; (b) Ohshita, K., Ishiyama, H., Takahashi, Y., Ito, J., Mikami, Y. and Kobayashi, J. (2007) Synthesis of penaresidin derivatives and its biological activity. *Bioorg. Med. Chem.*, **15**, 4910–4916.

21 (a) Takikawa, H., Maeda, T. and Mori, K. (1995) Synthesis of penaresidin A, an azetidine alkaloid with actomyosin atpase-activating property. *Tetrahedron Lett.*, **36**, 7689–7692; (b) Kobayashi, J., Tsuda, M., Cheng, J.-F., Ishibashi, M., Takikawa, H. and Mori, K. (1996) Absolute stereochemistry of penaresidins A and B. *Tetrahedron Lett.*, **37**, 6775–6776; (c) Takikawa, H., Maeda, T., Seki, M., Koshino, H. and Mori, K. (1997) Synthesis of sphingosine relatives. Part 19. Synthesis of penaresidin A and B, azetidine alkaloids with actomyosin ATPase-activating properties. *J. Chem. Soc. Perkin Trans. 1*, 97–111.

22 (a) Raghavan, S. and Krishnaiah, V. (2010) An efficient stereoselective synthesis of penaresidin A from (*E*)-2-protected amino-3,4-unsaturated sulfoxide. *J. Org. Chem.*, **75**, 748–761; (b) Liu, D.-G. and Lin, G.-Q. (1999) Novel enantioselective synthesis of penaresidin A and *Allo*-penaresidin A *via* the construction of a highly functionalized azetidine. *Tetrahedron Lett.*, **40**, 337–340;

(c) Knapp, S. and Dong, Y. (1997) Stereoselective synthesis of penaresidin A and related azetidine alkaloids. *Tetrahedron Lett.*, **38**, 3813–3816.

23 Yoda, H., Oguchi, T. and Takabe, K. (1997) Novel asymmetric synthesis of penaresidin B as a potent actomyosin ATPase activator. *Tetrahedron Lett.*, **38**, 3283–3284.

24 Hiraki, T., Yamagiwa, Y. and Kamikawa, T. (1995) Synthesis of a straight chain analog of penaresidins, azetidine alkaloids from marine sponge *Penares* sp. *Tetrahedron Lett.*, **36**, 4841–4844.

25 (a) Yoda, H., Uemura, T. and Takabe, K. (2003) Novel and practical asymmetric synthesis of an azetidine alkaloid, penaresidin B. *Tetrahedron Lett.*, **44**, 977–979; (b) Yoda, H. (2002) Recent advances in the synthesis of naturally occurring polyhydroxylated alkaloids. *Curr. Org. Chem.*, **6**, 223–243.

26 (a) Zafaralla, G. and Mobashery, S. (1993) Evidence for a new enzyme-catalyzed reaction other than β-lactam hydrolysis in turnover of a penem by the TEM-1 β-lactamase. *J. Am. Chem. Soc.*, **115**, 4962–4965; (b) Faraci, W.S. and Pratt, R.F. (1984) Elimination of a good leaving group from the 3'-position of a cephalosporin need not be concerted with β-lactam ring opening: TEM-2 β-lactamase-catalyzed hydrolysis of pyridine-2-azo-4'-(*N'*,*N'*-dimethylaniline) cephalosporin (PADAC) and of cephaloridine. *J. Am. Chem. Soc.*, **106**, 1489–1490; (c) Page, M.I. (1984) The mechanisms of reactions of β-lactam antibiotics. *Acc. Chem. Res.*, **17**, 144–151; (d) Cimarusti, C.M. (1984) Dependence of β-lactamase stability on substructures within β-lactam antibiotics. *J. Med. Chem.*, **27**, 247–253; (e) Knight, W.B., Green, B.G., Chabin, R.M., Gale, P., Maycock, A.L., Weston, H., Kuo, D.W., Westler, W.M. and Dorn, C.P. (1992) Specificity, stability and potency of monocyclic β-lactam inhibitors of human leukocyte elastase. *Biochemistry*, **31**, 8160–8170.

27 (a) Johnson, J.R., Woodward, R.B. and Robinson, R. (1949) The Constitution of the Penicillins, in *The Chemistry of Penicillin* (eds H.T. Clarke, J.R. Johnson

and R. Robinson), Princeton University Press, Princeton, NJ, USA, pp. 440–454; (b) Woodward, R.B. (1966) Recent advances in the chemistry of natural products. *Science*, **153**, 487–493; (c) Page, M.I. and Proctor, P. (1984) Mechanism of β-lactam ring opening in cephalosporins. *J. Am. Chem. Soc.*, **106**, 3820–3825; (d) Cohen, N.C. (1983) β-lactam antibiotics: geometrical requirements for antibacterial activities. *J. Med. Chem.*, **26**, 259–264.

28 (a) Knox, R. (1961) A survey of new penicillins. *Nature*, **192**, 492–496; (b) Singh, G.S. (2003) Recent progress in the synthesis and chemistry of azetidinones. *Tetrahedron*, **59**, 7631–7649; (c) Troisi, L., Granito, C. and Pindinelli, E. (2004) Synthesis and functionalizations of β-lactams, in *Targets in Heterocyclic Systems Volume 8: Chemistry and Properties* (eds O.A. Attanasi and D. Spinelli), Royal Society of Chemistry, Cambridge, UK, pp. 187–215.

29 (a) Staudinger, H. and der Ketene, Z.K. (1907) Diphenylketen. *Liebigs Ann. Chem.*, **356**, 51–123; (b) Cossio, F.P., Arrieta, A. and Sierra, M.A. (2008) The mechanism of the ketene–imine (staudinger) reaction in its centennial: still an unsolved problem? *Acc. Chem. Res.*, **41**, 925–936.

30 (a) Palomo, C., Aizpurua, J.M., Ganboa, I. and Oiarbide, M. (1999) Asymmetric synthesis of β-lactams by staudinger ketene-imine cycloaddition reaction. *Eur. J. Org. Chem.*, 3223–3235; (b) Palomo, C. and Aizpurua, J.M. (1999) Asymmetric synthesis of 3-amino-β-lactams via Staudinger ketene-imine cycloaddition reaction. *Chem. Heterocycl. Comp.*, **34**, 1222–1236.

31 (a) Brandi, A., Cicchi, S. and Cordero, F.M. (2008) Novel syntheses of azetidines and azetidinones. *Chem. Rev.*, **108**, 3988–4035; (b) Fletcher, S.R. and Trevor, K.I. (1978) Synthesis of α-methylene-β-lactams. *J. Chem. Soc. Chem. Commun.*, 903–904; (c) Easton, C.J. and Love, S.G. (1986) Direct introduction of a benzoyloxy substituent at the C-4 position of β-lactams. *Tetrahedron Lett.*, **27**, 2315–2318; (d) Miyachi, N., Kanda, F. and Shibasaki, M. (1989) Use of copper(I) trifluoromethanesulfonate in β-lactam synthesis. *J. Org. Chem.*, **54**, 3511–3513.

32 Tanaka, K., Yoda, H., Inoue, K. and Kaji, A. (1986) A new method for the synthesis of 3-methylene-2-azetidinones. *Synthesis*, 66–69.

33 Tanaka, K., Horiuchi, H. and Yoda, H. (1989) Dianion of N-phenyl-2-([phenylsulfonyl] methyl) propenamide as a versatile reagent for the preparation of α-methylene carbonyl compounds. *J. Org. Chem.*, **54**, 63–70.

34 Dürckheimer, W., Blumbach, J., Lattrell, R. and Scheunemann, K.H. (1985) Recent developments in the field of β-lactam antibiotics. *Angew. Chem. Int. Ed. Engl.*, **24**, 180–202.

35 Chain, E., Florey, H.W., Gardner, A.D., Heatley, N.G., Jennings, M.A., Orr-Ewing, J. and Sanders, A.G. (1940) Penicillin as a chemotherapeutic agent. *Lancet*, **236**, 226–228.

36 Wegman, M.A., Janssen, M.H.A., van Rantwijk, F. and Sheldon, R.A. (2001) Towards biocatalytic synthesis of β-lactam antibiotics. *Adv. Synth. Catal.*, **343**, 559–576.

37 Batchelor, F.R., Doyle, F.P., Nayler, J.H.C. and Rolinson, G.N. (1959) Synthesis of penicillin: 6-aminopenicillanic acid in penicillin fermentations. *Nature*, **183**, 257–258.

38 (a) Sheehan, J.C. and Henery-Logan, K.R. (1957) The total synthesis of penicillin V. *J. Am. Chem. Soc.*, **79**, 1262–1263; (b) Sheehan, J.C. and Henery-Logan, K.R. (1959) The total synthesis of penicillin V. *J. Am. Chem. Soc.*, **81**, 3089–3094; (c) Sheehan, J.C. (1996) Penicillin V, in *Classics in Total Synthesis: Targets, Strategies, Methods* (eds K.C. Nicolaou and E.J. Sorensen), Wiley-VCH Verlag GmbH, Weinheim, Germany, pp. 41–53.

39 Abraham, E.P., Newton, G.G.F., Crawford, K., Burton, H.S. and Hale, C.W. (1953) Cephalosporin N: a new type of penicillin. *Nature*, **171**, 343.

40 (a) Newton, G.G.F. and Abraham, E.P. (1955) Cephalosporin C, a new antibiotic containing sulfur and D-α-aminoadipic acid. *Nature*, **175**, 548; (b) Newton, G.G.F. and Abraham, E.P. (1956) Isolation of cephalosporin C, a penicillin-like antibiotic containing D-α-aminoadipic

acid. *Biochem. J.*, **62**, 651–658; (c) Abraham, E.P. and Newton, G.G.F. (1961) The structure of cephalosporin C. *Biochem. J.*, **79**, 377–393.

41 Crowfoot Hodgkin, D. and Maslen, E.N. (1961) The X-ray analysis of the structure of cephalosporin C. *Biochem. J.*, **79**, 393–402.

42 Morin, R.B., Jackson, B.G., Mueller, R.A., Lavagnino, E.R. and Scanlon, W.B. (1963) Chemistry of cephalosporin antibiotics. III. Chemical correlation of penicillin and cephalosporin antibiotics. *J. Am. Chem. Soc.*, **85**, 1896–1897.

43 Adrio, J.L. and Demain, A.L. (2002) Improvements in the formation of cephalosporins from penicillin G and other penicillins by bioconversion. *Org. Proc. Res. Dev.*, **6**, 427–433.

44 Woodward, R.B., Heusler, K., Gosteli, J., Naegeli, P., Oppolzer, W., Ramage, R., Ranganathan, S. and Vorbrüggen, H. (1966) The total synthesis of cephalosporin C. *J. Am. Chem. Soc.*, **88**, 852–853.

45 (a) Couty, F., Durrat, F. and Evano, G. (2005) Ring expansion and ring opening of azetidines, in *Targets in Heterocyclic Systems Volume 9: Chemistry and Properties* (eds O.A. Attanasi and D. Spinelli), Royal Society of Chemistry, Cambridge, UK, pp. 186–210; (b) Banik, B.K., Manhas, M.S. and Bose A.K. (1993) Versatile β-lactam synthons: enantiospecific synthesis of (−)-polyoxamic acid. *J. Org. Chem.*, **58**, 307–309; (c) Ma, S.-H., Yoon, D.H., Ha, H.-J. and Lee, W.K. (2007) Preparation of enantiopure 2-acylazetidines and their reactions with chloroformates. *Tetrahedron Lett.*, **48**, 269–271; (d) Ojima, I. (1995) Recent advances in the β-lactam synthon method. *Acc. Chem. Res.*, **28**, 383–389.

46 (a) Crépin, D., Dawick, J. and Aïssa, C. (2010) Combined rhodium-catalyzed carbon-hydrogen activation and β-carbon elimination to access eight-membered rings. *Angew. Chem. Int. Ed.*, **49**, 620–623; (b) Koya, S., Yamanoi, K., Yamasaki, R., Azumaya, I., Masu, H. and Saito, S. (2009) Selective synthesis of eight-membered cyclic ureas by the [6 + 2] cycloaddition reaction of 2-vinylazetidines and electron-deficient isocyanates. *Org. Lett.*, **11**, 5438–5441; (c) Kazi, B., Kiss, L., Forró, E. and Fülöp, F. (2010) Synthesis of orthogonally protected azepane β-amino ester enantiomers. *Tetrahedron Lett.*, **51**, 82–85; (d) Alcaide, B., Martin-Cantalejo, Y., Rodriguez-Lopez, J. and Sierra, M.A. (1993) New reactivity patterns of the β-lactam ring: tandem C3-C4 bond breakage-rearrangement of 4-acyl- or 4-imino-3,3-dimethoxy-2-azetidinones promoted by $SnCl_2 \cdot H_2O$. *J. Org. Chem.*, **58**, 4767–4770.

3
Epoxides and Oxetanes
Biswanath Das and Kongara Damodar

3.1
Introduction

The three-membered ring in epoxides or oxiranes [1] is highly strained. Epoxides are versatile building blocks in organic synthesis because the ring-opening of these compounds leads to the formation of alcohols of various structural patterns [2]. Oxetanes [3] can be used in synthesis as equivalents for the a^3-synthon, and undergo a wide range of chemical transformations. As with their three-membered ring analogs, the epoxides, oxetanes can be self-polymerized under Lewis acid conditions [4] or polymerized through anionic ring opening [5] to give a range of polyether architectures. They have also been converted into cage molecules for the selective complexation of cations [6]. Epoxides [7] and oxetanes [8] are important heterocyclic systems because they are present in several natural bioactive compounds. They are also biogenetic precursors of different natural products. In addition, they have played very important roles in the synthesis of various naturally occurring compounds and their analogs.

The preparations and reactions of epoxides and oxetanes have been described in some review articles [1, 2]. In the present review, we have included examples of some natural products containing epoxide and oxetane rings and mentioned the recent synthetic applications of these heterocycles to construct natural products (Table 3.1).

3.2
Epoxides in Natural Product Synthesis

Both the formation and opening of the epoxide ring are very important in the modern synthesis of natural products. The formation of the epoxide ring is useful to construct the natural products possessing the epoxide moiety, while the epoxide ring opening is used to generate the required oxygenated functions present in many natural products. Asymmetric epoxidation is a routine method to introduce the required chirality with oxygen-bearing functions.

Heterocycles in Natural Product Synthesis, First Edition. Edited by Krishna C. Majumdar and Shital K. Chattopadhyay.
© 2011 Wiley-VCH Verlag GmbH & Co. KGaA. Published 2011 by Wiley-VCH Verlag GmbH & Co. KGaA.

Table 3.1 Natural products containing epoxide and oxetane rings.

Serial No.	Trivial name	Structure	Source (Name of the plant/microbe/organism)	Isolation [Ref]	Biological activity	Synthesis [Ref]
1	(−)-Posticlure		*Orgyia postica*	[9]	Sex pheremone	[10]
2	(−)-Scyphostatine		*Desyseyphus mollissima*	[11]	Small molecule inhibitor of neutral sphingomyelinase (N-Smase)	[12]
3	Arenastatin A		Marine sponge *Dysidea arenaria*	[13]	Cytotoxic	[14]
4	(+)-(Z)-Laureatin		Red alga *Laurencia nipponica*	[15]		[16]
5	Manumycin A		*Streptomyces parvulus* (Tü 64)	[17]	Antibiotic	[18]

#	Name	Structure	Source	Ref	Activity	Ref
6	Taxol		Taxus species	[19]	Anticancer	[20]
7	Merrilactone A		Illicium merrillianum	[21]	Neuroprotective and neuritogenic	[22]
8	Ambuic acid		Pestalotiopsis sp. And Monochaetica sp.	[23]	Antifungal	[24]
9	Oxetin		Streptomyces sp. OM-2317	[25]	Antibiotic and herbicidal activity	[26]
10	Disparlure		Lymantria dispar (Gypsy moth)	[27]	Insect sex pheremone	[28]
11	Oleandomycin		Streptomyces antibioticus	[29]	Antibiotic	[30]
12	Coriolin		Coriolus Consors	[31]	Antibiotic	[32]

(Continued)

66 | *3 Epoxides and Oxetanes*

Table 3.1 (Continued)

Serial No.	Trivial name	Structure	Source (Name of the plant/microbe/organism)	Isolation [Ref]	Biological activity	Synthesis [Ref]
13	Epoxomicin		*Actinomycete* strain No. Q996-17	[33]	Proteasome inhibitor	[34]
14	Monocillin I		*Monocillium nordinii*	[35]	Antibiotic and antifungal	[36]
15	Multiplolide		*Xylaria multiplex*	[37]	Antifungal agent	[38]
16	Oximidine I		*Pseudomonas* sp. Q52002	[39]	Cytotoxic	[40]
17	Laulimalide		*Spongia mycofijiensis* and *Hyattela* sp.	[41]	Antimicrotubule agent	[42]

18	Cryptophycin 1		*Nostoc* sp. GSV 224	[43]	Cytotoxic	[44]
19	Epolactaene		*Penicillium* sp. BM-1689-P	[45]	Neurite out growth activity	[46]
20	(+)-11,12-epoxysarcophytol A		*Lobophytum* sp.	[47]	Anticancer	[48]
21	Palmarumycin JC1		*Jatropha curcus*	[49]	Antibacterial activity	Not reported
22	Epocarbazolin A		*Streptomyces anulatus*	[50]	Antisporiosis, antiasthmatic and antihypersensitivity	[51]
23	Oxetanocin		*Bacillus megaterium*	[52]	Antibacterial, antitumor, and antiviral activity	[53]

3.2.1
Synthesis of Natural Products Possessing an Epoxide Moiety

3.2.1.1 Synthesis of (−)-Posticlure [10]

(−)-Posticlure **1**, a *trans*-epoxide sex pheromone, was synthesized from diethyl L-tartarate **2** (Scheme 3.1). The latter was converted into a dihydroxy olefin **3** which afforded a *trans*-epoxide **4** of the required stereochemistry. The debenzylation of this compound yielded a hydroxyepoxide **5** which on oxidation followed by Wittig olefination generated (−)-posticlure **1**.

3.2.1.2 Synthesis of Natural Polyethers [54]

The epoxide cleavage cascade reaction is an important strategy for the construction of *trans-syn*-fused array of natural polyethers. As an example, the synthesis of some of the rings of gymnocin A is shown (Scheme 3.2).

3.2.1.3 Synthesis of (+)-11,12-Epoxysarcophytol A [48]

(+)-11,12-Epoxysarcophytol A **8**, an epoxy cembrane diterpenoid was isolated from marine soft coral, *Lobophytum* sp. The synthesis of the compound was achieved from a general triene precursor **9** by Sharpless epoxidation, followed by cyclization (Scheme 3.3).

3.2.1.4 Synthesis of (−)-Scyphostatine [12]

(−)-Scyphostatine **10**, a potent small molecule inhibitor of neutral spingomyelinase (N-SMase) was synthesized from L-tyrosine **11** (Scheme 3.4). The compound **11** was converted through a sequence of reactions into a cyclohexane derivative **12** which underwent epoxidation to form a β-epoxide **13** as a key intermediate. The epoxide was subsequently converted into (−)-scyphostatine **10**.

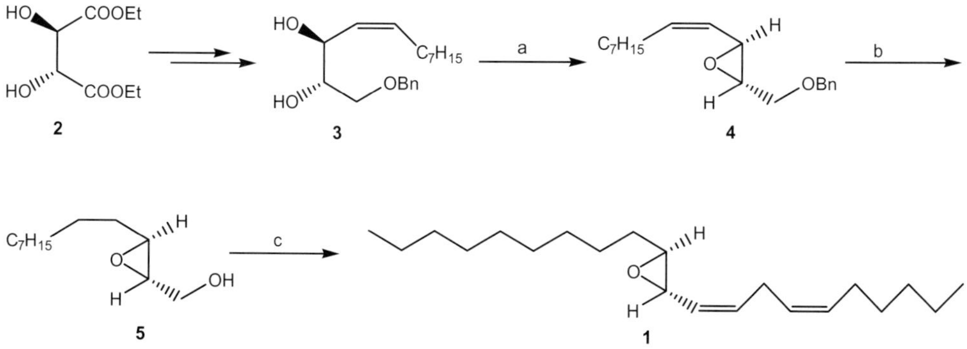

Scheme 3.1 Reagents and conditions: a) i) $CH_3C(OCH_3)_3$, *p*-TsOH, CH_2Cl_2, r.t., 30 min.; ii) CH_3COBr, CH_2Cl_2, r.t., 1.5 h; iii) K_2CO_3, MeOH, r.t., 2.5 h, 90%; b) $Pd(OH)_2/C$, H_2, MeOH/EtOAc, r.t., 12 h, 94%; c) i) PCC, $NaHCO_3$, CH_2Cl_2, 0 °C, 4 h; ii) (Z)-*n*-$C_5H_{11}C=CHCH_2CH_2P^+Ph_3I$, *n*-BuLi, r.t., 30 min., −80 °C, aldehydes from **5**, 1 h, r.t., overnight, 77%.

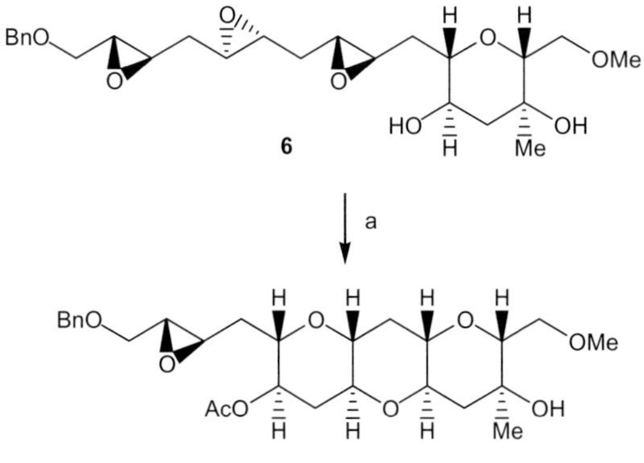

Scheme 3.2 Reagents and conditions: a) H₂O, 60°C, 5 days then Ac₂O, Et₃N, 23%.

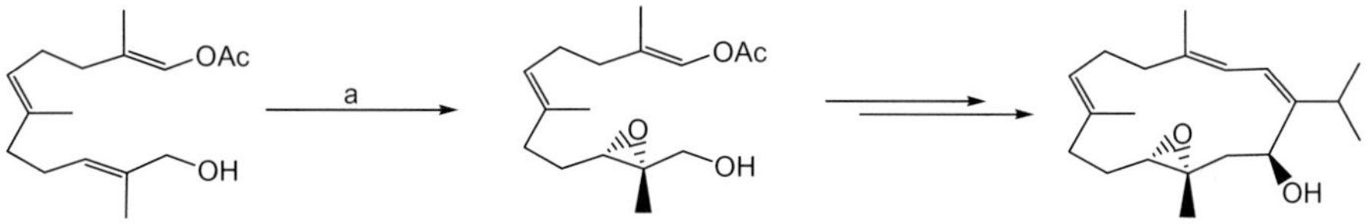

Scheme 3.3 Reagents and conditions: a) Ti(OⁱPr)₄, L-(+)-DET, t-BuO₂H, CH₂Cl₂, −40°C, 95%, 98% ee.

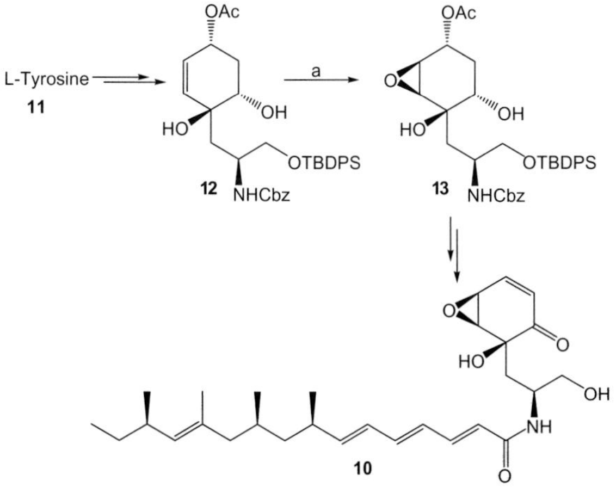

Scheme 3.4 Reagents and conditions: a) m-CPBA, CH₂Cl₂, 0°C, 84%.

3.2.1.5 Synthesis of Arenastatin A [14]

Arenastatin A **14**, a cyclic depsipeptide, was isolated from the marine sponge *Dysidea arenaria*. The compound possesses extremely potent antiproliferative activity. The cyclic peptide core **15** of the compound was synthesized starting from the aldehyde **16** (Scheme 3.5). The epoxide ring in **15** was introduced by an asymmetric Corey–Chaycovsky reaction mediated by D-camphor derived chiral sulfide **17**.

3.2.1.6 Synthesis of (+)-Ambuic Acid [24]

The bioactive epoxyquinol, ambuic acid **18** was isolated from *Pestalotiopsis*. The compound was synthesized from a simple starting material, 2,5-dimethoxybenzaldehyde **19** involving a multistep sequence (Scheme 3.6).

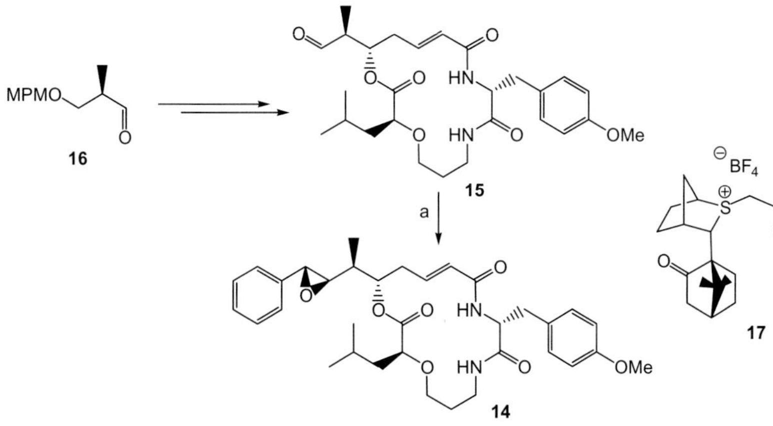

Scheme 3.5 Reagents and conditions: a) **17**, P$_2$-Et, CH$_2$Cl$_2$, −78 °C.

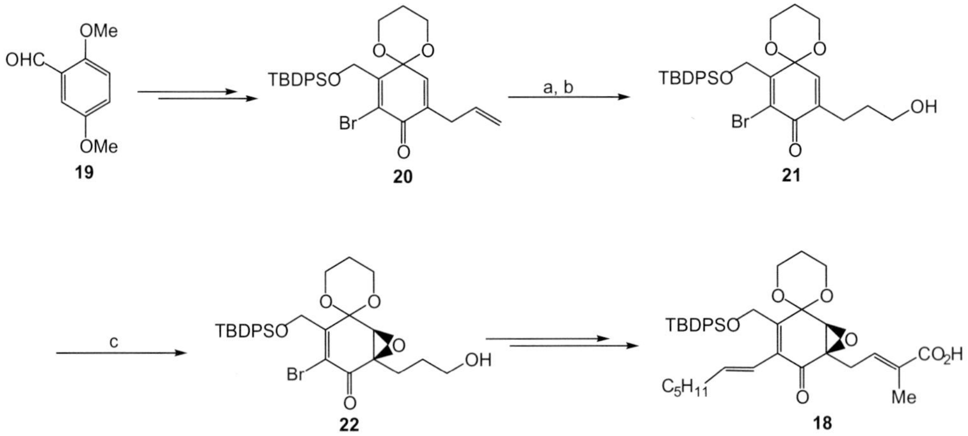

Scheme 3.6 Reagents and conditions: a) NaIO$_4$, OsO$_4$, THF/H$_2$O, 1.5 h, 62%; b) BH$_3$, *t*-uNH$_2$, MeOH/H$_2$O, THF, 0 °C, 20 min., 76%; c) Ph$_3$COOH, NaHMDS, L-DIPT, 4 A° MS, PhMe, −40 °C, 50 h, 91%, 91% ee.

3.2.1.7 Total Synthesis of Epocarbazolin A [51]

Epocarbazolin A **23**, a carbazole alkaloid, was isolated from *Streptomyces anulatus*. The synthesis of this compound in racemic form was achieved by a multicomponent reaction of an aryl triflate **24**, an aryl amine **25** and alkenylstannane **26**. The resulting carbazole compound **27** was then converted into the racemic epocarbazolin A **23** (Scheme 3.7).

3.2.1.8 Synthesis of Multiplolide A [38]

Multiplolide A **30** is an antifungal constituent of the culture broth of *Xylaria multiplex*. The molecule contains a 10-membered lactone having an epoxide ring. The synthesis of the compound was initiated with L-rhamnose **31** and D-mannitol **32**. The two fragments, **33** and **34**, derived from these two compounds were reacted and the resulting compound **35** was subjected to transformations to produce multiplolide A **30** (Scheme 3.8).

3.2.2 Synthesis of Natural Products Involving the Transformation of the Epoxide Moiety

3.2.2.1 Synthesis of Dodoneine [55]

The naturally occurring bioactive dihydropyranone dodoneine **38** was synthesized (Scheme 3.9) via an epoxide, (2S, 2S)-3-(4-(benzyloxy)phenethyl)oxirane-2-yl)

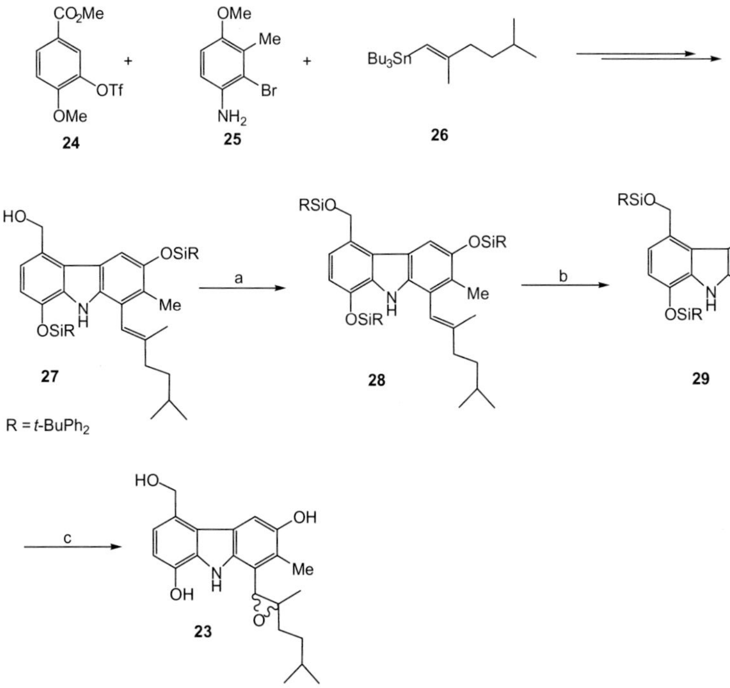

Scheme 3.7 Reagents and conditions: a) *t*-BuPh$_2$SiCl, DMF, DMAP, r.t., 4.5 h, 100%; b) dimehtyldioxirane, acetone/CH$_2$Cl$_2$, −20 °C, 24 h, 53%; c) TBAF, THF, r.t., 6 h, 24%.

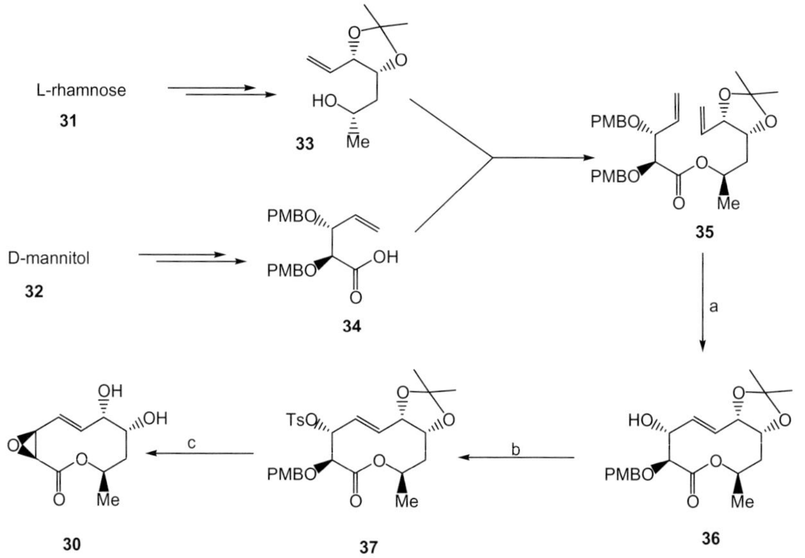

Scheme 3.8 Reagents and conditions: a) Grubb's second-generation catalyst, benzene, reflux, 6 h, 64%; b) p-TsCl, i-Pr$_2$NEt, DMAP, CH$_2$Cl$_2$, r.t., 20 h, 80%; c) i) TFA, CH$_2$Cl$_2$, r.t., 48 h; ii) NaH, THF, 0 °C, 1 h, 60% (in two steps).

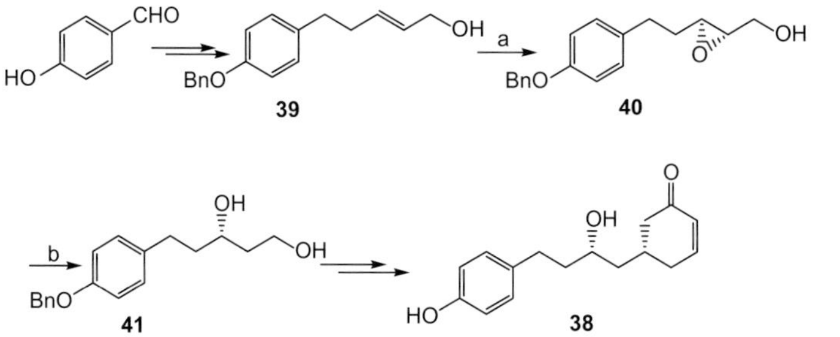

Scheme 3.9 Reagents and conditions: a) Ti(OiPr)$_4$, (+)-DIPT, TBHP, CH$_2$Cl$_2$, −20 °C, 12 h, 92%; b) Red-Al, THF, 0 °C, 0.5 h, 82%.

methanol **40**. The latter was prepared by epoxidation of the corresponding alkene **39** (generated from 4-hydroxy benzaldehyde) under Sharpless asymmetric epoxidation conditions using (+)-DIPT. The epoxide underwent reductive ring-opening to form (S)-5-(4-benzyloxy) phenyl) pentane-1,3-diol **41** which was subsequently converted into dodoneine **38**.

3.2.2.2 Synthesis of (−)-Pericosin B [56]

The highly functionalized C-7 cyclohexenoid natural products pericosines are cytotoxic metabolites of the fungus *Periconia byssoides*. The synthesis of (−)-pericosin B **42** was accomplished from commercially available (−)-quinic acid **43** through the formation and opening of epoxide strategy (Scheme 3.10).

3.2.2.3 Synthesis of (−)-Peucedanol [57]

Synthesis of (−)-peucedanol **48** was achieved from the commercially available esculetin **49** involving the formation of chiral epoxide (Scheme 3.11). The asymmetric epoxidation was carried out by using (La-[*R*]-BINOL) as catalyst. The epoxide underwent regioselective reduction to form the target molecule.

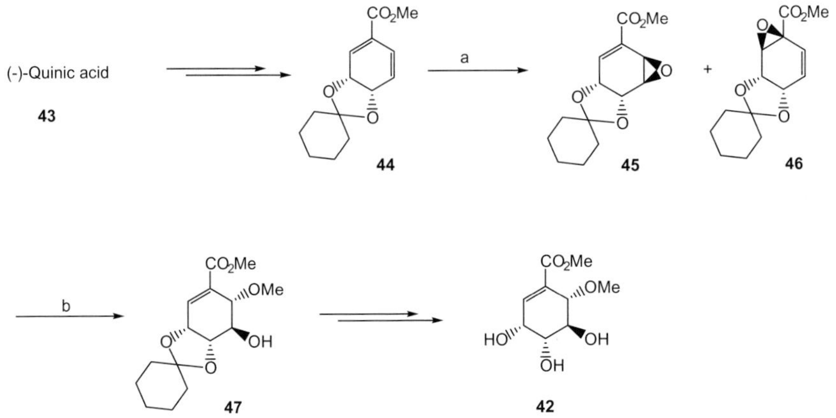

Scheme 3.10 Reagents and conditions: a) *m*-CPBA, 40 °C; b) Cat. HCl, MeOH, r.t., overnight, 54%.

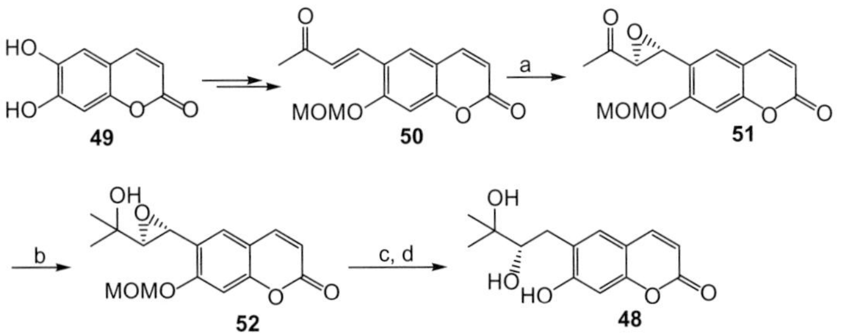

Scheme 3.11 Reagents and conditions: a) La-(R)-BINOL, Ph$_2$As=O, TBHP, Decane, 4A° MS, THF, r.t., 94%, 96% ee; b) MeMgBr, THF, −78 °C, 76%; c) NaBH$_4$, BH$_3$-THF, THF, 0 °C, 74%; d) Con. HCl–H$_2$O, THF, 40 °C, 92%.

3.2.2.4 Synthesis of (+)-Bourgeanic Acid [58]

The lichen metabolite, (+)-bourgeanic acid **53** was synthesized from the chiral 2-methyl butyl bromide **54** with 10% overall yield (Scheme 3.12). Compound **54** was converted into an olefin **55** which in turn was converted into an epoxide **56** by asymmetric epoxidation. The cleavage of the epoxide **56** generated the required chiral fragment **57** used in the synthesis of (+)-bourgeanic acid **53**.

3.2.2.5 Synthesis of (6S)-5,6-Dihydro-6([2R]-2-Hydroxy-6-Phenylhexyl)-2H-Pyran-2-one [59]

Synthesis of (6S)-5,6-dihydro-6([2R]-2-hydroxy-6-phenylhexyl)-2H-pyran-2-one **61**, a potent natural antifungal compound, was synthesized through the preparation of chiral epoxide followed by its ring opening (Scheme 3.13). The chiral epoxide **65** was prepared from an iodocarbonate **64** which was generated by diastereoselective iodine-induced electrophilic cyclization of a Boc-protected hydroxyalkene **63** formed from phenyl pentanol **62**. The hydroxyl group of the chiral epoxide **65** was protected and its ring opening was cleaved with Grignard reagent to produce a terminal alkene **66** which was used to synthesize the desired pyranone **61**.

3.2.2.6 Synthesis of Verbalactone [60]

Verbalactone **67**, a macrocyclic dimeric lactone having antibacterial activity was synthesized from hexanal **68** following similar sequences to those shown in

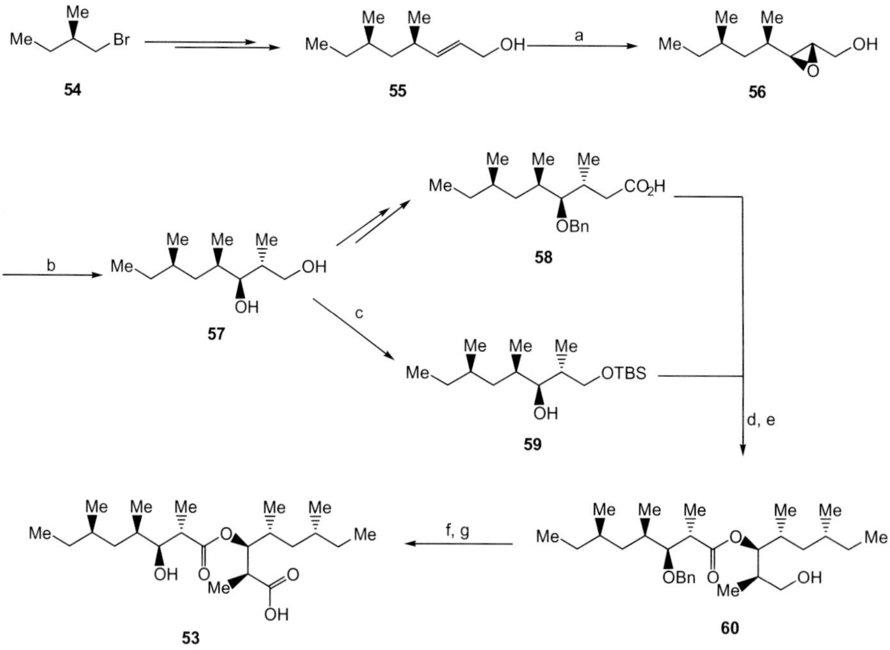

Scheme 3.12 Reagents and conditions: a) (+)-DET, Ti(O^{i}Pr)$_4$, TBHP, 99%; b) Me$_2$CuCNLi, 85%; c) TBSCl, imidazoles, 76%; d) TCBC, Et$_3$N, DMAP; e) TBAF (64% in two steps); f) pyridinium dichromate (PDC), 67%; g) H$_2$, Pd/C, 94%.

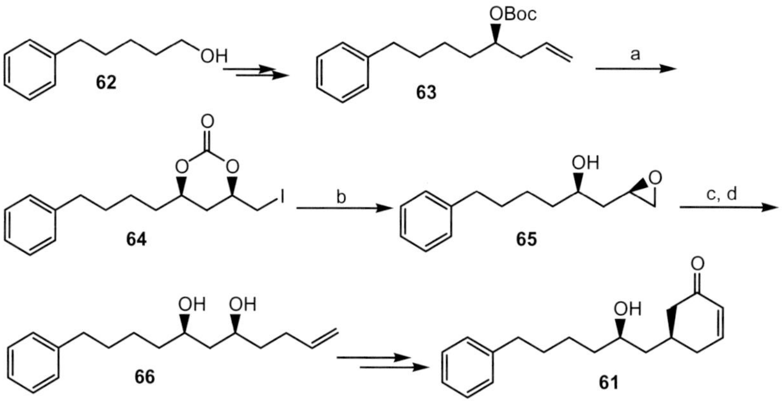

Scheme 3.13 Reagents and conditions: a) I$_2$, MeCN, −20°C, 6h, 72%; b) K$_2$CO$_3$, MeOH, r.t., 30 min., 81%; c) TBSCl, imidazole, CH$_2$Cl$_2$, 0°C–r.t., 5 h, 88%; d) vinyl magnesium bromide, CuBr, THF, −78−−40°C, 7 h, 80%.

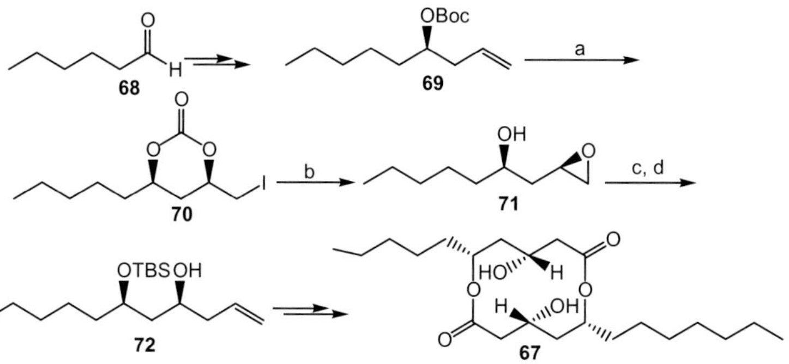

Scheme 3.14 Reagents and conditions: a) I$_2$, MeCN, −20°C, 6h, 72%; b) K$_2$CO$_3$, MeOH, 20°C, 30 min.; 82%; c) TBSCl, imidazole, CH$_2$Cl$_2$, 4h, 91%; d) vinyl magnesium bromide, CuCN, −20°C–r.t., 3 h, 85%.

Scheme 3.14. This synthesis also involved the preparation of a chiral epoxide **71** from an iodoacetate **70** produced from a Boc-protected hydroxyalkene **69**.

3.2.2.7 Synthesis of (2S, 3S)-(+)-Aziridine-2,3-Dicarboxylic Acid [61]
Naturally occurring (2S, 3S)-(+)-aziridine-2,3-dicarboxylic acid **73** was synthesized from diethyl (2R, 3R)-(−)-oxirane-2,3-dicarboxylate **74** via a ring-opening and ring-closing strategy (Scheme 3.15).

3.2.2.8 Synthesis of D-*erythro*-Sphingosine [62]
The enantioselective synthesis of D-*erythro*-sphingosine **77** which regulates fundamental and diverse cell process involved an asymmetric sulfur ylide reaction to produce a chiral epoxide intermediate **80** (Scheme 3.16). This epoxide **80** was subsequently converted into D-*erythro*-sphingosine **77**.

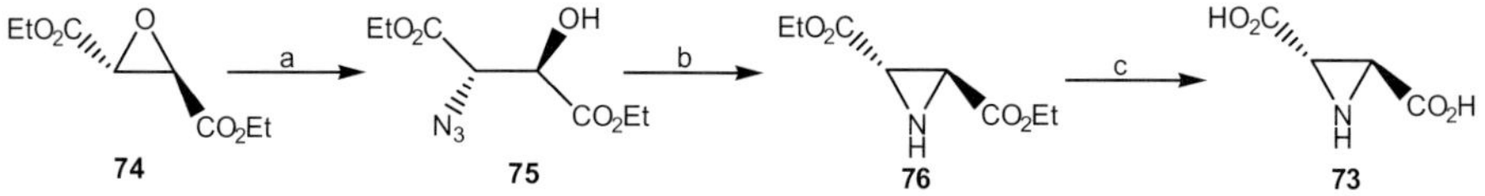

Scheme 3.15 Reagents and conditions: a) i) Me$_3$SiN$_3$, EtOH, DMF, 80 °C; ii. NH$_4$Cl, EtOH, H$_2$O, 72%; b) i) PPh$_3$, DMF; ii) heat, 3 h, 91%, 95% ee; c) i) LiOH, EtOH; ii) H$^+$, 69%.

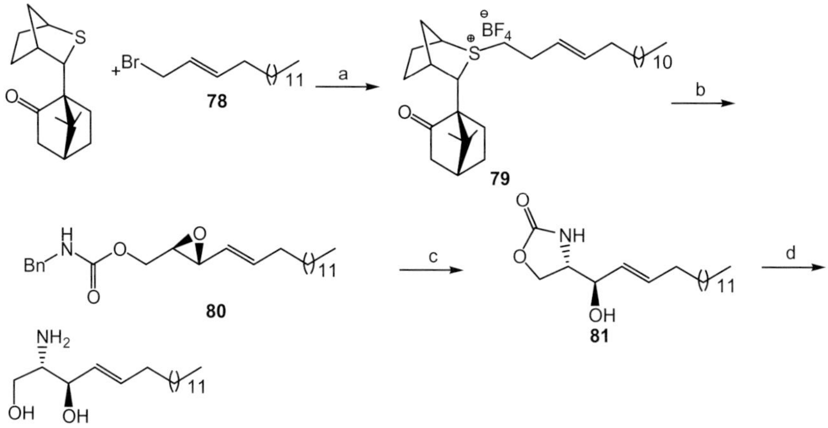

Scheme 3.16 Reagents and conditions: a) AgBF$_4$, CH$_2$Cl$_2$, r.t., 48 h, 80%; b) PhCH$_2$NHCOOCH$_2$CHO, EtP$_2$, CH$_2$Cl$_2$, −78 °C, 2 h, 80%; c) i) NaHMDS, THF, −15 °C, 5 h; ii) Li, EtNH$_2$, ^{t}BuOH, −78 °C, 4 h, 90%; d) KOH, EtOH, H$_2$O, reflux, 2.5 h, 100%.

3.2.2.9 Synthesis of (+)-L-733,060 [63]

(+)-L-733,060, **82** is a 2,3-disubstituted piperidine acting as a potent neurokinin substance P receptor antagonist. The synthesis of the compound was initiated from (+)-ethyl glycidate **83** (Scheme 3.17). The latter was treated with phenyllithium to afford an epoxyketone **84** which underwent diastereoselctive reductive amination to produce the corresponding anti-aminoepoxide **85**. Regioselctive intramolecular ring opening of **85** by heating in di-t-butyldicarbonate yielded the oxazolidinone **86** which was used for the synthesis of (+)-L-733,060, **82**.

3.2.2.10 Synthesis of (+)-Chelidonine [64]

(+)-Chelidonine **87** was synthesized from a polycyclic bromohydrin **89**, via conversion to an epoxy compound **90**. The epoxide ring was subsequently transformed to a hydroxyl group to give the desired alkaloid (Scheme 3.18).

3.2.2.11 Synthesis of (−)-Pironetin [65]

The novel unsaturated δ-lactone derivative, (−)-pironetin **91** showed immunosuppressive and plant growth regulator activities. The compound was synthesized (Scheme 3.19) from a chiral epoxyalcohol **92** which was derived from pent-2-ene-

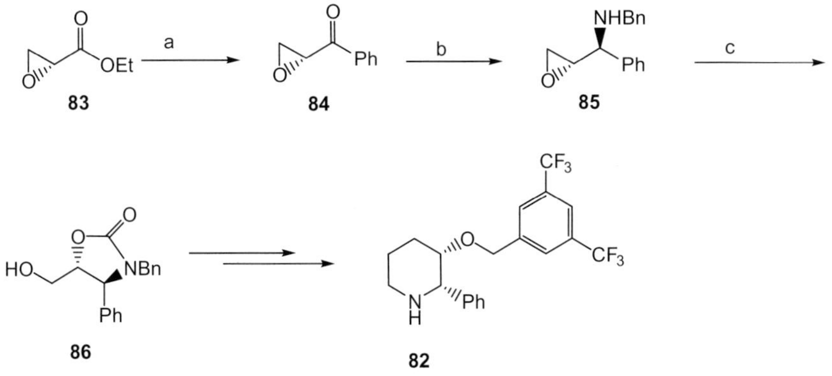

Scheme 3.17 Reagents and conditions: a) PhLi, Et$_2$O, −78 °C, 90%; b) BnNH$_2$, AcOH, Me$_4$NHB(OAc)$_3$, CH$_2$Cl$_2$, r.t., 70%; c) Boc$_2$O, 60–70 °C, 3M NaOH, r.t., 72%.

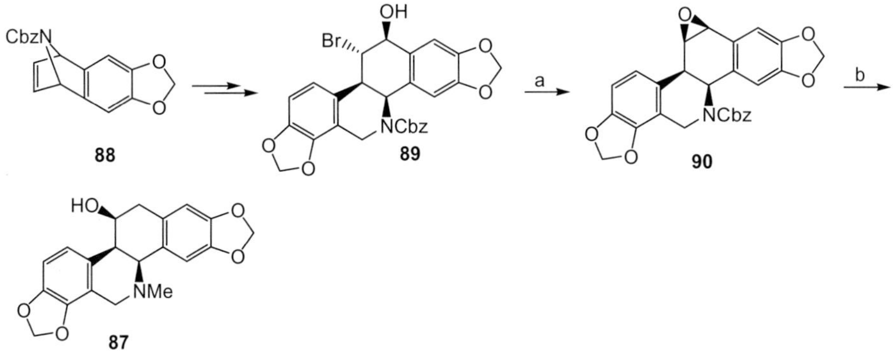

Scheme 3.18 Reagents and conditions: a) KO^{t}Bu, THF, −78 °C, 30 min., 95%; b) LAH, 1,4-dioxane, reflux, 18 h, 88%.

1-ol. The epoxyalcohol was benzylated and then the epoxide ring was regioselectively opened using acetylenic alanate followed by hydrogenation to afford a (Z)-homoallylic alcohol derivative **94**. The latter was converted into an epoxide **95** which was used for the synthesis of (−)-pironetin **91**.

3.2.2.12 Synthesis of (−)-Codonopsinine [66]
(−)-Codonopsinine **97** possesses antibiotic and hypotensive activities. It was synthesized through a stereoselective epoxidation/epoxide ring opening sequence (Scheme 3.20).

3.2.2.13 Synthesis of Sesamin and Dihydrosesamin [67]
Sesamin **101** and dihydrosesamin **102**, two naturally-occurring bioactive lignans were synthesized in racemic form applying the epoxide opening strategy (Scheme 3.21). The intramolecular radical cyclization of suitably substituted epoxy ethers

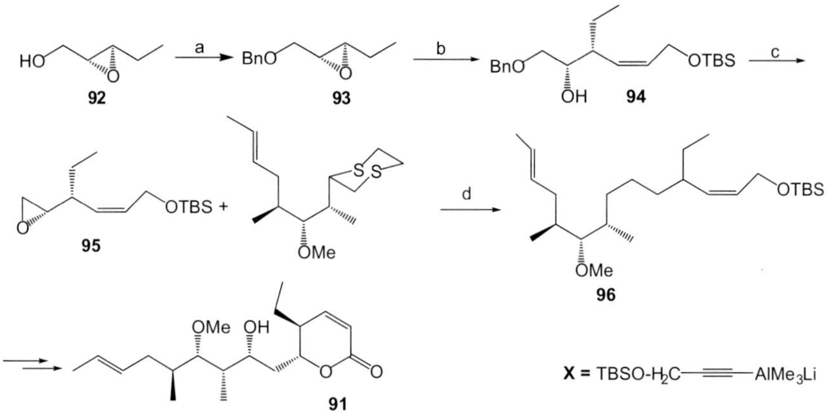

Scheme 3.19 Reagents and conditions: a) PhCOCl, Et$_3$N, CH$_2$Cl$_2$, 0°C, 100%; b) i) **X**, BF$_3$·OEt$_2$, ether, −78°C, 43%; ii) H$_2$, Lindlar Catalyst, MeOH, r.t., 100%; c) i) K$_2$CO$_3$, MeOH, r.t., 99%; ii) TsCl, Et$_3$N, DMAP, CH$_2$Cl$_2$, 0°C, 76%; iii) K$_2$CO$_3$, MeOH, r.t., 89%; d) n-BuLi, HMPA, THF, 0°C, 91%.

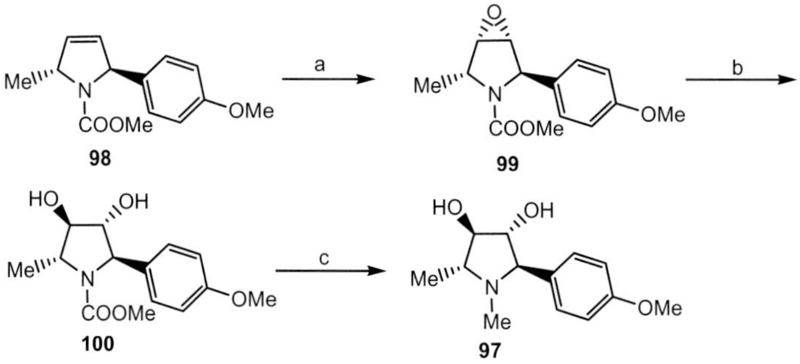

Scheme 3.20 Reagents and conditions: a) m-CPBA, toluene, r.t., 70% (α- and β-epoxide 93:7); b) dioxane/H$_2$O/H$_2$SO$_4$, 95°C, 9h, 55%; c) LiAlH$_4$, THF, 74%.

was carried out using bis(cyclopentadienyl)titanium(III) chloride as the radical source.

3.2.2.14 Synthesis of (9S, 12R, 13S)-Pinellic Acid [68]
(9S, 12R, 13S)-Pinellic acid **103**, a potentially useful oral adjuvant, was synthesized through an epoxide formation and cleavage strategy (Scheme 3.22). The epoxide **106** was prepared from an allyl alcohol **105**. The epoxide ring of **107** was cleaved with lithium in liquid NH$_3$ in the presence of Fe(NO$_3$)$_3$. The resulting alcohol **108** was then converted into (9S, 12R, 13S)-pinellic acid **103**.

3.2.2.15 Synthesis of (Z)-Nonenolide [69]
Nonenolide, an antiviral naturally occurring macrolide, contains a double bond with (E)-stereochemistry. Its (Z)-isomer **109** was synthesized from 4-penten-1-ol

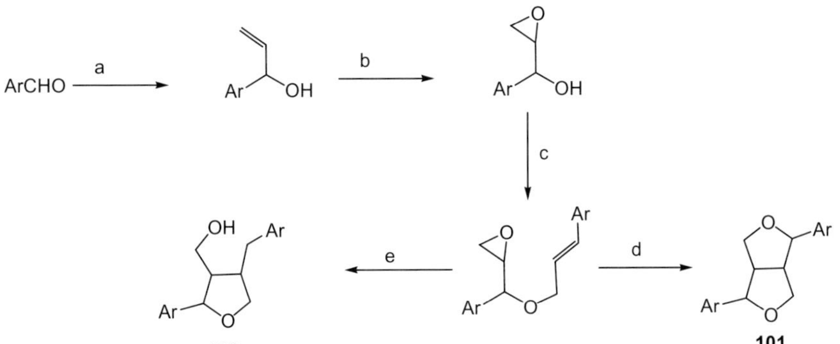

Scheme 3.21 Reagents and conditions: a) vinylmagnesium bromide, THF, r.t.; b) m-CPBA, CHCl$_3$; c) NaH, THF-DMSO, 0–5 °C, ArCH=CHCH$_2$Br; d) i) Cp$_2$TiCl, THF; ii) I$_2$, THF, 60 °C; e) i) Cp$_2$TiCl, THF, r.t.; ii) H$_3$O$^+$.

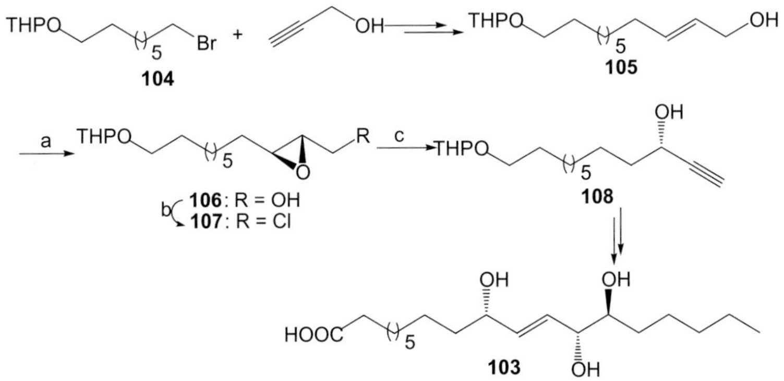

Scheme 3.22 Reagents and conditions: a) (+)-DET, Ti(O^{i}Pr)$_4$, dry CH$_2$Cl$_2$, TBHP, −24 °C, 4 h, 80%; b) TPP, dry CCl$_4$, reflux, 4 h, 90%; c) Li, Liq. NH$_3$, Fe (NO$_3$)$_3$, −33 °C, 1 h, 76%.

110 (Scheme 3.23). The latter was converted into a mono THP protected dihydroxy alkene **111**. This compound was transformed into its isomer **114** through an epoxide **112**. Compound **114** was subsequently used to synthesize the (Z)-isomer of nonenolide **109**.

3.2.2.16 Synthesis of (−)-Cubebol [70]
The synthesis of (−)-cubebol **115**, a major constituent of cubeb oil, was synthesized from (−)-menthone **116** (Scheme 3.24). The key steps are associated with the formation of an epoxide **118** (from a chlorohydrin derivative **117**) prepared from (−)-menthone) and its subsequent conversion into a cyclopropane derivative **119**.

3.2.2.17 Synthesis of (+)-Schweinfurthins B and E [71]
The synthesis of natural cytotoxic agents, (+)-schweinfurthins B **120** and E **121** was accomplished from vanillin **122** (Scheme 3.25). The key steps involve a

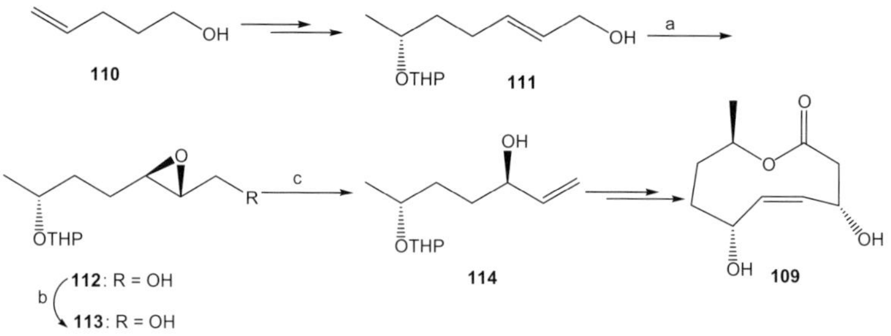

Scheme 3.23 Reagents and conditions: a) (−)-DET, Ti (O^{i}Pr)$_4$, cumene hydroperoxide, 4 A° MS, CH$_2$Cl$_2$, −20 °C, 5 h, 75%; b) I$_2$, Ph$_3$P, imidazole, ether/acetone (3:1), 0 °C to r.t., 1 h, 90%; c) activated Zn, EtOH, reflux, 1–2 h, 80%.

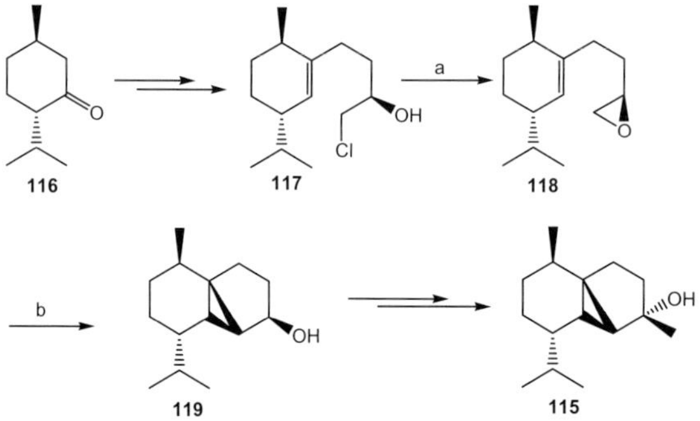

Scheme 3.24 Reagents and conditions: a) NaOH, MeOH, 82%; b) LTMP, ^{t}BuOMe, 0 °C to r.t., 20 h, 90%.

Shi epoxidation of an olefin **123** derived from vanillin and cascade cyclization initiated by treatment of the resulting epoxide **124** with BF$_3$·OEt$_2$. The cyclization product **125** was subsequently converted into (+)-schweinfurthins B **120** and E **121**.

3.2.2.18 Synthesis of (−)-Cleistenolide [72]
(−)-Cleistenolide **126**, a plant-derived antibacterial and antifungal compound, was synthesized from (R, R)-1,5-hexadiene-3,4-diol **128** which was in turn prepared from D-mannitol **127** (Scheme 3.26). The diol **128** was converted into a monoepoxide **129** which was used to synthesize the target molecule.

3.2.2.19 Synthesis of Decarestricine J [73]
Decarestricine J **132**, was isolated from a culture broth of *Penicillium simplicissimum* and was found to inhibit the biosynthesis of cholesterol. The synthesis of

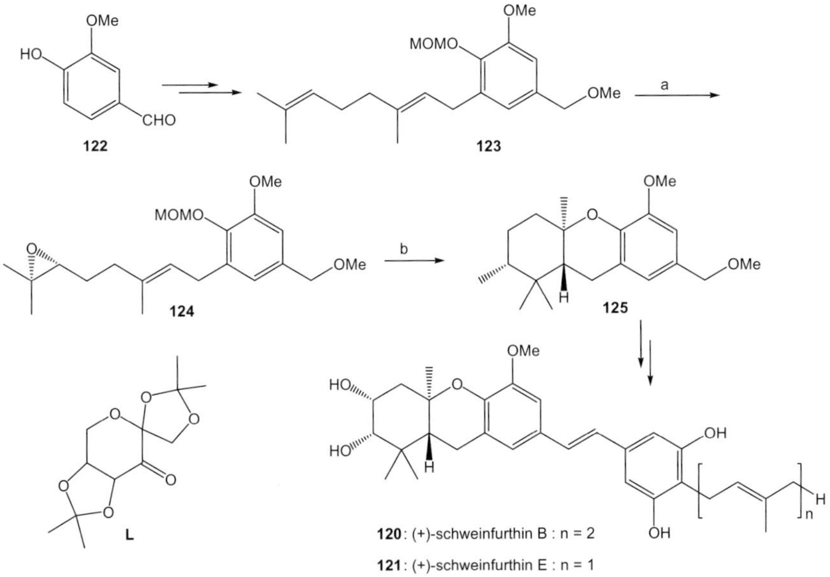

3.2 Epoxides in Natural Product Synthesis | 81

Scheme 3.25 Reagents and conditions: a) H$_2$O$_2$, L, aq. buffer, MeCN, 63%; b) BF$_3$.OEt$_2$, 69%.

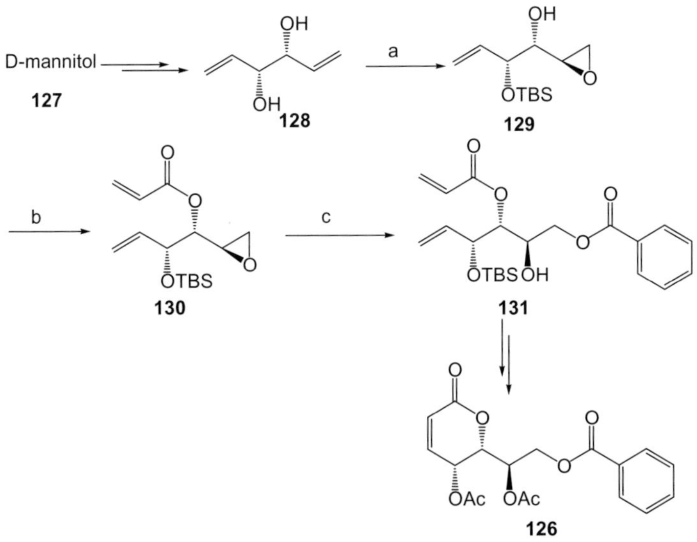

120: (+)-schweinfurthin B : n = 2
121: (+)-schweinfurthin E : n = 1

Scheme 3.26 Reagents and conditions: a) TBSCl, imidazole, CH$_2$Cl$_2$, 20 °C, 87%; Ti(O^{i}Pr)$_4$, L-(+)-DET, ^{t}BuOOH in decane, CH$_2$Cl$_2$, −30 °C, 28 d, 84%; b) CH$_2$=CHCOCl, ^{i}Pr$_2$NEt, CH$_2$Cl$_2$, 0 °C, 92%; c) C$_6$H$_5$COOH, ^{i}Pr$_2$NEt, 20 °C, 53%.

3 Epoxides and Oxetanes

the compound was accomplished using racemic propylene oxide **133** and by applying hydrolytic kinetic resolution (HKR) (Scheme 3.27). The HKR method involved readily available cobalt-based chiral salen complex as catalyst and water to resolve the racemic epoxide.

3.2.2.20 Synthesis of (−)-Gloeosporone [74]

The macrolide natural product, (−)-gloeosporone **143** was synthesized from 7-octen-1-ol **144** (Scheme 3.28). An epoxy alcohol **145** was prepared from this compound involving Jacobsen's hydrolytic kinetic resolution. The epoxy alcohol **145** was then converted into macrolide **149** which was subsequently transformed into (−)-gloeosporone **143**.

3.2.2.21 Synthesis of (S)-Dihydrokavain [75]

(S)-Dihydrokavain **150** isolated from kava plant, *Piper methystieum*, contains a lactone moiety which is responsible for various biological properties. Hydrolytic

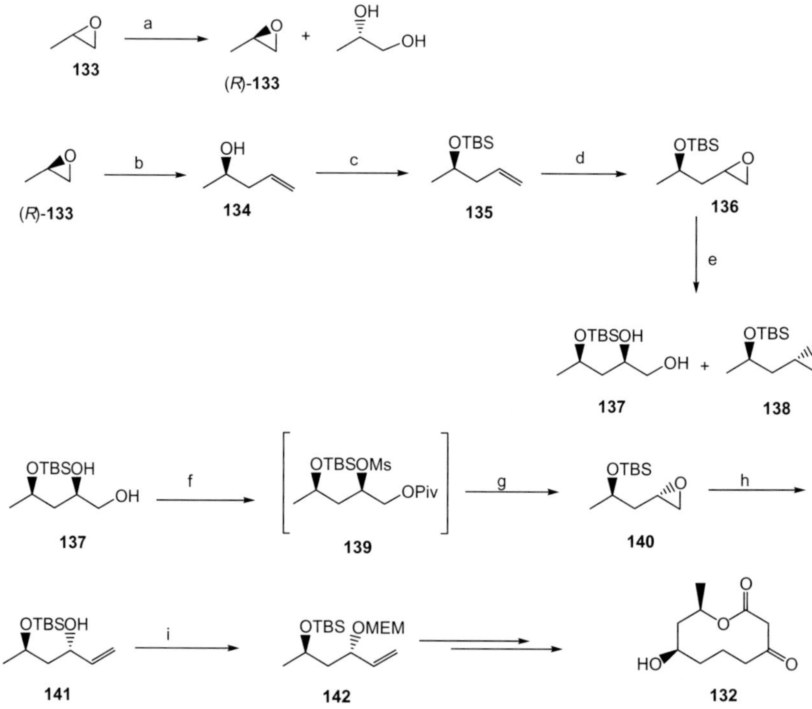

Scheme 3.27 Reagents and conditions: a) (R, R)-salen-Co-(OAc), H₂O, 0 °C, 14 h, (45%, for (R)-**115**); b) vinylmagnesium bromide, THF, CuI, −20 °C, 12 h, 90%; c) TBDMSCl, imidazoles, CH₂Cl₂, 0 °C to r.t., 2 h, 93%; e) (S, S)-salen-Co-(OAc), H₂O, 0 °C, 20 h, (70%, for **119**; 22% for **120**); f) i) PivCl, Et₃N, DMAP, r.t., 2 h; ii) MsCl, Et₃N, DMAP, 0 °C to r.t., overnight, 61% (for three steps); h) Me₃SiI, n-BuLi, THF, 2 h, 70%; i) i) DIPEA, MEMCl, CH₂Cl₂, 0 °C to r.t., 8 h; ii) TBAF, THF, 0 °C to r.t., 5 h, 80% (for two steps).

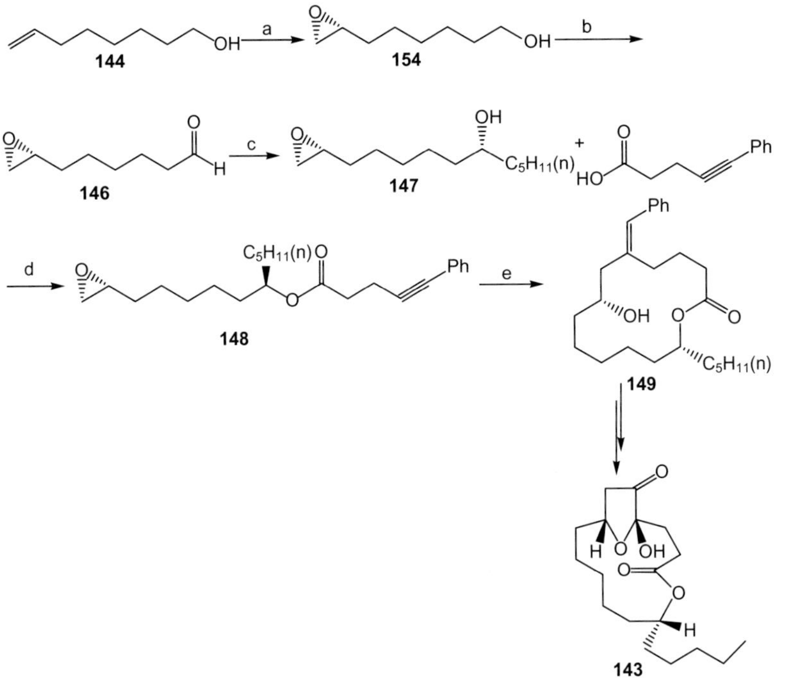

Scheme 3.28 Reagents and conditions: a) m-CPBA, CH$_2$Cl$_2$, (R, R)-(Co[salen][OAc]), H$_2$O, 35%; b) n-Pr$_4$NBnO$_4$, NMO, 3 A° MS, CH$_2$Cl$_2$, 95%; c) trans-cyclohexane diamine ditriflate, (n-C$_5$H$_{11}$)Zn, Ti(O^{i}Pr)$_4$, −20 °C, PhMe, 80%; d) DCC, DMAP, CH$_2$Cl$_2$, 85%; e) (Ni[cod]$_2$), Bu$_3$P, Et$_3$B, THF, 87%.

kinetic resolution of the racemic epoxide, phenyl butylenes oxide **151** yielded the desired chiral epoxide **152** (Scheme 3.29). Regioselective opening of the epoxide **152** with dimethylsulfonium methylide gave an allylic alcohol **153** which was converted into (S)-dihydrokavain **150**.

3.2.2.22 Synthesis of (−)-Phorocantholide-J [76]

Epoxides and oxetanes have been used for silicon-mediated, sequential ring expansions of 2-cycloalkenones into medium-sized lactones (Scheme 3.30).

The above method has been applied for the synthesis of (−)-phorocantholide-J **154**, a natural 10-membered ring olefinic lactone (Scheme 3.31). The compound is a part of the defense secretion of the insect eucalypt longicorn, *Phorocantha synonyma*. The reaction of cyclopentenone enol silyl ether **155** with t-butyl

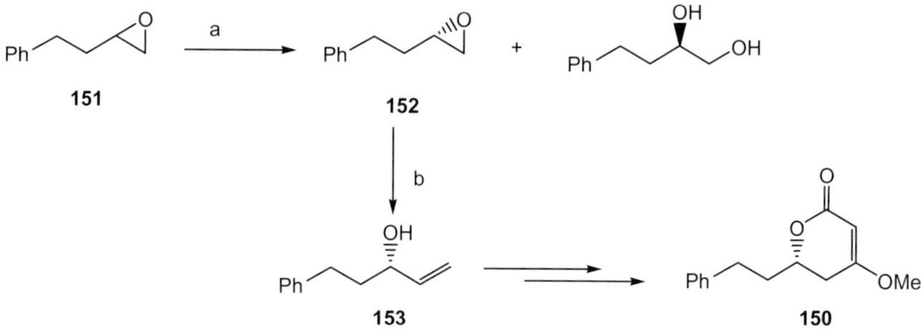

Scheme 3.29 Reagents and conditions: a) (S, S)-salen-Co-(OAc), H₂O, 25 °C, 46%; b) Me₃S⁺I⁻, n-BuLi, THF, −10 °C, 82%.

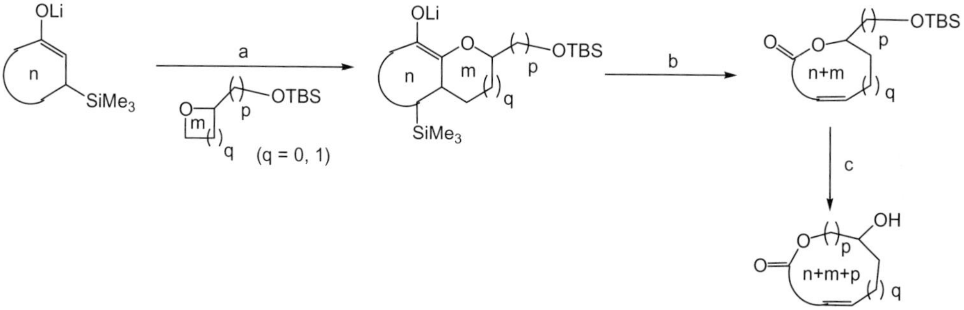

Scheme 3.30 Reagents and conditions: a) BF₃.OEt₂, THF; b) PhI(OAc)₂, I₂ or CAN; c) Bu₄NF, THF.

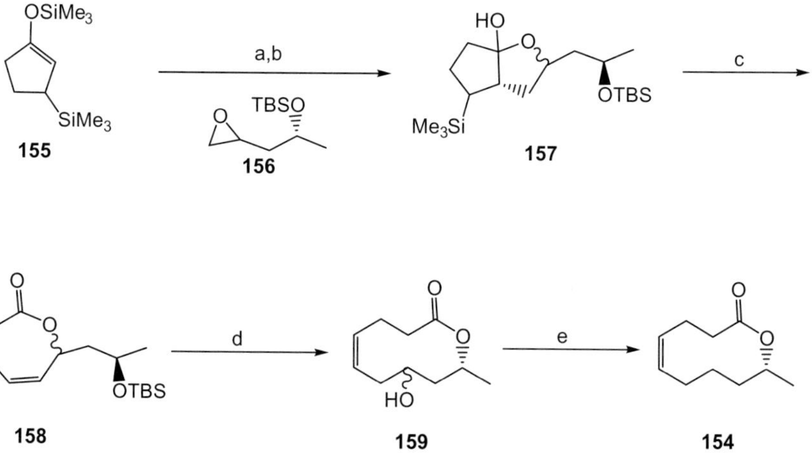

Scheme 3.31 Reagents and conditions: a) MeLi, THF, 0 °C; b) BF₃.OEt₂, −78 °C; c) PhI(OAc)₂, I₂, CH₂Cl₂; 58% in three steps; d) Bu₄NF, THF, 80%; e) i) ClC(S)OPh; ii) Bu₃SnH, AIBN, C₆H₆, 85%.

dimethylsilyl ether **156** of (*R*)-4-hydroxy-1-pentene oxide afforded the hemiketal **157**. Oxidative fragmentation of **157** yielded *cis*-olefinic ring lactone **158** as a diastereomeric mixture. Compound **158** on fluoride-induced desilylation underwent translactonization into the olefinic hydroxyllactone **159**. Finally the radical deoxygenation of **159** produced the natural (–)-phorocantholide-J **154** in an overall yield of 26%.

3.3
Oxetane in Natural Product Synthesis

Several natural products contain the oxetane moiety. The synthesis of this moiety is an important step in the construction of these molecules. Oxetanes are also important intermediates in the synthesis of various polyfunctional compounds.

3.3.1
Synthesis of Natural Products Possessing an Oxetane Moiety

3.3.1.1 Synthesis of Epi-oxetin [26]

Oxetin, a natural antibiotic and herbicide, contains an oxetane ring. With the goal of preparing this compound from L-serine **161**, Blauvett and Howell adopted a synthetic sequence which ultimately afforded epi-oxetin **160** (Scheme 3.32). The

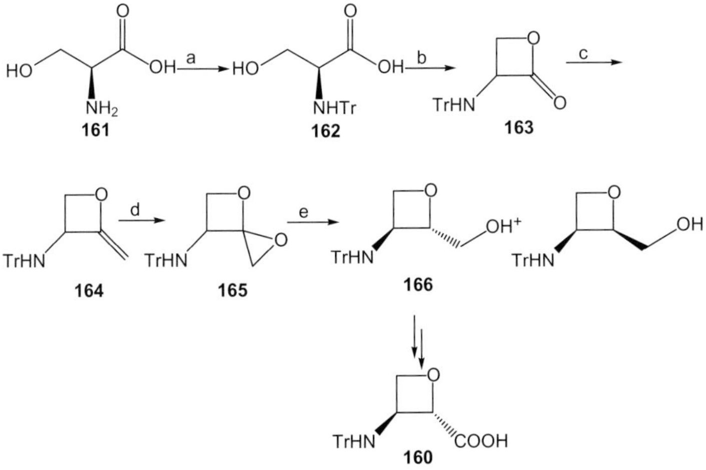

Scheme 3.32 Reagents and conditions: a) TMSCl, TEA; TEA, MeOH, TrCl, 66%; b) BOP, TEA, CH$_2$Cl$_2$, 73%; c) Cp$_2$TiMe$_2$, PhMe, 80°C, 60%; d) Dimethyldioxirane (DMDO), CH$_2$Cl$_2$, 90%; e) DIBAL-H, PhMe, −78°C, 65% (α:β isomers, 2:1).

oxetane core was derived from N-tritylserine and the acid group was generated through an epoxide formation.

3.3.1.2 Synthesis of Dioxatricyclic Segment of Dictyoxetane [77, 78]

Dictyoxetane **167**, a constituent of brown algae *Dictyota dichotoma*, is structurally related to the class of dolabellones which possess a wide range of biological properties. The dioxatricyclic core **168** of the molecule containing an oxetane ring was synthesized from 1,1-bis benzyloxy propanone **169** through the formation of an epoxide **174** (Scheme 3.33).

3.3.1.3 Synthesis of (−)-Merrilactone A [22]

(−)-Merrilactone A **175**, a sesquiterpenoid isolated from *Illicium merrillianum*, was found to exhibit neuroprotective and neuritogenic activities. The synthesis of this complex molecule required multistep transformations. The oxetane ring of this compound was prepared from a related epoxide **178** produced by epoxida-

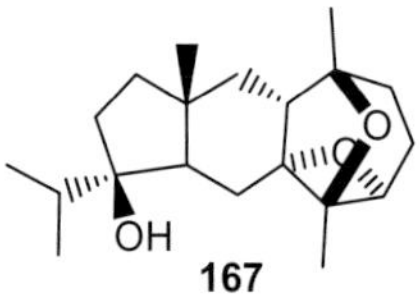

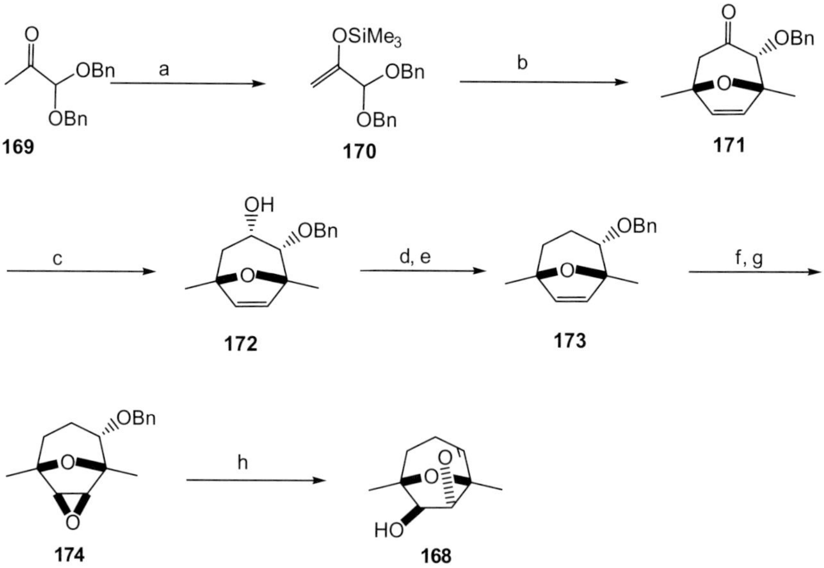

Scheme 3.33 Reagents and conditions: a) LDA, TMSCl, THF, −78 °C−r.t.; b) 2,5-dimethyl furan, TMSOTf, CH$_2$Cl$_2$, −78 °C, 53% (over two steps); c) DIBAL-H, THF, −78 °C, 94%; d) NaH, CS$_2$, CH$_3$I, THF, −0 °C−r.t. 77%; e) Bu$_3$SnH, AIBN, toluene, 95 °C, 92%; f) m-CPBA, CH$_2$Cl$_2$, 0 °C−r.t., 85%; g) H$_2$, Pd/C, MeOH, 85%; h) BF$_3$·OEt$_2$, CH$_2$Cl$_2$, 0 °C, 72%.

tion of the corresponding olefin **177**. The olefin **177** was prepared from a simple precursor, 2,3-dimethylmaleic anhydride **176** involving several conversions (Scheme 3.34).

3.3.1.4 Total Synthesis of (+)-(Z)-Laureatin [16]

(+)-(Z)-Laureatin **179**, an eight-membered cyclic ether, was isolated from red algae of the genus *Laurencia*. The compound contains an oxetane core which was generated from an epoxide **181**; this was in turn prepared from the corresponding olefinic compound **180** (Scheme 3.35).

3.3.1.5 Synthesis of Taxol [20]

The novel tetracyclic diterpenoid taxol (paclitaxel) **183** was originally obtained from the Pacific yew tree *Taxus brevifolia*. The compound is one of the most important

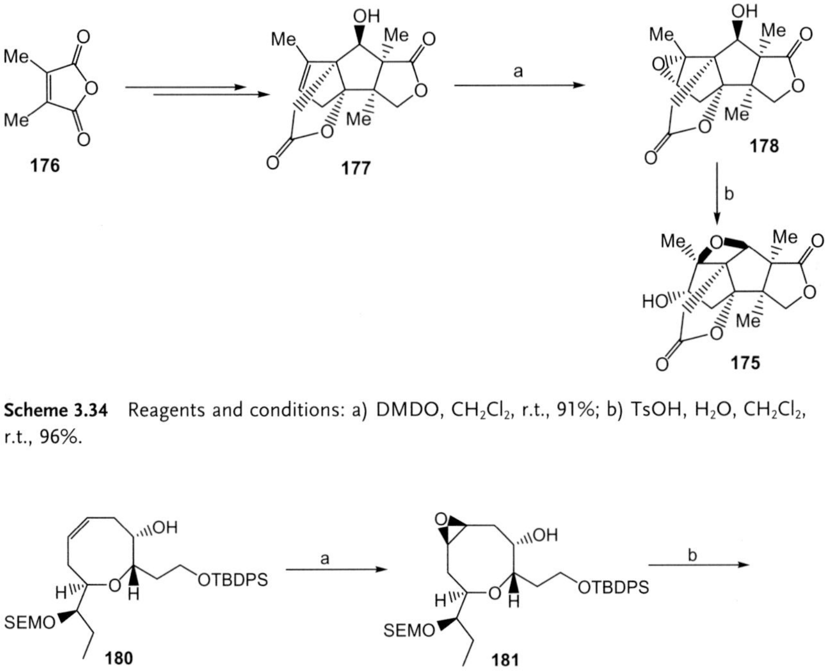

Scheme 3.34 Reagents and conditions: a) DMDO, CH$_2$Cl$_2$, r.t., 91%; b) TsOH, H$_2$O, CH$_2$Cl$_2$, r.t., 96%.

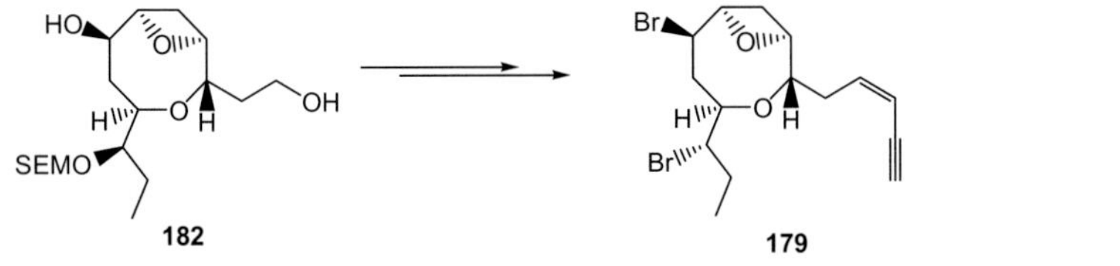

Scheme 3.35 Reagents and conditions: a) *m*-CPBA, CH$_2$Cl$_2$, 0 °C to r.t., 96%; b) aq. KOH, DMSO, 80 °C, 97%.

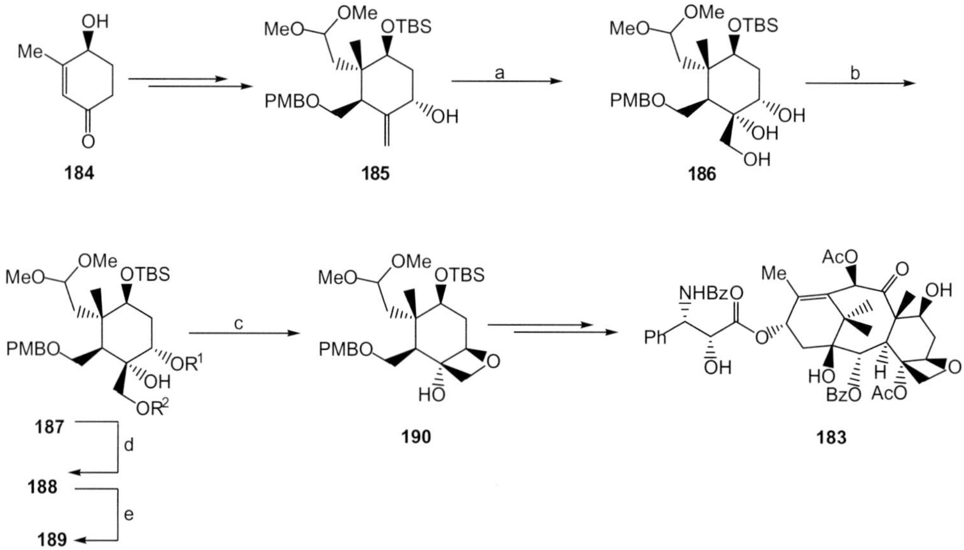

Scheme 3.36 Reagents and conditions: a) OsO$_4$, NMO, ^{t}BuOH. H$_2$O, r.t., 12h, 80%; b) Ac$_2$O, DMAP, pyridine, r.t., 2h; c) MsCl, r.t., 12h; d) K$_2$CO$_3$, MeOH, r.t., 1h; e) DBU, toluene, reflux, 1h, 65% (last four steps).

cancer therapeutic agents. It contains an oxetane ring which is the most vital part for its activity. The oxetane ring of taxol was synthesized from an allylic alcohol **184** (Scheme 3.36).

3.3.2
Synthesis of Natural Products Involving Transformation of the Oxetane Moiety

3.3.2.1 Synthesis of Erogorgiaene [79]

The synthesis of erogorgiaene **191**, a natural anti-tubercular agent, was initiated from *p*-methyl phenyl acetic acid **192** (Scheme 3.37). The latter was converted into a protected 1,3-diol **193** which was then converted into the oxetane **194**. A highly diastereoselctive intramolecular Friedel–Crafts reaction of this oxetane led to the formation of an alcohol **195** which was then transformed into erogorgiaene **191**.

3.3.2.2 Synthesis of *trans*-Whiskey Lactone [80]

trans-Whiskey lactone **196**, found in whiskey, brandy and wine stored in oak barrels, was synthesized starting from acetylinic aldehyde **197** (Scheme 3.38). The latter was converted into an oxetane derivative **199**. The enantiospecific ring expansion reaction of this oxetane **199** with a chiral copper catalyst bearing the bipyridine ligand as a catalyst afforded *trans*-whiskey lactone **196**.

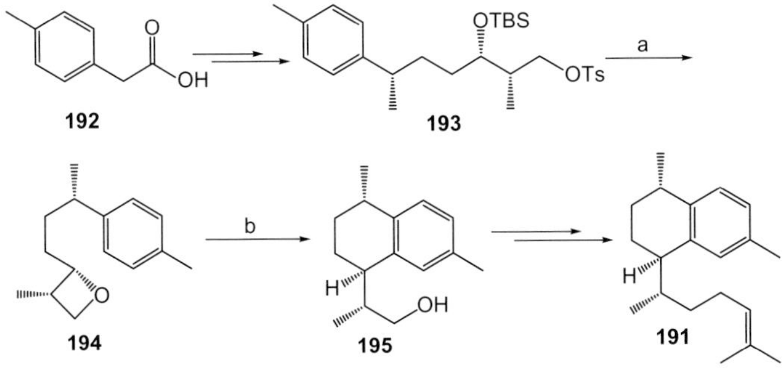

Scheme 3.37 Reagents and conditions: a) PTSA, MeOH, then NaH, THF, r.t., 12 h, 93%; b) BF$_3$·OEt$_2$, CH$_2$Cl$_2$, −78 °C to r.t., 3 h, 81%.

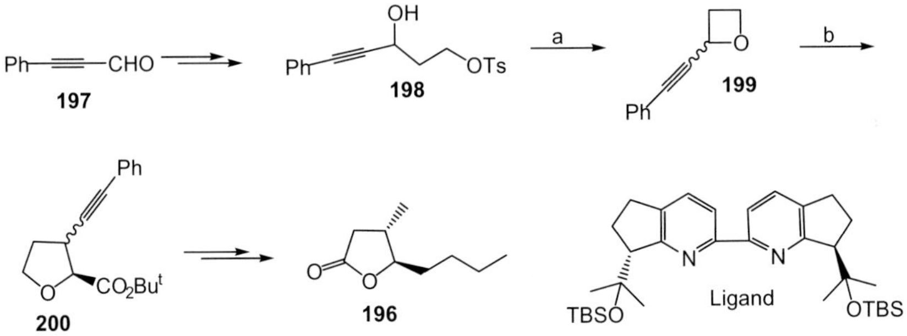

Scheme 3.38 Reagents and conditions: a) n-BuLi, 55%; b) CuOTf-ligand, 1%, N$_2$CHCOO^{t}Bu, 88%.

3.3.2.3 Synthesis of (±)-Sarracenin [81]

(±)-Sarracenin **201**, a tricyclic secoiridoid, was synthesized from the simple precursors, cyclopentadiene **202** and acetaldehyde **203** (Scheme 3.39). These two precursors underwent photocycloaddition to form two diastereomeric oxetanes, **204** and **205**. The oxetane ring of these compounds was opened with methanol and camphor sulfonic acid to form the alcohol **206** which was subsequently used to synthesize (±)-sarracenin **201**.

3.4 Conclusion

Epoxides and oxetanes are important heterocyclic units found in various naturally occurring molecules, some with potential bioactivities. They are also valuable

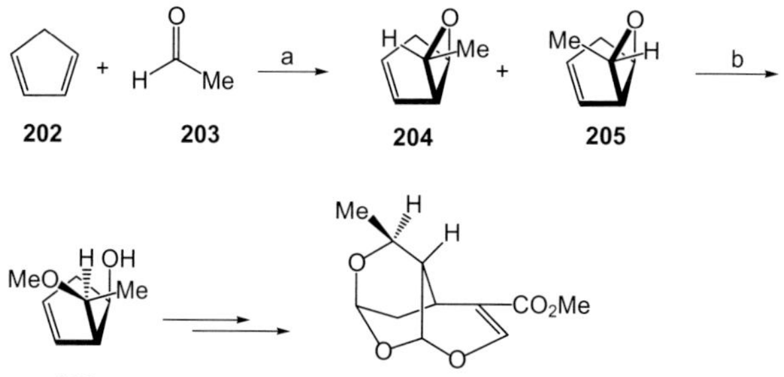

Scheme 3.39 Reagents and conditions: a) hv, <10°C, 7 h; b) MeOH, CSA, rt, 7 days.

intermediates in the synthesis of biologically active natural products. In the present article we have presented some bioactive natural molecules having epoxide and oxetane moieties in tabular form and also discussed the synthetic uses of these heterocycles in the synthesis of natural products.

Acknowledgment

The authors thank all the colleagues of their laboratories for their kind help and constructive suggestions.

References

1 (a) Padwa, A. and Murphree, S.S. (2006) Epoxides and aziridines – a mini review. ARKIVOC, 6–33; (b) Finar, I.L. (1973) Organic Chemistry: The Fundamental Principles, vol. 1, 6th edn, Longman Group Ltd., UK, pp. 111–112, 312–315; (c) Carruthers, W. and Coldham, I. (2005) Modern Methods of Organic Synthesis, 4th edn, Cambridge University Press, UK, pp. 331–346; (d) De Vos, D.E., Sels, B.F. and Jacobs, P.A. (2003) Practical heterogeneous catalysts for epoxide production. Adv. Synth. Catal., **345** (4), 457–473.

2 (a) Jacobsen, E.N. (2000) Asymmetric catalysis of epoxide ring-opening reactions. Acc. Chem. Res., **33** (6), 421–431; (b) Hodgson, D.M., Gibbs, A.R. and Lee, G.P. (1996) Enantioselective desymmetrisation of achiral epoxides. Tetrahedron, **52** (46), 14361–14384.

3 Lindeman, R.J. (1996) Comprehensive Heterocyclic Chemistry II (eds A.R. Katritzky and C.W. Rees), Pergamon, Oxford, UK, p. 1B.

4 (a) Magnusson, H., Malmström, E. and Hult, A. (1999) Synthesis of hyperbranched aliphatic polyethers via cationic ringopening polymerization of 3-ethyl-3- (hydroxymethyl) oxetanes. Macromol. Rapid Commun., **20** (8), 453–457; (b) Liaw, D., Liang, W. and Liaw, B. (1998) Preparation of new polyether with oxetane pendant group. J. Polym. Sci. [A1], **36**, 103–107.

5 Kudo, H., Morita, A. and Nishikubo, T. (2003) Synthesis of a hetero telechelic hyperbranched polyether. Anionic

ring-opening polymerization of 3-ethyl-3-(hydroxymethyl) oxetane using potassium tert-butoxide as an initiator. *Polym. J.*, **35** (1), 88–91.

6 Dale, J. and Fredriksen, S.B. (1992) Synthesis of branched polyether ligands designed for selective complexation of cations. *Acta Chem. Scand.*, **46**, 271–277.

7 Marco-Contelles, J., Molina, M.T. and Anjum, S. (2004) Naturally occuring cyclohexane epoxides: source, biological activities and synthesis. *Chem. Rev.*, **104** (6), 2857–2899.

8 (a) Nicolaou, K.C. and Guy, R.K. (1995) The conquest of taxol. *Angew. Chem. Int. Ed.*, **34** (19), 2079–2090; (b) Macias, F.A., Molinillo, J.M.G. and Massanet, G.M. (1993) First synthesis of two naturally occurring oxetane lactones: clementein and clementein B. *Tetrahedron*, **49** (12), 2499–2508.

9 Wakamura, S., Arakaki, N., Yamamoto, M., Hiradate, S., Yasui, H., Yasuda, T. and Ando, T. (2001) Posticlure: a novel *trans*-epoxide as a sex pheromone component of the tussock moth, *Orgyia postica* (Walker). *Tetrahedron Lett.*, **42** (4), 687–689.

10 Fernandes, R.A. (2007) An efficient synthesis of (−)-posticlure: the sex pheromone of *Orgyia postica*. *Eur. J. Org. Chem.*, 5064–5070.

11 Tanaka, M., Nara, F., Suzuki-Konagai, K., Hosoya, T. and Ogita, T. (1997) Structural elucidation of scyphostatin, an inhibitor of membrane-bound neutral sphingomyelinase. *J. Am. Chem. Soc.*, **119** (33), 7871–7872.

12 Takagi, R., Miyanaga, W., Tojo, K., Tsuyumine, S. and Ohkata, K. (2007) Stereoselective total synthesis of (+)-scyphostatin via a π-facially selective Diels–Alder reaction. *J. Org. Chem.*, **72** (11), 4117–4125.

13 Kobayashi, M., Aoki, S., Ohyabu, N., Kurosu, M., Wang, W. and Kitagawa, I. (1994) Arenastatin A, a potent cytotoxic depsipeptide from the Okinawan marine sponge *Dysidea arenaria*. *Tetrahedron Lett.*, **35** (43), 7969–7972.

14 (a) Kobayashi, M., Kurosu, M., Wang, W. and Kitagawa, I. (1994) A total synthesis of arenastatin-A, an extremely potent cytotoxic depsipeptide, from the Okinawan marine sponge Dysidea-arenaria. *Chem. Pharm. Bull.*, **42** (11), 2394–2396; (b) Kotoku, N., Narumi, F., Kato, T., Yamaguchi, M. and Kobayashi, M. (2007) Stereoselective total synthesis of arenastatin A, a spongean cytotoxic depsipeptide. *Tetrahedron Lett.*, **48** (40), 7147–7150.

15 (a) Irie, T., Izawa, M. and Kurosawa, E. (1970) Laureatin and isolaureatin, constituents of *Laurencia nipponica* Yamada. *Tetrahedron*, **26** (3), 851–870; (b) Irie, T., Izawa, M. and Kurosawa, E. (1968) Laureatin, a constituent from *Laurencia nipponica* Yamada. *Tetrahedron Lett.*, **9** (17), 2091–2096; (c) Irie, T., Izawa, M. and Kurosawa, E. (1968) Isolaureatin, a constituent from *Laurencia nipponica* Yamada. *Tetrahedron Lett.*, **9** (23), 2735–2738.

16 Sugimoto, M., Suzuki, T., Hagiwara, H. and Hoshi, T. (2007) The first total synthesis of (+)-(Z)-laureatin. *Tetrahedron Lett.*, **48** (7), 1109–1112.

17 Buzzetti, F., Gaumann, E., Hütter, R., Keller-Schierlein, W., Neipp, L., Prelog, V. and Zahner, H. (1963) Stopwechselprodkte von mikroorganismen. 41. Manumycin. *Pharm. Acta Helv.*, **38** (12), 871–874.

18 Alcaraz, L., Macdonald, G., Ragot, J., Lewis, N.J. and Taylor, R.J.K. (1999) Synthetic approaches to the manumycin A, B and C antibiotics: the first total synthesis of (+)-manumycin A. *Tetrahedron*, **55** (12), 3707–3716.

19 Wani, M.C., Taylor, H.L., Wall, M.E., Coggan, P. and Mcphail, A.T. (1971) Plant antitumor agents. VI. Isolation and structure of taxol, a novel antileukemic and antitumor agent from *Taxus brevifolia*. *J. Am. Chem. Soc.*, **93** (9), 2325–2327.

20 (a) Holton, R.A., Somoza, C., Kim, H.B., Liang, F., Biediger, R.J., Boatman, P.D., Shindo, M., Smith, C.C.H. and Kim, S. (1994) First total synthesis of taxol. 1. Fuctionalisation of the B ring; First total synthesis of taxol. 2. Completion of the C and D B rings. *J. Am. Chem. Soc.*, **116** (4), 1597–1598 & 1599–1600; (b) Nicolaou, K.C., Yang, Z., Liu, J.J., Ueno, H., Nantermet, P.G., Guy, R.K., Claiborne, C.F., Renaud, J., Couladouros, E.A.,

Paulvannan, K. and Sorensen, E.J. (1994) Total synthesis of taxol. *Nature*, **367**, 630–634.

21 Huang, J.-M., Yokoyama, R., Yang, C.-S. and Fukuyama, Y. (2000) Merrilactone A, a novel neurotrophic sesquiterpene dilactone from *Illicium merrillianum*. *Tetrahedron Lett.*, **41** (32), 6111–6114.

22 Inoue, M., Sato, T. and Hirama, M. (2006) Asymmetric total synthesis of (−)-merrilactone A: use of a bulky protecting group as long-range stereocontrolling element. *Angew. Chem. Int. Ed.*, **45** (29), 4843–4848.

23 Li, J.Y., Harpar, J.K., Grant, D.M., Tombe, B.O., Bashyal, B., Hess, W.M. and Strobel, G.A. (2001) Ambuic acid, a highly functionalized cyclohexenone with antifungal activity from *Pestalotiopsis spp.* and *Monochaetia sp. Phytochemistry*, **56** (5), 463–468.

24 Li, C.H., Johnson, R.P. and Porca, J.A., Jr. (2003) Total synthesis of the quinone epoxide dimer (+)-torreyanic acid: application of a biomimetic oxidation/electrocyclization/Diels-Alder dimerization cascade. *J. Am. Chem. Soc.*, **125** (17), 5095–5106.

25 Omura, S., Murata, M., Imamura, N., Iwai, Y., Tanaka, H., Furusaki, A. and Matsumoto, T. (1984) Oxetin, a new antimetabolite from an actinomycete-fermentation, isolation, structure and biological activity. *J. Antibiot.*, **37** (11), 1324–1332.

26 Blauvelt, M.L. and Howell, A.R. (2008) Synthesis of *epi*-oxetin via a serine-derived 2-methyleneoxetane. *J. Org. Chem.*, **73** (2), 517–521.

27 (a) Bierl, B.A., Beroza, M. and Collier, C.W. (1970) Potent sex attractant of gypsy moth: its isolation, identification and synthesis. *Science*, **170**, 87–89; (b) Iwaki, S., Marumo, S., Saito, T., Yamada, M. and Katagiri, K. (1974) Synthesis and activity of optically active disparlure. *J. Am. Chem. Soc.*, **96** (25), 7842–7844.

28 Prasad, K.R. and Anbarasan, P. (2007) Enantiodivergent synthesis of both enantiomers of Gypsy moth pheromone disparlure. *J. Org. Chem.*, **72** (8), 3155–3157.

29 Sobin, B.A., English, A.R. and Celmer, W.D. *Antibiotics Annual 1954-1955*, Medicinal Encyclopedia, Inc., New York, USA, p. 827.

30 Tatsuta, K., Ishiyama, T., Tajima, S., Koguchi, Y. and Gunji, H. (1990) The total synthesis of oleandomycin. *Tetrahedron Lett.*, **31** (5), 709–712.

31 (a) Takeuchi, T., Iinuma, H., Iwanaga, J., Takahashi, S., Takita, T. and Umezawa, H. (1969) Coriolin a new basidiomycetes antibiotic. *J. Antibiot.*, **22** (5), 215–217; (b) Nakamura, H., Takita, T., Umezawa, H., Kunishima, M., Nakayama, Y. and Iitaka, Y. (1974) Absolute configuration of coriolin, a new sesquiterpene antibiotic from *coriolus consors*. *J. Antibiot.*, **27** (4), 301–302 and references cited therein.

32 (a) Mizuno, H., Domon, K., Masuya, K., Tanino, K. and Kuwajima, I. (1999) Total synthesis of (−)-coriolin. *J. Org. Chem.*, **64** (8), 2648–2656; (b) Domon, K., Masuya, K., Tanino, K. and Kuwajima, I. (1997) Highly efficient method for coriolin synthesis. *Tetrahedron Lett.*, **38** (3), 465–468.

33 Hanada, M., Sugawara, K., Kaneta, K., Toda, S., Nishiyama, Y., Tomita, K., Yamamoto, H., Konishi, M. and Oki, T. (1992) Epoxomicin, a new antitumor agent of microbial origin. *J. Antibiot.*, **45** (11), 1746–1752.

34 (a) Katukojvala, S., Bartlett, K.N., Lotesta, S.D. and Williams, L.J. (2004) Spirodiepoxide in total synthesis: epoxomicin. *J. Am. Chem. Soc.*, **126** (47), 15348–15349; (b) Sin, N., Kim, K.B., Elofsson, M., Meng, L., Auth, H., Kwok, B.H.B. and Crews, C.M. (1999) Total synthesis of the potent proteasome inhibitor epoxomicin: a useful tool for understanding proteasome biology. *Bioorg. Med. Chem. Lett.*, **9** (15), 2283–2288.

35 Ayer, W.A., Lee, S.P., Tsuneda, A. and Hiratsuka, Y. (1980) The isolation, identification, and bioassay of the antifungal metabolites produced by *Monocillium nordinii*. *Can. J. Microbiol.*, **26** (7), 766–773.

36 Garbaccio, R.M., Stachel, S.J., Baeschlin, D.K. and Danishefsky, S.J. (2001) Concise asymmetric syntheses of radicicol and monocillin I. *J. Am. Chem. Soc.*, **123** (44), 10903–10908.

37 Bhoonphong, S., Kittakoop, P., Isaka, M., Pittayakhajonwut, D., Tanticharoen, M. and Thebtarononth, Y. (2001) Multiplolides A and B, new antifungal 10-membered lactones from *Xylaria multiplex*. *J. Nat. Prod.*, **64** (7), 965–967.

38 Ramana, C.V., Kaladkar, T.P., Chatterjee, S. and Gurjar, M.K. (2008) Total synthesis and determination of relative and absolute configuration of multiplolide A. *J. Org. Chem.*, **73** (10), 3817–3822.

39 Kim, J.W., Shin-ya, K., Furihata, K., Hayakawa, Y. and Seto, H. (1999) Oximidines I and II: novel antitumor macrolides from *Pseudomonas* sp. *J. Org. Chem.*, **64** (1), 153–155.

40 Harvey, J.E., Raw, S.A. and Taylor, R.J.K. (2003) The first synthesis of the epoxide-containing macrolactone nucleus of oximidine I. *Tetrahedron Lett.*, **44** (38), 7209–7212.

41 (a) Quinoa, E., Kakou, Y. and Crews, P. (1988) Fijianolides, polyketide heterocycles from a marine sponge. *J. Org. Chem.*, **53** (15), 3642–3644; (b) Corley, D.G., Herb, R., Moore, R.E., Scheuer, P.J. and Paul, V.J. (1988) Laulimalides. New potent cytotoxic macrolides from a marine sponge and a nudibranch predator. *J. Org. Chem.*, **53** (15), 3644–3646; (c) Jefford, C.W., Bernardinelli, G., Tanaka, J.-I. and Higa, T. (1996) Structures and absolute configurations of the marine toxins, latrunculin A and laulimalide. *Tetrahedron Lett.*, **37** (2), 159–162.

42 Ghosh, A.K. and Wang, Y. (2000) Total synthesis of (–)-laulimalide. *J. Am. Chem. Soc.*, **122** (44), 11027–11028.

43 Trimurtulu, G., Ohtani, I., Patterson, G.M.L., Moore, R.E., Corbett, T.H., Valeriote, F.A. and Demchik, L. (1994) Total structures of cryptophycins, potent antitumor depsipeptides from the blue-green alga *Nostoc sp.* Strain GSV 224. *J. Am. Chem. Soc.*, **116** (11), 4729–4737.

44 Li, L.-H. and Tius, M.A. (2002) Stereospecific synthesis of cryptophycin 1. *Org. Lett.*, **4** (10), 1637–1640.

45 (a) Kakeya, H., Takahashi, I., Okada, G., Isono, K. and Osada, H. (1995) Epolactaene, a novel neuritogenic compound in human neuroblastoma-cells, produced by a marine fungus. *J. Antibiot.*, **48** (7), 733–735; (b) Kakeya, H., Onozawa, C., Sato, M., Arai, K. and Osada, H. (1997) Neuritogenic effect of epolactaene derivatives on human neuroblastoma cells which lack high-affinity nerve growth factor receptors. *J. Med. Chem.*, **40** (4), 391–394.

46 (a) Hayashi, Y. and Narasaka, K. (1998) Asymmetric total synthesis of epolactaene. *Chem. Lett.*, **27** (4), 313–314; (b) Marumoto, S., Kogen, H. and Naruto, S. (1998) Absolute configuration and total synthesis of (+)-epolactaene, a neuritogenic agent from *Penicillium sp.* BM 1689-P active in human neuroblastoma cells. *J. Org. Chem.*, **63** (7), 2068–2069.

47 Bowden, B.F., Coll, J.C. and Tapiolas, D.M. (1983) Studies of Australian soft corals. XXXIII. New cembranoid diterpenes from a *Lobophytum* species. *Aust. J. Chem.*, **36** (11), 2289–2295.

48 Lan, J., Liu, Z., Yuan, H., Peng, L., Li, W.-D.Z., Li, Y., Li, Y. and Chan, A.S.C. (2000) First total synthesis and absolute configuration of marine cembrane diterpenoid (+)-11,12-epoxysarcophytol A. *Tetrahedron Lett.*, **41** (13), 2181–2184.

49 Ravindranath, N., Reddy, M.R., Mahender, G., Ramu, R., Kumar, K.R. and Das, B. (2004) Deoxypreussomerins from *Jatropha curcas*: are they also plant metabolites? *Phytochemistry*, **65** (16), 2387–2390.

50 Nihei, Y., Yamamoto, H., Hasegawa, M., Hanada, M., Fukagawa, Y. and Oki, T. (1993) Epocarbazolin A and epocarbazolin B, novel 5-lipoxygenase inhibitors-toxonomy, fermentation, isolation, structures and biological-activities. *J. Antibiot.*, **46** (1), 25–33.

51 Knöll, J. and Knölker, H.-J. (2006) First total synthesis of (±)-epocarbazolin A and epocarbazolin B, and asymmetric synthesis of (–)-epocarbazolin A via Shi epoxidation. *Tetrahedron Lett.*, **47** (34), 6079–6082.

52 Shimada, N., Hasegawa, S., Harada, T., Tomisawa, T., Fujii, A. and Takita, T. (1986) Oxetanocin, a novel nucleoside from bacteria. *J. Antibiot.*, **39** (11), 1623–1625.

53 Gumina, G. and Chu, C.K. (2002) Synthesis of L-oxetanocin. *Org. Lett.*, **4** (7), 1147–1149.

54 Van Dyke, A.R. and Jamison, T.F. (2009) Functionalized templates for the convergent assembly of polyethers: synthesis of the HIJK rings of Gymnocin A. *Angew. Chem. Int. Ed.*, **48** (24), 4430–4432.

55 Das, B., Suneel, K., Satyalakshmi, G. and Kumar, D.N. (2009) Stereoselective total synthesis of dodoneine. *Tetrahedron Asymmetry*, **20** (13), 1536–1540.

56 Usami, Y., Suzuki, K., Mizuki, K., Ichikawa, H. and Arimoto, M. (2009) Synthesis of (−)-pericosin B, the antipode of the cytotoxic marine natural product. *Org. Biomol. Chem.*, **7** (2), 315–318.

57 Nemoto, T., Ohshima, T. and Shibasaki, M. (2000) Enantioselective total syntheses of novel PKC activator (+)-decursin and its derivatives using catalytic asymmetric epoxidation of an enone. *Tetrahedron Lett.*, **41** (49), 9569–9574.

58 Reiss, T. and Breit, B. (2009) Total synthesis of (+)-bourgeanic acid utilizing o-DPPB-directed allylic substitution. *Org. Lett.*, **11** (15), 3286–3289.

59 Das, B., Laxminarayana, K., Krishnaiah, M. and Kumar, D.N. (2009) Stereoselective total synthesis of a potent natural antifungal compound (6S)-5,6,dihydro-6-([2R]-2-hydroxy-6-phenyl hexyl)-2H-pyran-2-one. *Bioorg. Med. Chem. Lett.*, **19** (22), 6396–6398.

60 Das, B., Laxminarayan, K., Krishnaiah, M. and Kumar, D.N. (2009) A stereoselective total synthesis of verbalactone. *Helv. Chim. Acta*, **92** (9), 1840–1844.

61 Legters, J., Thijs, L. and Zwanenburg, B. (1991) Synthesis of naturally occurring (2S,3S)-(+)-aziridine-2,3-dicarboxylic acid. *Tetrahedron*, **47** (28), 5287–5294.

62 Morales-Serna, J.A., Llaveria, J., Diaz, Y., Matheu, M.I. and Castillon, S. (2008) Asymmetric sulfur ylide based enantioselective synthesis of D-erythro-sphingosine. *Org. Biomol. Chem.*, **6** (24), 4502–4504.

63 Prevost, S., Phansavath, P. and Haddad, M. (2010) A stereoselective synthesis of (+)-L-733,060 from ethyl (R)-(+)-2,3-epoxypropanoate. *Tetrahedron Asymmetry*, **21** (1), 16–20.

64 Fleming, M.J., McManus, H.A., Rudolph, A., Chan, W.H., Ruiz, J., Dockendorff, C. and Lautens, M. (2008) Concise enantioselective total syntheses of (+)-homochelidonine, (+)-chelamidine, (+)-chelidonine, (+)-chelamine and (+)-norchelidonine by a Pd(II)-catalyzed ring-opening strategy. *Chem. A Eur. J.*, **14** (7), 2112–2124.

65 Watanabe, H., Watanabe, H. and Kitahara, T. (1998) Total synthesis of (−)-pironetin. *Tetrahedron Lett.*, **39** (45), 8313–8316.

66 Severino, E.A. and Correia, C.R.D. (2000) Heck arylation of endocyclic enecarbamates with diazonium salts. Improvements and a concise enantioselective synthesis of (−)-codonopsinine. *Org. Lett.*, **2** (20), 3039–3042.

67 Roy, S.C., Rana, K.K. and Guin, C. (2002) Short and stereoselective total synthesis of furano lignans (±)-dihydrosesamin, (±)-lariciresinol dimethyl ether, (±)-acuminatin methyl ether, (±)-sanshodiol methyl ether, (±)-lariciresinol, (±)-acuminatin, and (±)-lariciresinol monomethyl ether and furofuran lignans (±)-sesamin, (±)-eudesmin, (±)-piperitol methyl ether, (±)-pinoresinol, (±)-piperitol, and (±)-pinoresinol monomethyl ether by radical cyclization of epoxides using a transition-metal radical source. *J. Org. Chem.*, **67** (10), 3242–3248.

68 Sabitha, G., Reddy, E.V., Bhikshapathi, M. and Yadav, J.S. (2007) Toatal synthesis of (9S, 12R, 13S)-pinellic acid. *Tetrahedron Lett.*, **48** (2), 313–315.

69 Sabitha, G., Padmaja, P., Sudhakar, K. and Yadav, J.S. (2009) Total synthesis of the Z-isomers of nonenolide and desmethyl nonenolide. *Tetrahedron Asymmetry*, **20** (11), 1330–1336.

70 Hodgson, D.M., Salik, S. and Fox, D.J. (2010) Stereocontrolled syntheses of (−)-cubebol and (−)-10-epicubebol involving intramolecular cyclopropanation of α-lithiated epoxide. *J. Org. Chem.*, **75** (7), 2157–2168.

71 Topczewski, J.J., Neighbors, J.D. and Wiemer, D.F. (2009) Total synthesis of

(+)-schweinfurthins B and E. *J. Org. Chem.*, **74** (18), 6965–6972.

72 Schmidt, B., Kunz, O. and Biernat, A. (2010) Total synthesis of (–)-cleistenolide. *J. Org. Chem.*, **75** (7), 2389–2394.

73 Chowdhury, P.S., Gupta, P. and Kumar, P. (2009) Enantioselective synthesis of decarestrictine J. *Tetrahedron Lett.*, **50** (51), 7188–7190.

74 Trenkle, J.D. and Jamison, T.F. (2009) Macrocyclization by nickel-catalyzed, ester-promoted, epoxide-alkyne reductive coupling: total synthesis of (–)-gloeosporone. *Angew. Chem. Int. Ed.*, **48** (29), 5366–5368.

75 Raj, I.V.P. and Sudalai, A. (2008) Asymmetric synthesis of (S)-vigabatrin® and (S)-dihydrokavain via cobalt catalyzed hydrolytic kinetic resolution of epoxides. *Tetrahedron Lett.*, **49** (16), 2646–2648.

76 Posner, G.H., Hatcher, M.A. and Maio, W.A. (2005) New silicon-mediated, sequential ring expansions of *n*-sized 2-cycloalkenones into hydroxyolefinic $n + m + p$ medium-sized lactones: short synthesis of (–)-phoracantholide-J. *Org. Lett.*, **7** (19), 4301–4303.

77 Wittenberg, J., Beil, W. and Hoffmann, H.M.R. (1998) Synthesis of dioxatricyclic segments of dictyoxetane. oxygenated 6,8-dimethyl-2,7-dioxatricyclo[4.2.1.0^{3,8}] nonanes show antitumor activity. *Tetrahedron Lett.*, **39** (45), 8259–8262.

78 Proemmel, S., Wartchow, R. and Hoffmann, H.M.R. (2002) Synthesis and studies of marine natural products: the dictyoxetane core from 8-oxabicyclo[3.2.1] oct-6-en-3-ones. *Tetrahedron*, **58** (31), 6199–6206.

79 Yadav, J.S., Basak, A.K. and Srihari, P. (2007) An aldol approach to the synthesis of the anti-tubercular agent erogorgiaene. *Tetrahedron Lett.*, **48** (16), 2841–2843.

80 Ito, K., Yoshitake, M. and Katsuki, T. (1996) Chiral bipyrindine and biquinoline ligands: their asymmetric synthesis and application to the synthesis of *trans*-whisky lactone. *Tetrahedron*, **52** (11), 3905–3920.

81 Hoye, T.R. and Richardson, W.S. (1989) A short, oxetane-based synthesis of (±)-sarracenin. *J. Org. Chem.*, **54** (3), 688–693.

Part Two
Common Ring Heterocycles in Natural Product Synthesis

4
Furan and Its Derivatives
Alicia Boto and Laura Alvarez

4.1
Introduction

The furan ring is present in a variety of natural products, from acetogenins to terpenes and complex alkaloids [1–7]. Many of these products have a potent biological activity: some are sex pheromones, other are defensive compounds, and many have been clinically studied to treat a variety of ailments. For instance, the morphine-type alkaloids have been used from ancient times to treat pain; terpenes, such as salvinorin A or dysidiolide, are known for their hallucinogen and cytostatic properties; the furanocembranoid compounds have been studied for their potent cytotoxic activities, and the complex polyethers, such as brevitoxin, which have caused severe food poisoning in nature, are also used to understand the physiology of human ion channels. Some of these compounds are shown in Table 4.1 below, and as it can be seen, furan has been used by nature to create an amazing variety of structures, displaying many different biological activities.

This chapter is devoted to the applications of furan and its derivatives (hydrofurans, benzofurans, butenolides, etc) to prepare natural products, such as those recorded in Table 4.1 below. But the synthetic utility of these heterocycles is not limited to furan-containing products. Furan derivatives are also valuable precursors of sugars, indolizidine and strychnine alkaloids, polyene macrocycles, azulenes, terpenes, and so on [8, 9].

For this reason, the chapter is divided in two parts: the first describes furan-containing natural products, followed by a summary of synthetic methods to prepare the heterocycles (dehydration of sugar and lactols, metathesis, metal-catalyzed cycloadditions, reduction and oxidation reactions, etc). The second part describes the use of furan derivatives as synthetic intermediates, independently of whether the natural product target retains the furan system or not. This section is classified according to the chemical process: metallation, couplings, cycloadditions, nucleophilic additions, reduction and oxidation reactions, etc. Although furan chemistry is extensive, and only a selection is presented here, we will try to do justice to furan's versatility and potential in synthesis.

Heterocycles in Natural Product Synthesis, First Edition. Edited by Krishna C. Majumdar and Shital K. Chattopadhyay.
© 2011 Wiley-VCH Verlag GmbH & Co. KGaA. Published 2011 by Wiley-VCH Verlag GmbH & Co. KGaA.

4.2
Natural Products Containing the Furan Ring

The furan ring is present in a huge variety of natural products, such as polyketides, phenylpropanoids, alkaloids, and terpenes [1–7]. In general, the natural compounds containing furan rings and its derivatives (di-, tetrahydro, γ-lactone) have been found in all class of terrestrial (fungi, bacteria, insects, plants) and marine (bacteria, fungi, algae, mollusc, seaweed) organisms. The different structural types of natural products with furan rings cover the acetogenins (asimicin), morphinanes (morphine), cembranolides (deoxypukalide), manzamine and other alkaloids (nakadomarin A), macrolide antibiotics, polycyclic ethers, lignans (podophyllotoxin), gingkolides (ginkgolide B), quassinoids, limonoids, some taxane derivatives, furanflavonoids, furanoquinones, steroidal glycosydes (cephalostatins, ritterazines), macrodiolides (pamamycin), among others. A number of natural products belonging to these groups display a variety of biological activities.

4.2.1
Occurrence of Furan Rings in Natural Products

Table 4.1 shows representative examples of the families of natural products containing furans. More examples will be given in the second part of this review, and related structures can be found in recent reviews on natural products.

Table 4.1 Natural products containing furan rings.

Product	Natural source	Biological activity	Reference
Angelmarin (1)	Angelica pubescens	Cytotoxic	[10]
6-C-β-manno pyranosylapigenin (2)	Camellia sinensis	Anti-inflammatory, chemopreventive for colon cancer	[11]
Asiminocin (3)	Asirnina triloba	Antitumor, pesticidal, antimalarial, immunosuppressive and antifeedant	[12]

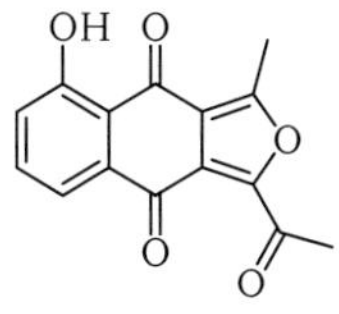

4.2 Natural Products Containing the Furan Ring

Bhimamycin B (4)

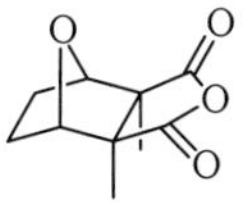

Cantharidin (5)

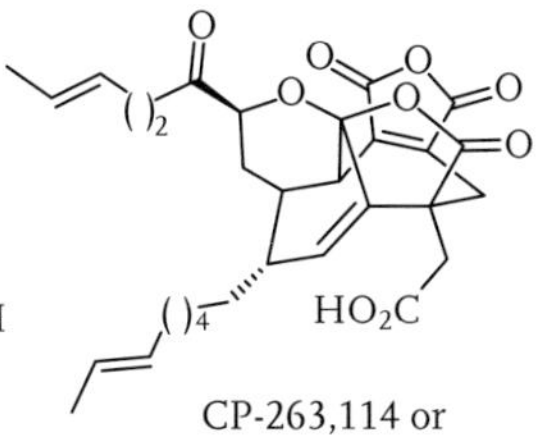

Cephalostatin (6)

CP-263,114 or phomoidride B (7)

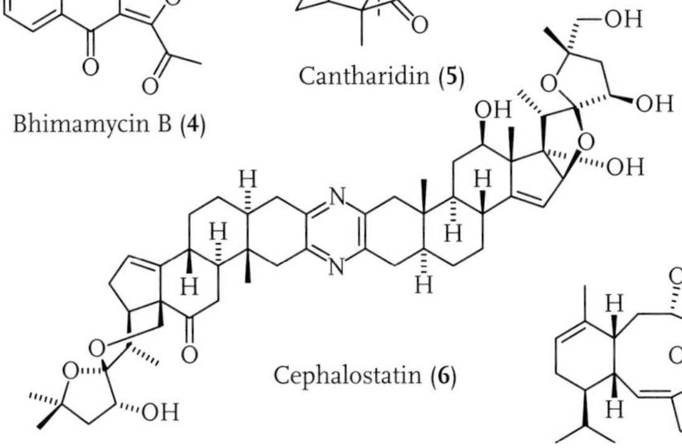

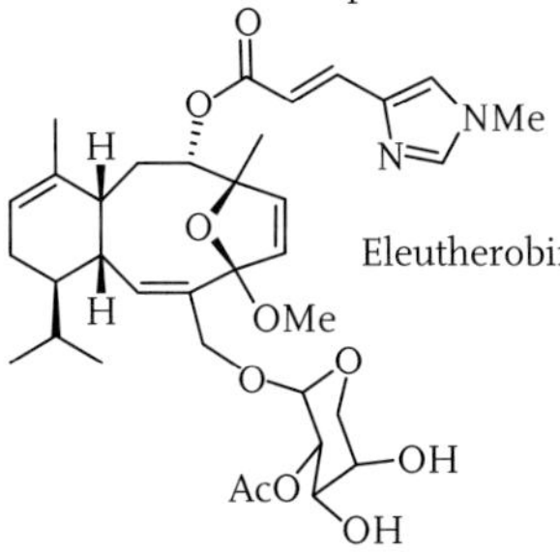

Eleutherobin (8)

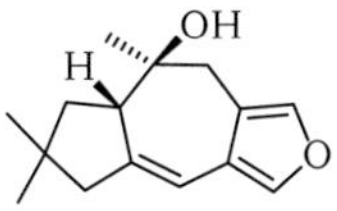

Furoscrobiculin B (9)

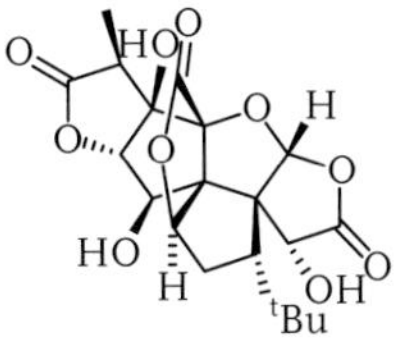

Ginkgolide B (10)

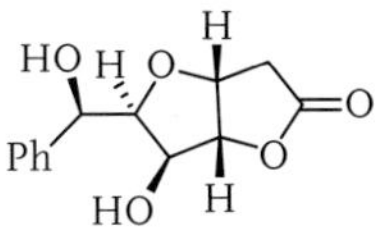

(+)-goniofufurone (11)

Table 4.1 (Continued)

Product	Natural source	Biological activity	Reference
Bhimamycin B (4)	*Streptomyces* sp. GW32/698	Antibacterial	[13]
Cantharidin (5)	*Mylabris phalerata* or *M. cichorii* (beetles)	Active against hepatocarcinoma	[14]
Cephalostatin (6)	*Cephalodiscus gilchristi* (marine tubeworm)	Anticancer agent	[15]
CP-263,114 (7)	Unidentified fungus	Squalene synthase inhibitor	[16]
Eleutherobin (8)	*Erythropodium caribaeorum*	Antitumor agent	[17]
Furoscrobiculin (9)	*Lactarius* and *Russula*	Mushroom's chemical defense	[18]
Ginkgolide B1 (10)	*Ginkgo biloba*	Preventing dementia and age-related cognitive decline	[19]
Goniofufurone (11)	*Goniothalamus giganteus*	Embryo toxic	[20]

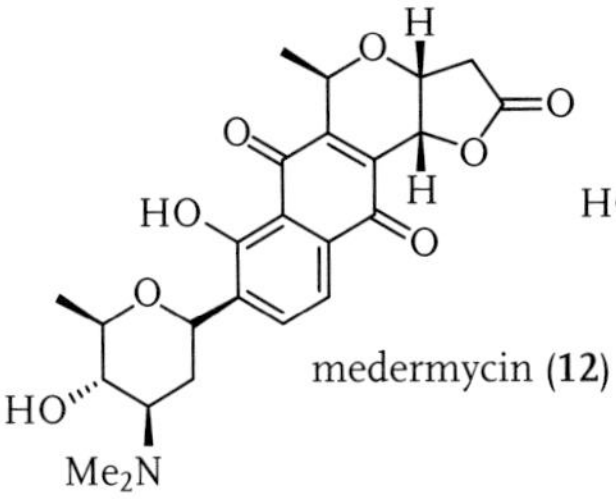

medermycin (**12**)

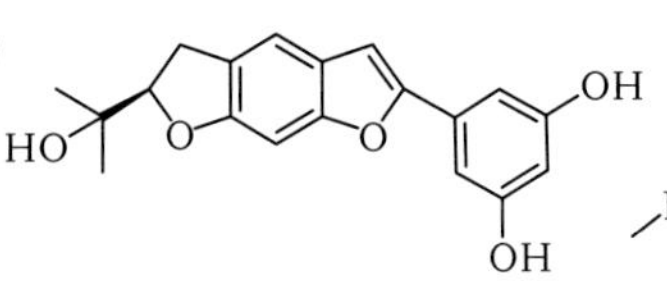

moracin O (**13**)

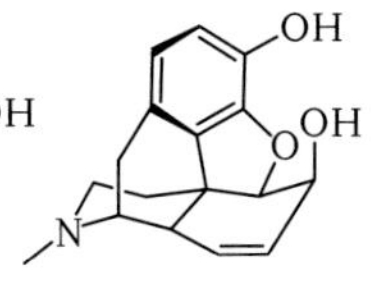

(−)-morphine (**14**)

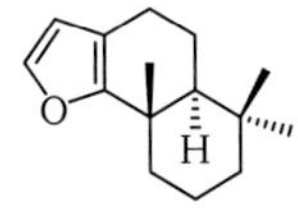

pallescensin A (**16**)

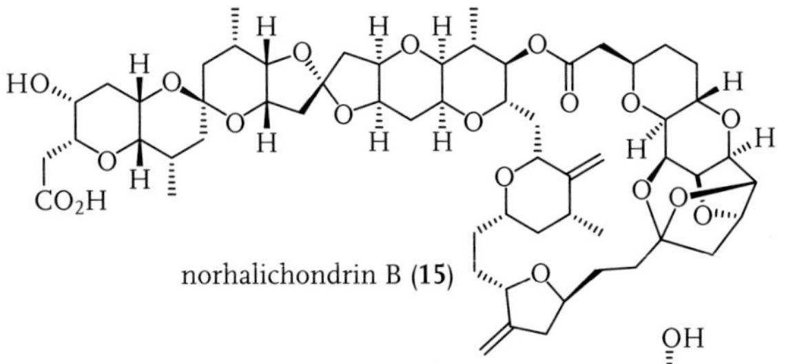

norhalichondrin B (**15**)

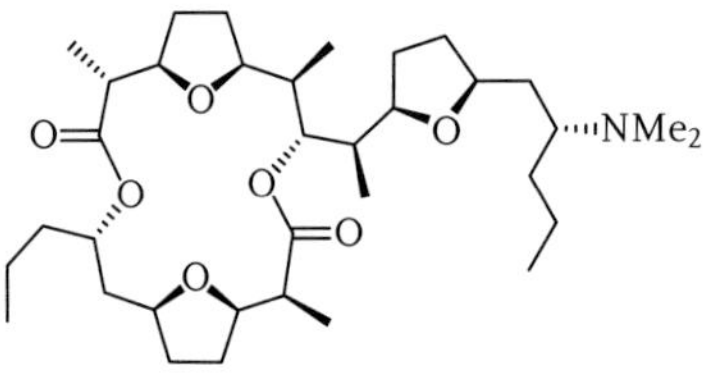

pamamycin (**17**)

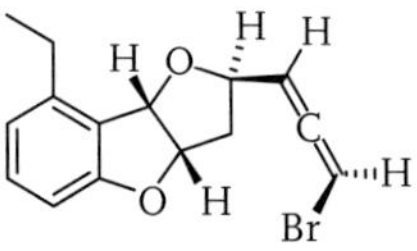

panacene (**18**)

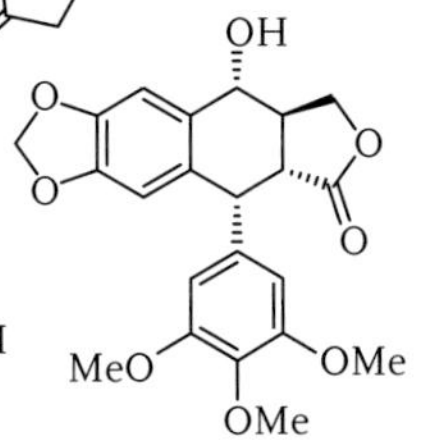

podophyllotoxin (**19**)

Table 4.1 (Continued)

Product	Natural source	Biological activity	Reference
Medermycin (**12**)	*Nocardiopsis* sp.	Selective inhibitor of the serine-threonine kinase AKT	[21]
Moracin O (**13**)	*Morus mesozygia*	Antitumoral	[22]
Morphine (**14**)	*Daphniphyllum. calycinum*	Enhance the mRNA expression of NGF	[23]
Norhalichondrin B (**15**)	*Halichondra okadaii* (marine sponge)	Antitumoral	[24]
Pallescensin A (**16**)	*Disidea pallescens* (marine sponge)	Involved in the defensive mechanisms employed by opisthobranch	[25]
Pamamycin (**17**)	*Streptomyces alboniger*	Antibiotic, antifungic	[26]
Panacene (**18**)	*Aplysia brasilina* (sea hare)	Shark antifeedant	[27]
Podophyllotoxin (**19**)	*Podophyllum* sp.	Cytotoxic, antiviral	[28]

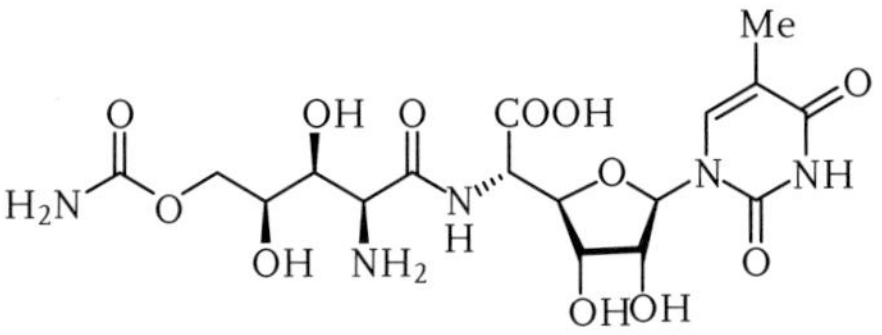

polyoxin J (20)

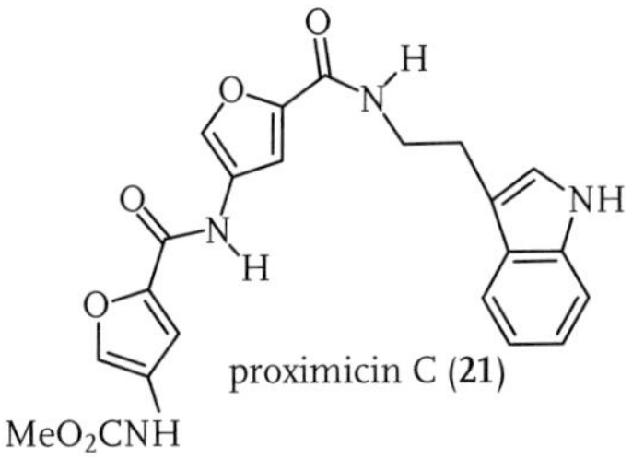

proximicin C (21)

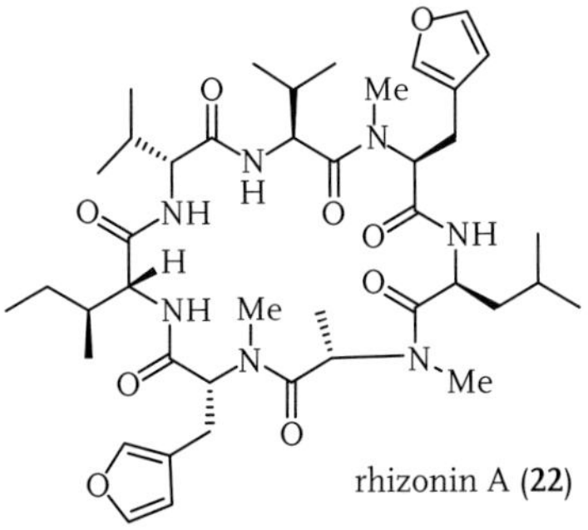

rhizonin A (22)

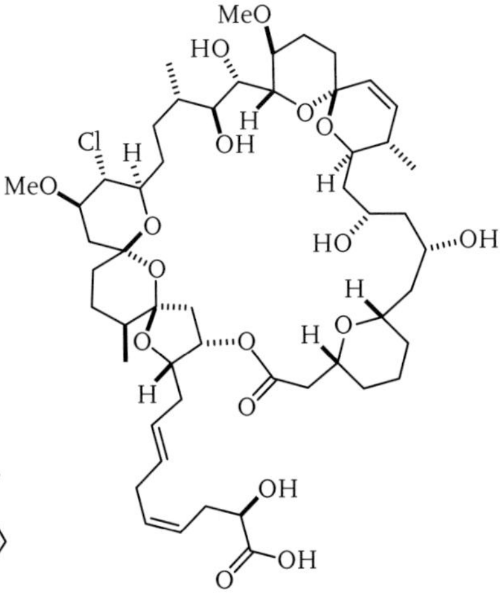

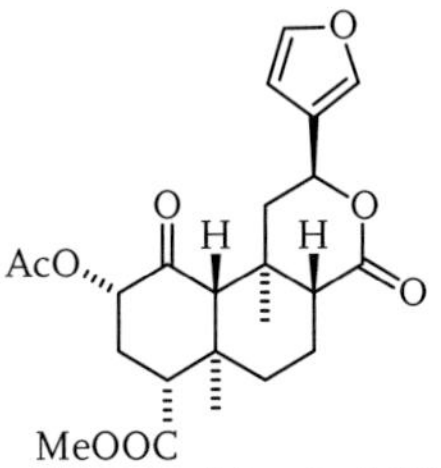

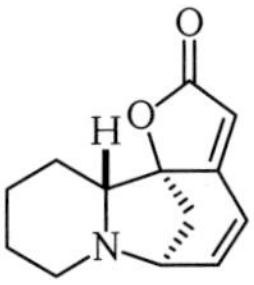

salvinorin A (23)

securinine (24)

spirastrellolide A (25)

Table 4.1 (Continued)

Product	Natural source	Biological activity	Reference
Polyoxin (20)	Streptomyces cacaoi var. asoensis	Antibiotic	[29]
Proximicin C (21)	Verrucosispora sp. (actinomycetes)	Cytotoxic	[30]
Rhizonin A (22)	Burkholderia rhizoxina, endofungal bacteria	Potent and non-specific hepatotoxin in rats	[31]
Salvinorin A (23)	Salvia divinorum	Psychoactive	[32]
Securinine (24)	Securinega suffructicosa	Specific GABA receptor antagonist	[33]
Spirastrellolide A (25)	Caribbean marine sponge Spirastrella coccinea	Ser/Thr protein phosphatase 2A inhibitor	[34]

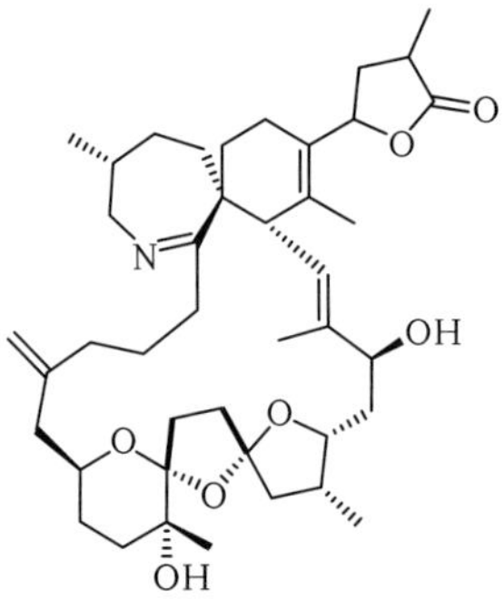

spirolide B (26)

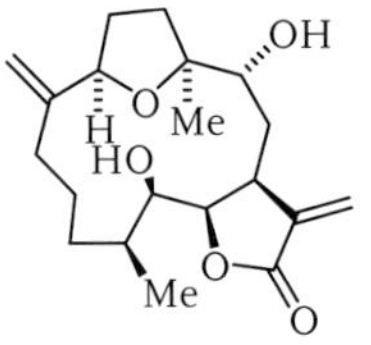

uprolide D (27)

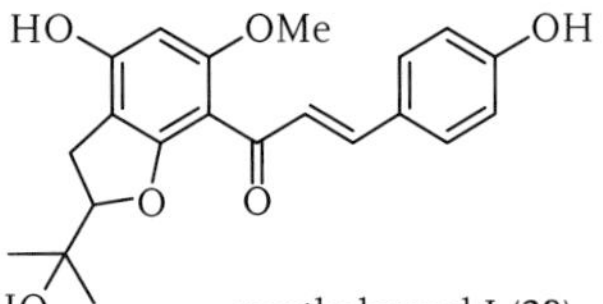

xanthohumol I (28)

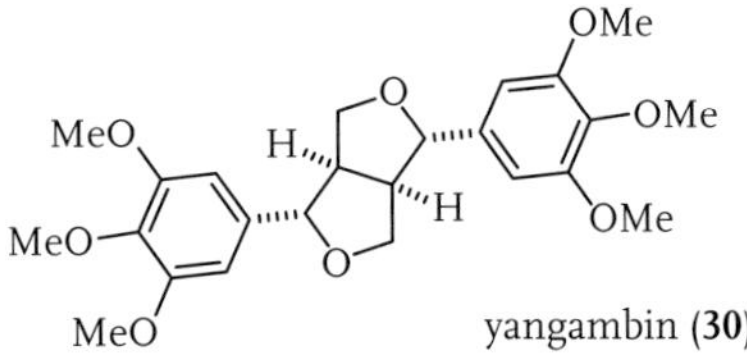

yangambin (30)

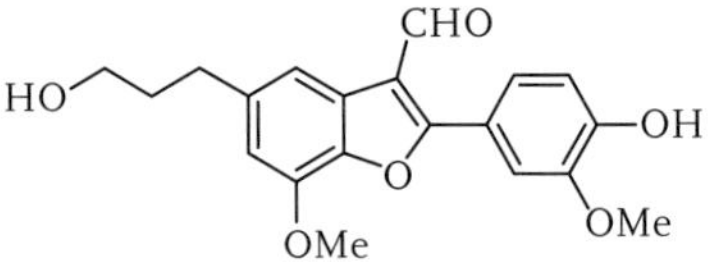

XH-14 (29)

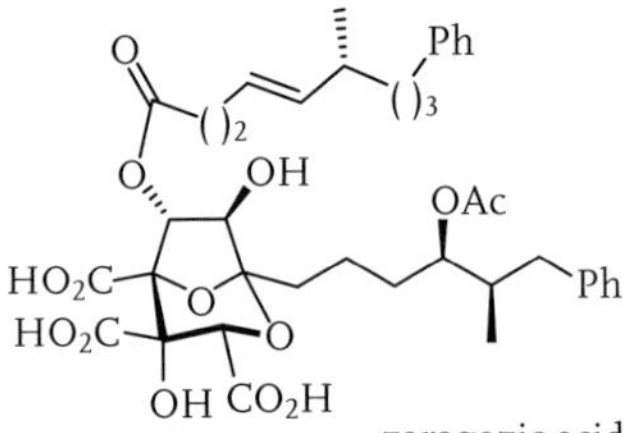

zaragozic acid C (31)

Table 4.1 (Continued)

Product	Natural source	Biological activity	Reference
Spirolide (26)	*Alexandrium ostenfeldii*	Marine toxin, food poisoning	[35]
Uprolide D (27)	*Eunicea mammosa*	Cytotoxic	[36]
Xanthohumol (28)	*Humulus lupulus*	Inhibit human stomach carcinoma cells	[7g, 37]
XH-14 (29)	*Salvia miltiorrhiza*	Antagonist A1 adenosine receptor	[38]
Yangambin (30)	*Liriodendron tulipifera* L.	Protective effects against cardiovascular collapse	[39]
Zaragozic acid (31)	*Sporormiella intermedia*	Inhibitor of sterol synthesis	[40]

4.2.2
Synthesis of Furans in Natural Products

A summary of synthetic methods used to prepare these heterocycles is given in the following diagrams.

4.2 Natural Products Containing the Furan Ring

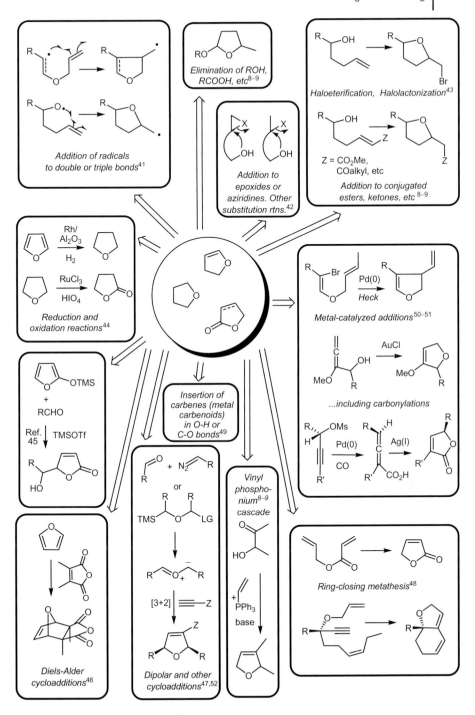

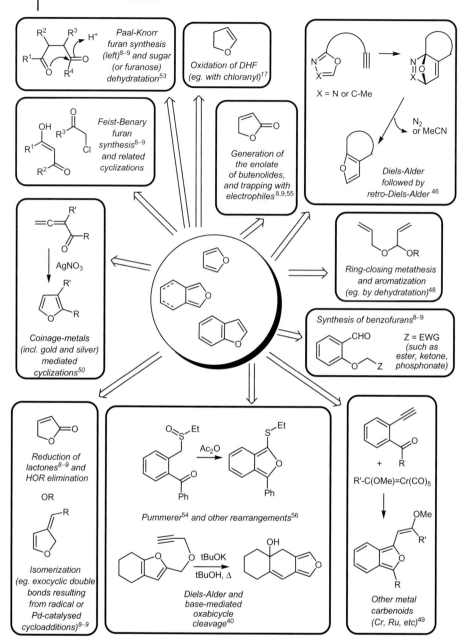

4.3
Furan Derivatives as Reagents in the Synthesis of Natural Products

Furan and its derivatives are valuable precursors of natural products. Furan itself is an aromatic compound, which gives Friedel–Crafts and other aromatic substitution reactions. Besides, it is easily metallated, and the resulting species can be alkylated or participate in sp^2-sp^n couplings. The benzofurans and isobenzofurans have also an aromatic character.

However, the aromaticity of furan is relatively small, in comparison to other heterocyclic systems, such as pyridine, pyrrol or thiophene. Therefore, furan can undergo Diels–Alder, cyclopropanation and other cycloaddition reactions. The aromaticity of the system can also be lost by reduction, addition of radicals to the heterocycle, addition of furan to electrophiles, rearrangements (e.g., Ireland-Claisen) etc. The isobenzofuran and the benzofuran systems also share this reactivity pattern.

Since furan and hydrofurans are electron-rich heterocycles, they are prone to react with electrophiles or electrophilic radicals. They are also readily oxidized to lactols, furanones, acids, etc. Besides, tetrahydrofurans can undergo ring-expansion reactions when the α–C is attached to a leaving group. Finally, enolates from butyrolactones can react with a variety of electrophiles.

Some furan derivatives can also react as electrophiles, such as 2-alkoxytetrahydrofurans or butyrolactones. In order to illustrate these synthetic possibilities, this section is classified according to the type of reaction, and not to the type of natural product target:

1) Metallation and reactions of metallated derivatives;
2) Reduction and oxidation reactions;
3) Other uses as nucleophiles (including Friedel–Crafts, TMSOF etc) or electrophiles;
4) Cycloadditions;
5) Furan in other reactions (radical reactions, ring expansion, cyclopropanation, etc);
6) Other uses of furans in synthesis (rigidity constraints, chiral auxiliaries, chelation).

4.3.1
Metallation

Metallated furans are often formed by reaction of halogenated furans with metals or organometallics [57]. For instance, 2-furyllithium can be generated by treatment of 2-bromofuran with butyllithium (Scheme 4.1) [58]. The metallated furan can react with electrophiles, such as the aldehyde **32**, to give alcohols (e.g., compounds **33a** and **33b**). This strategy was used by Marcos et al. to prepare analogs of the potent antitumoral dysidiolide **34** [58a].

Moreover, when furan is treated with strong bases, such as butyllithium, the 2-H is abstracted. In the following example, the resulting 2-furyllithium added to a chiral sulfinylimide **35**, affording the addition product **36** with high diastereoselectivity. Later oxidation of the furan ring to an acid completed the preparation of the core of polyoxins C and J **37** and of nikkomycin **38** [58b].

In the third example, the aldehyde of 3-methylfurfural **39** was protected with O,N-dimethyl hydroxylamine prior metallation with butyllithium [59a]. The lithiation took place at the unprotected 2-position, affording the intermediate **40**,

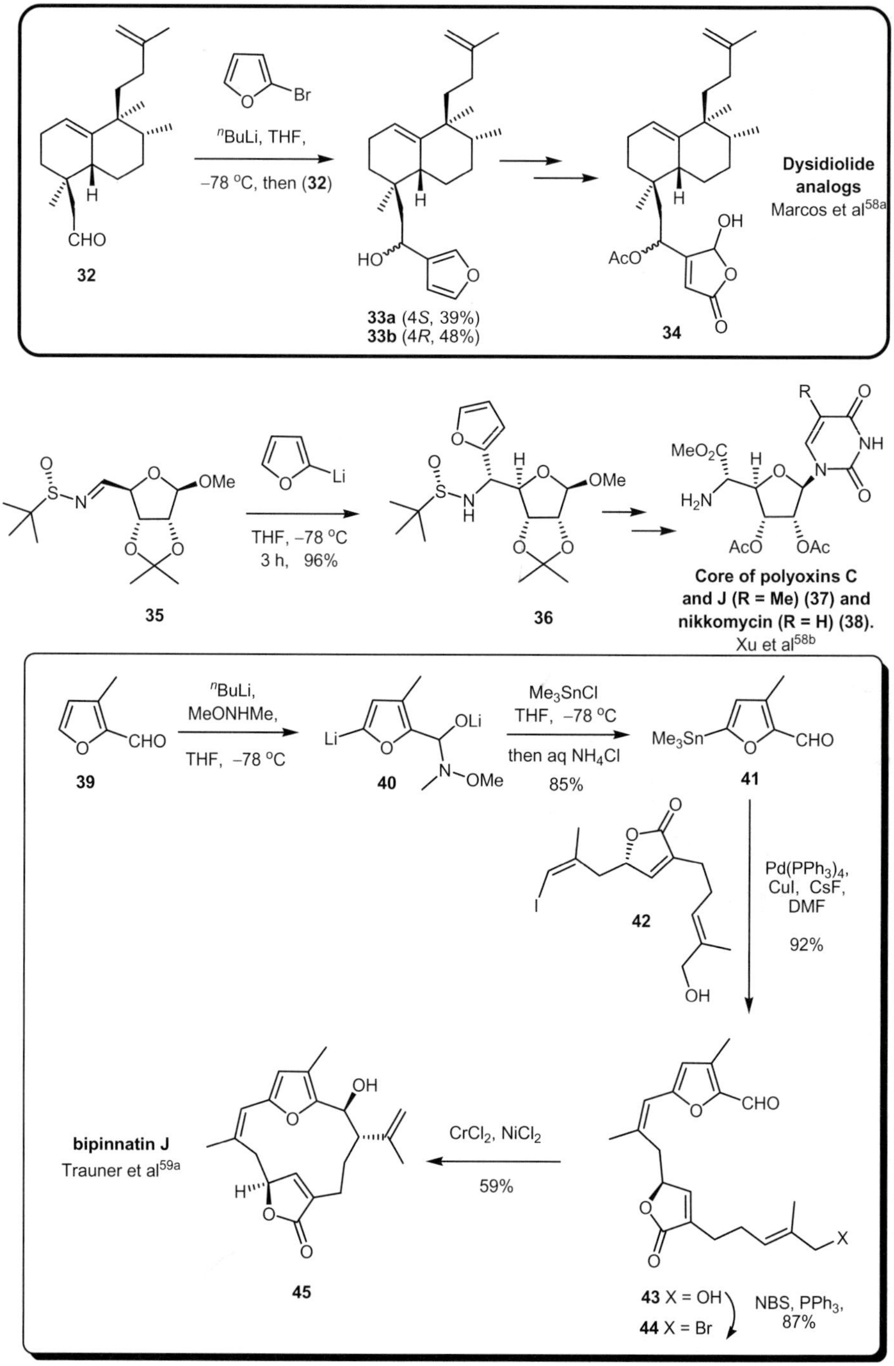

Scheme 4.1 Metallation reactions (I).

which reacted with Me$_3$SnCl to give the furanylstannane **41**. Furylstannanes are often used in synthesis as anion precursors or in sp^n-sp^n couplings. Thus, product **41** underwent a sp^2-sp^2 Stille coupling with the vinyliodide **42** to afford product **43**. After converting the hydroxyl into a bromo group (product **44**), a Nozaki-Hiyama-Kishi coupling was carried out, completing the synthesis of bipinnatin J **45**.

Although proton abstraction by bases usually takes place at C-2 (or C-5), the position of metallation can be controlled by chelation [59b]. When the protected furylcarbinol **46** (Scheme 4.2) was treated with BuLi, the usual abstraction of 2-H took place. The 2-lithiated intermediate reacted with the adjacent silyl ether (Brooks rearrangement) to generate the silylated product **47**. However, when the latter was treated with butyllithium, the metallation took place at C-4, since the resulting 4-furyllithium **48** was stabilized by chelation. The intermediate **48** reacted with tributyltin chloride to afford a stannylated derivative **49**. *Ipso*-substitution of tin by an iodo group afforded compound **50**, which served as precursor of the cyclohexenone **51**. This compound underwent an intramolecular Heck coupling, and after reduction of the ketone and protection of the resulting alcohol, the tricyclic product **52** was isolated. This compound was a key intermediate in the synthesis of phomoidride D **53**.

Among metallated furans, furyllithiums are the most usual in synthesis, but other organometallic reagents have been used as well [59–65]. For instance, the addition of 3-furyllithiums to aldehyde **54** proceeded in low yield; in order to overcome this problem, the titanium reagent **55** was used [60]. The resulting 2-silyloxy-3-alkylfuran underwent acid hydrolysis to a furanone, providing ricciocarpin B **56** in good yields.

In the case of the 2-alkylfuran **57**, the metallation with LDA and ZnBr$_2$ provided a furylzinc intermediate **58** [61a], which underwent a sp^2-sp^2 Negishi coupling with the vinyl iodide **59**, providing the 2,5-dialkyl furan **60**. The synthesis was completed by macrolactonization, followed by ring-closing metathesis, generating the butenolide in (−)-Z-deoxypukalide **61**.

In another remarkable example, the core of the potent antitumoral eleutherobin was assembled by an intramolecular Nozaki-Hiyama-Kishi reaction [62]. Thus, the 2-bromofuran **62** (Scheme 4.3) was transformed into the furylchromium derivative **63**. The organometallic moiety added to the aldehyde, afforded the tricyclic compound **64**, which was then converted into eleutherobin **65**. In this way, a highly complex system was generated from readily available precursors.

Some reduced furan derivatives have also been used in metallations: for instance, 2-lithium reagents derived from dihydrofurans have been used as masked carbonyl anions. In the example, coupling of the alkyliodide **66** [58c] and the lithium reagent **67** afforded the enol ether **68**. When this product was treated with MeMgBr in the presence of a nickel complex, a methyl group was introduced with concomitant opening of the enol ether, affording product **69** in good yields. This product was a key intermediate in the synthesis of the germacradienol **70**.

Finally, thebaine **71** was transformed into the silylated product **72** by treatment with butyllithium and trapping of the intermediate with Me$_3$SiCl [63]. The introduction of the silyl group was vital for the success of the backbone rearrangement.

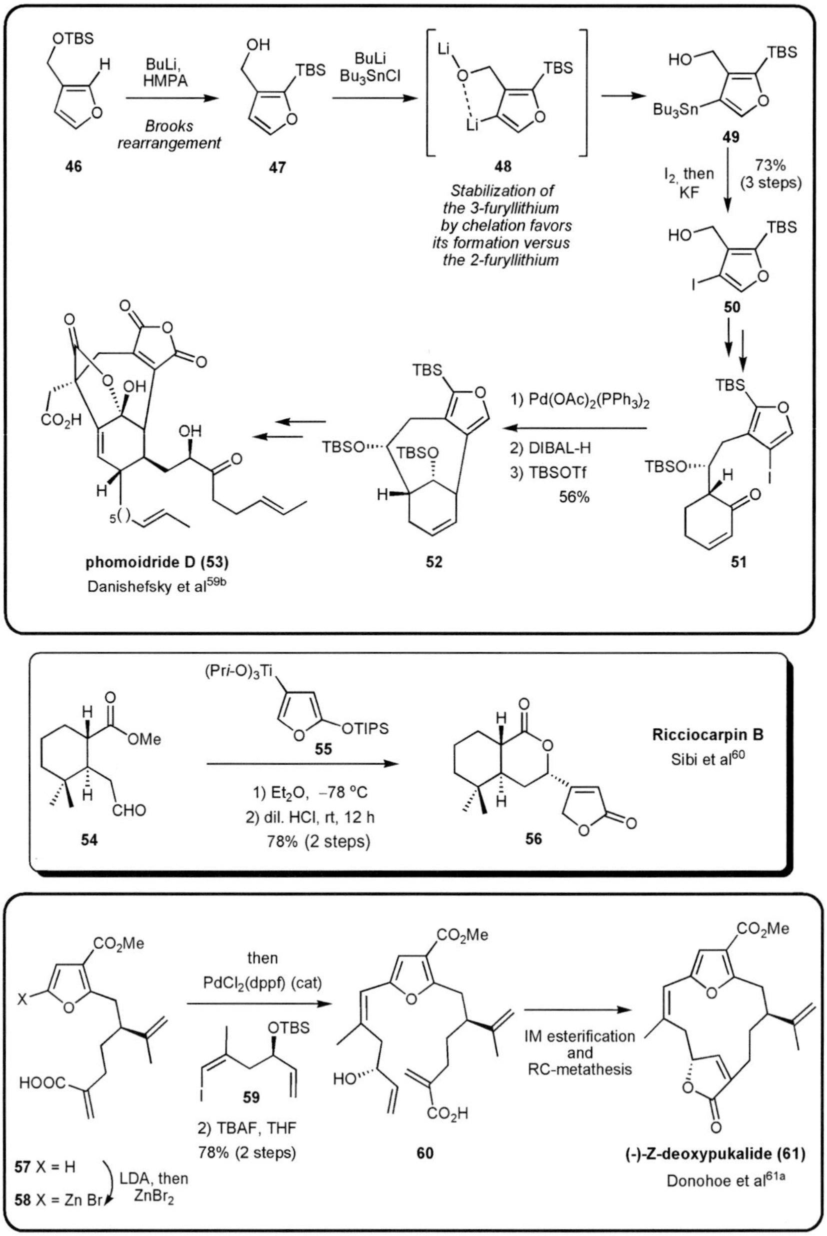

Scheme 4.2 Metallation reactions (II).

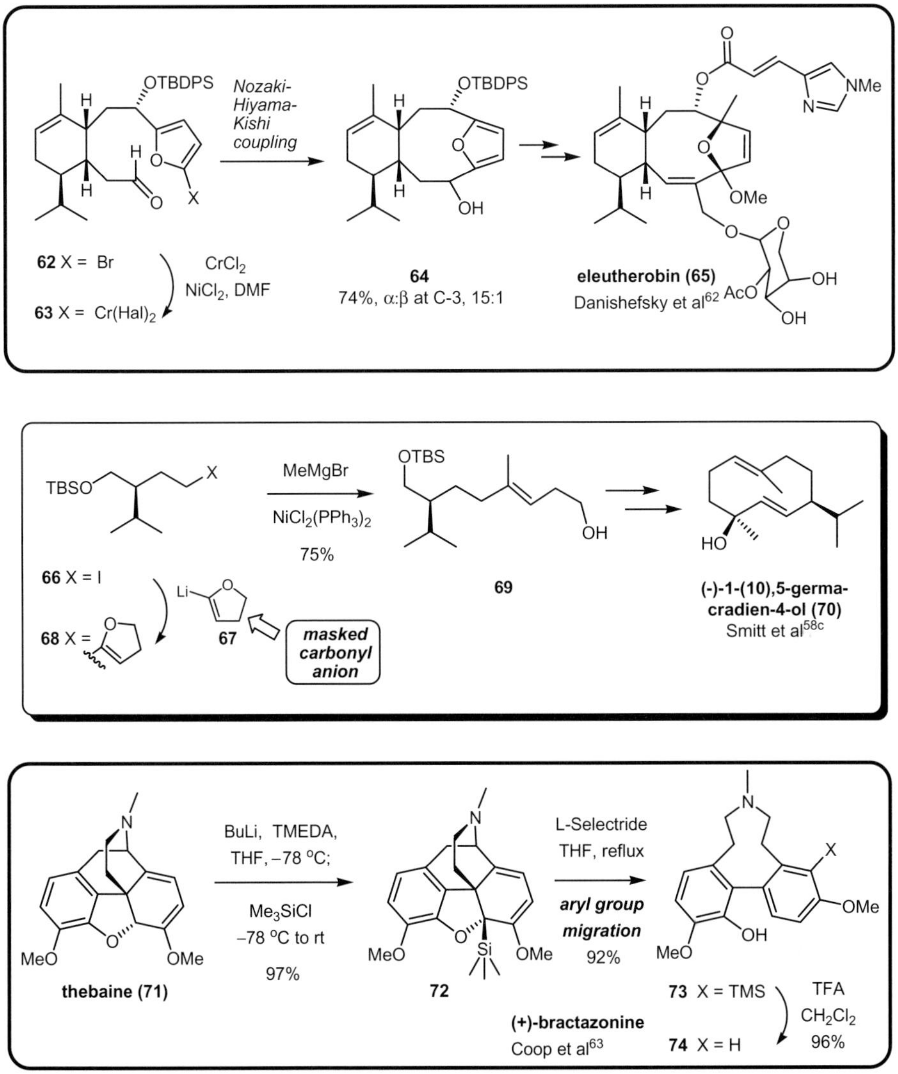

Scheme 4.3 Metallation reactions (III).

Thus, when compound **72** was treated with L-selectride, an aryl group migration took place, affording the tricyclic compound **73**. The silyl group was then removed by treatment with concentrated trifluoroacetic acid, yielding (+)-bractazonine **74**.

4.3.2
Reduction and Oxidation

The furan ring can be reduced by hydrogenation under pressure. For example, the furan ring in compound **75** (Scheme 4.4) [66a] was reduced with a Raney-Ni

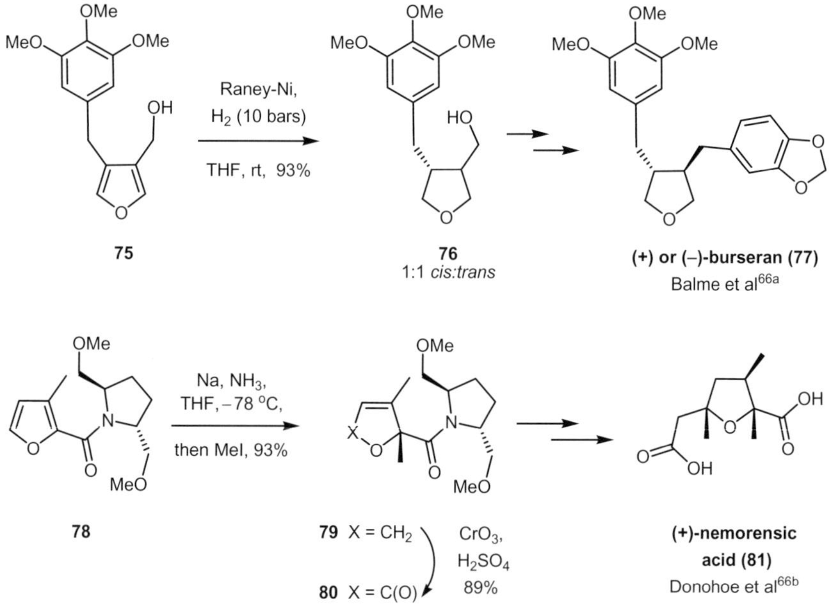

Scheme 4.4 Reduction of furan derivatives.

catalyst without affecting the benzene ring. The resulting tetrahydrofuran **76** was converted in a few steps into (+)-burseran **77**.

In the case of the substrate **78** [66b], the furan ring was reduced using Birch's conditions, and the anion intermediate was trapped with methyl iodide to give product **79**. The dihydrofuran ring was then oxidized to a butenolide using chromium trioxide, affording compound **80** in very good yields. This product was converted into (+)-nemorensic acid **81**, the upper part of nemorensine.

Since furan is an electron-rich heterocycle, it can be oxidized using peroxides, peracids, singlet oxygen, bromine, chromium reagents, etc. In the synthesis of the C40-C53 and C27-C38 domains of norhalichondrin **15**, (Scheme 4.5) [14b, 67a], the same starting furfural **82** was used. The addition of a chiral allyl(isopinocamphenyl)borane afforded the *anti*-furylcarbinol **83** with high enantio- and diastereoselectivity. Then the furan ring underwent an Achmatowicz oxidation, generating the lactol **84**. This product was reduced with triethylsilane, affording the pyranone **85** in high yields. Compound **85** was then transformed into norhalichondrin's C40-C53 domain **86**.

When the starting furfural **82** was treated with a chiral Z-allylborane, the *syn*-furylcarbinol **87** was obtained. The Achmatowicz oxidation, followed by reduction of the lactol **88**, provided product **89** in excellent yield. The silyl group was then removed, and the hydroxy group added *in situ* to the unsaturated ketone, generating a tetrahydrofuran ring. This ring was oxidized with Jones' reagent to a lactone, affording the bicyclic compound **90**. A few additional steps provided norhalichondrin's C27-C38 domain **91**.

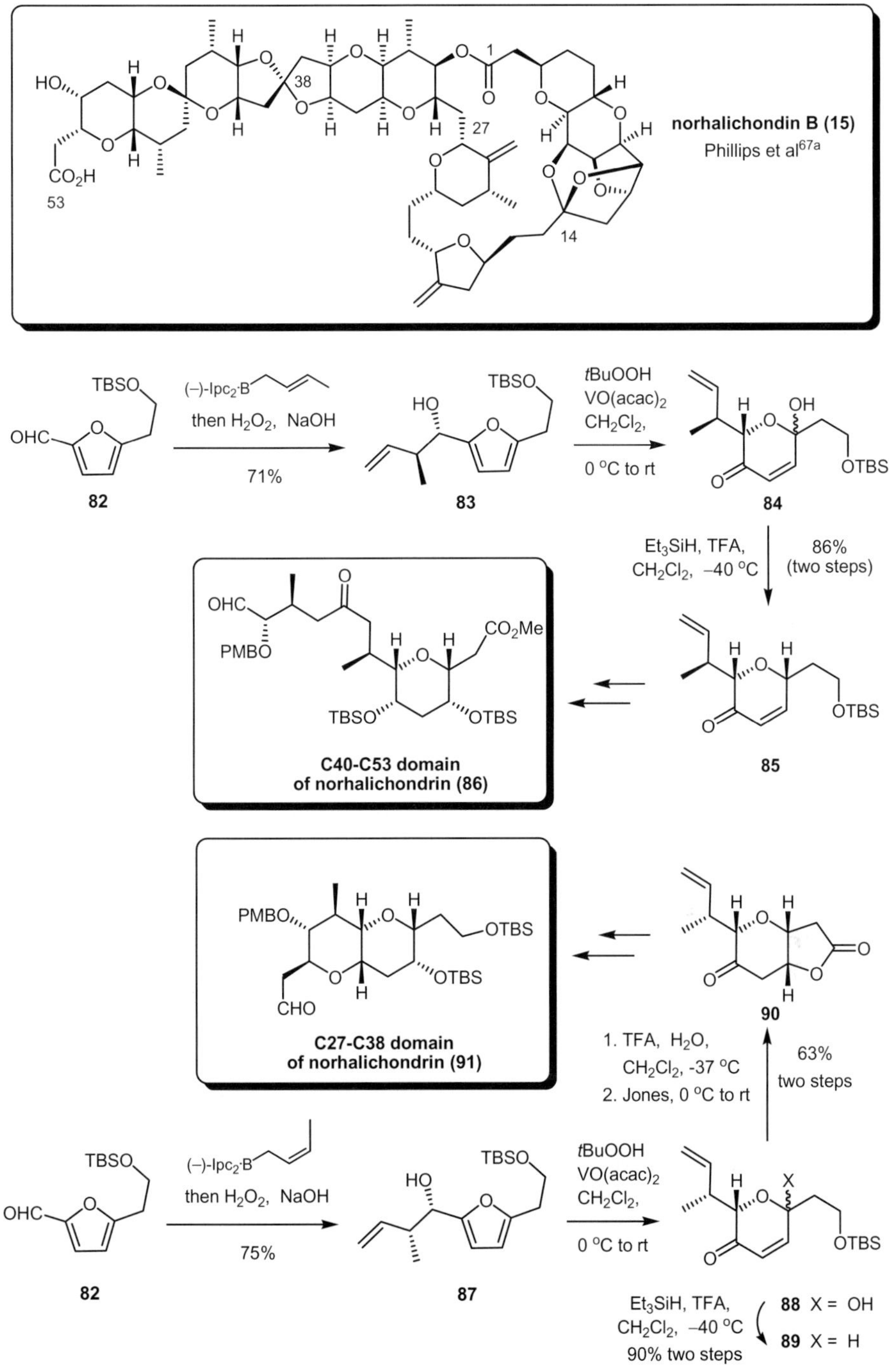

4.3 *Furan Derivatives as Reagents in the Synthesis of Natural Products* | 113

Scheme 4.5 Achmatowicz oxidation of furan rings in the synthesis of norhalichondrin fragments.

The Achmatowicz oxidation has been used by many groups to prepare sugars of unusual series. In O'Doherty's synthesis of anthrax tetrasaccharide **92** (Scheme 4.6) [67b], the furanone **93** serves as starting material for all the carbohydrate units. Thus, reduction with Noyori (*S*,*S*)-catalyst provided the (*S*)-furyl carbinol **94**, which was oxidized with aqueous bromine to the lactol **95**. The later was converted by ketone reduction, dihydroxylation, and protection of the hydroxy groups into the L-rhamnopyranose **96**. Then compound **96** was coupled to its precursor **95** using a palladium (0) catalyst, affording the disaccharide **97**. The synthetic cycle was repeated, yielding the trisaccharide **98**.

In a similar way, the starting furanone **93** was reduced with Noyori (*R*,*R*)-catalyst, giving the (*R*)-furyl carbinol **99**, which was oxidized to the lactol **100**. The latter was converted into the allylic azide **101**, and after dihydroxylation of the double bond, the azidosugar **102** was obtained. After some protection/deprotection steps, compound **103** was obtained and coupled to the trisaccharide **98**. The reduction of the azide group and some protection/deprotection steps completed the synthesis of the toxic anthrax tetrasaccharide **92**. A related strategy was used by Nicolaou in his synthesis of the complex marine polyether maitotoxin [1c, 67g–h].

The aza-Achmatowicz reaction has also been very useful in synthesis, particularly in the preparation of alkaloids and iminosugars. For example, the α-furylamine **104** (Scheme 4.7) was oxidized with *m*-chloroperbenzoic acid, and the resulting lactol protected as its methyl acetal. The *N*,*O*-acetal **105** was transformed into azimic acid **106** and into deoxocassine **107** [68a, b].

The oxidation of furyl carbinols to unsaturated dicarbonyl compounds has also been used in the synthesis of many bioactive natural products. Thus, the furyl carbinol **108** was oxidized with aqueous bromine to the ketoaldehyde **109**, which was then oxidized to the ketoacid **110**. Further manipulations generated a new furyl carbinol **111**, which also underwent oxidation, affording the ketoaldehyde **112**. The aldehyde was converted into the acid **113** and a macrolactonization was carried out. The resulting macrocycle **114** was deprotected to afford macrosphelide B **115** [69a].

Among the oxidizing agents, singlet oxygen deserves particular attention. This reactive species is prepared by bubbling oxygen, under light irradiation, into a solution containing a sensitizer such as Bengal Rose, methylene Blue, tetraphenylporphyrin (TPP), etc. The course of the reaction depends on the functional groups of the substrate [70]. Thus, when the furyl ketone **116** was treated with singlet oxygen, a [4 + 2] cycloaddition ensued to give the peroxide **117**. This intermediate rearranged *in situ* to the tricyclic compound **118**, which after reduction of the peroxy group with DMS and reketalization, afforded the bis-spiroketal unit of salinomycin **119** [70a].

In the case of furan substrate **120**, (Scheme 4.8) where no free hydroxy groups are present, the reaction took a different course. The initial cycloadduct **121** rearranged to the enol **122**, which underwent intramolecular condensation to give the litseaverticillols **123** [70b]. The method has been used to prepare a variety of structures, such as milbemycins, prunolides, dysidiolide analogs, etc [70].

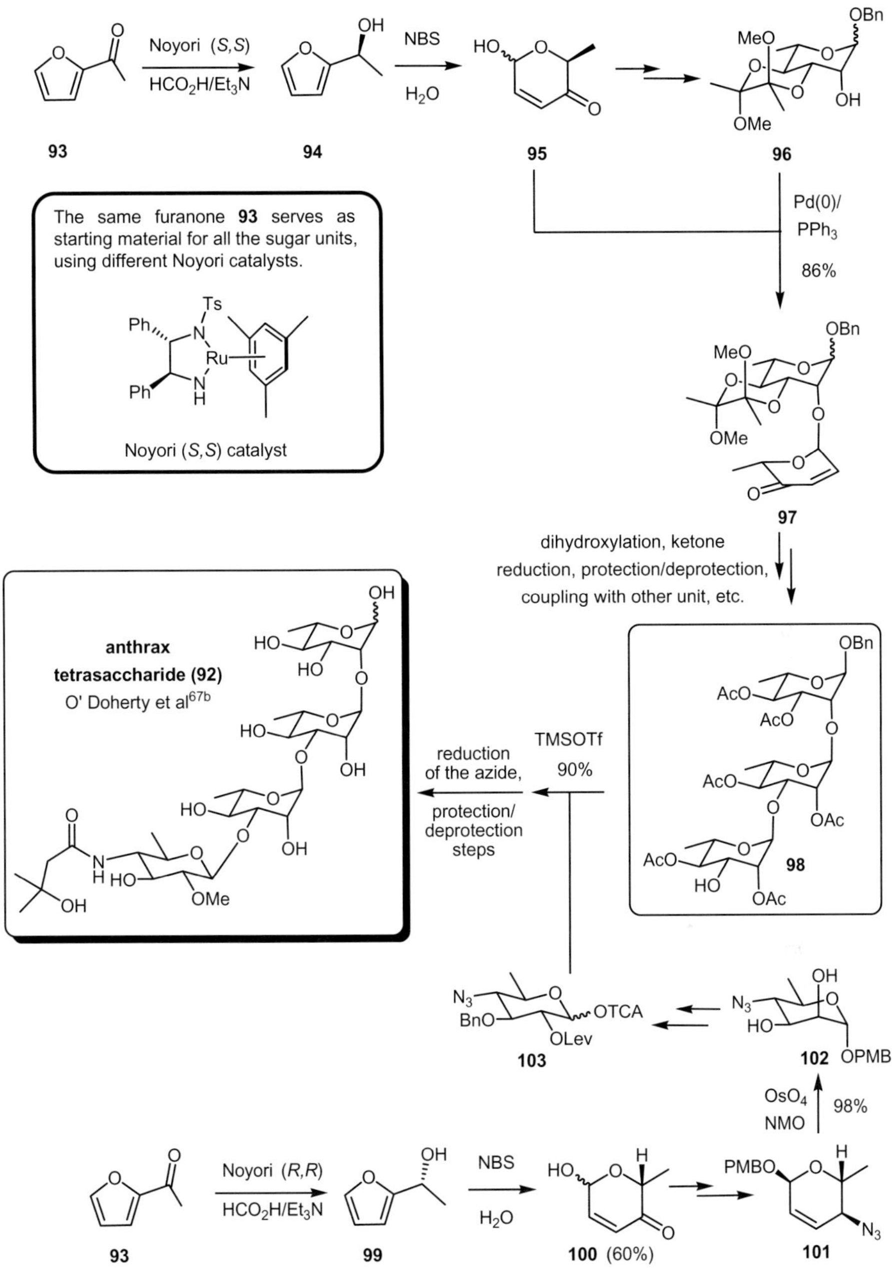

4.3 Furan Derivatives as Reagents in the Synthesis of Natural Products | 115

The same furanone **93** serves as starting material for all the sugar units, using different Noyori catalysts.

Noyori (S,S) catalyst

anthrax tetrasaccharide (92)
O' Doherty et al[67b]

Scheme 4.6 Achmatowicz oxidation of furan rings in the synthesis of anthrax tetrasaccharide.

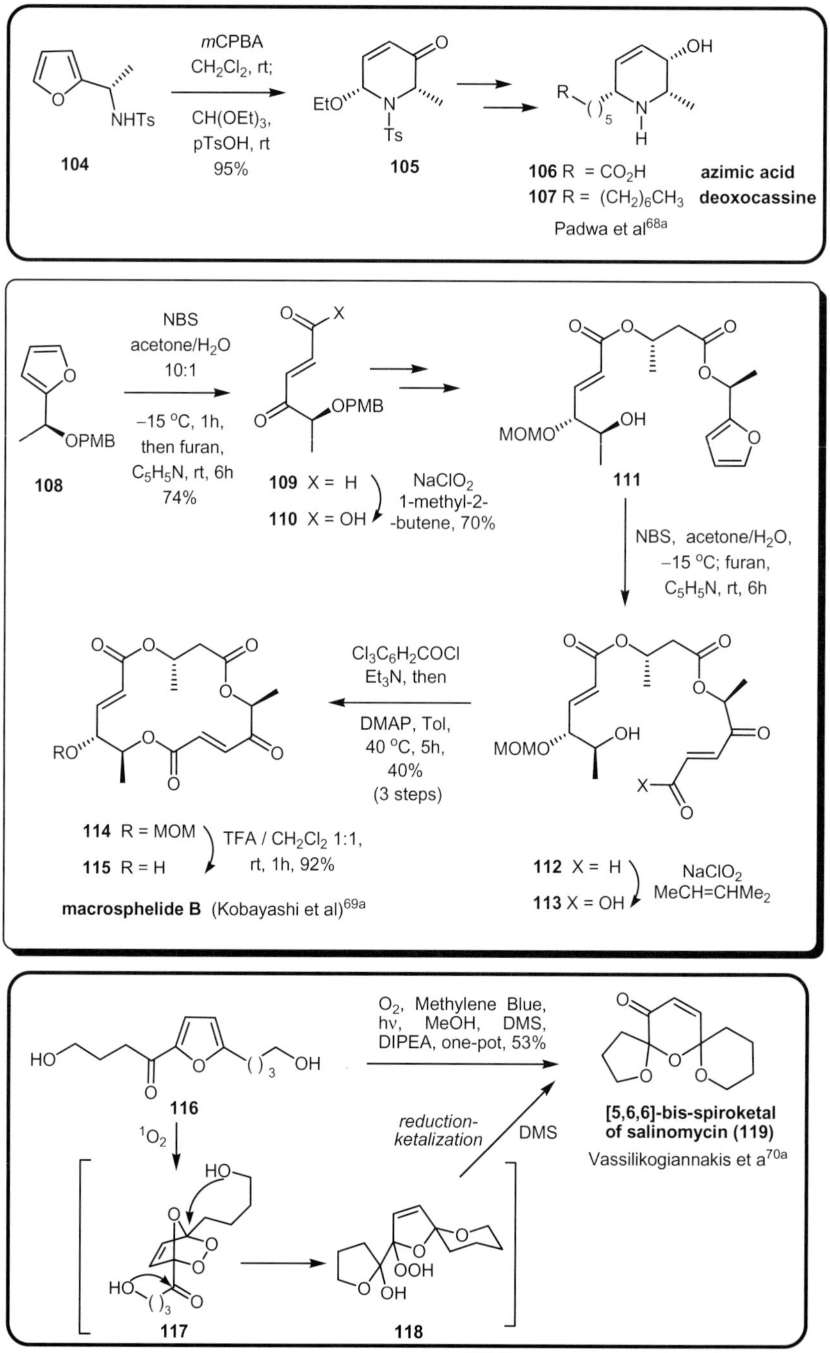

Scheme 4.7 Oxidation of furan rings in the synthesis of natural products.

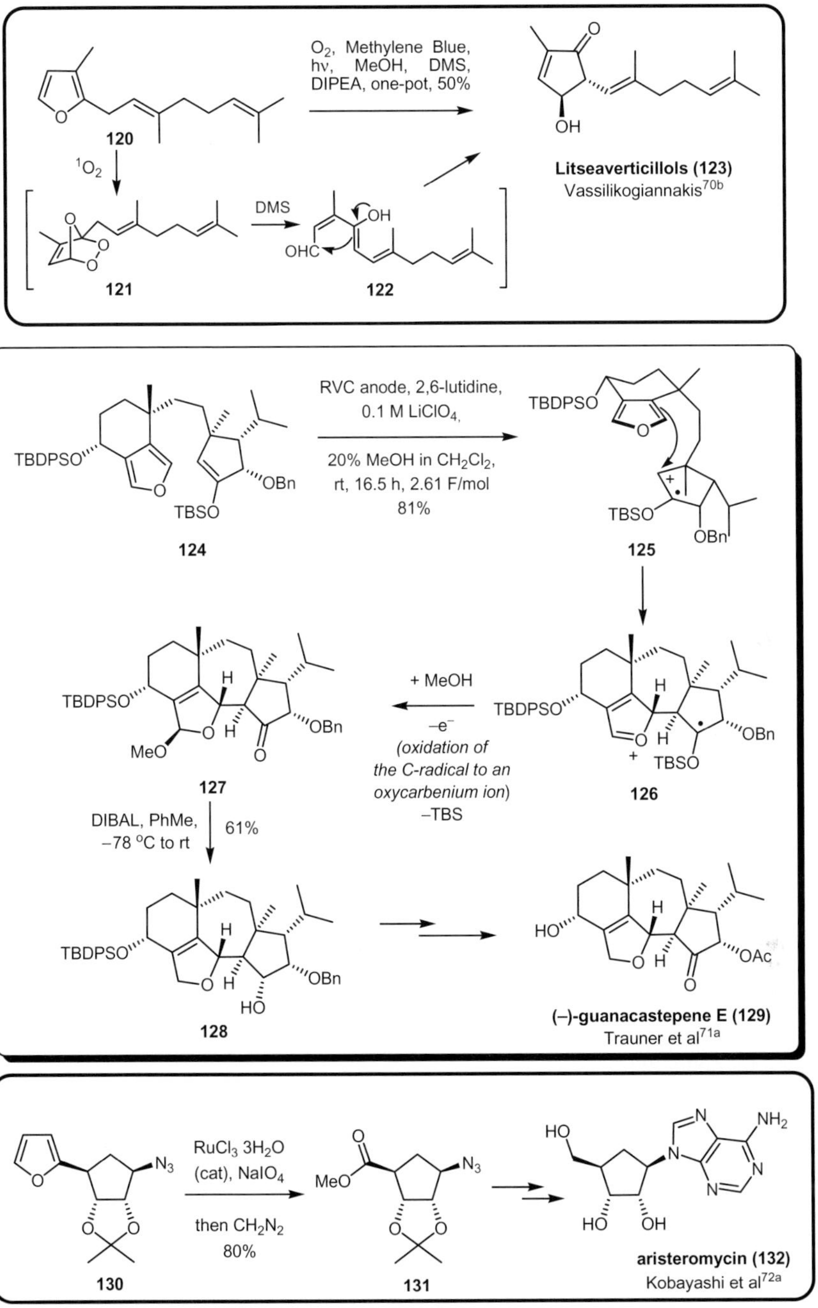

4.3 *Furan Derivatives as Reagents in the Synthesis of Natural Products* | 117

Scheme 4.8 Oxidation of furan rings in the synthesis of natural products.

The oxidation of the furan ring can also be achieved using electroorganic chemistry [71]. Thus, the electrooxidation of the enol ether in substrate **124** generated a cation radical **125** [71a]. The electron-rich furan added to the cation, affording the tetracyclic intermediate **126**. The oxycarbenium ion was trapped by methanol, and the C-radical was oxidized *in situ* to an oxycarbenium ion, which was desilylated to a ketone. The resulting product **127** was reduced with DIBAL to the dihydrofuran derivative **128**. After a few steps, the synthesis of (−)-guanacastepene E **129** was completed.

The furan ring can also be used as a masked acid, as shown in the synthesis of aristeromycin from substrate **130**. The furan ring was oxidized with RuO_4, generated from $RuCl_3$ and sodium periodate. The acid was methylated to give the ester **131**, which was readily converted into aristeromycin **132** [72a]. Other examples of oxidation of furans in the synthesis of natural products are shown in Scheme 4.9, and many others can be found in the recent literature.

4.3.3
Furan Derivatives as Electrophiles and Nucleophiles

The acetals derived from tetrahydro or dihydrofurans are useful electrophiles [74], as shown in the synthesis of dihydroclerodin (Scheme 4.10). Thus, the racemic methyl acetal **144**, in the presence of a Lewis acid, generated the oxycarbenium ion **145**, which reacted with the carvone-derived silyl enol ether **146**. The addition afforded two diastereomeric acetals **147** and **148**, which were readily separated by crystallization in iPr_2O. The major isomer **147** was then transformed into (+)-dihydroclerodin **149** [74a].

In a similar way, the allylsilane **150** reacted with the oxycarbenium ion derived from the acetal **151**. The resulting product **152** was converted into the aldehyde **153**, which reacted with a lithiated furanone **155** to give the addition product **156**. This compound was a key intermediate in Yoshii's synthesis of tetronomycin **157** [74b].

On the other hand, many furan derivatives have been used as nucleophiles [75–78]. Among them, the silyloxy furans derived from butenolides deserve special attention (Scheme 4.11) [76]. These reagents are often generated by treatment of the butenolide with a base and TMSOTf. Once formed, they react with a variety of electrophiles, as shown in the following examples.

Thus, the natural butenolide menisdaurilide **158** was converted into the silyloxy furan **159**, which added to the acyliminium ion derived from 2-hydroxypiperidine **160** [75a]. The resultant tricyclic product **161** was transformed into the tetracyclic alkaloid allosecurinine **162**.

In a similar way, the 2-hydroxypyrrolidine **163** generated an iminium ion on treatment with boron trifluoride, and this ionic intermediate was trapped by the silyloxy furan **164**, to give compound **165**. Then a ring-closing metathesis was used to generate the tricyclic compound **166**, a key precursor of (−)-norsecurinine **167** [75b].

The synthesis of rugulovasines by Martin *et al.* [75c], using two different approaches, also illustrates the scope of this methodology. Thus, the cyano deriva-

....Some other natural products obtained by oxidation of furan and furan derivatives

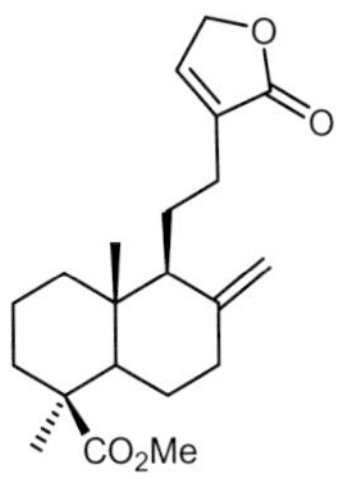

pinusolide (133)
Butenolide comes from oxidation of a furan ring with NBS
Prokop et al (2006)[67c]

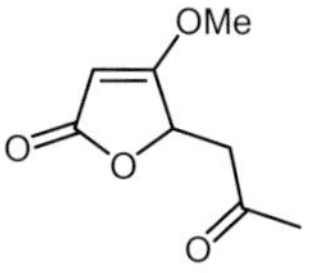

Annularin H (134)
Butenolide comes from oxidation of a dihydrofuran ring with Mn(OAc)$_3$/TBHP
Reissig et al (2007)[67d]

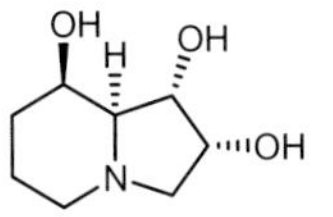

8a-epi-swainsonine (135)
A key step was the Achmatowicz oxidation of a furyl carbinol. The resulting lactone was later transformed into a lactam
O'Doherty et al (2008)[67e]

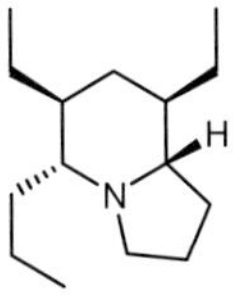

epi-indolizidine 223A (136) A key step was the aza-Achmatowicz oxidation of a 2-substituted furan which gave a lactamol ring
Padwa et al (2003)[68c]

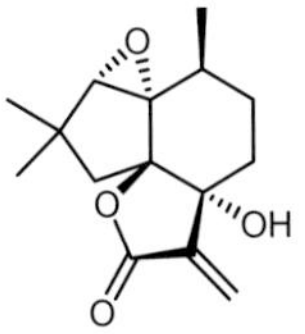

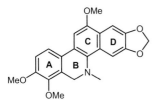

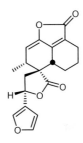

Alliacol (137)
The cyclohexane ring was created by electrooxidation of an enol ether, which reacted with a furan ring (from which the butenolide was formed)
Moeller et al (2004)[71c]

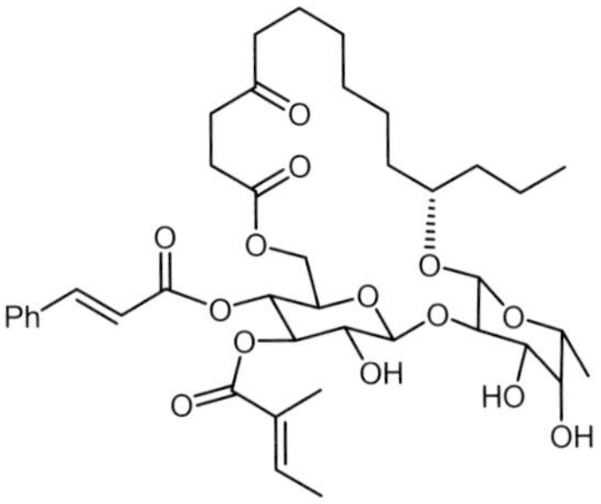

Ipomoeassin B (138)
Part of the macrocycle chain was built from a 2-substituted furan, by Achmatowicz oxidation with tBuOOH and VO(acac)2
Fürstner et al (2007)[67f]

Teuscorolide (139)
The upper butenolide comes from selective oxidation of a furan ring with PDC in DMF
Liu, Zhu et al (2008)[73a]

12-methoxydihydrochelerythrine (140)
The oxidation of a benzofuran ring was used to build the A-B ring system
Watanabe, Ishikaba et al (2003)[73b]

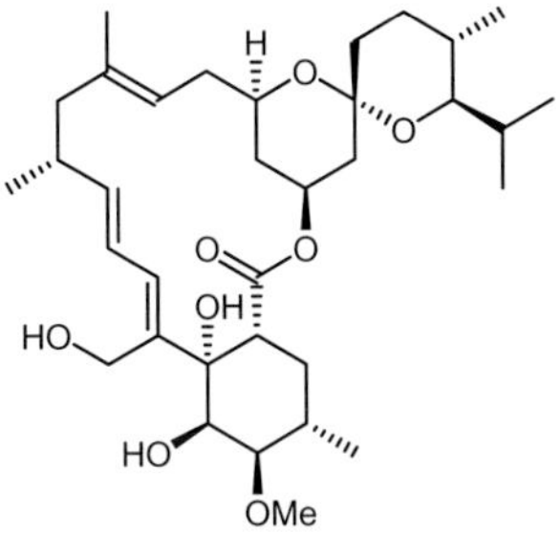

(6R)-6-hydroxy-3,4-dihydromilbemycin E (141)
The diene section of the macrocycle derives from a TMS-furan, by oxidation with singlet oxygen (O$_2$ and catalytic tetraphenylphorphirine)
Thomas et al (2005)[70c]

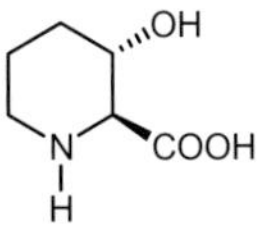

3-Hydroxy pipecolic acid (142)
The carboxyl group was formed by oxidation of a furan ring with RuCl$_3$/NaIO$_4$
Liu, Huang et al (2008)[72b]

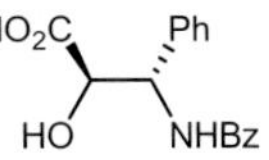

Taxol side chain (143)
The carboxyl group was formed from a furan, by oxidation with RuCl$_3$/NaIO$_4$
Aggarwal et al (2003)[72c]

Scheme 4.9 Oxidation of furan rings in the synthesis of natural products.

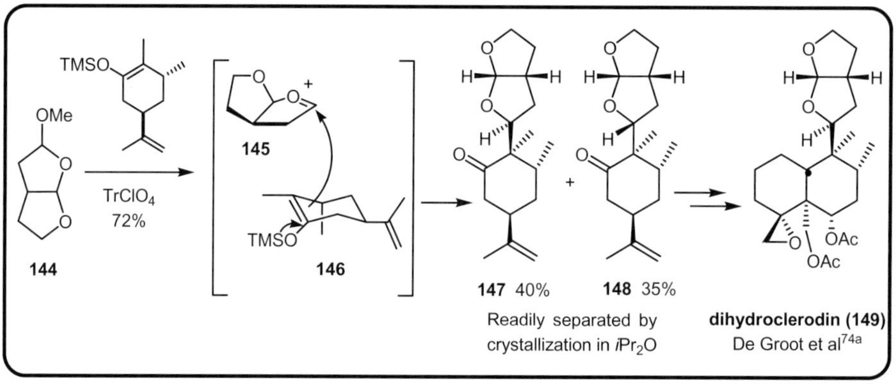

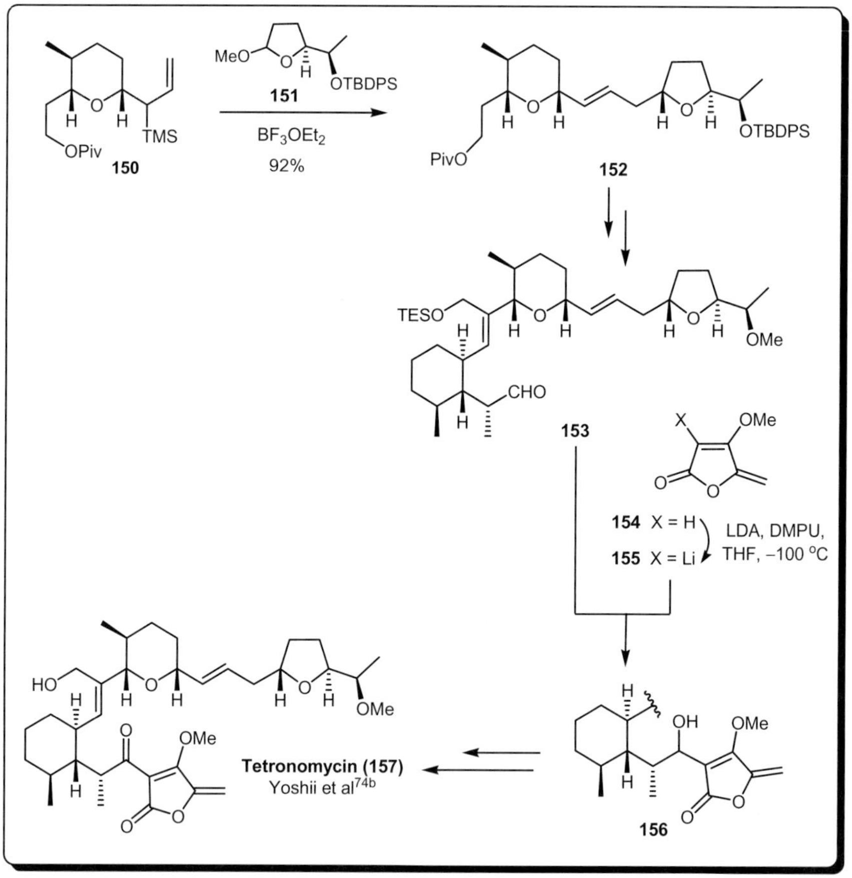

Scheme 4.10 Furan derivatives as electrophiles and nucleophiles.

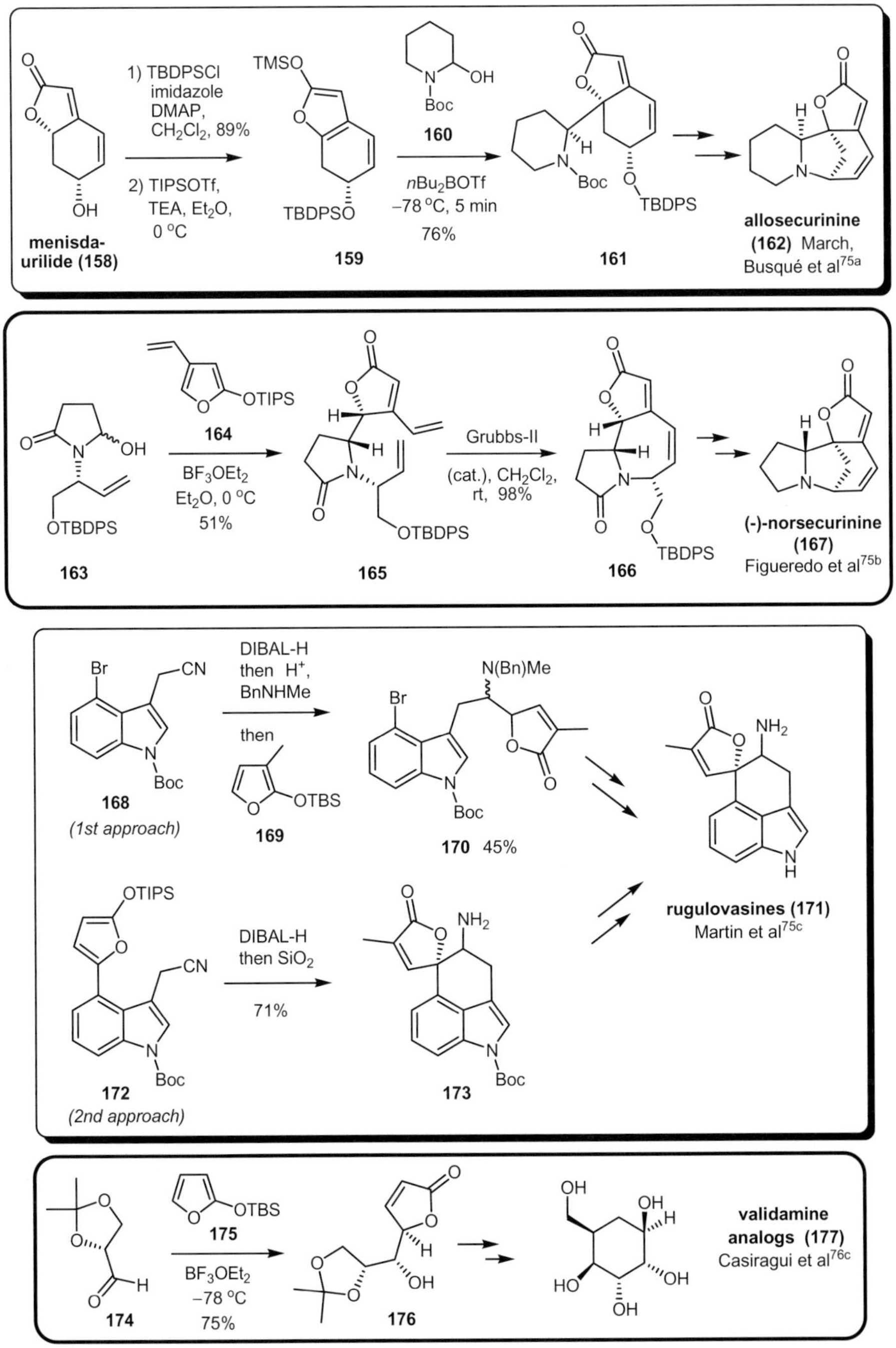

4.3 *Furan Derivatives as Reagents in the Synthesis of Natural Products* | 121

Scheme 4.11 Furan derivatives as nucleophiles in the synthesis of natural products.

tive **168** was converted into an imine, which was trapped by the silyloxy furan **169**. The addition product **170** was obtained in moderate yields, and then transformed into the target alkaloid **171**.

In order to improve the addition step yields, Martin explored the intramolecular addition of a silyloxy furan into an imine, generated *in situ* from a cyano group (conversion **172**→**173**). Remarkably, this new approach doubled the addition step yields, and also shortened the synthesis. Thus, the tetracyclic product **173** was converted in one step into rugulovasines **171** [75c].

The work of Casiraghi et al. on this methodology has also provided many interesting applications, such as the preparation of acetogenin analogs, iminosugars, and carbocycles [75d,76b–d]. Thus, the glyceraldehyde **174** reacted with the silyloxyfuran **175** to afford the butenolide **176**, precursor of product **177**. The latter is an analog of validamine, a component of several antibiotics [75d].

Furan silanes, such as compound **178** (Scheme 4.12) are also good nucleophiles [77a]. These furan derivatives are often prepared by metallation of furans followed by reaction with silyl chlorides. In the example given, when a Lewis acid was added, the furan ring added intramolecularly to the aldehyde, giving the tricyclic compound **179**. The TBS group was then removed by treatment with concentrated TFA, yielding (±)-6β-hydroxyeuropsin **180**.

Furan also reacted as a nucleophile in the Michael addition to compound **181**. The resultant product **182** was treated with dimethyldioxirane (DMDO) to oxidize the furan ring to a ketoaldehyde. The aldehyde underwent intramolecular condensation with the cyclic enol, affording compound **183**, a potential precursor of the agarofuran sesquiterpenes **184** [77b].

Many complex polycyclic systems can be readily assembled from simple furan substrates, as shown by Padwa et al. in the preparation of alkaloids of different families. Thus, alkylation of the ketone **185** (Scheme 4.13) with the α-iodoamide

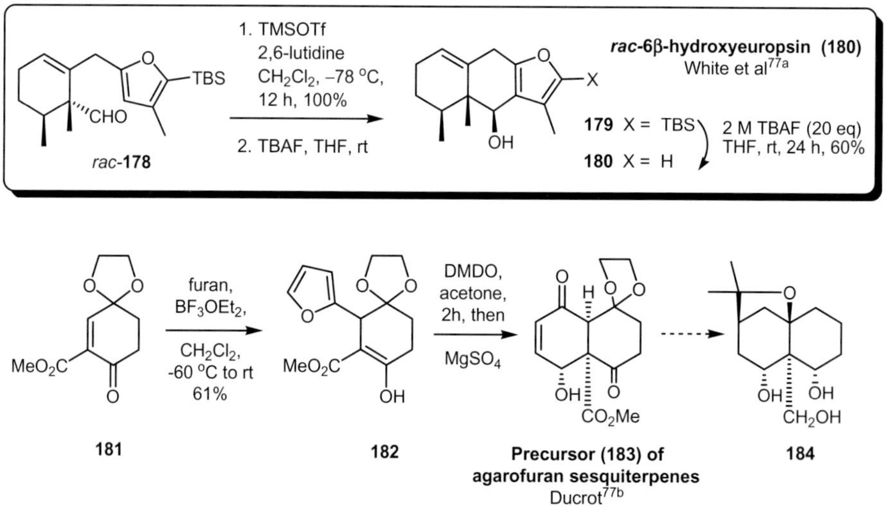

Scheme 4.12 Furan derivatives as nucleophiles in the synthesis of natural products.

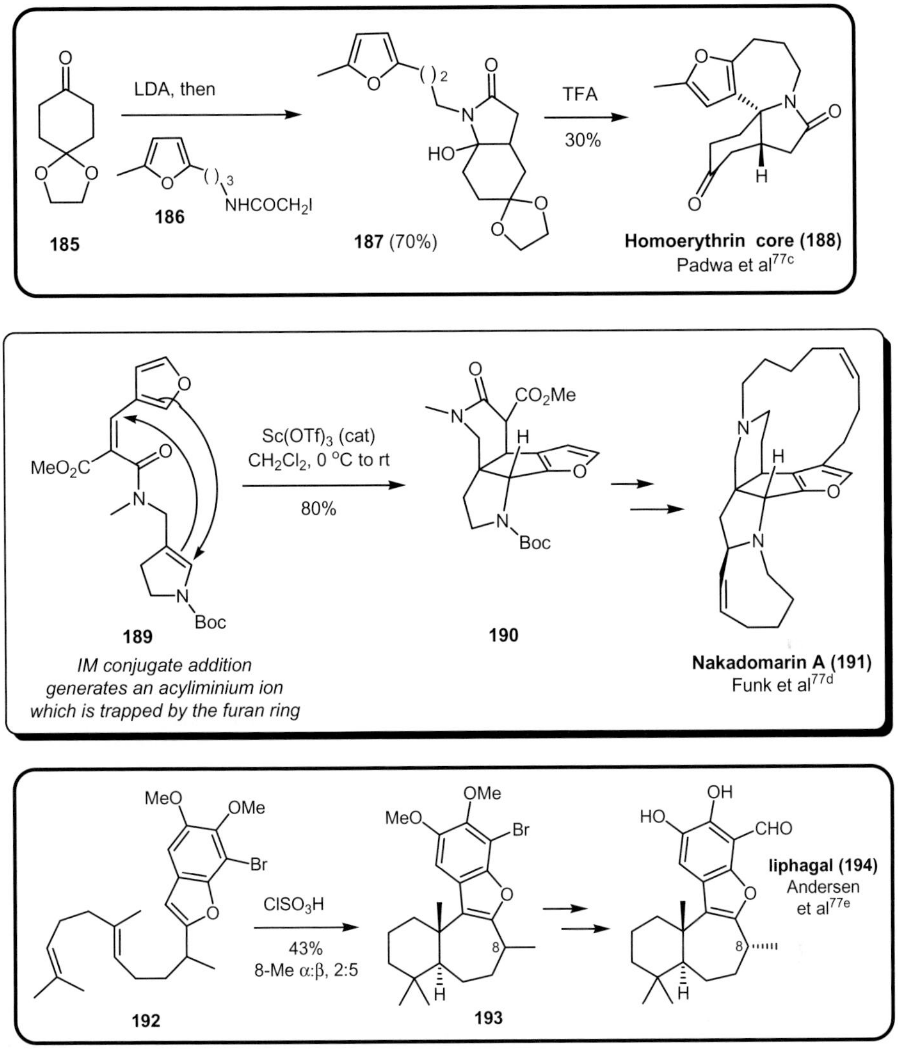

Scheme 4.13 Furan derivatives as nucleophiles in the synthesis of natural products.

186, followed by intramolecular formation of an N,O-acetal, afforded the tricyclic compound **187**. On acid treatment, the N,O-acetal generated an iminium ion, which was trapped by the furan ring, to give the tetracyclic homoerythrin core **188** [77c].

Another remarkable example was reported by Funk et al., using the encarbamate **189**. In the presence of a Lewis acid, the encarbamate added to the conjugated double bond, generating an acyliminium ion intermediate. This ion was trapped in situ by the furan ring, affording the tetracyclic compound **190** with high diastereoselectivity. Compound **190** was an advanced precursor of the marine product nakadomarin A **191** [77d].

Other aromatic furan derivatives have also been used as nucleophiles. For instance, on acid treatment the polyene **192** underwent a Johnson-type cyclization to generate the tetracyclic system **193**. This compound was readily deprotected to liphagal **194** [77e].

4.3.4
Furan in Cycloadditions

4.3.4.1 [2 + 1], [2 + 2] and [3 + 2] Cycloadditions

Due to the relatively low aromaticity of furan, cycloaddition reactions are much easier than with pyrrole, thiophene, and other aromatic systems. For instance, furans react readily with carbenes in cyclopropanation reactions (Scheme 4.14) [79]. Thus, furan **195** underwent a [2 + 1] cycloaddition with a rhodium carbenoid derived from ethyl α-diazoacetate, giving the intermediate **196**. This unstable cyclopropane underwent ring opening to give the enolate **197**, which generated a

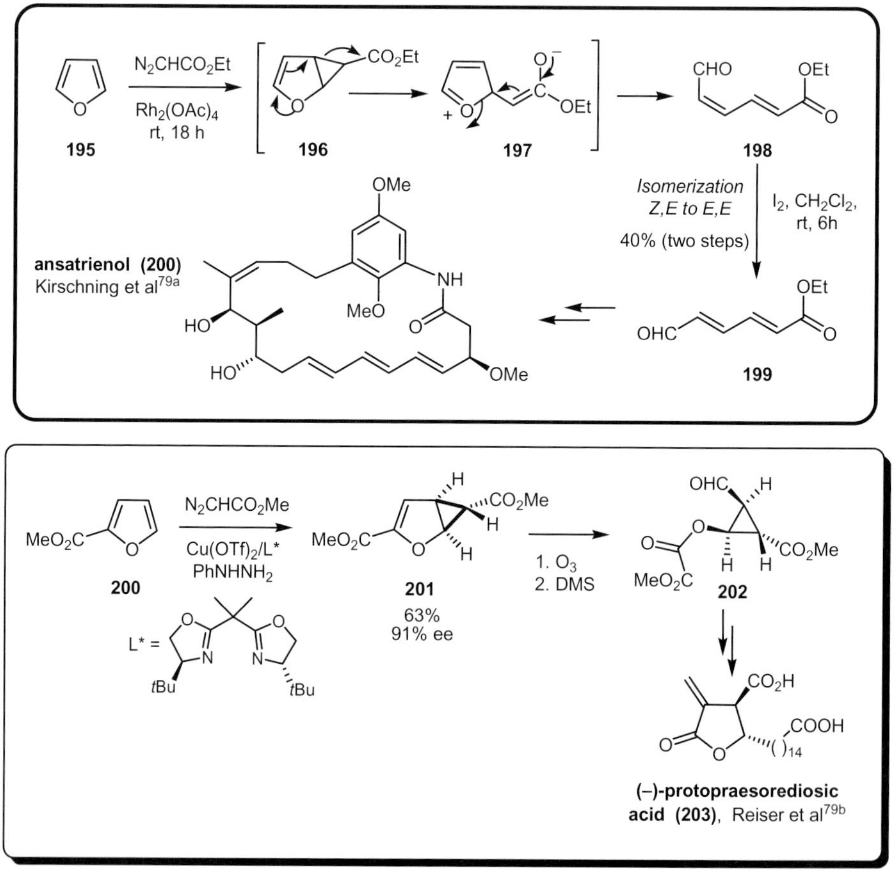

Scheme 4.14 [2 + 1] cycloadditions in the synthesis of natural products.

4.3 Furan Derivatives as Reagents in the Synthesis of Natural Products | 125

second ring cleavage, affording the Z,E-diene **198**. After iodine-promoted isomerization to the E,E-diene **198**, the chain was elaborated into the potent antibiotic ansatrienol **199** [79a].

In another example [79b], a chiral copper catalyst was used to convert the furan **200** into the cyclopropane **201**, with high enantioselectivity. Oxidative cleavage of the double bond gave the optically pure cyclopropane **202**, which served as precursor of (−)-protopraesorediosic acid **203**.

Finally, the cyclopropanation of furan rings can be followed by ring expansion, as occurred in the transformation of compound **204** (Scheme 4.15) into the cyclopropane intermediate **205** and finally, the guanacastepene derivative **206** [79c].

The [2 + 2] cycloadditions have also been used to prepare natural products (Scheme 4.16) [80]. The bis-butenolide **207** reacted with ethylene under photolytic conditions, and after removal of the silyl groups, the tetracyclic compound **208** was obtained in good yields and high diastereoselectivity. The diol was cleaved and then the molecule was elaborated into (+)-grandisol **209** [80a].

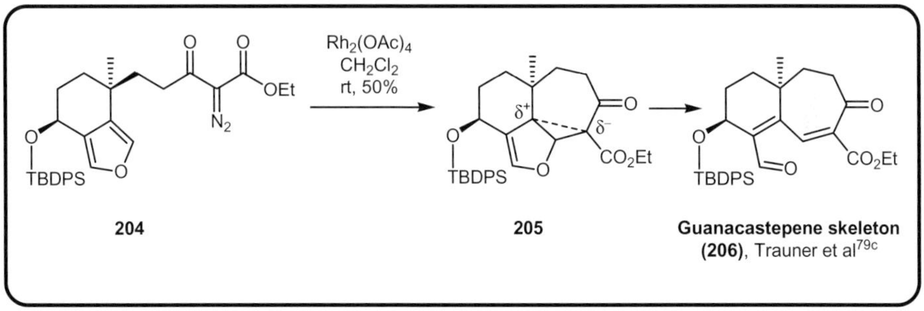

Scheme 4.15 [2 + 1] cycloadditions in the synthesis of natural products.

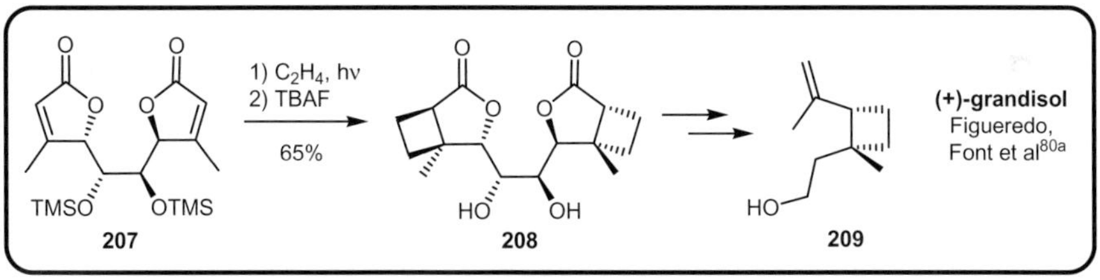

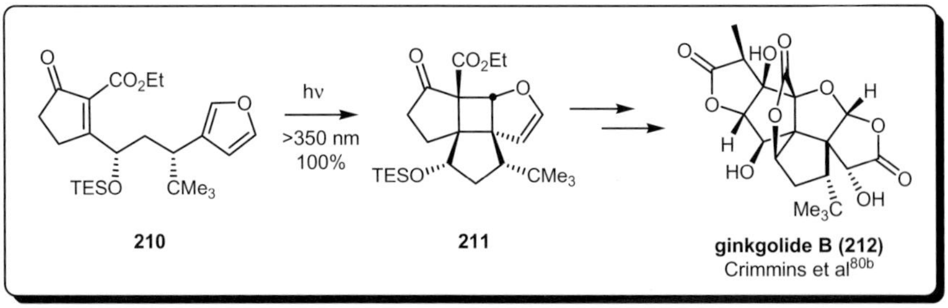

Scheme 4.16 [2 + 2] cycloadditions in the synthesis of natural products.

In the following example, compound **210** underwent an intramolecular [2 + 2] cycloaddition between the furan ring and the unsaturated ketoester, to give the cyclobutane derivative **211**. This compound was a key intermediate in the synthesis of ginkgolide B **212** [80b].

Furan has also been used in dipolar cycloadditions [81, 82], and its reaction with nitrones, such as compound **213** (Scheme 4.17), has provided hydrofuroisoxazols (e.g., product **214** and its derivative **215**). The N–O bond can be readily reduced, and the resulting aminoalcohols have been used to prepare iminosugars, such as the glycosidase inhibitor nojirimycin **216** [81a].

On the other hand, the addition of metallo-carbenoids to carbonyl groups generates carbonyl ylide dipoles (as in conversion **218**→**219**, Scheme 4.18). The [3 + 2] cycloaddition of dipole **219** provided the hexacyclic compound **220**, which is a precursor of Aspidosperma alkaloids [81b–d].

Push-pull dipoles such as **219** can be generated using different methodologies. Thus, the oxadiazole **221** (Scheme 4.19) underwent an intramolecular Diels–Alder cycloaddition to give the oxabicycle **222**. Loss of nitrogen followed, giving the carbonyl ylide dipole **223**. A second cycloaddition originated the polycyclic compound **224**, which was readily transformed into the Aspidosperma alkaloid vindorosine **225** [81e].

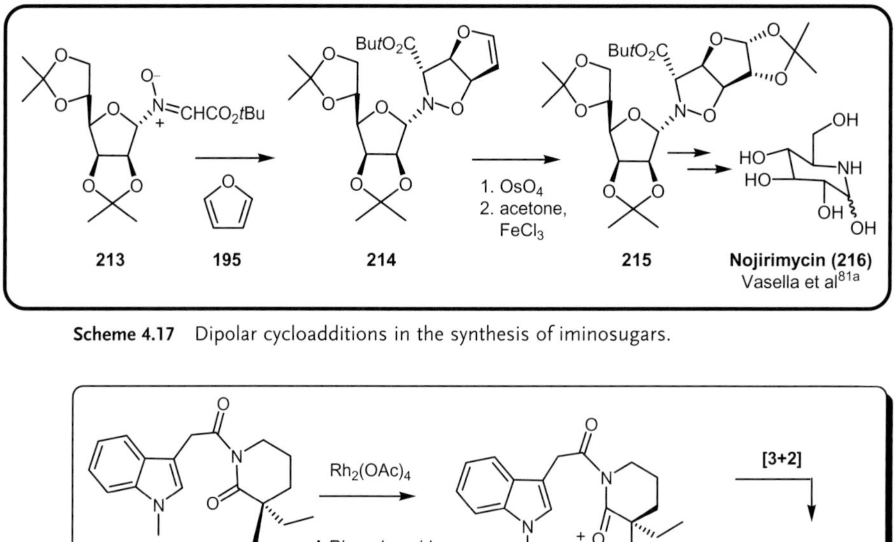

Scheme 4.17 Dipolar cycloadditions in the synthesis of iminosugars.

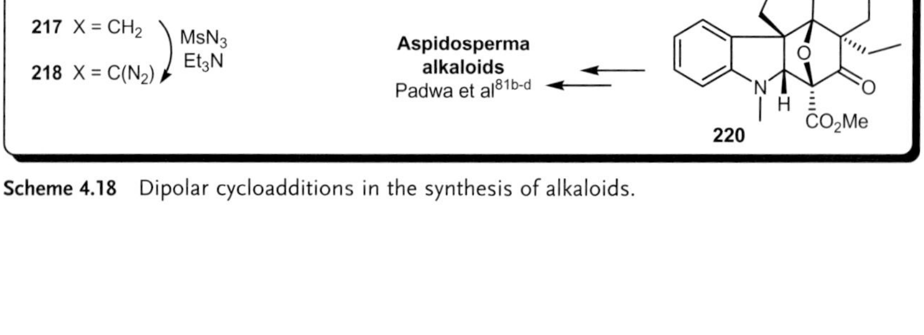

Scheme 4.18 Dipolar cycloadditions in the synthesis of alkaloids.

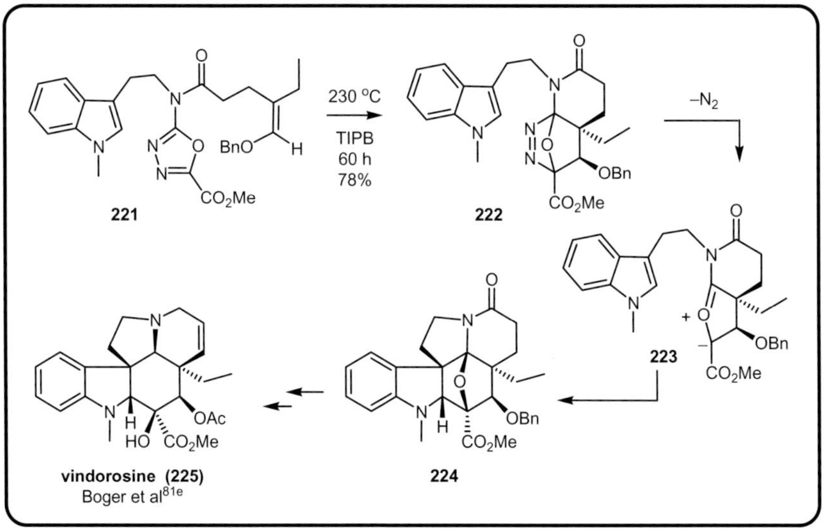

Scheme 4.19 Dipolar cycloadditions in the synthesis of alkaloids.

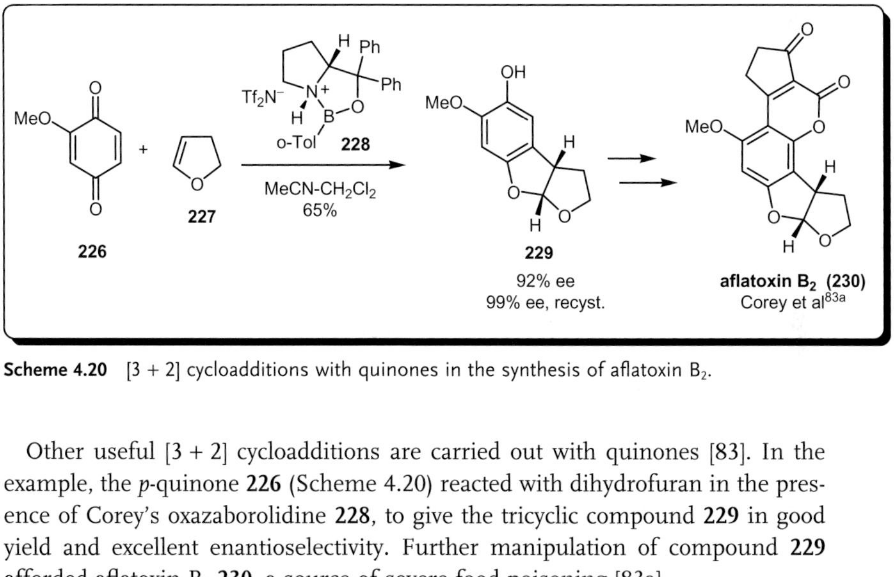

Scheme 4.20 [3 + 2] cycloadditions with quinones in the synthesis of aflatoxin B$_2$.

Other useful [3 + 2] cycloadditions are carried out with quinones [83]. In the example, the *p*-quinone **226** (Scheme 4.20) reacted with dihydrofuran in the presence of Corey's oxazaborolidine **228**, to give the tricyclic compound **229** in good yield and excellent enantioselectivity. Further manipulation of compound **229** afforded aflatoxin B$_2$ **230**, a source of severe food poisoning [83a].

4.3.4.2 Diels–Alder ([4 + 2] Cycloadditions)

The most important cycloadditions in natural product synthesis are the Diels–Alder reactions (4 + 2) [84–88]. For instance, the bromobenzene **231** (Scheme 4.21) was treated with sodium amide to generate a benzyne, which reacted with 2-methoxyfuran to give the Diels–Alder adduct **232**. Treatment with methanolic HCl cleaved the oxabicycle, yielding the naphthalene derivative **233**, which served

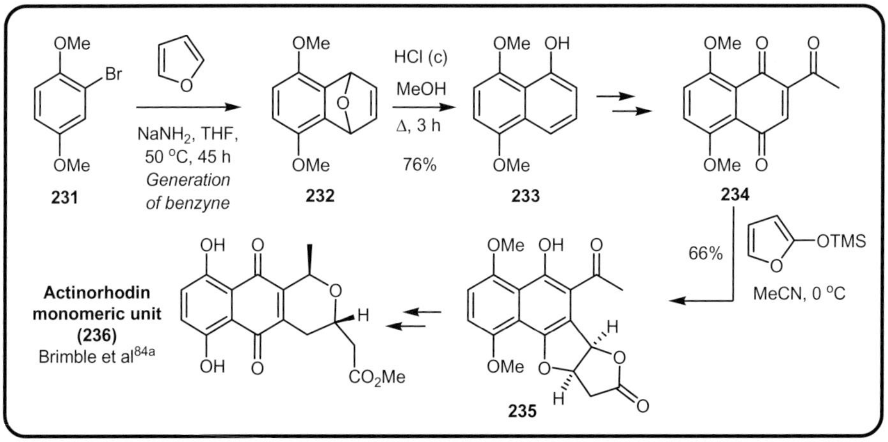

Scheme 4.21 Diels–Alder reaction in the synthesis of actinorhodin monomeric unit.

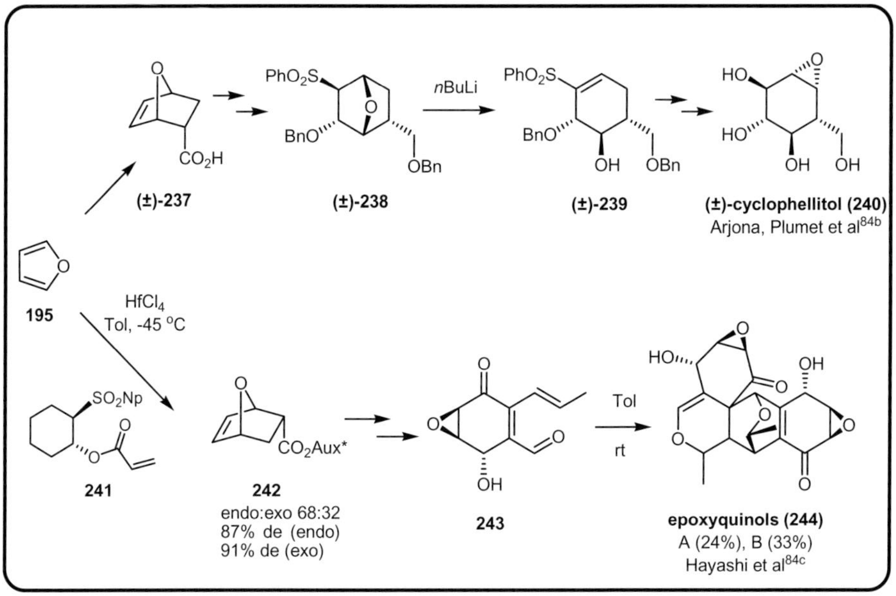

Scheme 4.22 Diels–Alder reactions in the synthesis of cyclophellitol and epoxyquinols.

as precursor of the naphthoquinone **234**. The latter underwent a [3 + 2] cycloaddition, giving the tetracyclic product **235**, with a fused furan-furanone ring system. This product was an advanced intermediate in the synthesis of actinorhodin monomeric unit **236** [84a].

The Diels–Alder reaction of furans with alkenes generates oxanorbornenes, which were used by Vogel et al. to prepare a variety of structures. Continuing with this work, Arjona and Plumet reported the synthesis of (±)-cyclophellitol (Scheme 4.22), using a Diels–Alder reaction between furan and an acrylate. The cycloadduct

237 was elaborated into the sulfone 238. The abstraction of its α-proton by BuLi, followed by cleavage of the β-ether, yielded the cyclohexenol 239. This product served as precursor of cyclophellitol 240 and other bioactive carbocycles [84b].

In a similar way, the Diels–Alder reaction between furan and the acrylate 241 yielded the oxanorbornene 242. This intermediate was epoxidized and then treated with LDA to generate an enolate, which evolved by β-ether cleavage. The resulting cyclohexenol (not shown) was converted in a few steps into product 243. A biomimetic cycloaddition was then carried out at room temperature, affording the epoxyquinols 244 [84c].

A number of methods have been developed to cleave oxabicycles [89]; among them, hydrogenation, reduction with SmI_2, acid-catalyzed cleavage (sometimes a Lewis acid is combined with a base), opening with Rh(I)-, Pd(0)- or related metal complexes, addition of nucleophiles to the double bond with concomitant ether cleavage, and in some cases, by simple heating. Thus, the Pummerer rearrangement of compound 245 (Scheme 4.23) afforded the furan 246, which underwent an intramolecular Diels–Alder reaction to give the oxabicycle (±)-247. On heating, this compound rearranged to the enamide (±)-248. After removing

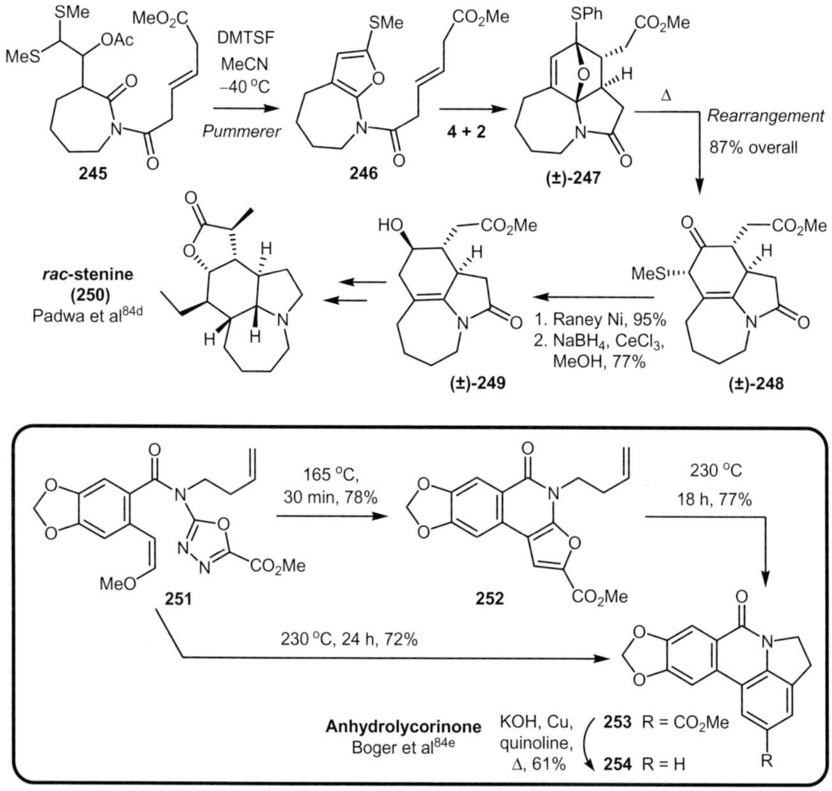

Scheme 4.23 Diels–Alder reactions in the synthesis of stenine and anhydrolycorinone.

the methylthio group and reducing the ketone, the tricyclic compound (±)-**249** was isolated. This product was converted in few steps into the alkaloid (±)-stenine **250** [84d].

In another example, on heating compound **251**, the oxadiazole ring and the enol ether group underwent an intramolecular Diels–Alder reaction, which was followed by a retro Diels–Alder with extrusion of nitrogen, generating the furan **252**. A second Diels–Alder reaction was carried out at high temperature, and the intermediate oxabicycle spontaneously underwent aromatization to afford product **253**. Remarkably, the direct conversion of compound **251** into the tetracyclic product **253** was also achieved in high yield. In the final step, hydrolysis of the ester group followed by decarboxylation gave anhydrolycorinone **254** [84e].

The reductive cleavage of the oxabicycle is illustrated in Winkler's synthesis of the eleutherobin core (Scheme 4.24) [84f]. A double Diels–Alder reaction was triggered by oxidation of the alcohol **255** in the presence of the furan-diene **256**. The first cycloaddition generated the oxabicyclic intermediate **257**, and after the second cycloaddition, the tetracyclic and highly functionalized product **258** was isolated. The free hydroxy group was protected as its silyl ether, giving compound **259** in satisfactory overall yields and high diastereoselectivity. This product was converted into ketone **260**, which under treatment with SmI$_2$ generated an anion-radical •C(R)$_2$–O⁻. The α-ether group was cleaved, yielding the tricyclic product **261**, a key precursor of the eleutherobin core **262**.

The acid-catalyzed cleavage of the oxabicycle was used in Padwa's remarkable synthesis of strychnine (Scheme 4.25) [84g]. Thus, on treatment of compound **263**

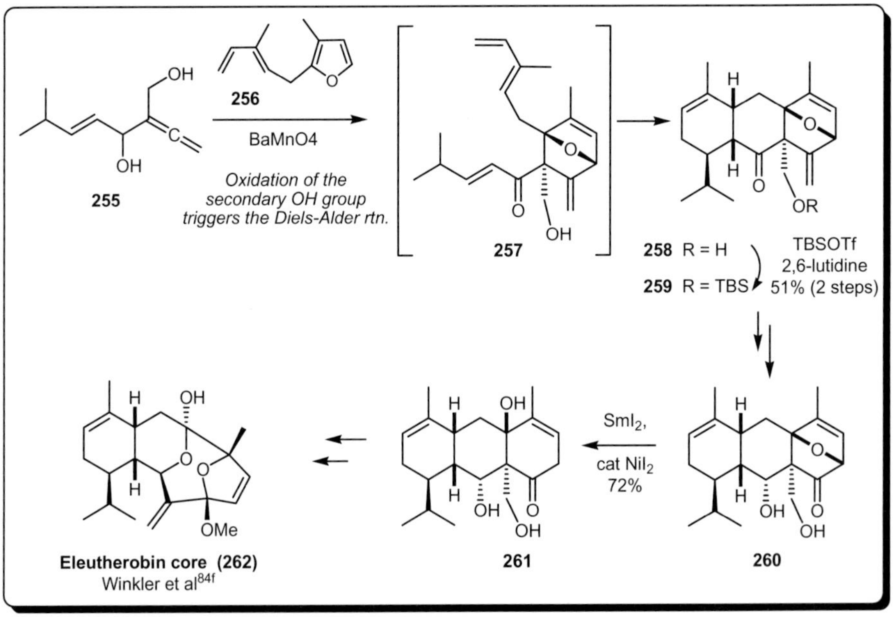

Scheme 4.24 Diels–Alder reactions in the synthesis of the eleutherobin core.

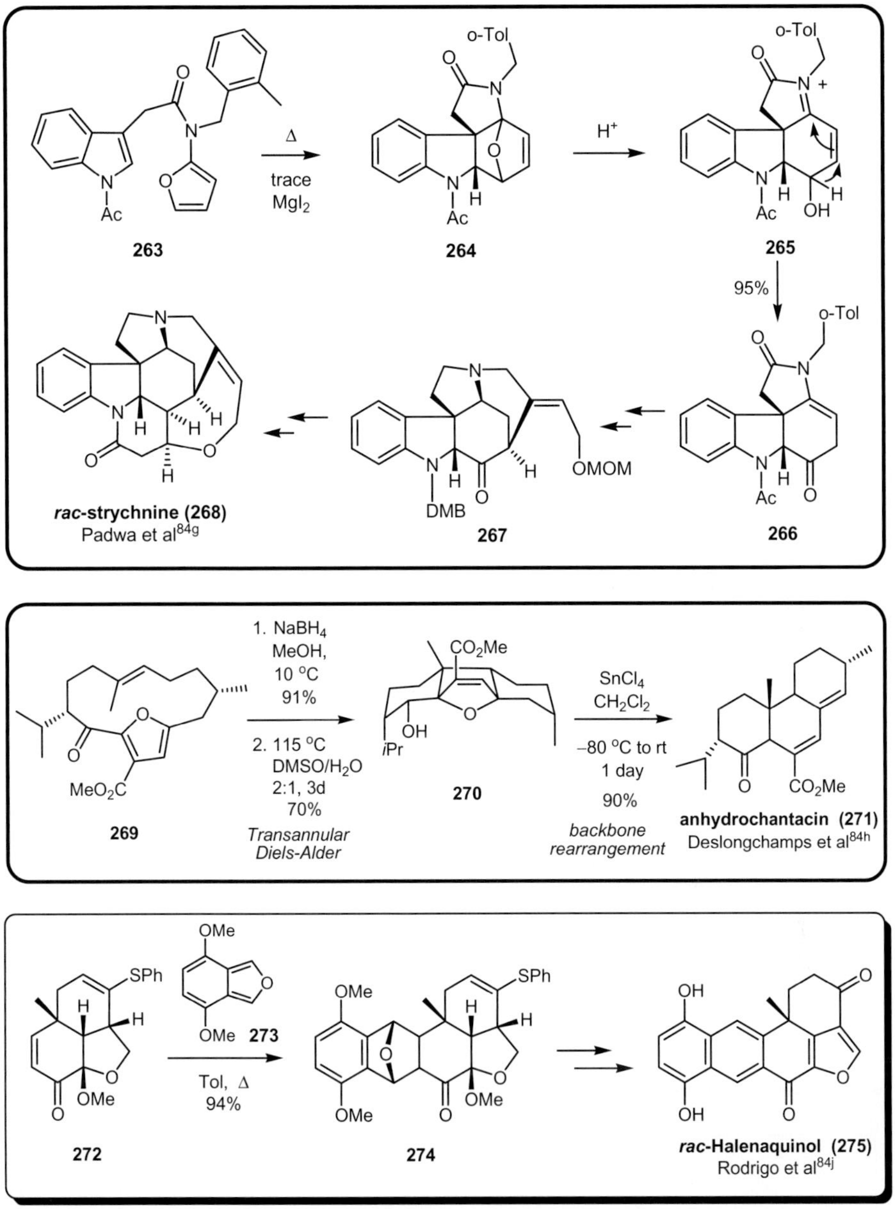

Scheme 4.25 Diels–Alder reactions in the synthesis of natural products.

with catalytic MgI$_2$, the α-amidofuran reacted with the indole unit, yielding the bicyclic intermediate **264**. The acid-catalyzed cleavage of the *N,O*-acetal generated an acyliminium ion **265**, from which was formed compound **266**. This compound was converted into the pentacyclic product **267**, a precursor of (±)-strychnine **268**.

Sometimes the cleavage of the oxabicycle under acid conditions can generate interesting skeletal rearrangements. In Deslongchamps's synthesis of anhydrochantacin [84h, i], a transannular Diels–Alder converted the macrocycle **269** into the tetracyclic compound **270**. On treatment with acid, the cleavage of the dihydrofuran ring was accompanied by backbone rearrangement, to give anhydrochantacin **271** in a very direct way.

Other heteroaromatic compounds have also been used in the synthesis of natural products. For instance, the isobenzofuran **272** reacted with the unsaturated ketone **273** to give the Diels–Alder adduct **274**, a key intermediate in the synthesis of (±)-halenaquinol **275** [84j].

Not only the Diels–Alder, but also the *retro*-Diels–Alder, are useful in synthesis (Scheme 4.26) [90]. Using a strategy that combines both reactions, furans can be used as protecting groups in the preparation of natural products. For example, the oxabicycle **276** was obtained by cycloaddition of furan and a maleimide, in order to protect the maleimide's double bond. The substitution of the mesylate by an azido group afforded compound **277**, which underwent a Staudinger reduction, generating the intermediate **278**. A retro-Diels–Alder ensued, releasing product **279**, which was transformed into phloeodictine A1 **280** [90b].

Similarly, the chiral oxanorbornene **281** was used to induce stereoselectivity in the addition of MeLi to the lactone and the addition of *i*PrMgBr to the resulting lactol. The second addition generated a diol which was oxidized to the lactone **282** with Jones' reagent. Then a *retro*-Diels–Alder was carried out, affording the chiral

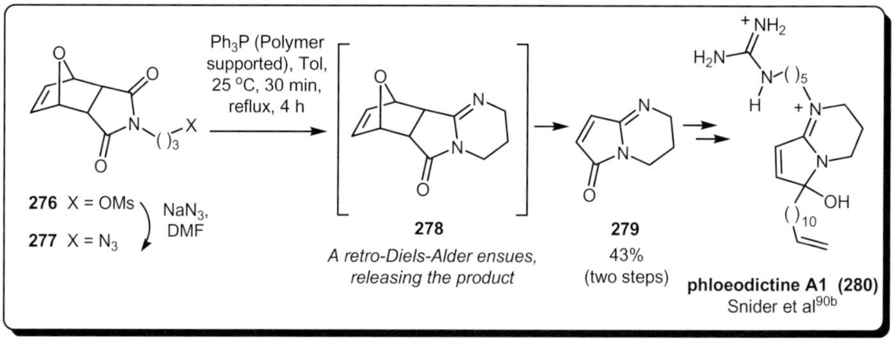

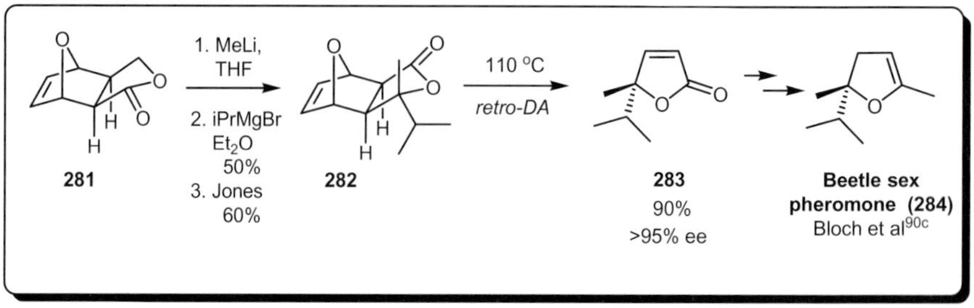

Scheme 4.26 *retro*-Diels–Alder reactions in the synthesis of natural products.

butenolide **283** with excellent enantioselectivity. This lactone is an advanced intermediate in the synthesis of the beetle sex pheromone **284** [90c, d].

4.3.4.3 [4 + 3], [6 + 4], [8 + 2] and [5 + 2] Cycloadditions

The [4 + 3] cycloadditions have also been used in the synthesis of different natural products containing seven-member rings (Scheme 4.27) [91]. Thus, when pentachloroacetone was treated with a base (NaTFE in TFE), the dipolar intermediate **285** was generated. This intermediate reacted with the furan **286** to give the bicyclic product **287**, which was dehalogenated to compound **288**. Further steps provided the tricyclic compound **289**. The oxabicycle underwent hydrogenolysis, affording the guaiane skeleton **290** [91a].

In a similar way, when the enol ether **291** was treated with a Lewis acid, a dipole **292** was formed. This reactive intermediate and the furan **293** underwent a [4 + 3] cycloaddition, giving the polycyclic compound **294**. In this case, the oxabicycle was cleaved by treatment with a Lewis acid/base system, affording the alkaloid imerubrine **295** [91b].

Two examples of less frequent cycloadditions [92] are shown in Scheme 4.28. Thus, when compound **296** was heated in benzene, an intramolecular [6 + 4] cycloaddition between the furan and the tropone rings took place. This process generated the ingenane core **297** which can be converted into ingenol **298** [92a].

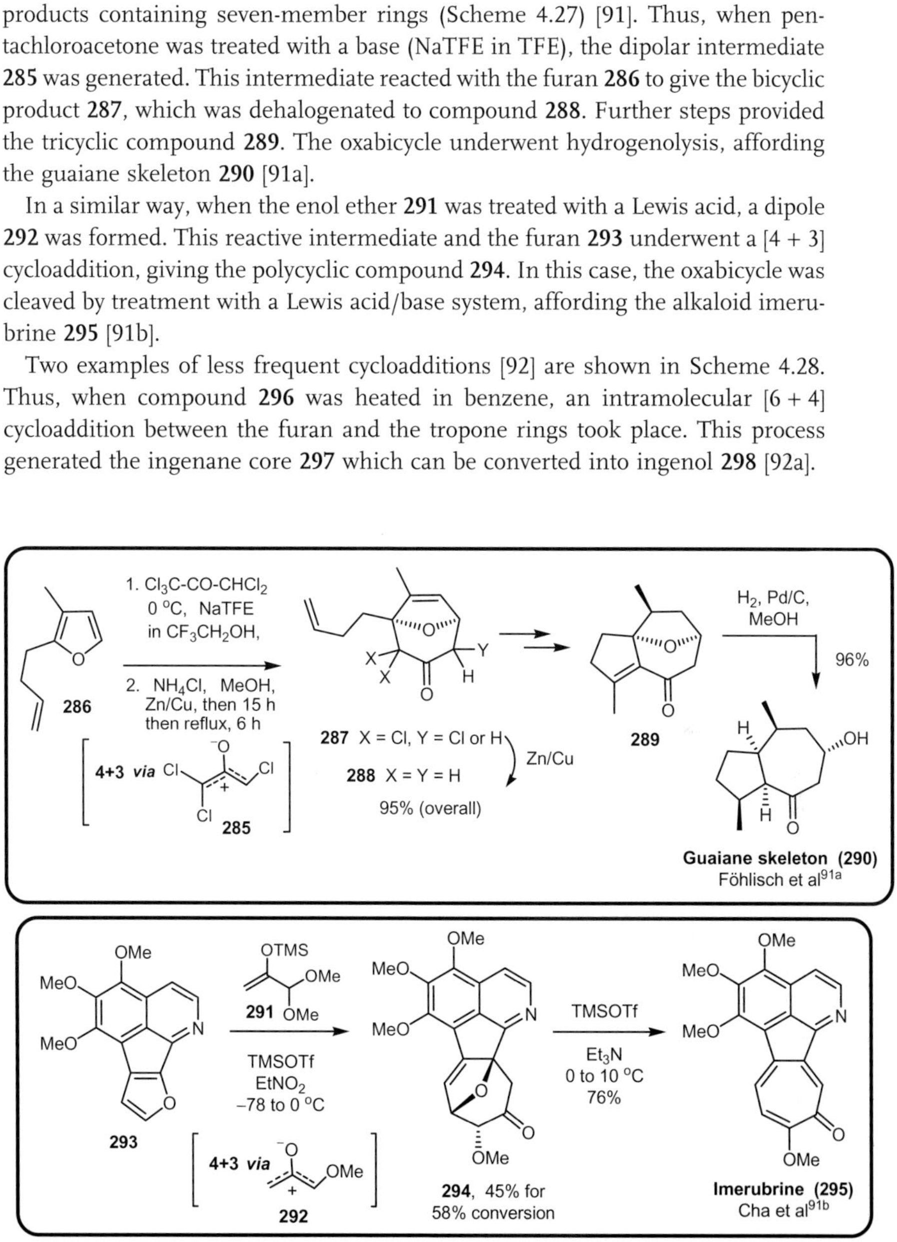

Scheme 4.27 [4 + 3] cycloadditions in the synthesis of natural products.

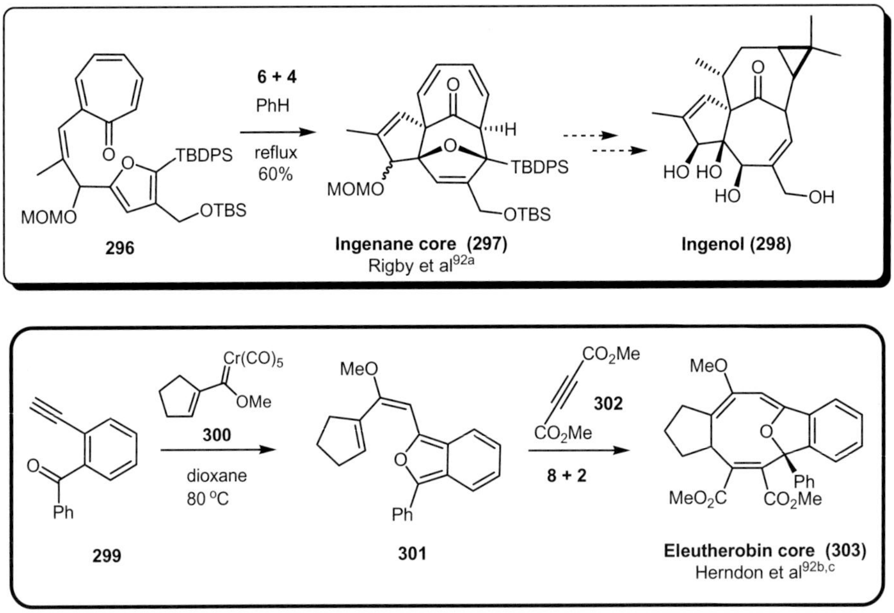

Scheme 4.28 [6 + 4] and [8 + 2] cycloadditions in the synthesis of natural products.

In the following example, the reaction of the alkyne **299** and the chromium carbenoid **300** formed the polyunsaturated system **301**. When this system was treated with the alkyne **302** and DMAD, an [8 + 2] cycloaddition took place, generating the eleutherobin core **303** [92b, c].

Finally, the furan ring has served as precursor of units which undergo cycloadditions [93]. An example is shown in Wender's remarkable synthesis of resiniferatoxin (Scheme 4.29). The furan **304** underwent an Achmatowicz oxidation to the acetal **305**, which was converted into its acetate **306**. On treatment with DBU, a pyrilium ion **307** was formed, which took part in a [5 + 2] cycloaddition, to afford the tricyclic compound **308**. This product was a key precursor in the synthesis of resiniferatoxin **309** [93a].

Pattenden postulated that this kind of transformation also happened in nature, and hence, bipinnatin J **310** would be the metabolic precursor of intricarene **311** [93b], via an *in vivo* oxidation and cycloaddition (Scheme 4.28). To check this hypothesis, bipinnatin J **310** was prepared from the furan **312** in good global yields. Then bipinnatin **310** was oxidized under classical Achmatowicz conditions and acetylated, affording the lactol **313**. This product underwent the [5 + 2] cycloaddition, via a pyrilium intermediate **314**, to afford intricarene **311**. Although the global oxidation–cycloaddition yields were low, the success in the preparation of target **311** strongly backed the biosynthetic hypothesis. In a parallel work, Trauner modified the oxidation and cycloaddition conditions, improving the yields of these two steps (81% and 26% respectively) [93c–d].

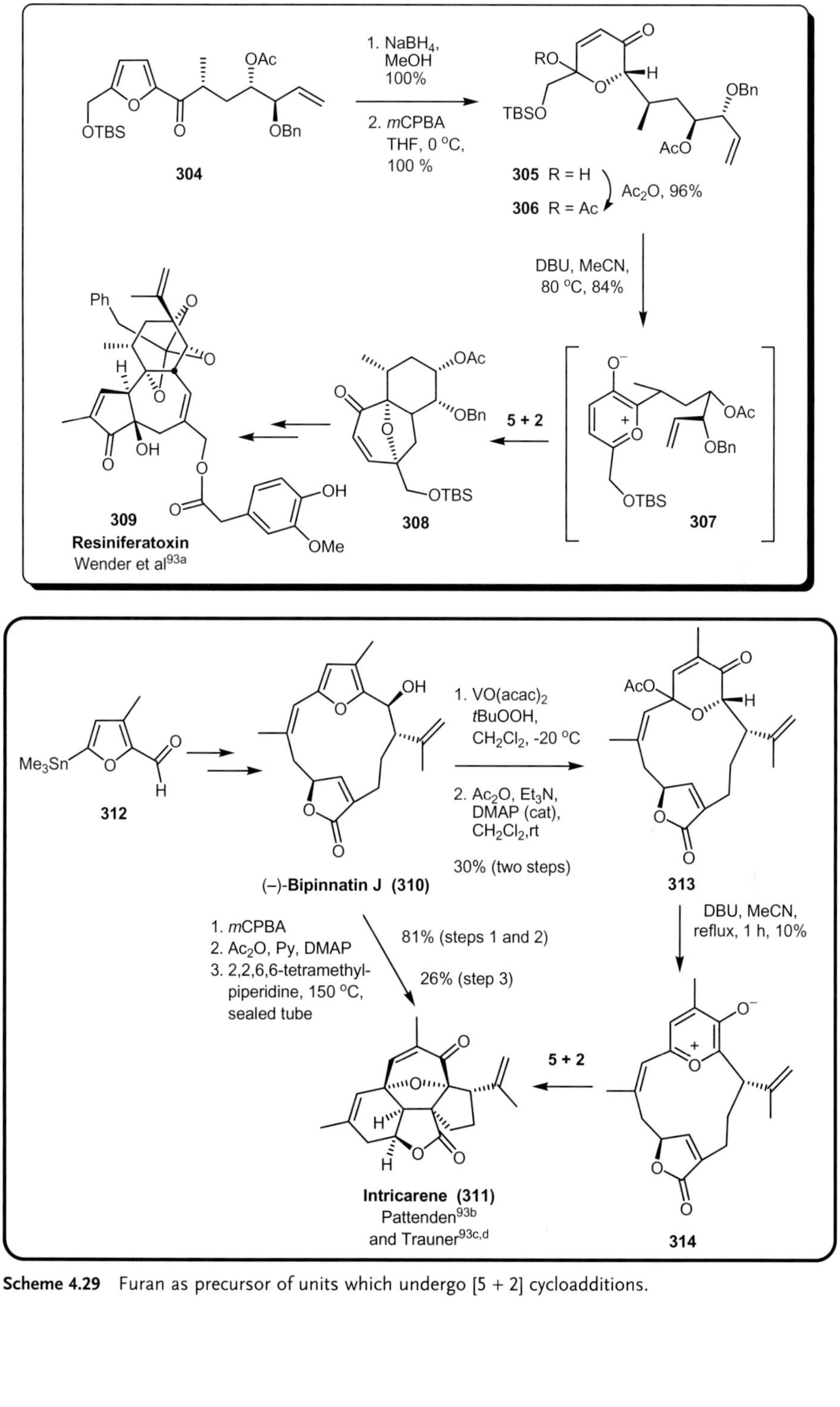

4.3 Furan Derivatives as Reagents in the Synthesis of Natural Products | 135

Scheme 4.29 Furan as precursor of units which undergo [5 + 2] cycloadditions.

4.3.5
Furan in Other Reactions

The addition of radicals to dihydrofuran or furan rings has allowed the formation of polycyclic systems (Scheme 4.30). Thus, the C-radical derived from the aryl iodide **315** added to the double bond of the dihydrofuran ring; the resulting radical was reduced, followed by elimination of the acetate. The reaction product **316** presented the tricyclic core of aflatoxin B1 **317** [94a].

In the case of substrate **318**, treatment with lauryl peroxide generated a primary C-radical which added to the furan ring (intermediate **319**). As commented before, furan has lower aromaticity than pyrrole, thiophene and other heteroaromatic rings. The 5-exo-cyclization destroyed the ring aromaticity, creating a spirocyclic system **320**. This methodology could be used in a future to prepare the tetracyclic alkaloid securinine **321** [94b].

The dearomatization of furan was also observed with compound **322** (Scheme 4.31), favored by the formation of a polyconjugated system **323**. Thus, acid-catalyzed dehydration generated a cationic intermediate which was stabilized by resonance with the adjacent furan ring. Addition of the side-chain hydroxy group generated the spirocyclic compound **323**, which is an analog of the antifeeding tonghaosu **324** [95a–c].

The dearomatization of furan has been observed in ionic cyclizations. Thus, when compound **325** was treated with acid, an acyliminium ion **326** was formed. Nucleophilic addition of furan generated an oxycarbenium ion which was trapped by water, forming a lactol **327** with excellent yield and high diastereoselectivity. This compound was transformed into the manzamine core **328** [95d].

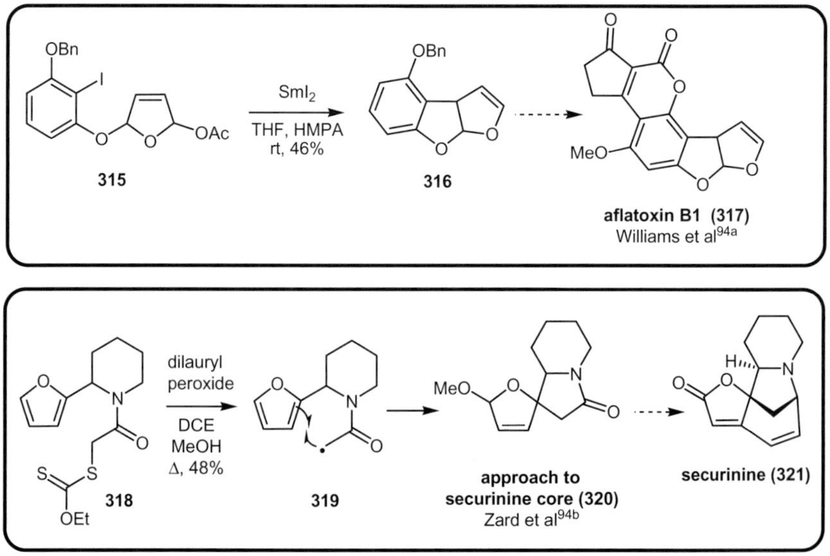

Scheme 4.30 Radical reactions in the synthesis of natural products or their analogs.

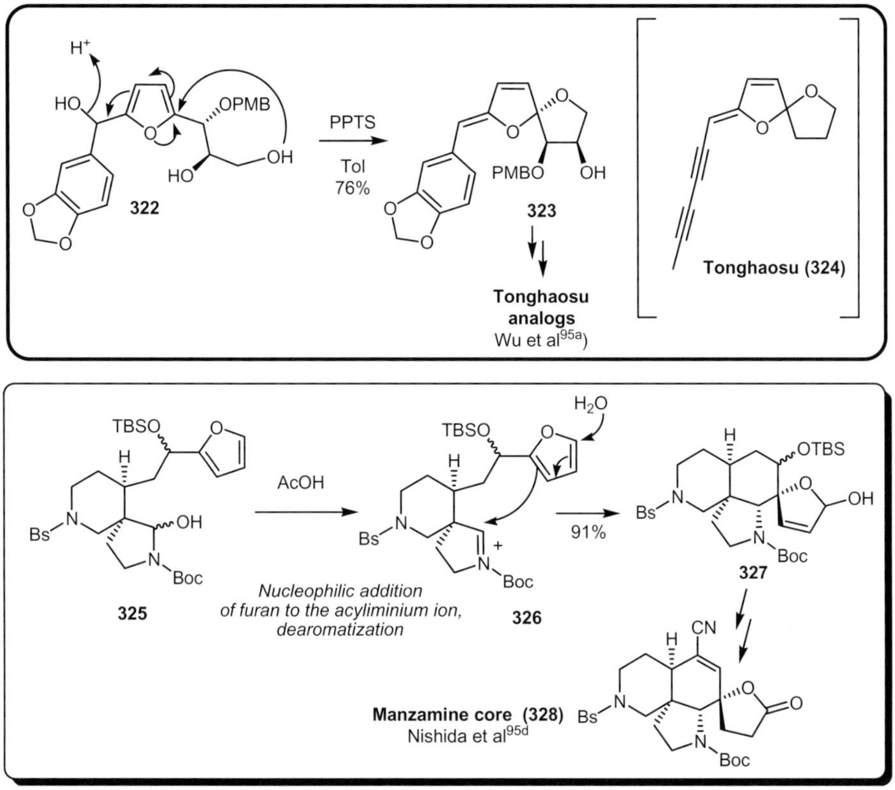

Scheme 4.31 Dearomatization reactions in the synthesis of natural products or their analogs.

The dihydrofuran-2,5-diols or the 5-hydroxyfuranones have been used in other interesting process developed by Trost: the extrusion of the 5-OR group with formation of an allylic cation stabilized by the palladium complex (e.g., conversion **329→330**, Scheme 4.32) [96]. The intermediate is trapped by nucleophiles, such as 2-naphtol, to give addition products (e.g., acetal **331**). The butenolide **331** was obtained in good yields and excellent enantioselectivity, and then was used in a palladium-catalyzed [2 + 3] dipolar cycloaddition, giving the bicyclic system **332**. This product served as precursor of brefeldin A **333** [96a].

Other interesting process is the Ireland-Claisen rearrangement [97], which has proven particularly useful in the synthesis of polyether compounds. Thus, treatment of compound **334** (Scheme 4.33) with LDA and TMSCl generated an enol silyl ether intermediate, which underwent the rearrangement, creating a C–C bond between the two furan units. The resulting acid was reduced to the alcohol **335**, and after some steps, another Ireland-Claisen rearrangement was carried out to join the B–C rings in monesin A **336** [97a, b].

The hydrofuran rings can also undergo ring-expansion reactions, as shown in the synthesis of salinomycin (Scheme 4.34). Thus, the mesylation of compound **337** was followed by extrusion of the mesylate and formation of an oxonium

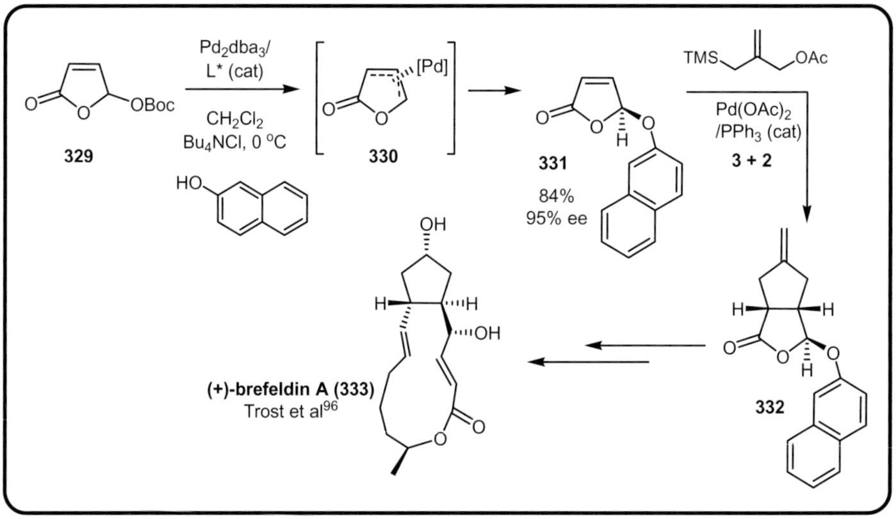

Scheme 4.32 Trost's synthesis of brefeldin A.

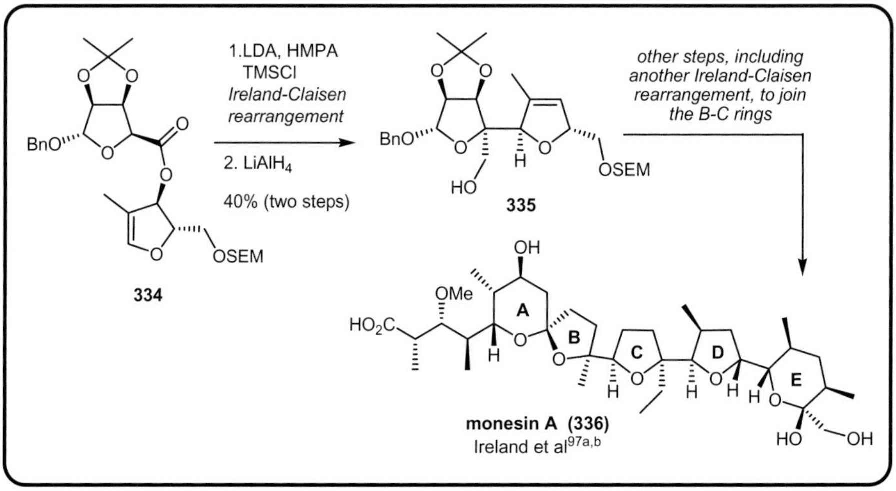

Scheme 4.33 Ireland-Claisen rearrangement in the synthesis of monesin A.

intermediate **338**. Addition of water produced the hydroxypyran unit in compound **339**. This compound was a valuable precursor of the antibiotic polyether salinomycin A **340** [98].

4.3.6
Other Uses of Furan in Synthesis

Sometimes the furan ring or its derivatives provide the rigidity needed for a reaction to occur. This is the case of compound **341** (Scheme 4.35), where the lactol

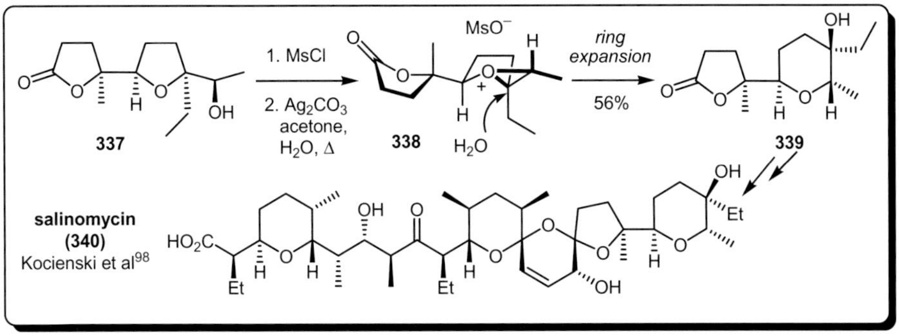

Scheme 4.34 Ring expansion in the synthesis of salinomycin.

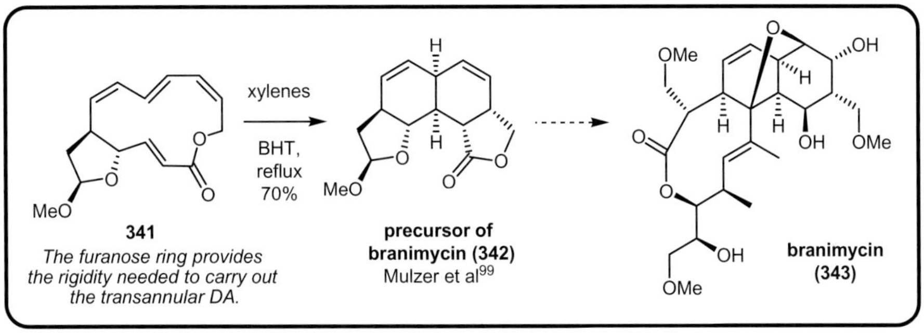

Scheme 4.35 Use of furan derivatives to achieve conformational constraint.

ring was introduced to achieve conformational constraint in the transannular Diels–Alder reaction. The cycloaddition failed with more flexible systems. The reaction product **342** is a precursor of branimycin **343** [99].

These rings have also been used as chelators, in order to approach reaction partners, or to control the stereochemistry of a process. Thus, in the synthesis of tetronasin by Ley (Scheme 4.36), compound **344** was treated with KHDMS, and an anionic cyclization cascade ensued. Chelation of K^+ by the furan ring and other groups (intermediate **345**) promoted the correct stereochemistry in this cascade, yielding the polycyclic system **346**, a key precursor of tetronasin sodium salt **347** [100].

Finally, furan derivatives have been used in catalysts (P[furyl]$_3$ ligands, Shi spiro catalysts)[101] or as chiral auxiliaries, as shown with compound **348** (Scheme 4.37). The samarium-promoted radical cyclization generated the spyrocyclic system **349** with high stereoselectivity. Then the furanose was removed by oxidation, and the product was used to prepare anastrephin **350** [102].

4.4 Summary

In summary, furan rings and their derivatives (hydrofurans, butenolides, benzofurans, etc) are versatile natural product precursors. The subject is extensive, and

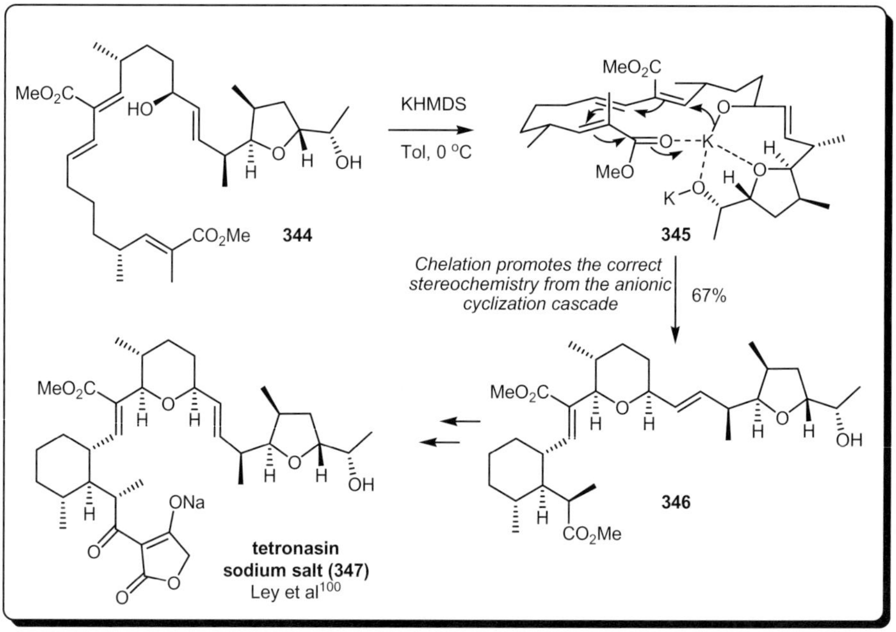

Scheme 4.36 Use of furan derivatives as chelators.

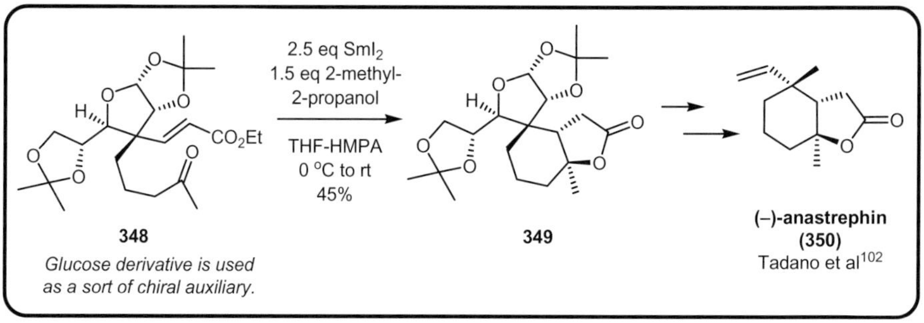

Scheme 4.37 Use of furan derivatives as chiral auxiliaries.

only a careful selection of their uses has been outlined here. Moreover, the advances in furan chemistry made in the recent years will probably yield interesting applications in natural products synthesis in the next future.

References

1 Furan in natural products, acetogenins, macrolides and other polyether derivatives: (a) Nicolaou, K.C. (2009) *J. Org. Chem.*, **74**, 951–972; (b) Paterson, I., and Dalby, S.M. (2009) *Nat. Prod. Rep.*, **26**, 865–873; (c) Nicolaou, K.C., Frederick, M.O. and Aversa, R.J. (2008) *Angew. Chem. Int. Ed.*, **47**, 7182–7225;

(d) Maezaki, N., Kojima, N. and Tanaka, T. (2006) *Synlett*, 993–1003; (e) Hale, K.J., Hummersone, M.G., Manaviazar, S. and Frigerio, M. (2002) *Nat. Prod. Rep.*, **19**, 413–453; (f) Faul, M.M. and Huff, B.E. (2000) *Chem. Rev.*, **100**, 2407–2473.

2 Furan in terpenes and steroids: (a) Gaich, T., Weinstabl, H. and Mulzer, J. (2009) *Synlett*, 1357–1366; (b) Yue, J.M., Liao, S.G. and Chen, H.D. (2009) *Chem. Rev.*, **109**, 1092–1140; (c) Prisinzano, T.E. and Rothman, R.B. (2008) *Chem. Rev.*, **108**, 1732–1743; (d) González, M.A. (2008) *Tetrahedron*, **64**, 445–467; (e) Fraga, B.M. (2008) *Nat. Prod. Rep.*, **25**, 1180–1209; (f) Busch, T. and Kirschning, A. (2008) *Nat. Prod. Rep.*, **25**, 318–341; (g) Keyzers, R.A., Northcote, P.T. and Davies-Coleman, M.T. (2006) *Nat. Prod. Rep.*, **23**, 321–334; (h) Liu, Y., Zhang, S. and Abreu, P.J.M. (2006) *Nat. Prod. Rep.*, **23**, 630–651; (i) Maifeld, S.V. and Lee, D. (2006) *Synlett*, 1623–1644; (j) Hanson, J.R. (2006) *Nat. Prod. Rep.*, **23**, 875–885; (k) Wipf, P. and Halter, R.J. (2005) *Org. Biomol. Chem.*, **3**, 2053–2061; (l) Connolly, J.D. and Hill, R.A. (2005) *Nat. Prod. Rep.*, **22**, 230–248; (m) Maurya, R. and Yadav, P.P. (2005) *Nat. Prod. Rep.*, **22**, 400–424; (n) Steyn, P.S. and van Heerden, F.R. (1998) *Nat. Prod. Rep.*, **15**, 397–413.

3 Furan in Lignans: Saleem, M., Kim, H.J., Ali, M.S. and Lee, Y.S. (2005) *Nat. Prod. Rep.*, **22**, 696–716.

4 Tetronic acids: Schobert, R. and Schlenk, A. (2008) *Bioorg. Med. Chem.*, **16**, 4203–4221.

5 Cembranes, cembranolides: (a) Roethle, P.A. and Trauner, D. (2008) *Nat. Prod. Rep.*, **25**, 298–317; (b) Tius, M.A. (1988) *Chem. Rev.*, **88**, 719–732.

6 Furan in alkaloids and as precursor of alkaloids: (a) Weinreb, S.M. (2009) *Nat. Prod. Rep.*, **26**, 758–775; (b) Jin, Z. (2009) *Nat. Prod. Rep.*, **26**, 363–381; (c) Michael, J.P. (2008) *Nat. Prod. Rep.*, **25**, 139–165; (d) Jin, Z. (2007) *Nat. Prod. Rep.*, **24**, 886–905; (e) Michael, J.P. (2005) *Nat. Prod. Rep.*, **22**, 603–626; (f) Pilli, R.A. and Ferreira de Oliveira, M.C. (2000) *Nat. Prod. Rep.*, **17**, 117–128; (g) Casiraghi, C., Zanardi, F., Rassu, G. and Spanu, P. (1995) *Chem. Rev.*, **95**, 1677–1716.

7 Furan in other natural products: (a) Nandy, J.P., Prakesch, M., Khadem, S., Reddy, P.T., Sharma, U. and Arya, P. (2009) *Chem. Rev.*, **109**, 1999–2060; (b) Bräse, S., Encinas, A., Keck, J. and Nising, C.F. (2009) *Chem. Rev.*, **109**, 3903–3990; (c) Blunt, J.W., Copp, B.R., Hu, W.P., Munro, M.H.G., Northcote, P.T. and Prinsep, M.R. (2009) *Nat. Prod. Rep.*, **26**, 170–245; (d) Lachia, M. and Moody, C.J. (2008) *Nat. Prod. Rep.*, **25**, 227–254; (e) Sperry, J., Bachu, P. and Brimble, M.A. (2008) *Nat. Prod. Rep.*, **25**, 376–400; (f) Cordier, C., Morton, D., Murrison, S., Nelson, A. and O'Leary-Steele, C. (2008) *Nat. Prod. Rep.*, **25**, 719–737; (g) Veitch, N.C. and Grayer, R.J. (2008) *Nat. Prod. Rep.*, **25**, 555–612; (h) Skoropeta, D. (2008) *Nat. Prod. Rep.*, **25**, 1131–1166; (i) Shoji, M. and Hayashi, Y. (2007) *Eur. J. Org. Chem.*, 3783–3800; (j) Blunt, J.W., Copp, B.R., Munro, M.H.G., Northcote, P.T. and Prinsep, M.R. (2003) *Nat. Prod. Rep.*, **20**, 1–48; (k) Faulkner, D.J. (2000) *Nat. Prod. Rep.*, **17**, 7–56; (l) Nicolaou, K.C., Vourloumis, D., Winssinger, N. and Baran, P.S. (2000) *Angew. Chem. Int. Ed.*, **39**, 44–122; (m) Mehta, G. and Singh, V.K. (1999) *Chem. Rev.*, **99**, 881–930; (n) Cereghetti, D.M. and Carreira, E.M. (2006) *Synthesis*, 914–942.

8 (a) Lipshutz, B.H. (1986) *Chem. Rev.*, **86**, 795–819; (b) Rodrigo, R. (1988) *Tetrahedron*, **44**, 2093–2135.

9 Synthesis of furan derivatives and their applications: (a) Brichacek, M. and Njardarson, J.T. (2009) *Org. Biomol. Chem.*, **7**, 1761–1770; (b) Wolfe, J.P. and Hay, M.B. (2007) *Tetrahedron*, **63**, 261–290; (c) Bellur, E., Feist, H. and Langer, P. (2007) *Tetrahedron*, **63**, 10865–10888; (d) Kirsch, S.F. (2006) *Org. Biomol. Chem.*, **4**, 2076–2080; (e) Zografos, A.L. and Georgiadis, D. (2006) *Synthesis*, 3157–3188; (f) Brown, R.C.D. (2005) *Angew. Chem. Int. Ed.*, **44**, 850–852; (g) Keay, B.A. (1999) *Chem. Soc. Rev.*, **28**, 209–215; (h) Wong, H.N.C. (1996) *Pure Appl. Chem.*, **68**, 335–344; (i) Hou, X.L., Cheung, H.Y., Hon, T.Y.,

Kwan, P.L., Lo, T.H., Tong, S.Y. and Wong, H.C.N. (1998) *Tetrahedron*, **54**, 1955–2020.

10 Angelmarin (1): (a) Jiang, H. and Hamada, Y. (2009) *Org. Biomol. Chem.*, **7**, 4173–4176; (b) Magolan, J. and Coster, M.J. (2009) *J. Org. Chem*, **74**, 5083–5086.

11 6-C-β-manno pyranosylapigenin (2): (a) Veitch, N.C. and Grayer, R.J. (2008) *Nat. Prod. Rep.*, **25**, 555–611; (b) Nakatsuka, T., Tomimori, Y., Fukuda, Y. and Nukaya, H. (2004) *Bioorg. Med. Chem. Lett.*, **14**, 3201–3203; (c) Furuta, T., Kimura, T., Kondo, S., Mihara, H., Wakimoto, T., Nukaya, H., Tsuji, K. and Tanaka, K. (2004) *Tetrahedron*, **60**, 9375–9379.

12 Asiminocin (3): (a) Marshall, J.A., Piettre, A., Paige, M.A. and Valeriote, F. (2003) *J. Org. Chem.*, **68**, 1771–1779; (b) Elliott, M.C. and Williams, E. (2001) *J. Chem. Soc. Perkin Trans. 1*, 2303–2340; (c) Elliott, M.C. (2000) *J. Chem. Soc. Perkin Trans. 1*, 1291–1318; (d) McLaughlin, J.L. (2008) *J. Nat. Prod.*, **71**, 1311–1321.

13 Bhimamycin B (4): Uno, H., Murakami, S., Fujimoto, A., Yamaoka, Y. (2005) *Tetrahedron Lett.*, **46**, 3997–4000.

14 Cantharidin (5): (a) Deng, L., Shen, L., Zhang, J., Yang, B., He, Q. and Hu, Y. (2007) *Can. J. Chem.*, **85**, 938–944; (b) For analogs, see:Baba, Y., Hirukawa, N. and Sodeoka, M. (2005) *Bioorg. Med. Chem.*, **13**, 5164–5170; (c) Palasonin:Dauben, W.G., Lam, J.Y.L. and Guo, Z.R. (1996) *J. Org. Chem.*, **61**, 4816–4819.

15 Cephalostatin (6): Lee, S., LaCour, T.G. and Fuchs, P.L. (2009) *Chem. Rev.*, **109**, 2275–2314.

16 CP-263,114 (7): Splegel, D.A., Njardarson, J.T., McDonald, I.M. and Wood, J.L. (2003) *Chem. Rev.*, **103**, 2691–2727.

17 Eleutherobin (8): (a) Castoldi, D., Caggiano, L., Panigada, L., Sharon, O., Costa, A.M. and Gennari, C. (2005) *Chemistry*, **12**, 51–62; (b) Sperry, J.B., Constanzo, J.R., Jasinski, J., Butcher, R.J. and Wright, D.L. (2005) *Tetrahedron Lett.*, **46**, 2789–2793; (c) Diederichsen, U. (2003) *Org. Synth. Highlights*, 317–325; (d) Chen, X.T., Bhattacharya, S.K., Zhou, B., Gutteridge, C.E., Pettus, T.R.R. and Danishefsky, S.J. (1999) *J. Am. Chem. Soc.*, **121**, 6563–6579.

18 Furoscrobiculin (9): Seki, M., Sakamoto, T., Suemune, H. and Kanematsu, K. (1997) *J. Chem. Soc. Perkin Trans. 1*, 1707–1714.

19 Ginkgolide B1 (10): (a) Gosh, A.K. (2009) *J. Med. Chem.*, **52**, 2163–2176; (b) Vogensen, S.B., Stromgaard, K., Shindou, H., Jaracz, S., Suehiro, M., Ishii, S., Shimizu, T. and Nakanishi, K. (2003) *J. Med. Chem.*, **46**, 601–608; (c) Crimmins, M.T., Pace, J.M., Nantermet, P.G., Kim-Meade, A.S., Thomas, J.B., Watterson, S.H. and Wagman, A.S. (2000) *J. Am. Chem. Soc.*, **122**, 8453–8463; (d) Corey, E.J., Kang, M., Desai, M.C., Ghosh, A.K. and Houpis, I.N. (1988) *J. Am. Chem. Soc.*, **110**, 649–651.

20 Goniofufurone (11): Prasad, K.R. and Gholap, S.I. (2008) *J. Org. Chem.*, **73**, 2–11.

21 Medermycin (12): (a) Brimble, M.A. and Brenstrum, T.J. (2001) *J. Chem. Soc. Perkin Trans. 1*, 1624–1634; (b) Brimble, M.A. (2000) *Pure Appl. Chem.*, **72**, 1635–1639; (c) For related griseusin, see:Brimble, M.A., Nairn, M.R. and Park, J.S.O. (1999) *Org. Lett.*, **1**, 1459–1462.

22 Moracin O (13): Kaur, N., Xia, Y., Dat, N.T., Gajulapati, K., Choi, Y., Hong, Y.S., Lee, J.J. and Lee, K. (2009) *Chem. Commun.*, 1879–1881.

23 Morphine (14): (a) Tanimoto, H., Saito, R. and Chida, N. (2008) *Tetrahedron Lett.*, **49**, 358–362; (b) Omori, A.T., Finn, K.J., Leisch, H., Carroll, R.J. and Hudlicky, T. (2007) *Synlett*, 2859–2862; (c) Uchida, K., Yokoshima, S., Kan, T. and Fukuyama, T. (2006) *Org. Lett.*, **8**, 5311–5313; (d) Parker, K.A. and Fokas, D. (2006) *J. Org. Chem.*, **71**, 449–455; (e) Trost, B.M. and Toste, F.D. (2005) *J. Am. Chem. Soc.*, **127**, 14785–14803; (f) Novak, B.H., Hudlicky, T., Reed, J.W., Mulzer, J. and Trauner, D. (2000) *Curr. Org. Chem.*, **4**, 343–362.

24 Norhalichondrin B (15): (a) Phillips, A.J., Jackson, K.L. and Henderson, J.A. (2009) *Chem. Rev.*, **109**, 3044–3079; (b)

Jackson, K.L., Henderson, J.A., Motoyoshi, H. and Phillips, A.J. (2009) *Angew. Chem. Int. Ed.*, **48**, 2346–2350.

25 Pallescensin A (16): Foot, J.S., Phillis, A.T. and Sharp, P.P. (2006) *Tetrahedron Lett.*, **47**, 6817–6820.

26 Pamamycin (17): (a) Wang, Y., Bernsmann, H., Gruner, M. and Metz, P. (2001) *Tetrahedron Lett.*, **42**, 7801–7804; (b) Metz, P. (2005) *Top. Curr. Chem.*, **244** (Natural Product Synthesis II), 215–249; (c) Kang, E.J. and Lee, E. (2005) *Chem. Rev.*, **105**, 4348–4378.

27 Panacene (18): (a) Sabot, C., Berard, D. and Canesi, S. (2008) *Org. Lett.*, **10**, 4629–4632; (b) Boukouvalas, J., Pouliot, M., Robichaud, J., McNeil, S. and Snieckus, V. (2006) *Org. Lett.*, **8**, 3597–3599.

28 Podophyllotoxin (19): (a) Wu, Y., Zhao, J., Chen, J., Pan, C., Li, L. and Zhang, H. (2009) *Org. Lett.*, **11**, 597–600; (b) Wu, Y., Zhang, H., Hao, Y., Zao, J., Chen, J. and Li, L. (2007) *Org. Lett.*, **9**, 1199–1202; (c) Reynolds, A.J., Scott, A.J., Turner, C.I. and Sherburn, M.S. (2003) *J. Am. Chem. Soc.*, **125**, 12108–12109; (d) Berkowitz, D.B., Choi, S. and Maeng, J.H. (2000) *J. Org. Chem.*, **65**, 847–860; (e) For epi-podophyllotoxin, see:Engelhardt, U., Sarkar, A., Linker, T. (2003) *Angew. Chem. Int. Ed.*, **42**, 2887–2889; (f) Ward, R.S. (1999) *Nat. Prod. Rep.*, **161**, 75–96; (g) Ward, R.S. (1992) *Synthesis*, 719–730.

29 Polyoxin (20): (a) Luo, Y.C., Zhang, H.H., Liu, Y.Z., Cheng, R.L. and Xu, P.F. (2009) *Tetrahedron Asymmetry*, **20**, 1174–1180; (b) Dondoni, A., Franco, S., Junquera, F., Merchan, F.L., Merino, P. and Tejero, T. (1997) *J. Org. Chem.*, **62**, 5497–5507; (c) For related nikkomycin, see:Hayashi, Y., Urushima, T., Shin, M. and Shoji, M. (2005) *Tetrahedron*, **61**, 11393–11404.

30 Proximicins (21): (a) Schneider, K., Keller, S., Wolter, F.E., Röglin, L., Beil, W., Seitz, O., Nicholson, G., Bruntner, C., Riedlinger, J., Fiedler, H.P. and Süssmuth, R.D. (2008) *Angew. Chem. Int. Ed.*, **47**, 3258–3261; (b) Wolter, F.E., Molinari, L., Socher, E.R., Schneider, K., Nicholson, G., Beil, W., Seitz, O. and Süssmuth, R.D. (2009) *Bioorg. Med. Chem. Lett.*, **19**, 3811–3815.

31 Rhizonin A (22): Nakatsuka, H., Shimokawa, K., Miwa, R., Yamada, K. and Uemura, D. (2009) *Tetrahedron Lett.*, **50**, 186–188.

32 Salvinorin (23): Simpson, D.S., Katavic, P.L., Lozama, A., Harding, W.W., Parrish, D., Deschamps, J.R., Dersch, C.M., Partilla, J.S., Rothman, R.B., Navarro, H. and Prisinzano, T.E. (2007) *J. Med. Chem.*, **50**, 3596–3603.

33 Securinine (24): (a) Weinreb, S.M. (2009) *Nat. Prod. Rep.*, **26**, 758–775; (b) Dhudshia, B., Cooper, B.F.T., MacDonald, C.L.B. and Thadani, A.N. (2009) *Chem. Commun.*, 463–465; (c) Leduc, A.B. and Kerr, M.A. (2008) *Angew. Chem. Int. Ed.*, **47**, 7945–7948; (d) Carson, C.A. and Kerr, M.A. (2006) *Angew. Chem. Int. Ed.*, **45**, 6560–6563; (e) Honda, T., Namiki, H., Watanabe, N. and Mizutani, H. (2004) *Tetrahedron Lett.*, **45**, 5211–5213; (f) Alibes, R., Ballbe, M., Busque, F., De March, P., Elias, L., Figueredo, M. and Font, J. (2004) *Org. Lett.*, **6**, 1813–1816 and references cited therein.

34 Spirastrellolide A (25): (a) Paterson, I., Anderson, E.A., Dalby, S.M., Lim, J.H., Genovino, J., Maltas, P. and Moessner, C. (2008) *Angew. Chem. Int. Ed.*, **47**, 3016–3020; (b) Paterson, I. Anderson, E.A., Dalby, S.M., Lim, J.H., Genovino, J., Maltas, P. and Moessner, C. (2008) *Angew. Chem. Int. Ed.*, **47**, 3021–3025; (c) Paterson, I. and Dalby, S.M. (2009) *Nat. Prod. Rep.*, **26**, 865–873.

35 Spirolide (26): (a) Meilert, K. and Brimble, M.A. (2006) *Org. Biomol. Chem.*, **4**, 2184–2192; (b) Ishihara, J., Ishizaka, T., Suzuki, T. and Hatakeyama, S. (2004) *Tetrahedron Lett.*, **45**, 7855–7858.

36 Uprolide D (27): (a) Marshall, J.A. and Hann, R. (2008) *J. Org. Chem.*, **73**, 6753-6757; (b) For the structurally related, neurotoxic pukalide, see:Donohoe, T.J., Ironmonger, A. and Kershaw, N.M. (2008) *Angew. Chem. Int. Ed.*, **47**, 7314–7316.

37 Xanthohumol (28): Wang, W.S., Zhou, Y.W., Ye, Y.H. and Li, M.L. (2004) *Chin. Chem. Lett.*, **15**, 1195–1196.

38 XH-14 (29): (a) Kao, C.L. and Chern, J.W. (2002) *J. Org. Chem.*, **67**, 6772–6787; (b) Lütjens, H. and Scammells, J. (1998) *Tetrahedron Lett.*, **39**, 6581–6584.

39 Yangambin (30): Mori, N., Watanabe, H. and Kitahara, T. (2006) *Synthesis*, 400–404.

40 Zaragozic acid (31): (a) Nakamura, S., Sato, H., Hirata, Y., Watanabe, N. and Hashimoto, S. (2005) *Tetrahedron*, **61**, 11078–11106; (b) Armstrong, A. and Blench, T.J. (2002) *Tetrahedron*, **58**, 9321–9349; (c) Schlessinger, R.H., Wu, X.H. and Pettus, T.R.R. (1995) *Synlett*, 536–538.

41 Preparation of furans using radical reactions: (a) Majumdar, K.C., Basu, P.K. and Mukhopadhyay, P.P. (2007) *Tetrahedron*, **63**, 793–826; (b) Majumdar, K.C., Basu, P.K. and Mukhopadhyay, P.P. (2005) *Tetrahedron*, **61**, 10603–10642; (c) Majumdar, K.C., Basu, P.K. and Mukhopadhyay, P.P. (2004) *Tetrahedron*, **60**, 6239–6279; (d) Procter, D.J., Edmons, D.J. and Johnston, D. (2004) *Chem. Rev.*, **104**, 3371–3403; (e) Hartung, J. (2001) *Eur. J. Org. Chem.*, 619–632; (f) Hartung, J., Gottwald, T. and Spehar, K. (2002) *Synthesis*, 1469–1498.

42 Preparation of furans using epoxide ring opening: (a) Marshall, J.A. and Hann, R.K. (2008) *J. Org. Chem.*, **73**, 6753–6757; (b) Berber, H., Delaye, P.O. and Mirand, C. (2008) *Synlett*, 94–96; (c) Gupta, S., Rajagopalan, M., Alhamadsheh, M.M., Tillekeratne, L.M.V. and Hudson, R.A. (2007) *Synthesis*, 3512–3518; (d) Curran, D.P., Zhang, Q., Richard, C., Lu, H., Gudipati, V. and Wilcox, C.S. (2006) *J. Am. Chem. Soc.*, **128**, 9561–9573; (e) Halim, R., Brimble, M.A. and Merten, J. (2005) *Org. Lett.*, **7**, 2659–2662.

43 Haloeterification and haloesterification: (a) Carley, S. and Brimble, M.A. (2009) *Org. Lett.*, **11**, 563–566; (b) Wang, C., Zhang, H., Liu, J., Shao, Z., Ji, Y. and Li, L. (2006) *Synlett*, 1051–1054; (c) Gu, Y. and Snider, B.B. (2003) *Org. Lett.*, **5**, 4385–4388; (d) Kinoshita, A. and Mori, M. (1996) *J. Org. Chem.*, **61**, 8356–8357.

44 Reduction and oxidations: (a) Zhou, Y.G. (2007) *Acc. Chem. Res.*, **40**, 1357–1366; (b) Vassilikogiannakis, G., Montagnon, T. and Tofi, M. (2008) *Acc. Chem. Res.*, **41**, 1001–1010; (c) Yoshida, J.-I., Kataoka, K., Horcajada, R. and Nagaki, A. (2008) *Chem. Rev.*, **108**, 2265–2299; (d) Hudlicky, T., Entwistle, D.A., Pitzer, K.K. and Thorpe, A.J. (1996) *Chem. Rev.*, **96**, 1195–1220; (e) Holder, N.L. (1982) *Chem. Rev.*, **82**, 267–332.

45 Oxycarbenium ions and furan-derived nucleophiles (such as TMSOF): (a) Harmata, M. and Rashatasakhon, P. (2003) *Tetrahedron*, **59**, 2371–2395; (b) Martin, S.F. (2002) *Acc. Chem. Res.*, **35**, 895–904; (c) Rassu, G., Zanardi, F., Battistini, L. and Casiraghi, G. (2000) *Chem. Soc. Rev.*, **29**, 109–118.

46 Diels–Alder reactions: (a) Clarke, P.A., Reeder, A.T. and Winn, J. (2009) *Synthesis*, 691–709; (b) Reymond, S. and Cossy, J. (2008) *Chem. Rev.*, **108**, 5359–5406; (c) Takao, K.I., Munakata, R. and Tadano, K.I. (2005) *Chem. Rev.*, **105**, 4779–4807; (d) Nicolaou, K.C., Snyder, S.A., Montagnon, T. and Vassilikogiannakis, G. (2002) *Angew. Chem. Int. Ed.*, **41**, 1668–1698; (e) Kappe, C.O., Murphree, S.S. and Padwa, A. (1997) *Tetrahedron*, **53**, 14179–14233; (f) Winkler, J.D. (1996) *Chem. Rev.*, **96**, 167–176; (g) Brieger, G. and Bennet, J.N. (1980) *Chem. Rev.*, **80**, 63–97; (h) See also: Jacobi, P.A. and Lee, K. (1997) *J. Am. Chem. Soc.*, **119**, 3409–3410 and references cited therein.

47 Dipolar cycloadditions: (a) Pellisier, H. (2007) *Tetrahedron*, **63**, 3235–3285; (b) Nair, V. and Suja, T.D. (2007) *Tetrahedron*, **63**, 12247–12275; (c) Padwa, A. and Weingarten, M.D. (1996) *Chem. Rev.*, **96**, 223–269; (d) Padwa, A. (1991) *Acc. Chem. Res.*, **24**, 22–28; (e) For other cycloaddition processes, see: Liao, C.C. and Peddinti, R.K. (2002) *Acc. Chem. Res.*, **35**, 856–866; (f) Harmata, M. (2001) *Acc. Chem. Res.*, **34**, 595–605.

48 (a) Nicolaou, K.C., Bulger, P.G. and Sarlah, D. (2005) *Angew. Chem. Int. Ed.*, **44**, 4490–4527; (b) Grubbs, R.H. (ed.) (2003) *Handbook of Metathesis*, Wiley-

VCH Verlag GmbH, Weinheim, Germany, ISBN: 3-527-30616-1.
49 Carbenes: (a) Dotz, K.H. and Stendel, J. (2009) *Chem. Rev.*, **109**, 3227–3274; (b) Warkentin, J. (2009) *Acc. Chem. Res.*, **42**, 205–212; (c) Cheng, Y. and Meth-Cohn, O. (2004) *Chem. Rev.*, **104**, 2507–2503; (d) Müller, P. (2004) *Acc. Chem. Res.*, **37**, 243–251; (e) Reissig, H.U. and Zimmer, R. (2003) *Chem. Rev.*, **103**, 1151–1196; (f) Davies, H.M.L. and Beckwith, R.E.J. (2003) *Chem Rev.*, **103**, 2861–2903; (g) Sierra, M.A. (2000) *Chem. Rev.*, **100**, 3591–3637; (h) Doyle, M.P. and Forbes, D.C. (1998) *Chem. Rev.*, **98**, 911–935; (i) Ye, T. and McKervey, M.A. (1994) *Chem. Rev.*, **94**, 1091–1160.
50 Other metal-catalyzed furan synthesis: (a) Alvarez-Corral, M., Muñoz-Dorado, M. and Rodriguez-Garcia, I. (2008) *Chem. Rev.*, **108**, 3174–3198; (b) Yamamoto, Y. and Patil, N.T. (2008) *Chem. Rev.*, **108**, 3395-3442; (c) Hashmi, A.S.K. (2007) *Chem. Rev.*, **107**, 3180–3211; (d) Ma, S. (2005) *Chem. Rev.*, **105**, 2829–2871; (e) Schröter, S., Stock, C. and Bach, T. (2005) *Tetrahedron*, **61**, 2245–2267; (f) Zeni, G. and Larock, R.C. (2004) *Chem. Rev.*, **104**, 2285–2309; (g) Trost, B.M. and Crawley, M.L. (2003) *Chem. Rev.*, **103**, 2921–2943; (h) Yet, L. (2000) *Chem. Rev.*, **100**, 2963–3007.
51 For recent examples of metal-catalyzed furan synthesis, see: (a) Brasholz, M. and Reissig, H.U. (2009) *Eur. J. Org. Chem.*, 3595–3604; (b) Istrate, F.M. and Gagosz, F.L. (2008) *J. Org. Chem.*, **73**, 730–733; (c) Marshall, J.A. and Van Devender, E.A. (2001) *J. Org. Chem.*, **66**, 8037–8041; (d) Hiroya, K., Suzuki, N., Yasuhara, A., Egawa, Y., Kasano, A. and Sakamoto, T. (2000) *J. Chem. Soc. Perkin Trans. 1*, 4339–4346; (e) Lutjens, H. and Scammells, P.J. (1998) *Tetrahedron Lett.*, **39**, 6581–6584; (f) Marshall, J.A. and Sehan, C.A. (1997) *J. Org. Chem.*, **62**, 4313–4320.
52 Furans using biomimetic reactions, multicomponent processes and cascade reactions: (a) Touré, B.B. and Hall, D.G. (2009) *Chem. Rev.*, **109**, 4439–4486; (b) Bulger, P.G., Bagal, S.K. and Marquez, R. (2008) *Nat. Prod. Rep.*, **25**, 254–297; (c) Padwa, A. and Bur, S.K. (2007) *Tetrahedron*, **63**, 5341–5378; (d) Nicolaou, K.C., Edmons, D.J. and Bulger, P.G. (2006) *Angew. Chem. Int. Ed.*, **45**, 7134–7186; (e) Parsons, P.J., Penkett, C.S. and Shell, A.J. (1996) *Chem. Rev.*, **96**, 195–206; (f) Parsons, P.J., Penkett, C.S. and Shell, A.J. (1996) *Chem. Rev.*, **96**, 195–206.
53 Ketalization, furans from ketals: (a) Boto, A., Hernández, D. and Hernández, R. (2007) *Org. Lett.*, **9**, 1721–1724; (b) Jung, M.E. and Min, S.J. (2004) *Tetrahedron Lett.*, **45**, 6753–6755; (c) Fenlon, T.W., Schwaebisch, D., Mayweg, A.V.W., Lee, V., Adlington, R.M. and Baldwin, J.E. (2007) *Synlett*, 2679–2682; (d) Gennari, C., Castoldi, D. and Sharon, O. (2007) *Pure Appl. Chem.*, **79**, 173–180; (e) For some synthesis of furans from sugars, see:Mereyala, H.B., Gadikota, R.R. and Krishnan, R. (1997) *J. Chem. Soc. Perkin Trans. 1*, 3567–3571; (f) Dixon, D.J., Ley, S.V., Gracza, T. and Szolcsanyi, P. (1999) *J. Chem. Soc. Perkin Trans. 1*, 839–841; (g) Yoon, S.H., Moon, H.S., Hwang, S.K., Choi, S. and Kang, S.K. (1998) *Bioorg. Med. Chem.*, **6**, 1043–1049; (h) Mereyala, H.B., Gadikota, R.R., Joe, M., Arora, S.K., Dastidar, S.G. and Agarwal, S. (1999) *Bioorg. Med. Chem.*, **7**, 2095–2103.
54 Rearrangements: (a) Bur, S.K. and Padwa, A. (2004) *Chem. Rev.*, **104**, 2401–2432; (b) Tsubuki, M., Ohinata, A., Tanaka, T., Takahashi, K. and Honda, T. (2005) *Tetrahedron*, **61**, 1095–1100; (c) Caruana, P.A. and Frontier, A.J. (2004) *Tetrahedron*, **60**, 10921–10926.
55 For some synthesis of lactones and their reactions, see: (a) Prasad, K.R. and Gholap, S.L. (2008) *J. Org. Chem.*, **73**, 2–11; (b) Huang, Y. and Pettus, T.R.R. (2008) *Synlett*, 1353–1356; (c) Lee, C.L., Lin, C.F., Lin, W.R., Wang, K.S., Chang, Y., Lin, S.R., Wu, Y.C. and Wu, M.J. (2005) *Bioorg. Med. Chem.*, **13**, 5864–5872; (d) Krohn, K., Riaz, M. and Flörke, U. (2004) *Eur. J. Org. Chem.*, 1261–1270.
56 For the preparation of benzofuran and isobenzofuran rings and their reactivity, see: (a) McAllister, G.D., Hartley, R.C., Dawson, M.J. and Knaggs, A.R. (1998) *J. Chem. Soc. Perkin Trans. 1*, 3453–3457;

(b) Traulsen, T. and Friedrichsen, W. (2000) *J. Chem. Soc. Perkin Trans. 1*, 1387–1398; (c) Uno, H., Murakami, S., Fujimoto, A. and Yamaoka, Y. (2005) *Tetrahedron Lett.*, **46**, 3997–4000; (d) Dixit, M., Sharon, A., Maulik, P.R. and Goel, A. (2006) *Synlett*, 1497–1502; (e) Slamet, R. and Wege, D. (2007) *Tetrahedron*, **63**, 12621–12628; (f) Yue, D., Yao, T. and Larock, R.C. (2005) *J. Org. Chem.*, **70**, 10292–10296.

57 Metallation: (a) Chinchilla, R., Nájera, C. and Yus, M. (2005) *Tetrahedron*, **61**, 3139–3176; (b) Chinchilla, R., Nájera, C. and Yus, M. (2004) *Chem. Rev.*, **104**, 2667–2722.

58 Lithiation: (a) Marcos, I.S., Escola, M.A., Moro, R.F., Basabe, P., Diez, D., Sanz, F., Mollinedo, F., de la Iglesia-Vicente, J., Sierra, B.G. and Urones, J.G. (2007) *Bioorg. Med. Chem.*, **15**, 5719–5737; (b) Xu, P.F., Luo, Y.C. and Zhang, H.H. (2009) *Synlett*, 833–837; (c) Smitt, O. and Högberg, H. (2002) *Synlett*, 1273–1276; (d) See also:Nakatsuka, H., Shimokawa, K., Miwa, R., Yamada, K. and Uemura, D. (2009) *Tetrahedron Lett.*, **50**, 186–188; (e) Gogoi, S. and Argade, N.P. (2008) *Synthesis*, 1455–1459; (f) Bruyère, H., Dos Reis, C., Samaritani, S., Ballereau, S. and Royer, J. (2006) *Synthesis*, 1673–1681; (g) Oh, S., Jeong, I.H., Shin, W.S. and Lee, S. (2003) *Bioorg. Med. Chem. Lett.*, **13**, 2009–2012; (h) Marshall, J.A., Bartley, G.S. and Wallace, E.M. (1996) *J. Org. Chem.*, **61**, 5729–5735; (i) Shen, C.C., Chou, S.C., Chou, C.J. and Ho, L.K. (1996) *Tetrahedron Asymmetry*, **7**, 3141–3146.

59 Stannylated derivatives: (a) Roethle, P.A. and Trauner, D. (2006) *Org. Lett.*, **8**, 345–347; (b) Tan, Q. and Danishefsky, S.J. (2000) *Angew. Chem. Int. Ed.*, **39**, 4509–4511; (c) Meng, D.F., Tan, Q. and Danishefsky, S.J. (1999) *Angew. Chem. Int. Ed.*, **38**, 3197–3201; (d) Meng, D.F. and Danishefsky, S.J. (1999) *Angew. Chem. Int. Ed.*, **38**, 1485–1488; See also: (e) Clark, J.S., Northall, J.M., Marlin, F., Nay, B., Wilson, C., Blake, A.J. and Waring, M.J. (2008) *Org. Biomol. Chem.*, **6**, 4012–4025; (f) Cases, M., González-López de Turiso, F. and Pattenden, G. (2001) *Synlett*, 1869–1872; (g) Pattenden, G. and Sinclair, D.J. (2002) *J. Organomet. Chem.*, **653**, 261–268; (h) Paterson, I., Brown, R.E. and Urch, C. (1999) *Tetrahedron Lett.*, **40**, 5807–5810; (i) Yu, P., Yang, Y., Zhang, Z.Y., Mak, T.C.W. and Wong, H.N.C. (1997) *J. Org. Chem.*, **62**, 6359–6366; (j) Paterson, I., Brown, R.E. and Urch, C. (1999) *Tetrahedron Lett.*, **40**, 5807–5810.

60 Titanium reagents: Sibi, M.P. and He, L. (2004) *Org. Lett.*, **6**, 1749–1752.

61 Furyl zincs: (a) Donohoe, T.J., Ironmonger, A. and Kershaw, N.M. (2008) *Angew. Chem. Int. Ed.*, **47**, 7314–7316; (b) Mukaiyama, T., Suzuki, K., Yamada, T. and Tabusa, F. (1990) *Tetrahedron*, **46**, 265–276.

62 Chromium derivatives: Chen, X.T., Bhattacharya, S.K., Zhou, B., Gutteridge, C.E., Pettus, T.R.R. and Danishefsky, S.J. (1999) *J. Am. Chem. Soc.*, **121**, 6563–6579.

63 Silyl derivatives: (a) Chen, W., Wu, H., Bernard, D., Metcalf, M.D., Deschamps, J.R., Flippen-Anderson, J.L., MacKerell, A.D., Jr. and Coop, A. (2003) *J. Org. Chem.*, **68**, 1929–1932; (b) Tan, Z. and Negishi, E.-I. (2006) *Org. Lett.*, **8**, 2783–2785.

64 Seleno derivatives: Brecht-Forster, A., Fitremann, J. and Renaud, P. (2002) *Helv. Chim. Acta*, **85**, 3965–3974.

65 Boranes, boronates, boronic acids: Yick, C.Y., Tsang, T.K. and Wong, H.N.C. (2003) *Tetrahedron*, **59**, 325–333.

66 Reductions: (a) Garçon, S., Vassiliou, S., Cavicchioli, M., Hartmann, B., Monteiro, N. and Balme, G. (2001) *J. Org. Chem.*, **66**, 4069–4073; (b) Donohoe, T.J., Guillermin, J.B., Frampton, C. and Walter, D.S. (2000) *Chem. Commun.*, 465–466.

67 Achmatowicz to lactones: (a) Henderson, J.A., Jackson, K.L. and Phillips, A.J. (2007) *Org. Lett.*, **9**, 5299–5302; (b) Guo, H. and O'Doherty, G.A. (2008) *J. Org. Chem.*, **73**, 5211–5220; (c) Shults, E.E., Velder, J., Schmalz, H.G., Chernov, S.V., Rubalova, T.V., Gatilov, Y.V., Henze, G., Tolstikov, G.A. and Prokop, A. (2006) *Bioorg. Med. Chem. Lett.*, **16**, 4228–4232; (d) Brasholz, M. and Reissig, H.U. (2007) *Synlett*, 1294–1298; (e) Abrams, J.N., Babu, R.S., Guo, H., Le, D., Le, J., Osbourn, J.M.

and O'Doherty, G.A. (2008) *J. Org. Chem.*, **73**, 1935–1940; (f) Fürstner, A. and Nagano, T. (2007) *J. Am. Chem. Soc.*, **129**, 1906–1907; (g) Nicolaou, K.C., Frederick, M.O., Burtoloso, A., Denton, R.M., Rivas, F., Cole, K.P., Aversa, R.J., Gibe, R., Umezawa, T. and Suzuki, T. (2008) *J. Am. Chem. Soc.*, **130**, 7466–7476; (h) Nicolaou, K.C., Cole, K.P., Frederick, M.O., Aversa, R.J. and Denton, R.M. (2007) *Angew. Chem. Int. Ed.*, **46**, 8875–8879; (i) See also: Singh, V., Singh, V. (2009) *Tetrahedron Lett.*, **50**, 3092–3094; (j) Robertson, J., Stevens, K. and Naud, S. (2008) *Synlett*, 2083–2086; (k) Akaike, H., Horie, H., Kato, K. and Akita, H. (2008) *Tetrahedron Asymmetry*, **19**, 1100–1105; (l) Guo, H. and O'Doherty, G.A. (2006) *Org. Lett.*, **8**, 1609–1612; (m) Crawford, C., Nelson, A. and Patel, I. (2006) *Org. Lett.*, **8**, 4231–4234; (n) Robertson, J., Meo, P., Dallimore, J.W.P., Doyle, B.M. and Hoarau, C. (2004) *Org. Lett.*, **6**, 3861–3863; (o) Li, M., Scott, J. and O'Doherty, G.A. (2004) *Tetrahedron Lett.*, **45**, 1005–1009; (p) Li, M. and O'Doherty, G.A. (2004) *Tetrahedron Lett.*, **45**, 6407–6411; (q) Haukas, M.H. and O'Doherty, G.A. (2002) *Org. Lett.*, **4**, 1771–1774; (r) Harris, J.M. and O'Doherty, G.A. (2000) *Org. Lett.*, **2**, 2983–2986; (s) Haukaas, M.H. and O'Doherty, G.A. (2001) *Org. Lett.*, **3**, 401–404; (t) Balachari, D. and O'Doherty, G.A. (2000) *Org. Lett.*, **2**, 4033–4036; (u) Taniguchi, T., Takeuchi, M. and Ogasawara, K. (1998) *Tetrahedron Asymmetry*, **9**, 1451–1456; (v) De Mico, A., Margarita, R. and Piancatelli, G. (1995) *Tetrahedron Lett.*, **33**, 3553–3556; (w) Martin, S.F., Lee, W.C., Pacofsky, G.J., Gist, R.P. and Mulhern, T.A. (1994) *J. Am. Chem. Soc.*, **116**, 4674–4688; (x) Martin, S.F. and Zinke, P.W. (1991) *J. Org. Chem.*, **56**, 6606–6611; (y) Martin, S.F., Pacofsky, G.J., Gist, R.P. and Lee, W.C. (1989) *J. Am. Chem. Soc.*, **111**, 7634–7636.

68 Aza-Achmatowicz to lactams: (a) Leverett, C.A., Cassidy, M.P. and Padwa, A. (2006) *J. Org. Chem.*, **71**, 8591–8601; (b) Cassidy, M.P. and Padwa, A. (2004) *Org. Lett.*, **6**, 4029–4031; (c) Harris, J.M. and Padwa, A. (2003) *J. Org. Chem.*, **68**, 4371–4381; (d) See also: Zhang, H.X., Xia, P. and Zhou, W.S. (2003) *Tetrahedron*, **59**, 2015–2020; (e) Yang, C., Liao, L., Xu, Y., Zhang, H., Xia, P. and Zhou, W. (1999) *Tetrahedron Asymmetry*, **10**, 2311–2318.

69 Achmatowicz to polyenes: (a) Kobayashi, Y., Kumar, G.B., Kurachi, T., Acharya, H.P., Yamazaki, T. and Kitazume, T. (2001) *J. Org. Chem.*, **66**, 2011–2018; (b) See also: Hayashi, Y., Shoji, M., Ishikawa, H., Yamaguchi, J., Tamura, T., Imai, H., Nishigaya, Y., Takabe, K., Kakeya, H. and Osada, H. (2008) *Angew. Chem. Int. Ed.*, **47**, 6657–6660; (c) Raczko, J. (2003) *Tetrahedron*, **59**, 10181–10186; (d) Kobayashi, Y. and Wang, Y.G. (2002) *Tetrahedron Lett.*, **43**, 4381–4384; (e) Lee, W.W., Shin, H.J. and Chang, S. (2001) *Tetrahedron Asymmetry*, **12**, 29–31; (f) Kobayashi, Y., Kumar, B.G. and Kurachi, T. (2000) *Tetrahedron Lett.*, **41**, 1559–1563

70 Singlet oxygen: (a) Tofi, M., Montagnon, T., Georgiou, T. and Vassilikogiannakis, G. (2007) *Org. Biomol. Chem.*, **5**, 772–777; (b) Vassilikogiannakis, G., Margaros, I., Montagnon, T. and Stratakis, M. (2005) *Chem. Eur. J.*, **11**, 5899–5907; (c) Helliwell, M., Karim, S., Parmee, E.R. and Thomas, E.J. (2005) *Org. Biomol. Chem.*, **3**, 3636–3653; (d) See also: Georgiou, T., Tofi, M., Montagnon, T. and Vassilikogiannakis, G. (2006) *Org. Lett.*, **8**, 1945–1948; (e) Sofikiti, N., Tofi, M., Montagnon, T., Vassilikogiannakis, G. and Stratakis, M. (2005) *Org. Lett.*, **7**, 2357–2359; (f) García, I., Gómez, G., Teijeira, M., Terán, C. and Fall, Y. (2006) *Tetrahedron Lett.*, **47**, 1333–1335; (g) Teijeira, M., Suárez, P.L., Gómez, G., Terán, C. and Fall, Y. (2005) *Tetrahedron Lett.*, **46**, 5889–5892; (h) Pérez, M., Canoa, P., Gómez, G., Terán, C. and Fall, Y. (2004) *Tetrahedron Lett.*, **45**, 5207–5209; (i) Shimazawa, R., Suzuki, T., Dodo, K. and Shirai, R. (2004) *Bioorg. Med. Chem.*, **14**, 3291–3294; (j) de la Torre, M.C., García, I. and Sierra, M.A. (2002) *J. Nat. Prod.*, **65**, 661–668; (k) Imamura, P.M. and Costa, M. (2000) *J. Nat. Prod.*, **63**, 1623–1625; (l) Shiraki, R., Sumino, A.,

Tadano, K.I. and Ogawa, S. (1996) *J. Org. Chem.*, **61**, 2845–2852; (m) Parmee, E.R., Mortlock, S.V., Stacey, N.A., Thomas, E.J. and Mills, O.S. (1997) *J. Chem. Soc. Perkin Trans. 1*, 381–390

71 Electrooxidation: (a) Miller, A.K., Hughes, C.C., Kennedy-Smith, J.J., Gradi, S.N. and Trauner, D. (2006) *J. Am. Chem. Soc.*, **128**, 17057–17062; (b) Mihelcic, J. and Moeller, K.D. (2003) *J. Am. Chem. Soc.*, **125**, 36–37; (c) See also:Wu, H. and Moeller, K.D. (2007) *Org. Lett.*, **9**, 4599–4602; (d) Nieto-Mendoza, E., Guevara-Salazar, J.A., Ramírez-Apan, M.T., Frontana-Uribe, B.A., Cogordan, J.A. and Cárdenas, J. (2005) *J. Org. Chem.*, **70**, 4538–4541; (e) Sperry, J.B., Constanzo, J.R., Jasinski, J., Butcher, R.J. and Wright, D.L. (2005) *Tetrahedron Lett.*, **46**, 2789–2793.

72 Oxidation to acid with RuO_4: (a) Ainai, T., Wang, Y.G., Tokoro, Y. and Kobayashi, Y. (2004) *J. Org. Chem.*, **69**, 655–659; (b) See also:Liu, L.X., Peng, Q.L. and Huang, P.Q. (2008) *Tetrahedron Asymmetry*, **19**, 1200–1203; (c) Aggarwal, V.A. and Vasse, J.L. (2003) *Org. Lett.*, **5**, 3987–3990; (d) See also:Luo, Y.C., Zhang, H.H., Liu, Y.Z., Cheng, R.L. and Xu, P.F. (2009) *Tetrahedron Asymmetry*, **20**, 1174–1180; (e) Krüger, J. and Carreira, E.M. (1998) *J. Am. Chem. Soc.*, **120**, 837–838; (f) Pagenkopf, B.L., Krüger, J., Stojanovic, A. and Carreira, E.M. (1998) *Angew. Chem. Int. Ed. Engl.*, **37**, 3124–3126; (g) Dondoni, A., Franco, S., Junquera, F., Merchán, F.L., Merino, P. and Tejero, T. (1997) *J. Org. Chem.*, **62**, 5497–5507; (h) Kobayashi, Y., Nakano, M. and Okui, H. (1997) *Tetrahedron Lett.*, **38**, 8883–8886

73 Other oxidations: (a) Chen, I.C., Wu, Y.K., Liu, H.J. and Zhu, J.L. (2008) *Chem. Commun.*, 4720–4722; (b) Watanabe, T., Ohashi, Y., Yoshino, R., Komano, N., Eguchi, M., Maruyama, S. and Ishikawa, T. (2003) *Org. Biomol. Chem.*, **1**, 3024–3032.

74 Furans as electrophiles: (a) Meulemans, T.M., Stork, G.A., Macaev, F.Z., Jansen, B.J.M. and De Groot, A. (1999) *J. Org. Chem.*, **64**, 9178–9188; (b) Hori, K., Hikage, N., Inagaki, A., Mori, S., Nomura, K. and Yoshii, E. (1992) *J. Org. Chem.*, **57**, 2888–2902; (c) Hori, K., Kazuno, H., Nomura, K. and Yoshii, E. (1993) *Tetrahedron Lett.*, **34**, 2183–2186.

75 Furans as nucleophiles (TMSOF): (a) Bardají, G.G., Cantó, M., Alibés, R., Bayón, P., Busqué, F., de March, P., Figueredo, M. and Font, J. (2008) *J. Org. Chem.*, **73**, 7657–7662; (b) Alibes, R., Bayon, P., de March, P., Figueredo, M., Font, J., García-García, E. and González-Gálvez, D. (2005) *Org. Lett.*, **7**, 5107–5109; (c) Liras, S., Lynch, C.L., Fryer, A.M., Vu, B.T. and Martin, S.F. (2001) *J. Am. Chem. Soc.*, **123**, 5918–5924; (d) Rassu, G., Auzzas, L., Zambrano, V., Burreddu, P., Pinna, L., Battistini, L., Zanardi, F. and Casiraghi, G. (2004) *J. Org. Chem.*, **69**, 1625–1628.

76 For other examples of TMSOF as nucleophiles see: (a) Snider, B.B. and Che, Q. (2004) *Org. Lett.*, **6**, 2877–2880; (b) Zanardi, F., Battistini, L., Rassu, G., Auzzas, L., Pinna, L., Marzocchi, L., Acquotti, D. and Casiraghi, G. (2000) *J. Org. Chem.*, **65**, 2048–2064; (c) Rassu, G., Auzzas, L., Pinna, L., Battistini, L., Zanardi, F., Marzocchi, L., Acquotti, D. and Casiraghi, G. (2000) *J. Org. Chem.*, **65**, 6307–6318; (d) Rassu, G., Pinna, L., Spanu, P., Zanardi, F., Battistini, L. and Casiraghi, G. (1997) *J. Org. Chem.*, **62**, 4513–4517; (e) Martin, S.F., Barr, K.J., Smith, D.W. and Bur, S.K. (1999) *J. Am. Chem. Soc.*, **121**, 6990–6997; (f) Mukai, C., Hirai, S., Kim, I.J., Kido, M. and Hanaoka, M. (1996) *Tetrahedron*, **52**, 6547–6565

77 Furan as nucleophile: (a) Shanmugham, M.S. and White, J.D. (2004) *Chem. Commun.*, 44–45; (b) Boyer, F.D., Descoins, C.L., Thanh, G.V., Descoins, C., Prangé, T. and Ducrot, P.H. (2003) *Eur. J. Org. Chem.*, 1172–1183; (c) Rose, M.D., Cassidy, M.P., Rashatasakhon, P. and Padwa, A. (2007) *J. Org. Chem.*, **72**, 538–549; (d) Nilson, M.G. and Funk, R.L. (2006) *Org. Lett.*, **8**, 3833–3836; (e) Marion, F., Williams, D.E., Patrick, B.O., Hollander, I., Mallon, R., Kim, S.C., Roll, D.M., Fieldberg, L., Van Soest, R. and Andersen, R.J. (2006) *Org. Lett.*, **8**, 321–324.

78 For other examples of furans as nucleophiles see: (a) Cassidy, M.P.,

Özdemir, A.D. and Padwa, A. (2005) *Org. Lett.*, **7**, 1339–1342; (b) Mal, S.K., Kar, G.K. and Ray, J.K. (2004) *Tetrahedron*, **60**, 2805–2811; (c) Padwa, A., Heidelbaugh, T.M., Kuethe, J.T., McClure, M.S. and Wang, Q. (2002) *J. Org. Chem.*, **67**, 5928–5937; (d) Inone, M., Carson, M.W., Frontier, A.J. and Danishefsky, S.J. (2001) *J. Am. Chem. Soc.*, **123**, 1878–1889; (e) Sasaki, S., Hamada, Y. and Shiori, T. (1997) *Tetrahedron Lett.*, **38**, 3013–3016; (f) Jung, M.E. and Siedem, C.S. (1993) *J. Am. Chem. Soc.*, **115**, 3822–3823.

79 Cyclopropanations: (a) Kashin, D., Meyer, A., Wittenberg, R., Schöning, K.U., Kamlage, S. and Kirschning, A. (2007) *Synthesis*, 304–319; (b) Chhor, R.B., Nosse, B., Sörgel, S., Böhm, C., Seitz, M. and Reiser, O. (2003) *Chem. Eur. J.*, **9**, 260–270; (c) Hughes, C.C., Kennedy-Smith, J.J. and Trauner, D. (2003) *Org. Lett.*, **5**, 4113–4115; (d) Foot, J.S., Phillis, A.T., Sharp, P.P., Willis, A.C. and Banwell, M.G. (2006) *Tetrahedron Lett.*, **47**, 6817–6820.

80 [2 + 2] cycloadditions: (a) de March, P., Figueredo, M., Font, J., Raya, J., Alvarez-Larena, A. and Piniella, J.F. (2003) *J. Org. Chem.*, **68**, 2437–2447; (b) Crimmins, M.T., Pace, J.M., Nantermet, P.G., Kim-Meade, A.S., Thomas, J.B., Watterson, S.H. and Wagman, A.S. (2000) *J. Am. Chem. Soc.*, **122**, 8453–8463.

81 (a) Vasella, A. and Voeffray, R. (1982) *Helv. Chim. Acta*, **65**, 1134–1144; (b) Padwa, A. and Price, A.T. (1998) *J. Org. Chem.*, **63**, 556–565; (c) Padwa, A. and Price, A.T. (1995) *J. Org. Chem.*, **60**, 6258–6259; (d) Padwa, A., Lynch, S.M., Mejía-Oneto, J.M. and Zhang, H. (2005) *J. Org. Chem.*, **70**, 2206–2218; (e) Elliott, G.I., Velcicky, J., Ishikawa, H., Li, Y.K. and Boger, D.L. (2006) *Angew. Chem. Int. Ed.*, **45**, 620–622.

82 (a) Padwa, A. and Pearson, W.H. (2003) *Synthetic Applications of 1,3-Dipolar Cycloaddition Chemistry: Towards Heterocycles and Natural Products*, Series: Chemistry of Heterocyclic Compounds, vol. 59, John Wiley & Sons, Inc., Hoboken, USA. ISBN: 0-471-28061-5; (b) Sessions, E.H., O'Connor, R.T., Jr. and Jacobi, P.A. (2007) *Org. Lett.*, **9**, 3221–3224; (c) Jacobi, P.A. and Lee, K. (2000) *J. Am. Chem. Soc.*, **122**, 4295–4303; (d) Sessions, E.H. and Jacobi, P.A. (2006) *Org. Lett.*, **8**, 4125–4128; (e) Hodgson, D.M., Le Strat, F., Avery, T.D., Donohue, A.C. and Bruckl, T. (2004) *J. Org. Chem.*, **69**, 8796–8803; (f) Ou, L., Hu, Y., Song, G. and Bai, D. (1999) *Tetrahedron*, **55**, 13999–14004; (g) Padwa, A., Brodney, M.A., Marino, J.P., Jr., Osterhout, M.H. and Price, A.T. (1997) *J. Org. Chem.*, **62**, 67–77; (h) Padwa, A., Brodney, M.A., Marino, J.P., Jr. and Sheenan, S.M. (1997) *J. Org. Chem.*, **62**, 78–87; (i) Sheenan, S.M. and Padwa, A. (1997) *J. Org. Chem.*, **62**, 438–439; (j) Kuethe, J.T. and Padwa, A. (1997) *J. Org. Chem.*, **62**, 774–775; (k) Cochran, J.E. and Padwa, A. (1995) *J. Org. Chem.*, **60**, 3938–3939.

83 [3 + 2] cycloadditions with quinones: (a) Zhou, G. and Corey, E.J. (2005) *J. Am. Chem. Soc.*, **127**, 11958–11959; (b) See also: Souza, F.E. and Rodrigo, R. (1999) *Chem. Commun.*, 1947–1948; (c) Carlini, R., Higgs, K., Older, C., Randhawa, S. and Rodrigo, R. (1997) *J. Org. Chem.*, **62**, 2330–2331; (d) Brimble, M.A., Nairn, M.R. and Park, J.S.O. (2000) *J. Chem. Soc. Perkin Trans. 1*, 697–709; (e) Brimble, M.A. and Brenstrum, T.J. (2001) *J. Chem. Soc. Perkin Trans. 1*, 1624–1634; (f) Berand, D., Jean, A. and Canesi, S. (2007) *Tetrahedron Lett.*, **48**, 8238–8241.

84 Diels–Alder: (a) Brimble, M.A., Duncalf, L.J. and Phythian, S.J. (1997) *J. Chem. Soc. Perkin Trans. 1*, 1399–1403; (b) Aceña, J.L., Arjona, O. and Plumet, J. (1997) *J. Org. Chem.*, **20**, 3360–3364; (c) Shoji, M., Imai, H., Mukaida, M., Sakai, K., Kakeya, H., Osada, H. and Hayashi, Y. (2005) *J. Org. Chem.*, **70**, 79–91; (d) Padwa, A. and Ginn, J.D. (2005) *J. Org. Chem.*, **70**, 5197–5206; (e) Wolkenberg, S.E. and Boger, D.L. (2002) *J. Org. Chem.*, **67**, 7361–7364; (f) Winkler, J.D., Quinn, K.J., MacKinnon, C.H., Hiscock, S.D. and McLaughlin, E.C. (2003) *Org. Lett.*, **5**, 1805–1808; (g) Zhang, H., Boonsombat, J. and Padwa, A. (2007) *Org. Lett.*, **9**, 279–282; (h) Toró, A. and Deslongchamps, P. (2003) *J. Org. Chem.*,

68, 6847–6852; (i) Toró, A., L'Heureux, A. and Deslongchamps, P. (2000) *Org. Lett.*, **2**, 2737–2740; (j) Sutherland, H.S., Souza, F.E.S. and Rodrigo, R.G.A. (2001) *J. Org. Chem.*, **66**, 3639–3641.

85 Other Diels–Alder, with 2-amino or 2-thiofurans as substrates: (a) Padwa, A. and Wang, Q. (2006) *J. Org. Chem.*, **71**, 7391–7402; (b) Wang, Q. and Padwa, A. (2006) *Org. Lett.*, **8**, 601–604; (c) Padwa, A., Bur, S.K. and Zhang, H. (2005) *J. Org. Chem.*, **70**, 6833–6841; (d) Padwa, A., Brodney, M.A., Lynch, S.M., Rashatasakhon, P., Wang, Q. and Zhang, H. (2004) *J. Org. Chem.*, **69**, 3735–3745; (e) Wang, Q. and Padwa, A. (2004) *Org. Lett.*, **6**, 2189–2192; (f) Takadoi, M., Yamaguchi, K. and Terashima, S. (2003) *Bioorg. Med. Chem.*, **11**, 1169–1186; (g) Padwa, A., Ginn, J.D., Bur, S.K., Eidell, C.K. and Lynch, S.M. (2002) *J. Org. Chem.*, **67**, 3412–3424; (h) Padwa, A., Dimitroff, M. and Liu, B. (2000) *Org. Lett.*, **2**, 3233–3235.

86 Cyclotrimerizations: Anderson, E.A., Alexanian, E.J. and Sorensen, E.J. (2004) *Angew. Chem. Int. Ed.*, **43**, 1998–2001.

87 Recent examples of IMDAF and related cycloadditions: (a) Zubkov, F.I., Ershova, J.D., Orlova, A.A., Zaytsev, V.P., Nikitina, E.V., Peregudov, A.S., Gurbanov, A.V., Borisov, R.S., Khrustalev, V.N., Maharramov, A.M. and Varlamov, A.V. (2009) *Tetrahedron*, **65**, 3789–3803; (b) Nielsen, L.B., Slamet, R. and Wege, D. (2009) *Tetrahedron*, **65**, 4569–4577; (c) Gao, S., Wang, Q. and Chen, C. (2009) *J. Am. Chem. Soc.*, **131**, 1410–1412; (d) Quiroz-Florentino, H., Aguilar, R., Santoyo, B.M., Díaz, F. and Tamariz, J. (2008) *Synlett*, 1023–1028; (e) Sarang, P.S., Yadav, A.A., Patil, P.S., Krishna, U.M., Trivedi, G.K. and Salunkhe, M.M. (2007) *Synthesis*, 1091–1095; (f) Brubaker, J.D. and Myers, A.G. (2007) *Org. Lett.*, **9**, 3523–3525; (g) Chittiboyina, A.G., Kumar, G.M., Carvalho, P.D., Liu, Y., Zhou, Y.D., Nagle, D.G. and Avery, M.A. (2007) *J. Med. Chem.*, **50**, 6299–6302; (h) Li, C.C., Wang, C.H., Liang, B., Zhang, X.H., Deng, L.J., Liang, S., Chen, J.H., Wu, Y.D. and Yang, Z. (2006) *J. Org. Chem.*, **71**, 6892–6897; (i) Merten, J., Hennig, A., Schwab, P., Fröhlich, R., Tokalov, S.V., Gutzeit, H.O. and Metz, P. (2006) *Eur. J. Org. Chem.*, 1144–1161; (j) Parsons, P.J., Board, J., Waters, A.J., Hitchcock, P.B., Wakenhut, F. and Walter, D.S. (2006) *Synlett*, 3243–3246; (k) Li, C.C., Liang, S., Zhang, X.H., Xie, Z.X., Chen, J.H., Wu, Y.D. and Yang, Z. (2005) *Org. Lett.*, **7**, 3709–3712; (l) Baba, Y., Hirukawa, N. and Sodeoka, M. (2005) *Bioorg. Med. Chem.*, **13**, 5164–5170; (m) Pedrosa, R., Sayalero, S., Vicente, M. and Casado, B. (2005) *J. Org. Chem.*, **70**, 7273–7278; (n) Du, X., Chu, H.V. and Kwon, O. (2004) *Tetrahedron Lett.*, **45**, 8843–8846; (o) Du, X., Chu, H.V. and Kwon, O. (2003) *Org. Lett.*, **5**, 1923–1926; (p) Baran, A., Kazaz, C., Seçen, H. and Sütbeyaz, Y. (2003) *Tetrahedron*, **59**, 3643–3648; (q) Fokas, D., Patterson, J.E., Slobodkin, G. and Baldino, C.M. (2003) *Tetrahedron Lett.*, **44**, 5137–5140.

88 For other interesting examples of IMDAF and related cycloadditions: (a) Claeys, S., Van Haver, D., De Clercq, P.J., Milanesio, M. and Viterbo, D. (2002) *Eur. J. Org. Chem.*, 1051–1062; (b) Lautens, M., Colucci, J.T., Hiebert, S., Smith, N.D. and Bouchain, G. (2002) *Org. Lett.*, **4**, 1879–1882; (c) Richter, F., Bauer, M., Pérez, C., Maichle-Mössmer, C. and Maier, M.E. (2002) *J. Org. Chem.*, **67**, 2474–2480; (d) Arjona, O., Menchaca, R. and Plumet, J. (2001) *Org. Lett.*, **3**, 107–109; (e) Arjona, O., Menchaca, R. and Plumet, J. (2001) *J. Org. Chem.*, **66**, 2400–2413; (f) Aceña, J.L., Arjona, O., León, M.L. and Plumet, J. (2000) *Org. Lett.*, **2**, 3683–3686; (g) Trembleau, L., Patiny, L. and Ghosez, L. (2000) *Tetrahedron Lett.*, **41**, 6377–6381; (h) Gilbert, A.M., Miller, R. and Wulff, W.D. (1999) *Tetrahedron*, **55**, 1607–1630; (i) Brickwood, A.C., Drew, M.G.B., Harwood, L.M., Ishikawa, T., Marais, P. and Morisson, V. (1999) *J. Chem. Soc. Perkin Trans. 1*, 913–921; (j) Kurosu, M., Marcin, L.R., Grinsteiner, T.J. and Kishi, Y. (1998) *J. Am. Chem. Soc.*, **120**, 6627–6628; (k) Chen, C.H., Rao, P.D. and Liao, C.C. (1998) *J. Am. Chem. Soc.*, **120**, 13254–13255; (l) Baba, Y., Sakamoto, T., Soejima, S. and

Kanematsu, K. (1994) *Tetrahedron*, **50**, 5645–5658; (m) Kanematsu, K., Soejima, S. and Weng, G. (1991) *Tetrahedron Letters*, **32**, 4761–4764.

89 Oxabicycle ring opening: (a) Lautens, M., Faghou, K. and Hiebert, S. (2003) *Acc. Chem. Res.*, **36**, 48–58; (b) Padwa, A. (2009) *J. Org. Chem.*, **74**, 6421–6441.

90 Retro Diels–Alder: (a) Klunder, A.J.H., Zhu, J. and Zwanenburg, B. (1999) Transient chirality in retro Diels–alder. *Chem. Rev.*, **99**, 1163–1198; (b) Neubert, B.J. and Snider, B.B. (2003) *Org. Lett.*, **5**, 765–768; (c) Bloch, R. and Brillet, C. (1991) *Tetrahedron Asymmetry*, **2**, 797–800; (d) Bloch, R., Bortolussi, M., Girard, C. and Seck, M. (1992) *Tetrahedron*, **48**, 453–462.

91 [4 + 3] cycloadditions: (a) Föhlisch, B., Flogaus, R., Henle, G.H., Sendelbach, S. and Henkel, S. (2006) *Eur. J. Org. Chem.*, 2160–2173; (b) Lee, J.C. and Cha, J.K. (2001) *J. Am. Chem. Soc.*, **123**, 3243–3246; (c) See also:Kim, H. and Hoffmann, H.M.R. (2000) *Eur. J. Org. Chem.*, 2195–2201; (d) Harmata, M. and Kahraman, M. (1998) *Tetrahedron Lett.*, **39**, 3421–3424.

92 [6 + 4] cycloaddition: (a) Rigby, J.H. and Chouraqui, G. (2005) *Synlett*, 2501–2503; (b) [8 + 2] cycloaddition:Zhang, L., Buckingham, C. and Herndon, J.W. (2005) *Org. Lett.*, **7**, 1665–1667; (c) Luo, Y., Herndon, J.W. and Cervantes-Lee, F. (2003) *J. Am. Chem. Soc.*, **125**, 12720–12721.

93 [5 + 2] cycloadditions with pyrilium ions derived from furans: (a) Wender, P.A., Jesudason, C.D., Nakahira, H., Tamura, N., Tebbe, A.L. and Ueno, Y. (1997) *J. Am. Chem. Soc.*, **119**, 12976–12977; (b) Tang, B., Bray, C.D. and Pattenden, G. (2006) *Tetrahedron Lett.*, **47**, 6401–6404; (c) Roethle, P.A. and Trauner, D. (2006) *Org. Lett.*, **8**, 345–347; (d) Roethle, P.A., Hernandez, P.T. and Trauner, D. (2006) *Org. Lett.*, **8**, 5901–5904; (e) See also:Wender, P.A., Rice, K.D. and Schnute, M.E. (1997) *J. Am. Chem. Soc.*, **119**, 7897–7898; (f) Wender, P.A., Bi, F.C., Buschmann, N., Gosselin, F., Kan, C., Kee, J.M. and Ohmura, H. (2006) *Org. Lett.*, **8**, 5373–5376; (g) Magnus, P. and Ollivier, C. (2002) *Tetrahedron Lett.*, **43**, 9605–9609; (h) Magnus, P., Waring, M.J., Ollivier, C. and Lynch, V. (2001) *Tetrahedron Lett.*, **42**, 4947–4950.

94 Radicals: (a) Holzapfel, C.W. and Williams, D.B.G. (1995) *Tetrahedron*, **51**, 8555–8564; (b) Guindeuil, S. and Zard, S.Z. (2006) *Chem. Commun.*, 665–667; (c) Crich, D., Hwang, J.T. and Yuan, H. (1996) *J. Org. Chem.*, **61**, 6189–6198.

95 Dearomatization via ionic reactions: (a) Yin, B.L., Hu, T.S., Yue, H.J., Gao, Y., Wu, W.M. and Wu, Y.L. (2004) *Synlett*, 306–310; (b) Gao, Y., Wu, W.L., Wu, Y.L., Ye, B. and Zhou, R. (1998) *Tetrahedron*, **54**, 12523–12538; (c) Tokumaru, K., Arai, S. and Nishida, A. (2006) *Org. Lett.*, **8**, 27–30.

96 (a) Trost, B.M. and Crawley, M.L. (2002) *J. Am. Chem. Soc.*, **124**, 9328–9329; (b) See also:Meira, P.R.R., Moro, A.V. and Correia, C.R.D. (2007) *Synthesis*, 2279–2286; (c) Trost, B.M. and Kallander, L.S. (1999) *J. Org. Chem.*, **64**, 5427–5435.

97 Ireland-Claisen: (a) Ireland, R.E., Armstrong, J.D., III, Lebreton, J., Meissner, R.S. and Rizzacasa, M.A. (1993) *J. Am. Chem. Soc.*, **115**, 7152–7165; (b) Ireland, R.E., Meissner, R.S. and Rizzacasa, M.A. (1993) *J. Am. Chem. Soc.*, **115**, 7166–7172; (c) See also:Kraus, G.A. and Wei, J. (2004) *J. Nat. Prod.*, **67**, 1039–1040.

98 Ring expansion: Kocienski, P., Brown, R.C.D., Pommier, A., Procter, M. and Schmidt, B. (1998) *J. Chem. Soc. Perkin Trans. 1*, 9–39.

99 Rigidifiers: Felzmann, W., Arion, V.B., Mieusset, J.-L. and Mulzer, J. (2006) *Org. Lett.*, **8**, 3849–3851.

100 (a) Ley, S.V., Brown, D.S., Clase, J.A., Fairbanks, A.J., Lennon, I.C., Osborn, H.M.I., Stokes, E.S.E. and Wadsworth, D.J. (1998) *J. Chem. Soc. Perkin Trans. 1*, 2259–2276; (b) See also:Boons, G.-J., Lennon, I.C., Ley, S.V., Owen, E.S.E., Staunton, J. and Wadsworth, D.J. (1994) *Tetrahedron Lett.*, **35**, 323–326; (c) Hori, K., Kazuno, H., Nomura, K. and Yoshii, E. (1993) *Tetrahedron Lett.*, **34**, 2183–2186.

101 (a) Andersen, N.G. and Keay, B.A. (2001) *Chem. Rev.*, **101**, 997–1030; (b)

Wong, O.A. and Shi, Y. (2008) *Chem. Rev.*, **108**, 3968–3987; (c) See also:Spescha, M. and Rihs, G. (1993) *Helv. Chim. Acta*, **76**, 1219–1230; (d) Tuttle, J.B., Ouellet, S.G. and MacMillan, D.W.C. (2006) *J. Am. Chem. Soc.*, **128**, 12662–12663; (e) Northrup, A.B. and MacMillan, D.W.C. (2002) *J. Am. Chem. Soc.*, **124**, 2458–2460.

102 (a) Tadano, K., Isshiki, Y., Minami, M. and Ogawa, S. (1993) *J. Org. Chem.*, **58**, 6266–6279; (b) Tadano, K., Isshiki, Y., Minami, M. and Ogawa, S. (1992) *Tetrahedron Lett.*, **33**, 7899–7902.

5
Pyran and Its Derivatives

Hideto Miyabe, Okiko Miyata and Takeaki Naito

5.1
Introduction

Simple and unsaturated pyran ring systems seen in 2H-pyran and 4H-pyran are not so stable that their production via biosynthetic processes is not so common. However, six-membered oxygen heterocyclic ring derivatives including pyran compounds are found in an abundant range of naturally occurring compounds. Fusion of additional rings to the basic pyran skeleton results in a substantially more stable heterocyclic system. Therefore, condensed pyrans are widely distributed in nature and have quite an extensive chemistry, both in synthesis and applications. The abundance of pyran ring-containing heterocycles in bioactive natural products continues to encourage the development of novel and improved syntheses. A large number of new references dedicated to the synthesis of pyrans and their derivatives have been found, however, many traditional approaches are still of great value.

The syntheses of six-membered oxygen heterocycles and their benzo derivatives are subject to annual review [1]. Numerous total syntheses of complex pyran-containing natural products are found in the literature; specific examples include syntheses of actin-binding marine macrolides [2], total synthesis of oxacyclic macrodiolide natural products [3], total synthesis of marine polycyclic ethers [4, 5], advances in the total synthesis of biologically important marine macrolides [6], and convergent strategies for syntheses of *trans*-fused polycyclic ethers [7, 8].

This chapter covers selected works, mainly on the total syntheses of pyran natural products, published in the past three years. Additionally, the text of this chapter focuses on synthetic methodology employed for the construction of pyran rings, excluding many examples of synthetic works which employed very interesting reactions but under conventional reaction conditions, in their elegant total and formal syntheses of pyran natural products. Many natural pyran derivatives with different oxidation levels are known, including glycoside, δ-lactone, pyrone, chroman, coumarin, flavone, and others which are excluded from this chapter because of the page limit.

In Table 5.1 relatively small-sized compounds are incorporated by describing their chemical structure, isolated source, isolation, biological activity, and synthesis

Heterocycles in Natural Product Synthesis, First Edition. Edited by Krishna C. Majumdar
and Shital K. Chattopadhyay.
© 2011 Wiley-VCH Verlag GmbH & Co. KGaA. Published 2011 by Wiley-VCH Verlag GmbH & Co. KGaA.

Table 5.1 Pyran-based natural products.

Serial No.	Trivial name	Structure	Source	Isolation [Ref]	Biological activity	Synthesis [Ref]
1	Aspergillide B, Aspergillide C		*Aspergillus ostianus* (marine-derived fungus)	[9]	Cytotoxicity [9]	[10, 11]
2	Cardinalin 3		*Dermocybe cardinalis* (New Zealand toadstool), *Ventilago goughii*	[12, 13]	Cytotoxicity [12]	[14]
3	Centrolobine		*Centrolobium robustum* (porcupine tree), Stem of *Brosinium potabile*	[14–17]	Activity against *Leishmania amazonensis promastigotes* [14–17]	[18–25]
4	Diospongin B		Rhizomes of *Dioscorea spongiosa*	[26]	Inhibitory activities on bone resorption [26]	[27, 28]

5	Dysiherbaine, Neodysiherbaine A	*Dysidea herbacea* (Micronesian sponge)	[29, 30]	Convulsant activity [31]; Agonist for ionotropic glutamate receptors [32, 33]	[34–42]
6	FR901464	Culture broth of a bacterium of *Pseudomonas* sp. No. 2663	[43–45]	Antitumor activity [43, 44]	[46–51]
7	GEX1A (Herboxidiene, TAN-1609)	*Streptomyces* sp. A7847	[52, 53]	Herbicidal activity [52]; Antitumor activity [54]; Activation of the LDL receptor [55]	[56–59]
8	Hongoquercin A	Unidentified terrestrial fungus	[60]	Antibacterial property [60, 61]	[61, 62]

(Continued)

Table 5.1 (Continued)

Serial No.	Trivial name	Structure	Source	Isolation [Ref]	Biological activity	Synthesis [Ref]
9	Jerangolid D		Secondary metabolite produced by myxobacterium *Sorangium cellulosum*	[63]	Antifungal property [63]	[64]
10	Jimenezin		Seeds of *Rollinia mucosa*	[65]	Cytotoxic activity [65]	[66–68]
11	Lasonolide A		*Forcepia* sp. (Caribbean marine sponge)	[69]	Antitumor activity [69]	[70–75]
12	Methyl sarcophytoate		*Sarcophyton glaucum* (Okinawan soft coral)	[76, 77]	Cytotoxic activity [76]	[78, 79]

13	Napyradiomycin A1	Strains of *Chainia rubra* and *Streptomyces* sp	[80, 81]	Antibacterial activity [80]; Nonsteroidal estrogen antagonists [81]	[82, 83]
14	Spirastrellolide A	*Spirastrella coccinea* (marine sponge)	[84, 85]	Protein phosphatase 2A inhibitor [84, 85]	[86, 87]
15	Theopederin D	*Theonella* genus (marine sponge)	[88]	Cytotoxic activity [88]	[89, 90]
16	Zincophorin	Cultured strains of *Streptomyces griseus*	[91, 92]	Antiviral activity [93]	[94–99]

with their references cited. In the text, sections are arranged in the order of the substituted pattern on the pyran skeleton, commencing from (i) 2,6-disubstituted pyran natural products; followed by (ii) 2,6-cyclic pyran compounds; (iii) complexly fused pyran compounds with additional rings; (iv) fused pyran compounds with aromatic rings; and (v) fused pyran compounds with aliphatic rings.

5.2
Application of Pyran Moieties in the Synthesis of Natural Products

5.2.1
2,6-Disubstituted Pyran Natural Products

The first total synthesis of the marine metabolite bistramide A was achieved by Kozmin's group [100]. The structurally related compound was synthesized by Wipf's group [101]. In 2006, Crimmins' group also completed the total synthesis of bistramide A via the diastereoselective formation of tetrahydropyran ring by the reaction of acetal **1** and 3-penten-2-one in the presence of TMSOTf (Scheme 5.1) [102]. For the preparation of the tetrahydropyran fragment, [4 + 2] annulation of crotylsilane **3** with aldehydes was developed by Panek's group [103a]. In their total synthesis of bistramide A, the spiroketal fragment was also synthesized by [4 + 2] annulation strategy [103b, c].

Pederin displays nanomolar toxicity against a broad range of cancer cell lines. Therefore, the chemical synthesis of pederin has been pursued by many groups. In 2007, Rawal's group reported the concise total synthesis of this natural product (Scheme 5.2) [104]. In this study, the pyranone **7** was formed by diastereoselective Mukaiyama–Michael reaction of **5** and **6**. The conjugated addition of silyl ketene

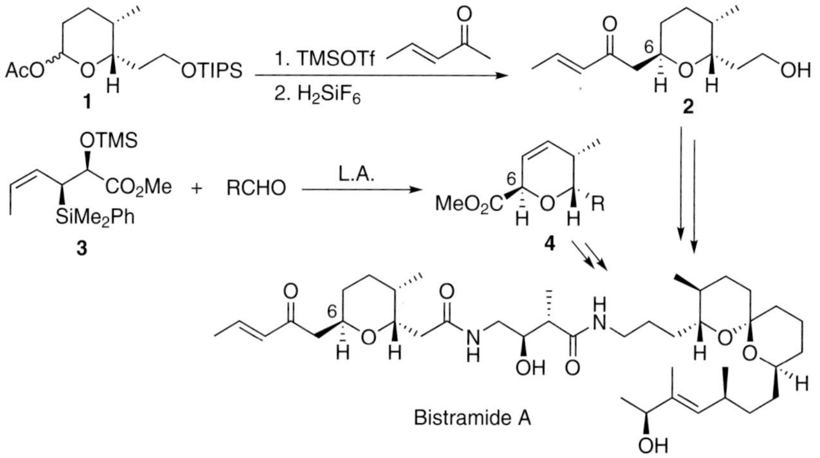

Scheme 5.1

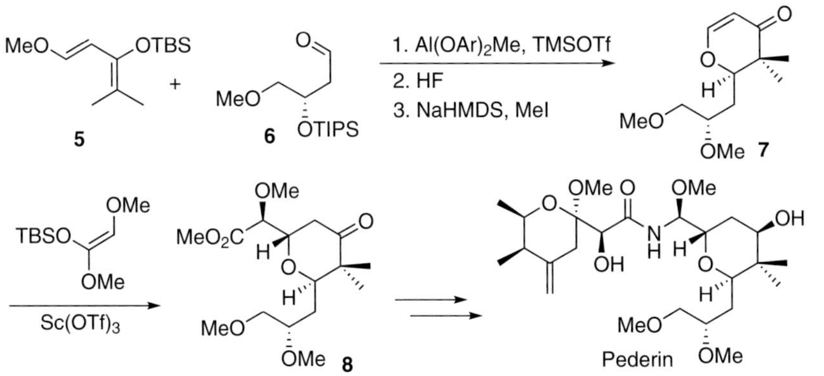

5.2 Application of Pyran Moieties in the Synthesis of Natural Products | 159

Scheme 5.2

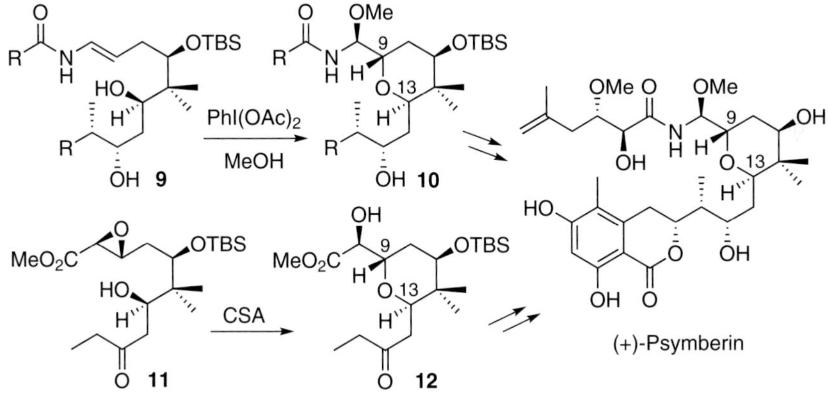

Scheme 5.3

acetal onto pyranone **7** produced ester **8** in excellent diastereoselectivity, which was converted into the final natural product.

Total synthesis of the marine sponge cytotoxin psymberin, a new member of the pederin family, has drawn much attention from the synthetic chemistry community. The first total synthesis was reported by De Brabander's group [105]. A second total synthesis of this natural product was accomplished by Huang's group (Scheme 5.3) [106]. The tetrahydropyran ring was effectively constructed by PhI(OAc)$_2$-mediated oxidative cyclization of enamide **9**. Total synthesis of (+)-psymberin was also achieved by Smith III's group via the significant formation of a tetrahydropyran ring, by cyclization of epoxide **11** using camphor sulfonic acid [107]. Williams' group reported the formal synthesis of psymberin via preparation of the tetrahydropyran ring using oxidation of allene [108].

The synthesis of pyranicin is the subject of current interest, because of its impressive biological activity against a number of cancer cell lines. First, Takahashi and Nakata synthesized pyranicin via SmI$_2$-induced reductive cyclization of acrylate [109]. Second total synthesis was achieved by Rein's group via an asymmetric Horner–Emmons reaction [110]. In 2008, Phillips reported concise total synthesis based on the Sharpless asymmetric kinetic resolution of furfuryl alcohol **13** [111]. The resulting intermediate hemiacetal **14** was immediately reduced with i-Pr$_3$SiH to give the pyran derivative **15** (Scheme 5.4). Total synthesis was also achieved by Makabe's group. A key feature of this synthesis is the use of palladium-catalyzed diastereoselective cyclization of the allylic ester **16** [112].

A stepwise [3 + 3] annulation reaction leading to functionalized pyrans was successfully applied to the total synthesis of (+)-rhopaloic acid B having a 2,5-disubstituted pyran moiety by Harrity's group (Scheme 5.5) [113]. This process involves the reaction of epoxide with allylmagnesium reagent **19**, prepared from allyl alchohol **18**, and the palladium-catalyzed cyclodehydration reaction of **20**. The structurally related rhopaloic acid A has also been synthesized by several groups [114].

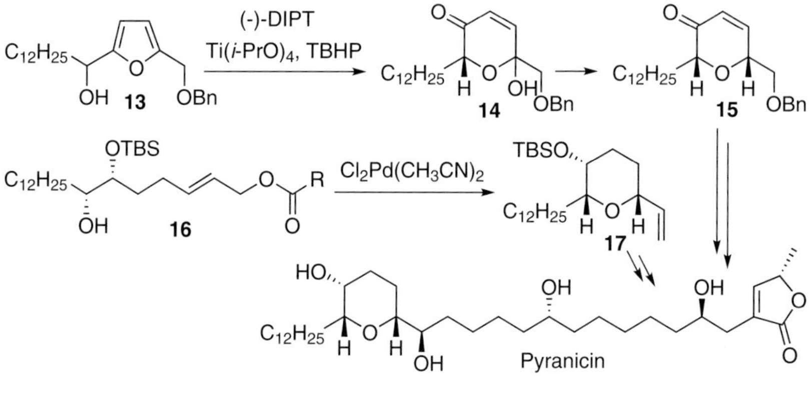

Scheme 5.4

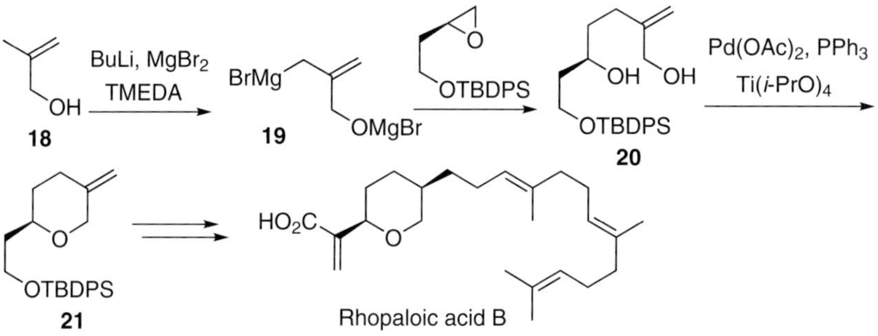

Scheme 5.5

5.2.2
2,6-Cyclic Pyran Compounds

The synthesis of bryostatins has been challenging chemists. Until recently, only three total syntheses had been accomplished [115], although many synthetic studies have been reported, including synthesis of bryostatin analogs. In 2008, Trost's group reported an elegant total synthesis of bryostatin 16 by a rare palladium-catalyzed alkyne-ynoate coupling macrocyclization (Scheme 5.6) [116].

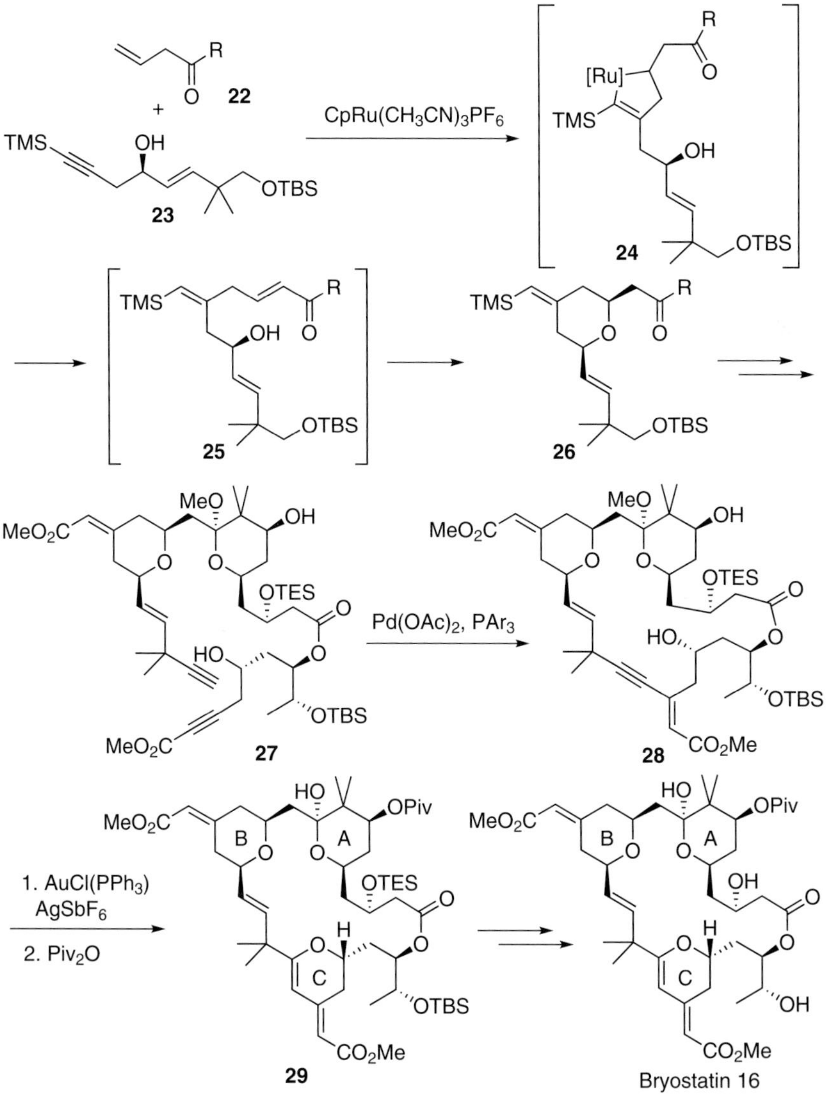

Scheme 5.6

The tetrahydropyran ring B fragment **26** was prepared by a ruthenium-catalyzed coupling reaction of alkene **22** with enantiomerically enriched alkyne **23** (90% ee) via ruthenacycle **24** and hydroxyenone **25**. The alkyne-ynoate coupling reaction of **27** was developed as a key macrocyclization step. Treatment of **27** with Pd(OAc)$_2$ and phosphine ligand successfully provided the macrocycle **28**. The tetrahydropyran ring C was prepared by cationic gold-catalyzed 6-*endo* cyclization of **28**. The desired 6-*endo* product **29** was obtained after the reaction of the hindered secondary alcohol moiety with Piv$_2$O. Finally, global deprotection led to total synthesis of the final natural product.

Kendomycin is a 16-membered conformationally restricted macrocycle comprising a functionalized pyran ring and unique quinine methide chromophore (Scheme 5.7). The first total synthesis of kendomycin was accomplished by Lee's group via the significant formation of macrocycle using a macro-C-glycosidation approach [117]. The macroglycosidation of phenol **30** occurred smoothly through

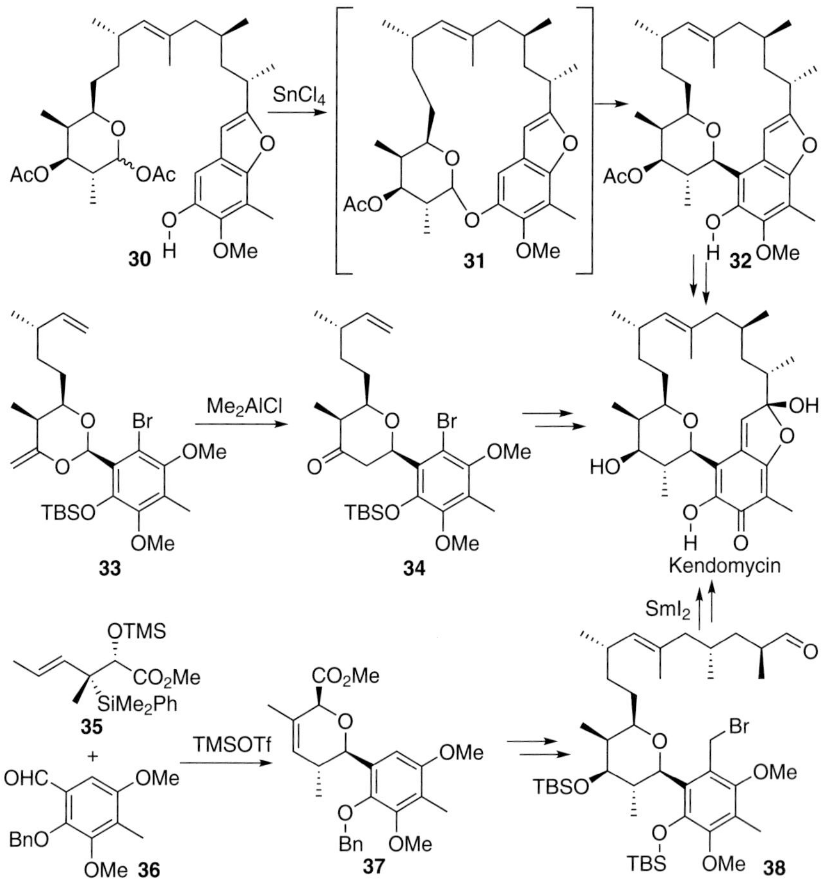

Scheme 5.7

5.2 Application of Pyran Moieties in the Synthesis of Natural Products | 163

facile formation of O-glycoside **31** and subsequent rearrangement to C-glycoside **32**. Smith III's group reported the total synthesis via a ring-closing metathesis to generate the macrocycle [118]. In this synthesis, the sterically encumbered pyran ring fragment **34** was constructed by the Petasis–Ferrier rearrangement of the unstable enol-acetal **33** which was prepared by Petasis–Tebbe methylidenation. Total synthesis was also reported by Panek's group [119]. The pyran ring was constructed by [4 + 2] annulation reaction of crotylsilane **35** and aldehyde **36**. As the significant formation of macrocycle, SmI$_2$-assisted cyclization of **38** was employed. Formal synthesis was achieved by using a Prins cyclization [120].

The synthesis of leucascandrolide A has generated substantial interest, resulting in several total and formal syntheses [121]. The first total synthesis was reported by Leighton's group [122]. In 2007, Floreancig's group reported a concise formal synthesis using electron-transfer-initiated cyclization (ETIC) method (Scheme 5.8) [123]. The hydroformylation of alcohol **39** gave lactol **40** which led to **41**. The key ETIC reaction of **41** progressed quite smoothly by treatment with ceric ammonium nitrate (CAN) to form tetrahydropyranone **43** as a single stereoisomer. This reaction proceeded through oxidative cleavage of the benzylic C–C bond to form oxocarbenium ion **42**. Evans' group investigated the one-pot diastereoselective

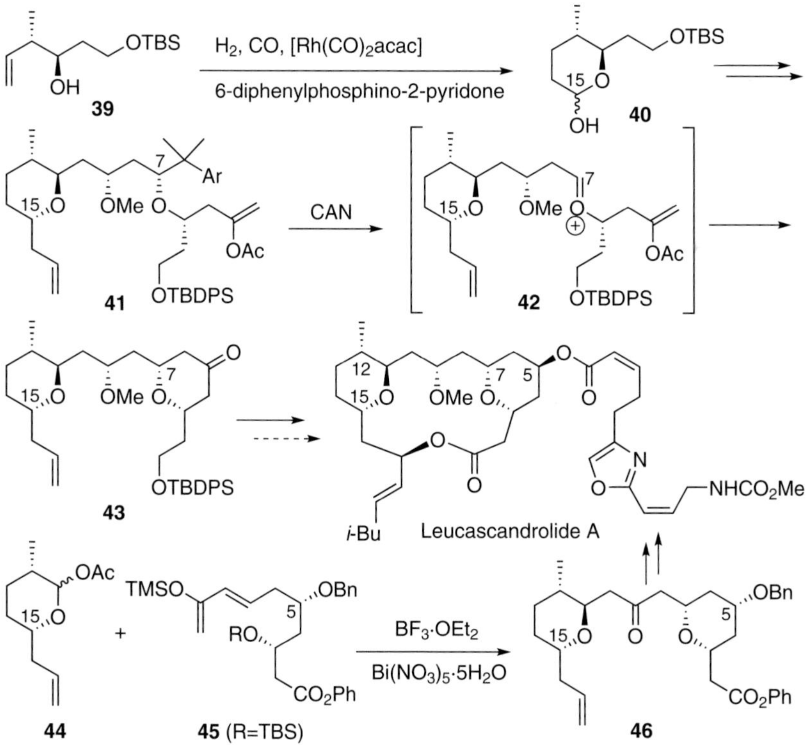

Scheme 5.8

sequential two-component etherification/oxa-conjugate addition reaction for the construction of the non-adjacent bis(tetrahydropyran) core of the natural product [124]. The reaction of acetate **44** with the diene **45** afforded the bis(tetrahydropyran) core **46** which led to total synthesis of the final natural product.

The cytotoxic natural product neopeltolide has recently stimulated a flurry of synthetic interest. The stereochemistry of neopeltolide was revised from the original structural assignment by two total syntheses (Scheme 5.9) [125, 126]. The first total synthesis and stereochemical reassignment were reported by Panek's group [125]. In their synthesis, the triflic acid-promoted [4 + 2] annulation of allylsilane

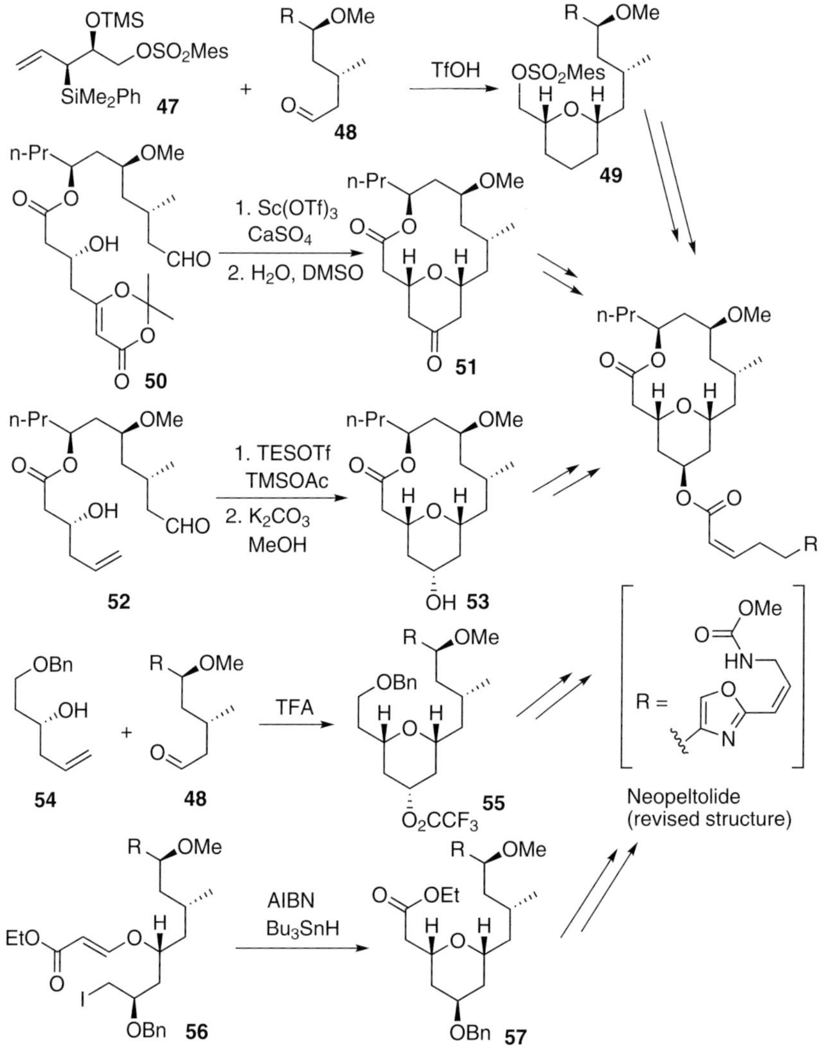

Scheme 5.9

47 and aldehyde 48 afforded functionalized pyran 49 with good diastereoselectivity. Scheidt's group investigated the Lewis acid-catalyzed macrocyclization, which was successfully applied to the total synthesis [126]. In the key step, scandium(III) triflate promoted macrocyclization of 50 to produce cyclic product 51 after heating in wet DMSO. Lee's group reported the total synthesis by macrocyclization using a Prins cyclization strategy [127]. The Prins cyclization of homoallyl alcohol 52 proceeded efficiently and hydrolysis of the Prins product gave macrolide 53. In this reaction, macrolactone 53 was constructed with complete stereocontrol at the two new stereogenic centers. Maier's group also reported the synthesis using a TFA-mediated Prins cyclization between homoallyl alcohol 54 and aldehyde 48 [128]. The radical reaction was successfully applied to the total synthesis by Taylor's group [129]. This synthesis was highlighted by a radical cyclization of 56 to establish the requisite stereochemistry of the tetrahydropyran core. The product 57 was obtained as nearly a single diastereomer. Other reports on total synthesis of neopeltolide have appeared to date [130].

5.2.3
Complex Pyran Natural Products

(+)-Cortistatin A is a novel steroidal alkaloid boasts a heptacyclic skeleton featuring an oxabicyclo[3.2.1]octene and some an isoquinoline structural motif. (+)-Cortistatin A was totally synthesized by Nicolaou's group who employed two key reactions involving intramolecular 1,4-addition/aldol/dehydration sequence for the construction of oxabicyclo[3.2.1]octene unit and Sonogashira coupling of the isoquinoline part with the pentacyclic part (Scheme 5.10) [131]. For the formation of pyran ring via cascade reaction, the precursor 58 was prepared. Upon heating hydroxy enone-enal 58 at reflux in dioxane in the presence of K_2CO_3, a 1,4-hydroxy enone addition/aldol/dehydration cascade proceeded smoothly to afford pentacyclic dienone 60. Finally, 60 was coupled with the isoquinoline part via Sonogashira coupling and other functionalization to afford the natural product. Related interesting studies on the construction of the oxabicyclo[3.2.1]octene unit have been

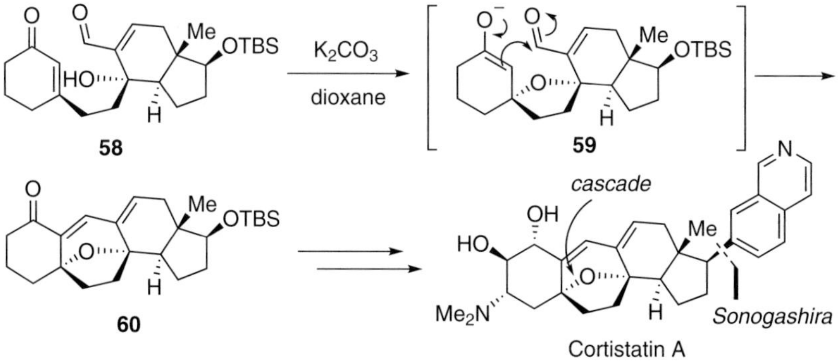

Scheme 5.10

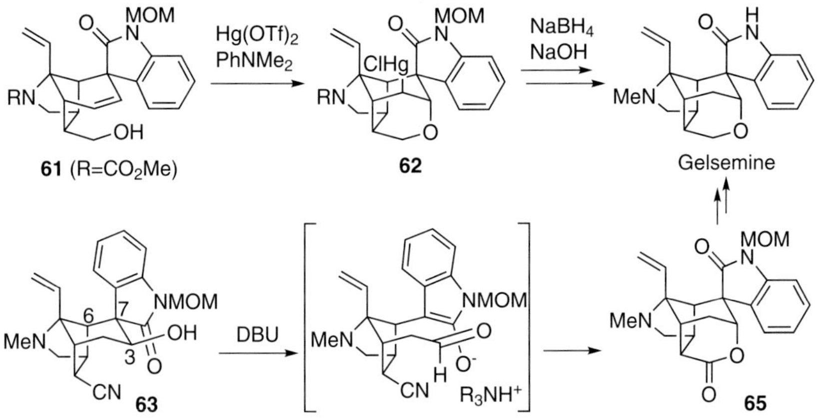

Scheme 5.11

reported [132, 133] and an interesting review covered total synthesis versus semisynthesis [134].

Many total syntheses of gelsemine with highly complex structure have been published since 1994 and recently two groups accomplished the total synthesis by respective characteristic strategies (Scheme 5.11) [135, 136]. Particularly, construction of the pyran ring of the natural product was established independently by different ways. Danishefsky's group employed mercuric cyclization procedure [135]. Treatment of alcohol **61** with Hg(OTf)$_2$ N,N-dimethylaniline complex afforded pyran ring **62** from which the mercury part was removed by reduction with NaBH$_4$ under basic conditions to afford the natural product. Overman's group employed an interesting reaction of hydroxyl nitrile **63** which was simply heated with DBU to afford hexacyclic lactone **65** via three steps involving epimerization of the spirooxindole stereocenter C_7, epimerization of the C_3 alcohol, and intramolecular condensation of the C_3 alcohol with the *endo* cyanide [136]. The lactone **65** was derived into pyran ring of the final natural product via two-step reductions. A tricyclic core structure related to gelsemine was prepared by a very interesting three-step bridge swapping strategy involving the bridging ether oxygen and intramolecular Michael addition of a tethered cyanoactamide unit [137].

The first new class of antibiotic discovered in more than four decades, platensimycin, has an intriguing structure which features a hydrophilic aromatic unit and a lipophilic tetracyclic unit, linked together by an amide bond. It was totally synthesized by several groups including synthesis of the analog [138–149]. The first synthesis of racemic platensimycin was completed by the Nicolaou's group about four months after its isolation (Scheme 5.12) [138]. In his synthesis, tetracyclic core **67** was constructed by ketyl radical cyclization of the precursor **66** with SmI$_2$. This racemic synthesis of **67** was later upgraded to an asymmetric one. In the total synthesis of platensimycin, much attention was focused by synthetic chemists on the stereoselective preparation of the tetracyclic core **67**. Yamamoto's group published a very elegant enantioselective synthesis of **67** which is based

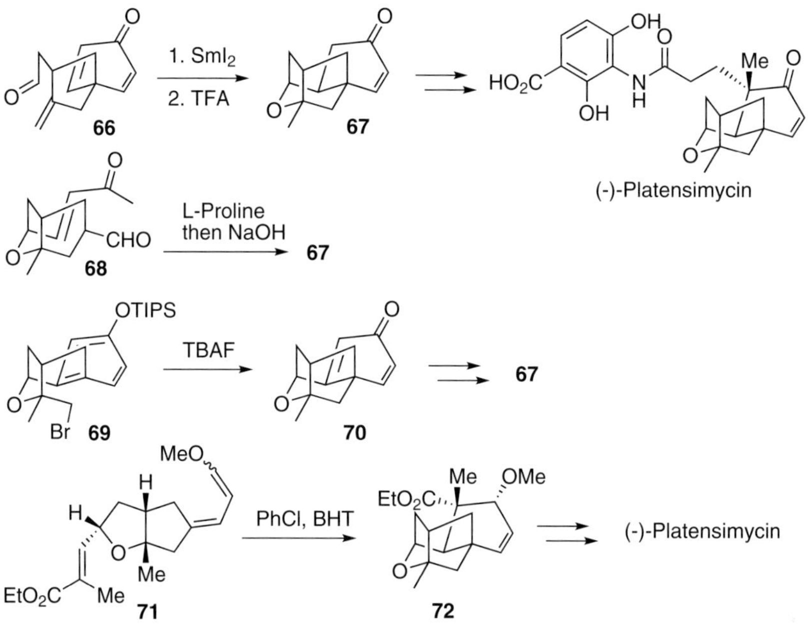

5.2 Application of Pyran Moieties in the Synthesis of Natural Products

Scheme 5.12

on Robinson annulation, to construct the quaternary center [141]. Treatment of formyl ketone **68** with L-proline followed by NaOH gave desired **67** via organocatalyst-mediated Michael addition and aldol condensation. In the enantioselective synthesis of **67** by Corey's group, **69** was converted into the tetracyclic structure **70** which is the precursor to **67** by treatment with TBAF [143]. Ghosh's group achieved total synthesis of (−)-platensimycin via strategy for the preparation of the oxatetracyclic core involving an intramolecular Diels–Alder reaction which constructed three chiral centers including two all-carbon quaternary chiral centers [147]. Natural (+)-carvone was converted into the precursor **71** for tetracyclic core carrying pyran ring. Triene **71** was subjected to a thermal Diels–Alder reaction at 270 °C in a sealed tube to afford the desired oxatetracyclic core **72** which was coupled with the anilide part to furnish the natural product.

The fascinating structural motif of the phomactin family has attracted a number of synthetic efforts, including Yamada's pioneering synthesis of a phomactin member [150] and elegant total syntheses of phomactin A reported by Pattenden [151] and Halcomb [152]. Recently, a total synthesis of racemic phomactin A was reported which highlights the final completion of a complex natural product target that had commenced with an intramolecular oxa [3+3] annulation strategy [153] in the construction of the ABC-tricycle (Scheme 5.13) [154]. The precursor **74** was formed *in situ* from the corresponding aldehyde **73**. Treatment of **74** with either piperidinium acetate, or with firstly TMS$_2$NH and then piperidine and Ac$_2$O, gave ABD tricycle **75**, which was converted into the final natural product.

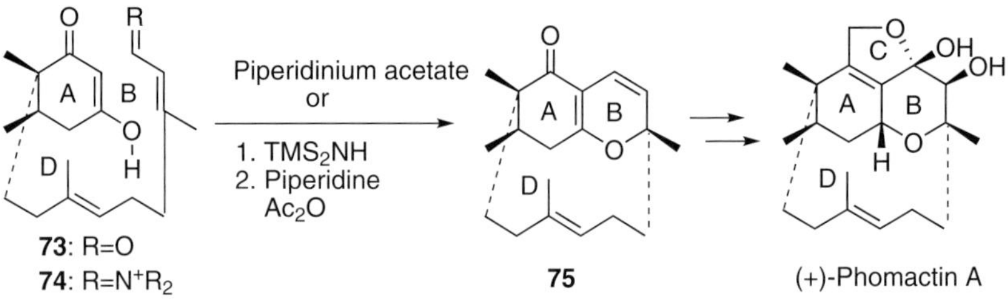

168 | *5 Pyran and Its Derivatives*

Scheme 5.13

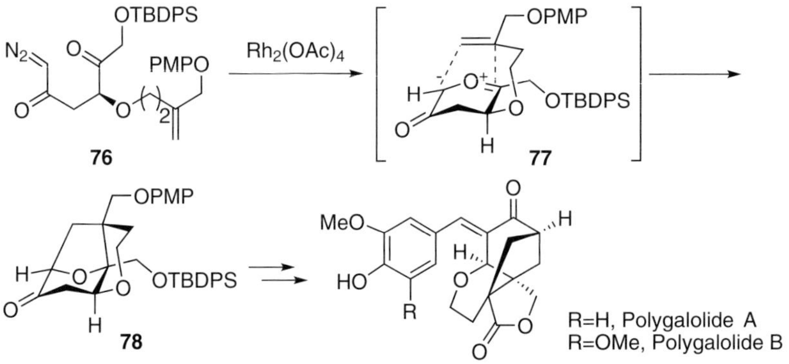

Scheme 5.14

The first total synthesis of (−)-polygalolides A and B was achieved by Hashimoto's group (Scheme 5.14) [155]. The critical tandem carbonyl ylide formation /1,3-dipolar cycloaddition of **76** proceeded efficiently in the presence of Rh$_2$(OAc)$_4$ in the manner as shown in **77**. Finally condensation with aryl aldehyde and functionalization completed the total synthesis of natural products. Later, a short, formal, and biomimetic synthesis of two natural products via stereospecific and regiospecific [5 + 2] cycloaddition reaction was reported by the same group collaborating with Snider's group [156].

5.2.4
Fused Pyran Compounds with Aromatic Rings

Following a previous model study on efficient biomimetic routes to the tetracyclic core of natural product [157], (−)-berkelic acid was totally synthesized for the first time via oxa-Pictet–Spengler reaction and the relative configuration at C18 and C19 and the absolute stereochemistry were reassigned (Scheme 5.15) [158]. Ketal aldehyde **79** and chiral alcohol **80** were condensed by treatment with Dowex 50QWX8-400-H$^+$ to give the tetracyclic core **81** after allylation. Functionalization of the side chain completed the total synthesis of natural product, (−)-berkelic acid and its 22-epimer. As this work was completed, Fürstner's group reported a syn-

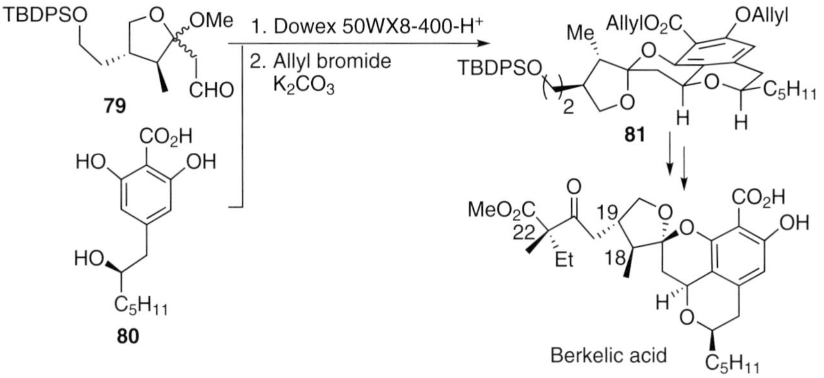

Scheme 5.15

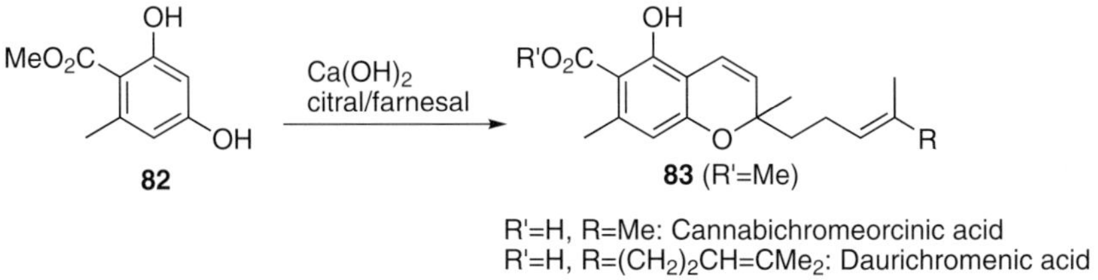

Scheme 5.16

thesis of the enantiomer of the methyl ester of berkelic acid and reassigned the stereochemistry at C18 and C19 [159]. The synthesis not only resolved the open discussion but also suggested that the original assignment needs to be revised.

The complex bioactive natural and non-natural benzopyran congeners were synthesized using one- or two-step approaches in very good yields from the reactions of resorcyclic acid derivatives with citral and/or farnesal [160]. Treatment of a natural product **82** with citral/farnesal in the presence of Ca(OH)$_2$ gave the desired benzopyrans **83** (Scheme 5.16). The observed regioselectivity could be the result of a complexation of Ca^{2+} ion with both the phenolic groups, thus activating the 3-position of **82** for the condensation reaction. Saponification of **83** gave cannabichromeorcinic acid and daurichromenic acid.

Three members of the cannabinoid class, cannabinol, were synthesized using a microwave-mediated [2 + 2 + 2] cyclotrimerization reaction [161] as the key step [162]. The diyne **84** was subjected to Ru-catalyzed [2 + 2 + 2] cyclotrimerization reaction under microwave irradiation to deliver the pyran **85** as a single regioisomer (Scheme 5.17). A few steps involving oxidation of B-ring in **85** and introduction of methyl group, and deprotection completed total synthesis of natural products including B-ring opened cannabinodiol.

Fused pyran-γ-lactones are common structural motifs among many classes of natural products, such as a family of potent pyranonaphthoquinone antibiotics

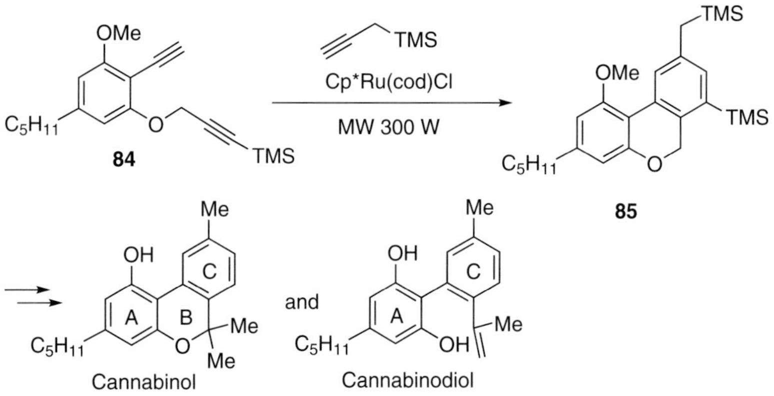

Scheme 5.17

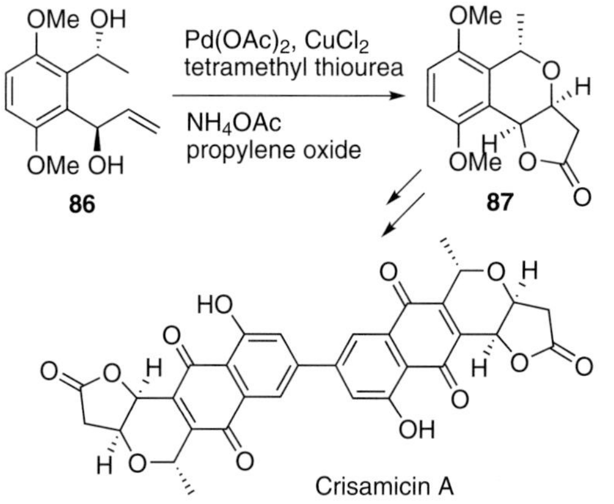

Scheme 5.18

such as crisamicin A and frenolicin B. A diversity-oriented synthesis of a range of fused pyran-γ-lactones was reported through a versatile Pd-thiourea complex-catalyzed intramolecular alkoxycarbonylative annulation which was successively applied to the first total synthesis of crisamicin A (Scheme 5.18) [163]. The key precursor **86** was annulated under optimized conditions involving 10% Pd(OAc)$_2$/tetramethyl thiourea, CuCl$_2$, propylene oxide, and NH$_4$OAc to give the key intermediate **87** which was finally derived to crisamicin A via Pd-catalyzed homocoupling.

Proposing that calyxin F regioisomer, enone **90**, is a possible structure of epicalyxin F, Rychnovsky et al., synthesized enone **90** in a highly convergent strategy from three simple pieces (Scheme 5.19) [164]. Several catalysts were evaluated to construct pyran ring **89** and eventually the use of TMSOTf gave the benzopyran **89** as a separable 5 : 1 mixture of diastereomers. **89** was converted into alcohol **90**,

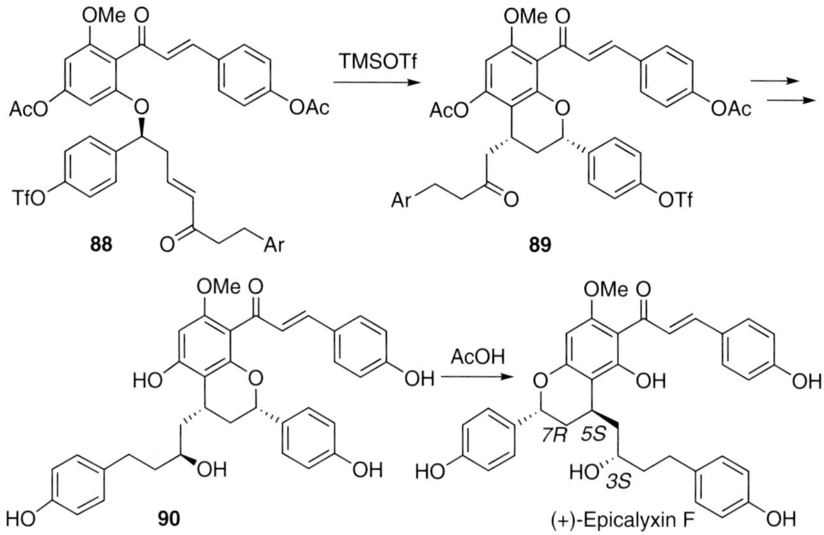

Scheme 5.19

which however was not identical with epicalyxin F which was shown to rearrange to epicalyxin F on treatment with acid.

5.2.5
Fused Pyran Compounds with Aliphatic Rings

Rubioncolin B was totally synthesized via an intramolecular Diels–Alder reaction involving an *ortho*-quinone methide **92** as the diene and naphthofuran as the dienophile (Scheme 5.20) [165]. Exposure of precursor **91** to 2 equiv of TASF in the presence of PhI(OAc)$_2$ directly provided rubioncolin B methyl ether. Probably, this oxidation/tautomerization/Diels–Alder cascade begins by TASF-mediated desilylation providing a phenoxide followed by oxidation to oxonium ion, and re-desilylation leading to the formation of *ortho*-quinone methide tautomer **92** which cyclized, as required, to the methyl ether of natural product via the *endo* transition state.

Two types of Prins cyclization were successfully applied to the total synthesis of (−)-blepharocalyxin D, a cytotoxic dimeric diarylheptanoid (Scheme 5.21) [166]. Stereoselective Prins cyclization of alcohol **93** with *p*-anisaldehyde for constructing the pyran ring was conducted by modifying Willis' conditions [167]. A second Prins reaction of methylidene-substiututed 7-membered cyclic acetals **95** is unprecedented but proceeded effectively in the presence of Lewis acid to afford axial aldehyde **96** which led to total synthesis of the final natural product.

The halichondrin family with impressive levels of cytotoxicity has attracted significant attention, highlighted by the total syntheses of halichondrin B and norhalichondrin B by Kishi's group [168]. In 2009, Phillips' group reported the total

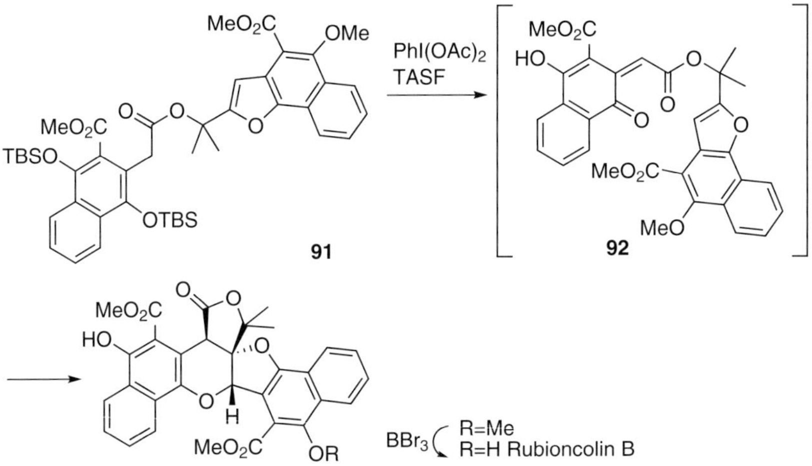

Scheme 5.20

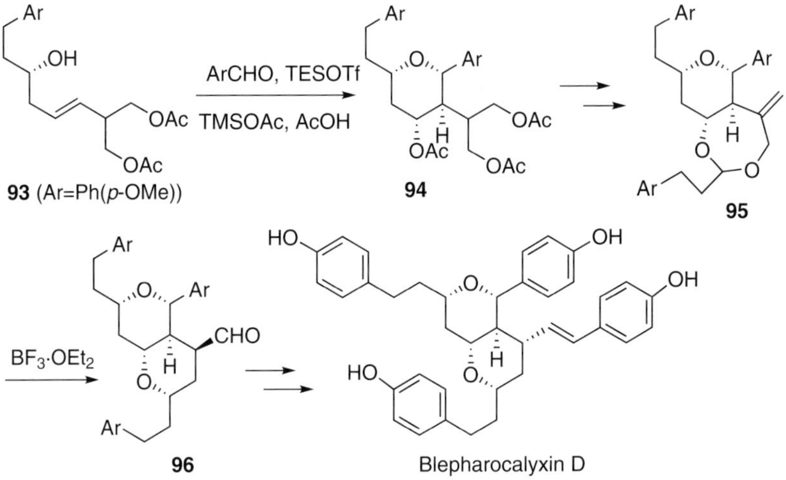

Scheme 5.21

synthesis of norhalichondrin B (Scheme 5.22) [169]. A key feature of this total synthesis is the construction of C40–C53 domain **99** and C27–C38 domain **102** by the conversion of furans into pyranones. The Achmatowicz oxidation of **97** produced an intermediate pyranone hemiacetal, which was immediately subjected to ionic hydrogenation using Et_3SiH to give the pyranone **98** as a single diastereomer. Next, the pyranone **98** was successfully led to C40–C53 domain **99**. The C27–C38 domain **102** was also prepared via a similar two-step protocol.

An enantioselective total synthesis of pseudodehydrothyrsiferol was accomplished via the significant sequence formation of the highly strained tetrahydro-

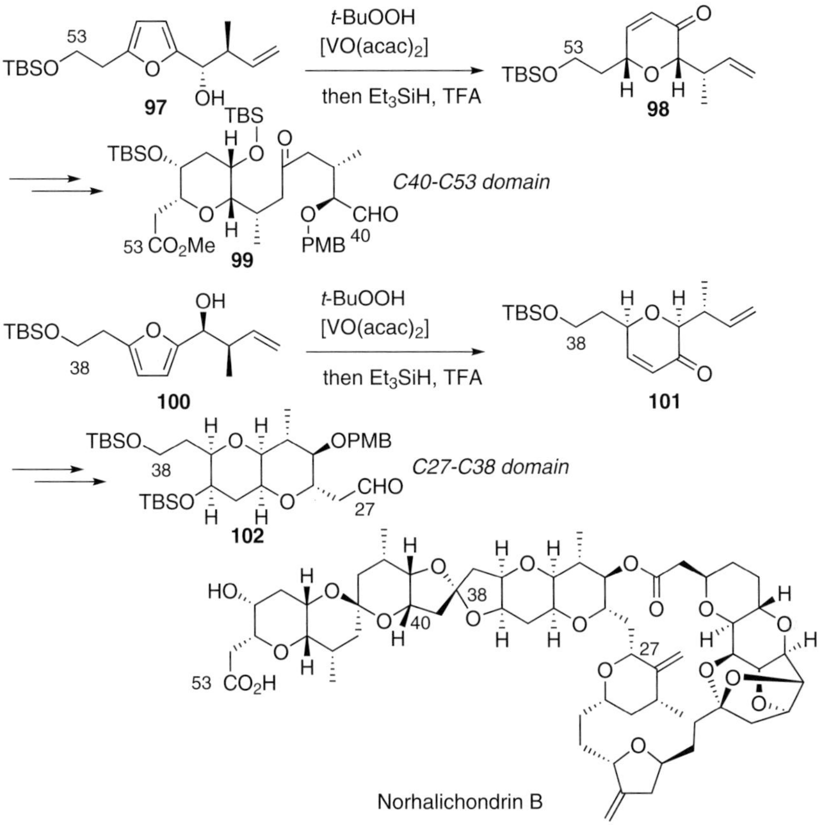

5.2 Application of Pyran Moieties in the Synthesis of Natural Products | 173

Scheme 5.22

pyran by a Mitsunobu-type Sn2 reaction (Scheme 5.23) [170]. The left-hand pyran ring was constructed by 6-*endo* cyclization of known allylic epoxide **103** under acidic conditions, which is favored over 5-*exo* mode because the terminal group assisted cleavage of next C–O bond. The second pyran ring was provided by a Mitsunobu-type Sn2 reaction of fully functionalized glycol **105** with CMMP, the most reactive Tsunoda reagent [171] for the total synthesis of natural pseudodehydrothyrsiferol.

(−)-FR182877 which is a unique hexacyclic structure with twelve contiguous stereogenic centers and also known to exhibit a potent cytotoxic activity comparable to that of taxol was totally synthesized by four groups [172, 173]. Sorensen's [172a] and Evans' [173b] groups employed closely related strategies, involving consecutive transannular cycloadditions. Nakata's group developed tandem intramolecular Diels–Alder (IMDA)-intramolecular hetero-Diels–Alder (IMHDA) reactions of polyene **106** for the stereoselective preparation of the AB-ring and the CD-ring moieties of (−)-FR182877 (Scheme 5.24) [173]. The first IMDA reaction produced a mixture of diastereomers, but the undesired product did not undergo

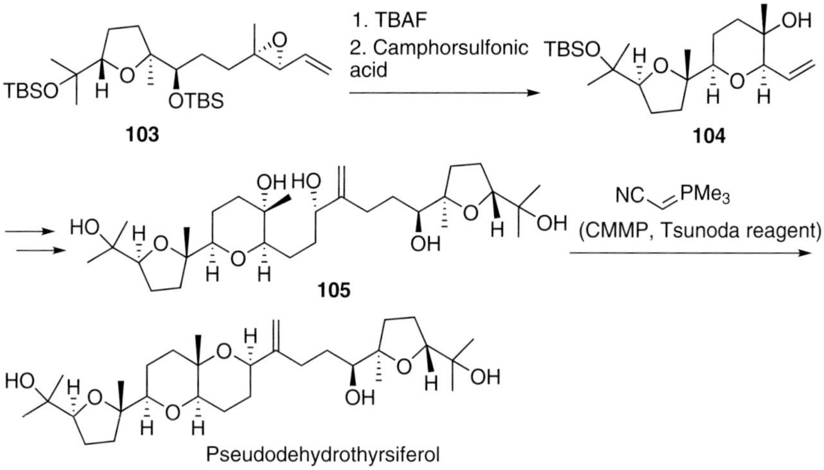

103

104

105

Pseudodehydrothyrsiferol

Scheme 5.23

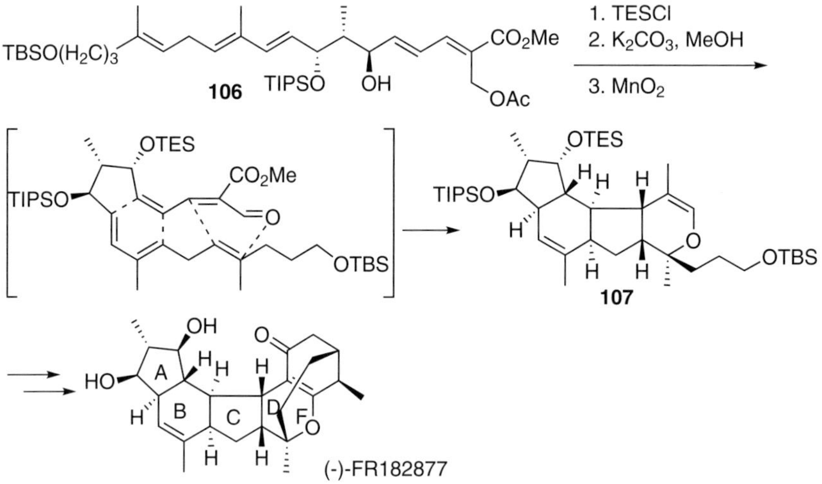

106

107

(-)-FR182877

Scheme 5.24

the subsequent IMHDA reaction. It is speculated that the reason for the trend in stereoselectivity is the role played by the stereoelectronic effect induced by the allylic substituent.

Conformational locking through allylic strain as a device for stereocontrol was successfully applied to the total synthesis of grandisine A as a small molecule natural product binding to opioid receptors [174]. Lewis acid-catalyzed diene-aldehyde cyclocondensation of diene **108** with acetaldehyde occurred efficiently to afford exclusively *endo*-product **109** which was derived to chiral hydroxyketone **110** via optical resolution of the intermediate (Scheme 5.25). Hydroxyketone **110** was

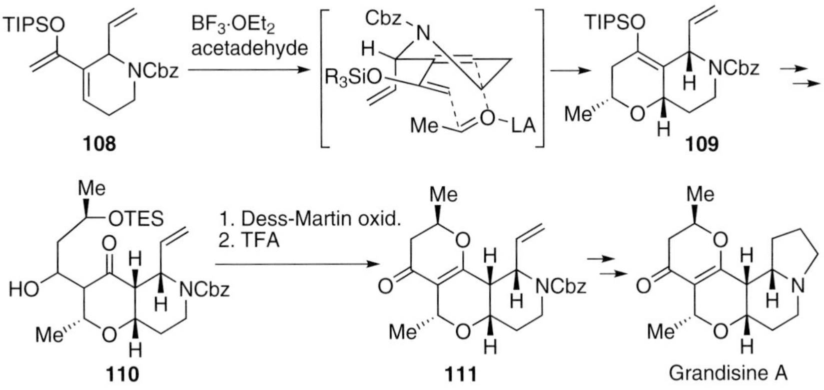

Scheme 5.25

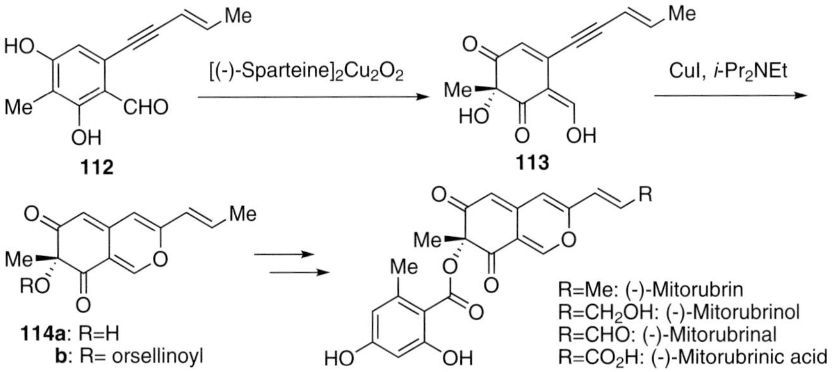

Scheme 5.26

oxidized and subsequent acid-catalyzed deprotection of the silyl group led to the formation of second pyran ring **111** which was converted into the final natural product.

After a synthesis of racemic mitorubrinic acid reported by Pettus' group [175], asymmetric syntheses of (−)-mitorubin and related azaphilone natural products were reported [176]. Though Pettus' group [175] employed conventional method for the construction of bicyclic pyran part from the corresponding lactone, key steps involve copper-mediated, enantioselective oxidative dearomatization to prepare the azaphilone core and olefin cross-metathesis for side-chain installation in asymmetric synthesis by Porco's group (Scheme 5.26) [176]. o-Alkynylbenzaldehyde **112** was submitted to [(−)-Sparteine]$_2$Cu$_2$O$_2$-mediated oxidative dearomatization to give vinylogous acid **113** which, after workup, was directly submitted to Cu(I)-catalyzed cycloisomerization to afford mitorubin core structure **114a**. After esterification with orsellinic acid fragment to **114b**, olefin cross-metathesis was successfully applied to install the requisite side chains for natural product synthesis.

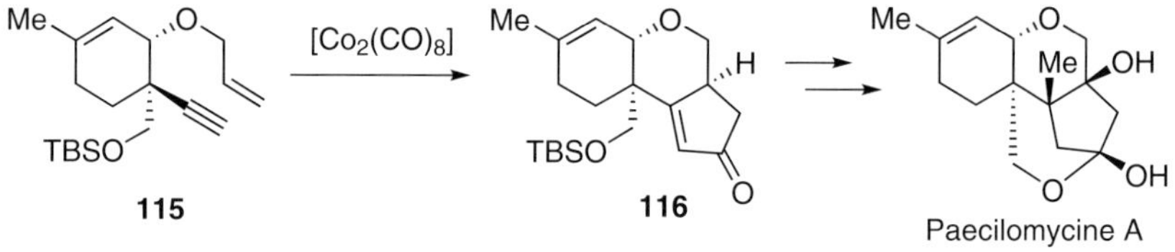

Scheme 5.27

In the course of studies on small molecule natural product-derived agents in the treatment of neurological diseases, total synthesis of paecilomycine A was achieved via the crucial pyran ring construction using intramolecular Pauson–Khand reaction (Scheme 5.27) [177]. The yield of Pauson–Khand product **116** from ene-yne **115** was moderate but the reaction was scalable and **116** was readily purified and advanced to reach the final natural product via further functionalizations involving carbon enlargement and cyclic hemiacetal formation.

5.3
Conclusion

As described above, different synthetic strategies and methodologies have been applied to the synthesis of various types of pyran natural products. The successful total syntheses highlight the level of progress and the state of the art in contemporary organic synthesis. However, there still remains a pressing demand for realizing even more efficient and scaleable synthetic pathways to provide environmentally benign and cost-effective routes to more complex pyran compounds and their analogs for biologically studies and also (pre)clinical evaluation.

References

1. (a) Hepworth, J.D. and Heron, B.M. (2009) Six-membered ring systems: with O and/or S atoms. *Prog. Heterocycl. Chem.*, **20**, 399–431, 365–398; (b) Hepworth, J.D. and Heron, B.M. (2007) Six-membered ring systems: with O and/or S atoms. *Prog. Heterocycl. Chem.*, **18**, 376–401; (c) Hepworth, J.D. and Heron, M.B. (2005) Six-membered ring systems: with O and/or S atoms. *Prog. Heterocycl. Chem.*, **17**, 362–388.
2. Yeung, K.S. and Paterson, I. (2002) Actin-binding marine macrolides: total synthesis and biological importance. *Angew. Chem. Int. Ed.*, **41** (24), 4632–4653.
3. Kang, E.J. and Lee, E. (2005) Total synthesis of oxacyclic macrodiolide natural products. *Chem. Rev.*, **105** (12), 4348–4378.
4. Nakata, T. (2005) Total synthesis of marine polycyclic ethers. *Chem. Rev.*, **105** (12), 4314–4347.
5. Nicolaou, K.C., Frederick, M.O. and Aversa, R.J. (2008) The continuing saga of the marine polyether biotoxins. *Angew. Chem. Int. Ed.*, **47** (38), 7182–7225.

6 Yeung, K.S. and Paterson, I. (2005) Advances in the total synthesis of biologically important marine macrolides. *Chem. Rev.*, **105** (12), 4237–4313.

7 Inoue, M. (2005) Convergent strategies for syntheses of trans-fused polycyclic ethers. *Chem. Rev.*, **105** (12), 4379–4405.

8 Clark, J.S. (2006) Construction of fused polycyclic ethers by strategies involving ring-closing metathesis. *Chem. Commun.*, (34), 3571–3571.

9 Kito, K., Ookura, R., Yoshida, S., Namikoshi, M., Ooi, T. and Kusumi, T. (2008) New cytotoxic 14-membered macrolides from marine-derived fungus *Aspergillus ostianus*. *Org. Lett.*, **10** (2), 225–228.

10 Hande, S.M. and Uenishi, J. (2009) Total synthesis of aspergillide B and structural discrepancy of aspergillide A. *Tetrahedron Lett.*, **50** (2), 189–192.

11 Nagasawa, T. and Kuwahara, S. (2009) Enantioselective total synthesis of aspergillide C. *Org. Lett.*, **11** (3), 761–764.

12 Buchanan, M.S., Gill, M. and Yu, J. (1997) Pigments of fungi. Part 43: 1,2 Cardinalins 1–6, novel pyranonaphthoquinones from the fungus *Dermocybe cardinalis* Horak. *J. Chem. Soc. Perkin Trans.*, **1** (6), 919–926.

13 Jammula, S.R., Pepalla, S.B., Telikepalli, H., Rao, K.V.J. and Thomson, R.H. (1991) Benzisochromanquinones from *Ventilago goughii*. *Phytochemistry*, **30** (11), 3741–3744.

14 Govender, S., Mmutlane, E.M., van Otterlo, W.A.L. and de Koning, C.B. (2007) Bidirectional racemic synthesis of the biologically active quinone cardinalin 3. *Org. Biomol. Chem.*, **5** (15), 2433–2440.

15 Galeffi, C., Casinovi, C.G. and Marini-Bettolo, G.B. (1965) Synthesis of centrolobine. *Gazz. Chim. Ital.*, **95** (1–2), 95–100.

16 Craveiro, A.A., Prado da Costa, A., Gottlieb, O.R. and Welerson de Albuquerque, P.C. (1970) Diarylheptanoids of *Centrolobium* species. *Phytochemistry*, **9** (8), 1869–1875.

17 Alcantara, A.F., Souza, M.R. and Pilo-Veloso, D. (2000) Constituents of *Brosimum potabile*. *Fitoterapia*, **71** (5), 613–615.

18 Colobert, F., Mazery, R.D., Solladié, G. and Carreño, M.C. (2002) First enantioselective total synthesis of (–)-centrolobine. *Org. Lett.*, **4** (10), 1723–1725.

19 Marumoto, S., Jaber, J.J., Vitale, J.P. and Rychnovsky, S.D. (2002) Synthesis of (–)-centrolobine by Prins cyclizations that avoid racemization. *Org. Lett.*, **4** (22), 3919–3922.

20 Evans, P.A., Cui, J. and Gharpure, S.J. (2003) Stereoselective construction of cis-2,6-disubstituted tetrahydropyrans via the reductive etherification of δ-trialkylsilyloxy substituted ketones: total synthesis of (–)-centrolobine. *Org. Lett.*, **5** (21), 3883–3885.

21 Boulard, L., BouzBouz, S., Cossy, J., Franck, X. and Fifadère, B. (2004) Two successive one-pot reactions leading to the expeditious synthesis of (–)-centrolobine. *Tetrahedron Lett.*, **45** (35), 6603–6605.

22 Chandrasekhar, S., Prakash, S.J. and Shyamsunder, T. (2005) Asymmetric synthesis of the pyran antibiotic (–)-centrolobine. *Tetrahedron Lett.*, **46** (39), 6651–6653.

23 Sabitha, G., Reddy, K.B., Reddy, G.S.K.K., Fatima, N. and Yadav, J.S. (2005) TMSI-mediated prins cyclization: diastereoselective synthesis of 4-iodo-2,6-disubstituted tetrahydropyrans and synthesis of (±)-centrolobine. *Synlett*, (15), 2347–2351.

24 Chan, K.-P. and Loh, T.-P. (2005) Prins cyclizations in silyl additives with suppression of epimerization: versatile tool in the synthesis of the tetrahydropyran backbone of natural products. *Org. Lett.*, **7** (20), 4491–4494.

25 Böhrsch, V. and Blechert, S. (2006) A concise synthesis of (–)-centrolobine via a diastereoselective ring rearrangement metathesis–isomerisation sequence. *Chem. Commun.*, (18), 1968–1970.

26 Yin, J., Kouda, K., Tezuka, Y., Le Tran, Q., Miyahara, T., Chen, Y. and Kadota, S. (2004) New diarylheptanoids from the rhizomes of *Dioscorea spongiosa* and their antiosteoporotic activity. *Planta Med.*, **70** (1), 54–58.

27 Sawant, K.B. and Jennings, M.P. (2006) Efficient Total syntheses and structural verification of both diospongins A and B via a common δ-lactone intermediate. *J. Org. Chem.*, **71** (20), 7911–7914.

28 Chandrasekhar, S., Shyamsunder, T., Prakash, S.J., Prabhakar, A. and Jagadeesh, B. (2006) First total synthesis of (–)-diospongin B. *Tetrahedron Lett.*, **47** (1), 47–49.

29 Sakai, R., Kamiya, H., Murata, M. and Shimamoto, K. (1997) Dysiherbaine: a new neurotoxic amino acid from the micronesian marine sponge dysidea herbacea. *J. Am. Chem. Soc.*, **119** (18), 4112–4116.

30 Sakai, R., Koike, T., Sasaki, M., Shimamoto, K., Oiwa, C., Yano, A., Suzuki, K., Tachibana, K. and Kamiya, H. (2001) Isolation, structure determination, and synthesis of neodysiherbaine A, a new excitatory amino acid from a marine sponge. *Org. Lett.*, **3** (10), 1479–1482.

31 Sakai, R., Swanson, G.T., Sasaki, M., Shimamoto, K. and Kamiya, H. (2006) Dysiherbaine: a new generation of excitatory amino acids of marine origin. *Cent. Nerv. Syst. Agents Med. Chem*, **6** (2), 83–108.

32 Swanson, G.T., Green, T., Sakai, R., Contractor, A., Che, W., Kamiya, H. and Heinemann, S.F. (2002) Differential activation of individual subunits in heteromeric kainate receptors. *Neuron*, **34** (4), 589–598.

33 Sanders, J.M., Pentikainen, O.T., Settimo, L., Pentikainen, U., Shoji, M., Sasaki, M., Sakai, R., Johnson, M.S. and Swanson, G.T. (2006) Determination of binding site residues responsible for the subunit selectivity of novel marine-derived compounds on kainate receptors. *Mol. Pharmacol.*, **69** (6), 1849–1860.

34 Snider, B.B. and Hawryluk, N.A. (2000) Synthesis of (–)-dysiherbaine. *Org. Lett.*, **2** (5), 635–638.

35 Sasaki, M., Koike, T., Sakai, R. and Tachibana, K. (2000) Total synthesis of (–)-dysiherbaine, a novel neuroexcitotoxic amino acid. *Tetrahedron Lett.*, **41** (20), 3923–3926.

36 Masaki, H., Maeyama, J., Kamada, K., Esumi, T., Iwabuchi, Y. and Hatakeyama, S. (2000) Total synthesis of (–)-dysiherbaine. *J. Am. Chem. Soc.*, **122** (21), 5216–5217.

37 Phillips, D. and Chamberlin, A.R. (2002) Total synthesis of dysiherbaine. *J. Org. Chem.*, **67** (10), 3194–3201.

38 Takahashi, K., Matsumura, T., Ishihara, J. and Hatakeyama, S. (2007) A highly stereocontrolled total synthesis of dysiherbaine. *Chem. Commun.*, (40), 4158–4160.

39 Sasaki, M., Tsubone, K., Aoki, K., Akiyama, N., Shoji, M., Oikawa, M., Sakai, R. and Shimamoto, K. (2008) Rapid and efficient synthesis of dysiherbaine and analogues to explore structure–activity relationships. *J. Org. Chem.*, **73** (1), 264–273.

40 Lygo, B., Slack, D. and Wilson, C. (2005) Synthesis of neodysiherbaine. *Tetrahedron Lett.*, **46** (39), 6629–6632.

41 Takahashi, K., Matsumura, T., Corbin, G.R.M., Ishihara, J. and Hatakeyama, S. (2006) A highly stereocontrolled total synthesis of (–)-neodysiherbaine A. *J. Org. Chem.*, **71** (11), 4227–4231.

42 Shoji, M., Akiyama, N., Tsubone, K., Lash, L.L., Sanders, J.M., Swanson, G.T., Sakai, R., Shimamoto, K., Oikawa, M. and Sasaki, M. (2006) Total synthesis and biological evaluation of neodysiherbaine A and analogues. *J. Org. Chem.*, **71** (14), 5208–5220.

43 Nakajima, H., Sato, B., Fujita, T., Takase, S., Terano, H. and Okuhara, M. (1996) New antitumor substances, FR901463, FR901464 and FR901465. I. Taxonomy, fermentation, isolation, physico-chemical properties and biological activities. *J. Antibiot.*, **49** (12), 1196–1203.

44 Nakajima, H., Hori, Y., Terano, H., Okuhara, M., Manda, T., Matsumoto, S. and Shimomura, K. (1996) New antitumor substances, FR901463, FR901464 and FR901465. *J. Antibiot.*, **49** (12), 1204–1211.

45 Nakajima, H., Takase, S., Terano, H. and Tanaka, H. (1997) New antitumor substances, FR901463, FR901464 and FR901465. III. Structures of FR901463,

FR901464 and FR901465. *J. Antibiot.*, **50** (1), 96–99.

46 Thompson, C.F., Jamison, T.F. and Jacobsen, E.N. (2000) Total synthesis of FR901464. convergent assembly of chiral components prepared by asymmetric catalysis. *J. Am. Chem. Soc.*, **122** (42), 10482–10483.

47 Thompson, C.F., Jamison, T.F. and Jacobsen, E.N. (2001) FR901464: total synthesis, proof of structure, and evaluation of synthetic analogues. *J. Am. Chem. Soc.*, **123** (41), 9974–9983.

48 Horigome, M., Motoyoshi, H., Watanabe, H. and Kitahara, T. (2001) A synthesis of FR901464. *Tetrahedron Lett.*, **42** (46), 8207–8210.

49 Motoyoshi, H., Horigome, M., Watanabe, H. and Kitahara, T. (2006) Total synthesis of FR901464: second generation. *Tetrahedron*, **62** (7), 1378–1389.

50 Albert, B.J., Sivaramakrishnan, A., Naka, T. and Koide, K. (2006) Total synthesis of FR901464, an antitumor agent that regulates the transcription of oncogenes and tumor suppressor genes. *J. Am. Chem. Soc.*, **128** (9), 2792–2793.

51 Albert, B.J., Sivaramakrishnan, A., Naka, T., Czaicki, N.L. and Koide, K. (2007) Total syntheses, fragmentation studies, and antitumor/antiproliferative activities of FR901464 and its low picomolar analogue. *J. Am. Chem. Soc.*, **129** (9), 2648–2659.

52 Isaac, B.G., Ayer, S.W., Elliott, R.C. and Stonard, R.J. (1992) Herboxidiene: a potent phytotoxic polyketide from streptomyces sp. A7847 herboxidiene: a potent phytotoxic polyketide from streptomyces sp. a7847. *J. Org. Chem.*, **57** (26), 7220–7226.

53 Sakai, Y., Yoshida, T., Ochiai, K., Uosaki, Y., Saitoh, Y., Tanaka, F., Akiyama, T., Akinaga, S. and Mizukami, T. (2002) GEX1 compounds, novel antitumor antibiotics related to herboxidiene, produced by streptomyces sp. I. taxonomy, production, isolation, physicochemical properties and biological activities. *J. Antibiot.*, **55** (10), 855–862.

54 Sakai, Y., Tsujita, T., Akiyama, T., Yoshida, T., Mizukami, T., Akinaga, S., Horinouchi, S., Yoshida, M. and Yoshida, T. (2002) GEX1 compounds, novel antitumor antibiotics related to herboxidiene, produced by streptomyces sp. II. The effects on cell cycle progresion and gene expression. *J. Antibiot.*, **55** (10), 863–872.

55 Koguchi, Y., Nishio, M., Kotera, J., Omori, K., Ohnuki, T. and Komatsubara, S. (1997) Trichostatin A and herboxidiene up-regulate the gene expression of low density lipoprotein receptor. *J. Antibiot.*, **50** (11), 970–971.

56 Blakemore, P.R., Kocieński, P.J., Morley, A. and Muir, K. (1999) A synthesis of herboxidiene. *J. Chem. Soc., Perkin Trans.*, **1** (8), 955–968.

57 Banwell, M., McLeod, M., Premraj, R. and Simpson, G. (2000) Total synthesis of herboxidiene, a complex polyketide from *Streptomyces* species A7847. *Pure Appl. Chem.*, **72** (9), 1631–1634.

58 Zhang, Y. and Panek, J.S. (2007) Total synthesis of herboxidiene/GEX 1A. *Org. Lett.*, **9** (16), 3141–3143.

59 Murray, T.J. and Forsyth, C.J. (2008) Total synthesis of GEX1A. *Org. Lett.*, **10** (16), 3429–3431.

60 Roll, D.M., Manning, J.K. and Carter, G.T. (1998) Hongoquercins A and B, new sesquiterpenoid antibiotics. *J. Antibiot.*, **51** (7), 635–639.

61 Tsujimori, H., Bando, M. and Mori, K. (2000) Synthetic microbial chemistry, XXXII synthesis and absolute configuration of hongoquercin A, an antibacterial sesquiterpene-substituted orsellinic acid isolated as a fungal metabolite. *Eur. J. Org. Chem.*, **2**, 297–302.

62 Kurdyumov, A.V. and Hsung, R.P. (2006) An unusual cationic [2 + 2] cycloaddition in a divergent total synthesis of hongoquercin a and rhododaurichromanic acid A. *J. Am. Chem. Soc.*, **128** (19), 6272–6273.

63 Gerth, K., Washausen, P., Höfle, G., Irschik, H. and Reichenbach, H. (1996) Antibiotics from gliding bacteria. The jerangolids: a family of new antifungal compounds from *Sorangium cellulosum* (Myxobacteria): production, physico-chemical and biological properties of jerangolid. *J. Antibiot.*, **49** (1), 71–75.

64 Pospíšil, J. and Markó, I.E. (2007) Total synthesis of jerangolid D. *J. Am. Chem. Soc.*, **129** (12), 3516–3517.

65 Chavez, D., Acevedo, L.A. and Mata, R. (1998) Jimenezin, a novel annonaceous acetogenin from the seeds of rollinia mucosa containing adjacent tetrahydrofuran–tetrahydropyran ring systems. *J. Nat. Prod.*, **61** (4), 419–421.

66 Takahashi, S., Maeda, K., Hirota, S. and Nakata, T. (1999) Total synthesis of a new cytotoxic acetogenin, jimenezin, and the revised structure. *Org. Lett.*, **1** (12), 2025–2028.

67 Hwang, C.H., Keum, G., Sohn, K.I., Lee, D.H. and Lee, E. (2005) Stereoselective synthesis of (−)-jimenezin. *Tetrahedron Lett.*, **46** (39), 6621–6623.

68 Bandur, N.G., Brückner, D., Hoffmann, R.W. and Koert, U. (2006) Total synthesis of jimenezin via an intramolecular allylboration. *Org. Lett.*, **8** (17), 3829–3831.

69 Horton, P.A., Koehn, F.E., Longley, R.E. and McConnell, O.J. (1994) Lasonolide A, a new cytotoxic macrolide from the marine sponge forcepia sp. *J. Am. Chem. Soc.*, **116** (13), 6015–6016.

70 Lee, E., Song, H.Y., Kang, J.W., Kim, D.-S., Jung, C.-K. and Joo, J.M. (2002) Lasonolide A: structural revision and synthesis of the unnatural (−)-enantiomer. *J. Am. Chem. Soc.*, **124** (3), 384–385.

71 Song, H.Y., Joo, J.M., Kang, J.W., Kim, D.S., Jung, C.K., Kwak, H.S., Park, J.H., Lee, E., Hong, C.Y., Jeong, S., Jeon, K. and Park, J.H. (2003) Lasonolide A: structural revision and total synthesis. *J. Org. Chem.*, **68** (21), 8080–8087.

72 Kang, S.H., Kang, S.Y., Kim, C.M., Choi, H., Jun, H., Lee, B.M., Park, C.M. and Jeong, J.W. (2003) Total synthesis of natural (+)-lasonolide A. *Angew. Chem. Int. Ed.*, **42** (39), 4779–4782.

73 Kang, S.H., Kang, S.Y., Choi, H.W., Kim, C.M., Jun, H.S. and Youn, J.H. (2004) Stereoselective total synthesis of the natural (+)-lasonolide A. *Synthesis*, **7**, 1102–1114.

74 Yoshimura, T., Yakushiji, F., Kondo, S., Wu, X., Shindo, M. and Shishido, K. (2006) Total synthesis of (+)-lasonolide A. *Org. Lett.*, **8** (3), 475–478.

75 Ghosh, A.K. and Gong, G. (2007) Enantioselective total synthesis of macrolide antitumor agent (−)-lasonolide A. *Org. Lett.*, **9** (8), 1437–1440.

76 Kusumi, T., Igari, M., Ishitsuka, M.O., Ichikawa, A., Itezono, Y., Nakayama, N. and Kakisawa, H. (1990) A novel chlorinated biscembranoid from the marine soft coral *Sarcophyton glaucum*. *J. Org. Chem.*, **55** (26), 6286–6289.

77 Ishitsuka, M.O., Kusumi, T. and Kakisawa, H. (1991) Isolation of a monomeric counterpart of the marine biscembranoids, a biogenetic Diels–Alder precursor. *Tetrahedron Lett.*, **32** (25), 2917–2918.

78 Ichige, T., Okano, Y., Kanoh, N. and Nakata, M. (2007) Total synthesis of methyl sarcophytoate. *J. Am. Chem. Soc.*, **129** (32), 9862–9863.

79 Ichige, T., Okano, Y., Kanoh, N. and Nakata, M. (2009) Total synthesis of methyl sarcophytoate, a marine natural biscembranoid. *J. Org. Chem.*, **74** (1), 230–243.

80 Shiomi, K., Iinuma, H., Hamada, M., Naganawa, H., Manabe, M., Matsuki, C., Takeuchi, T. and Umezawa, H. (1986) Novel antibiotics napyradiomycins production, isolation, physico-chemical properties and biological activity. *J. Antibiot.*, **39** (4), 487–493.

81 Hori, Y., Abe, Y., Shigematsu, N., Goto, T., Okuhara, M. and Kohsaka, M. (1993) Napyradiomycins A and B1: Non-steroidal estrogen-receptor antagonists produced by a *Streptomyces*. *J. Antibiot.*, **46** (12), 1890–1893.

82 Tatsuta, K., Tanaka, Y., Kojima, M. and Ikegami, H. (2002) The first total synthesis of (±)-napyradiomycin A1. *Chem. Lett*, **2002** (1), 14–15.

83 Snyder, S.A., Tang, Z.-Y. and Gupta, R. (2009) Enantioselective total synthesis of (−)-napyradiomycin A1 via asymmetric chlorination of an isolated olefin. *J. Am. Chem. Soc.*, **131** (16), 5744–5745.

84 Williams, D.E., Roberge, M., Van Soest, R. and Andersen, R.J. (2003) Spirastrellolide A, an antimitotic macrolide isolated from the Caribbean marine sponge spirastrella coccinea. *J. Am. Chem. Soc.*, **125** (18), 5296–5297.

85 Williams, D.E., Lapawa, M., Feng, X., Tarling, T., Roberge, M. and Andersen, R.J. (2004) Spirastrellolide A: revised structure, progress toward the relative configuration, and inhibition of protein phosphatase 2A. *Org. Lett.*, **6** (15), 2607–2610.

86 Paterson, I., Anderson, E.A., Dalby, S.M., Lim, J.H., Genovino, J., Maltas, P. and Moessner, C. (2008) Total synthesis of spirastrellolide A methyl ester – Part 1: synthesis of an advanced C17-C40 bis-spiroacetal subunit. *Angew. Chem. Int. Ed.*, **47** (16), 3016–3020.

87 Paterson, I., Anderson, E.A., Dalby, S.M., Lim, J.H., Genovino, J., Maltas, P. and Moessner, C. (2008) Total synthesis of spirastrellolide A methyl ester – Part 2: subunit union and completion of the synthesis. *Angew. Chem. Int. Ed.*, **47** (16), 3021–3025.

88 Fusetani, N., Sugawara, T. and Matsunaga, S. (1992) Bioactive marine metabolites. 41. Theopederins A–E, potent antitumor metabolites from a marine sponge, Theonella sp. *J. Org. Chem*, **57** (14), 3828–3832.

89 Kocienski, P., Narquizian, R., Raubo, P., Smith, C., Farrugia, L.J., Muir, K. and Boyle, F.T. (2000) Synthetic studies on the pederin family of antitumour agents. Syntheses of mycalamide B, theopederin D and pederin. *J. Chem. Soc., Perkin Trans.*, **1** (15), 2357–2384.

90 Green, M.E., Rech, J.C. and Floreancig, P.E. (2008) Total synthesis of theopederin D. *Angew. Chem. Int. Ed.*, **47** (38), 7317–7320.

91 Grafe, U., Schade, W., Roth, M., Radics, L., Incze, M. and Ujszaszy, K. (1984) Griseochelin, a novel carboxylic acid antibiotic from Streptomyces griseus. *J. Antibiot.*, **37** (8), 836–846.

92 Brooks, H.A., Gardner, D., Poyser, J.P. and King, T.J. (1984) The structure and absolute stereochemistry of zincophorin (antibiotic M144255): a monobasic carboxylic acid ionophore having a remarkable specificity for divalent cations. *J. Antibiot.*, **37** (11), 1501–1504.

93 Tonew, E., Tonew, M., Gräfe, U. and Zopel, P. (1988) Griseochelin methyl ester, a new polyether derivative with antiviral activity. *Pharmazie*, **43** (10), 717–719.

94 Danishefsky, S.J., Selnick, H.G., DeNinno, M.P. and Zelle, R.E. (1987) The total synthesis of zincophorin. *J. Am. Chem. Soc.*, **109** (5), 1572–1574.

95 Danishefsky, S.J., Selnick, H.G., Zelle, R.E. and DeNinno, M.P. (1988) Total synthesis of zincophorin. *J. Am. Chem. Soc.*, **110** (13), 4368–4378.

96 Defosseux, M., Blanchard, N., Meyer, C. and Cossy, J. (2003) Total synthesis of zincophorin methyl ester. *Org. Lett.*, **5** (22), 4037–4040.

97 Defosseux, M., Blanchard, N., Meyer, C. and Cossy, J. (2004) Total synthesis of zincophorin and its methyl ester. *J. Org. Chem*, **69** (14), 4626–4647.

98 Komatsu, K., Tanino, K. and Miyashita, M. (2004) Stereoselective total synthesis of the ionophore antibiotic zincophorin. *Angew. Chem. Int. Ed.*, **43** (33), 4341–4345.

99 Song, Z. and Hsung, P.R. (2007) A formal total synthesis of (+)-zincophorin. Observation of an unusual urea-directed stork–crabtree hydrogenation. *Org. Lett.*, **9** (11), 2199–2202.

100 Statsuk, A.V., Liu, D. and Kozmin, S.A. (2004) Synthesis of bistramide A. *J. Am. Chem. Soc.*, **126** (31), 9546–9547.

101 Wipf, P. and Hopkins, T.D. (2005) Total synthesis and structure validation of (+)-bistramide C. *Chem. Commun.*, (27), 3421–3423.

102 Crimmins, M.T. and DeBaillie, A.C. (2006) Enantioselective total synthesis of bistramide A. *J. Am. Chem. Soc.*, **128** (15), 4936–4937.

103 (a) Lowe, J.T. and Panek, J.S. (2005) Synthesis and [4 + 2]-annulation of enantioenriched (Z)-crotylsilanes: preparation of the C1–C13 fragment of bistramide A. *Org. Lett.*, **7** (15), 3231–3234; (b) Lowe, J.T., Wrona, I.E. and Panek, J.S. (2007) Total synthesis of bistramide A. *Org. Lett.*, **9** (2), 327–330; (c) Wrona, I.E., Lowe, J.T., Turbyville, T.J., Johnson, T.R., Beignet, J., Beutler, J.A. and Panek, J.S. (2009) Synthesis of a 35-member stereoisomer library of bistramide A: evaluation of effects on actin state, cell cycle and tumor cell

growth. *J. Org. Chem.*, **74** (5), 1897–1916.
104 Jewett, J.C. and Rawal, V.H. (2007) Total synthesis of pederin. *Angew. Chem. Int. Ed.*, **46** (34), 6502–6504.
105 (a) Jiang, X., García-Fortanet, J. and De Brabander, J.K. (2005) Synthesis and complete stereochemical assignment of psymberin/irciniastatin A. *J. Am. Chem. Soc.*, **127** (32), 11254–11255; (b) Jiang, X., Williams, N. and De Brabander, J.K. (2007) Synthesis of psymberin analogues: probing a functional correlation with the pederin/mycalamide family of natural products. *Org. Lett.*, **9** (2), 227–230.
106 (a) Huang, X., Shao, N., Palani, A., Aslanian, R. and Buevich, A. (2007) The Total Synthesis of Psymberin. *Org. Lett.*, **9** (13), 2597–2600; (b) Huang, X., Shao, N., Huryk, R., Palani, A., Aslanian, R. and Seidel-Dugan, C. (2009) The discovery of potent antitumor agent C11-deoxypsymberin/irciniastatin A: total synthesis and biology of advanced psymberin analogs. *Org. Lett.*, **11** (4), 867–870.
107 Smith, A.B., III, Jurica, J.A. and Walsh, S.P. (2008) Total synthesis of (+)-psymberin (irciniastatin A): catalytic reagent control as the strategic cornerstone. *Org. Lett.*, **10** (24), 5625–5628.
108 Shangguan, N., Kiren, S. and Williams, L.J. (2007) A formal synthesis of psymberin. *Org. Lett.*, **9** (6), 1093–1096.
109 Takahashi, S., Kubota, A. and Nakata, T. (2003) Total synthesis of a cytotoxic acetogenin, pyranicin. *Org. Lett.*, **5** (8), 1353–1356.
110 (a) Strand, D. and Rein, T. (2005) Total synthesis of pyranicin. *Org. Lett.*, **7** (2), 199–202; (b) Strand, D., Norrby, P.-O. and Rein, T. (2006) Divergence en route to nonclassical annonaceous acetogenins. Synthesis of pyranicin and pyragonicin. *J. Org. Chem.*, **71** (5), 1879–1891.
111 Griggs, N.D. and Phillips, A.J. (2008) A concise and modular synthesis of pyranicin. *Org. Lett.*, **10** (21), 4955–4957.
112 (a) Hattori, Y., Furuhata, S., Okajima, M., Konno, H., Abe, M., Miyoshi, H., Goto, T. and Makabe, H. (2008) Synthesis of pyranicin and its inhibitory action with bovine heart mitochondrial complex I. *Org. Lett.*, **10** (5), 717–720; (b) Furuhata, S., Hattori, Y., Okajima, M., Konno, H., Abe, M., Miyoshi, H., Goto, T. and Makabe, H. (2008) Synthesis of pyranicin and its deoxygenated analogues and their inhibitory action with bovine heart mitochondrial complex I. *Tetrahedron*, **64** (33), 7695–7703.
113 (a) Brioche, J.C.R., Goodenough, K.M., Whatrup, D.J. and Harrity, J.P.A. (2007) A [3 + 3] annelation approach to (+)-rhopaloic acid B. *Org. Lett.*, **9** (20), 3941–3943; (b) Brioche, J.C.R., Goodenough, K.M., Whatrup, D.J. and Harrity, J.P.A. (2008) Investigation of an organomagnesium-based [3 + 3] annelation to pyrans and its application in the synthesis of rhopaloic acid A. *J. Org. Chem.*, **73** (5), 1946–1953.
114 (a) Snider, B.B. and He, F. (1997) Total synthesis of (±)-rhopaloic acid A. *Tetrahedron Lett.*, **38** (31), 5453–5454; (b) Takagi, R., Sasaoka, A., Kojima, S. and Ohkata, K. (1997) Synthesis of norsesterterpene rac- and ent-rhopaloic acid A. *Chem. Commun.*, (17), 1887–1888; (c) Kadota, K. and Ogasawara, K. (2003) A stereocontrolled synthesis of (+)-rhopaloic acid A using a dioxabicyclo[3.2.1]octane chiral building block. *Heterocycles*, **59** (2), 485–490.
115 (a) Kageyama, M., Tamura, T., Nantz, M.H., Roberts, J.C., Somfai, P., Whritenour, D.C. and Masamune, S. (1990) Synthesis of bryostatin 7. *J. Am. Chem. Soc.*, **112** (20), 7407–7408; (b) Evans, D.A., Carter, P.H., Carreira, E.M., Charette, A.B., Prunet, J.A. and Lautens, M. (1999) Total synthesis of Bryostatin 2. *J. Am. Chem. Soc.*, **121** (33), 7540–7552; (c) Ohmori, K., Ogawa, Y., Obitsu, T., Ishikawa, Y., Nishiyama, S., Yamamura and Trost (2000) Total synthesis of Bryostatin 3. *Angew. Chem. Int. Ed.*, **39** (13), 2290–2294.
116 (a) Trost, B.M. and Dong, G. (2008) Total synthesis of bryostatin 16 using atom-economical and chemoselective approaches. *Nature*, **456** (7221), 485–488; (b) Miller, A.K. (2009) Catalysis in the

117 Yuan, Y., Men, H. and Lee, C. (2004) Total synthesis of kendomycin: a macro-C-glycosidation approach. *J. Am. Chem. Soc.*, **126** (45), 14720–14721.

118 (a) Smith, A.B., III, Mesaros, E.F. and Meyer, E.A. (2005) Total synthesis of (–)-kendomycin exploiting a Petasis–Ferrier rearrangement/ring-closing olefin metathesis synthetic strategy. *J. Am. Chem. Soc.*, **127** (19), 6948–6949; (b) Smith, A.B., III, Mesaros, E.F. and Meyer, E.A. (2006) Evolution of a total synthesis of (–)-kendomycin exploiting a Petasis–Ferrier rearrangement/ring-closing olefin metathesis strategy. *J. Am. Chem. Soc.*, **128** (15), 5292–5299.

119 Lowe, J.T. and Panek, J.S. (2008) Total synthesis of (–)-kendomycin. *Org. Lett.*, **10** (17), 3813–3816.

120 (a) Bahnck, K.B. and Rychnovsky, S.D. (2006) Rapid stereocontrolled assembly of the fully substituted C-aryl glycoside of kendomycin with a Prins cyclization: a formal synthesis. *Chem. Commun.*, (22), 2388–2390; (b) Bahnck, K.B. and Rychnovsky, S.D. (2008) Formal synthesis of (–)-kendomycin featuring a Prins-cyclization to construct the macrocycle. *J. Am. Chem. Soc.*, **130** (39), 13177–13181.

121 (a) Kopecky, D.J. and Rychnovsky, S.D. (2001) Mukaiyama aldol–Prins cyclization cascade reaction: a formal total synthesis of leucascandrolide A. *J. Am. Chem. Soc.*, **123** (34), 8420–8421; (b) Wang, Y., Janjic, J. and Kozmin, S.A. (2002) Synthesis of leucascandrolide A via a spontaneous macrolactolization. *J. Am. Chem. Soc.*, **124** (46), 13670–13671; (c) Wipf, P. and Reeves, J.T. (2002) A formal total synthesis of leucascandrolide A. *Chem. Commun.*, (18), 2066–2067; (d) Fettes, A. and Carreira, E.M. (2002) Total synthesis of leucascandrolide A. *Angew. Chem. Int. Ed.*, **41** (21), 4098–4101; (e) Paterson, I. and Tudge, M. (2003) Stereocontrolled total synthesis of (+)-leucascandrolide A. *Angew. Chem. Int. Ed.*, **42** (3), 343–347; (f) Williams, D.R., Plummer, S.V. and Patnaik, S. (2003) *Angew. Chem. Int. Ed.*, **42** (33), 3934–3938; (g) Crimmins, M.T. and Siliphaivanh, P. (2003) Enantioselective total synthesis of (+)-leucascandrolide A macrolactone. *Org. Lett.*, **5** (24), 4641–4644; (h) Williams, D.R., Patnaik, S. and Plummer, S.V. (2003) Leucascandrolide A: a second generation formal synthesis. *Org. Lett.*, **5** (26), 5035–5038; (i) Su, Q. and Panek, J.S. (2005) Total synthesis of (+)-leucascandrolide A. *Angew. Chem. Int. Ed.*, **44** (8), 1223–1225; (j) Ferrié, L., Reymond, S., Capdevielle, P. and Cossy, J. (2007) Formal chemoselective synthesis of leucascandrolide A. *Org. Lett.*, **9** (13), 2461–2464; (k) Van Orden, L.J., Patterson, B.D. and Rychnovsky, S.D. (2007) Total synthesis of leucascandrolide A: a new application of the Mukaiyama aldol–Prins reaction. *J. Org. Chem.*, **72** (15), 5784–5793.

122 Hornberger, K.R., Hamblett, C.L. and Leighton, J.L. (2000) Total synthesis of leucascandrolide A. *J. Am. Chem. Soc.*, **122** (51), 12894–12895.

123 Jung, H.H., Seiders, J.R., II and Floreancig, P.E. (2007) Oxidative cleavage in the construction of complex molecules: synthesis of the leucascandrolide A macrolactone. *Angew. Chem. Int. Ed.*, **46** (44), 8464–8467.

124 Evans, P.A. and Andrews, W.J. (2008) A sequential two-component etherification/oxa-conjugate addition reaction: asymmetric synthesis of (+)-leucascandrolide A macrolactone. *Angew. Chem. Int. Ed.*, **47** (29), 5426–5429.

125 Youngsaye, W., Lowe, J.T., Pohlki, F., Ralifo, P. and Panek, J.S. (2007) Total synthesis and stereochemical reassignment of (+)-neopeltolide. *Angew. Chem. Int. Ed.*, **46** (48), 9211–9214.

126 Custar, D.W., Zabawa, T.P. and Scheidt, K.A. (2008) Total synthesis and structural revision of the marine macrolide neopeltolide. *J. Am. Chem. Soc.*, **130** (3), 804–805.

127 Woo, S.K., Kwon, M.S. and Lee, E. (2008) Total synthesis of (+)-neopeltolide by a Prins macrocyclization. *Angew. Chem. Int. Ed.*, **47** (17), 3242–3244.

128 (a) Vintonyak, V.V. and Maier, M.E. (2008) Formal total synthesis of

neopeltolide. *Org. Lett.*, **10** (6), 1239–1242; (b) Vintonyak, V.V., Kunze, B., Sasse, F. and Maier, M.E. (2008) Total synthesis and biological activity of neopeltolide and analogues. *Chem. Eur. J.*, **14** (35), 11132–11140.

129 Kartika, R., Gruffi, T.R. and Taylor, R.E. (2008) Concise enantioselective total synthesis of neopeltolide macrolactone highlighted by ether transfer. *Org. Lett.*, **10** (21), 5047–5050.

130 (a) Fuwa, H., Naito, S., Goto, T. and Sasaki, M. (2008) Total synthesis of (+)-neopeltolide. *Angew. Chem. Int. Ed.*, **47** (25), 4737–4739; (b) Paterson, I. and Miller, N.A. (2008) Total synthesis of the marine macrolide (+)-neopeltolide. *Chem. Commun.*, (39), 4708–4710.

131 Nicolaou, K.C., Sun, Y.P., Peng, X.S., Polet, D. and Chen, D.Y.K. (2008) Total synthesis of (+)-cortistatin A. *Angew. Chem. Int. Ed.*, **47** (38), 7310–7313.

132 Liu, L., Gao, Y., Che, C., Wu, N., Wang, D.Z., Li, C.C. and Yang, Z. (2009) A model study for the concise construction of the oxapentacyclic core of cortistatins through intramolecular Diels–Alder and oxidative dearomatization–cyclization reactions. *Chem. Commun.*, (6), 662–664.

133 Yamashita, S., Iso, K. and Hirama, M. (2008) A concise synthesis of the pentacyclic framework of cortistatins. *Org. Lett.*, **10** (16), 3413–3415.

134 Nising, C.F. and Brase, S. (2008) Highlights in steroid chemistry: total synthesis versus semisynthesis. *Angew. Chem. Int. Ed.*, **47** (49), 9389–9391.

135 Ng, F.W., Lin, H. and Danishefsky, S.J. (2002) Explorations in organic chemistry leading to the total synthesis of (±)-gelsemine. *J. Am. Chem. Soc.*, **124** (33), 9812–9824.

136 Madin, A., O'Donnell, C.J., Oh, T., Old, D.W., Overman, L.E. and Sharp, M.J. (2005) Use of the intramolecular heck reaction for forming congested quaternary carbon stereocenters. Stereocontrolled total synthesis of (±)-gelsemine. *J. Am. Chem. Soc.*, **127** (51), 18054–18065.

137 Tchabanenko, K., Simpkins, N.S. and Male, L. (2008) A concise approach to a gelsemine core structure using an oxygen to carbon bridge swapping strategy. *Org. Lett.*, **10** (21), 4747–4750.

138 Nicolaou, K.C., Li, A. and Edmonds, D.J. (2006) Total synthesis of platensimycin. *Angew. Chem. Int. Ed.*, **45** (42), 7086–7090.

139 Zou, Y., Chen, C.H., Taylor, C.D., Foxman, B.M. and Snider, B.B. (2007) Formal synthesis of (±)-platensimycin. *Org. Lett.*, **9** (9), 1825–1828.

140 Nicolaou, K.C., Tang, Y. and Wang, J. (2007) Formal synthesis of (±)-platensimycin. *Chem. Commun.*, (19), 1922–1923.

141 Li, P., Payette, J.N. and Yamamoto, H. (2007) Enantioselective route to platensimycin: an intramolecular robinson annulation approach. *J. Am. Chem. Soc*, **129** (31), 9534–9535.

142 Tiefenbacher, K. and Mulzer, J. (2008) Synthesis of platensimycin. *Angew. Chem. Int. Ed.*, **47** (14), 2548–2555.

143 Lalic, G. and Corey, E.J. (2007) An effective enantioselective route to the platensimycin core. *Org. Lett.*, **9** (23), 4921–4923.

144 Matsuo, J., Takeuchi, K. and Ishibashi, H. (2008) Stereocontrolled formal synthesis of (±)-platensimycin. *Org. Lett.*, **10** (18), 4049–4052.

145 Kim, C.H., Jang, K.P., Choi, S.Y., Chung, Y.K. and Lee, E. (2008) A carbonyl ylide cycloaddition approach to platensimycin. *Angew. Chem. Int. Ed.*, **47** (21), 4009–4011.

146 Kaliappan, K.P. and Ravikumar, V. (2007) An expedient enantioselective strategy for the oxatetracyclic core of platensimycin. *Org. Lett.*, **9** (12), 2417–2419.

147 Ghosh, A.K. and Xi, K. (2009) Total synthesis of (–)-platensimycin, a novel antibacterial agent. *J. Org. Chem.*, **74** (3), 1163–1170.

148 Wang, J., Lee, V. and Sintim, H.O. (2009) Efforts towards the identification of simpler platensimycin analogues – the total synthesis of oxazinidinyl platensimycin. *Chem. Eur. J.*, **15** (12), 2747–2750.

149 Nicolaou, K.C., Stepan, A.F., Lister, T., Li, A., Montero, A., Tria, G.S., Turner, C.I., Tang, Y., Wang, J., Denton, R.M. and Edmonds, D.J. (2008) Design,

synthesis, and biological evaluation of platensimycin analogues with varying degrees of molecular complexity. *J. Am. Chem. Soc.*, **130** (39), 13110–13119.

150 Miyaoka, H., Saka, Y., Miura, S. and Yamada, Y. (1996) Total synthesis of phomactin D. *Tetrahedron Lett.*, **37** (39), 7107–7110.

151 Foote, K.M., Hayes, C.J., John, M.P. and Pattenden, G. (2003) Synthetic studies towards the phomactins. Concise syntheses of the tricyclic furanochroman and the oxygenated bicyclo[9.3.1] pentadecane ring systems in phomactin A. *Org. Biomol. Chem.*, **1** (27), 3917–3948.

152 Mohr, P.J. and Halcomb, R.L. (2003) Total synthesis of (+)-phomactin A using a B-alkyl suzuki macrocyclization. *J. Am. Chem. Soc.*, **125** (7), 1712–1713.

153 Cole, K.P. and Hsung, R.P. (2005) Unique structural topology and reactivities of the ABD tricycle in phomactin A. *Chem. Commun.*, (46), 5784–5786.

154 Tang, Y., Cole, K.P., Buchanan, G.S., Li, G. and Hsung, R.P. (2009) Total synthesis of Phomactin A. *Org. Lett.*, **11** (7), 1591–1594.

155 Nakamura, S., Sugano, Y., Kikuchi, F. and Hashimoto, S. (2006) Total synthesis and absolute stereochemistry of polygalolides A and B. *Angew. Chem. Int. Ed.*, **45** (39), 6532–6535.

156 Snider, B.B., Wu, X., Nakamura, S. and Hashimoto, S. (2007) A short, formal, biomimetic synthesis of (±)-polygalolides A and B. *Org. Lett.*, **9** (5), 873–874.

157 Zhou, J. and Snider, B.B. (2007) Biomimetic synthesis of the tetracyclic core of berkelic acid. *Org. Lett.*, **9** (10), 2071–2074.

158 Wu, X., Zhou, J. and Snider, B.B. (2009) Synthesis of (−)-berkelic acid. *Angew. Chem. Int. Ed.*, **48** (7), 1283–1286.

159 Buchgraber, P., Snaddon, T.N., Wirtz, C., Mynott, R., Goddard, R. and Fürstner, A. (2008) A synthesis-driven structure revision of berkelic acid methyl ester. *Angew. Chem. Int. Ed.*, **47** (44), 8450–8454.

160 Mondal, M., Puranik, V.G. and Argade, N.P. (2007) A facile phenol-driven intramolecular diastereoselective thermal/base-catalyzed dipolar [2 + 2] annulation reactions: an easy access to complex bioactive natural and unnatural benzopyran congeners. *J. Org. Chem.*, **72** (6), 2068–2076.

161 Gandon, V., Aubert, C. and Malacria, M. (2006) Recent progress in cobalt-mediated [2 + 2 + 2] cycloaddition reactions. *Chem. Commun.*, (21), 2209–2217.

162 Teske, J.A. and Deiters, A. (2008) A cyclotrimerization route to cannabinoids. *Org. Lett.*, **10** (11), 2195–2198.

163 (a) Li, Z., Gao, Y., Jiao, Z., Tang, Y., Dai, M., Wang, G., Wang, Z. and Yang, Z. (2008) Total synthesis of Crisamicin A. *Org. Lett.*, **10** (14), 3017–3020; (b) Li, Z., Gao, Y., Jiao, Z., Wu, N., Wang, D.Z. and Yang, Z. (2008) Diversity-oriented synthesis of fused pyran γ-lactones via an efficient Pd–thiourea-catalyzed alkoxycarbonylative annulation. *Org. Lett.*, **10** (22), 5163–5166.

164 Tian, X. and Rychnovsky, S.D. (2007) Synthesis and structural reassignment of (+)-Epicalyxin F. *Org. Lett.*, **9** (24), 4955–4958.

165 Lumb, J.P., Choong, K.C. and Trauner, D. (2008) ortho-quinone methides from para-quinones: total synthesis of Rubioncolin B. *J. Am. Chem. Soc.*, **130** (29), 9230–9231.

166 Ko, M.H., Lee, D.G., Kim, M.A., Kim, H.J., Park, J., Lah, M.S. and Lee, E. (2007) Total synthesis of (−)-Blepharocalyxin D. *Org. Lett.*, **9** (1), 141–144.

167 Barry, C.S., Crosby, S.R., Harding, J.R., Hughes, R.A., King, C.D., Parker, G.D. and Willis, C.L. (2003) Stereoselective synthesis of 4-Hydroxy-2,3,6-trisubstituted tetrahydropyrans. *Org. Lett.*, **5** (14), 2429–2432.

168 (a) Aicher, T.D., Buszek, K.R., Fang, F.G., Forsyth, C.J., Jung, S.H., Kishi, Y., Matelich, M.C., Scola, P.M., Spero, D.M. and Yoon, S.K. (1992) Total synthesis of halichondrin B and norhalichondrin B. *J. Am. Chem. Soc.*, **114** (8), 3162–3164; (b) Namba, K., Jun, H.-S. and Kishi, Y. (2004) A simple but remarkably effective device for forming the C8–C14 polycyclic ring system of halichondrin B. *J. Am. Chem. Soc.*, **126** (25), 7770–7771.

169 Jackson, K.L., Henderson, J.A., Motoyoshi, H. and Phillips, A.J. (2009) A total synthesis of norhalichondrin B. *Angew. Chem. Int. Ed.*, **48** (13), 2346–2350.

170 Hioki, H., Motosue, M., Mizutani, Y., Noda, A., Shimoda, T., Kubo, M., Harada, K., Fukuyama, Y. and Kodama, M. (2009) Total synthesis of pseudodehydrothyrsiferol. *Org. Lett.*, **11** (3), 579–582.

171 Sakamoto, I., Kaku, H. and Tsunoda, T. (2003) Preparation of (Cyanomethylene) trimethylphosphorane as a new mitsunobu-type reagent. *Chem. Pharm. Bull.*, **51** (4), 474–476.

172 (a) Vanderwal, C.D., Vosburg, S., Weiler, S. and Sorensen, E.J. (2003) An enantioselective synthesis of FR182877 provides a chemical rationalization of its structure and affords multigram quantities of its direct precursor. *J. Am. Chem. Soc.*, **125** (18), 5393–5407; (b) Evans, D.A. and Starr, J.T. (2003) A cycloaddition cascade approach to the total synthesis of (−)-FR182877. *J. Am. Chem. Soc.*, **125** (44), 13531–13540; (c) Clarke, P.A., Davie, R.L. and Peace, S. (2005) Synthesis of the B-ring of FR182877. Investigation of the reactions of 6-fumaryl 1,3,8-nonatrienes. *Tetrahedron*, **61** (9), 2335–2351.

173 Tanaka, N., Suzuki, T., Matsumura, T., Hosoya, Y. and Nakata, M. (2009) Total synthesis of (−)-FR182877 through tandem IMDA-IMHDA reactions and stereoselective transition-metal-mediated transformations. *Angew. Chem. Int. Ed.*, **48** (14), 2580–2583.

174 Maloney, D.J. and Danishefsky, S.J. (2007) Conformational locking through allylic strain as a device for stereocontrol – total synthesis of Grandisine A. *Angew. Chem. Int. Ed.*, **46** (41), 7789–7792.

175 Marsini, M.A., Gowin, K.M. and Pettus, T.R.R. (2006) Total synthesis of (±)-mitorubrinic acid. *Org. Lett.*, **8** (16), 3481–3483.

176 Zhu, J. and Porco, J.A., Jr. (2006) Asymmetric syntheses of (−)-mitorubrin and related azaphilone natural products. *Org. Lett.*, **8** (22), 5169–5171.

177 Min, S.J. and Danishefsky, S.J. (2007) Total synthesis of Paecilomycine A. *Angew. Chem. Int. Ed.*, **46** (13), 2199–2202.

6
Pyrrole and Its Derivatives

Dipakranjan Mal, Brateen Shome and Bidyut Kumar Dinda

6.1
Introduction

The five-membered and aromatic nitrogen heterocycle "pyrrole" had been known in the literature as a component of coal tar since 1834 [1]. However, the seminal synthetic work of Paal and Knorr [2, 3], set the beginning of pyrrole chemistry, which centered on the exploration of their reactivities as aromatics [4], and the synthetic chemistry of tetrapyrroles [5] embodying natural products like chlorophyll, heme and vitamin B12. Now, pyrrole natural products are subdivided into two groups: (i) tetrapyrroles and (ii) pyrrole-containing small molecules [5]. Since the tetrapyrroles are well documented in the literature, we have excluded them from the present chapter.

The second subclass "pyrrole-containing small molecules" is rarely encountered in natural sources because of the exceptional reactivity of pyrrole moieties to electrophilic substitution reactions. The chemistry of pyrrole-containing natural products dates back to the isolation and the total synthesis of tripyrrole prodigiosin by Rapoport and Holden in the 1960s [6].

The number of pyrrole natural products is constantly growing and at present is so large that it is difficult to include all the members in a limited number of pages, so only representative members are described in Table 6.1. In this chapter, we aim to highlight (i) synthesis of the pyrrole natural products and (ii) the use of pyrrole building blocks in the synthesis of allied natural products. Related reviews already exist [7], so we chose to focus on recent developments with brief and contextual mention of the older literature. Since the chemistry of pyrrolidine-, pyrrolizidine- and indolizidine-containing natural products is established and periodically reviewed, their discussion in this article is made only in passing.

Table 6.1 Structures and bioactivity of pyrrole-containing natural products.

Serial No.	Trivial name	Structure	Source	Isolation [Ref]	Bioactivity	Synthesis [Ref]
1	Acanthamide A. Five similar compounds are known.		*Acanthostylotella* sp.	[8]	NR	†NR
2	Cyclopentadiene[c]pyrrole-1,3-diol		*Castanea mollissima* Blume	[9]	NR	NR
3	Cylindradine A		*Axinella cylindratus*	[10]	Active against P388 cell line	NR
4	Daminin. Bromo derivative is also known.		Marine sponge *Axinella damicornis*	[11]	Neuroprotective activity	[11]
5	Funebrine. Three more similar compounds are known.		*Quararibea funebris*	[12]	Inhibitor of GlyT1	[13]
6	Glaciapyrrole A. Two more congeners are known.		Alaskan Marine sediment	[14]	NR	NR
7	Hymenialdisine. Its debromo analog is also known.		Marine sponge *Stylissa carteri*	[15]	Inhibitor of tumor necrosis factor-α	[16]

8	Isobatzelline E	Marine sponge *Zyzzya fuliginosa*	[17]	Active to pancreatic cancer	[18]
9	Lamellarin A	Indian ascidian *Didemnum obscurum*	[19]	Cytotoxic	[20]
10	Longamide B. Few more congeners are known.	Marine sponge *Agelas dispar*	[21]	Anti-microbial activity	[22]
11	Manzacidin A	Mediterranean sponge *Axinella verrucosa*	[23]	Activator of ATPase	[24]
12	Marineosin	Marine-Derived *Actinomycete*	[25]	Active against colon carcinoma	
13	Marinopyrrole B. Five more congeners are reported.	Marine *Streptomyces* sp.	[26]	Antibacterial activity	[26]
14	Mukanadin A. Two more congeners are reported.	Marine sponge *Agelas nakamurai*	[27]	NR	NR

(Continued)

Table 6.1 (Continued)

Serial No.	Trivial name	Structure	Source	Isolation [Ref]	Bioactivity	Synthesis [Ref]
15	Mycalazal. Seven more members are known.		*Mycale tenuispiculata*	[28]	Inhibitors of tumor cell lines	NR
16	Nagelamide A. Eight more members are reported.		Marine sponge *Agelas* sp.	[29]	Antibacterial activity and phosphatase inhibition	NR
17	Neolamellarin A		Sponge *Dendrilla nigra*	[30]	Antitumor activity	[31]
18	Neopyrrolomycin B. Three more N-aryl congeners are known.		*Streptomyces* sp.	[32]	Antibacterial activity	[33]
19	Phenylpyrrole		*Rapidithrix thailandica*	[34]	Inactive to acetylcholinestertase	[34]
20	Polycitone B		Marine ascidian *Polycitor africanus*	[35]	Antitumor activity	[36]

21	Prodigiosin R1		*Streptomyces griseo-viridis*	[37]	Active against cervical carcinoma	[38]
22	Pyranigrin D		Sponge Fungus *Aspergillus niger*	[39]	NR	NR
23	Pyrrolomycin B. Ten more congeners are known.		*Actinomycete* strain SF-208	[40]	Antibacterial activity	NR
24	Pyrrolo[2,1-c] oxazine-carboxaldeyde		*Celastrus orbiculatus* Thunb	[41]	NR	NR
25	Pyrrolo[1,2-a] pyrrolone		*Cynoglossum gansuense*	[42]	NR	NR
26	Pyrrolostatin		*Streptomyces chrestomyceticus*	[43]	Peroxidation inhibitory activity	[44]
27	Rhazinicine		*Kopsia dasyrachis*	[45]	Cytotoxic to vincristine-resistant cells	[46]

(*Continued*)

Table 6.1 (Continued)

Serial No.	Trivial name	Structure	Source	Isolation [Ref]	Bioactivity	Synthesis [Ref]
28	Rigidin		*Cystodytes sp.*	[47]	Weakly active to cancer cell lines	[48]
29	Rumbrin		*Auxarthron umbrinum*	[49]	Antioxidant activities	[50]
30	Solsodomine A		*Solanum sodomaeum L*	[51]	Active against mycobacterium sp.	NR
31	Sventrin		Caribbean Sponge *Agelas sventres*	[52]	Anti-feedant against reef fish	NR
32	Tambjamine A. Eleven congeners are also known.		*Tambje abdere, T. eliora,* and *Roboastra tigris*	[53]	Immunosuppressive activity	[54]
33	Zyzzyanones B. Two more congeners are also known.		Marine Sponge *Zyzzya fuliginosa*	[55]	Active against carcinoma	NR

† NR = not reported.

6.2
Synthesis of Pyrrole Natural Products

A pyrrole nucleus is constructed by one of the established routes which include: (i) Paal–Knorr synthesis; (ii) Knorr synthesis; (iii) Hantzsch synthesis; (iv) van Leusen synthesis; (v) Barton–Zard synthesis; (vi) Gupton synthesis; (vii) Kenner synthesis and many others as found in the standard text books [4]. These methods, particularly Paal–Knorr synthesis, have been successfully applied to fabricate pyrrole nuclei of the natural products over the years. However, their applicability is reliant on the nature of N-pyrrole protection and the presence of other N-containing sensitive functionalities. One of the useful but subsidiary strategies is the installation of an ester group in the nuclei to prevent polymerization of the pyrrole rings and moderate their reactivity. This strategy is not devoid of drawbacks because the decarboxylation of the 2-carboxy groups becomes an additional problem [56]. With the advent of pyrrole natural products, the chemistry of aromatic pyrrole rings has undergone many innovations involving transition metal-catalyzed ring closures and cross-couplings. In the following sections we present recent examples to illustrate total syntheses in categories based on the number of pyrrole units, that is, monopyrroles, dipyrroles and tripyrroles.

6.2.1
Monopyrrolic Natural Products

Mononuclear pyrrole natural products are typically found to contain an electron-deficient aromatic ring and an amide function in C2 of the pyrrole unit. Additionally, they often contain a bromine and/or chlorine, especially when isolated from the marine sponges. Arylpyrroles represented by pyrrolinitrins (e.g., **7**) and pseudilins (e.g., **15**) have attracted considerable attention for their antibacterial activities [57]. Pratt synthesis of pyrrolinitrin **7** used the effect of triisopropylsilyl (TIPS) group in directing halogenation at C3 and C4 positions rather than more electrophilic C2 and C5 positions (Scheme 6.1) [58]. The use of the halogenated derivatives in Suzuki coupling thus culminated in efficient and general synthesis of arylpyrroles. Bromo compound **2** was obtained by selective monobromination of N-protected pyrrole **1** with NBS. Lithiation of bromopyrrole **2** with t-BuLi followed by quenching of the lithio salt with C_2Cl_6 gave chloropyrrole **3**. Direct chlorination of **1** to **3** was not achievable. Borylation of **4**, obtained by treatment of **3** with NBS, with pinacolboron gave Suzuki substrate **5**, which was reacted with 2-bromo-6-chloronitrobenzene in the presence of $Pd(OAc)_2$ to give aryl pyrrole **6**. Desilylation of **6** with tetrabutylammonium fluoride (TBAF) furnished natural product pyrrolinitrin **7**.

The Knölker approach to halogenated arylpyrroles entailed an entirely new synthesis of pyrroles. As an example, the synthesis of pentabromopseudilin **15** is presented (Scheme 6.2) [59]. The pyrrole ring of **15** was constructed by silver(I) catalyzed cyclization of aryl substituted homopropargylamine **11**, which was prepared by addition of trimethylsilylpropargylmagnesium bromide **9** to

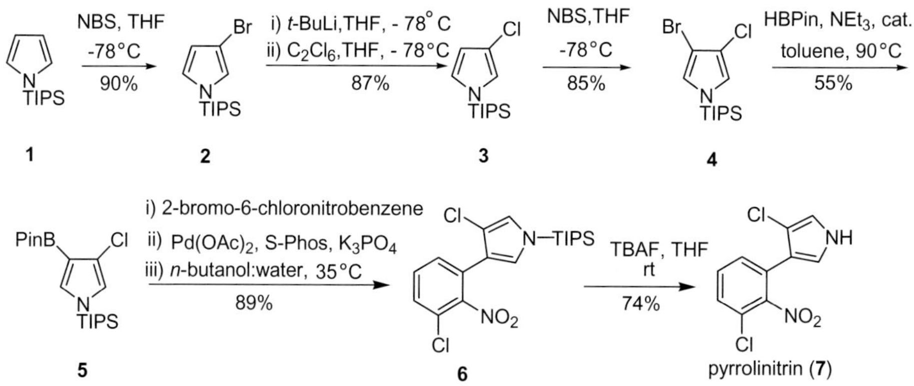

Scheme 6.1 Pratt synthesis of pyrrolinitrin.

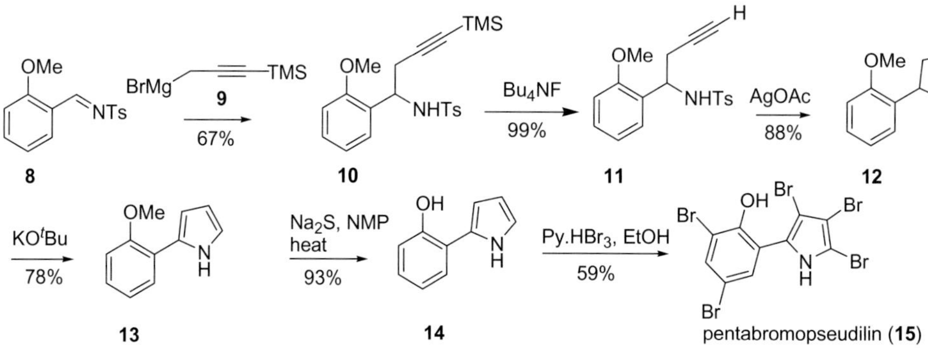

Scheme 6.2 Knölker synthesis of pentabromopseudilin.

N-tosylaldimine **8**, followed by treatment of the resulting acetylene desilylation with TBAF. Silver(I) acetate-catalyzed ring closure of **11** in acetone or dichloromethane under reflux temperature provided pyrroline **12** in excellent yield. Treatment of **12** with potassium *t*-butoxide (*t*-BuOK) in DMSO promoted elimination of *p*-toluenesulfinic acid to furnish aryl pyrrole **13**. Cleavage of the methyl ether **13** with sodium sulfide afforded pseudilin **14**. Subsequent bromination of **14** using pyridinium tribromide afforded pentabromopseudilin **15**. In contrast, the synthesis of pentachloropseudilin began with a prechlorinated Schiff base in the place of **8** and involved a sequence similar to that for pentabromopseudilin.

Fürstner platinum(II)-catalyzed enyne metathesis is an unconventional route for synthesizing the *meta* pyrrolophane unit of streptorubin B [60]. The metathesis involves cleavage of a C=C bond leading to the ring expansion of pre-existing ring with a new C=C double bond on a bridge atom as depicted in Scheme 6.3. Although it is suggested that the "enyne metathesis" proceeds through a "nonclas-

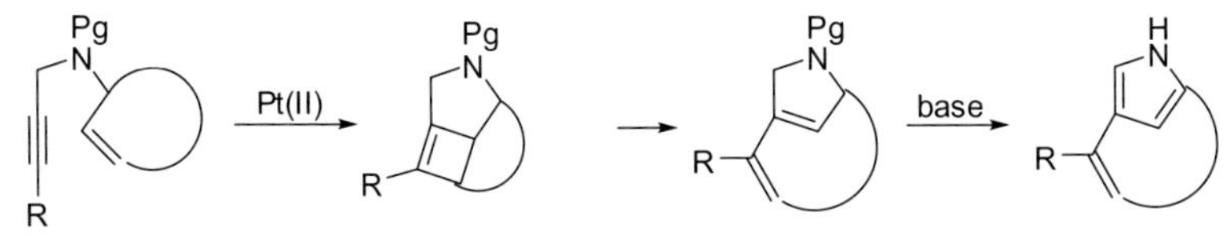

Scheme 6.3 Fürstner pyrrole synthesis.

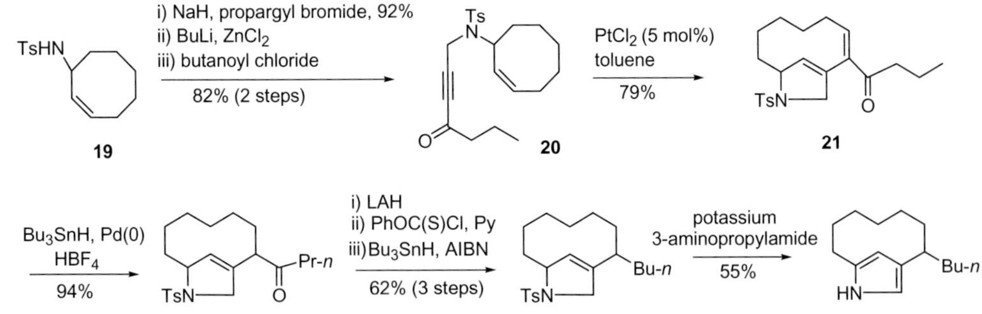

Scheme 6.4 Fürstner synthesis of streptorubin B.

sical" carbocation intermediate, the formalism involving intermediates **17** and **17'** explains the transformation. Base-catalyzed elimination of the protecting group (Pg) results in the aromatization of the pyrroline ring of **17** and gives *meta*-pyrrolophane **18**.

The actual route for the total synthesis of streptorubin B began with cyclooctene **19**, which was obtained by allylic amination of cyclooctene with chloramine-T and selenium (Scheme 6.4) [60]. *N*-Propargylation of **19** with propargyl bromide followed by acylation of the resulting alkyne with butyryl chloride furnished enyne **20**. Treatment of **20** with $PtCl_2$ (5 mol%) in toluene delivered metathetic product **21** in 79% yield. Conjugate reduction of **21** with Bu_3SnH gave **22**. A three-step (LAH reduction, xanthate formation and BTH reduction) sequence applied to the pyrroline **22** gave side-chain deoxygenated pyrroline **23**. Treatment of **23** with potassium 3-aminopropylamide furnished metapyrrolophane **24** via elimination of *p*-toluenesulfinic acid. The *m*-pyrrolophane **24** yielded streptorubin B, on condensation with the appropriate dipyrrolic aldehyde.

Lamellarins, first isolated in 1985, represent pyrrolic natural products with a carboxy function at C2 and highly oxygenated aryl groups at C3, C4 and/or C5. This polycyclic group of natural products has received enormous importance by virtue of their biological activities and structural diversity. They are closely related to the pyrrole natural products, lukianols, polycitrins, polycitones, ningalins and sorniamides. The general structural features of these natural products are shown in Scheme 6.5.

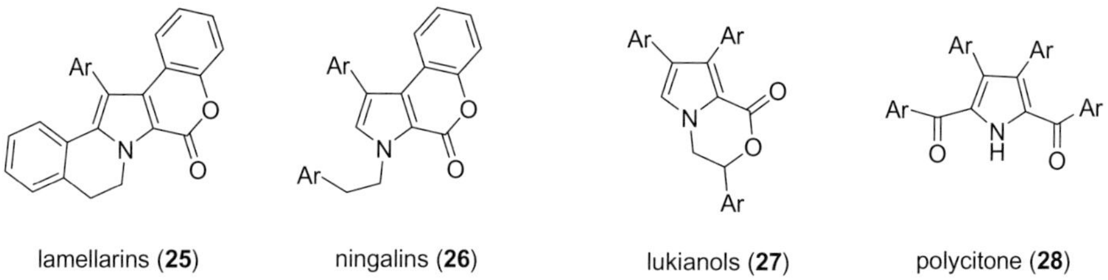

lamellarins (25)   ningalins (26)   lukianols (27)   polycitone (28)

Scheme 6.5 Structural patterns of lamellarins and related compounds.

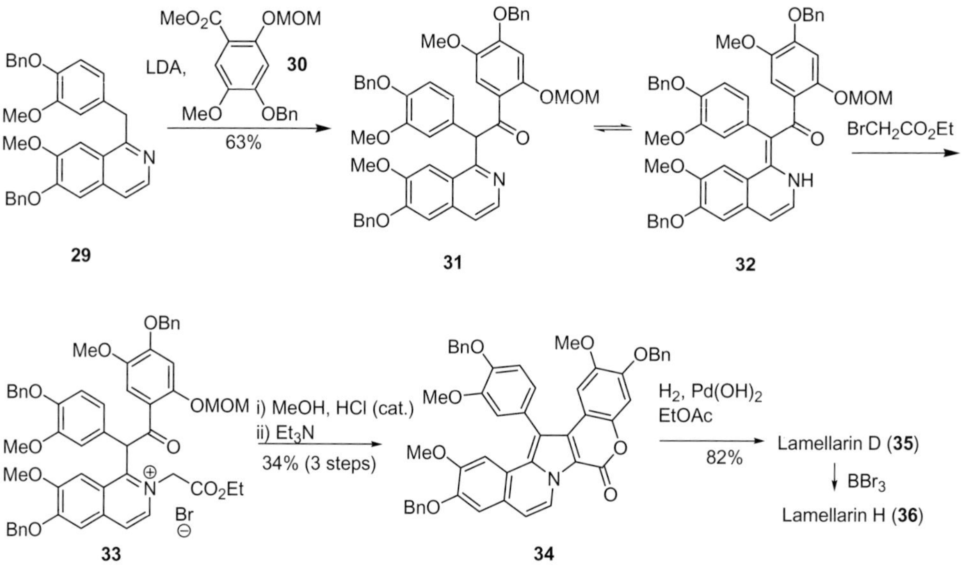

Scheme 6.6 Ishibashi synthesis of lamellarin D and H.

The chemistry of this group has recently been reviewed [20, 61]. The synthetic studies on the natural products have revalidated the importance of Paal–Knorr pyrrole synthesis and many text book syntheses. In addition, they demonstrated utility of cross-couplings and inverse electron demand Diels–Alder reactions. The first total syntheses of lamellarin D **35** and H **36** were accomplished by Ishibashi group in five steps, starting from readily accessible benzylisoquinoline **29** and the oxygenated benzoate **30** (Scheme 6.6) [62]. Condensation of the substrates via acylation of the lateral benzylic anion afforded **31**/ **32**. The mixture of **31** and **32** was alkylated with ethyl bromoacetate to produce iminium salt **33**. Cyclization of the salt **33** with triethylamine gave pentacycle **34**. Debenzylation of **34** with H_2/Parlman catalyst yielded lamellarin D **35**, which was demethylated with BBr_3 to provide lamellarin H **36**.

The synthesis of lamellarin H **36** was also achieved by the Faulkner group and the Banwell group [63]. The key reactions were Sonogashira coupling and intramo-

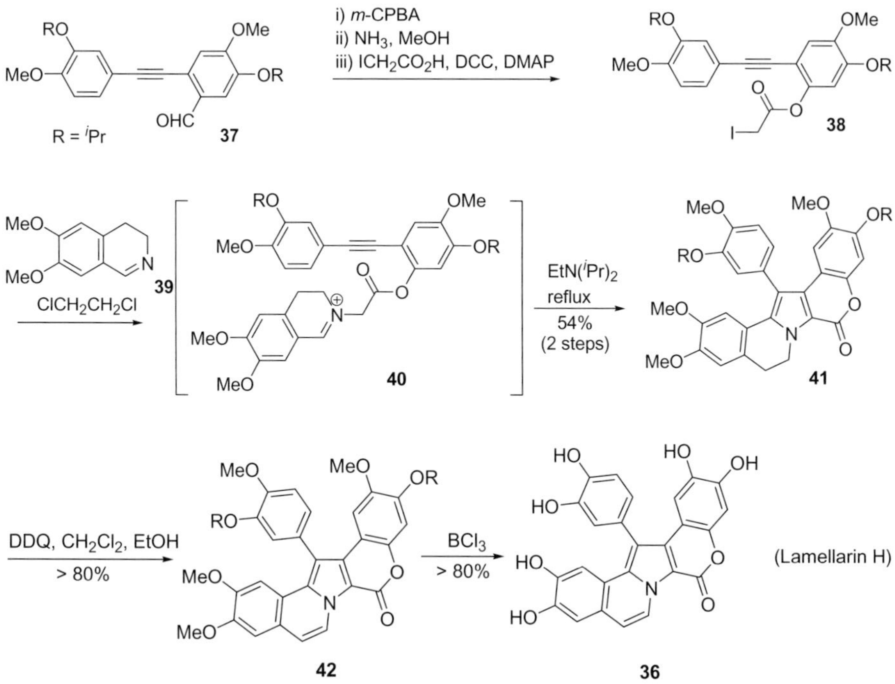

Scheme 6.7 Faulkner synthesis of lamellarin H.

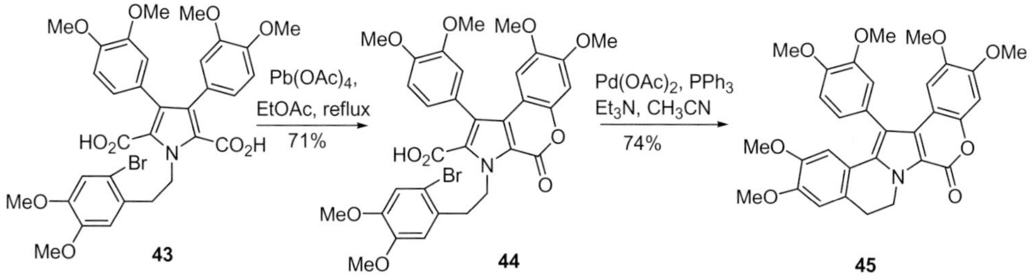

Scheme 6.8 Steglich synthesis of lamellarin G trimethyl ether.

lecular 1,3-dipolar cycloaddition of an unactivated alkyne. The strategy allowed construction of two heterocyclic rings in one operation. Alkyne **37**, assembled by a Sonogashira coupling was elaborated to **38** through a Baeyer–Villiger reaction and an esterification with iodoacetic acid (Scheme 6.7). Quanternization of **39** with **38** followed by *in situ* treatment with Hunig base promoted intramolecular [3 + 2] cycloaddition of **40** and furnished **41** with entire framework of lamellarin H **36**. Sequential reactions of **41** with DDQ and BCl₃ yielded lamellarin H **36** in good yield.

Steglich synthesis of lamellarin G trimethyl ether **45** (Scheme 6.8) demonstrates that a formal substitution at C2 of a pyrrole ring can be achieved via decarboxylative intramolecular Heck reaction (**44**→**45**) [64]. The precursor **44** was constructed via lead tetraacetate mediated oxidative lactonization of symmetric pyrrole dicarboxylic acid **43**, which, in turn, was prepared by a Paal–Knorr synthesis with a homobenzyl amine.

Gupton synthesis (Scheme 6.9) of lamellarins involves construction of a 2,4-disubstituted pyrrole motif by condensation of an α-amino ketone with a vinaminidium salt [65]. In the total synthesis of polycitone, phenylacetic acid **46** was treated with POCl$_3$-DMF to give vinaminidium salt **47**, which was reacted with amino ketone **48** to produce 2-aroylpyrrole **49**. The second aroyl group of the target was installed by Friedel–Crafts (FC) reaction of **49** with *p*-anisic acid **50** in the presence of trifluoroacetic anhydride (TFAA). The aryl group at C4 was then incorporated by a Suzuki coupling reaction between **51** and **52**. The resulting tetrasubstituted pyrrole **53** was then transformed to polycitone via an AlI$_3$ mediated O-demethylation and a Mitsunobu N-alkylation of the NH group. Although 3-iodopyrrole **51** actively underwent the Suzuki coupling reaction, the corresponding bromo analog was inert to the reaction with the boronic acid **52**.

In early investigations on the synthesis of pyrrolic natural products, the classical methods of pyrrole formation have been used. These methods are now being complemented by one or more of the cross-couplings such as Heck, Sonogashira, Stille, Suzuki, etc. It is not an exaggeration to say that modern total syntheses often take recourse to at least one cross-coupling. Handy synthesis of lamellarin G (Scheme 6.10) is an example where three different Suzuki couplings under the same set of reaction conditions have been utilized to complete the total synthesis [66]. The product of the first Suzuki coupling **56**, obtained from **54** and **55**, was selectively brominated with NBS to give substrate **57**. The second Suzuki coupling between **57** and **58** provided pyrrole **59**. A further 3-step sequence applied to **59** furnished bromopyrrole **60**. The last Suzuki coupling, that is, the coupling between **60** and **61** resulted in the formation of lamellarin G trimethyl ether **62**. Although this step was problematic, the yield improved to 46% under optimized conditions.

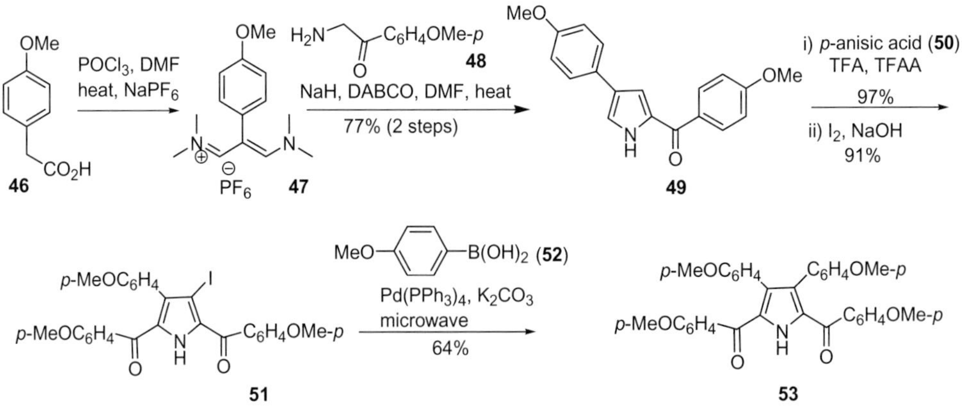

Scheme 6.9 Gupton polycitone synthesis.

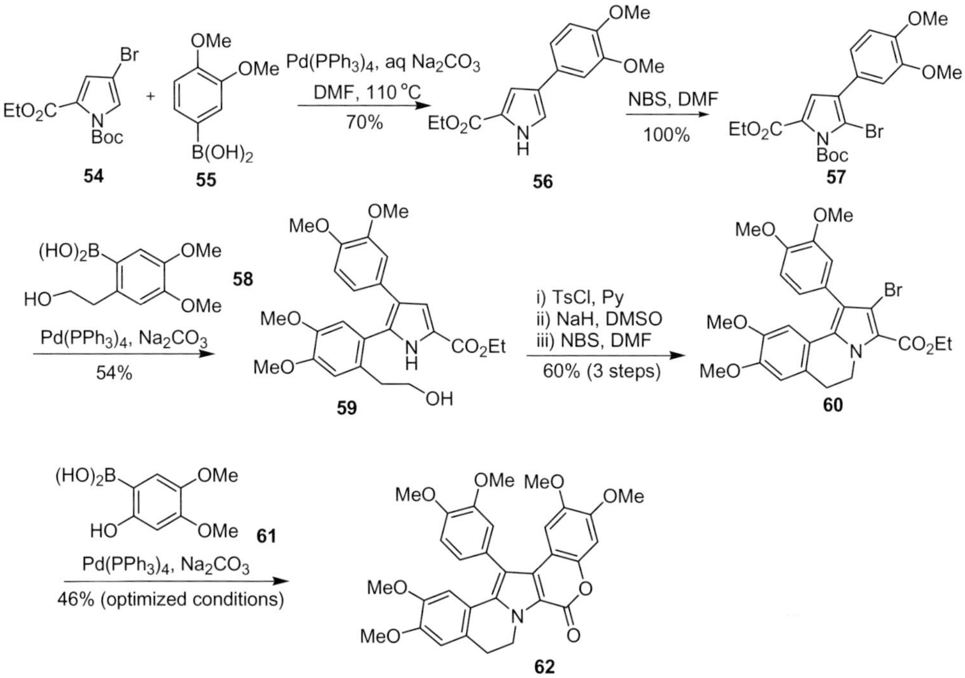

Scheme 6.10 Handy synthesis of lamellarin G trimethyl ether.

It is interesting to note that the yields of the couplings progressively decrease with increasing steric bulk around the pyrrole nuclei. In the Handy synthesis, they are 70%, 54% and 46%, although the same set of experimental conditions were used.

The oroidin (pyrrole-imidazole) group is now a well-known class of alkaloids. Since the first isolation of an oroidin in 1971, a few hundreds of such compounds have been reported. These alkaloids are viewed as a pyrrole derivative conjugated to an imidazole moiety through an amide linker. Almost all the members contain a pyrrole unit and vary widely in the imidazole segments. The structural diversity arises from the intramolecular interaction between the pyrrole units and the imidazole units as is apparent from the structural overview of the sub-class (Scheme 6.11). The majority of the members feature bromo substituted pyrrole-2-carboxy derivatives.

The total synthesis of oroidin natural products is generally accomplished in two stages: (i) coupling of pyrrole units to imidazole units and (ii) pyrrole annulation or vice-versa. The chemistry of the alkaloids has been comprehensively reviewed by the groups led by Papeo and Weinreb [67, 68]. Herein, we highlight selected methods of conjugation of pyrrole carboxamide in Schemes 6.12–6.14. The Birman group used conjugation (Scheme 6.12) through selective amidation with a trichloroacylpyrrole derivative in the late stage of their synthesis of sceptrin [69]. Trichloroacetyl derivative **71** smoothly reacted with diamine **72** to give diamide

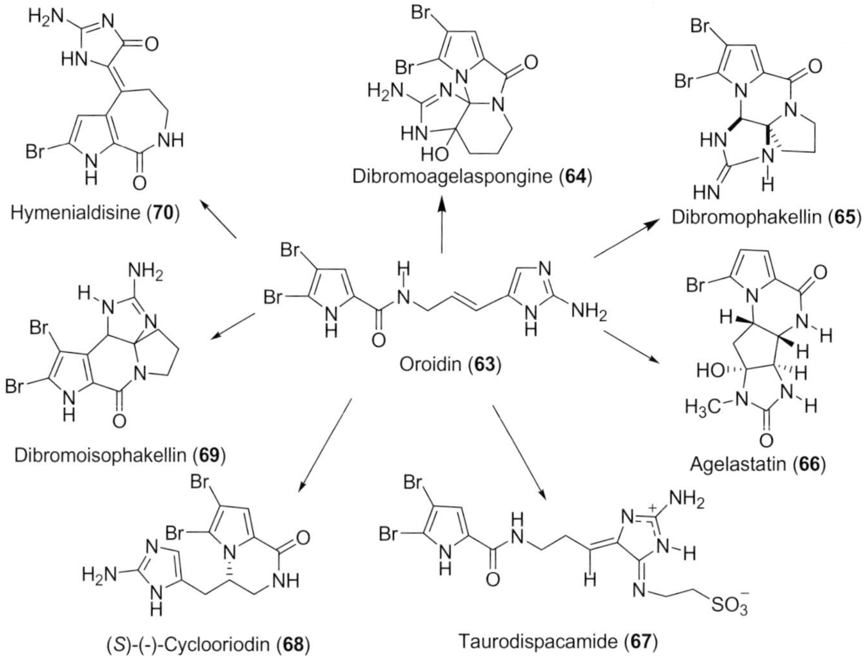

Scheme 6.11 A schematic view of the oroidin skeletons.

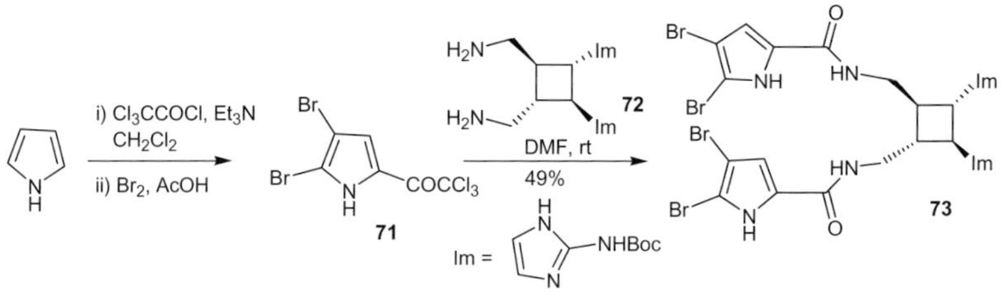

Scheme 6.12 Birman approach.

73. Such formation of an amide bond is a popular method in pyrrole chemistry [70, 71].

Weinreb et al. used different tactics to conjugate the amides. For example, primary amide **75** was reacted with cis-iodoalkene **74** under Buchwald conditions to produce cis enamide **76** in their synthesis of ageladin A (Scheme 6.13) [72]. Bromine substituents of **75** remained intact in the reaction.

In the synthesis of nagelamide D, Lovely and co-workers adopted a double Mitsunobu reaction between **77** and **78** which produced N-alkylated product **79** (Scheme 6.14) [73]. The pyrrolohydantoin rings of **79** were then selectively hydro-

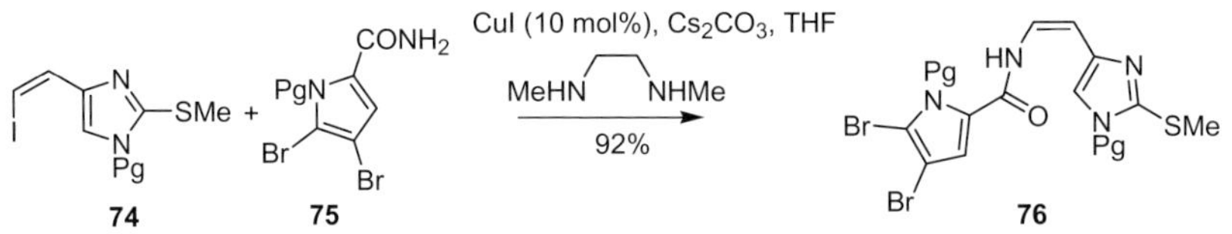

6.2 Synthesis of Pyrrole Natural Products | 201

Scheme 6.13 Weinreb approach.

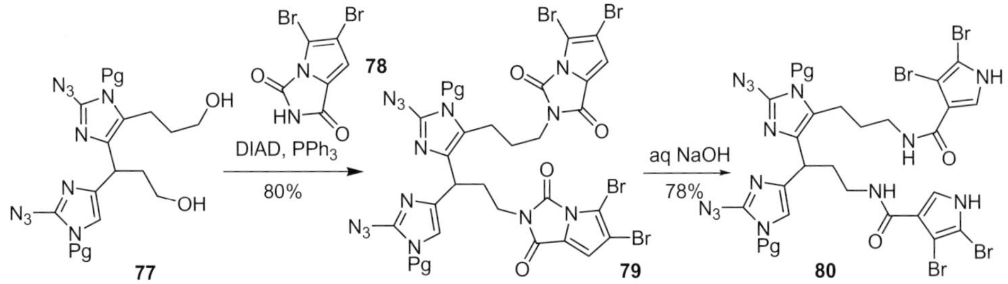

Scheme 6.14 Lovely's conjugation.

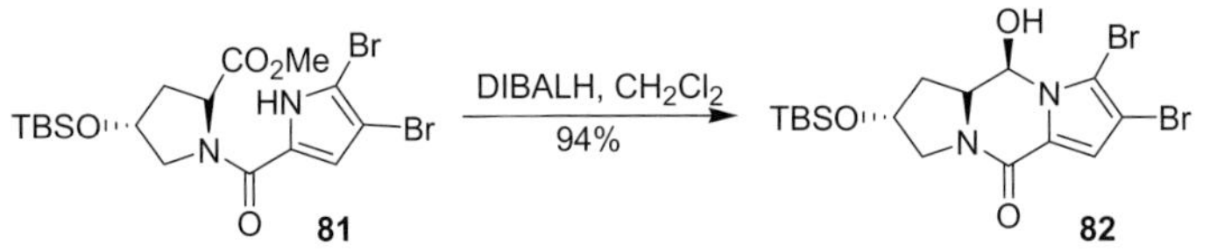

Scheme 6.15 Lindel cyclization.

lyzed to give the conjugate **80**. PPh$_3$ was reacted with DIAD to prevent the competitive formation of the iminophosphorane which is normally formed in the reaction between PPh$_3$ and the azide moiety.

For the synthesis of annulated pyrrole structures, various methods of cyclization of the linear conjugates of pyrrole-imidazoles are now available. As illustration, we cite selected reactions in Schemes 6.15–6.19. As in Scheme 6.15, the Lindel group constructed the annulated piperazinone **82** via intramolecular addition of the pyrrole "N" to an aldehyde produced in situ by DIBALH reduction of ester function [74].

Papeo synthesis of longamide B methyl ester **85** was performed in four steps from **83** (Scheme 6.16) [75]. The stable pyrrolyl aminol **84** was prepared in two steps from trichloroacetyl derivative **83** via amidation with 2,2-dimethoxyethanamine followed by acid-catalyzed deketalization. Reaction of **84** with the anion of trimethylphosphonoacetate followed by LiOH treatment yielded piperazinone annulated pyrrole **85**. The formation of **85** proceeded through Wittig–Horner reaction followed by intramolecular aza-Michael addition.

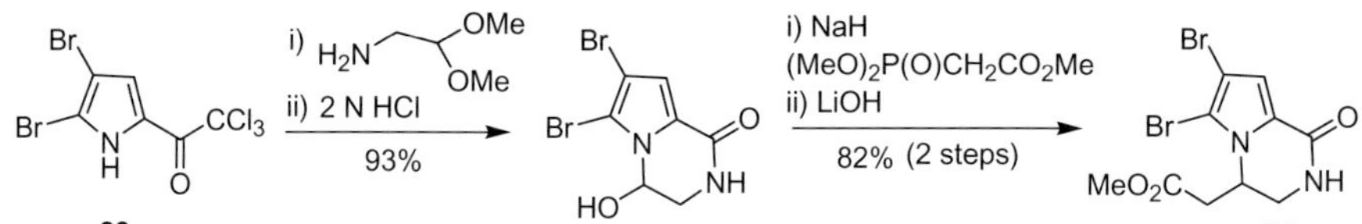

83 84 85

Scheme 6.16 Papeo synthesis of longamide B.

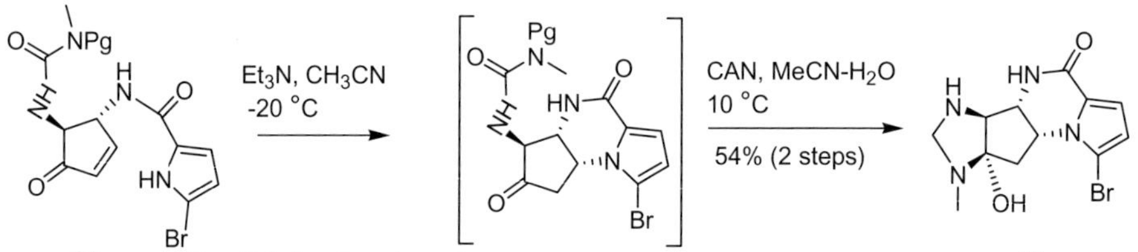

86 Pg = 2,4-dimethoxybenzyl 87 88

Scheme 6.17 Chida cyclization.

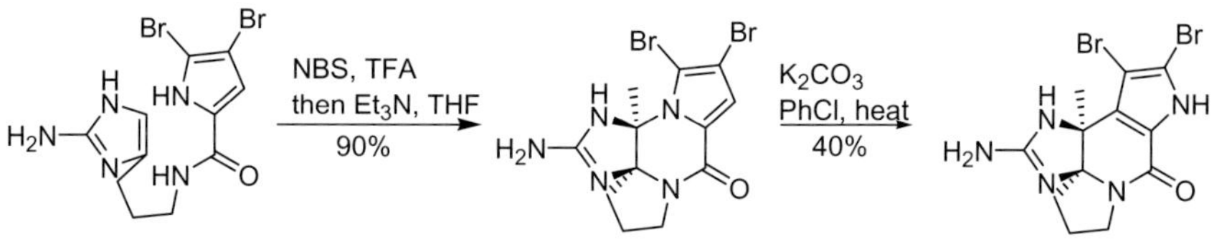

Scheme 6.18 Horne annulation to isophakellin.

Aza-Michael addition is one of the commonly pursued strategies for *N*-alkylation of the pyrrole nucleus. It constitutes one of the late steps in the synthesis of agelastatin **66** (Scheme 6.17) [76a]. The product of the addition **87** was not isolated. It was directly converted to **88**. The successes depend on the reaction conditions [76b].

Horne group employed bromine induced annulation that is, **89**→**65** (Scheme 6.18). The second step of the approach is noteworthy. Isomerization of **65** took place under basic conditions to give **69** with a new C–C bond.

Extension of the Pummerer reaction is an effective way of annulating piperazinone (Scheme 6.19) [77]. In the presence of Stang reagent, PhI(CN)OTf, the pyrrole-imidazole conjugate **90** underwent annulation to give the structural core of **93**. One of the possible mechanisms could be sequential formation of intermediates **91** and **92**.

Trost and Dong constructed the pyrrolopiparazinone framework of agelastatin A **66** by two successive Pd-catalyzed alkylations (Scheme 6.20) [78, 79]. Pyrrole nitrogen of **95** was sufficiently nucleophilic to undergo Tsuji–Trost alkylation with

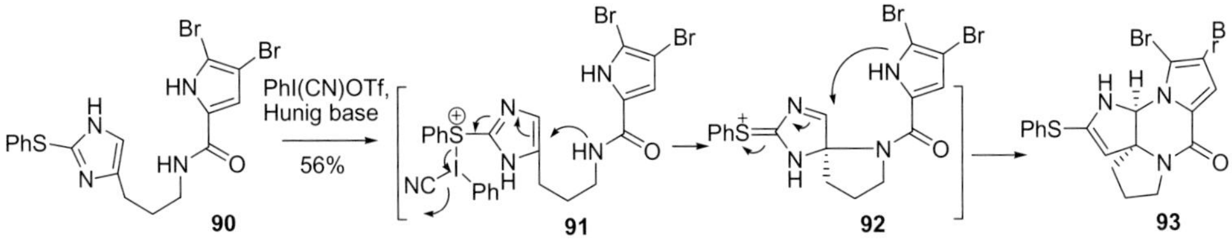

Scheme 6.19 Feldman annulation.

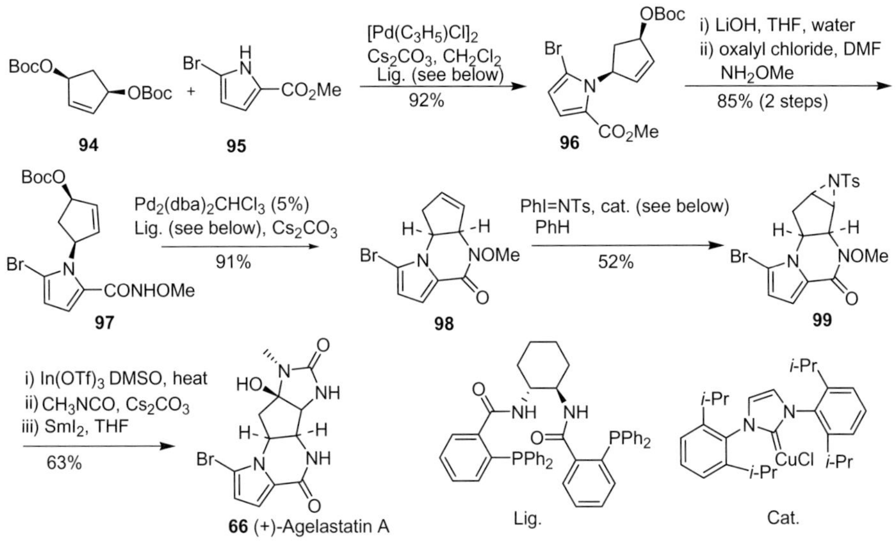

Scheme 6.20 Trost synthesis of agelastatin A.

Boc-protected derivative **94**, giving **96** in enantiomerically pure form. Similarly, the side chain N-methoxyamide nitrogen of **97** underwent intramolecular alkylation to give tricyclic heterocycle **98**. Further elaboration of **99** through aziridination, oxidative cleavage of aziridine ring and urea formation with methylisocyanate resulted in the total synthesis of (−)-agelastatin **66**.

6.2.2
Dipyrrolic Natural Products

Natural products featuring a bipyrrole unit are frequently found in marine organisms. Most of them are either brominated or chlorinated or both. The linkage between the pyrrole units is either 1,2′ or 2,2′. The Gribble group brought bipyrroles to the forefront of scientific interest and contributed significantly to the newer aspects of pyrrole chemistry. In the synthesis of bipyrrole **103**, reaction of N-methylpyrrole with N-chlorosuccinimide (NCS) **100** was exploited to join two pyrrole units in 1,2′ fashion giving **101** (Scheme 6.21) [80]. It is also demonstrated that the succinimide **101** transforms to bipyrrole **102** on reduction with triethoxysilane in the presence of Ti(O^{i}Pr)$_4$. Introduction of the nuclear chlorine atoms was performed by the use of SO_2Cl_2 albeit in only 16%. 2,2′-Bipyrrole **107** was synthesized by exploiting the nucleophilic character of 1-methylpyrrole at C2 toward an iminium chloride (Scheme 6.22). The reaction of methylpyrrole with **104** in the presence of $POCl_3$ gave **105**, which was aromatized with Pd-C to give **106**. Perchlorination of **106** with SO_2Cl_2 provided only detectable amount of **107**.

Synthesis of mixed halogenated 2,2′-bipyrroles has not been reported so far. 1,3′-Bipyrroles have been synthesized from 3-nitropyrrole derivatives via Paal–Knorr condensation. In the synthesis of the core of marinopyrroles (Scheme 6.23), the Paal–Knorr condensation of **108** with **109** afforded bipyrrole **110**, which on bromination with NBS gave **111**.

2,2′-Bipyrroles are also embodied in the tambjamine family which numbers around a dozen. They are isolated from both marine and terrestrial sources. The synthesis developed by the Banwell group display varied aspects of chemistry of 2-substituted pyrroles, which includes an organosilicon promoted ipso-substitution (Scheme 6.24) [57]. Boc-protected pyrrole was efficiently dibrominated at C2 and

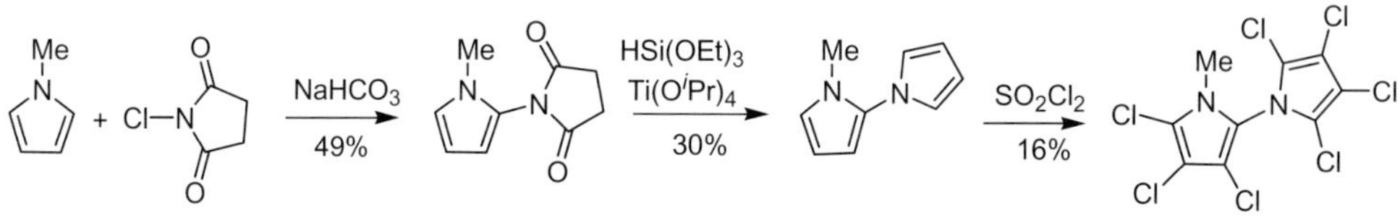

100 **101** **103**

Scheme 6.21 Gribble synthesis of bipyrrole **103**.

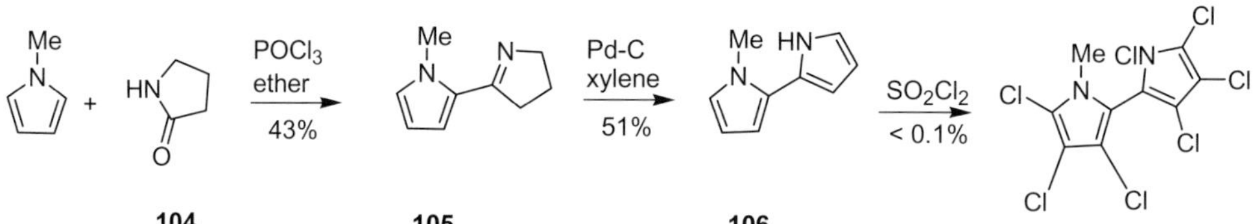

104 **105** **106** **107**

Scheme 6.22 Gribble synthesis of 2,2'-bipyrrole **107**.

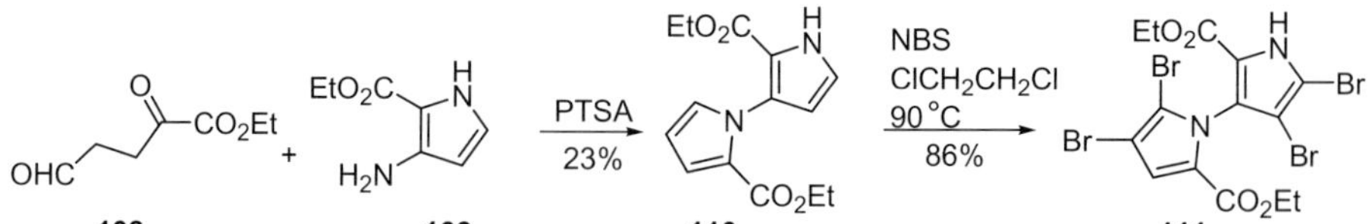

108 **109** **110** **111**

Scheme 6.23 Synthesis of the marinopyrrole core.

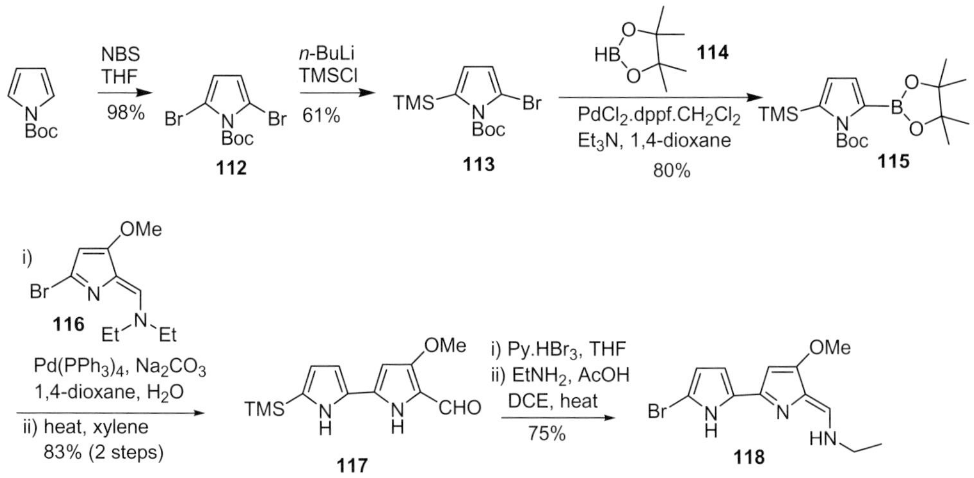

Scheme 6.24 Banwell synthesis of tembjjamine G.

C5 positions to give **112**. Through lithiation, one of the bromine atoms was replaced by a silyl group to give **113**. The remaining bromine was substituted by pinacolboron to furnish **115**. Suzuki coupling between **115** and azafulvene **116**, obtained in one step from the corresponding pyrrolinone, followed by thermal removal of the Boc group provided the desired bipyrrole **117**. Ipso substitution of the silyl group in **117** with bromine by treatment with pyridinium hydrobromide perbromide resulted in the corresponding bromo compound. The synthesis of tambjamine **118** was completed by enamination of the intermediate aldehyde **117** with ethylamine. This strategy was broadly applicable to the synthesis of seven similar natural products.

6.2.3
Tripyrrolic Natural Products: Prodigiosins

Prodigiosins, the naturally occurring pigments, are characterized as dipyrroles linked to a third pyrrole unit by a methane linker. For the sake of distinction, we have classified them as tripyrroles. The central pyrrole rings have azafulvene structures and are proposed to exist in two geometrical forms in solution. Their pharmacological profiles qualify them as potential immunosuppressants. With the synthesis of dipyrroles discussed in Section 6.2.2, it would suffice to include only how the prodigiosin frameworks are assembled from a dipyrrole and a monopyrrole in three different modes [81]. The transformation **119→120** represents a FC type hydroxyalkylation of a monopyrrole with a bipyrrole. Transposition of the aldehyde group between the pyrrolic units and their reaction offer another approach. The third one, that is **121→120**, involves a Suzuki coupling (Scheme 6.25).

The earliest synthesis of a prodigiosin is due to Rapoport and Holden (Scheme 6.26) [6]. It features many interesting facets of pyrrole chemistry and functional

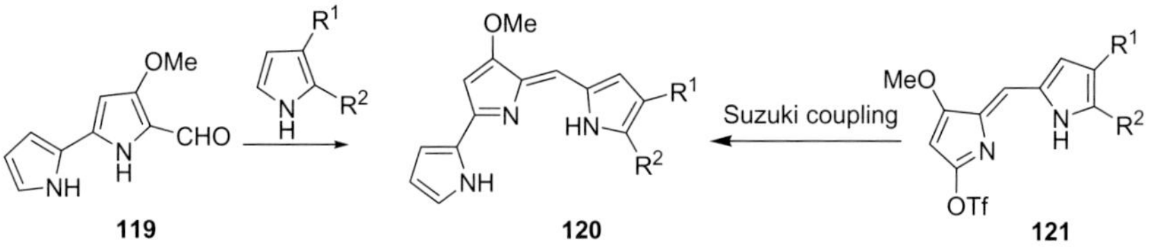

Scheme 6.25 Approaches to prodogiosins.

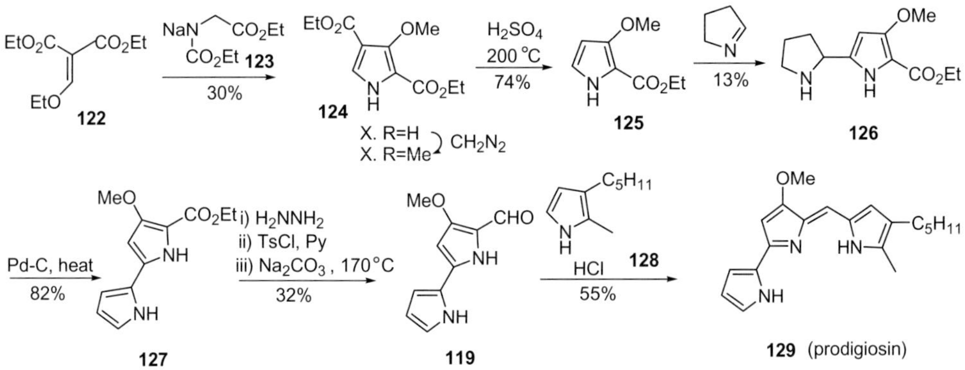

Scheme 6.26 Rapoport synthesis of prodigiosin.

group transformations. Uncommon condensation of glycinate **123** with ethoxymethylene malonate **122** was adopted to construct the pyrrole of **124**. This was then O-methylated in 44% yield by treatment with CH_2N_2 and subsequently subjected to reaction with H_2SO_4 at 200 °C for selective deethoxycarbonylation to give **125**. The resulting mono ester underwent an addition reaction with 1-pyrroline to give bipyrrole derivative **126**, which was duly dehydrogenated to give bipyrrole **127**. Application of McFayden–Stevens reduction protocol (acyl hydrazide formation, tosylation and base treatment) to **127** afforded dipyrrolic aldehyde **119**. Condensation of **128** with dialkyl pyrrole **128** in the presence of HCl led to prodigiosin **129**. The only drawback of the approach is the poor yields. However, this approach laid the foundation for newer approaches.

Boger synthesis (Scheme 6.27) of prodigiosin relied on the use of the Rapoport aldehyde [82]. It was synthesized from diazine **132**, which was prepared by DA-RDA of tetrazine **130** and 1,1-dimethoxyethene **131**. A circuitous route to decarboxylation was performed on **133**. Reaction of **133** with $I_2/NaHCO_3$ gave the diiodo derivative **134**, hydrogenation of which furnished **135**. Necessary bipyrrole carbonyl **137** was prepared by condensation of **135** with **136** in the presence of NaH. Pd-mediated oxidative cyclization of **137** in the presence of **138** resulted in the tricyclic intermediate **139**, which was selectively cleaved to give 2,2′-bipyrrole aldehyde **119** through hydrolysis and reduction.

Thomson synthesis of metacycloprodigiosin **147** shows the versatility of the Paal–Knorr pyrrole synthesis (Scheme 6.28) [38]. Macrocyclic acyl ketone **140**

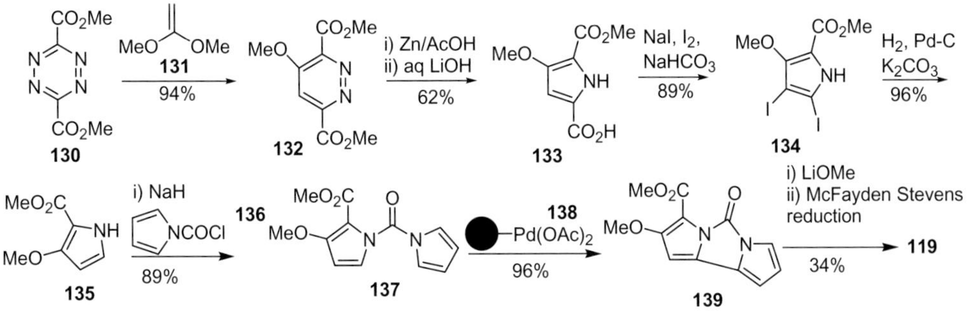

Scheme 6.27 Boger and Patel route to bipyrrole aldehyde.

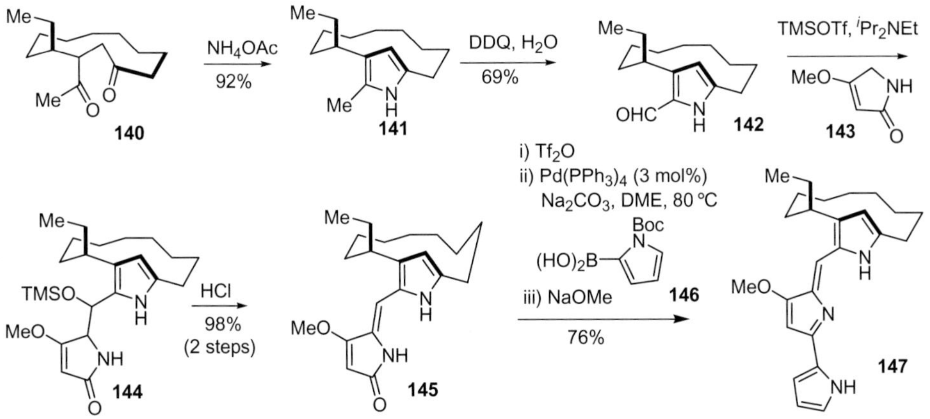

Scheme 6.28 Thomson synthesis of metacycloprodigiosin.

underwent Paal–Knorr reaction with ammonium acetate to yield meta-pyrrolophane **141** in 92% yield. DDQ oxidation of **141** in aqueous acetonitrile caused oxidation of the pyrrolic methyl group to give carboxaldehyde **142**. The second pyrrole ring was conjugated to the aldehyde **142** through aldol type condensation of lactam **143**. Third pyrrole unit was installed via a Suzuki coupling of Boc-protected boronic acid **146** with the triflate derivative of pyrrolylmethinepyrrole **145**.

Recently, a novel synthesis of 4,5-disubstituted 2-formylpyrroles from 4-formyloaxazoles has been reported by Senanayaka et al [83]. This is considered an extension of the Cornforth rearrangement on to a vinylogous system. As illustrated in Scheme 6.29, the aldol **148** was dehydrated to unsaturated system **149** with an oxazole ring, which underwent base-catalyzed rearrangement with aqueous NaOH to give pyrrole-2-carboxaldehyde **151** through **150**.

Reeves' synthesis of butylcycloheptylprodigiosin **156** by the above method of pyrrole formation is remarkably brief and innovative (Scheme 6.30) [84]. In five steps from hydroxyketone **152**, the prodigiosin **156** was regiospecifically synthesized in 23% overall yield. The synthetic strategy seems to be quite versatile. The

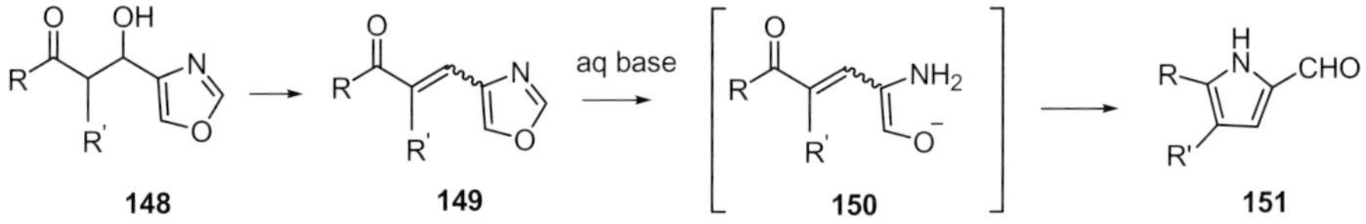

Scheme 6.29 Synthesis of pyrrolic aldehydes.

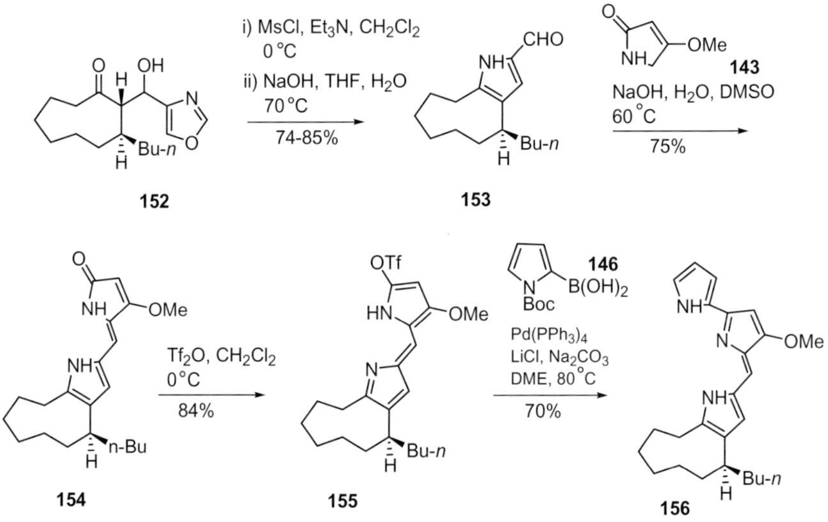

Scheme 6.30 Reeves' synthesis of butylcycloheptyl prodigiosin.

aldol **152** was converted to **153** in a two-step sequence, which was condensed with **143** to furnish **154**. Triflation of **154** with Tf$_2$O followed by Suzuki coupling of the resulting triflate **155** with boronic acid **146** provided prodigiosin **156**.

Although roseophilin **167** is not considered a tripyrrolic natural product, its structural resemblance with prodigiosin alkaloids warrants brief discussion. The major difference between a prodigiosin and a roseophilin is a methoxyfuran ring rather than a methoxypyrrole ring in the center of its backbone. Fürstner synthesis entails a new pyrrole synthesis, which is based on Pd-catalyzed pyrrole formation in [4 + 1] mode (Scheme 6.31) [85]. The macrocyclic ketone **157**, featuring a γ-carbonyloxy α,β-unsaturated moiety on reaction with BnNH$_2$ gave pyrrole derivative **158** via initial N-alkylation at the γ-position of the moiety. Intramolecular FC reaction of **158** in the presence of chloroenamine **159** gave tricycle **160**. Quite interestingly, treatment of **160** with t-BuOK generated bridge-head double bond towards pyrrole ring, en route to installation of the isopropyl group in the arylic position. Eventually through alteration of the protecting groups, that is, Bn to SEM group, the pyrrole moiety **164** was linked to binuclear unit **165** to furnish **166**. Desilylation of **166** with TBAF gave roseophilin **167**.

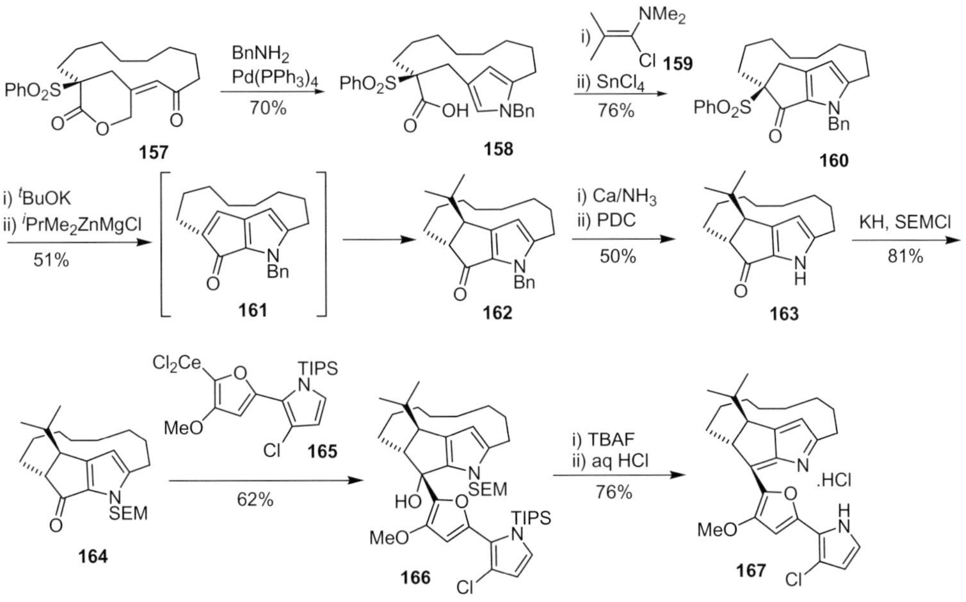

Scheme 6.31 Fürstner synthesis of roseophilin.

6.3
Synthesis of Non-pyrrole Natural Products from Pyrrole Derivatives

There are two or three major strategies that are prevalent for the synthesis of pyrrole-related natural products. Perhydrogenation of pyrroles to pyrrolidines is one of them. In recent years, Birch reduction to pyrroline derivatives has been a popular pathway for the synthesis of pyrrolidine and indolizidine natural products [86].

In the synthesis of monomorine **173**, unusual N-alkylation of the potassium salt of pyrrole, with a γ-butyrolactone has been exercised to fabricate the indolizidine framework (Scheme 6.32) [87]. The γ-pyrrolic acid, thus obtained, was methylated to give **169**. A BBr$_3$-promoted intramolecular FC reaction afforded indolizidine **170** in 90% yield. The keto function in **170** was then reduced to a methylene group with sodium cyanoborohydride in the presence of ZnI$_2$. The butyl chain of monomorine **173** was introduced by FC reaction with butyryl chloride with AgOTf in 20% yield. Perhydrogenation of **172** with H$_2$ in the presence of Pd-C, H$_2$SO$_4$, furnished (±)-monomorine **173**.

Donohoe *et al.* have recently shown that Birch reduction of pyrrole-2,5-dicarboxylate **174** is stereodivergent and stereoselective. Pyrrole **174** was readily prepared by dilithiation with lithium tetramethyl piperidide followed by quenching with methyl chloroformate (Scheme 6.33). Quenching of the reaction of **174** and Li in liquid ammonia with saturated ammonium chloride exclusively gives

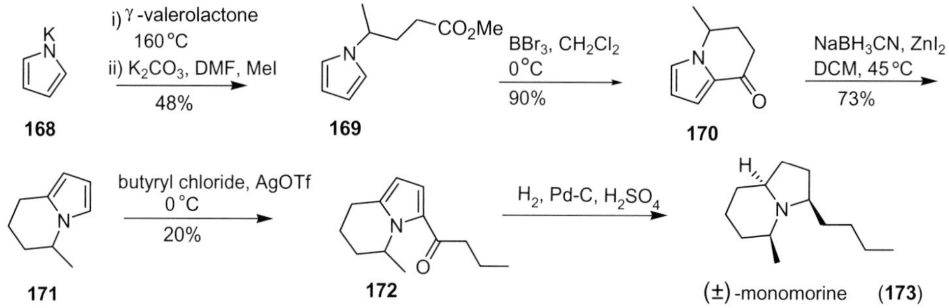

Scheme 6.32 Synthesis of monomorine.

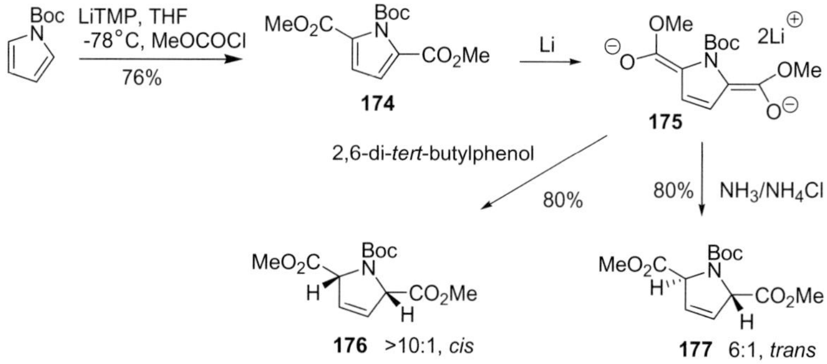

Scheme 6.33 Stereodivergent Birch reduction of pyrrole dicarboxylates.

trans pyrroline isomer **177** and that with 2,6-di-t-butylphenol gives cis-pyrroline isomer **176** as the major product [88].

The trans isomer **177** was converted to differentially functionalized pyrroline derivative **180** in five steps [89]. Polydihydroxylation of **177** with OsO$_4$ followed by protection of the resulting diol with dimethoxypropane afforded **178**. NaBH$_4$ reduction of **178** in a mixture of THF and methanol, followed by protection of the resulting alcohol afforded mono ester **185**. This was then further reduced to the aldehyde **180** by treatment with DIBALH. Aldehyde **180** was converted in several steps to (±)-1-epiaustraline **181**. (Scheme 6.34)

Robins' synthesis of isoretronecanol **187** exemplifies an extraordinary way of assembling an pyrrolizidine nucleus [90]. Regiospecific 1,3-dipolar cycloaddition of ethyl propiolate to in situ generated azomethine ylide **183** from hydroxyproline **182** afforded **184** in 84% yield. Routine functional group manipulations and perhydrogenation thereafter, resulted in the synthesis of enantiomerically pure (+)-isoretronecanol **187** through **185** and **186**. (Scheme 6.35)

The fact that N-Boc protected pyrrole undergoes Diels–Alder reaction with ethynyl tolyl sulfone was used as the pivotal step in a concise synthesis of

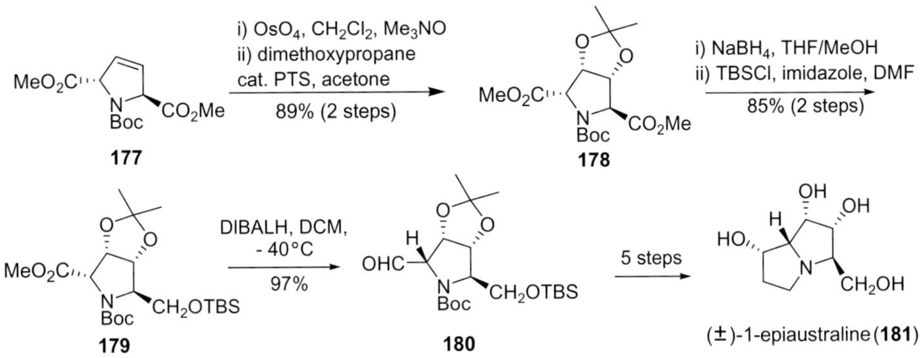

Scheme 6.34 Total synthesis of epiaustraline.

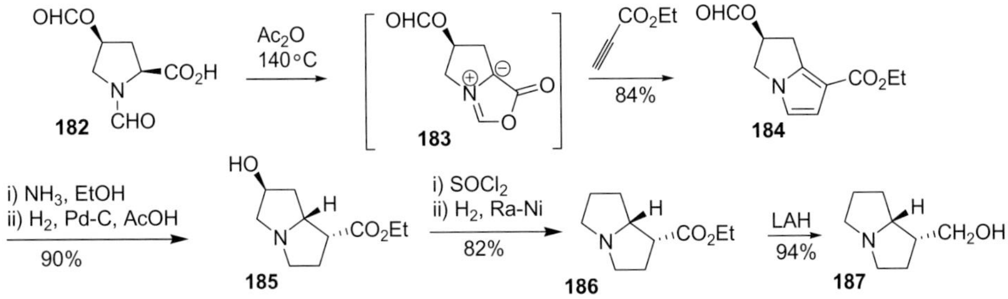

Scheme 6.35 Robins' synthesis of pyrrolizidine (+)-isoretronecanol.

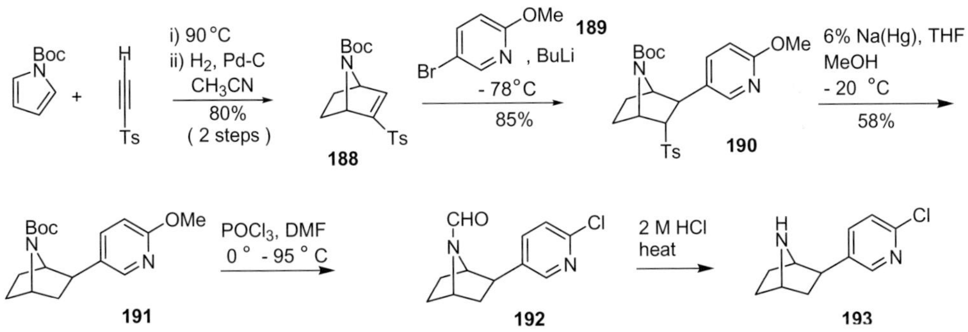

Scheme 6.36 Simpkins' total synthesis of epibatidine.

epibatidine **193** [91]. The cycloadduct **188** underwent conjugate addition with lithiopyridine derivative **189** to give [2.2.1] azabicycloheptane **190**. Removal of the sulfone group in **190**, substitution of aromatic OMe group by chlorine atom and deformylation afforded epibatidine **193** via **191** and **192** in good overall yield. (Scheme 6.36)

Unlike typical cycloaddition of a nitrone with alkenes and alkynes, the interaction of the cyclic nitrone **195** with the *N*-TIPS-protected pyrrole **194** underwent substitution at C3 position [92]. The resulting product **196** was converted to the glycine appended pyrrole **197** via hydrogenolysis. For the introduction of the methoxycarbonyl at C5, commonly used trichloroacetylation followed by methanolysis was adopted to furnish **199**. Perhydrogenation of **199** and subsequent selective RuO_4 oxidation gave pyrrolidinone **201**, which, on deprotection of the Boc group gave penmacric acid **202**. (Scheme 6.37)

The susceptibility of a pyrrole ring to oxidation has been successfully exploited in the synthesis of non-pyrrole natural products. Steglich synthesis demonstrated that pyrrole diacids could be oxidatively degraded to non-pyrrole derivatives (Scheme 6.38) [93]. Diacid **204**, prepared from **203** through an oxidative dimerization was degraded with NaOCl to give imide **205**, which was then converted to polycitrin A **208** in three steps through the intermediates **206** and **207**.

In the synthesis (Scheme 6.39) of isochrysohermidine **212**, the bipyrrolic nucleus **209**, obtained by a double Diels–Alder-retro Diels–Alder strategy followed by Boger contraction as the key step, was converted to pyrrole dicarboxylic acid **210**. It was oxidatively degraded with singlet oxygen to give isochrysohermidine **212** through fragmentation of the intermediate endoperoxide adduct **211** [94].

Aziridinyl and pyrrolic aminols are fairly stable and survive many organic transformations [95]. In the total synthesis of tarchonanthuslactone **220** (Scheme 6.40), accomplished by the Dixon group, a pyrrolic carbinol was utilized as an efficient stereocontrolling element for the synthesis of a 1,3-diol. Use of carbonyl

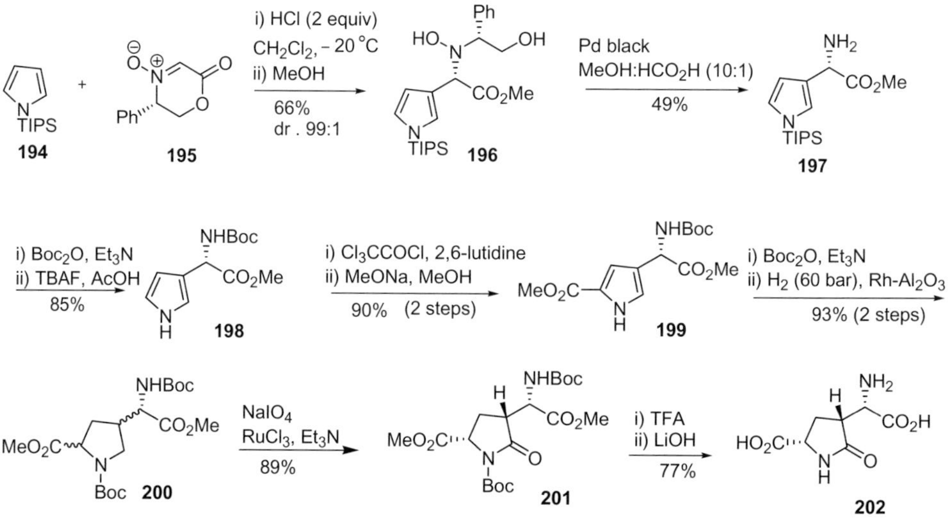

Scheme 6.37 Synthesis of penmacric acid.

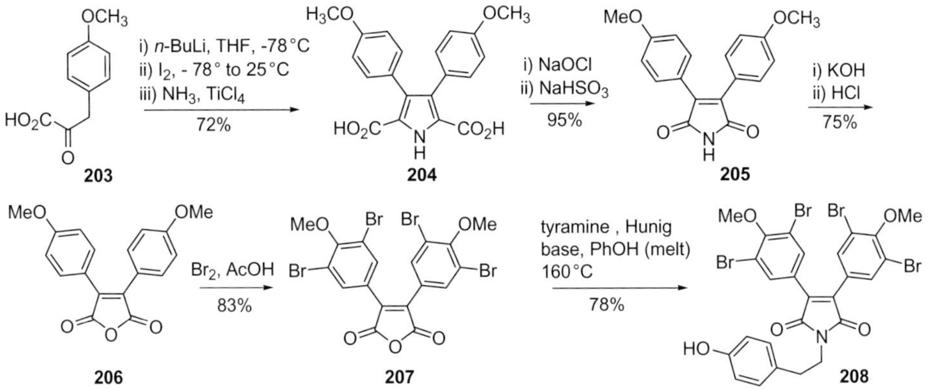

6.3 Synthesis of Non-pyrrole Natural Products from Pyrrole Derivatives | 213

Scheme 6.38 Steglich synthesis of polycitrin.

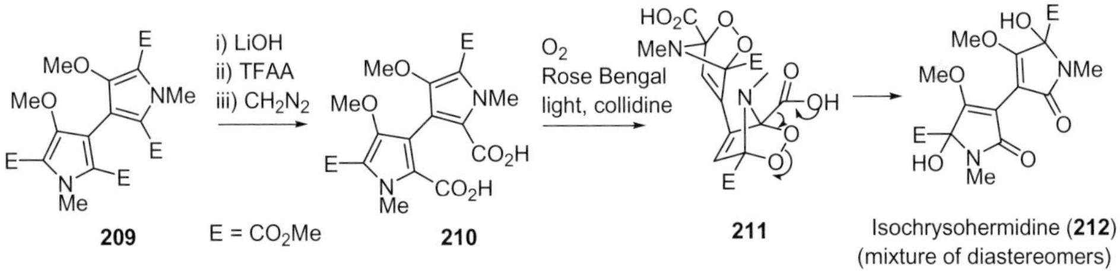

Scheme 6.39 Boger synthesis of isochrysohermidine.

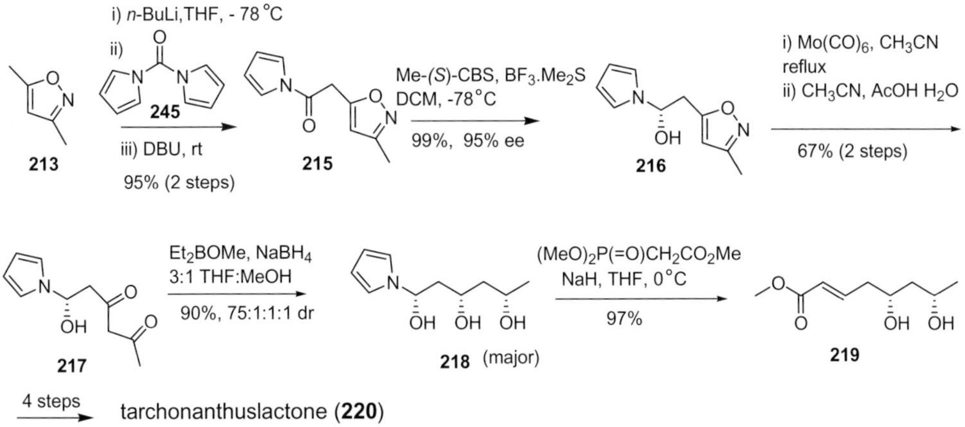

Scheme 6.40 Synthesis of tarchonanthuslactone.

dipyrrole allowed introduction of a masked aldehyde group in the side chain of 3,5-dimethylisoxazole **213**. The catalytic reduction of **215** with Corey-Bakshi-Sibata reagent afforded α-aminol **216** in enantiomerically pure form. Reductive cleavage of the isoxazole ring in **216** with $Mo(CO)_6$ gave the diketo substrate **217**. Due to stereoinduction of the carbinol chiral center, the stereoisomeric triol **218** was produced as the major product. HWE reaction of the triol with trimethylphosphonoacetate in the presence of NaH gave pyrrole-free triol **219**, en route to tarchonanthuslactone **220** in four steps.

6.4
Conclusion

The emergence of pyrroles as natural products has triggered reinvestigation of the established chemistry of pyrroles as well as the development of new chemistry. For example, the scope of Birch reduction as a means to synthesize pyrrole derivatives has enormously increased. The chemistry delineated by the cross-coupling, transition metal-catalyzed cyclizations and Diels–Alder reactions has greatly expanded in their scope. With newer methods of pyrrole synthesis [96] and newer members isolated [97], the chemistry of pyrrole natural products is likely to develop as a more attractive class of alkaloids.

Notes added in proofs

Three different groups of pyrrole natural products have recently been isolated. Nitropyrrolines A-E, characterized as 2-nitropyrrole derivatives with a farnesyl group at C4, have been reported by Fenical *et al.* Several of them were found to be cytotoxic to HCT-116 human carcinoma cells [98]. Heronapyrroles A-C, biosynthetically related to the nitropyrrolins, have been reported by Capon *et al.* They display cytotoxicity to Gram-positive bacteria. Heronapyrrole C contains a chain with two tetrahydrofuran units [99]. Keropsamide, a new class of pyrrole pigments, have been isolated from marine ciliates. Their structural skeleton seems to be derived from condensation of 3-pyrrolepropenoic acid with an unsaturated bromotyramine [100]. In a recent synthetic endeavor, Jia *et al.* reported synthesis of lamellarins D and H, and ningalin B featuring three different oxidative steps [101].

Acknowledgments

The authors are grateful to IIT, Kharagpur, DST and CSIR for the uninterrupted financial support. Mr. Prithiba Mitra is specially thanked for his literary assistance in the preparation of the manuscript.

References

1. Joule, J.A. and Mills, K. (2010) *Heterocyclic Chemistry*, John Wiley & Sons, Ltd, Chichester, UK.
2. Paal, C. (1885) Synthese von Thiophen- und Pyrrolderivaten. *Ber. Dtsch. Chem. Ges.*, **18**, 367–371.
3. Knorr, L. (1885) Einwirkung des Diacetbernsteinsäureesters auf Ammoniak und primäre Aminbasen. *Ber. Dtsch. Chem. Ges.*, **18**, 299–311.
4. (a) Taylor, E.C. and Jones, R.A. (1990) *Pyrroles*, John Wiley & Sons, Inc., New York, USA; (b) Gilchrist, T.L. (1997) *Heterocyclic Chemistry*, 3rd edn, Addison Wesley Longman, Essex, UK.
5. Walsh, C.T., Garneau-Tsodikova, S. and Howard-Jones, A.R. (2006) Biological formation of pyrroles: nature's logic and enzymatic machinery. *Nat. Prod. Rep.*, **23**, 517–531.
6. Rapoport, H. and Holden, K.G. (1962) The synthesis of prodigiosin. *J. Am. Chem. Soc.*, **84** (4), 635–642.
7. (a) O'Hagan, D. (2000) Pyrrole, pyrrolidine, pyridine, piperidine and tropane alkaloids. *Nat. Prod. Rep.*, **17**, 435–446; (b) Liddell, J.R. (2002) Pyrrolizidine alkaloids. *Nat. Prod. Rep.*, **19**, 773–781.
8. Ebada, S.S., Edrada-Ebel, R., de Voogd, N.J., Wray, V. and Proksch, P. (2009) Dibromopyrrole alkaloids from the marine sponge Acanthostylotella sp. *Nat. Prod. Commun.*, **4** (1), 47–52.
9. Zhang, D.S., Gao, H.Y., Song, X.M., Tang, Z.S. and Wu, L.J. (2008) A new alkaloid from *Castanea mollissima* Blume. *Chin. Chem. Lett.*, **19** (7), 832–834.
10. Kuramoto, M., Miyake, N., Ishimaru, Y., Ono, N. and Uno, H. (2008) Cylindradines A and B: novel bromopyrrole alkaloids from the marine sponge *Axinella cylindratus*. *Org. Lett.*, **10** (23), 5465–5468.
11. Aiello, A., D'Esposito, M., Fattorusso, E., Menna, M., Müeller, W.E.G., Perovi _ UNDEFINED-Ottstadt, S., Tsuruta, H., Gulder, T.A.M. and Bringmann, G. (2005) Daminin, a bioactive pyrrole alkaloid from the mediterranean sponge *Axinella damicornis*. *Tetrahedron*, **61** (30), 7266–7270.
12. Zennie, T.M. and Cassady, J.M. (1990) Funebradiol, a new pyrrole lactone alkaloid from *Quararibea funebris* flowers. *J. Nat. Prod.*, **53** (6), 1611–1614.
13. Dong, Y., Pai, N.N., Ablaza, S.L., Yu, S.X., Bolvig, S., Forsyth, D.A. and Le Quesne, P.W. (1999) Quararibea metabolites. 4.Total synthesis and conformational studies of (±)-funebrine and (±)-funebral. *J. Org. Chem.*, **64** (8), 2657–2666.
14. Macherla, V.R., Liu, J., Bellows, C., Teisan, S., Nicholson, B., Lam, K.S. and Potts, B.C.M. (2005) Glaciapyrroles A, B, and C, pyrrolosesquiterpenes from a *Streptomyces* sp. isolated from an Alaskan marine sediment. *J. Nat. Prod.*, **68** (5), 780–783.
15. Eder, C., Proksch, P., Wray, V., Steube, K., Bringmann, G., van Soest, R.W.M., Sudarsono, S., Ferdinandus, E., Pattisina, L.A., Wiryowidagdo, S. and Moka, W. (1999) New alkaloids from the Indopacific sponge *Stylissa carteri*. *J. Nat. Prod.*, **62** (1), 184–187.
16. Mangu, N., Spannenberg, A., Beller, M. and Tse, M.K. (2010) Synthesis of novel annulated hymenialdisine analogues via palladium-catalyzed cross-coupling reactions with arylboronic acids. *Synlett*, (2), 211–214.
17. Chang, L.C., Otero-Quintero, S., Hooper, J.N.A. and Bewley, C.A. (2002) Batzelline D and isobatzelline E from the Indopacific sponge *Zyzzya fuliginosa*. *J. Nat. Prod.*, **65** (5), 776–778.
18. Tao, X.L., Cheng, J.F., Nishiyama, S. and Yamamura, S. (1994) Synthetic studies on tetrahydropyrroloquinoline-containing natural products: syntheses of discorhabdin C, batzelline C and isobatzelline C. *Tetrahedron*, **50** (7), 2017–2028.
19. Krishnaiah, P., Reddy, V.L.N., Venkataramana, G., Ravinder, K., Srinivasulu, M., Raju, T.V., Ravikumar,

K., Chandrasekar, D., Ramakrishna, S. and Venkateswarlu, Y. (2004) New lamellarin alkaloids from the Indian ascidian *Didemnum obscurum* and their antioxidant properties. *J. Nat. Prod.*, **67** (7), 1168–1171.

20 Fan, H., Peng, J., Hamann, M.T. and Hu, J.F. (2008) Lamellarins and related pyrrole-derived alkaloids from marine organisms. *Chem. Rev.*, **108** (1), 264–287.

21 Cafieri, F., Fattorusso, E. and Taglialatela-Scafati, O. (1998) Novel bromopyrrole alkaloids from the sponge *Agelas dispar*. *J. Nat. Prod.*, **61** (1), 122–125.

22 Sun, X.T. and Chen, A. (2007) Total synthesis of rac-longamide B. *Tetrahedron Lett.*, **48** (19), 3459–3461.

23 Aiello, A., D'Esposito, M., Fattorusso, E., Menna, M., Müller, W.E.G., Perović-Ottstadt, S. and Schroeder, H.C. (2006) Novel bioactive bromopyrrole alkaloids from the mediterranean sponge *Axinella verrucosa*. *Biol. Med. Chem.*, **14** (1), 17–24.

24 Woo, J.C.S. and MacKay, D.B. (2003) Diastereoselective isothiourea iodocyclization for manzacidin synthesis. *Tetrahedron Lett.*, **44** (14), 2881–2883.

25 Boonlarppradab, C., Kauffman, C.A., Jensen, P.R. and Fenical, W. (2008) Marineosins A and B, cytotoxic piroaminals from a marine-derived actinomycete. *Org. Lett.*, **10** (24), 5505–5508.

26 (a) Hughes, C.C., Prieto-Davo, A., Jensen, P.R. and Fenical, W. (2008) The marinopyrroles, antibiotics of an unprecedented structure class from a marine *Streptomyces* sp. *Org. Lett.*, **10** (4), 629–631; (b) Hughes, C.C., Kauffman, C.A., Jensen, P.R. and Femical, W. (2010) Structures, reactivities, and antibiotic properties of the marinoprroles A-F. *J. Org. Chem.*, **75** (10), 3240–3250.

27 Uemoto, H., Tsuda, M. and Kobayashi, J. (1999) Mukanadins A–C, new bromopyrrole alkaloids from marine sponge *Agelas nakamurai*. *J. Nat. Prod.*, **62** (11), 1581–1583.

28 Crestia, D., Demuynck, C. and Bolte, J. (2004) Transketolase and fructose-1,6-bis-phosphate aldolase, complementary tools for access to new ulosonic acid analogues. *Tetrahedron*, **60** (10), 2417–2425.

29 Endo, T., Tsuda, M., Okada, T., Mitsuhashi, S., Shima, H., Kikuchi, K., Mikami, Y., Fromont, J. and Kobayashi, J. (2004) Nagelamides A–H, new dimeric bromopyrrole alkaloids from marine sponge *Agelas* species. *J. Nat. Prod.*, **67** (8), 1262–1267.

30 Liu, R., Liu, Y., Zhou, Y.D. and Nagle, D.G. (2007) Molecular-targeted antitumor agents. 15. neolamellarins from the marine sponge *Dendrilla nigra* inhibit hypoxia-inducible factor-1 activation and secreted vascular endothelial growth factor production in breast tumor cells. *J. Nat. Prod.*, **70** (11), 1741–1745.

31 Arafeh, K.M. and Ullah, N. (2009) Synthesis of neolamellarin A, an inhibitor of hypoxia-inducible factor-1. *Nat. Prod. Commun.*, **4** (7), 925–926.

32 Hopp, D.C., Rhea, J., Jacobsen, D., Romari, K., Smith, C., Rabenstein, J., Irigoyen, M., Clarke, M., Francis, L., Luche, M., Carr, G.J. and Mocek, U. (2009) Neopyrrolomycins with broad spectrum antibacterial activity. *J. Nat. Prod.*, **72** (2), 276–279.

33 Tatsuta, K. and Itoh, M. (1994) Synthesis and biological evaluation of neopyrrolomycin analogs. *J. Antibiot.*, **47** (2), 262–265.

34 Sangnoi, Y., Sakulkeo, O., Yuenyongsawad, S., Kanjana-opas, A., Ingkaninan, K., Plubrukarn, A. and Suwanborirux, K. (2008) Acetylcholinesterase-inhibiting activity of pyrrole derivatives from a novel marine gliding bacterium, *Rapidithrix thailandica*. *Mar. Drugs*, **6** (4), 578–586.

35 Rudi, A., Evan, T., Aknin, M. and Kashman, Y. (2000) Polycitone B and prepolycitrin A: two novel alkaloids from the marine ascidian *Polycitor africanus*. *J. Nat. Prod.*, **63** (6), 832–833.

36 Kreipl, A.T., Reid, C. and Steglich, W. (2002) Total syntheses of the marine

pyrrole alkaloids polycitone A and B. *Org. Lett.*, **4** (19), 3287–3288.
37 Kawasaki, T., Sakurai, F. and Hayakawa, Y. (2008) A prodigiosin from the roseophilin producer *Streptomyces griseoviridis. J. Nat. Prod.*, **71** (7), 1265–1267.
38 Clift, M.D. and Thomson, R.J. (2009) Development of a merged conjugate addition/oxidative coupling sequence. Application to the enantioselective total synthesis of metacycloprodigiosin and prodigiosin R1. *J. Am. Chem. Soc.*, **131** (40), 14579–14583.
39 Hiort, J., Maksimenka, K., Reichert, M., Perovic-Ottstadt, S., Lin, W.H., Wray, V., Steube, K., Schaumann, K., Weber, H., Proksch, P., Ebel, R., Müller, W.E.G. and Bringmann, G. (2004) New natural products from the sponge-derived fungus *Aspergillus niger. J. Nat. Prod.*, **67** (9), 1532–1543.
40 Ezaki, N., Shomura, T., Koyama, M., Niwa, T., Kojima, M., Inouye, S., Ito, T. and Niida, T. (1981) New chlorinated nitropyrrole antibiotics, pyrrolomycin A and B (SF-2080 A and B). *J. Antibiot.*, **34** (10), 1363–1365.
41 Guo, Y., Li, X., Wang, J., Xu, J. and Li, N. (2005) A new alkaloid from the fruits of *Celastrus orbiculatus. Fitoterapia*, **76** (2), 273–275.
42 Jin, Y.P., Wei, X.N. and Shi, Y.P. (2007) Chemical constituents from *Cynoglossum gansuense. Helv. Chim. Acta*, **90** (4), 776–782.
43 Kato, S., Shindo, K., Kawai, H., Odagawa, A., Matsuoka, M. and Mochizuki, J. (1993) Pyrrolostatin, a novel lipid peroxidation inhibitor from *Streptomyces chrestomyceticus*: taxonomy, fermentation, isolation, structure elucidation and biological properties. *J. Antibiot.*, **46** (6), 892–899.
44 Fumoto, Y., Eguchi, T., Uno, H. and Ono, N. (1999) Synthesis of pyrrolostatin and its analogues. *J. Org. Chem.*, **64** (17), 6518–6521.
45 Kam, T.S., Subramaniam, G. and Chen, W. (1999) Alkaloids from *Kopsia dasyrachis. Phytochemistry*, **51** (1), 159–169.
46 Beck, E.M., Hatley, R. and Gaunt, M.J. (2008) Synthesis of rhazinicine by a metal-catalyzed C–H bond functionalization strategy. *Angew. Chem. Int. Ed.*, **47** (16), 3004–3007.
47 Tsuda, M., Nozawa, K., Shimbo, K. and Kobayashi, J. (2003) Rigidins B-D, new pyrrolopyrimidine alkaloids from a tunicate *Cystodytes* species. *J. Nat. Prod.*, **66** (2), 292–294.
48 Gupton, J.T., Banner, E.J., Scharf, A.B., Norwood, B.K., Kanters, R.P.F., Dominey, R.N., Hempel, J.E., Kharlamova, A., Bluhn-Chertudi, I. and Hickenboth, C.R. (2006) The application of vinylogous iminium salt derivatives to an efficient synthesis of the pyrrole containing alkaloids rigidin and rigidin E. *Tetrahedron*, **62** (35), 8243–8255.
49 Yamagishi, Y., Matsuoka, M., Odagawa, A., Kato, S., Shindo, K. and Mochizuki, J. (1993) Rumbrin, a new cytoprotective substance produced by *Auxarthron umbrinum*. I. Taxonomy, production, isolation and biological activities. *J. Antibiot.*, **46** (6), 884–887.
50 Clark, B.R. and Murphy, C.D. (2009) Biosynthesis of pyrrolylpolyenes in *Auxarthron umbrinum. Org. Biomol. Chem.*, **7** (1), 111–116.
51 El Sayed, K.A., Hamann, M.T., Abd El-Rahman, H.A. and Zaghloul, A.M. (1998) New pyrrole alkaloids from *Solanum sodomaeum. J. Nat. Prod.*, **61** (6), 848–850.
52 Assmann, M., Zea, S. and Köck, M. (2001) Sventrin, a new bromopyrrole alkaloid from the Caribbean sponge *Agelas sventres. J. Nat. Prod.*, **64** (12), 1593–1595.
53 Carte, B. and Faulkner, D.J. (1983) Defensive metabolites from three nembrothid nudibranchs. *J. Org. Chem.*, **48** (14), 2314–2318.
54 Pinkerton, D.M., Banwell, M.G. and Willis, A.C. (2007) Total syntheses of tambjamines C, E, F, G, H, I and J, BE-18591, and a related alkaloid from the marine bacterium *Pseudoalteromonas tunicata. Org. Lett.*, **9** (24), 5127–5130.
55 Utkina, N.K., Makarchenko, A.E. and Denisenko, V.A. (2005) Zyzzyanones B-D, dipyrroloquinones from the marine sponge *Zyzzya fuliginosa. J. Nat. Prod.*, **68** (9), 1424–1427.

56 Jolicoeur, B., Chapman, E.E., Thompson, A. and Lubell, W.D. (2006) Pyrrole protection. *Tetrahedron*, **62** (50), 11531–11563.

57 Hanefeld, U. and Laatsch, H. (1991) Synthesis of isopentabromopseudilin. *Liebigs Ann. Chem.*, 865–869.

58 Morrison, M.D., Hanthorn, J.J. and Pratt, D.A. (2009) Synthesis of pyrrolnitrin and related halogenated phenylpyrroles. *Org. Lett.*, **11** (5), 1051–1054.

59 Martin, R., Jaeger, A., Boehl, M., Richter, S., Fedorov, R., Manstein, D.J., Gutzeit, H.O. and Knölker, H.J. (2009) Total synthesis of pentabromo- and pentachloropseudilin, and synthetic analogues-allosteric inhibitors of myosin ATPase. *Angew. Chem. Int. Ed.*, **48** (43), 8042–8046.

60 Fürstner, A., Szillat, H., Gabor, B. and Mynott, R. (1998) Platinum- and acid-catalyzed enyne metathesis reactions: mechanistic studies and applications to the syntheses of streptorubin B and metacycloprodigiosin. *J. Am. Chem. Soc.*, **120** (33), 8305–8314.

61 Ridley, C.P., Reddy, M.V., Rocha, G., Bushman, F.D. and Faulkner, D.J. (2002) Total synthesis and evaluation of lamellarin α 20-sulfate analogues. *Bioorg. Med. Chem.*, **10** (10), 3285–3290.

62 Ishibashi, F., Miyazaki, Y. and Iwao, M. (1997) Total syntheses of lamellarin D and H. The first synthesis of lamellarin-class marine alkaloids. *Tetrahedron*, **53** (17), 5951–5962.

63 Banwell, M. and Hockless, D. (1997) Convergent total synthesis of lamellarin K. *Chem. Commun.*, (23), 2259–2260.

64 Heim, A., Terpin, A. and Steglich, W. (1997) Biomimetic synthesis of lamellarin G trimethyl ether. *Angew. Chem. Int. Ed.*, **36** (1/2), 155–156.

65 Gupton, J.T., Miller, R.B., Krumpe, K.E., Clough, S.C., Banner, E.J., Kanters, R.P.F., Du, K.X., Keertikar, K.M., Lauerman, N.E. and Solano, J.M. (2005) The application of vinylogous iminium salt derivatives to an efficient relay synthesis of the pyrrole containing alkaloids polycitone A and B. *Tetrahedron*, **61** (7), 1845–1854.

66 Handy, S.T., Zhang, Y. and Bregman, H. (2004) A modular synthesis of the lamellarins: total synthesis of lamellarin G trimethyl ether. *J. Org. Chem.*, **69** (7), 2362–2366.

67 Weinreb, S.M. (2007) Some recent advances in the synthesis of polycyclic imidazole-containing marine natural products. *Nat. Prod. Rep.*, **24** (5), 931–948.

68 Forte, B., Malgesini, B., Piutti, C., Quartieri, F., Scolaro, A. and Papeo, G. (2009) A submarine journey. The pyrrole-imidazole alkaloids. *Mar. Drugs*, **7** (4), 705–753.

69 Birman, V.B. and Jiang, X.T. (2004) Synthesis of sceptrin alkaloids. *Org. Lett.*, **6** (14), 2369–2371.

70 Su, S., Seiple, I.B., Young, I.S. and Baran, P.S. (2008) Total syntheses of (±)-massadine and massadine chloride. *J. Am. Chem. Soc.*, **130** (49), 16490–16491.

71 Ando, N. and Terashima, S. (2006) A novel synthesis of the 2-aminoimidazol-4-carbaldehyde derivatives, versatile synthetic intermediates for 2-aminoimidazole alkaloids. *Synlett*, (17), 2836–2840.

72 Meketa, M.L. and Weinreb, S.M. (2007) A new total synthesis of the zinc matrixmetalloproteinase inhibitor ageladine a featuring a biogenetically patterned 6π-2-azatriene electrocyclization. *Org. Lett.*, **9** (5), 853–855.

73 Bhandari, M.R., Sivappa, R. and Lovely, C.J. (2009) Total synthesis of the putative structure of nagelamide D. *Org. Lett.*, **11** (7), 1535–1538.

74 Zoellinger, M., Mayer, P. and Lindel, T. (2007) Enantioselective total synthesis of (−)-dibromophakellstatin. *Synlett*, (17), 2756–2758.

75 Papeo, G., Gomez-Zurita, M.A., Borghi, D. and Varasi, M. (2005) Total synthesis of (±)-cyclooroidin. *Tetrahedron Lett.*, **46** (50), 8635–8638.

76 (a) Hama, N., Matsuda, T., Sato, T. and Chida, N. (2009) Total synthesis of (−)-agelastatin A: the application of a sequential sigmatropic rearrangement. *Org. Lett.*, **11** (12), 2687–2690; (b) Domostoj, M.M., Irving, E., Scheinmann, F. and Hale, K.J. (2004)

New total synthesis of the marine antitumor alkaloid (−)-agelastatin A. *Org. Lett.*, **6** (15), 2615–2618.
77 Feldman, K.S. and Skoumbourdis, A.P. (2005) Extending Pummerer reaction chemistry. Synthesis of (±)-dibromophakellstatin by oxidative cyclization of an imidazole derivative. *Org. Lett.*, **7** (5), 929–931.
78 Trost, B.M. and Dong, G. (2006) New class of nucleophiles for palladium-catalyzed asymmetric allylic alkylation. Total synthesis of agelastatin A. *J. Am. Chem. Soc.*, **128** (18), 6054–6055.
79 Trost, B.M. and Dong, G.B. (2009) A stereodivergent strategy to both product enantiomers from the same enantiomer of a stereoinducing catalyst: agelastatin A. *Chem. Eur. J.*, **15** (28), 6910–6919.
80 Wu, J., Vetter, W., Gribble, G.W., Schneekloth, J.S. Jr., Blank, D.H. and Görls, H. (2002) Structure and synthesis of the natural heptachloro-1′-methyl-1,2′-bipyrrole (Q1). *Angew. Chem. Int. Ed.*, **41** (10), 1740–1743.
81 Fürstner, A. (2003) Chemistry and biology of roseophilin and the prodigiosin alkaloids: a survey of the last 2500 years. *Angew. Chem. Int. Ed.*, **42** (31), 3582–3603.
82 Boger, D.L. and Patel, M. (1988) Total synthesis of prodigiosin, prodigiosene, and desmethoxyprodigiosin: Diels–Alder reactions of heterocyclic azadienes and development of an effective palladium(II)-promoted 2,2′-bipyrrole coupling procedure. *J. Org. Chem.*, **53** (7), 1405–1415.
83 Reeves, J.T., Song, J.J., Tan, Z., Lee, H., Yee, N.K. and Senanayake, C.H. (2007) A general synthesis of substituted formylpyrroles from ketones and 4-formyloxazole. *Org. Lett.*, **9** (10), 1875–1878.
84 Reeves, J.T. (2007) A concise synthesis of butylcycloheptylprodigiosin. *Org. Lett.*, **9** (10), 1879–1881.
85 Fürstner, A. and Weintritt, H. (1998) Total synthesis of roseophilin. *J. Am. Chem. Soc.*, **120** (12), 2817–2825.
86 Donohoe, T.J. and Thomas, R.E. (2007) Partial reduction of pyrroles: application to natural product synthesis. *Chem. Rec.*, **7** (3), 180–190.
87 Amos, R.I.J., Gourlay, B.S., Molesworth, P.P., Smith, J.A. and Sprod, O.R. (2005) Annulation of pyrrole: application to the synthesis of indolizidine alkaloids. *Tetrahedron*, **61** (34), 8226–8230.
88 Donohoe, T.J. and Sintim, H.O. (2004) A concise total synthesis of (±)-1-epiaustraline. *Org. Lett.*, **6** (12), 2003–2006.
89 Donohoe, T.J., Sintim, H.O. and Hollinshead, J. (2005) A noncarbohydrate based approach to polyhydroxylated pyrrolidizines: total syntheses of the natural products hyacinthacine A1 and 1-epiaustraline. *J. Org. Chem.*, **70** (18), 7297–7304.
90 Robins, D.J. and Sakdarat, S. (1979) Synthesis of the 8β-pyrrolizidine bases (+)-isoretronecanol, (+)-laburnine, and (+)-supinidine. *J. Chem. Soc. Chem. Commun.*, (24), 1181–1182.
91 Giblin, G.M.P., Jones, C.D. and Simpkins, N.S. (1998) The total synthesis of the analgesic alkaloid epibatidine. *J. Chem. Soc. Perkin Trans. I*, (22), 3689–3698.
92 Berini, C., Pelloux-Leon, N., Minassian, F. and Denis, J.N. (2009) From N-triisopropylsilylpyrrole to an optically active C-4 substituted pyroglutamic acid: total synthesis of penmacric acid. *Org. Biomol. Chem.*, **7** (21), 4512–4516.
93 Terpin, A., Polborn, K. and Steglich, W. (1995) Biomimetic total synthesis of polycitrin A. *Tetrahedron*, **51** (36), 9941–9946.
94 Boger, D.L. and Baldino, C.M. (1993) d,l- and meso-Isochrysohermidin: total synthesis and interstrand DNA crosslinking. *J. Am. Chem. Soc.*, **115** (24), 11418–11425.
95 Scott, M.S., Luckhurst, C.A. and Dixon, D.J. (2005) A total synthesis of tarchonanthuslactone exploiting N-pyrrole carbinols as efficient stereocontrolling elements. *Org. Lett.*, **7** (26), 5813–5816.
96 Pahadi, N.K., Paley, M., Jana, R., Waetzig, S.R. and Tunge, J.A. (2009)

Formation of *N*-alkylpyrroles via intermolecular redox amination. *J. Am. Chem. Soc.*, **131** (46), 16626–16627.

97 Hertiani, T., Edrada-Ebel, R., Ortlepp, S., van Soest, R.W.M., de Voogd, N.J., Wray, V., Hentschel, U., Kozytska, S., Mueller, W.E.G. and Proksch, P. (2010) From anti-fouling to biofilm inhibition: new cytotoxic secondary metabolites from two Indonesian agelas sponges. *Bioorg. Med. Chem.*, **18** (3), 1297–1311.

98 Kwon, H.C., Espindola, A.P.D.M., Park, J.S., Davo-Prieto, A., Rose, M., Jensen, P.R. and Fenical, W. J. (2010) Nitropyrrolins A-E, cytotoxic farnesyl-α-nitropyrroles from a marine-derived bacterium within the actinomycete family Streptomycetaceae. *J. Nat. Prod.*, **73**, 2047–2052.

99 Raju, R, Piggott, A.M., Diaz, L.X.B., Khalil, Z. and Capon, R.J. (2010) Heronapyrroles A-C: farnesylated 2-nitropyrroles from an Australian marine-derived *Streptomyces* sp. *Org. Lett.*, **12** (22), 5158–5161.

100 Guella, G., Frassanito, R., Mancini, I., Sandron, T., Modeo, L., Verni, F., Dini, F. and Petroni, G. (2010) Keronopsamides, a new class of pigments from marine ciliates. *Eur. J. Org. Chem.*, (3) 427–434.

101 Li, Q., Jiang, J., Fan, A., Cu, Y. and Jia, Y. (2011) Total synthesis of lamellarins D, H, and R and ningalin B. *Org. Lett.*, **13** (2), 312–315.

7
Indoles and Indolizidines

Sarah M. Bronner, G.-Yoon J. Im and Neil K. Garg

7.1
Introduction

Alkaloids, as defined by Hesse in *Alkaloids: Nature's Curse or Blessing?*, are nitrogen-containing organic molecules of natural origin with a greater or lesser degree of basic character [1]. Since the first description of the isolation and characterization of (−)-morphine in the early 1800s by Sertürner [2] and the synthesis of (+)-coniine by Ladenberg in 1886 [3], alkaloids have served as platforms for the development of powerful new methodologies in organic chemistry. Two classes of alkaloids that have received significant attention are those that contain indole or indolizidine heterocycles (Figure 7.1). Each of these motifs comprises fused 5,6-ring systems and contains one nitrogen atom, but vary in the placement of nitrogen and the degree of ring unsaturation.

The indole heterocycle is ubiquitous among bioactive molecules and natural products. More than 10 000 biologically active indole derivatives have been discovered to date, of which over 200 are currently marked as drugs or undergoing clinical trials [4]. Not surprisingly, indole alkaloids derived from terrestrial and marine environments have captured significant interest from scientists aiming to discover new therapeutic agents. The fascination with indole alkaloids can also be traced to the structural complexity embedded in countless indole-containing natural products. These alkaloids often possess challenging structural features that cannot be tackled by conventional chemistry and, therefore, require the discovery of new chemical innovations. This phenomenon has held true for many decades, with some of the most impressive examples being reported over 50 years ago, such as Woodward's classic total syntheses of strychnine [5] and reserpine [6].

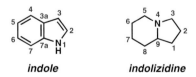

Figure 7.1 Indole and indolizidine.

Heterocycles in Natural Product Synthesis, First Edition. Edited by Krishna C. Majumdar and Shital K. Chattopadhyay.
© 2011 Wiley-VCH Verlag GmbH & Co. KGaA. Published 2011 by Wiley-VCH Verlag GmbH & Co. KGaA.

Although not as pervasive as indole, the indolizidine heterocycle is also observed in many natural products. Similar to indole alkaloids, indolizidines are thought to have great potential as leads for the discovery of novel therapeutics [7]. Interestingly, naturally occurring molecules that contain the indolizidine heterocycle are commonly found in extracts from the skins of amphibians or arthropods, although indolizidine alkaloids have also been discovered in plant and marine organisms [7, 8].

Several excellent reviews and books describing synthetic achievements in indole and indolizidine alkaloid total synthesis are available [1, 9, 10]. This chapter describes a small sampling of the most impressive syntheses involving indoles and indolizidines that have been reported within the last decade (Table 7.1). Several groundbreaking studies, such as the total syntheses of diazonamide A [11, 12], vinblastine [13], quadrigemine C [14], psycholeine [14], aspidophytine [15], okaramine N [16], and indolizomycin [17] have not been included here, but have been reviewed elsewhere [10a,b].

7.2
Applications of Indoles and Indolizidines in the Synthesis of Natural Products

7.2.1
Indoles and Oxindoles

7.2.1.1 Total Synthesis of Actinophyllic Acid (Overman)

Actinophyllic acid **1**, (Scheme 7.1) is a structurally unique indole alkaloid isolated from the leaves of the *Alstonia actinophylla* tree collected on the Cape York Peninsula of Australia by Quinn, Carroll, and co-workers [18]. It was discovered in 2005 as part of an ongoing effort to unearth natural products that would inspire the development of new therapeutics for cardiovascular disease [18]. Actinophyllic acid **1** was shown to inhibit carboxypeptidase U (CPU), an enzyme that has been demonstrated to attenuate the rate of blood clot removal, a process known as fibrinolysis [43].

Actinophyllic acid **1** possesses an unprecedented skeleton comprised of 1-azabicyclo[4.4.2]dodecane and 1-azabicyclo[4.2.1]nonane fragments, which together are not found in any other indole alkaloid [18]. The challenges associated with assembling the multiply bridging polycyclic structure and installing five contiguous stereocenters render **1** an intriguing target for total synthesis. Whereas Wood and co-workers have recently reported significant progress toward the synthesis of actinophyllic acid **1** [44], the Overman laboratory described the only completed synthesis of this molecule in 2008 [19].

Overman and co-workers envisioned deriving (±)-actinophyllic acid **1** from intermediate **13** by late-stage installation of the C20 hemiketal (Scheme 7.1). The acyl pyrrolidine motif embedded in **13** would be assembled via an aza-Cope–Mannich reaction [45] of an *in-situ* generated iminium ion species **14**. The details of this crucial complexity-generating transformation will be described below. The key

Table 7.1 Natural products containing the indole or indolizidine units.

Serial No.	Trivial name	Structure	Source	Isolation [Ref]	Biological activity	Synthesis [Ref]
1	Actinophyllic acid		*Alstonia actinophylla* (tree leaves)	[18]	Carboxypeptidase U (CPU) inhibition	[19]
1	Dragmacidin F		marine sponge of the genus *Halicorte*	[20]	Antiviral activity against HSV-1 and HIV-1	[21]
3	Penitrem D		*Penicillium crustosum* (fungus)	[22]	Tremorgen (causes extended tremors and limb weakness)	[23]
4	Welwitindolinone A isonitrile		*Hapalosiphon welwitschii* (cyanobacteria)	[24]	Antifungal activity	[25, 26]
5	11,11′-Dideoxyverticillin A		*Penicillium* fungus (from the surface of green alga *Avruinvillea longicuulis*)	[27]	Exhibits toxicity against HCT-116 human colon carcinoma	[28]

(*Continued*)

Table 7.1 (Continued)

Serial No.	Trivial name	Structure	Source	Isolation [Ref]	Biological activity	Synthesis [Ref]
6	Minfiensine	(structure 6)	Strychnos minfiensis (African plant)	[29]	None reported	[30–33]
7	Norfluorocurarine	(structure 7)	Diplorrhynchus condylocarpon subsp. Mossambicensis (African plant)	[34, 35]	CNS stimulant; convulsant (10× potency of strychnine)	[36]
8	Psychotrimine	(structure 8)	Psychotria rostrata (leaves of plant)	[37]	Treatment of constipation	[38]
9, 10, 11	Myrmicarins 215A, 215B, and 217	(structures 9 (Z), 10 (E), 11)	Myrmicaria opaciventris (an African ant)	[39]	Biological toxin	[40]
12	Serratezomine A	(structure 12)	Lycopodium serratum (club moss)	[41]	None reported	[42]

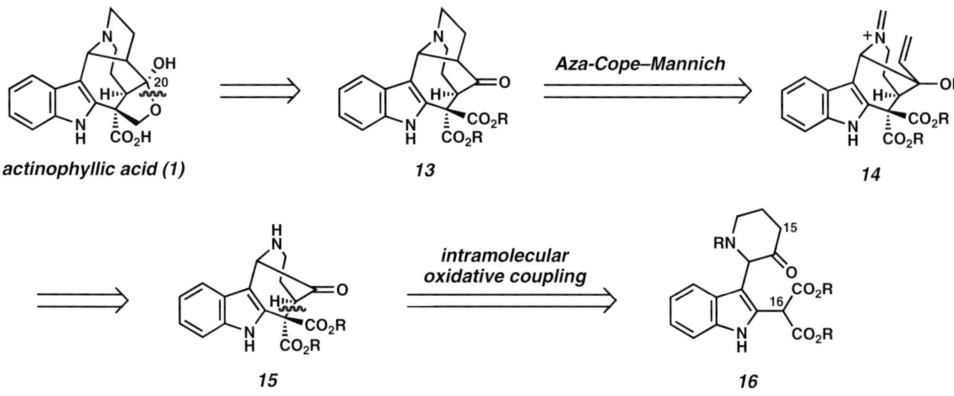

Scheme 7.1 Overman's strategy for the total synthesis of (±)-actinophyllic acid **1**.

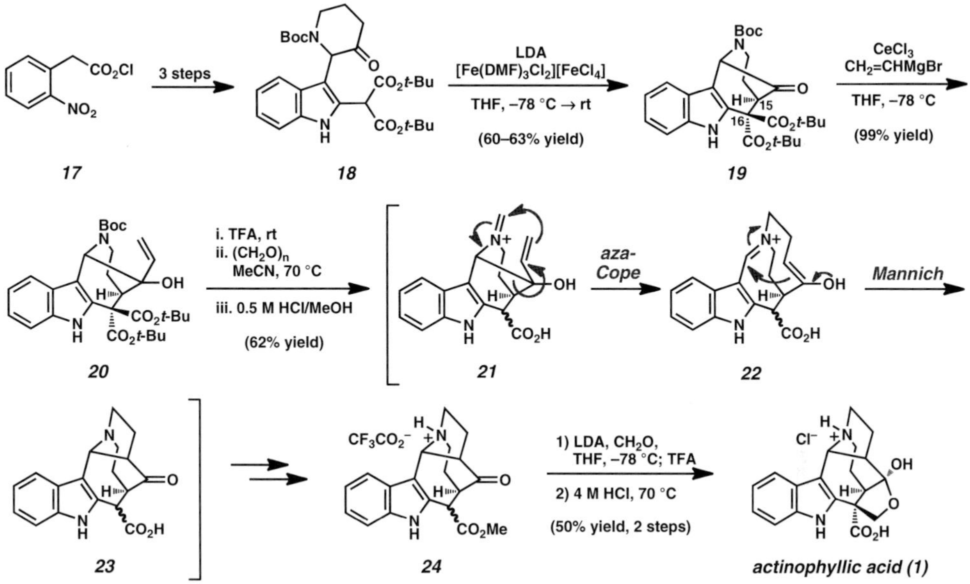

Scheme 7.2 The total synthesis of (±)-actinophyllic acid **1**.

substrate **14** could be derived from ketone **15**, which in turn would arise from an intramolecular oxidative coupling of the dienolate generated from indole-2-malonate derivative **16**. In the forward sense, this ambitious coupling would fashion the C15–C16 bond linkage and provide a significant portion of the actinophyllic acid skeleton.

The forward synthesis of actinophyllic acid **1** commences with the assembly of the C3-substituted indole-2-malonate **18** from acid chloride **17** over a three-step sequence (Scheme 7.2) [46]. Treatment of ketodiester **18** with LDA, followed by introduction of the Fe(III) oxidant [Fe(DMF)$_3$Cl$_2$][FeCl$_4$] [47], yields the desired bicycle **19** in 60–63% yield. Impressively, this transformation can be carried out on scales up to ten grams and represents the first example described in the literature of an oxidative intramolecular coupling of a ketone and a malonic ester enolate [48]. It is noteworthy that the assembly of the complex keto-bridged bicycle under these oxidative conditions proceeds in the absence of an indole *N*-protecting group.

Having established the necessary C15–C16 linkage, the next goal is to execute the critical aza-Cope–Mannich reaction sequence. This requires elaboration of bicycle **19** to the key rearrangement substrate **20**, which is readily achieved upon diastereoselective addition of vinylmagnesium bromide to the ketone of **19** in the presence of $CeCl_3$ [49]. After some experimentation, conditions were discovered that enable the conversion of allylic alcohol **20** to the desired pentacyclic product **24** through a one-pot, three-step reaction sequence. The remarkable transformation proceeds by: (i) treatment of **20** with TFA at room temperature, which leads to cleavage of the *N*-Boc and *t*-butyl esters, with subsequent decarboxylation; (ii) removal of TFA and treatment with paraformaldehyde, which promotes the formation of an intermediate formaldiminium ion. This species readily undergoes aza-Cope–Mannich rearrangement as planned (see transition structures **21** and **22**) to provide ketoacid **23**; (iii) exposure of **23** to HCl/MeOH to furnish the corresponding methyl ester, which is isolated as the TFA salt **24**. With **24** readily available, the total synthesis of actinophyllic acid **1** is achieved in two additional steps involving stereoselective aldol reaction between formaldehyde and the enolate [50] derived from ester **24**, followed by acid-mediated hydrolysis of the methyl ester.

The total synthesis of (±)-actinophyllic acid **1** described by Overman proceeds in only eight steps from acid chloride **17** and is achieved in 8% overall yield. Of note, seven of the eight steps in the synthesis are used to construct C–C or C–N bonds. The most striking features of this synthetic achievement include the scalable, intramolecular oxidative enolate coupling of substrate **18** to access ketodiester **19** and the aza-Cope–Mannich reaction to assemble the unprecedented actinophyllic acid ring system.

7.2.1.2 Total Synthesis of Dragmacidin F (Stoltz)

The dragmacidins are a family of novel marine alkaloids that have piqued the interest of the chemical community due to their structural complexity and promising biological profiles [21, 51–53]. The most exquisite member of this family is dragmacidin F **2**, (Scheme 7.3), which was isolated in 2000 from a marine sponge of the genus *Halicorte* collected off the southern coast of Ustica Island [20]. Alkaloid **2** was found to display promising *in vitro* antiviral activity against HSV-1 (EC_{50} = 95.8 µM) and HIV-1 (EC_{50} = 0.91 µM) [20]. Dragmacidin F **2** is the only dragmacidin alkaloid that does not possess a bis(indole) framework, although the western portion of the molecule is probably assembled biosynthetically from a bis(indole) precursor [20, 51]. Distinguishing structural features of **2** include a differentially substituted pyrazinone, a pyrrole-fused-[3.3.1]-bicycle, an aminoimidazole, and a 6-bromoindole unit. These rings provide a daunting total of seven

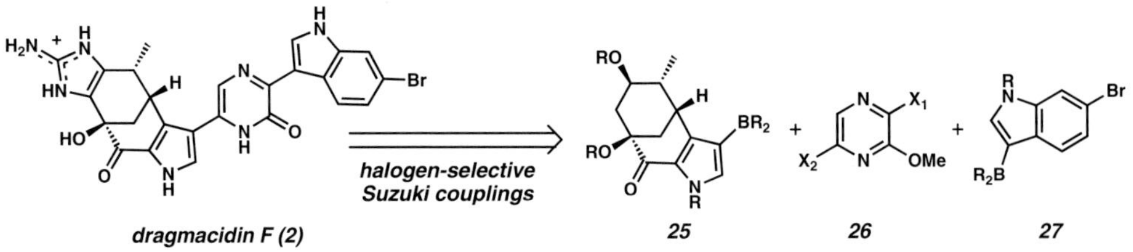

Scheme 7.3 Stoltz's strategy for the synthesis of dragmacidin F **2**.

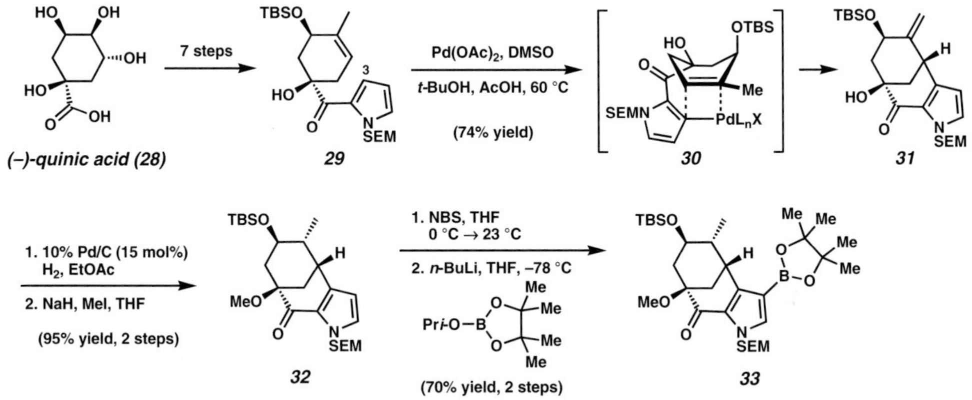

Scheme 7.4 Synthesis of fragment **33**.

nitrogen atoms that are contained within the polycyclic core of the natural product. This section highlights the total synthesis of (+)-dragmacidin F **2** reported by Stoltz and co-workers in 2004 [21], which remains the only total synthesis of **2** to date. It should be noted that prior to this achievement, Stoltz also accomplished the total synthesis of the related alkaloid dragmacidin D [52].

Stoltz's strategy for the preparation of dragmacidin F **2** is depicted retrosynthetically in Scheme 7.3. It was envisioned that the underlying framework of the natural product could be constructed via a series of cross-coupling reactions. The Suzuki coupling [54] was deemed the cross-coupling reaction of choice because of its widespread use in the synthesis and functionalization of heterocycles [55]. Thus, **2** would be derived from three fragments: a pyrroloboronate **25**, a differentially substituted pyrazine **26**, and a boronylated indole **27**. A key challenge in executing this cross-coupling sequence would be to maintain the 6-bromoindole unit, while also obtaining selectivity for sequential coupling of the two halogens (X_1 and X_2) present in pyrazine fragment **26**. Stoltz had previously reported rapid syntheses of suitable fragments **26** and **27** [52], so assembly of bicycle **25** was considered the first critical goal en route to dragmacidin F **2**.

Stoltz's total synthesis of dragmacidin F **2** commences with elaboration of commercially available (−)-quinic acid **28**, (Scheme 7.4) [56] to pyrrole-appended

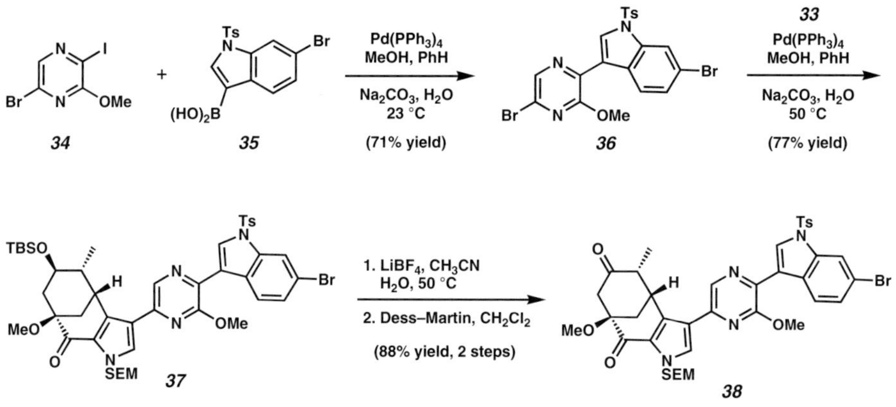

Scheme 7.5 Assembly of the dragmacidin F framework.

cyclohexene **29** over seven steps. In a key complexity-generating step, **29** undergoes Pd(II)-mediated oxidative carbocyclization [57] to construct the [3.3.1]-bicyclic framework of the natural product. Presumably, this transformation proceeds by initial C3 palladation of the pyrrole ring, followed by olefin insertion (see transition structure **30**) and β-hydride elimination to furnish bicycle **31**. Next, diastereoselective olefin reduction of bicycle **31** and methyl protection of the tertiary alcohol yields the intermediate **32**. Introduction of the key boronate motif is achieved in two additional steps by regioselective bromination of pyrrole **32**, followed by borylation using a metal-halogen exchange/quenching protocol. This sequence provides efficient access to boronic ester **33**, the key fragment needed for the Suzuki cross-coupling sequence.

The assembly of the dragmacidin F skeleton is highlighted in Scheme 7.5. In the first cross-coupling, pyrazine **34** and boronic acid **35** react under standard Suzuki coupling conditions at 23 °C to deliver indolopyrazine **36**. Here, selective coupling of the pyrazinyl iodide occurs, leaving both aryl bromides intact. Subsequent coupling of **36** with boronic ester **33** also proceeds smoothly under standard Suzuki-coupling conditions to furnish triaryl species **37**. In this case, the reaction is carried out at 50 °C to ensure complete selectivity for the coupling of the activated pyrazinyl bromide, without disturbing the bromoindole. Recent computational studies by Houk and co-workers suggest that selectivity in this process is determined by both the energy required to distort the carbon–halogen bond to the transition-state geometry and the interaction of the heterocycle π* molecular orbitals with the metal catalyst [58]. Silyl ether **37** is then elaborated to ketone **38** through a two-step sequence involving selective cleavage of the TBS protecting group followed by oxidation of the resulting secondary alcohol.

Having constructed the core framework of dragmacidin F **2**, the remaining synthetic challenges include the installation of the aminoimidazole and the removal of all protecting groups. To address the former of these goals, the authors

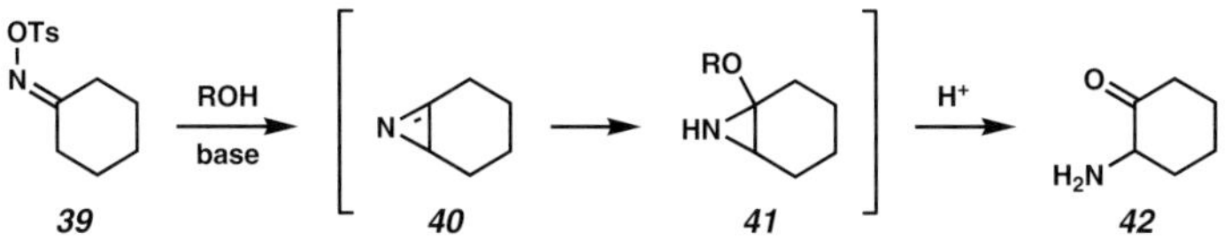

Scheme 7.6 Neber rearrangement.

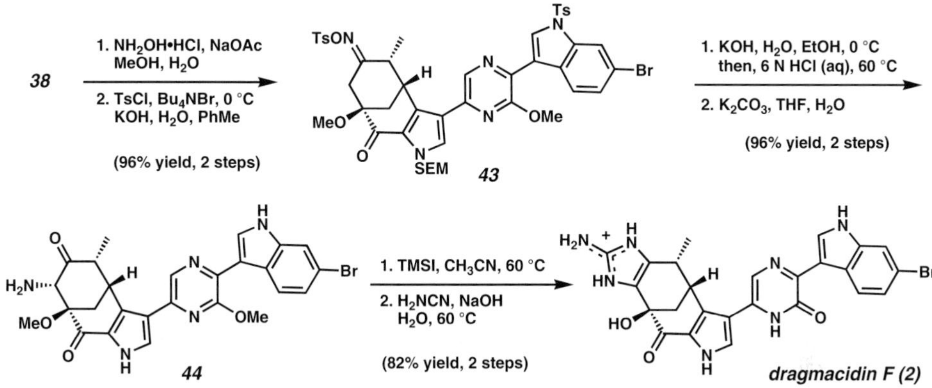

Scheme 7.7 Total synthesis of (+)-dragmacidin F **2**.

sought to install a nitrogen substituent α to the ketone via a Neber rearrangement [59]. In a typical Neber rearrangement (Scheme 7.6), a tosyl oxime **39** is treated with base in the presence of an alcohol to generate an intermediate azirene **40**. Attack of **40** by alkoxide furnishes an alkoxy aziridine **41**, which upon hydrolysis gives rise to the amino ketone product **42**. The necessary tosyl oxime derivative **43** is readily prepared from ketone **38** (Scheme 7.7). Upon sequential treatment of tosyl oxime **43** with KOH, HCl, and finally K$_2$CO$_3$ the targeted amino ketone **44** is obtained in 96% yield via the key Neber rearrangement sequence. During this process, which represents the first example of a Neber rearrangement in complex molecule total synthesis, both the SEM and Ts protecting groups are also removed. With bis(ether) **44** in hand, the stage is set for completion of the total synthesis. Bis(demethylation) of **44** with TMSI furnishes the corresponding deprotected aminoketone. Finally, reaction of this intermediate with cyanamide under basic conditions provides (+)-dragmacidin F **2**.

The enantiospecific synthesis of (+)-dragmacidin F **2** by Stoltz and co-workers is accomplished in 21 steps with an overall yield of 8% beginning from (−)-quinic acid. The route features a number of key steps, including an oxidative carbocyclization to build the [3.3.1]-bicycle, a sequence of halogen-selective cross-couplings to convergently assemble the carbon skeleton, and a late-stage Neber rearrangement for the eventual installation of the aminoimidazole unit. This effort remains the only total synthesis of dragmacidin F **2** to date, with the exception of Stoltz's related

230 | *7 Indoles and Indolizidines*

synthesis of (−)-**2**, which interestingly also begins from the (−)-enantiomer of quinic acid [21b].

7.2.1.3 Total Synthesis of Penitrem D (Smith)

Wilson and co-workers isolated the first penitrem alkaloid in 1968 from the ergot fungus *Penicillium cyclopium* [22]. In the years that followed, the Steyn group disclosed structures of penitrems A–F, isolated from the fungus *Penicillium crustosum* [60]. Penitrems A–F, which are the six penitrems known to date, possess unusual bioactivity as they are among the few known tremogens compounds capable of causing prolonged tremoring [22, 60b]. The partial resolution studies by Horeau have allowed for the absolute configuration of penitrem A to be established [61].

The penitrems are the most complex members of the indole-diterpene family of alkaloids. Their intricate structural framework, represented by penitrem D **3**, (Scheme 7.8), is composed of nine rings, including a highly substituted indole (rings D and E), a cyclobutanone moiety (ring B), and an eight-membered cyclic ether (ring A). Penitrem D **3** possesses eleven stereocenters, including vicinal all-carbon quaternary centers and two allylic hydroxyl groups.

The complexity of the penitrems has captured the attention of synthetic chemists, with extensive studies reported by the Curran and Smith laboratories. Curran and co-workers disclosed their radical and palladium-catalyzed cyclization strategy to access the BCD ring system of penitrem D in 2004 [62]. However, the enantiospecific approach described in 2000 by Smith represents the only total synthesis of penitrem D to date [23, 63].

The synthetic approach by Smith is shown in Scheme 7.8. The authors anticipated that the natural product **3** could be obtained from intermediate **45** by the

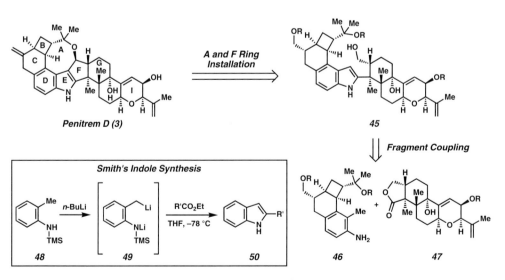

Scheme 7.8 Smith's strategy for the synthesis of (−)-penitrem D **3**.

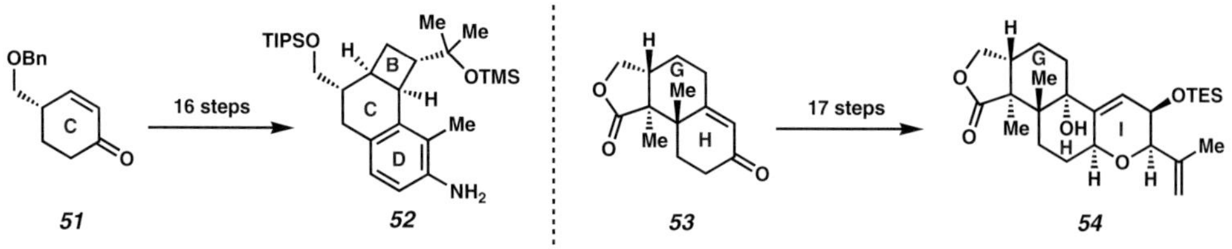

Scheme 7.9 Synthesis of fragments **52** and **54**.

late-stage installation of the A and F rings. In turn, it was envisioned that **45** could arise by the coupling of fragments **46** and **47**, which would proceed with the concomitant formation of ring E and the indole heterocycle. Although this disconnection was somewhat unconventional, the Smith group had previously validated this coupling methodology on simpler systems in an effort to gain entry into tremorgenic indole diterpenoid alkaloids [64]. Generally, this indole synthesis proceeds by treatment of a *N*-TMS-*o*-toluidine **48** with *n*-BuLi to afford an intermediate dilithiated species **49**, which then reacts with either an ester or lactone to ultimately furnish an indole product **50**.

After much experimentation, robust routes to the two coupling fragments were developed (Scheme 7.9). The western portion **52**, possessing the BCD ring system of the natural product, is prepared in enantioenriched form in 16 steps from enone **51**. The eastern portion **54**, which contains the GHI rings of penitrem D, is synthesized in 17 steps from enone **53** (prepared in 15 steps from the Wieland–Miescher ketone [65]).

With fragments **52** and **54** accessible, attention was directed toward the critical fragment-coupling step (Scheme 7.10). Impressively, under optimized reaction conditions, concomitant coupling/indole synthesis proceeds in 81% yield to furnish indole **56**, presumably via cyclization of an intermediate keto-silylamide (see transition structure **55**). This achievement sets the stage for construction of the F and A rings. Oxidation under Parikh–Doering conditions [66], followed by selective deprotection of the TMS and TES protecting groups provides aldehyde **57**. Of note, **57** exists as an inconsequential mixture of the depicted aldehyde and two hemiaminal isomers generated by attack of the indole nitrogen onto the aldehyde. Nonetheless, treatment of **57** with Sc(OTf)$_3$ leads to the formation of the A and F rings and delivers nonacycle **58** in 62% yield. Nonacycle **58** is elaborated to the targeted natural product over five steps, thus completing the total synthesis of penitrem D **3**.

The synthesis of penitrem D **3** proceeds in 43 steps from the Wieland–Miescher ketone (longest linear sequence) and features a number of elegant transformations. Key complexity generating steps include the coupling of fragments **52** and **54** with simultaneous indole formation, as well as the late-stage construction of the A and F rings using a Lewis-acid promoted cyclization cascade. Smith's striking achievement remains the only total synthesis of any penitrem alkaloid to date.

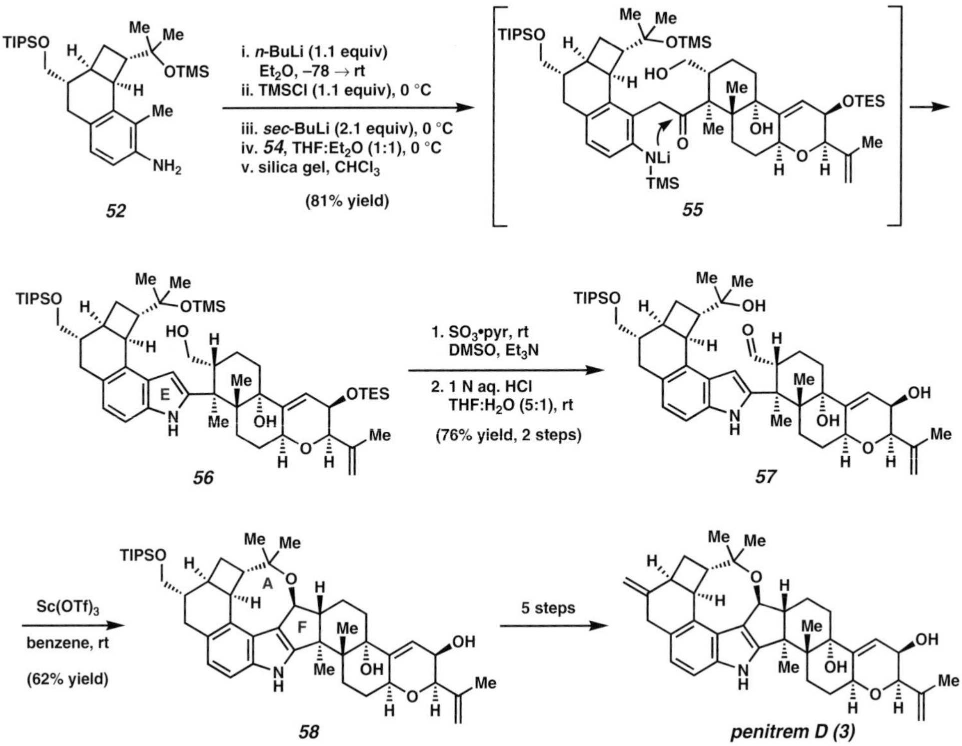

Scheme 7.10 Total synthesis of (−)-penitrem D **3**.

7.2.1.4 Total Synthesis of Welwitindolinone A Isonitrile (Baran, Wood)

In 1994, Moore and co-workers isolated several unusual oxindole alkaloids from the cyanobacteria *Hapalosiphon welwitschii* and *Westiella intricata* termed the welwitindolinones [24]. Ten welwitindolinones have been isolated to date [24, 67], but over 60 related natural products have been discovered including the fischerindoles and hapalindoles [68]. Of the known welwitindolinones, welwitindolinone A isonitrile **4**, (Scheme 7.11) is structurally unique, possessing a spirocycle at C3, which is fused to a bicyclo[4.2.0]octane core. This core, although compact, is densely functionalized and features two quaternary carbon centers, a tertiary center, an isonitrile, and an alkyl chloride.

Welwitindolinone A isonitrile **4** has attracted considerable attention due largely to its interesting carbon skeleton as well as its biological profile (i.e., antifungal activity) [69]. The pioneering synthetic studies by the Wood and Baran laboratories have recently culminated in total syntheses of this molecule. The Baran laboratory disclosed an enantiospecific synthesis of (+)-welwitindolinone A isonitrile **4** in 2005 [25], which was immediately followed by Wood's synthesis of (±)-**4** in early 2006 [26]. This section will cover both syntheses of **4**, in addition to Baran's synthesis of (−)-12-*epi*-fischerindole I **76** en route to (+)-**4**.

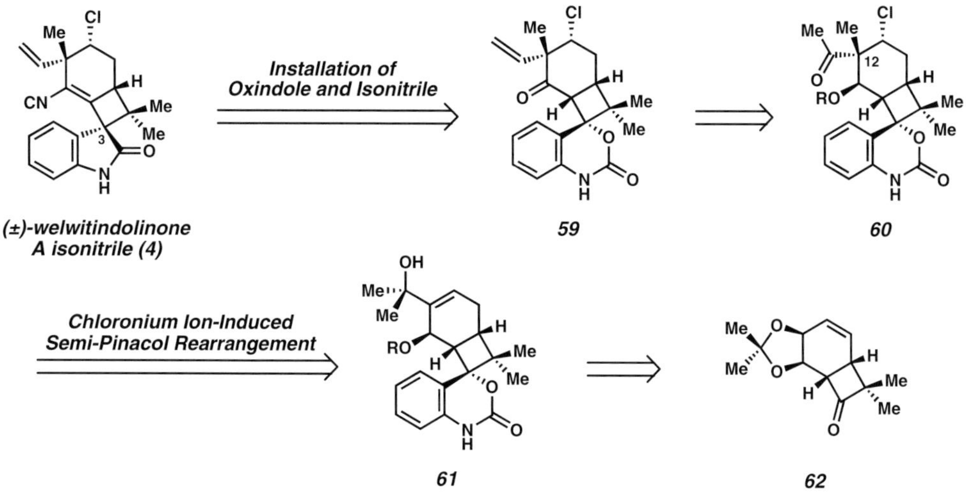

Scheme 7.11 Wood's strategy for the synthesis of welwitindolinone A isonitrile 4.

Wood's strategy for the synthesis of **4** is depicted retrosynthetically in Scheme 7.11, which shows that **4** would arise from a compound of the type **59** via late-stage installation of the isonitrile and oxindole moieties. In turn, it was envisioned that alkene **59** could be accessed from ketone **60**. In a key disconnection, the C12 quaternary center would be installed through a chloronium ion-induced semi-pinacol rearrangement [70] of precursor **61**, the details of which will be described below. Finally, alcohol **61** would be derived from readily available ketone **62**.

Wood's total synthesis initiates with a ketene [2 + 2] cycloaddition of **63** and **64** to provide cyclobutanone **62**, which possesses the bicyclo[4.2.0]-octane core of the natural product (Scheme 7.12). Over five steps, ketone **62** is elaborated to carbamate **65** in 64% yield. Subsequently, a four-step sequence is used to convert **65** to tertiary alcohol **66**, the substrate to be employed in the critical chloronium ion-induced semi-pinacol rearrangement mentioned above. Upon treatment of tertiary alcohol **66** with NaOCl and $CeCl_3 \cdot 7H_2O$ [71], product **68**, possessing the key quaternary carbon stereocenter at C12, is obtained in 78% yield. The reaction is thought to proceed via formation of an intermediate chloronium ion, which is prone to undergo rearrangement (see transition structure **67**). The methyl ketone of **68** is converted to terminal olefin **70** through a sequence involving desilylation [72] and diastereoselective ketone reduction (**68**→**69**) [73], followed by dehydration with Martin's sulfurane [74] (**69**→**70**). Dess–Martin oxidation of alcohol **70** provides ketone **59**.

With access to **59**, the remaining hurdles to complete the synthesis involve installation of the oxindole and isonitrile units. To this end, **59** is subjected to $(Boc)_2O$, DMAP, and DBU in dichloromethane to effect concomitant Boc-protection/CO_2-elimination (Scheme 7.13). Subsequent treatment of the intermediate enone with methoxyamine delivers oxime **71**. This sequence notably

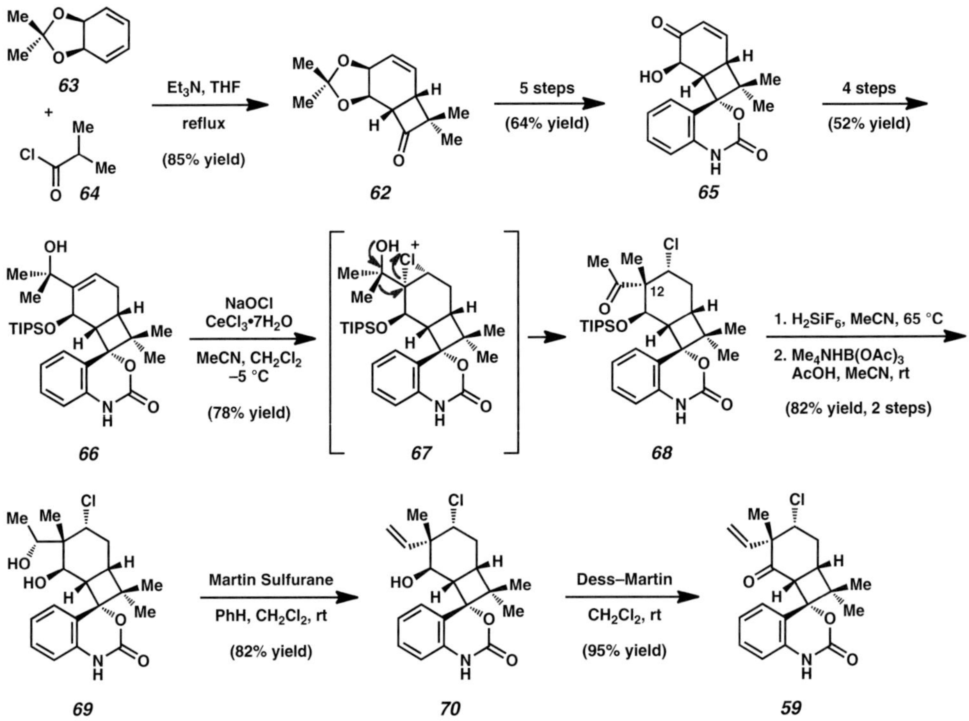

Scheme 7.12 Wood's synthesis of intermediate **59**.

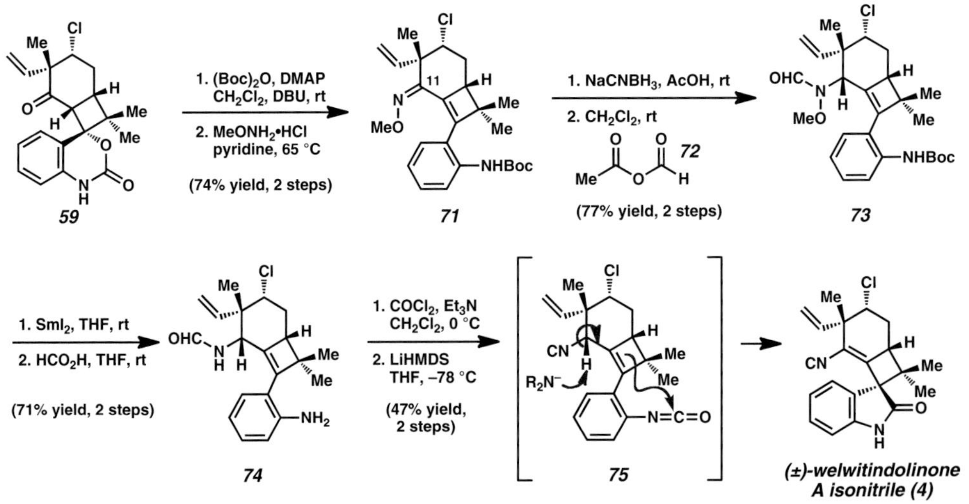

Scheme 7.13 Wood's total synthesis of welwitindolinone A isonitrile **4**.

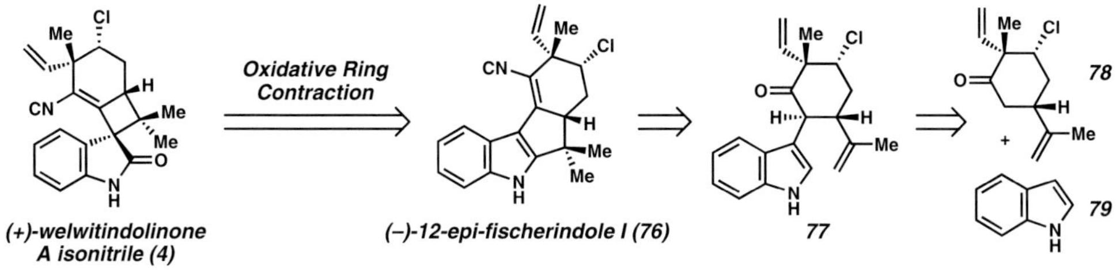

Scheme 7.14 Baran's strategy for the synthesis of welwitindolinone A isonitrile 4.

introduces a C11 nitrogen substituent that is required for the eventual installation of the isonitrile. Imine reduction of **71** and subsequent formylation [75] with **72** furnishes methoxyamide **73**. Next, **73** undergoes reductive N–O bond cleavage, followed by Boc-deprotection to give formamide **74**. Finally, to complete the total synthesis, **74** is subjected to phosgene/Et$_3$N to affect dehydration and isocyanate formation. The crude product is subsequently treated with LiHMDS to induce cyclization (see transition structure **75**) to afford **4** in 47% yield over the two steps.

Baran's strategy for the synthesis of **4** is quite different from Wood's and is inspired by biosynthetic considerations (Scheme 7.14). Baran proposes that a related natural product, 12-*epi*-fischerindole I **76** likely serves as the biosynthetic precursor to **4** [25]. Thus, the authors hoped to access (+)-welwitindolinone A isonitrile **4** through an oxidative ring contraction of (−)-12-*epi*-fisherindole I (**76**). **76** would be derived from indolyl cyclohexanone **77**, which, in turn would arise by the coupling of two fragments: carvone derivative **78** and indole **79**. The Baran laboratory had previously discovered conditions for such couplings, en route to various fischerindole and hapalindole natural products, thus providing considerable support for this ambitious disconnection [76].

Baran's rapid total synthesis of (−)-12-*epi*-fischerindole I **76** is shown in Scheme 7.15. The approach commences with the sequential treatment of (*S*)-carvone oxide **80** with LiHMDS and vinylmagnesium bromide. This presumably leads to the formation of an intermediate magnesium alkoxide, which facilitates intramolecular delivery of the vinyl group (see transition structure **81**) to furnish **82**, with the C12 quaternary center in place [77]. Chlorination of alcohol **82** with NCS/PPh$_3$ occurs with stereochemical inversion to provide alkyl chloride **78**. In a key step, ketone **78** is allowed to react with indole **79** in the presence of LiHMDS and Cu(II)-2-ethylhexanoate, thus delivering ketoindole **77** in 62% yield. It should be noted that the indole-ketone coupling process was originally designed to proceed by the coupling of two *in-situ* generated neutral radical intermediates, but other mechanistic pathways may be operative [76].

With ketone **77** in hand, the stage is now set to form the final ring and complete the synthesis of 12-*epi*-fischerindole I **76**. To assemble the framework of **76**, indolyl alkene **77** undergoes acid-mediated cyclization to provide tetracycle **83**. Reductive amination of **83** installs the C11 nitrogen substituent and furnishes secondary

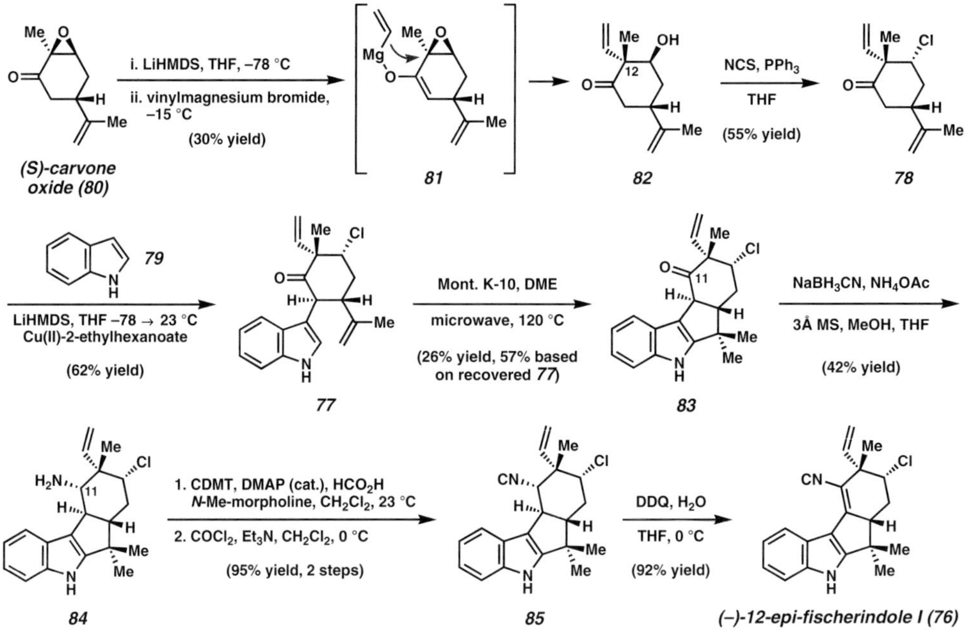

Scheme 7.15 Baran's total synthesis of 12-*epi*-fischerindole I **76**.

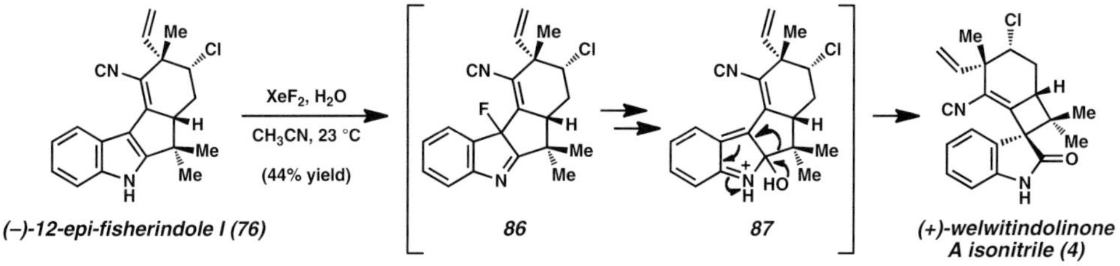

Scheme 7.16 Baran's total synthesis of welwitindolinone A isonitrile **4**.

amine **84**, which, in turn is converted to isonitrile **85** in 95% yield using a two-step protocol [78, 79]. The synthesis of (−)-12-*epi*-fischerindole I **76** is completed upon DDQ-promoted oxidation of late-stage intermediate **85**.

With access to 12-*epi*-fischerindole I **76** in eight steps, the proposed biomimetic oxidative ring contraction was investigated as a means to install the fused spirocyclic framework of **4** (Scheme 7.16). The desired ring contraction was originally achieved using *t*-BuOCl, but was hampered by competitive isonitrile decomposition. Ultimately, a milder variant was developed whereby **76** was readily converted to **4** upon exposure to XeF$_2$ [80] in wet acetonitrile at ambient temperature. Presumably, the transformation proceeds by initial fluorination to provide **86**, which undergoes attack by water and loss of fluoride to produce an intermediate hydroxy-iminium ion species. Ring contraction of this intermediate (see transition structure **87**) furnishes the natural product **4**.

The Wood and Baran syntheses of welwitindolinone A isonitrile **4** provide excellent examples of how challenges encountered in complex molecule synthesis can lead to innovations in synthetic chemistry. Impressive steps in Wood's approach to (±)-**4** include a chloronium-ion induced semi-pinacol rearrangement to install the alkyl chloride and quaternary center, as well as an isocyanate/isonitrile cyclization to introduce the oxindole (23 steps, 2.5% overall yield). Baran's synthesis features a number of novel transformations including an oxidative coupling of indole **79** and ketone **78**, as well as a biomimetic oxidative ring contraction of (−)-12-*epi*-fischerindole I **76** to give (+)-**4** (nine steps, 1% overall yield).

7.2.2
Indolines

7.2.2.1 Total Synthesis of 11,11′-Dideoxyverticillin A (Movassaghi)

Pyrrolidinoindoline natural products have fascinated the chemical community on the basis of their complex structural frameworks and interesting biological profiles [81]. Dimeric pyrrolidinoindolines, in which two such fragments are connected via a C3–C3′ linkage, are a special subset of this family of natural products that have been isolated from a variety of sources. Representative members of this class include (+)-chimonanthine **88** [82], (+)-WIN 64821, **89** [83], and (+)-11,11′-dideoxyverticillin A **5** [27] shown in Figure 7.2. Significant synthetic work has been described toward the synthesis of dimeric pyrrolidinoindolines, yet construction of the vicinal quaternary centers still remains a formidable challenge. Successful approaches to this sterically congested motif include Overman's asymmetric Heck reaction and bis(alkylation) strategies, which have been reviewed [84].

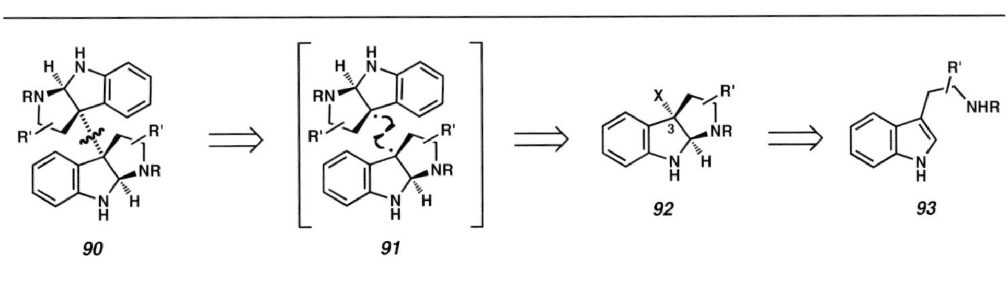

Figure 7.2 Movassaghi's approach to dimeric pyrrolidinoindoline alkaloids.

Recently, Movassaghi and co-workers reported the elegant total syntheses of several dimeric pyrrolidinoindoline alkaloids [28, 85, 86]. Their biosynthetic approach to dimers **90** [87], summarized retrosynthetically in Figure 7.2, relies on the reductive dimerization of an *in-situ* generated pyrrolinoindoline radical species (see transition structure **91**). A C3-halogenated pyrrolidinoindoline such as **92** was envisioned to be the ideal starting material for such a coupling sequence. In turn, **92** would be obtained from the tryptamine precursor **93**.

Although several syntheses of dimeric pyrrolidinoindolines have now been disclosed using the dimerization strategy described above [85, 86], this section will focus on Movassaghi's total synthesis of (+)-11,11′-dideoxyverticillin A **5** [28], which is arguably the most complex target prepared to-date by this approach. (+)-**5** was isolated in 1996 from the marine-derived fungus of the genus *Penicillium*, and was found to exhibit *in vitro* toxicity against HCT-116 human colon carcinoma (IC_{50} = 30 ng mL^{-1}) [27]. In addition to possessing the dimeric pyrrolidinoindoline moiety, **5** contains an intricate epidithiodiketopiperazine motif. Although natural products bearing these synthetically challenging units have been known for nearly four decades, none had been prepared synthetically prior to Movassaghi's achievement in 2009.

Movassaghi's biomimetic synthesis of (+)-11,11′-dideoxyverticillin A **5** commences via the rapid assembly of the dimeric pyrrolidinoindoline diketopiperazines (Scheme 7.17). Boc-deprotection of amide **94** (prepared from commercially available amino acid derivatives), followed by morpholine-catalyzed cyclization provides *cis*-diketopiperazine **95**. Next, **95** is treated with bromine to afford tetracyclic bromide **96**, which upon exposure to methyl iodide and potassium carbonate yields indoline **97**. Reductive dimerization of bromide **97** proceeds in the presence of tris(triphenylphosphine)cobalt(I) chloride in acetone to afford dimeric diketopiperazine **98**. This key step is notable as the vicinal quaternary centers are assembled with ease, while the desired *cis*-fusion on the 5,5-ring systems remains undisturbed [86].

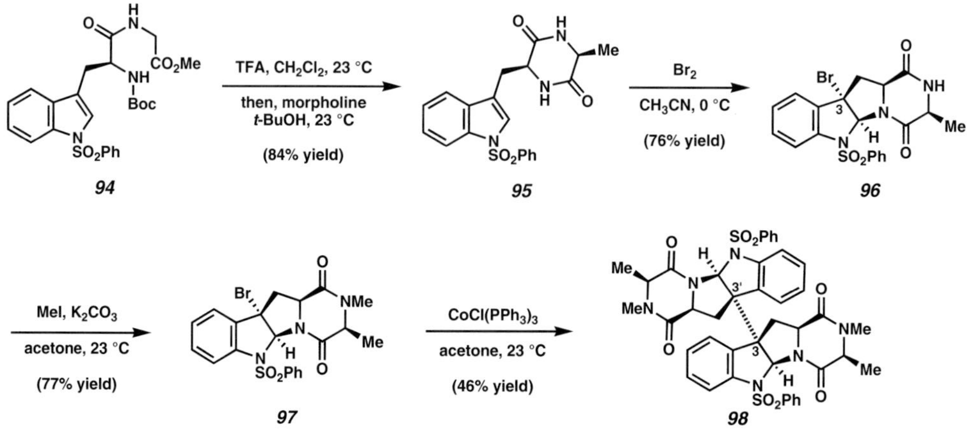

Scheme 7.17 Dimerization of 3-bromoindoline **97**.

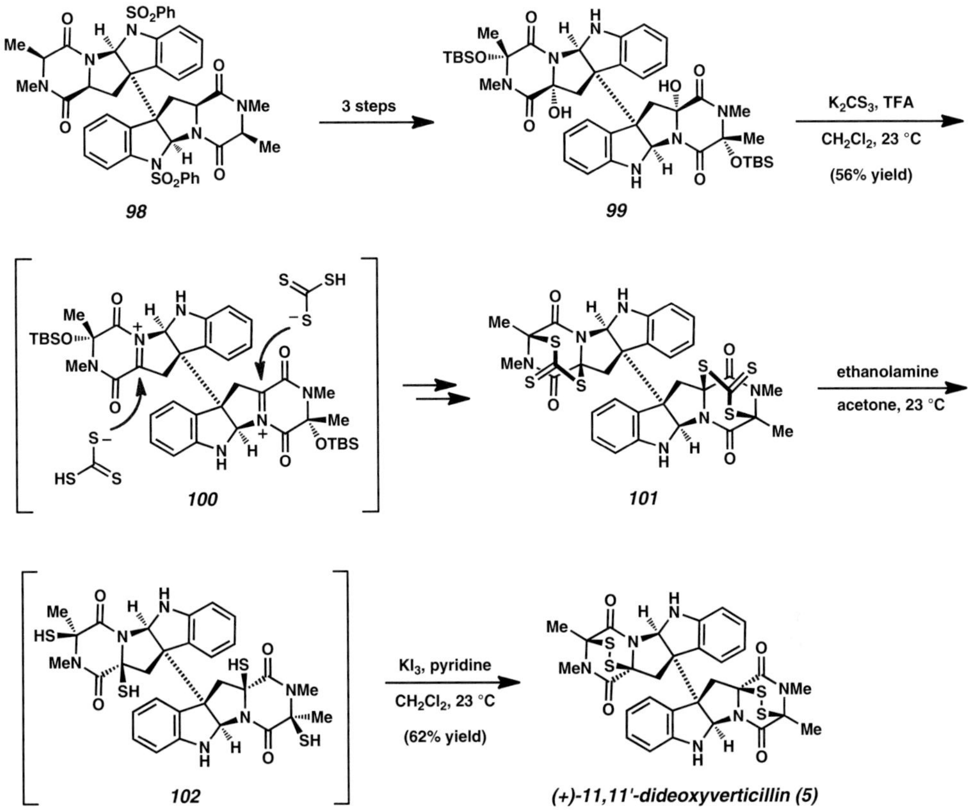

Scheme 7.18 Installation of the epidisulfide moieties and completion of the total synthesis of (+)-11,11′-dideoxyverticillin **5**.

Having constructed the key dimeric diketopiperazine intermediate, the focus shifted to the introduction of the epidisulfide moieties (Scheme 7.18). **98** is first elaborated in a three-step sequence to bis(hemiaminal) **99**. Recognizing that epidithiodiketopiperazines are extremely sensitive functional groups [88], mild conditions for their installation would be necessary. After extensive experimentation it was found that treatment of bis(hemiaminal) **99** with potassium trithiocarbonate and trifluoroacetic acid in dichloromethane delivers bis(dithiepanethione) **101**. This transformation likely proceeds via iminium ion generation and kinetic trapping with the trithiocarbonate nucleophile (see transition structure **100**), followed by intramolecular cyclization to form the (bis)dithiepanethione. It is notable that the conversion of **99** to **101** facilitates the replacement of four C–O bonds with C–S bonds, a process that occurs with complete stereocontrol.

In order to assemble the final epidisulfide linkages, the selective removal of the central thiocarbonyl functionality would be required. To achieve this, bis(dithiepanethione) **101** is first treated with ethanolamine. The resulting reaction mixture is subsequently partitioned between aqueous hydrochloric acid and dichloromethane to afford a solution of intermediate **102** in dichloromethane. Finally, addition of potassium triiodide induces oxidative dithiane formation, thus affording (+)-11,11′-dideoxyverticillin A **5**.

The stereo- and chemoselective total synthesis by Movassaghi allows for the enantiospecific construction of (+)-11,11′-dideoxyverticillin A **5** in only nine steps from amide **94**. This biosynthetic strategy represents the first synthesis of a dimeric epidithioketopiperazine natural product, and also provides an approach for the construction of other members of the fascinating family of dimeric pyrrolidinoindoline alkaloids.

7.2.2.2 Total Synthesis of Minfiensine (Overman, Qin, MacMillan)

Minfiensine **6**, (Figure 7.3) was isolated in 1989 by Massiot and co-workers from the African plant *Strychnos minfiensis* [29]. Although minfiensine is a Strychnos alkaloid, its structure is more reminiscent of the akuammiline alkaloids [81a,89] by virtue of its 1,2,3,4-tetrahydro-9a-4a-(iminoethano)-9H-carbazole core [90]. Salient structural features of **6** include a pentacyclic ring system and three stereogenic centers, one of which is quaternary.

Minfiensine's daunting molecular scaffold has inspired several elegant syntheses. Overman reported first and second-generation routes to (+)-minfiensine in 2005 [30] and 2008 [31], respectively. Qin also disclosed a synthesis of (±)-minfiensine in 2008 [32], and subsequently, MacMillan unveiled a total synthesis of (+)-minfiensine in 2009 [33]. Although all three laboratories assemble the minfiensine core using different approaches, each ultimately complete the total synthesis of **6** by accessing a tetracyclic precursor of the type **103** shown in Figure 7.3. This section will cover Overman's second-generation strategy, followed by descriptions of the Qin and MacMillan syntheses.

Overman's preparation of the tetracyclic core of minfiensine is shown in Scheme 7.19. Aryl triflate **106** is assembled from anisidine **104** and morpholine enamine **105** using a robust five-step sequence that proceeds in 46% overall yield. Upon exposure of **106** to Pd(0) [91] and subsequently TFA, a domino catalytic asymmetric Heck–N-acyliminium ion cyclization takes place to furnish the targeted tetracycle pyrrolidinoindoline **110** in 75–87% yield and 99% ee. This critical transformation likely proceeds by oxidative addition, coordination/insertion (see transition structure **107**), followed by β-hydride elimination of intermediate **108** to ultimately afford enamine **109**. A final TFA induced tautomerization and cyclization of the crude product **109** provides access to the desired tetracycle **110**. This cascade not

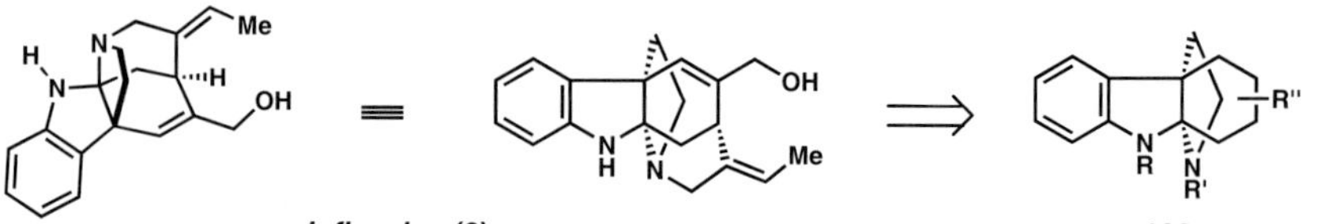

Figure 7.3 Minfiensine **6** and general synthetic strategy.

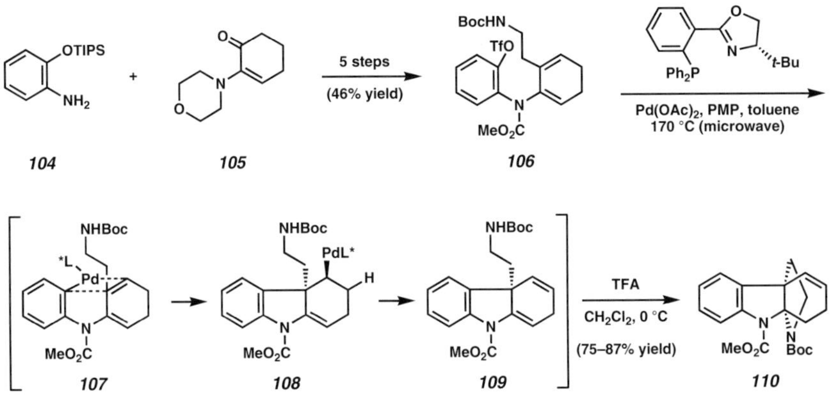

Scheme 7.19 Overman's synthesis of the minfiensine core.

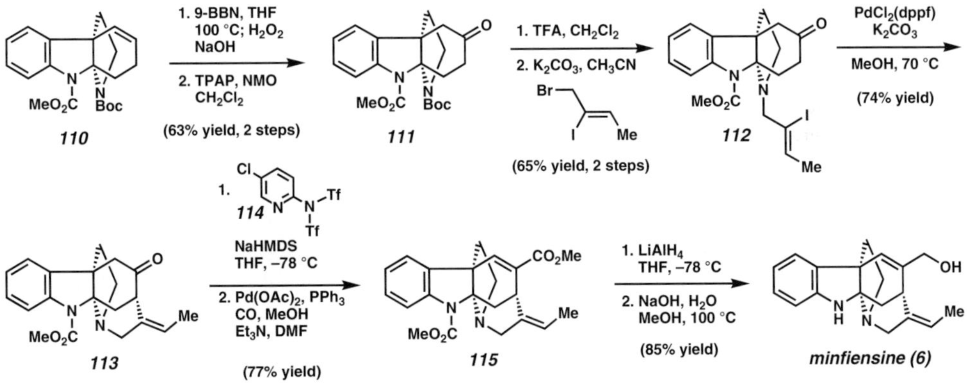

Scheme 7.20 Completion of the total synthesis of **6** by Overman.

only provides access to the key late-stage intermediate, but also demonstrates the utility of the asymmetric Heck cyclization reaction for the synthesis of sterically congested ring systems.

Overman's 2nd generation route to install the piperidine ring and complete the total synthesis of 6 is shown in Scheme 7.20. Hydroboration of tetracycle **110** with 9-borabicyclo[3.3.1]nonane followed by Ley oxidation [92] yields ketone **111**. Subsequently, Boc-deprotection and alkylation introduces the vinyl iodide functionality required to assemble the final ring. This task is ultimately achieved using a palladium-mediated enolate coupling reaction using a variation of Cook's conditions [93], thus allowing for the conversion of **112** to pentacycle **113** in 74% yield. Treatment of cyclohexanone **113** with Comins' reagent **114** [94] furnishes the corresponding vinyl triflate, which is then subjected to palladium-catalyzed carbonylation to yield α,β-unsaturated ester **115**. Finally, reduction of the ester moiety of **115** with LiAlH$_4$, followed by deprotection of the carbamate protecting group under

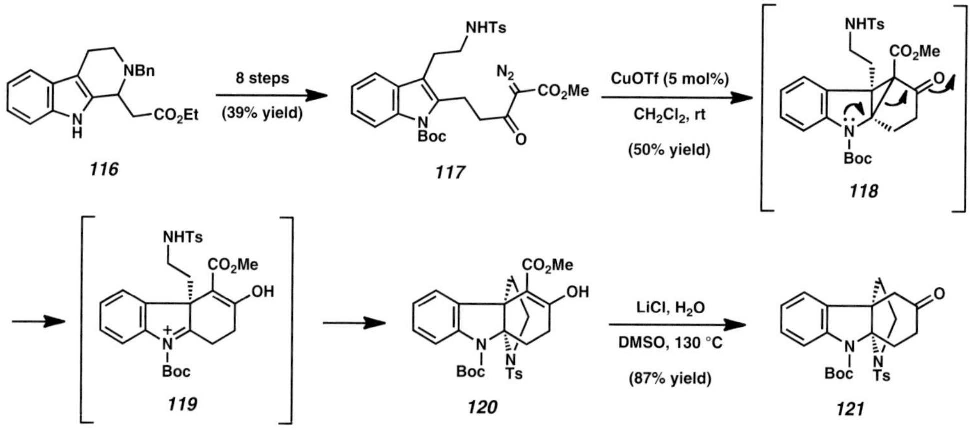

Scheme 7.21 Qin's synthesis of the minfiensine core.

basic conditions affords (+)-minfiensine **6**. Overman's second-generation total synthesis of (+)-minfiensine proceeds in 15 steps from morpholine enamine **105** and with an overall yield of 6.5%. Of note, each C–C bond-forming step employed in the synthesis uses palladium catalysis.

The approach employed by Qin to synthesize minfiensine **6** is also quite elegant, while differing significantly from the route described by Overman. Qin's strategy for assembling minfiensine's tetracyclic core involves an efficient one-pot cyclopropanation/ring fragmentation/cyclization cascade using an indole-derived started material (Scheme 7.21). To construct the cascade precursor, N-benzyl tetrahydrocarboline **116** (prepared in two steps from tryptamine) [95] is elaborated to diazo compound **117** using a robust eight-step sequence, which proceeds in 39% overall yield. In the key complexity-generating step of the synthesis, treatment of diazo species **117** with catalytic CuOTf at room temperature affords tetracycle **120** in 50% yield. The transformation is believed to involve decomposition of the diazo functionality to provide an intermediate cyclopropane, which in turn fragments to indolenium cation **119** (see transition structure **118**). Subsequent cyclization by the proximal sulfonamide furnishes the tetracyclic product **120**. Krapcho decarboxylation of **120** then affords ketone **121** [96].

In order to next assemble the piperidine ring, Qin and co-workers envisioned elaborating ketone **121** to vinyl iodide **124** by the removal of the tosyl group followed by alkylation of the nitrogen (Scheme 7.22). However, selective detosylation of **121** proved problematic. As a workaround, a stepwise route for the conversion of **121** to **124** was developed. Reduction of cyclohexanone **121** with NaBH$_4$, followed by tosyl deprotection yields **122**. Subsequent alkylation with **123** and alcohol oxidation delivers ketone **124**. Akin to Overman's route, a palladium-catalyzed intramolecular α-ketone vinylation [93] reaction of **124** assembles the piperidine ring to deliver pentacycle **125**. In three additional steps, ketone **125** is elaborated to minfiensine **6** in a manner similar to Overman's endgame strategy (Scheme

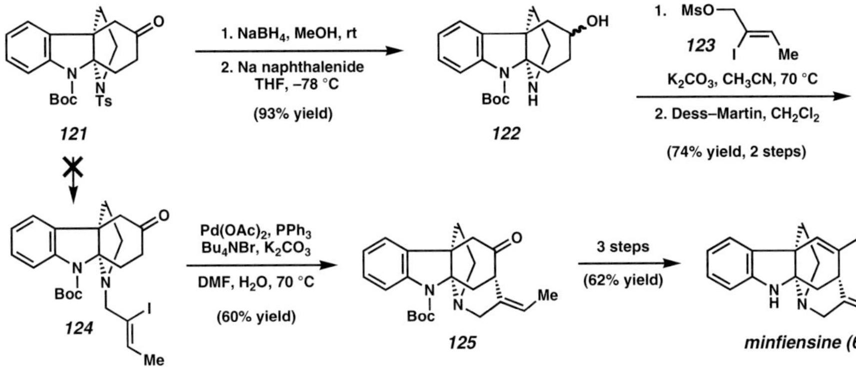

Scheme 7.22 Completion of the total synthesis of **6** by Qin.

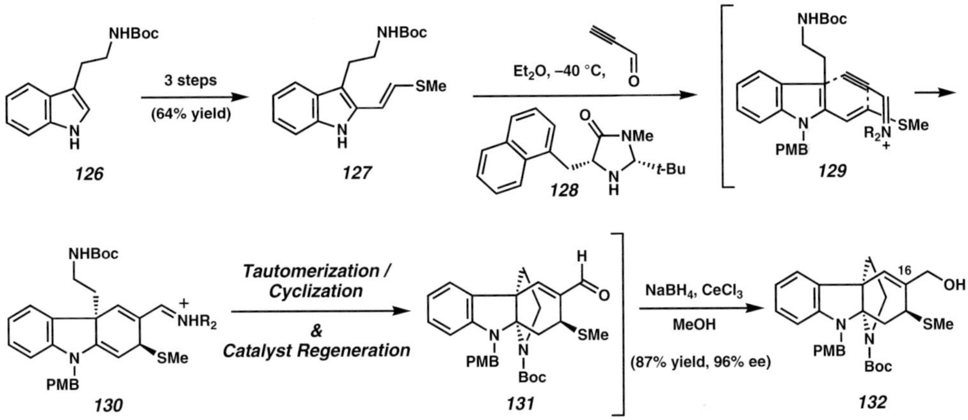

Scheme 7.23 MacMillan's synthesis of the minfiensine core.

7.20). The synthesis of (±)-**6** by Qin is 18 steps from tetrahydrocarboline **116**, and most notably, uses a unique one-pot cyclopropanation/ring fragmentation/cyclization cascade reaction to assemble the key tetracylic core of the natural product. The strategy has also proven amenable to the synthesis of the structurally related alkaloid vincorine [97].

Macmillan's rapid and enantiospecific synthesis of (+)-minfiensine **6** was disclosed in 2009, shortly after the report by Qin. The synthesis begins with the elaboration of tryptamine **126** to vinyl sulfide **127**, a substrate to be used in an organocatalytic [4 + 2] cycloaddition/tautomerization/cyclization cascade reaction (Scheme 7.23). In the event, treatment of 2-vinylindole **127** with propynal in the presence of organocatalyst **128** (TBA salt, 15 mol%), followed by reductive workup, leads to the formation of tetracycle **132** in 87% yield and 96% ee. This process is thought to occur by condensation of the secondary amine catalyst with propynal

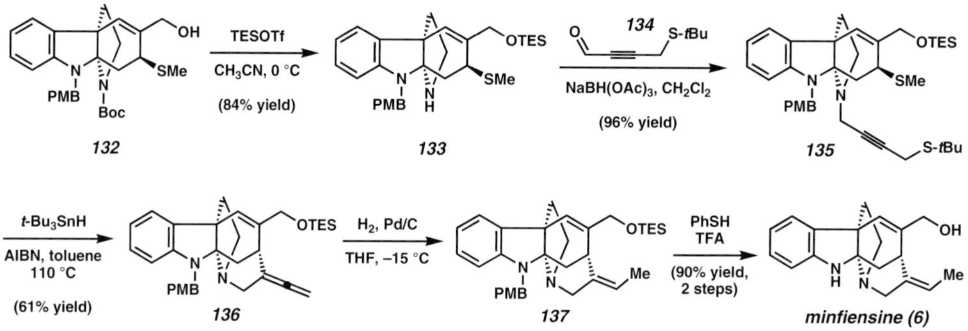

Scheme 7.24 Completion of the total synthesis of **6** by MacMillan.

to afford an intermediate iminium ion, which then reacts with **127** in an *endo*-selective Diels–Alder cycloaddition (see transition state structure **129**). Subsequent tautomerization of the PMB-substituted enamine in **130** to the corresponding iminum ion, followed by attack by the pendant *N*-Boc-protected amine (with catalyst regeneration) delivers the tetracyclic pyrroloindoline **131**. Reductive workup using Luche conditions [98] furnishes allylic alcohol **132**. It should be noted that **132** possesses much of the natural product framework, including the C16 hydroxymethyl group.

Similar to the Overman and Qin syntheses, the remaining challenge involves assembly of the piperidine ring. Whereas both Overman and Qin employed Pd-catalyzed transformations to achieve this goal, MacMillan uses a radical cyclization strategy that allows for the completion of the total synthesis in just five additional steps (Scheme 7.24). Treatment of **132** with TESOTf facilitates concomitant protection of the allylic alcohol and cleavage of the *N*-Boc-protecting group to furnish secondary amine **133**. Reductive amination of **133** with aldehyde **134** in the presence of NaBH(OAc)$_3$ provides the key cyclization substrate **135** in 96% yield. Although classic conditions for radical generation failed, the desired radical cyclization could be achieved using *t*-Bu$_3$SnH [99] and AIBN in refluxing toluene to yield allene **136** in an impressive 61% yield. To complete the total synthesis, allene **136** is treated with 10% Pd/C and H$_2$ in THF at −15 °C to effect regio- and diastereoselective reduction to trans-ethylidene **137**. Finally, global deprotection of **137** with TFA and thiophenol yields (+)-minfiensine **6**. Macmilllan's enantiospecific total synthesis of (+)-minfiensine is accomplished in just nine steps with an impressive overall yield of 21%. Key features of this synthesis include a novel asymmetric organocatalytic Diels–Alder/amine cyclization sequence to rapidly construct a key tetracyclic intermediate and a 6-*exo*-dig radical cyclization to install the final ring.

The Overman, Qin, and MacMillan syntheses of minfiensine **6** highlight the ability of synthetic chemists to develop creative strategies for the assembly of complex molecular architectures. Specifically, each of the routes features remarkable cascade transformations to build minfiensine's pyrrolidinoindoline core that

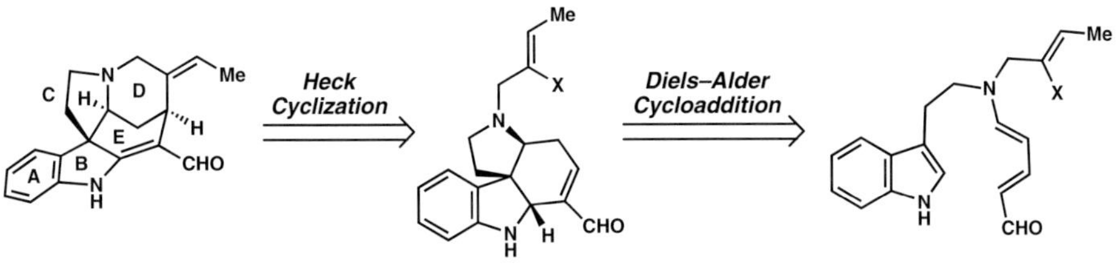

Scheme 7.25 Vanderwal's retrosynthesis of norfluorocurarine **7**.

rely on either Pd-catalysis, carbene chemistry, or organocatalysis. It is certain that the elegant strategies used to prepare minfiensine **6** will have applications in other synthetic endeavors.

7.2.2.3 Total Synthesis of Norfluorocurarine (Vanderwal)

The *Strychnos* alkaloid norfluorocurarine **7**, (Scheme 7.25) was first isolated by Stauffacher from the West African plant *Diplorrhynchus condylocarpon* subsp. *Mossambicensis* [34]. Others have since disclosed the isolation of norfluorocurarine from various plant sources [35]. Norfluorocurarine **7** contains the ABCE tetracyclic framework that is characteristic of the *Strychnos*, *Aspidosperma*, and *Iboga* families [1, 100]. Members of the *Strychnos* family have attracted much interest by the synthetic community [101], and three total syntheses of norfluorocurarine have been reported to-date [36, 102, 103]. The rapid synthesis of norfluorocurarine **7** by Vanderwal and co-workers will be the focus of this section [36].

In Vanderwal's retrosynthesis of norfluorocurarine **7**, shown in Scheme 7.25, the D ring would be constructed via an intramolecular Heck cyclization of vinyl halide **138**, a strategy also used in the synthesis of *Strychnos* alkaloids by Rawal [101b,104]. In turn, it was proposed that vinyl halide **138** could be formed from the intramolecular Diels–Alder cycloaddition of tryptamine-derived Zincke aldehyde **139** [105], a particularly ambitious disconnection since electron-rich indoles are generally considered poor dienophiles. The authors envisioned circumventing potentially unfavorable reactivity by using a Diels–Alder precursor bearing an electron-withdrawing group on the terminus of the diene. Such inverse electron-demand intramolecular Diels–Alder reactions had not previously been used in the syntheses of *Strychnos* alkaloids [101d].

Vanderwal's strategy is executed in the forward direction beginning with tryptamine **140** [106], which is converted to amine **141** by a straightforward alkylation reaction (Scheme 7.26). Next, Zincke aldehyde **145** is generated upon reaction of **141** with pyridinium salt **142** and subsequent hydrolysis with NaOH. This reaction presumably proceeds through intermediate **143**, which undergoes 6π electrocyclic pyridine ring opening to form imine intermediate **144**. A final basic workup facilitates imine hydrolysis and yields the key intermediate **145**.

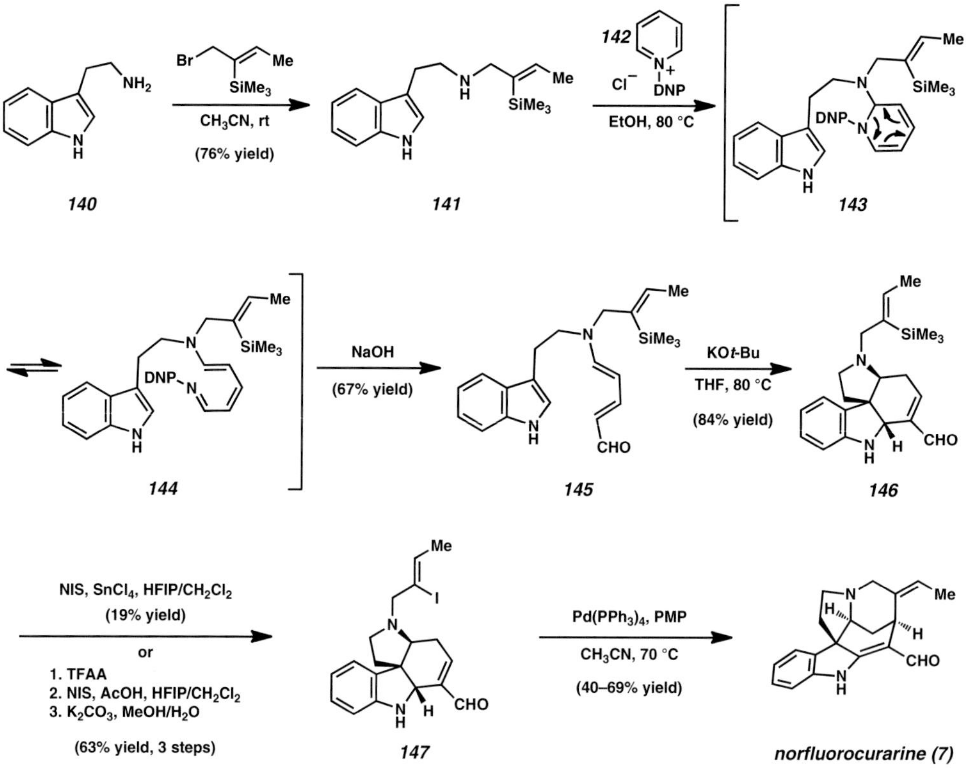

Scheme 7.26 The total synthesis of norfluorocurarine **7**.

With Zincke aldehyde **145** in hand, the next targets for construction were rings C and E of norfluorocurarine via the intramolecular [4 + 2] cycloaddition reaction. On model systems, it was observed that elevated temperatures led to undesired thermal Zincke aldehyde rearrangements [107]. Furthermore, use of Lewis acids and protic acids effected degradation or yielded Pictet–Spengler type products, rather than the desired cycloaddition. Ultimately, it was discovered that treatment of precursor **145** with *t*-BuOK in THF at 80 °C results in a successful and stereoselective anionic bicyclization reaction to afford tetracycle **146** in 84% yield.

The final steps of the synthesis involve the conversion of tetracycle **146** to vinyl iodide **147**. Although **147** could be obtained in one step by employing NIS and SnCl$_4$, this provided the desired product in prohibitively low yields. An alternate three-step approach was developed to access substrate **147**, which proceeds in 63% overall yield. In this sequence, the indoline nitrogen is protected with TFAA prior to iodination. Subsequent cleavage of the trifluoroacetamide furnishes late-stage intermediate **147**. Finally, Heck cyclization of **147** constructs the E ring, thus completing the total synthesis of norfluorocurarine **7**.

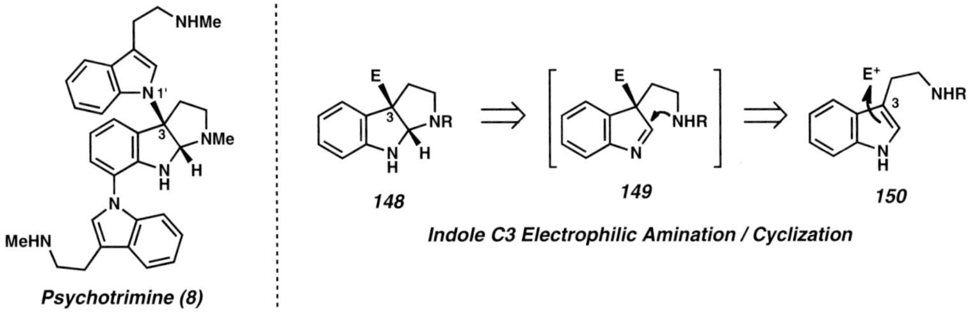

Scheme 7.27 Baran's strategy for the synthesis of psychotrimine **8**.

Remarkably, Vanderwal's concise total synthesis of **7** proceeds in five steps from tryptophan (or seven steps using the three-step iodination sequence). The brevity of the synthesis stems from the strikingly rapid construction of the ABCE strychnos core via a base-promoted anionic bicyclization of a tryptamine-based Zincke aldehyde.

7.2.2.4 Total Synthesis of Psychotrimine (Baran)

Psychotrimine **8**, (Scheme 7.27) was isolated by Takayama and co-workers in 2004 from the leaves of the plant *Psychotria rostrata*, which is used in Malaysia for the treatment of constipation [37]. Structural elucidation through spectroscopic analysis revealed psychotrimine to be a polymeric indole alkaloid [37, 84] whose biosynthesis presumably employs a tryptophan dimerization reaction. Psychotrimine **8** contains two stereocenters, one of which is quaternary at C3. Although numerous indoline natural products are known, psychotrimine is unique as it contains a rare C3–N1' tryptamine linkage.

The unusual structure of psychotrimine **8** has intrigued a number of synthetic groups. A few years after isolation, Takayama disclosed the first total synthesis of psychotrimine [108], which proceeds in 16 steps via the stepwise assembly of the pyrrolidinoindole motif. Around the same time, Rainer described a method to construct C3–N1' heterodimeric indolines from 3-bromopyrroloindoline precursors [109]. This section, however, will highlight the inventive and concise synthesis of psychotrimine described by Baran in 2008 [38].

Baran's strategy for accessing the pyrrolidinoindole core of psychotrimine **8** is shown in Scheme 7.27. It was envisioned that the indoline framework **148** could be assembled from the reaction of a tryptamine starting material with an electrophilic nitrogen source (see **150**) via intermediate indolenine **149**. Although analogous strategies have been used to access C3 halogenated [110] or alkylated [111] indolines, the analogous transformation for installation of a C3-nitrogen substituent was unknown.

The implementation of this unique approach to psychotrimine **8** is shown in Scheme 7.28. As extensive experimentation to achieve the direct coupling of tryptamine derivatives with indole subunits proved unsuccessful, the authors turned

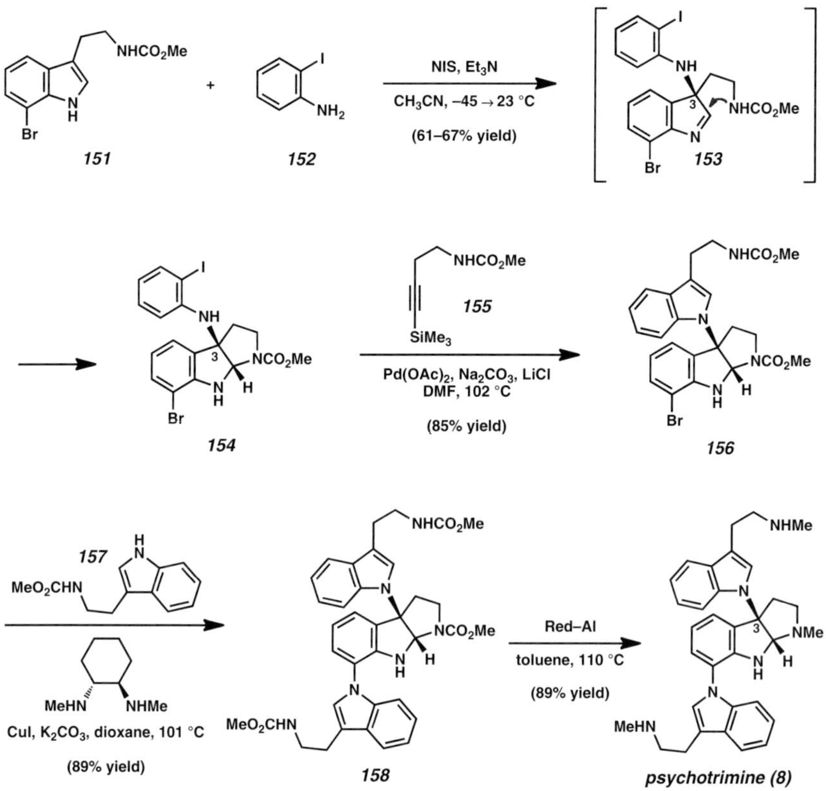

Scheme 7.28 The total synthesis of psychotrimine **8**.

to *o*-iodoaniline **152** as a substitute for the indole coupling partner. After optimization of reaction conditions, the authors discovered that tryptamine **151** and *o*-iodoaniline **152** undergo coupling in the presence of NIS and Et₃N to furnish indoline **154** in 61–67% yield. The transformation is believed to proceed via an initial *N*-halogenation [112] of iodoaniline **152**, followed by nucleophilic attack by **151** to deliver indolenine intermediate **153**. Subsequent trapping by the proximal carbamate furnishes the indoline product **154**.

Following installation of the requisite C3-substituent, attention was directed toward the completion of the total synthesis. Larock annulation [113] of *o*-iodoaniline **154** with alkyne **155** proceeds chemoselectively to assemble the indole motif present in **156**. In turn, **156** is coupled with tryptamine derivative **157** under Buchwald–Goldberg–Ullman conditions [114] to provide bis(indole) **158**, which possesses the full carbon skeleton of the targeted natural product. Finally, a triple reduction of the carbamate functionalities found in **158** with Red-Al yields psychotrimine **8**.

The rapid synthesis described by Baran proceeds in only four steps from tryptamine **151**, and delivers racemic psychotrimine **8** in 41–45% overall yield. The route is scalable and has facilitated the preparation of over two grams of the natural product. The synthesis features a novel oxidative C–N bond-forming reaction that not only installs the C3-nitrogen substituent, but also provides the psychotrimine indoline core. The generality of this method was more recently demonstrated in the enantiospecific syntheses of the related C3-N1' indoline-containing natural products kapakahines B and F [115].

7.2.3
Indolizidines

7.2.3.1 Total Synthesis of Myrmicarins 215A, 215B and 217 (Movassaghi)

The myrmicarins are a family of alkaloids isolated from the poison gland secretions of the African ant, *Myrmicaria opaciventris* [39]. Numerous members belong to this family of alkaloids, including M430A **159**, and simpler tricyclic compounds M215A **9**, M215B **10**, and M217 **11** (Figure 7.4). The indolizidine framework, often seen as a pyrroloindolizine core, is a structural feature shared by the myrmicarins that have made these alkaloids popular targets for total synthesis. Notable synthetic achievements in this area have been reviewed [116], and include the racemic synthesis of M217 by Schröder and Francke [117], the synthesis of (+)-M217 by Vallée [118], the formal synthesis of (−)-M217 by Lazzaroni [119], as well as the syntheses of M215A and M215B by Vallée [120]. Although several approaches to the myrmicarin alkaloids have been reported, the focus of this section will be on Movassaghi's syntheses of (−)-M215A **9**, (+)-M215B **10** and (+)-M217 **11** [40].

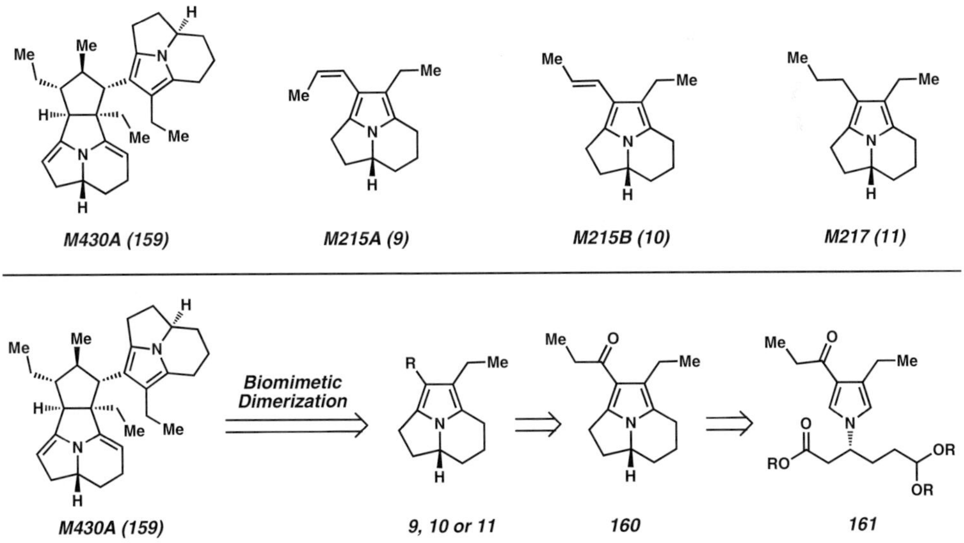

Figure 7.4 Myrmicarin alkaloids and Movassaghi's strategy for the synthesis of **9**, **10** and **11**.

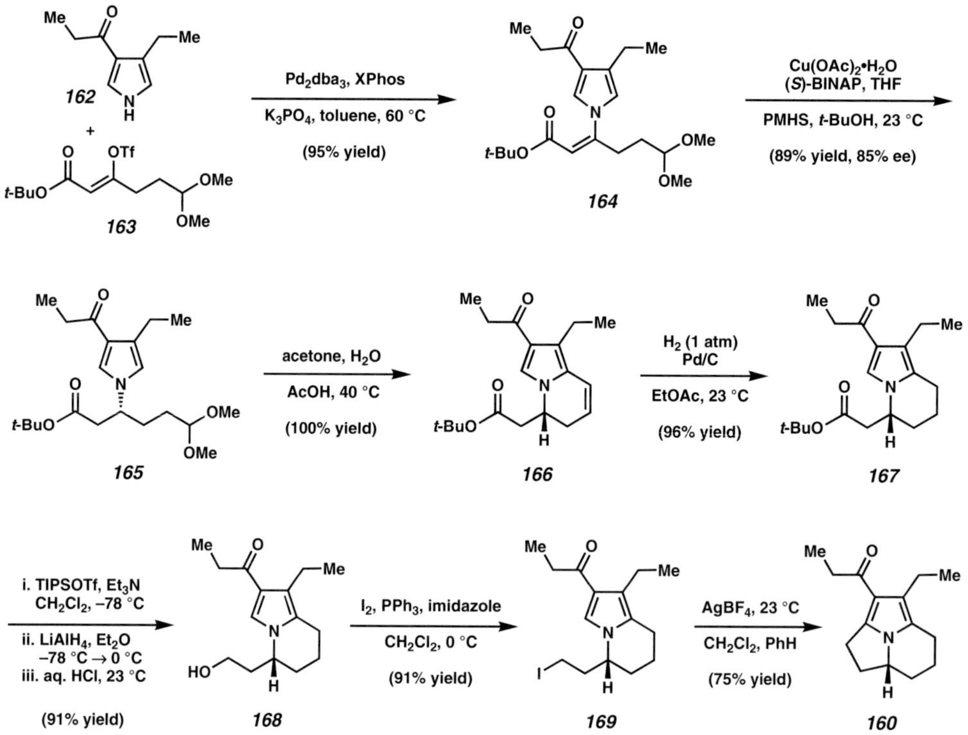

Scheme 7.29 Synthesis of tricycle **160**.

Movassaghi hypothesized that higher order myrmicarins such as M430A **159** could arise from a biomimetic dimerization of **9**, **10**, **11**, or some derivative thereof (Figure 7.4) [121]. It was proposed that all three natural products could be accessed from tricycle **160**. Tricycle **160**, in turn, would be synthesized from N-substituted pyrrole **161**. It was hypothesized that **161** could be prepared in enantioenriched form beginning from readily available starting materials.

Scheme 7.29 shows Movassaghi's synthesis of tricycle **160**, a late-stage intermediate that would serve as the common precursor towards the synthesis of **9**, **10** and **11**. Coupling of pyrrole **162** and vinyl triflate **163** (each readily accessible) under palladium catalysis affords N-vinyl pyrrole **164** [122]. This transformation represents the first example of a palladium-catalyzed N-vinylation of an azaheterocycle with a vinyl triflate and, notably, proceeds well on multigram scale. Enantioselective conjugate reduction of vinylogous amide **164** under Cu(OAc)$_2$/BINAP/PMHS conditions described by Buchwald [123] affords the optically active pyrrole **165** in 89% yield and 85% ee. Exposure of pyrrole **165** to AcOH in acetone and H$_2$O leads to cyclization with loss of methanol to deliver intermediate **166**, which contains the 6-membered ring of the natural product framework. Reduction of the olefin present in bicyclic pyrrole **166** under standard conditions affords

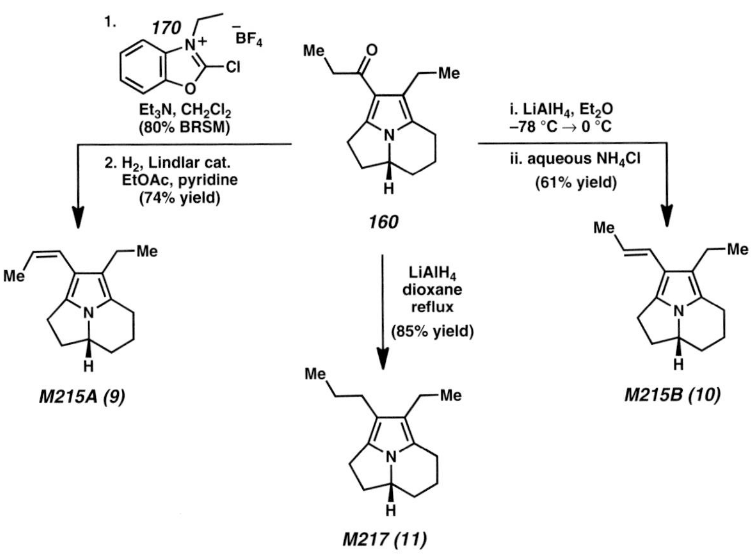

Scheme 7.30 Total synthesis of myrmicarins M215A, **9**, (+)-M215B, **10**, and (+)-M217, **11**.

indolizidine derivative **167** in 96% yield. To assemble the remaining 5-membered ring of the myrmicarin scaffold, **167** is converted to tricycle **160** through a sequence involving ester reduction with *in-situ* ketone protection (**167**→**168**), iodination (**168**→**169**), and Ag-mediated cyclization (**169**→**160**). Through this route, tricycle **160** is accessed in enantioenriched form in only seven steps from **162** and **163**.

As shown in Scheme 7.30, tricycle **160** serves as the common intermediate en route to (−)-M215A **9**, (+)-M215B **10**, and (+)-M217 **11**. Reduction of **160** with LiAlH$_4$ under refluxing conditions provides M217 **11** in a straightforward manner. However, accessing M215A **9** and M215B **10** as discrete olefin isomers was deemed more challenging due, in part, to their extreme sensitivity to air oxidation. Ultimately, it was found that reduction of **160** with LiAlH$_4$, followed by hydroxyl elimination in the presence of ammonium chloride furnishes M215B **10** with the desired *E* olefin geometry. To access M215A **9**, key intermediate **160** is first dehydrated with 2-chloro-3-ethylbenzoxazolium tetrafluoroborate **170** [124] to provide a particularly unstable intermediate alkyne. This species is then subjected to partial reduction under Lindlar conditions to afford M215A **9**, possessing the necessary *Z* olefin configuration.

Movassaghi's biomimetic approach to myrmicarin alkaloids M215A **9**, M215B **10**, and M217 **11** features several innovative transformations, such as a palladium-catalyzed pyrrole vinylation, an asymmetric conjugate reduction of a vinylogous amide, and sequential Friedel–Crafts cyclizations to access the key pentasubstituted pyrrole scaffold. The success of these steps ultimately provides access to late-stage intermediate **160**, which then serves as a common intermediate to three

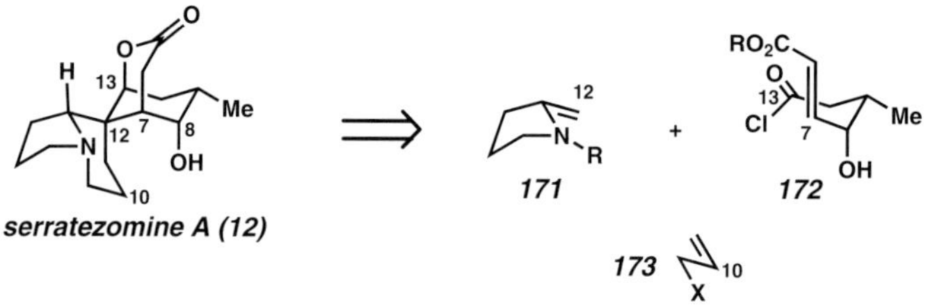

Scheme 7.31 Johnston's strategy for the synthesis (+)-serratezomine A **12**.

natural products. Movassaghi's concise enantiospecific syntheses of M215A **9**, (+)-M215B **10**, and (+)-M217 **11** also lay the foundation for the syntheses of higher order myrmicarin alkaloids.

7.2.3.2 Total Synthesis of Serratezomine A (Johnston)

Serratezomine A **12**, (Scheme 7.31), a spirocyclic indolizidine, was isolated from the club moss *L. serratum* collected from Sapporo, Japan in 2000 [41]. **12** is a member of the structurally intriguing *Lycopodium* family of natural products, which has received much attention from the synthetic community [125–130]. However, serratezomine A **12** differs structurally from most *Lycopodium* alkaloids by virtue of its indolizidine framework. Beyond the indolizidine ring, **12** possesses a number of features that render it a formidable target for synthesis, including the C12 spirocenter embedded within the [3.3.1]-bicyclic lactone. The natural product contains six stereogenic centers, all of which are contiguous. The challenge in synthesizing serratezomine A **12** is further compounded by an observation reported by Morita and Kobayashi, who noted that **12** is prone to undergo a thermodynamically favored lactone isomerization from O13 to O8 in the presence of acid [131]. Despite these hurdles, Johnston and co-workers were able to complete a total synthesis of (+)-serratezomine A in 2009 [42].

Johnston's retrosynthesis of serratezomine A **12** is shown in Scheme 7.31 and is based on the conceptual disconnection of (+)-serratezomine A into three fragments: **171**, **172**, and **173**. It was envisioned that an unsaturated 2-methylpyrrolidine **171** could provide a doubly nucleophilic C12 site that would enable the assembly of the cyclohexyl ring when reacted with **172**, a compound that is electrophilic at both C7 and C13. The three-carbon fragment **173** would serve as a building block for late-stage installation of the piperidine ring.

The convergent synthesis of (+)-serratezomine A **12** is shown in Scheme 7.32. Commercially available aldehyde **174** is readily converted over five steps to carboxylic acid **175**. Upon conversion of acid **175** to the corresponding acid chloride, the key doubly electrophilic fragment (see **172**, Scheme 7.31) is available. Subsequent trapping of the acid chloride with β-stannyl enamine **176** (prepared by a free radical-mediated vinyl amination reaction [132]) affords the vinylogous amide **177** in 65% yield. Oxidative dealkylation of pyrrolidine **177** in the presence of Ce(NH$_4$)$_4$(NO$_2$)$_6$ (CAN) promotes cyclization of the deprotected vinylogous amide onto the unsaturated ester to afford unsaturated cyclohexanone **178**.

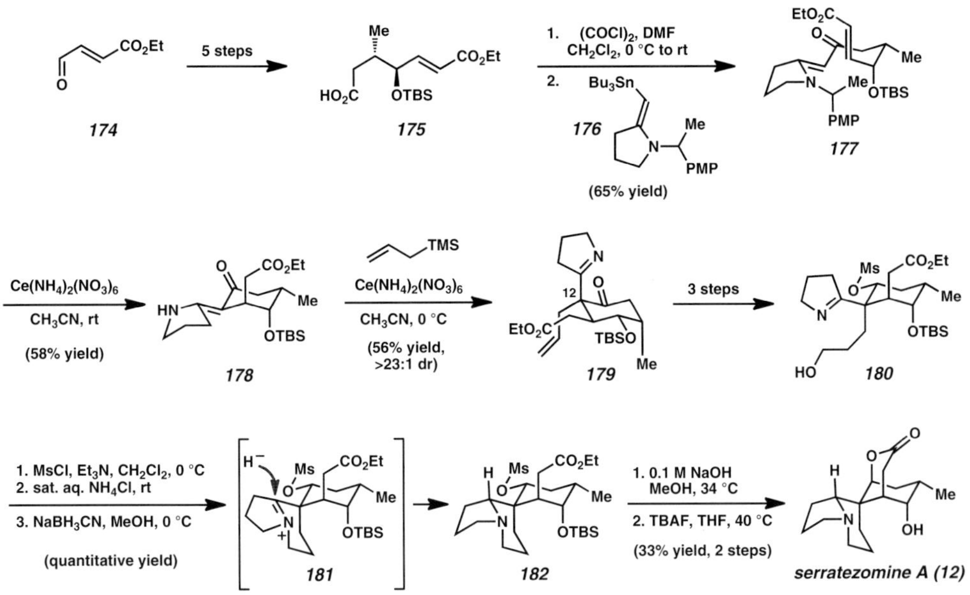

7.2 Applications of Indoles and Indolizidines in the Synthesis of Natural Products | 253

Scheme 7.32 Total synthesis of (+)-serratezomine A **12**.

With the cyclohexyl ring of the natural product in place, the principle remaining challenges include installation of the piperidine and lactone rings. Toward the assembly of the piperidine ring, cyclohexanone **178** is subjected to a Ce(IV)-mediated oxidative allylation [133] to yield β-imino ketone **179** in 56% yield and 23:1 dr. It should be noted that **179** possesses the key C12 quaternary stereocenter and all of the carbon atoms needed to complete the natural product synthesis. β-Imino ketone **179** is further elaborated to alcohol **180** via a three-step sequence. Upon activation of the primary alcohol to the corresponding mesylate, cyclization occurs to form an intermediate iminium ion, which is stabilized upon addition of ammonium chloride. Reduction of the iminium ion with sodium cyanoborohydride (see transition structure **181**) provides the pyrrolidine ring of indolizidine **182** in quantitative yield. Saponification of the ester moiety in **182** with NaOH followed by displacement of the secondary mesylate installs the [3.3.1]-bicyclic lactone [134]. Finally, TBAF deprotection furnishes (+)-serratezomine A **12**.

The convergent, enantiospecific total synthesis of (+)-serratezomine A **12** described by Johnston is achieved in just 15 steps (longest linear sequence) from aldehyde **174**. Notable features of this synthesis include the use of a β-stannyl enamine as a doubly nucleophilic synthetic lynchpin and a Ce(IV)-induced oxidative allylation [133] to introduce the C12 all-carbon quaternary stereocenter en route to **12**. Johnston's strategy provides an innovative means to assemble this challenging and unusual indolizidine-containing *Lycopodium* alkaloid.

7.3
Conclusion

Indole and indolizidine alkaloids have been a tremendous source of scientific fascination over the past century. The structural diversity encompassed by these molecules, coupled with their promising biological profiles, has inspired numerous studies in the area of chemical synthesis. In the last decade, hundreds of indole and indolizidine alkaloids have been prepared, often accompanied by the invention of new synthetic strategies and methodologies. The syntheses described in this chapter capture only some of the most impressive achievements in this critical area of study. The rich arena of indole and indolizidine alkaloid total syntheses will surely continue to flourish, thus stimulating further discoveries in both chemistry and human medicine.

Acknowledgment

This chapter is dedicated to Professor Larry Overman for his inspiring achievements in alkaloid total synthesis.

References

1 Hesse, M. (2002) *Alkaloids: Nature's Curse Or Blessing?* Wiley-VCH Verlag GmbH, Weinheim, Germany, p. 5.
2 (a) Sertürner, F.W. (1817) Ueber das Morphium, eine neue salzfähige Grundlage, und die Mekonsäure, als Hauptbestandtheile des Opiums. *Ann. Der Physik (Berlin)*, **55** (1), 56–90; (b) Schmitz, R. and Kuhlen, F.-J. (1989) Schmerz- und Betäubungsmittel vor 1600. Ein fast unbekanntes Kapitel der Arzneimittelgeschichte. *Pharm. Unserer Zeit*, **18** (1), 11–19; (c) Huxtable, R.J. and Schwartz, S.K.W. (2001) The isolation of morphine–first principles in science and ethics. *Mol. Interv.*, **1** (4), 189–191.
3 Ladenburg, A. (1886) Versuche zur Synthese des Coniin. *Ber. Dtsch. Chem. Ges.*, **19** (1), 439–441.
4 MDL Information Systems Inc., MDL Drug Data Report. MDL Information Systems Inc., San Leandro, CA.
5 (a) Woodward, R.B., Cava, M.P., Ollis, W.D., Hunger, A., Daeniker, H.U. and Schenker, K. (1954) The total synthesis of strychnine. *J. Am. Chem. Soc.*, **76** (18), 4749–4751; (b) Woodward, R.B., Cava, M.P., Ollis, W.D., Hunger, A., Daeniker, H.U. and Schenker, K. (1963) The total synthesis of strychnine. *Tetrahedron*, **19** (2), 247–288.
6 (a) Woodward, R.B., Bader, F.E., Bickel, H., Frey, A.J. and Kierstead, R.W. (1956) The total synthesis of reserpine. *J. Am. Chem. Soc.*, **78** (9), 2023–2025; (b) Woodward, R.B., Bader, F.E., Bickel, H., Frey, A.J. and Kierstead, R.W. (1958) The total synthesis of reserpine. *Tetrahedron*, **2** (1), 1–57.
7 Daly, J.W., Garraffo, H.M. and Spande, T.F. (1999) Alkaloids from amphibian skin, in *Alkaloids: Chemical and Biological Perspectives*, vol. 13, 1st edn (ed. S.W. Pelletier), Pergamon, New York, USA, pp. 1–161.
8 Daly, J.W., Spande, T.F. and Garraffo, H.M. (2005) Alkaloids from amphibian skin: a tabulation of over eight-hundred compounds. *J. Nat. Prod.*, **68** (10), 1556–1575.
9 (a) Horton, D.A., Bourne, G.T. and Smyth, M.L. (2003) The combinatorial

synthesis of bicyclic privileged structures or privileged substructures. *Chem. Rev.*, **103** (3), 893–930; (b) Nicolaou, K.C. and Snyder, S.A. (2004) The essence of total synthesis. *Proc. Nat. Acad. Sci. U.S.A.*, **101** (33), 11929–11936; (c) Nicolaou, K.C., Vourloumis, D., Winssinger, N. and Baran, P.S. (2000) The art and science of total synthesis at the dawn of the twenty-first century. *Angew. Chem. Int. Ed.*, **39** (1), 44–122.

10 (a) Nicolaou, K.C. and Sorensen, E.J. (1996) *Classics in Total Synthesis*, Wiley-VCH Verlag GmbH, Weinheim, Germany; (b) Nicolaou, K.C. and Snyder, S.A. (2003) *Classics in Total Synthesis II*, Wiley-VCH Verlag GmbH, Weinheim, Germany; (c) Hudlicky, T. and Reed, J.W. *The Way of Synthesis: Evolution of Design and Methods for Natural Products*, Wiley-VCH Verlag GmbH, Weinheim, Germany; (d) Corey, E.J. and Cheng, X.-M. (1989) *The Logic of Chemical Synthesis*, John Wiley & Sons, Inc., New York, USA.

11 (a) Burgett, A.W.G., Li, Q., Wei, Q. and Harran, P.G. (2003) A concise and flexible total synthesis of (−)-diazonamide A. *Angew. Chem. Int. Ed.*, **42** (40), 4961–4966; (b) Li, J., Jeong, S., Esser, L. and Harran, P.G. (2001) Total synthesis of nominal diazonamides – part 1: convergent preparation of the structure proposed for (−)-diazonamide A. *Angew. Chem. Int. Ed.*, **40** (24), 4765–4769; (c) Li, J., Burgett, A.W.G., Esser, L., Amezcua, C. and Harran, P.G. (2001) Total synthesis of nominal diazonamides – part 2: on the true structure and origin of natural isolates. *Angew. Chem. Int. Ed.*, **40** (24), 4770–4773.

12 (a) Nicolaou, K.C., Bella, M., Chen, D.Y.-K., Huang, X., Ling, T. and Snyder, S.A. (2002) Total synthesis of diazonamide A. *Angew. Chem. Int. Ed.*, **41** (18), 3495–3499; (b) Nicolaou, K.C., Bheema Rao, P., Hao, J., Reddy, M.V., Rassias, G., Huang, X., Chen, D.Y.-K. and Synder, S.A. (2003) The second total synthesis of diazonamide A. *Angew. Chem. Int. Ed.*, **42** (15), 1753–1758.

13 Yokoshima, S., Uedo, T., Kobayashi, S., Sato, A., Kuboyama, T., Tokuyama, H. and Fukuyama, T. (2002) Stereocontrolled total synthesis of (+)-vinblastine. *J. Am. Chem. Soc.*, **124** (10), 2137–2139.

14 Lebsack, A.D., Link, J.T., Overman, L.E. and Stearns, B.A. (2002) Enantioselective total synthesis of quadrigemine C and psycholeine. *J. Am. Chem. Soc.*, **124** (31), 9008–9009.

15 He, F., Bo, Y., Altom, J.D. and Corey, E.J. (1999) Enantioselective total synthesis of aspidophytine. *J. Am. Chem. Soc.*, **121** (28), 6771–6772.

16 Baran, P.S., Guerrero, C.A. and Corey, E.J. (2003) Short, enantioselective total synthesis of okaramine N. *J. Am. Chem. Soc.*, **125** (19), 5628–5629.

17 (a) Kim, G., Chu-Moyer, M.Y. and Danishefsky, S.J. (1990) The total synthesis of dl-indolizomycin. *J. Am. Chem. Soc.*, **112** (5), 2003–2005; (b) Kim, G., Chu-Moyer, M.Y., Danishefsky, S.J. and Schulte, G.K. (1993) The total synthesis of indolizomycin. *J. Am. Chem. Soc.*, **115** (1), 30–39.

18 Carroll, A.R., Hyde, E., Smith, J., Quinn, R.J., Guymer, G. and Forster, P.I. (2005) Actinophyllic acid, a potent indole alkaloid inhibitor of the coupled enzyme assay carboxypeptidase U/hippuricase from the leaves of *Alstonia actinophylla* (apocynaceae). *J. Org. Chem.*, **70** (3), 1096–1099.

19 (a) Martin, C.L., Overman, L.E. and Rohde, J.M. (2008) Total synthesis of (±)-actinophyllic acid. *J. Am. Chem. Soc.*, **130** (24), 7568–7569; (b) Taniguchi, T., Martin, C.L., Monde, K., Nakanishi, K., Berova, N. and Overman, L.E. (2009) Absolute configuration of actinophyllic acid as determined through chiroptical data. *J. Nat. Prod.*, **72** (3), 430–432.

20 Cutignano, A., Bifulco, G., Bruno, I., Casapullo, A., Gomez-Paloma, L. and Riccio, R. (2000) Dragmacidin F: a new antiviral bromoindole alkaloid from the Mediterranean sponge *Halicortex* sp. *Tetrahedron*, **56** (23), 3743–3748.

21 (a) Garg, N.K., Caspi, D.D. and Stoltz, B.M. (2004) The total synthesis of (+)-dragmacidin F. *J. Am. Chem. Soc.*, **126** (31), 9552–9553; (b) Garg, N.K., Caspi, D.D. and Stoltz, B.M. (2005) Development of an enantiodivergent

strategy for the total synthesis of (+)- and (−)-dragmacidin F from a single enantiomer of quinic acid. *J. Am. Chem. Soc.*, **127** (16), 5970–5978.

22 Wilson, B.J., Wilson, C.H. and Hayes, A.W. (1968) Tremorgenic toxin from *Penicillium cyclopium* grown on food materials. *Nature*, **220** (5162), 77–78.

23 (a) Smith, A.B., III, Kanoh, N., Ishiyama, H. and Hartz, R.A. (2000) Total synthesis of (−)-penitrem D. *J. Am. Chem. Soc.*, **122** (45), 11254–11255; (b) Smith, A.B., III, Kanoh, N., Ishiyama, H., Minakawa, N., Rainier, J.D., Hartz, R.A., Cho, Y.S., Cui, H. and Moser, W.H. (2003) Tremorgenic indole alkaloids. The total synthesis of (−)-penitrem D. *J. Am. Chem. Soc.*, **125** (27), 8228–8237.

24 Stratmann, K., Moore, R.E., Bonjouklian, R., Deeter, J.B., Patterson, G.M.L., Shaffer, S., Smith, C.D. and Smitka, T.A. (1994) Welwitindolinones, unusual alkaloids from the blue-green algae *Hapalosiphon welwitschii* and *Westiella intricata*. Relationship to fischerindoles and hapalinodoles. *J. Am. Chem. Soc.*, **116** (22), 9935–9942.

25 (a) Baran, P.S. and Richter, J.M. (2005) Enantioselective total syntheses of welwitindolinone A and fischerindoles I and G. *J. Am. Chem. Soc.*, **127** (44), 15394–15396; (b) Baran, P.S., Maimone, T.J. and Richter, J.M. (2007) Total synthesis of marine natural products without using protecting groups. *Nature*, **446** (7134), 404–408; (c) Richter, J.M., Ishihara, Y., Masuda, T., Whitefield, B.W., Llamas, T., Pohjakallio, A. and Baran, P.S. (2008) Enantiospecific total synthesis of the hapalindoles, fischerindoles, and welwitindolinones via a redox economic approach. *J. Am. Chem. Soc.*, **130** (52), 17938–17954.

26 (a) Reisman, S.E., Ready, J.M., Hasuoka, A., Smith, C.J. and Wood, J.L. (2006) Total synthesis of (±)-welwitindolinone A isonitrile. *J. Am. Chem. Soc.*, **128** (5), 1448–1449; (b) Reisman, S.E., Ready, J.M., Weiss, M.M., Hasuoka, A., Hirata, M., Tamaki, K., Ovaska, T.V., Smith, C.J. and Wood, J.L. (2008) Evolution of a synthetic strategy: total synthesis of (±)-welwitindolinone A isonitrile. *J. Am. Chem. Soc.*, **130** (6), 2087–2100.

27 Son, B.W., Jensen, P.R., Kauffman, C.A. and Fenical, W. (1999) New cytotoxic epidithiodioxopiperazines related to verticillin A from a marine isolate of the fungus *Penicillium*. *Nat. Prod. Lett.*, **13** (3), 213–222.

28 Kim, J., Ashenhurst, J.A. and Movassaghi, M. (2009) Total synthesis of (+)-11,11′-dideoxyverticillin A. *Science*, **324** (5924), 238–241.

29 Massiot, G., Thépenier, P., Jacquier, M.-J., Le Men-Olivier, L. and Delaude, C. (1989) Normavacurine and minfiensine, two new alkaloids with $C_{19}H_{22}N_2O$ formula from *Strychnos* species. *Heterocycles*, **29** (8), 1435–1438.

30 Dounay, A.B., Overman, L.E. and Wrobleski, A.D. (2005) Sequential catalytic asymmetric Heck–iminium ion cyclization: enantioselective total synthesis of the *Strychnos* alkaloid minfiensine. *J. Am. Chem. Soc.*, **127** (29), 10186–10187.

31 Dounay, A.B., Humphreys, P.G., Overman, L.E. and Wrobleski, A.D. (2008) Total synthesis of the *Strychnos* alkaloid (+)-minfiensine: Tandem enantioselective intramolecular Heck–iminium ion cyclization. *J. Am. Chem. Soc.*, **130** (15), 5368–5377.

32 Shen, L., Zhang, M., Wu, Y. and Qin, Y. (2008) Efficient assembly of an indole alkaloid skeleton by cyclopropanation: concise total synthesis of (±)-minfiensine. *Angew. Chem. Int. Ed.*, **47** (19), 3618–3621.

33 Jones, S.B., Simmons, B. and MacMillan, D.W.C. (2009) Nine-step enantioselective total synthesis of (+)-minfiensine. *J. Am. Chem. Soc.*, **131** (38), 13606–13607.

34 Stauffacher, D. (1961) Alkaloide aus *Diplorrhynchus condylocarpon* (MUELL. ARG.) Pichon ssp. *mossambicensis* (BENTH.) Duvign. *Helv. Chim. Acta*, **44** (7), 2006–2015.

35 (a) Rakhimov, D.A., Malikov, V.M. and Yusunov, C.Y. (1969) Structure of vincanicine. *Khim. Prir. Soedin.*, **5** (5), 461–462; (b) Clivio, P., Richard, B., Deverre, J.-R., Sevenet, T., Zeches, M. and Le Men-Oliver, L. (1991) Alkaloids

from leaves and root bark of *Ervatamia hirta*. *Phytochemistry*, **30** (11), 3785–3792.

36 Martin, D.B.C. and Vanderwal, C.D. (2009) Efficient access to the core of the *Strychnos*, *Aspidosperma* and *Iboga* alkaloids. A short synthesis of norfluorocurarine. *J. Am. Chem. Soc.*, **131** (10), 3472–3473.

37 Takayama, H., Mori, I., Kitajima, M., Aimi, N. and Lajis, N.H. (2004) New type of trimeric and pentameric indole alkaloids from *Psychotria rostrata*. *Org. Lett.*, **6** (17), 2945–2948.

38 Newhouse, T. and Baran, P.S. (2008) Total synthesis of (±)-psychotrimine. *J. Am. Chem. Soc.*, **130** (33), 10886–10887.

39 (a) Francke, W., Schröder, F., Walter, F., Sinnwell, V., Baumann, H. and Kaib, M. (1995) New alkaloids from ants: identification and synthesis of (3R,5S,9R)-3-butyl-5-(1-oxopropyl) indolizidine and (3R,5R,9R)-3-butyl-5-(1-oxopropyl)indolizidine, constituents of the poison gland secretion in *Myrmicaria eumenoides* (hymenoptera, formicidae). *Liebigs Ann.*, (6), 965–977; (b) Schröder, F., Franke, S., Francke, W., Baumann, H., Kaib, M., Pasteels, J.M. and Daloze, D. (1996) A new family of tricyclic alkaloids from *Myrmicaria* ants. *Tetrahedron*, **52** (43), 13539–13546; (c) Schröder, F., Sinnwell, V., Baumann, H. and Kaib, M. (1996) Myrmicarin 430A: a new heptacyclic alkaloid from *Myrmicaria* ants. *Chem. Commun.*, (18), 2139–2140; (d) Schröder, F., Sinnwell, V., Baumann, H., Kaib, M. and Francke, W. (1997) Myrmicarin 663: a new decacylic alkaloid from ants. *Angew. Chem. Int. Ed.*, **36** (1), 77–80.

40 Movassaghi, M. and Ondrus, A.E. (2005) Enantioselective total synthesis of tricyclic myrmicarin alkaloids. *Org. Lett.*, **7** (20), 4423–4426.

41 Morita, H., Arisaka, M., Yoshida, N. and Kobayashi, J. (2000) Serratezomines A–C, new alkaloids from *Lycopodium serratum* var. *serratum*. *J. Org. Chem.*, **65** (19), 6241–6245.

42 Chandra, A., Pigza, J.A., Han, J.-S., Mutnick, D. and Johnston, J.N. (2009) Total synthesis of the *Lycopodium* alkaloid (+)-serratezomine A. *J. Am. Chem. Soc.*, **131** (10), 3470–3471.

43 Leurs, J. and Hendriks, D. (2005) Carboxypeptidase U (TAFIa): a metallocarboxypeptidase with a distinct role in haemostasis and a possible risk factor for thrombotic disease. *Thromb. Haemost.*, **94** (3), 471–487.

44 Vaswani, R.G., Day, J.J. and Wood, J.L. (2009) Progress toward the total synthesis of (±)-actinophyllic acid. *Org. Lett.*, **11** (20), 4532–4535.

45 (a) Overman, L.E. (1992) Charge as a key component in reaction design. The invention of cationic cyclization reactions of importance in synthesis. *Acc. Chem. Res.*, **25** (8), 352–359; (b) Royer, J., Bonin, M. and Micouin, L. (2004) Chiral heterocycles by iminium ion cyclization. *Chem. Rev.*, **104** (5), 2311–2352.

46 Mahboobi, S. and Bernauer, K. (1988) Synthesis of esters of 3-(2-aminoethyl)-1H-indole-2-acetic acid and 3-(2-aminoethyl)-1H-indole-2-malonic acid (= 2-[3-(2-aminoethyl)-1H-indol-2-yl] propanedioic acid). 4th communication on indoles, indolenines, and indolines. *Helv. Chim. Acta*, **71** (8), 2034–2041.

47 (a) Tobinaga, S. and Kotani, E. (1972) Intramolecular and intermolecular oxidative coupling reactions by a new iron complex [Fe(DMF)$_3$Cl$_2$][FeCl$_4$]. *J. Am. Chem. Soc.*, **94** (1), 309–310; (b) Frazier, R.H., Jr. and Harlow, R.L. (1980) Oxidative coupling of ketone enolates by ferric chloride. *J. Org. Chem.*, **45** (26), 5408–5411.

48 (a) Cohen, T., McNamara, K., Kuzemko, M.A., Ramig, K., Landi, J.J., Jr. and Dong, Y. (1993) A novel one-flask cyclopentannulation involving a dilithiomethane equivalent as a β-connector of two enones. A highly efficient total synthesis of (±)-hirsutene. *Tetrahedron*, **49** (36), 7931–7942; (b) Paquette, L.A., Bzowej, E.I., Branan, B.M. and Stanton, K.J. (1995) Oxidative coupling of the enolate anion of (1R)-(+)-verbenone with Fe(III) and Cu(II) salts. Two modes of conjoining this bicyclic ketone across a benzene ring. *J. Org. Chem.*, **60** (22), 7277–7283; (c) Baran, P.S., Hafensteiner, B.D., Ambhaikar, N.B., Guerrero, C.A. and Gallagher, J.D. (2006) Enantioselective

total synthesis of avrainvillamide and the stephacidins. *J. Am. Chem. Soc.*, **128** (26), 8678–8693.

49 Imamoto, T., Takiyama, N., Nakamura, K., Hatajima, T. and Kamiya, Y. (1989) Reactions of carbonyl compounds with Grignard reagents in the presence of cerium chloride. *J. Am. Chem. Soc.*, **111** (12), 4392–4398.

50 Schlosser, M. and Coffinet, D. (1971) SCOOPY-reaktionen: stereoselektivität der allyl-alkohol-synthese via betain-ylide. *Synthesis*, (7), 380–381.

51 Garg, N.K. and Stoltz, B.M. (2006) A unified synthetic approach to the pyrazinone dragmacidins. *Chem. Commun.*, (36), 3769–3779.

52 Garg, N.K., Sarpong, R. and Stoltz, B.M. (2002) The first total synthesis of dragmacidin D. *J. Am. Chem. Soc.*, **124** (44), 13179–13184.

53 (a) Jiang, B. and Gu, X.-H. (2000) Syntheses and cytotoxicity evaluation of bis(indolyl)thiazole, bis(indolyl) pyrazinone and bis(indolyl)pyrazine: analogues of cytotoxic marine bis(indole) alkaloid. *Bioorg. Med. Chem.*, **8** (2), 363–371; (b) Jiang, B. and Gu, X.-H. (2000) Syntheses of bis(3′-indolyl)-2(1H)-pyrazinones. *Heterocycles*, **53** (7), 1559–1568; (c) Yang, C.-G., Wang, J. and Jiang, B. (2002) Enantioselective synthesis of the aminoimidazole segment of dragmacidin D. *Tetrahedron Lett.*, **43** (6), 1063–1066; (d) Miyake, F.Y., Yakushijin, K. and Horne, D.A. (2002) Synthesis of marine sponge bisindole alkaloids dihydrohamacanthins. *Org. Lett.*, **4** (6), 941–943; (e) Yang, C.-G., Liu, G. and Jiang, B. (2002) Preparing functional bis(indole) pyrazine by stepwise cross-coupling reactions: an efficient method to construct the skeleton of dragmacidin D. *J. Org. Chem.*, **67** (26), 9392–9396; (f) Feldman, K.S. and Ngernmeesri, P. (2005) Dragmacidin E synthesis studies. Preparation of a model cycloheptannelated indole fragment. *Org. Lett.*, **7** (24), 5449–5452; (g) Huntley, R.J. and Funk, R.L. (2006) A strategy for the total synthesis of dragmacidin E. Construction of the core ring system. *Org. Lett.*, **8** (21), 4775–4778.

54 (a) Miyaura, N. and Suzuki, A. (1995) Palladium-catalyzed cross-coupling reactions of organoboron compounds. *Chem. Rev.*, **95** (7), 2457–2483; (b) Suzuki, A. (2005) Carbon–carbon bonding made easy. *Chem. Commun.*, (38), 4759–4763; (c) Doucet, H. (2008) Suzuki–Miyaura cross-coupling reactions of alkylboronic acid derivatives or alkyltrifluoroborates with aryl, alkenyl or alkyl halides and triflates. *Eur. J. Org. Chem.*, (12), 2013–2030.

55 Li, J.J. and Gribble, G.W. (2000) *Palladium in Heterocyclic Chemistry: A Guide for the Synthetic Chemist*, vol. 20, 1st edn, Pergamon, Amsterdam, The Netherlands, pp. 355–373.

56 (a) Barco, A., Benetti, S., De Risi, C., Marchetti, P., Pollini, G.P. and Zanirato, V. (1997) D-(–)-Quinic acid: a chiron store for natural product synthesis. *Tetrahedron: Asymmetry*, **8** (21), 3515–3545; (b) Huang, P.-Q. (1999) D-Quinic acid, a versatile chiron in organic synthesis. *Youji Huaxue*, **19** (4), 364–373; (c) Hanessian, S., Pan, J., Carnell, A., Bouchard, H. and Lesage, L. (1997) Total synthesis of (–)-reserpine using the chiron approach. *J. Org. Chem.*, **62** (3), 465–473; (d) Hanessian, S. (1983) *Total Synthesis of Natural Products: The 'Chiron' Approach*, E.J. Baldwin (ed.) Pergamon Press, Oxford, UK, pp. 206–208.

57 (a) Stoltz, B.M. (2004) Palladium catalyzed aerobic dehydrogenation: from alcohols to indoles and asymmetric catalysis. *Chem. Lett.*, **33** (4), 362–367; (b) Trend, R.M., Ramtohul, Y.K., Ferreira, E.M. and Stoltz, B.M. (2003) Palladium-catalyzed oxidative Wacker cyclizations in nonpolar organic solvents with molecular oxygen: a stepping stone to asymmetric aerobic cyclizations. *Angew. Chem. Int. Ed.*, **42** (25), 2892–2895; (c) Ferreira, E.M. and Stoltz, B.M. (2003) Catalytic C–H Bond functionalization with palladium(II): aerobic oxidative annulations of indoles. *J. Am. Chem. Soc.*, **125** (32), 9578–9579; (d) Zhang, H., Ferreira, E.M. and Stoltz, B.M. (2004) Direct oxidative Heck cyclizations:

intramolecular Fujiwara–Moritani arylations for the synthesis of functionalized benzofurans and dihydrobenzofurans. *Angew. Chem. Int. Ed.*, **43** (45), 6144–6148.

58 (a) Legault, C.Y., Garcia, Y., Merlic, C.A. and Houk, K.N. (2007) Origin of regioselectivity in palladium-catalyzed cross-coupling reactions of polyhalogenated heterocycles. *J. Am. Chem. Soc.*, **129** (42), 12664–12665; (b) Garcia, Y., Schoenebeck, F., Legault, C.Y., Merlic, C.A. and Houk, K.N. (2009) Theoretical bond dissociation energies of halo-heterocycles: trends and relationships to regioselectivity in palladium-catalyzed cross-coupling reactions. *J. Am. Chem. Soc.*, **131** (18), 6632–6639.

59 (a) Neber, P.W. and Friedolsheim, A.V. (1926) Über eine neue art der umlagerung von oximen (New kind of rearrangement of oximes). *Justus Liebigs Ann. Chem.*, **449** (1), 109–134; (b) O'Brien, C. (1964) The rearrangement of ketoxime O-sulfonates to amino ketones (the Neber rearrangement). *Chem. Rev.*, **64** (2), 81–89.

60 (a) De Jesus, A.E., Hull, W.E., Steyn, P.S., van Heerden, F.R., Vleggaar, R. and Wessels, P.L. (1982) High-field ^{13}C N.M.R. evidence for the formation of [1,2-^{13}C]acetate from [2-^{13}C]acetate during the biosynthesis of penitrem A by *Penicillium crustosum*. *J. Chem. Soc., Chem. Commun.*, (15), 837–838; (b) De Jesus, A.E., Gorst-Allman, C.P., Steyn, P.S., van Heerden, F.R., Vleggaar, R., Wessels, P.L. and Hull, W.E. (1983) Tremorgenic mycotoxins from *Penicillium crustosum*. Biosynthesis of penitrem A. *J. Chem. Soc., Perkin Trans. 1*, (8), 1863–1868; (c) De Jesus, A.E., Steyn, P.S., van Heerden, F.R., Vleggaar, R., Wessels, P.L. and Hull, W.E. (1981) Structure and biosynthesis of the penitrems A–F, six novel tremorgenic mycotoxins from *Penicillium crustosum*. *J. Chem. Soc., Chem. Commun.*, (6), 289–291; (d) De Jesus, A.E., Steyn, P.S., van Heerden, F.R., Vleggaar, R., Wessels, P.L. and Hull, W.E. (1983) Tremorgenic mycotoxins from *Penicillium crustosum*: isolation of penitrems A–F and the structure elucidation and absolute configuration of penitrem A. *J. Chem. Soc., Perkin Trans. 1*, (8), 1847–1856.

61 (a) Horeau, A. (1961) Principe et applications d'une nouvelle methode de determination des configurations dite 'par dedoublement partiel.' *Tetrahedron Lett.*, **2** (15), 506–512; (b) Horeau, A. and Sutherland, J.K. (1966) The absolute configuration of some caryophyllene derivatives. *J. Chem. Soc. C*, 247–248; (c) Herz, W. and Kagan, H.B. (1967) Determination of the absolute configuration of hydroxylated sesquiterpene lactones by Horeau's method of asymmetric esterification. *J. Org. Chem.*, **32** (1), 216–218.

62 Rivkin, A., González-López de Turiso, F., Nagashima, T. and Curran, D.P. (2004) Radical and palladium-catalyzed cyclizations to cyclobutenes: an entry to the BCD ring system of penitrem D. *J. Org. Chem.*, **69** (11), 3719–3725.

63 (a) Smith, A.B., III, Haseltine, J.N. and Visnick, M. (1989) Tremorgenic indole alkaloid studies. 6. Preparation of an advanced intermediate for the synthesis of penitrem D. Synthesis of an indole oxocane. *Tetrahedron*, **45** (8), 2431–2449; (b) Smith, A.B., III, Nolen, E.G., Shirai, R., Blase, F.R., Ohta, M., Chida, N., Hartz, R.A., Fitch, D.M., Clark, W.M. and Sprengeler, P.A. (1995) Tremorgenic indole alkaloids. 9. Asymmetric construction of an advanced F-G-H-ring lactone precursor for the synthesis of penitrem D. *J. Org. Chem.*, **60** (24), 7837–7848.

64 (a) Smith, A.B., III and Visnick, M. (1985) An expedient synthesis of substituted indoles. *Tetrahedron Lett.*, **26** (32), 3757–3760; (b) Smith, A.B., III, Visnick, M., Haseltine, J.N. and Sprengeler, P.A. (1986) Organometallic reagents in synthesis: a new protocol for construction of the indole nucleus. *Tetrahedron*, **42** (11), 2957–2969.

65 Gutzwiller, J., Buchschacher, P. and Fürst, A. (1977) A procedure for the preparation of (S)-8a-methyl-3,4,8,8a-tetrahydro-1,6(2H,7H)-naphthalenedione. *Synthesis*, (3), 167–168.

66 Parikh, J.R. and Doering, W.v.E. (1967) Sulfur trioxide in the oxidation of alcohols by dimethyl sulfoxide. *J. Am. Chem. Soc.*, **89** (21), 5505–5507.

67 Jimenez, J.I., Huber, U., Moore, R.E. and Patterson, G.M.L. (1999) Oxidized welwitindolinones from terrestrial *Fischerella* spp. *J. Nat. Prod.*, **62** (4), 569–572.

68 (a) Moore, R.E., Cheuk, C. and Patterson, G.M.L. (1984) Hapalindoles: new alkaloids from the blue-green alga *Hapalosiphon fontinalis*. *J. Am. Chem. Soc.*, **106** (21), 6456–6457; (b) Moore, R.E., Cheuk, C., Yang, X.-Q.G., Patterson, G.M.L., Bonjouklian, R., Smitka, T.A., Mynderse, J.S., Foster, R.S., Jones, N.D., Swartzendruber, J.K. and Deeter, J.B. (1987) Hapalindoles, antibacterial and antimycotic alkaloids from the cyanophyte *Hapalosiphon fontinalis*. *J. Org. Chem.*, **52** (6), 1036–1043; (c) Schwartz, R.E., Hirsch, C.F., Springer, J.P., Pettibone, D.J. and Zink, D.L. (1987) Unusual cyclopropane-containing hapalindolinones from a cultured cyanobacterium. *J. Org. Chem.*, **52** (16), 3704–3706; (d) Moore, R.E., Yang, X.-Q.G. and Patterson, G.M.L. (1987) Fontonamide and anhydrohapaloxindole A, two new alkaloids from the blue-green alga *Hapalosiphon fontinalis*. *J. Org. Chem.*, **52** (17), 3773–3777; (e) Smitka, T.A., Bonjouklian, R., Doolin, L., Jones, N.D., Deeter, J.B., Yoshida, W.Y., Prinsep, M.R., Moore, R.E. and Patterson, G.M.L. (1992) Ambiguine isonitriles, fungicidal hapalindole-type alkaloids from three genera of blue-green algae belonging to the Stigonemataceae. *J. Org. Chem.*, **57** (3), 857–861; (f) Park, A., Moore, R.E. and Patterson, G.M.L. (1992) Fischerindole L, a new isonitrile from the terrestrial blue-green alga *Fischerella muscicola*. *Tetrahedron Lett.*, **33** (23), 3257–3260; (g) Huber, U., Moore, R.E. and Patterson, G.M.L. (1998) Isolation of a nitrile-containing indole alkaloid from the terrestrial blue-green alga *Hapalosiphon delicatulus*. *J. Nat. Prod.*, **61** (10), 1304–1306; (h) Klein, D., Daloze, D., Braekman, J.C., Hoffmann, L. and Demoulin, V. (1995) New hapalindoles from the cyanophyte *Hapalosiphon laingii*. *J. Nat. Prod.*, **58** (11), 1781–1785; (i) Moore, R.E., Yang, X.-Q.G., Patterson, G.M.L., Bonjouklian, R. and Smitka, T.A. (1989) Hapalonamides and other oxidized hapalindoles from *Hapalosiphon fontinalis*. *Phytochemistry*, **28** (5), 1565–1567; (j) Raveh, A. and Carmeli, S. (2007) Antimicrobial ambiguines from the cyanobacterium *Fischerella* sp. collected in Israel. *J. Nat. Prod.*, **70** (2), 196–201; (k) Becher, P.G., Keller, S., Jung, G., Süssmuth, R.D. and Jüttner, F. (2007) Insecticidal activity of 12-*epi*-hapalindole J isonitrile. *Phytochemistry*, **68** (19), 2493–2497; (l) Asthana, R.K., Srivastava, A., Singh, A.P., Deepali, Singh, S.P., Nath, G., Srivastava, R. and Srivastava, B.S. (2006) Identification of an antimicrobial entity from the cyanobacterium *Fischerella* sp. isolated from bark of *Azadirachta indica* (Neem) tree. *J. Appl. Phycol.*, **18** (1), 33–39.

69 Smith, C.D., Zilfou, J.T., Stratmann, K., Patterson, G.M.L. and Moore, R.E. (1995) Welwitindolinone analogs that reverse P-glycoprotein-mediated multiple drug resistance. *Mol. Pharmacol.*, **47** (2), 241–247.

70 Wang, B.M., Song, Z.L., Fan, C.A., Tu, Y.Q. and Chen, W.M. (2003) Halogen cation induced stereoselective semipinacol-type rearrangement of allylic alcohols: a highly efficient approach to α-quaternary β-haloketo compounds. *Synlett*, (10), 1497–1499. See also references therein.

71 Moreno-Dorado, F.J., Guerra, F.M., Manzano, F.L., Aladro, F.J., Jorge, Z.D. and Massanet, G.M. (2003) $CeCl_3$/NaClO: a safe and efficient reagent for the allylic chlorination of terminal olefins. *Tetrahedron Lett.*, **44** (35), 6691–6693.

72 Pilcher, A.S. and DeShong, P. (1993) Improved protocols for the selective deprotection of trialkylsilyl ethers using fluorosilicic acid. *J. Org. Chem.*, **58** (19), 5130–5134.

73 (a) Evans, D.A. and Chapman, K.T. (1986) The directed reduction of β-hydroxy ketones employing $Me_4NHB(OAc)_3$. *Tetrahedron Lett.*, **27**

(49), 5939–5942; (b) Evans, D.A., Chapman, K.T. and Carreira, E.M. (1988) Directed reduction of β-hydroxy ketones employing tetramethylammonium triacetoxyborohydride. *J. Am. Chem. Soc.*, **110** (11), 3560–3578.
74. Martin, J.C. and Arhart, R.J. (1971) Sulfuranes. III. Reagent for the dehydration of alcohols. *J. Am. Chem. Soc.*, **93** (17), 4327–4239.
75. Keck, G.E., Wager, T.T. and McHardy, S.F. (1999) Reductive cleavage of N–O bonds in hydroxylamines and hydroxamic acid derivatives using samarium diiodide. *Tetrahedron*, **55** (40), 11755–11772.
76. Baran, P.S. and Richter, J.M. (2004) Direct coupling of indoles with carbonyl compounds: short, enantioselective, gram-scale synthetic entry into the hapalindole and fischerindole alkaloid families. *J. Am. Chem. Soc.*, **126** (24), 7450–7451.
77. Wender, P.A., Erhardt, J.M. and Letendre, L.J. (1981) Reaction of allylically substituted enolates with organometallic reagents: a convenient source of enolonium ion equivalents. *J. Am. Chem. Soc.*, **103** (8), 2114–2116.
78. De Luca, L., Giacomelli, G., Porcheddu, A. and Salaris, M. (2004) A new, simple procedure for the synthesis of formyl amides. *Synlett*, (14), 2570–2572.
79. (a) Ugi, I., Fetzer, U., Eholzer, U., Knupfer, H. and Offermann, K. (1965) Isonitrile syntheses. *Angew. Chem. Int. Ed.*, **4** (6), 472–484; (b) Ugi, I., Betz, W., Fetzer, U. and Offermann, K. (1961) Isonitrile, X. Notiz zur Darstellung von Isonitrilen aus monosubstituierten Formamiden durch Wasserabspaltung mittels Phosgen und Trialkylaminen. *Chem. Ber.*, **94** (10), 2814–2816.
80. Shellhamer, D.F., Carter, D.L., Chiaco, M.C., Harris, T.E., Henderson, R.D., Low, W.S.C., Metcalf, B.T., Willis, M.C., Heasley, V.L. and Chapman, R.D. (1991) Reaction of xenon difluoride with indene in aqueous 1,2-dimethoxyethane and tetrahydrofuran. *J. Chem. Soc., Perkin Trans. 2*, (3), 401–403.
81. (a) Anthoni, U., Christophersen, C. and Nielson, P.H. (1999) Naturally occurring cyclotryptophans and cyclotryptamines, in *Alkaloids: Chemical and Biological Perspectives*, vol. 13, 1st edn (ed. S.W. Pelletier), John Wiley & Sons, Inc., New York, USA, pp. 163–236; (b) Takano, S. and Ogasawara, K. (1989) Alkaloids of the calabar bean, in *The Alkaloids: Chemistry and Pharmacology*, vol. 36, (ed. A. Brossi), Academic Press, San Diego, CA, USA, pp. 225–251.
82. (a) Tokuyama, T. and Daly, J.W. (1983) Steroidal alkaloids (batrachotoxins and 4β-hydroxybatrachotoxins), 'indole alkaloids' (calycanthine and chimonanthine) and a piperidinyldipyridine alkaloid (noranabasamine) in skin extracts from the Colombian poison-dart frog *Phyllobates terribilis* (Dendrobatidae). *Tetrahedron*, **39** (1), 41–47; (b) Lajis, N.H., Mahmud, Z. and Toia, R.F. (1993) The alkaloids of *Psychotria rostrata*. *Planta Med.*, **59** (4), 383–384; (c) Verotta, L., Pilati, T., Tatò, M., Elisabetsky, E., Amador, T.A. and Nunes, D.S. (1998) Pyrrolidinoindoline alkaloids from *Psychotria colorata*. *J. Nat. Prod.*, **61** (3), 392–396.
83. Barrow, C.J., Ping, C., Snyder, J.K., Sedlock, D.M., Sun, H.H. and Cooper, R. (1993) WIN 64821, a new competitive antagonist to substance P, isolated from an *Aspergillus* species: structure determination and solution conformation. *J. Org. Chem.*, **58** (22), 6016–6021.
84. Steven, A. and Overman, L.E. (2007) Total synthesis of complex cyclotryptamine alkaloids: stereocontrolled construction of quaternary carbon stereocenters. *Angew. Chem. Int. Ed.*, **46** (29), 5488–5508.
85. Movassaghi, M., Schmidt, M.A. and Ashenhurst, J.A. (2008) Concise total synthesis of (+)-WIN 64821 and (−)-ditryptophenaline. *Angew. Chem. Int. Ed.*, **47** (8), 1485–1487.
86. Movassaghi, M. and Schmidt, M.A. (2007) Concise total synthesis of (−)-calycanthine, (+)-chimonanthine, and (+)-folicanthine. *Angew. Chem. Int. Ed.*, **46** (20), 3725–3728.
87. Kirby, G.W. and Robins, D.J. (1980) The biosynthesis of gliotoxin and related

epipolythiodioxopiperazines, in *The Biosynthesis of Mycotoxins: A Study in Secondary Metabolism* (ed. P.S. Steyn), Academic Press, New York, USA, pp. 301–326.

88 Fukuyama, T., Nakatsuka, S.-I. and Kishi, Y. (1981) Total synthesis of gliotoxin, dehydrogliotoxin and hyalodendrin. *Tetrahedron*, **37** (11), 2045–2078.

89 Ramírez, A. and García-Rubio, S. (2003) Current progress in the chemistry and pharmacology of akuammiline alkaloids. *Curr. Med. Chem.*, **10** (18), 1891–1915.

90 Higuchi, K. and Kawasaki, T. (2007) Simple indole alkaloids and those with a nonrearranged monoterpenoid unit. *Nat. Prod. Rep.*, **24** (4), 843–868.

91 Loiseleur, O., Meier, P. and Pfaltz, A. (1996) Chiral phosphanyldihydrooxazoles in asymmetric catalysis: enantioselective Heck reactions. *Angew. Chem. Int. Ed.*, **35** (2), 200–202.

92 Ley, S.V., Norman, J., Griffith, W.P. and Marsden, S.P. (1994) Tetrapropylammonium perruthenate, $Pr_4N^+RuO_4^-$, TPAP: a catalytic oxidant for organic synthesis. *Synthesis*, (7), 639–666.

93 (a) Wang, T. and Cook, J.M. (2000) General approach for the synthesis of sarpagine/ajmaline indole alkaloids. Stereospecific total synthesis of the sarpagine alkaloid (+)-vellosimine. *Org. Lett.*, **2** (14), 2057–2059; (b) Zhao, S., Liao, X. and Cook, J.M. (2002) Enantiospecific, stereospecific total synthesis of (+)-majvinine, (+)-10-methoxyaffinisine, and (+)-Na-methylsarpagine as well as the total synthesis of the Alstonia bisindole macralstonidine. *Org. Lett.*, **4** (5), 687–690; (c) Cao, H., Yu, J., Wearing, X.Z., Zhang, C., Liu, X., Deschamps, J. and Cook, J.M. (2003) The first enantiospecific synthesis of (−)-koumidine via the intramolecular palladium-catalyzed enolate driven cross coupling reaction. The stereospecific introduction of the 19-(Z) ethylidene side chain. *Tetrahedron Lett.*, **44** (43), 8013–8017; (d) Yu, J., Wang, T., Liu, X., Deschamps, J., Flippen-Anderson, J., Liao, X. and Cook, J.M. (2003) General approach for the synthesis of sarpagine indole alkaloids. Enantiospecific total synthesis of (+)-vellosimine, (+)-normacusine B, (−)-alkaloid Q_3, (−)-panarine, (+)-N_a-methylvellosimine, and (+)-N_a-methyl-16-epipericyclivine. *J. Org. Chem.*, **68** (20), 7565–7581; (e) Zhou, H., Liao, X. and Cook, J.M. (2004) Regiospecific, enantiospecific total synthesis of the 12-alkoxy-substituted indole alkaloids, (+)-12-methoxy-N_a-methylvellosimine, (+)-12-methoxyaffinisine, and (−)-fuchsiaefoline. *Org. Lett.*, **6** (2), 249–252; (f) Yu, J., Wearing, X.Z. and Cook, J.M. (2005) A general strategy for the synthesis of vincamajine-related indole alkaloids: stereocontrolled total synthesis of (+)-dehydrovoachalotine, (−)-vincamajinine, and (−)-11-methoxy-17-epivincamajine as well as the related quebrachidine diol, vincamajine diol, and vincarinol. *J. Org. Chem.*, **70** (10), 3963–3979.

94 Comins, D.L. and Dehghani, A. (1992) Pyridine-derived triflating reagents: an improved preparation of vinyl triflates from metallo enolates. *Tetrahedron Lett.*, **33** (42), 6299–6302.

95 (a) Bailey, P.D. and Hollinshead, S.P. (1987) Synthesis of 3-(indol-2-yl) propenoate derivatives of tryptamine, valuable intermediates for the preparation of indole alkaloids. *Tetrahedron Lett.*, **28** (25), 2879–2882; (b) Chavan, S.P., Sharma, P., Sivappa, R. and Kalkote, U.R. (2006) A ring closing metathesis approach to the indole alkaloid mitralactonine. *Tetrahedron Lett.*, **47** (52), 9301–9303.

96 Krapcho, A.P., Weimaster, J.F., Eldridge, J.M., Jahngen, E.G.E., Jr., Lovey, A.J. and Stephens, W.P. (1978) Synthetic applications and mechanism studies of the decarbalkoxylations of geminal diesters and related systems effected in dimethyl sulfoxide by water and/or by water with added salts. *J. Org. Chem.*, **43** (1), 138–147.

97 Zhang, M., Huang, X., Shen, L. and Qin, Y. (2009) Total synthesis of the akuammiline alkaloid (±)-vincorine. *J. Am. Chem. Soc.*, **131** (16), 6013–6020.

98 (a) Luche, J.-L. (1978) Lanthanides in organic chemistry. 1. Selective 1,2 reductions of conjugated ketones. *J. Am. Chem. Soc.*, **100** (7), 2226–2227; (b) Luche, J.-L., Rodriguez-Hahn, L. and Crabbé, P. (1978) Reduction of natural enones in the presence of cerium trichloride. *J. Chem. Soc., Commun.*, (14), 601–602; (c) Gemal, A.L. and Luche, J.-L. (1981) Lanthanoids in organic synthesis. 6. The reduction of α-enones by sodium borohydride in the presence of lanthanoid chlorides: synthetic and mechanistic aspects. *J. Am. Chem. Soc.*, **103** (18), 5454–5459.

99 Pike, P.W., Gilliatt, V., Ridenour, M. and Hershberger, J.W. (1988) Substituent effects upon the kinetics of hydrogen transfer from triorganotin hydrides to the 5-hexen-1-yl radical. *Organometallics*, **7** (10), 2220–2223.

100 Southon, I.W. and Buckingham, J. (1989) *Dictionary of Alkaloids*, Chapman and Hall Ltd, New York, USA.

101 (a) Fevig, J.M., Marquis, R.W., Jr. and Overman, L.E. (1991) New approach to *Strychnos* alkaloids. Stereocontrolled total synthesis of (±)-dehydrotubifoline. *J. Am. Chem. Soc.*, **113** (13), 5085–5086; (b) Rawal, V.H., Michoud, C. and Monestel, R. (1993) General strategy for the stereocontrolled synthesis of *Strychnos* alkaloids: a concise synthesis of (±)-dehydrotubifoline. *J. Am. Chem. Soc.*, **115** (7), 3030–3031; (c) Bonjoch, J., Solé, D. and Bosch, J. (1993) A new, general synthetic pathway to *Strychnos* indole alkaloids. First total synthesis of (±)-echitamidine. *J. Am. Chem. Soc.*, **115** (5), 2064–2065; (d) Boonsombat, J., Zhang, H., Chughtai, M.J., Hartung, J. and Padwa, A. (2008) A general synthetic entry to the pentacyclic *Strychnos* alkaloid family, using a [4 + 2]-cycloaddition/rearrangement cascade sequence. *J. Org. Chem.*, **73** (9), 3539–3550.

102 Crawley, G.C. and Harley-Mason, J. (1971) Total synthesis of (±)-fluorocurarine, the racemate of a calabash-curare alkaloid. *J. Chem. Soc. D*, (13), 685–686.

103 (a) Solé, D., Bonjoch, J. and Bosch, J. (1996) Total synthesis of the *Strychnos* alkaloids (±)-akuammicine and (±)-norfluorocurarine from 3a-(o-nitrophenyl)hexahydroindol-4-ones by nickel(0)-promoted double cyclization. *J. Org. Chem.*, **61** (13), 4194–4195; (b) Bonjoch, J., Solé, D., García-Rubio, S. and Bosch, J. (1997) A general synthetic entry to *Strychnos* alkaloids of the curan type via a common 3a-(2-nitrophenyl)hexahydroindol-4-one intermediate. Total syntheses of (±)- and (–)-tubifolidine, (±)-akuammicine, (±)-19,20-dihydroakuammicine, (±)-norfluorocurarine, (±)-echitamidine, and (±)-20-epilochneridine. *J. Am. Chem. Soc.*, **119** (31), 7230–7240.

104 (a) Rawal, V.H. and Iwasa, S. (1994) A short, stereocontrolled synthesis of strychnine. *J. Org. Chem.*, **59** (10), 2685–2686; (b) Bonjoch, J. and Solé, D. (2000) Synthesis of strychnine. *Chem. Rev.*, **100** (9), 3455–3482.

105 (a) Zincke, T. (1904) Ueber Dinitrophenylpyridiniumchlorid und dessen Umwandlungsprodukte. *Liebigs Ann. Chem.*, **330** (2), 361–374; (b) Zincke, T. (1904) I. Ueber Dinitrophenylpyridiniumchlorid und dessen Umwandlungsprodukte. *Liebigs Ann. Chem.*, **333** (2), 296–345; (c) Zincke, T. and Wurker, W. (1904) Ueber Dinitrophenylpyridiniumchlorid und dessen Umwandlungsprodukte (2. Mittheilung.) *Liebigs Ann. Chem.*, **338** (1), 107–141; (d) König, W. (1904) Über eine neue, vom Pyridin derivierende Klasse von Farbstoffen. *J. Prakt. Chem.*, **69** (1), 105–137.

106 Metz, P. and Linz, C. (1994) Claisen rearrangement of *N*-silyl ketene *N,O*-acetals generated from allyl *N*-phenylimidates. *Tetrahedron*, **50** (13), 3951–3966.

107 Michels, T.D., Rhee, J.U. and Vanderwal, C.D. (2008) Synthesis of δ-tributylstannyl-α,β,γ,δ-unsaturated aldehydes from pyridines. *Org. Lett.*, **10** (21), 4787–4790.

108 Matsuda, Y., Kitajima, M. and Takayama, H. (2008) First total synthesis of trimeric indole alkaloid, psychotrimine. *Org. Lett.*, **10** (1), 125–128.

109 Espejo, V.R. and Rainier, J.D. (2008) An expeditious synthesis of C(3)–N(1') heterodimeric indolines. *J. Am. Chem. Soc.*, **130** (39), 12894–12895.

110 López, C.S., Pérez-Balado, C., Rodríguez-Graña, P. and de Lera, A.R. (2008) Mechanistic insights into the stereocontrolled synthesis of hexahydropyrrolo[2,3-b]indoles by electrophilic activation of tryptophan derivatives. *Org. Lett.*, **10** (1), 77–80.

111 (a) Nakagawa, M., Ma, J. and Hino, T. (1990) The Nicholas reaction of indoles. Propargylation of indoles with (propargyl)dicobalt hexacarbonyl cations. *Heterocycles*, **30** (1 Spec. Issue), 451–462; (b) Hino, T., Hasumi, K., Yamaguchi, H., Taniguchi, M. and Nakagawa, M. (1985) Inverted prenylation of N_b-methoxycarbonyltryptamine. Synthesis of 3a-(1,1-dimethylallyl)pyrrolo[2,3-b]indole and 2-(1,1-dimethylallyl) tryptamine derivatives. *Chem. Pharm. Bull.*, **33** (12), 5202–5206; (c) Muthusubramanian, P., Carlé, J.S. and Christophersen, C. (1983) Marine alkaloids. 7. Synthesis of debromoflustramine B and related compounds. *Acta Chem. Scand. B*, **B37**, 803–807; (d) Nakagawa, M., Matsuki, K. and Hino, T. (1983) A new synthesis of beta-carboline. *Tetrahedron Lett.*, **24** (21), 2171–2174; (e) Bocchi, V., Casanati, G. and Marchelli, R. (1978) Insertion of isoprene units into indole and 3-substituted indoles in aqueous systems. *Tetrahedron*, **34** (7), 929–932.

112 Gassman, P.G., van Bergen, T.J., Gilbert, D.P. and Cue, B.W. (1974) Azasulfonium salts. A general method for the synthesis of indoles. *J. Am. Chem. Soc.*, **96** (17), 5495–5508.

113 (a) Larock, R.C. and Yum, E.K. (1991) Synthesis of indoles via palladium-catalyzed heteroannulation of internal alkynes. *J. Am. Chem. Soc.*, **113** (17), 6689–6690; (b) Ma, C., Liu, X., Li, X., Flippen-Anderson, J., Yu, S. and Cook, J.M. (2001) Efficient asymmetric synthesis of biologically important tryptophan analogues via a palladium-mediated heteroannulation reaction. *J. Org. Chem.*, **66** (13), 4525–4542.

114 (a) Klapars, A., Antilla, J.C., Huang, X. and Buchwald, S.L. (2001) A general and efficient copper catalyst for the amidation of aryl halides and the N-arylation of nitrogen heterocycles. *J. Am. Chem. Soc.*, **123** (31), 7727–7729; (b) Antilla, J.C., Klapars, A. and Buchwald, S.L. (2002) The copper-catalyzed N-arylation of indoles. *J. Am. Chem. Soc.*, **124** (39), 11684–11688.

115 Newhouse, T., Lewis, C.A. and Baran, P.S. (2009) Enantiospecific total syntheses of kapakahines B and F. *J. Am. Chem. Soc.*, **131** (18), 6360–6361.

116 Ondrus, A.E. and Movassaghi, M. (2009) Total synthesis and study of myrmicarin alkaloids. *Chem. Commun.*, (28), 4151–4165.

117 Schröder, F. and Francke, W. (1998) Synthesis of myrmicarin 217, a pyrrole[2,1,5-cd]indolizine from ants. *Tetrahedron*, **54** (20), 5259–5264.

118 Sayah, B., Pelloux-Léon, N. and Vallée, Y. (2000) First synthesis of nonracemic (R)-(+)-myrmicarin 217. *J. Org. Chem.*, **65** (9), 2824–2826.

119 Settambolo, R., Guazzelli, G. and Lazzaroni, R. (2003) Intramolecular cyclodehydration of (4S)-(+)-4-carboxyethyl-4-(pyrrol-1-yl)butanal as the key step in the formal synthesis of (S)-(−)-myrmicarin 217. *Tetrahedron Asymmetry*, **14** (11), 1447–1449.

120 Sayah, B., Pelloux-Léon, N., Milet, A., Pardillos-Guindet, J. and Vallée, Y. (2001) Highly regioselective Vilsmeier–Haack acylation of hexahydropyrroloindolizine. *J. Org. Chem.*, **66** (7), 2522–2525.

121 (a) Ondrus, A.E. and Movassaghi, M. (2006) Dimerization of (+)-myrmicarin 215B. A potential biomimetic approach to complex myrmicarin alkaloids. *Tetrahedron*, **62** (22), 5287–5297; (b) Movassaghi, M., Ondrus, A.E. and Chen, B. (2007) Efficient and stereoselective dimerization of pyrroloindolizine derivatives inspired by a hypothesis for the biosynthesis of complex myrmicarin alkaloids. *J. Org. Chem.*, **72** (26), 10065–10074.

122 Tomori, H., Fox, J.M. and Buchwald, S.L. (2000) An improved synthesis of

functionalized biphenyl-based phosphine ligands. *J. Org. Chem.*, **65** (17), 5334–5341.
123 Rainka, M.P., Aye, Y. and Buchwald, S.L. (2004) Copper-catalyzed asymmetric conjugate reduction as a route to novel β-azaheterocyclic acid derivatives. *Proc. Natl. Acad. Sci. U.S.A.*, **101** (16), 5821–5823.
124 Tsuji, T., Watanabe, Y. and Mukaiyama, T. (1979) One-step formation of alkynes from aryl ketones. *Chem. Lett.*, (5), 481–482.
125 (a) Ayer, W.A. (1991) The lycopodium alkaloids. *Nat. Prod. Rep.*, **8** (5), 455–463; (b) Ma, X. and Gang, D.R. (2004) The lycopodium alkaloids. *Nat. Prod. Rep.*, **21** (6), 752–772; (c) Hirasawa, Y., Kobayashi, J. and Morita, H. (2009) The lycopodium alkaloids. *Heterocycles*, **77** (2), 679–729.
126 Linghu, X., Kennedy-Smith, J.J. and Toste, F.D. (2007) Total synthesis of (+)-fawcettimine. *Angew. Chem. Int. Ed.*, **46** (40), 7671–7673.
127 Kozak, J.A. and Dake, G.R. (2008) Total synthesis of (+)-fawcettidine. *Angew. Chem. Int. Ed.*, **47** (22), 4221–4223.
128 Yang, H., Carter, R.G. and Zakharov, L.N. (2008) Enantioselective total synthesis of lycopodine. *J. Am. Chem. Soc.*, **130** (29), 9238–9239.
129 (a) Beshore, D.C. and Smith, A.B., III (2007) Total syntheses of (+)-lyconadin A and (−)-lyconadin B. *J. Am. Chem. Soc.*, **129** (14), 4148–4149; (b) Beshore, D.C. and Smith, A.B., III (2008) The lyconadins: enantioselective total syntheses of (+)-lyconadin A and (−)-lyconadin B. *J. Am. Chem. Soc.*, **130** (41), 13778–13789; (c) Bisai, A., West, S.P. and Sarpong, R. (2008) Unified strategy for the synthesis of the 'miscellaneous' *Lycopodium* alkaloids: total synthesis of (±)-lyconadin A. *J. Am. Chem. Soc.*, **130** (23), 7222–7223.
130 Nishikawa, Y., Kitajima, M. and Takayama, H. (2008) First asymmetric total syntheses of cernuane-type *Lycopodium* alkaloids, cernuine, and cermizine D. *Org. Lett.*, **10** (10), 1987–1990.
131 Morita, H. and Kobayashi, J. (2002) A biomimetic transformation of serratinine into serratezomine A through a modified Polonovski reaction. *J. Org. Chem.*, **67** (15), 5378–5381.
132 (a) Nugent, B.M., Williams, A.L., Prabhakaran, E.N. and Johnston, J.N. (2003) Free radical-mediated vinyl amination: a mild, general pyrrolidinyl enamine synthesis. *Tetrahedron*, **59** (45), 8877–8888; (b) Prabhakaran, E.N., Nugent, B.M., Williams, A.L., Nailor, K.E. and Johnston, J.N. (2002) Free radical-mediated vinyl amination: access to N,N-dialkyl enamines and their β-stannyl and β-thio derivatives. *Org. Lett.*, **4** (24), 4197–4200.
133 (a) Hwu, J.R., Chen, C.N. and Shiao, S.-S. (1995) Silicon-controlled allylation of 1,3-dioxo compounds by use of allyltrimethylsilane and ceric ammonium nitrate. *J. Org. Chem.*, **60** (4), 856–862; (b) Zhang, Y., Raines, A.J. and Flowers, R.A., II (2003) Solvent-dependent chemoselectivities in Ce(IV)-mediated oxidative coupling reactions. *Org. Lett.*, **5** (13), 2363–2365; (c) Zhang, Y., Raines, A.J. and Flowers, R.A., II (2004) Solvent-dependent chemoselectivities in additions of β-carbonyl imines to allyltrimethylsilane with CTAN. *J. Org. Chem.*, **69** (19), 6267–6272.
134 Ziegler, F.E., Klein, S.I., Pati, U.K. and Wang, T.F. (1985) Cyclic diastereoselection as a synthetic route to quassinoids: a Claisen rearrangement based strategy for bruceantin. *J. Am. Chem. Soc.*, **107** (9), 2730–2737.

8
Pyridine and Its Derivatives

Paula Kiuru and Jari Yli-Kauhaluoma

8.1
Introduction

Pyridine-containing natural products consist of a variety of intriguing chemical structures that originate from the kingdoms of Animalia, Plantae, Fungi, Protista, and Prokaryota. The most commonly known compounds with an aromatic 6π electron pyridine moiety are nicotine, niacin (vitamin B_3 or nicotinic acid) and pyridoxine (vitamin B_6) (Figure 8.1). Pyridine-containing structures are more abundant in natural products than the structures having other pyridine oxidation states, such as those containing tetrahydropyridine, dihydropyridine, piperidine or pyridone moieties. At a first glance, pyridine-containing natural compounds appear to comprise a very complex group; however, many structural similarities can be observed. Long-chain 3-alkylpyridines are the largest structurally coherent group of compounds, originating mainly from marine sponges of the order Haplosclerida. In addition, compounds that act as nicotinic acetylcholine receptor agonists often possess bipyridine or piperidinylpyridine structures, making them of special biological interest.

In this chapter we have given priority to natural compounds that have had their syntheses published during the period from 2000 to 2009. We have intentionally excluded pyridine-containing macrocyclic peptides and sesquiterpenes from this chapter due to space requirements. Several syntheses have been published for the majority of compounds presented in Table 8.1; we refer only to the most recent publications here. For some compounds, there is only one published synthesis; additionally, some complex compounds, such as arenosclerins and halichlonacyclamines, lack a published synthesis. Novel pyridine-containing natural products are continuously being isolated and structurally characterized. Thus, some recently found compounds are also included in Table 8.1. Several thorough reviews have been published for some compounds, including nicotine, epibatidine and streptonigrin [1]. Therefore, their syntheses are not discussed in the present work.

Heterocycles in Natural Product Synthesis, First Edition. Edited by Krishna C. Majumdar and Shital K. Chattopadhyay.
© 2011 Wiley-VCH Verlag GmbH & Co. KGaA. Published 2011 by Wiley-VCH Verlag GmbH & Co. KGaA.

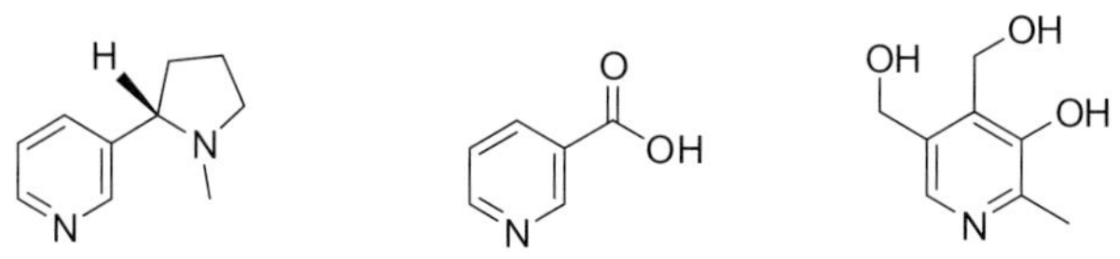

Figure 8.1 Nicotine, niacin (vitamin B₃ or nicotinic acid) and pyridoxine (vitamin B₆).

8.2
Application of the Pyridine Moiety in the Synthesis of Natural Products

8.2.1
Pyridines

8.2.1.1 Synthesis of Noranabasamine Enantiomers

Noranabasamine **35** is an alkaloid that has been isolated from the Dendrobatidae amphibian *Phyllobates terribilis* [56]. Noranabasamine is structurally related to the corresponding plant alkaloid anabasamine, which is known to inhibit acetylcholine esterase and possess anti-inflammatory activity [88].

The retrosynthesis of **35** is centered on the C–C disconnection of the terminal pyridine ring to yield 2-substituted piperidines as the initial synthesis targets [57]. The Trudell group chose the enantioselective iridium-catalyzed N-heterocyclization reaction with either enantiomer of 1-phenylethylamine and 1-(5-methoxypyridin-3-yl)-1,5-pentanediol **56** as a synthetic strategy for obtaining the 2-substituted piperidines.

The pentanediol intermediate **56** was prepared by lithiation of the commercially available 5-bromo-2-methoxypyridine **55** and the subsequent regioselective addition of δ-valerolactone (Scheme 8.1). This reaction sequence gave the ring-opened hydroxy ketone intermediate, which was reduced to the corresponding diol intermediate **56** in 88% yield. The Ir-catalyzed [1.5 mol% of (Cp*IrCl₂)₂] N-heterocyclization reaction between **56** and (R)- and (S)-1-phenylethylamine gave the corresponding 2-substituted piperidines **57** with a diastereomeric ratio of 95:5 in 72% and 69% yield, respectively.

The N-1-phenylethyl substituent of the chromatographically separated diastereomers **57** was removed by hydrogenolysis, and the obtained debenzylated products were treated with POCl₃ to yield the corresponding 2-chloro-substituted analogs. The 2-chloro analog **58** was subjected to the Pd(0)-catalyzed Suzuki–Miyaura coupling with 3-pyridineboronic acid in the presence of Pd₂(dba)₃, PCy₃ and K₃PO₄ in 1,4-dioxane and water to give (S)-(−)-noranabasamine **35** in 84% yield. The enantiopurity of **35** was determined by NMR study and was shown to be >86% e.e. The corresponding (R)-enantiomer was prepared in an analogous manner.

8.2.1.2 Synthesis of Quaterpyridine Nemertelline

Nemertelline **29** is one example of a structurally very interesting class of neurotoxic pyridine-containing natural compounds. Nemertelline (3,2′:3′,2″:4″,3‴-quaterpy-

Table 8.1 Natural products containing the pyridine moiety. Please note that literature references are given only from the compounds with structures presented in the table.

Serial No.	Trivial name	Structure	Source	Isolation [Ref]	Biological activity	Synthesis [Ref]
1	Amphimedoside A		Sponge *Amphimedon* sp.	[2]	Cytotoxic	No
2	Anabaseine		Hoplonemertine *Paranemertes peregrina*	[3]	Nicotinic acetylcholine receptors (nAChRs)	[4]
3	Anabasine		*Anabasis aphylla*, *Nicotiana glauca*	[5]	nAChRs	[6]
4	Arenosclerin A (-E)		Sponge *Arenosclera brasiliensis*	[7]	Antimicrobial, cytotoxic	No
5	Caerulomycin C (A-E)		*Streptomyces caeruleus*	[8]	Antifungal	[9]
6	Coniine		Poison hemlock *Conium maculatum*	[10]	Neurotoxic	[11]
7	Cortamidine oxide		Basidiomycete *Cortinarius* sp.	[12]	Antimicrobial	No
8	Cribochaline A (-B)		Sponge *Cribochalina* sp.	[13]	Antifungal	No

(*Continued*)

Table 8.1 (Continued)

Serial No.	Trivial name	Structure	Source	Isolation [Ref]	Biological activity	Synthesis [Ref]
9	Cyclohaliclonamine A (-E)		Sponge *Haliclona* sp.	[14]	Not reported	No
10	Cyclostellettamine A (-L)		Sponge *Haliclona* sp., *Stelletta maxima*, *Pachychalina* sp.	[15]	Histone deacetylase, muscarinic acetylcholine receptors (mAChRs)	[16]
11	Cytisine		Legume *Laburnum anagyroides*	[17]	nAChRs	[18]
12	Epibatidine		Frog *Epipedobates anthonyi*	[19]	Analgesic nAChRs	[1b–d, 20]
13	Fuzanin C (-D)		*Kitasatospora* sp. IFM10917	[21]	Cytotoxic	No
14	Hachijodine F (A-G)		Sponge *Amphimedon* sp.	[22]	Cytotoxic	[23]
15	Haliclamine A (-F)		Sponge *Haliclona* sp.	[24]	Cytotoxic	[25]
16	Haliclonacyclamine A (-E)		Sponge *Haliclona* and *Pachychalina alcaloidifera*	[26]	Antibacterial, cytotoxic, antifungal	No

17	Haminol-1 (-8)		Marine mollusk *Haminoea*	[27]	Antibacterial	[28]
18	3-Hydroxy-11-norcytisine		Legume *Laburnum anagyroides*	[29]	nAChRs	[30]
19	Ikimine A (-D)		Unidentified Micronesian sponge	[31]	Cytotoxic	[32]
20	Iromycin A (-F)		*Streptomyces* sp.	[33]	Nitric oxide synthase (NOS) inhibitor	[34]
21	Jussiaeiine A		Gorse *Ulex jussiaei*	[35]	Not reported	[36]
22	Kuraramine		*Sophora flavescens*	[37]	Not reported	[36]
23	Malloapeltine		Roots of *Mallotus apelta*	[38]	Anti-HIV	No
24	2-[(Methylthiomethyl)dithio]pyridine-N-oxide		*Allium stipitatum*	[39]	Antimicrobial, *Mycobacterium tuberculosis*	No

(*Continued*)

Table 8.1 (Continued)

Serial No.	Trivial name	Structure	Source	Isolation [Ref]	Biological activity	Synthesis [Ref]
25	Montipyridine		Stony coral *Montipora* sp.	[40]	Cytotoxic	[41]
26	Muscopyridine		Male musk deer *Moschus moschiferus*	[42]	Musk	[43]
27	Nakinadine A (-F)		Sponge *Amphimedon* sp.	[44]	Cytotoxic	No
28	Navenone B (A,C)		Sea slug *Navanax inermis*	[45]	Alarm pheromone	[46]
29	Nemertelline		Marine worm *Amphiporus angulatus*	[47]	Neurotoxin	[48]
30	Nicotine		Tobacco *Nicotiana tabacum*	[49]	nAChRs	[1a]
31	Niphatesine A (-H)		Sponge *Niphates* sp.	[50]	Antimicrobial	[32b, 51]

32	Niphatoxin A (-C)		Sponge *Niphates* sp.	[52]	Cytotoxic	[53]
33	Niphatyne A (-B)		Sponge *Niphates* sp.	[54]	Cytotoxic	No
34	Njaoaminium A (-C)		Sponge *Reniera* sp.	[55]	Cytotoxic	No
35	Noranabasamine		Poison dart frog *Phyllobates terribilis*	[56]	nAChRs and ion channels	[57]
36	Patungensin		*Lysimachia patungensis*	[58]	Not reported	No
37	Phenylpyridinylbutenol		*Streptomyces* sp.	[59]	Antiproliferative	No
38	Phormidinine A (-B)		Cyanobacterium *Phormidium* sp.	[60]	Not reported	No
39	Piericidin A1 (-B1)		*Streptomyces mobaraensis* and *S. pactam*	[61]	Mitochondrial electron transport chain protein	[62]
40	Pipecoline		Gray pine *Pinus sabiana*	[63]	—	[11]

(Continued)

274 | 8 Pyridine and Its Derivatives

Table 8.1 (Continued)

Serial No.	Trivial name	Structure	Source	Isolation [Ref]	Biological activity	Synthesis [Ref]
41	Pyrinadine A (-G)		Sponge *Cribrochalina* sp.	[64]	Cytotoxic	[65]
42	Pulo'upone		Mollusk *Philinopsis speciosa*	[66]	Antimicrobial	[67]
43	Pyrinodemin A (-F)		Sponge *Amphimedon* sp.	[68]	Cytotoxic	[69]
44	Simplakidine A		Caribbean sponge *Plakortis simplex*	[70]	Cytotoxic	No
45	Simplexidine		Sponge *Plakortis simplex*	[71]	Not reported	No
46	Solenopsin A (-C)		Fire ant *Solenopsis saevissima, S. invicta*	[72]	Angiogenesis, cytotoxic	[73]
47	Spongidine A (-D)		Vanuatu sponge *Spongia* sp.	[74]	Anti-inflammatory (phospholipase A2: PLA2)	[75]

48	Stenusine	(structure)	Rove beetle *Stenus* sp.	[76]	Antimicrobial	[77]
49	Streptonigrin	(structure)	*Streptomyces flocculus*	[78]	Anticancer, antiviral	[1e, 79]
50	Theonelladine A (-D)	(structure)	Sponge *Theonella swinhoei*	[80]	Cytotoxic	[32b, 81]
51	Untenine A (-C)	(structure)	Sponge *Callyspongia* sp.	[82]	Antifouling	[82]
52	Viscosaline	(structure)	Arctic sponge *Haliclona viscosa*	[83]	Antibacterial	[81a]
53	Viscosamine	(structure)	Arctic sponge *Haliclona viscosa*	[84]	Not reported	[85]
54	Xestamine A (-H)	(structure)	Sponge *Xestospongia wiedenmayeri*	[86]	Antimicrobial	[32b, 87]

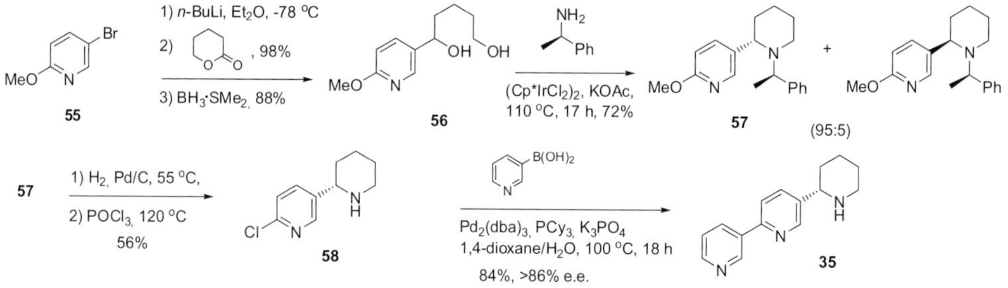

Scheme 8.1 Synthesis of (S)-(−)-noranabasamine [57].

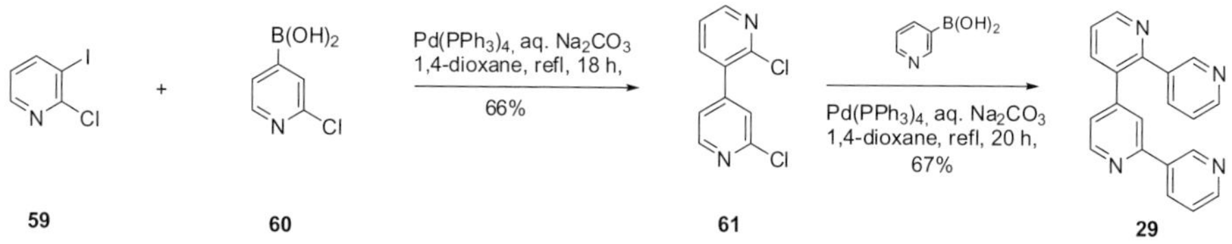

Scheme 8.2 Synthesis of nemertelline [48a].

ridine) was isolated from the Hoplonemertine marine worm *Amphiporus angulatus* [47]. The correct structure of nemertelline was reported in 1995 [89].

Rault and co-workers presented a regioselective two-step total synthesis of nemertelline (Scheme 8.2) that commenced with the highly selective Suzuki cross-coupling reaction between the commercially available 2-chloro-3-iodopyridine **59** and (2-chloropyridin-4-yl)boronic acid **60** to give 2,2′-dichloro-3,4′-bipyridine **61** in 66% yield [48a]. It is notable that no coupling was observed to take place on the chlorine-substituted sp² carbon atoms of the starting materials. The second Suzuki coupling between **61** and (pyridin-3-yl)boronic acid gave nemertelline **29** in 67% yield.

8.2.1.3 Synthesis of Caerulomycin C

Bipyridinic caerulomycins B–C were isolated from the cultures of actinomycetes *Streptomyces caeruleus* [8]. Synthesis of caerulomycin C **5** uses the so-called halogen dance reaction (Scheme 8.3) [9]. The synthesis was started from the commercially available picolinic acid **62** with the electrophilic 4-chlorination and the subsequent formation of N,N-diisopropyl amide **63**. Chlorine was substituted with the less carbanion-stabilizing methoxy group, as the halogen dance of iodine was found to produce a mixture of products. The 1,3-migration of iodine was conducted with the treatment of **64** with lithium diisopropylamide (LDA). Nucleophilic aromatic substitution at C-5 proved difficult, so the modified Ullmann coupling was used. Bromine was introduced to the 3-position, and 1,4-migration of bromine transferred it to the 6-position **65**, where the Negishi cross-coupling of another pyridine ring was carried out using $Pd_2(dba)_3$-Ph_3P as a catalyst. Finally,

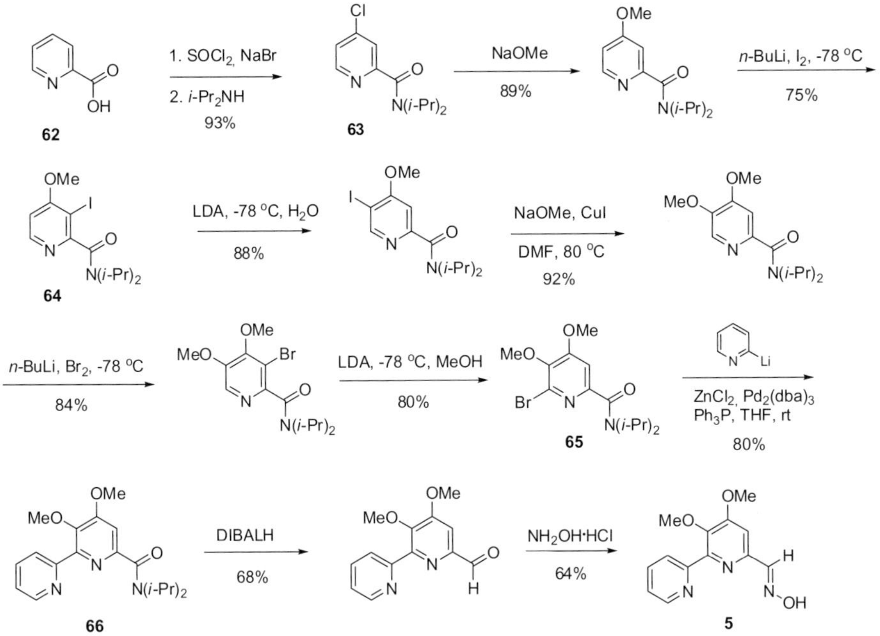

Scheme 8.3 Synthesis of caerulomycin C [9].

the bipyridylic amide **66** was converted to the oxime moiety of caerulomycin C **5** in two steps.

8.2.1.4 Synthesis of the Spongidine Isomer

The attempted synthesis of tetracyclic spongidine **47**, an anti-inflammatory pyridine alkaloid isolated from marine sponges belonging to the genus *Spongia* sp. [74], did not give the desired target compound. Instead, it yielded the isomeric spongidine **67** [75].

First, commercially available geranyl acetate **68** was subjected to allylic oxidation with SeO_2 and reduction with $NaBH_4$ (Scheme 8.4). The resulting alcohol **69** was converted to the corresponding bromide **70** using LiBr via the mesylate intermediate. The dianion of methyl acetoacetate was alkylated with the bromodiene **70** to give the allylic alcohol **71** in 65% yield. The subsequent conversion, via the mesylate intermediate, to the *E,Z*-bromide **72** proceeded in 70% yield. The allylic *E,Z*-bromide was coupled with the lithium anion of picoline to give *E,E*-(pyridin-3-yl) tridecadienoate **73** as the precursor for the subsequent radical cyclization reaction.

Mn(III)-promoted oxidative free-radical cyclization of the nucleophilic carbon-centered radicals with the pyridin-3-yl ring proceeded stereospecifically, giving the fused pyridine tetracycle **74** in 40% yield. Thorough spectroscopic characterization of the reaction product confirmed the formation of the tetracyclic β-keto ester via radical cyclization to the *ortho* position of the pyridine ring instead of the desired *para* position.

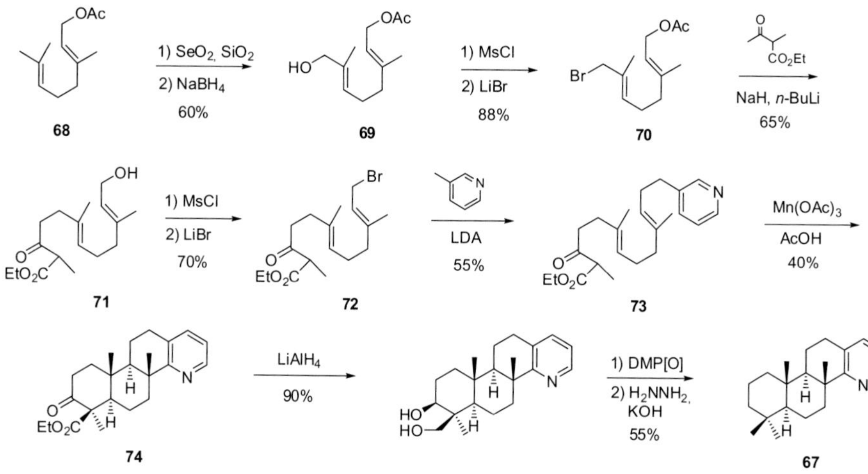

Scheme 8.4 Synthesis of the spongidine isomer [75].

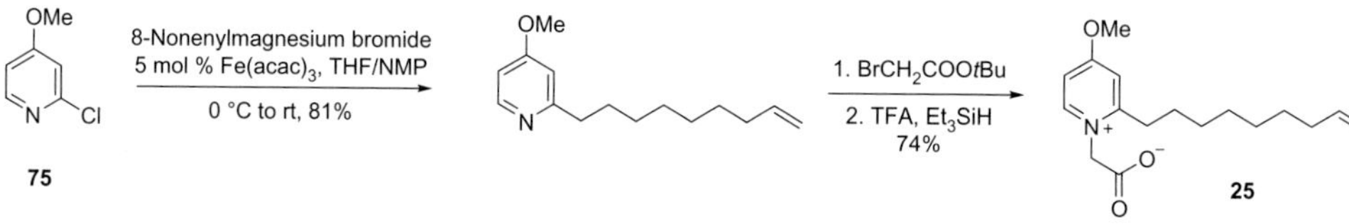

Scheme 8.5 Synthesis of montipyridine [41].

To convert the obtained tetracycle **74** to the isomeric spongidine, its β-keto ester functionality was reduced to the corresponding diol, oxidized to the β-keto aldehyde using Dess–Martin periodinane and finally subjected to the classical Wolff–Kishner conditions to give the *gem*-dimethyl moiety of the isomeric spongidine **67** in 55% yield.

8.2.2
2-Alkylpyridines

8.2.2.1 Synthesis of Montipyridine
An efficient method to prepare 2-alkylpyridinium montipyridine **25**, which is isolated from stony coral *Montipora* sp. [40], was reported by the Fürstner group (Scheme 8.5) [41]. Iron-catalyzed cross-coupling was used to attach the unsaturated nonenyl side chain to the pyridine ring **75**. The coupling reaction is based on the formation of [Fe(MgX)$_2$], the so-called "inorganic Grignard reagent".

8.2.2.2 Synthesis of Piericidin A1
Piericidins A1 **39** and B1 are isolated from the Actinobacteria *Streptomyces mobaraensis* and *S. pactam* [61]. They are potent inhibitors of the mitochondrial

8.2 Application of the Pyridine Moiety in the Synthesis of Natural Products

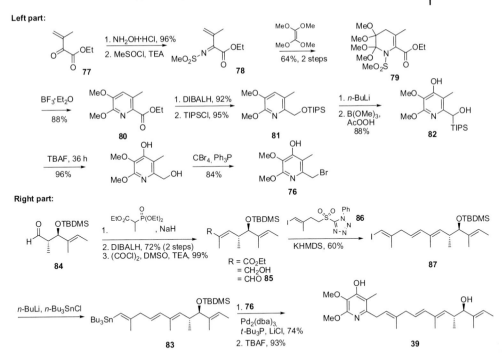

Scheme 8.6 Synthesis of piericidin A1 by the Boger group [62a].

electron transport chain NADH–ubiquinone oxidoreductase (Complex I). Two total syntheses have been reported for piericidin A1 **39** [62]. Boger and co-workers presented the total synthesis for **39** in 2006 (Scheme 8.6) [62a]. They based their approach on the inverse electron demand hetero Diels–Alder reaction, between N-sulfonyl-1-aza-1,3-butadiene and tetramethoxyethene, and the subsequent Lewis acid-promoted aromatization of the hetero Diels–Alder adduct. The synthesis strategy of the **39** side-chain comprises an asymmetric anti-aldol reaction, the Wadsworth–Horner–Emmons olefination and the Julia–Kocienski olefination. Finally, the heterobenzylic palladium-catalyzed Stille cross-coupling reaction was expected to provide the desired target compound.

Heterobenzylic bromide **76** was synthesized via a 10-step linear sequence for the final Stille cross-coupling reaction. The synthesis of **76** commenced with the oxime formation of the α-keto ester **77** and the reaction between the oxime and methylsulfinyl chloride to give the initial O-methanesulfinate and its subsequent rearrangement product, sulfonylimine **78**. The inverse electron demand hetero Diels–Alder reaction, between the electron-deficient N-sulfonyl-1-aza-1,3-butadiene **78** and tetramethoxyethene, provided the chromatographically stable cycloadduct **79** in good yield. Its chemical structure was confirmed through a single-crystal X-ray diffraction study.

The Lewis acid-promoted (BF$_3$·Et$_2$O) aromatization of **79** gave the picolinate **80** in good yield. The subsequent reduction of its ethyl ester moiety gave the corresponding alcohol, which was protected as a triisopropylsilyl (TIPS) ether to produce **81**. The treatment of **81** with excess *n*-BuLi followed by trimethylborate, with final oxidative cleavage of the aryl boronate in the presence of peracetic acid, gave the desired *C*-silylated intermediate **82**. Removal of the TIPS group by tetra-*n*-butylammonium fluoride (TBAF) and subsequent bromination with a CBr_4-Ph_3P system gave the heterobenzylic bromide **76** in good yield.

The tributyltin derivative of the side chain **83** was synthesized from the *t*-butyldimethylsilyl (TBDMS)-protected aldehyde **84**. The reaction sequence of the Wadsworth–Horner–Emmons olefination, diisobutylaluminium hydride (DIBALH) reduction of the ester moiety and the Swern oxidation provided the aldehyde part **85** for the Julia–Kocienski coupling. This modified Julia coupling between **85** and the iodo sulfone **86** in the presence of potassium hexamethyldisilazide (KHMDS) gave the desired *trans* alkene **87** in 60% yield. The lithium–iodine exchange reaction of **87** with *n*-BuLi, and subsequent treatment with *n*-Bu_3SnCl gave the vinyl stannane **83**. Finally, the Pd-catalyzed Stille cross-coupling reaction between stannane **83** and the heterobenzylic bromide **76**, with subsequent removal of the TBDMS protective group, gave the target piericidin A1 **39**.

Recently, the Akita group presented a new synthesis of **39** based on the Julia olefination, combining two halves (Scheme 8.7) [62b]. The left part, stannylpyridine **88**, was synthesized according to Keaton and Phillips, who used it for the

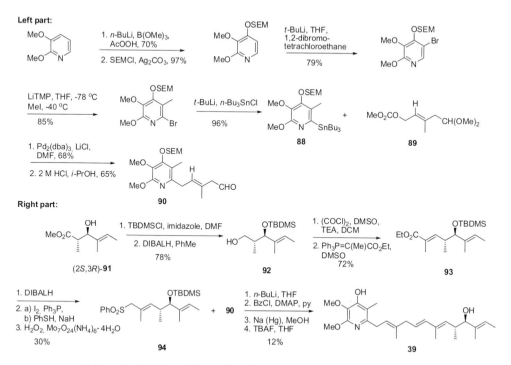

Scheme 8.7 Synthesis of piericidin A1 by the Akita group [62b].

synthesis of (−)-7-demethylpiericidin A1 by titanium(II)-mediated cyclization of siloxy(enynes) [90]. The Stille cross-coupling of **88** and **89** afforded the E/Z mixture of **90** in 2.2:1 ratio. Treatment with HCl and separation gave **90** in 65% yield.

The methyl hexenoate (2S,3R)-**91** was silylated, and the ester moiety was reduced with DIBALH to the corresponding 4-hexen-1-ol **92**. The alcohol **92** was subjected to the Swern oxidation, and the resulting aldehyde underwent the Wittig reaction with [1-(ethoxycarbonyl)ethylidene]triphenylphosphorane to give the homologated (4R,5R)-octadienoate **93** as a 5:1 mixture of geometric isomers (2E:2Z) in 72% yield (two steps). The subsequent reduction of **93** with DIBALH, treatment of the obtained alcohol with I_2/Ph_3P and conversion of the resulting allyl iodide to the corresponding allyl phenyl sulfide proceeded smoothly. Oxidation of the sulfide with H_2O_2 in the presence of ammonium molybdate gave the sulfone **94** for the subsequent Julia olefination. The reaction of the left part **90** and the sulfone **94** in the presence of n-BuLi produced the diastereomeric hydroxysulfones in 68% yield. Their subsequent conversion into benzoates and treatment with a 5% Na amalgam gave a 3:1 mixture of piericidin A1 geometric isomers (E:Z) as silyl ethers. Finally, desilylation with TBAF and the chromatographic separation of products gave piericidin A1 **39** in 69% yield.

8.2.3
3-Alkylpyridine, 3-Alkylpyridinium and 3-Alkyltetrahydropyridine Compounds

3-Alkylpyridines are the major class of pyridine-containing natural products. Approximately one hundred monomeric, cyclic or polymeric 3-alkylpyridines have been isolated from marine sponges, mainly from the order Haplosclerida at present. Their broad spectrum of bioactivities (e.g., cytotoxic, antimicrobial and antifouling properties) has recently been reviewed by Turk et al. [91] and Sepčić [92]. Owing to the large number of 3-alkylpyridines, only a few examples of their syntheses are presented here.

8.2.3.1 Synthesis of Xestamines

Ikimines **19**, cribochalines **8**, hachijodines **14**, niphatesines **31**, niphatynes **33**, theonelladines **50** and xestamines **54** are structured in a relatively similar manner to monomeric 3-alkylpyridines. They have differences in the substitution pattern and length of the alkyl chain, but contain nitrogen functionality at the chain end. Haminols **17**, and navenones **28** contain an oxygen moiety at the end of the chain. In 2003, Larock and co-workers used efficient palladium-catalyzed coupling in a three-component reaction with 3-iodopyridine, terminal long chain dienes and amines to synthesize nitrogen-ended ikimines **19**, niphatesines **31**, theonelladines **50** and xestamines **54** [32b]. In Scheme 8.8, the synthesis of xestamine D **95** is presented. The disadvantage of this otherwise expedient multicomponent procedure is that it produced two regioisomers **95** and **96** in a 91:9 ratio.

A different approach to the synthesis of xestamines (C **97**, E **98** and H) was presented by the Zard group (Scheme 8.9) [87]. Synthesis of the side chain relies on β-keto γ-xanthyl phosphonate **99**, which was subjected to the xanthate transfer

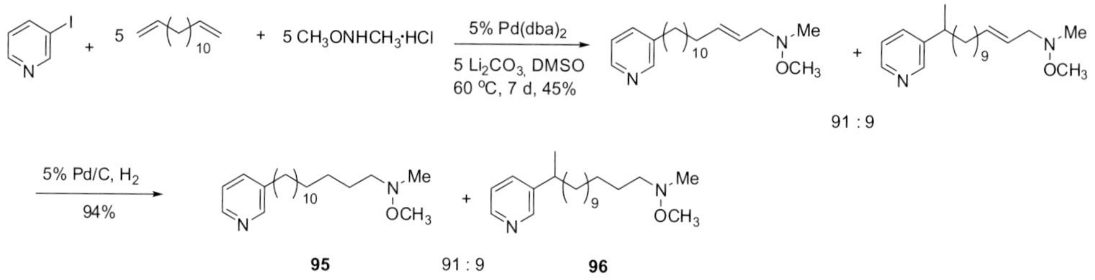

Scheme 8.8 Synthesis of xestamine D [32b].

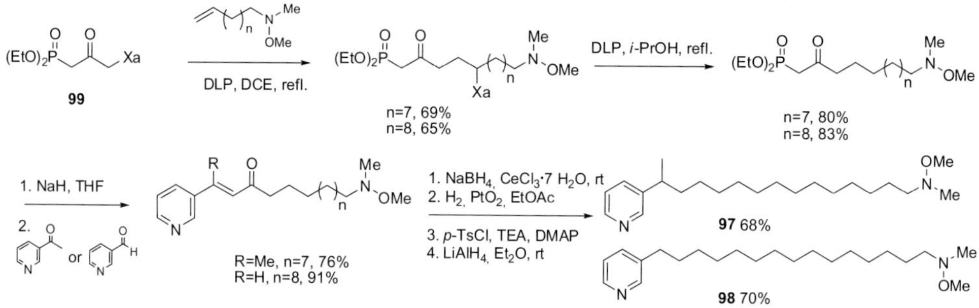

Scheme 8.9 Synthesis of xestamines C and E [87] (Xa = Xanthate).

radical reaction with dilauroyl peroxide. The xanthate group was removed, and the side chain was attached to the pyridine moiety using a Horner–Wadsworth–Emmons olefination. The corresponding α,β-unsaturated carbonyl compounds were reduced to obtain xestamine C **97** or E **98**.

8.2.3.2 Synthesis of Pyrinadine A

Pyrinadines A–G, which are bicyclic 3-alkylpyridines, were isolated from the Okinawan marine sponge *Cribrochalina* sp. They differ in their alkyl chain length and number of double bonds. Anwar and Lee [65] presented a biomimetic synthesis of pyrinadine A **41** in 50% total yield (Scheme 8.10). Pyridine-3-propanal was coupled to the phosphonium salt **100** by means of the Wittig reaction [16c]. The 3-pyridinyl alcohol **101** was oxidized with 2-iodoxybenzoic acid to the corresponding aldehyde **102**, converted to oxime **103** and reduced to hydroxylamine **104** in the presence of NaBH$_3$CN. The oxidative dimerization (air, open reaction vessel, 2 d) of **104** to pyrinadine A **41** proceeded in 76% yield.

8.2.3.3 Synthesis of Pyrinodemin A

Pyrinodemin A **43**, a bicyclic 3-alkylpyridine, was isolated from Okinawan sponge *Amphimedon* sp. in 1999 by Kobayashi [68a]. This was followed by the isolation of related pyrinodemins B–D [68b]. Original assignment of the olefin (16′–17′) in pyrinodemin A caused some controversy. Thus, several isomers, including (14′–

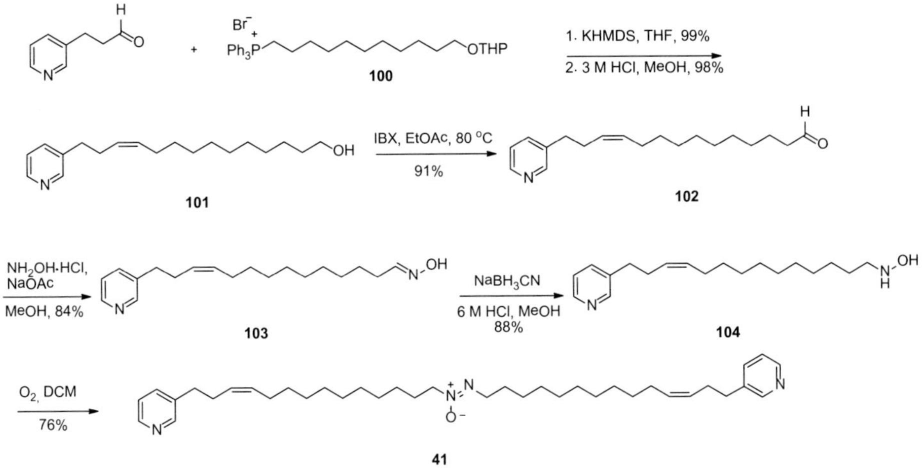

Scheme 8.10 Synthesis of pyrinadine A [65].

15′) and (15′–16′), were synthesized to help to solve the structure [69b–f]. Kobayashi revised the structure of pyrinodemin A in 2005 [69b], and indicated the double bond regiochemistry to be (15′–16′). This was confirmed by HPLC analysis with the authentic natural product. The most recent synthesis of pyrinodemin A by Pouilhès et al. is presented in Scheme 8.11 [69a].

The synthesis of pyrinodemin A **43** commenced with the construction of its 3,7-dioxa-2-azabicyclo[3.3.0]octane core by an asymmetric intramolecular 1,3-dipolar cycloaddition reaction via the dipolar oxazoline N-oxide intermediate. The hetero cycloadduct **105** was obtained in good yield and with complete stereoselectivity. The chiral auxiliary was removed in a three-step process (including a modified Agami–Couty N-debenzylation) due to the instability of the N–O σ bond towards catalytic hydrogenation [i.e., reduction with $Zn(BH_4)_2$, conversion of the hydroxyl group to the mesylate, and the cyanide-mediated β-elimination reaction]. After removal of the 4-methoxybenzyl protecting group of the oxazoline and the N-protection with the t-butoxycarbonyl (t-Boc) group, the primary hydroxyl group was converted to the iodide via the mesylate intermediate in excellent yield. The side chain of pyrinodemin A was obtained from the Cu(I)-mediated coupling reaction between the iodide **106** and the alkenyl Grignard reagent in 59% yield. The subsequent B-alkyl Suzuki reaction between the 9-borabicyclo[3.3.1]nonane (9-BBN) adduct of the oxazolidine **107** and 3-bromopyridine, as well as the following N-deprotection with trifluoroacetic acid (TFA), gave the bicyclic oxazolidine side chain **108**. In the final step, a reductive amination was used to couple **108** and the aldehyde **109** (synthesized according to Baldwin et al. [93]) in 30% yield. An alternative approach to the synthesis was to introduce both pyridine moieties simultaneously using the double B-alkyl Suzuki cross-coupling reaction. This worked, but with very low yield.

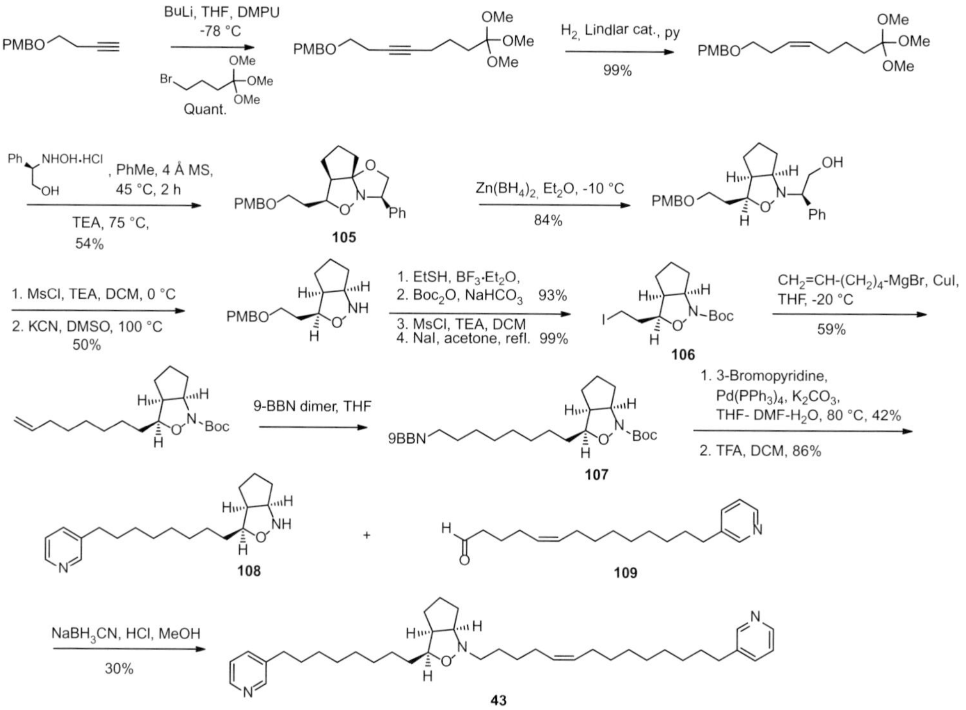

Scheme 8.11 Synthesis of pyrinodemin A [69a].

8.2.3.4 Synthesis of Haliclamine A

The Haplosclerida sponge genus *Haliclona* produces haliclamines, which are cyclic dimeric 3-alkyltetrahydropyridines. Haliclamines A **15** and B are found in the Hawaiian species [24a], and haliclamines C–E originate from the Arctic sponge *H. viscosa* [24b, c]. Structurally these compounds are related to cyclohaliclonamines **9**, which are pyridinium analogues of haliclamines. Synthesis and biosynthetic pathways of several polycyclic alkylpyridines and alkylpiperidines have been reviewed by Berlinck [94]. Haliclamine A **15** has been synthesized by Michelliza et al. [25a], Baldwin et al. [25b], and Morimoto et al. [25c]. The former synthesis is presented below.

For syntheses of both 3-alkylpyridine chains the thiophene derivative **110** is used (Scheme 8.12). To access the longer chain, the pyridine moiety is attached to thiophene **110** using the Wittig reaction. In the synthesis of the shorter nine-carbon chain, the lithium anion of thiophene is allowed to react with nicotinaldehyde (synthesis not shown). The thiophene ring of both chains is then reduced with Raney nickel, the *t*-butyldimethylsilyl (TBDMS) ether is deprotected and the resulting cyclopropyl carbinol **111** is rearranged into homoallylic bromides **112** in the presence of ZnBr$_2$ and trimethylsilylbromide at −20 °C. The bromine substituents

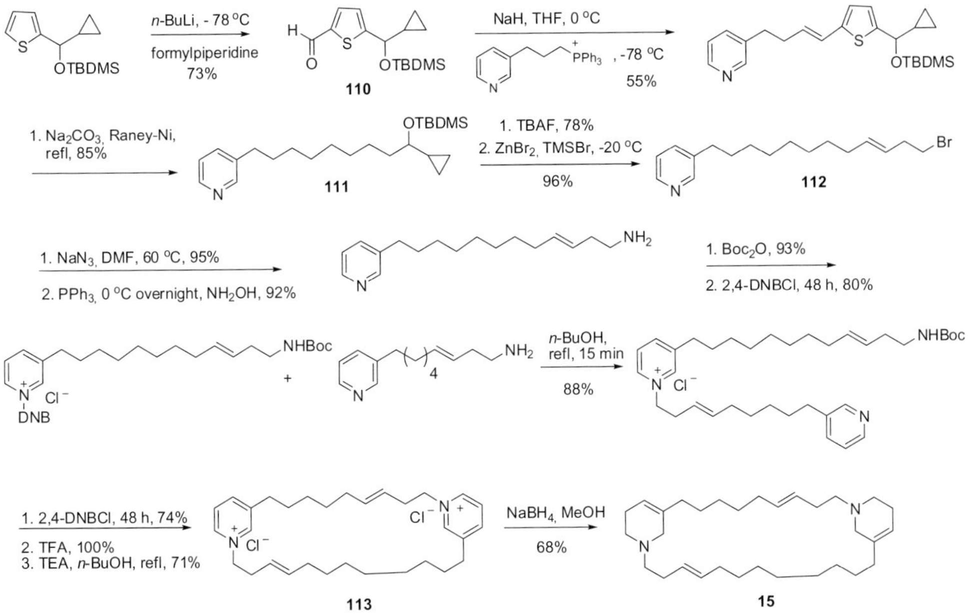

Scheme 8.12 Synthesis of haliclamine A [25a].

were transformed to the corresponding amino groups via azide intermediates. The Zincke procedure was used for the attachment of chains and the subsequent formation of the asymmetric pyridinium macrocycle **113**, which was eventually reduced to the target compound haliclamine A **15**.

Viscosamine **53** is an interesting trimeric cyclic 3-alkylpyridine isolated from the Arctic sponge *Haliclona viscosa* [84]. Another similar but unsaturated trimeric compound has been isolated from the Pacific *Halichlona* sp. [95]. Other oligomeric 3-alkylpyridinium metabolites with anti-acetylcholinesterase activity have been isolated from the sponges *Amphimedon compressa* (amphitoxin) [96] and *Reniera sarai* [97]. Linear oligomeric 3-alkylpyridinium metabolites from *R. sarai* have been synthesized by the Sepčić group [98].

8.2.4
Piperidines

8.2.4.1 Synthesis of Coniine and Pipecoline

2-Alkylpiperidine coniine **6** is a neurotoxic alkaloid from the poison hemlock (*Conium maculatum*) in the family Apiaceae [10]. Coniine **6** is also found in *Aloe* species, as well as in the yellow pitcher plant (*Sarracenia flava*) [99]. The compound 2-methylpiperidine pipecoline **40** is found in *C. maculatum* and in pine *Pinus sabiana* [63]. *Conium* species also contain other neurotoxic piperidine alkaloids, such as γ-coniceine, *N*-methylconiine and conhydrine. The hemlock alkaloids have been reviewed in 2005 by Reynolds [99]. The synthesis of coniine **6** in 1886 was

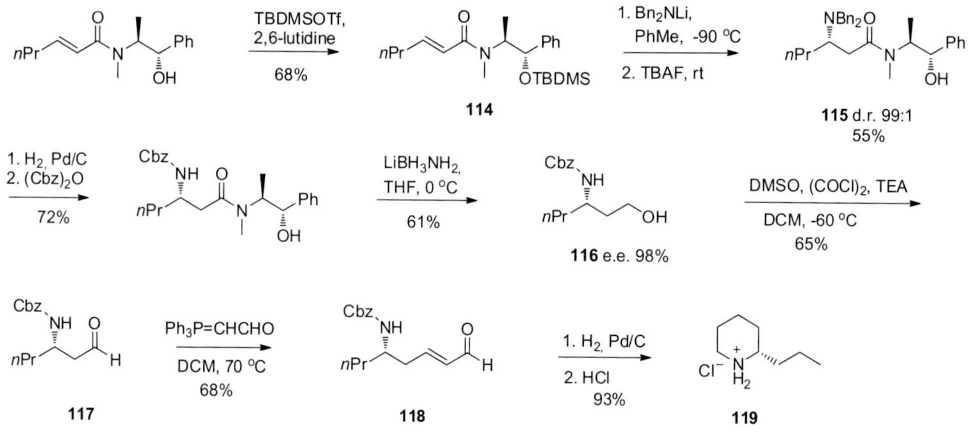

Scheme 8.13 Asymmetric synthesis of (S)-coniine [11a].

one of the first alkaloid syntheses performed. Since then numerous syntheses have been published.

Etxebarria et al. used (S,S)-(+)-pseudoephedrine as a chiral auxiliary in the synthesis of the alkaloids (S)-coniine hydrochloride **119** (Scheme 8.13) and (R)-pipecoline **40** [11a]. The silylated precursor enamide **114** was subjected to the aza-Michael reaction with lithium N,N-dibenzylamide. After desilylation with TBAF, the alcohol **115** was obtained with a diastereomeric ratio of 61 : 39. However, the chromatographic (SiO$_2$) separation gave the desired diastereomer in a 99 : 1 ratio. After a switch from benzyl to carboxybenzyl protection groups, the chiral auxiliary was removed reductively with lithium aminoborohydride to give the alcohol **116** (98% e.e.). This γ-amino alcohol **116** was subjected to the Swern oxidation and the Wittig reaction to produce β-amino aldehyde **117** and α,β-unsaturated δ-aldehyde **118**, respectively. The final reaction cascade of catalytic hydrogenation (Pd/C), deprotection and reductive amination furnished (S)-coniine hydrochloride **119** in 93% yield.

Another approach to the synthesis of (R)-coniine **128** and (R)-pipecoline **40** is the ring-closing metathesis used by the groups of Chang [11b], Gramain [11c] and Vankar [11d]. Here we present the synthesis by Vankar et al. (Scheme 8.14) [11d]. Imines **120** and **121**, having (R)-α-methylbenzylamine as a chiral auxiliary, were allylated with allyl bromide and zinc. The bis-allyl compounds **122** and **123** were obtained, in turn, with allyl bromide in the presence of NaH and n-butylammonium iodide. This step can also be done in one pot by quenching the first step with allyl bromide. The ring-closing metathesis using the Grubbs first generation catalyst gave tetrahydropyridines **124** : **125** and **126** : **127** in 85 : 15 d.r.. Finally, catalytic hydrogenation (Pd/C) gave (R)- (−)-coniine **128** and (R)- (−)-pipecoline **40** in 90% yields.

8.2.4.2 Synthesis of Stenusine

The alkaloid stenusine **48** is isolated from a rove beetle of the genus *Stenus* [76]. This compound acts as an enhancer to spread surface-active fluid, which the beetle

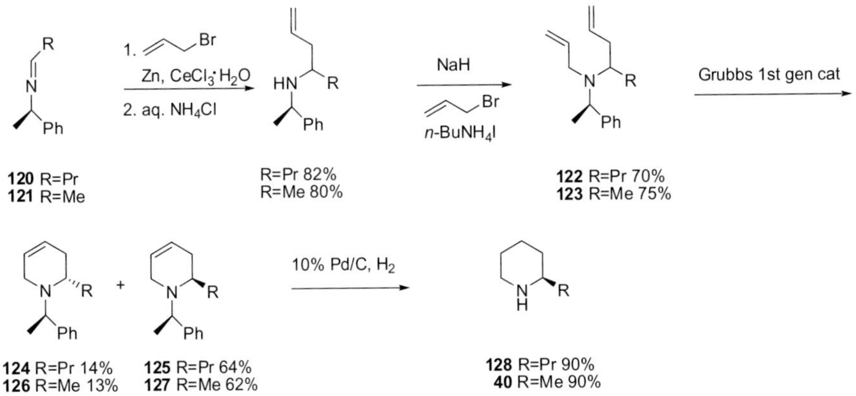

Scheme 8.14 Synthesis of (R)- (−)-coniine and (R)- (−)-pipecoline by Vankar [11d].

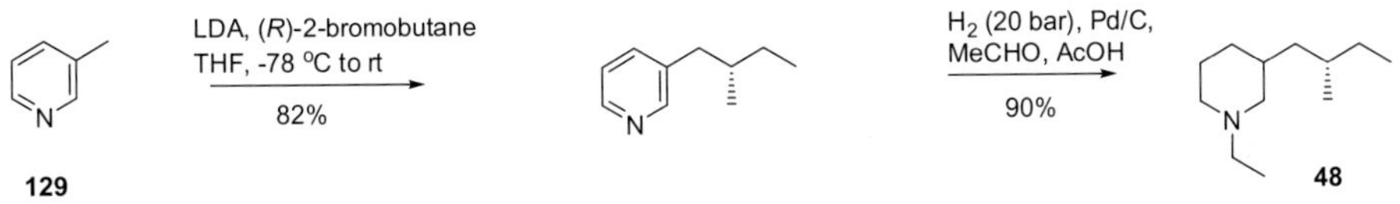

Scheme 8.15 Synthesis of stenusine [77].

excretes to prevent it from drowning. It also enables the beetle to propel over the water surface with speeds up to 75 cm/s [77]. Natural stenusine from *S. comma* is a mixture consisting of four stereoisomers: (2′R,3R):(2′S,3R):(2′S,3S):(2′R,3S) in the ratio 13%:40%:43%:4%. Gedig et al. reported a two-step synthesis for **48** with a stereoisomeric ratio that was similar to the natural stenusine (10%:29%:54%:7%) [77]. The methyl group of 3-picoline **129** is alkylated to give the (S)-enantiomer in 66% e.e. and the (R)-enantiomer in 17% e.e (Scheme 8.15). Catalytic hydrogenation (20 bar) in the presence of acetaldehyde afforded **48** in 74% yield (two steps). This method can also be used for the synthesis of norstenusine, a related compound with 3-isobutyl chain.

8.2.5
Pyridones

8.2.5.1 Synthesis of (±)-Cytisine
The compound (−)-cytisine **11** is a natural product isolated from *Laburnum anagyroides* and other plants of the Fabaceae family of angiosperms [17]. It has been shown that **11** acts as a partial agonist at neuronal nicotinic acetylcholine receptors [100]. The van Tamelen group prepared (−)-cytisine by an 11-step synthesis in the 1950s [101], and several total syntheses have been reported since that time [18a].

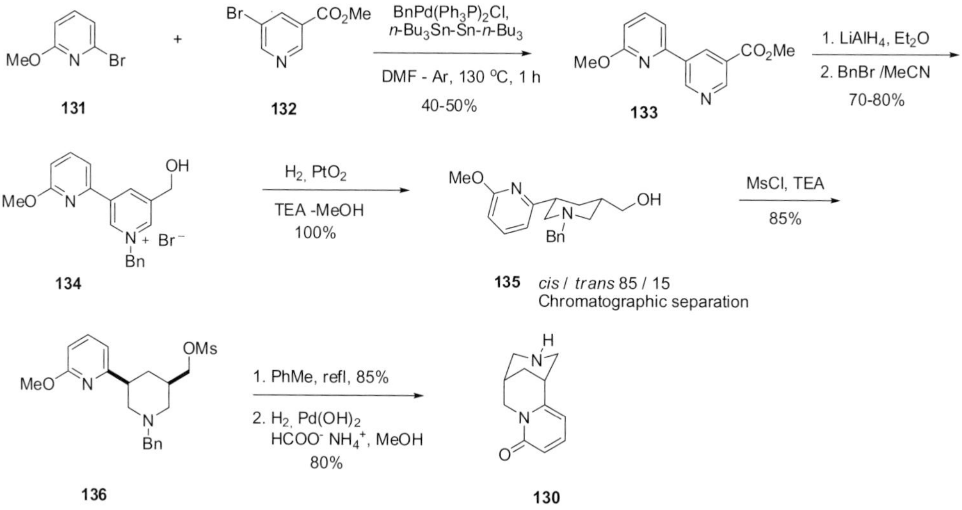

Scheme 8.16 Synthesis of (±)-cytisine [18b].

Herein we want to highlight the total synthesis of (±)-cytisine **130** via a five-step route with *in situ* Stille biaryl pyridine coupling reaction, which has been presented by the O'Neill group (Scheme 8.16) [18b].

First, the C–C bond between 2-bromo-6-methoxypyridine **131** and methyl 5-bromonicotinate **132** was formed in the *in situ* Stille coupling. This gave the methoxybipyridinyl ester **133** in 40–50% yield, with a small amount of the homocoupling product between the **132** aryls. The selective reduction of ester moiety of **133** and the subsequent selective *N*-benzylation provided the pyridinium salt **134** in good yield. Catalytic hydrogenation (PtO$_2$) gave the 3,5-*cis*-piperidine **135** quantitatively (an 85:15 mixture of *cis* and *trans* diastereomers).

Finally, the cyclization of the 3,5-*cis*-piperidine **135** to the diazabicyclo[3.3.1]nonane framework was achieved through a three-step process. The hydroxy group of **135** was converted to mesylate **136**, which was refluxed in toluene to give *N*-benzylcytisine in 85% yield. The removal of the *N*-benzyl substituent was accomplished by Pd-catalyzed hydrogenolysis to give (±)-cytisine **130**.

8.2.5.2 Synthesis of Iromycin A

Iromycins are microbial polyketide metabolites from the Actinobacteria *Streptomyces* sp. [33]. A close structural similarity can be found between these compounds and piericidin A **39** (*cf.* Section 8.2.2.2). Shojaei *et al.* [34] combined alane **137**, in the presence of *n*-BuLi, with the pyrone ring fragment **138** and obtained **139** in 83% yield (Scheme 8.17). The alane **137** was prepared from a 1:4 mixture of allyl bromides **140**. The Cu(I)-catalyzed alkynylation with (trimethylsilyl)acetylene gave the single enyne **141**. After desilylation, **141** was carboaluminated by the Negishi procedure with zirconocene dichloride as a catalyst. Conversion of the oxygen-containing pyrone **139** to the corresponding pyri-

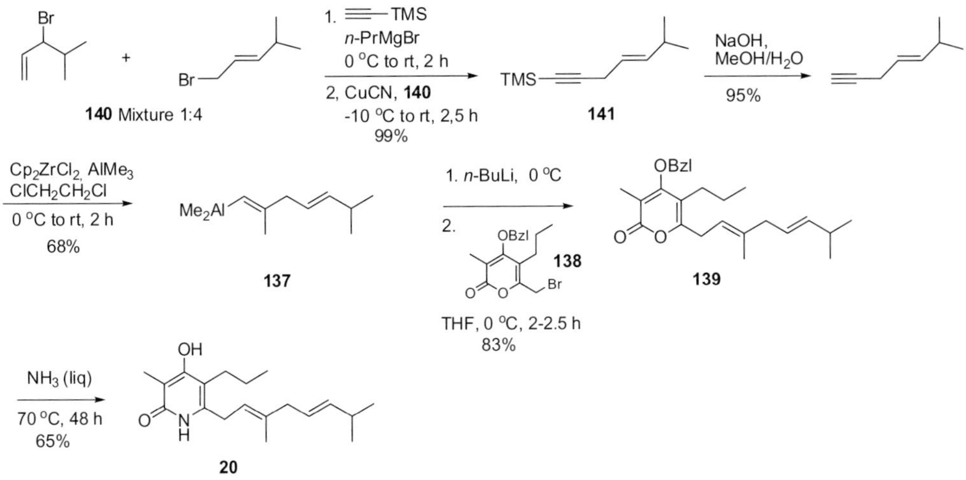

Scheme 8.17 Synthesis of iromycin A [34].

done **20** was carried out in 65% yield by heating **139** in liquid ammonia in an autoclave.

8.3
Conclusion

In this chapter we have highlighted a collection of syntheses towards pyridine-containing natural products. The target molecules range from relatively simple (coniine **6**) to highly complex (piericidin A1 **39**) structures. A representative array of different synthetic strategies was used to construct these natural compounds. Of special interest is the observation that the 4-substitution in pyridine-containing natural products is rare. No 4-alkylpyrines have been characterized at this time, to our knowledge, indicating the lack of a suitable biosynthetic pathway. As the pyridine moiety occurs in a wide variety of structures in nature, many of which have been shown to possess significant biological and pharmacological activities, it is likely that we will see several new syntheses of pyridine-containing natural compounds in the future. Also, as most of the animal phyla are exclusive to the sea environment, it is also conceivable that several natural compounds with pyridine moieties embedded in various carbon skeletons will be isolated from marine ecosystems.

Acknowledgment

Financial support from the Academy of Finland (Grant 120975), European Union Seventh Framework Programme (Grant Agreement KBBE-2009-3-245137) and the University of Helsinki Research Funds (Grant 07991) is greatly acknowledged.

References

1. (a) Wagner, F.F. and Comins, D.L. (2007) Recent advances in the synthesis of nicotine and its derivatives. *Tetrahedron*, **63** (34), 8065–8082; (b) Garraffo, H.M., Spande, T.F. and Williams, M. (2009) Epibatidine: from frog alkaloid to analgesic clinical candidates. A testimonial to 'true grit'! *Heterocycles*, **79** (1), 207–217; (c) Ivy, C.F. (2009) Epibatidine analogs synthesized for characterization of nicotinic pharmacophores – A review. *Heterocycles*, **79** (1), 99–120; (d) Olivo, H.F. and Hemenway, M.S. (2002) Recent syntheses of epibatidine. A review. *Org. Prep. Proced. Int.*, **34** (1), 1–26; (e) Bringmann, G., Reichert, Y. and Kane, V.V. (2004) The total synthesis of streptonigrin and related antitumor antibiotic natural products. *Tetrahedron*, **60** (16), 3539–3574.
2. Takekawa, Y., Matsunaga, S., van Soest, R.W.M. and Fusetani, N. (2006) Amphimedosides, 3-alkylpyridine glycosides from a marine sponge *Amphimedon* sp. *J. Nat. Prod.*, **69** (10), 1503–1505.
3. Kem, W.R., Abbott, B.C. and Coates, R.M. (1971) Isolation and structure of a hoplonemertine toxin. *Toxicon*, **9** (1), 15–22.
4. Wang, J., Papke, R.L. and Horenstein, N.A. (2009) Synthesis of H-bonding probes of α7 nAChR agonist selectivity. *Bioorg. Med. Chem. Lett.*, **19** (2), 474–476, and several other reports.
5. (a) Orekhov, A.P. (1929) The alkaloids of *Anabasis aphylla*. *C. R. Acad. Sci.*, **189**, 945; (b) Feinstein, L., Hannan, P.J. and McCabe, E.T. (1951) Extraction of alkaloids from tree tobacco. *J. Ind. Eng. Chem.*, **43** (6), 1402–1403.
6. Numerous reports, the most recent: Giera, D.S., Sickert, M. and Schneider, C. (2009) A straightforward synthesis of (S)-anabasine via the catalytic, enantioselective vinylogous Mukaiyama-Mannich reaction. *Synthesis*, **22**, 3797–3802.
7. Torres, Y.R., Berlinck, R.G.S., Magalhães, A., Schefer, A.B., Ferreira, A.G., Hajdu, E. and Muricy, G. (2000) Arenosclerins A-C and haliclonacyclamine E, new tetracyclic alkaloids from a Brazilian endemic Haplosclerid sponge *Arenosclera brasiliensis*. *J. Nat. Prod.*, **63** (8), 1098–1105.
8. McInnes, A.G., Smith, D.G., Wright, J.L.C. and Vining, L.C. (1977) Caerulomycins B and C, new 2,2′-dipyridyl derivatives from Streptomyces. *Can. J. Chem.*, **55** (24), 4159–4165.
9. Sammakia, T., Stangeland, E.L. and Whitcomb, M.C. (2002) Total synthesis of caerulomycin C via the halogen dance reaction. *Org. Lett.*, **4** (14), 2385–2388.
10. Wertheim, T. (1856) Ueber ein neues Alkaloïd in *Conium maculatum*. *Liebigs Ann. Chem.*, **100** (3), 328–339.
11. (a) Etxebarria, J., Vicario, J.L., Badia, D. and Carrillo, L. (2007) A general and enantiodivergent method for the asymmetric synthesis of piperidine alkaloids: concise synthesis of (R)-pipecoline, (S)-coniine and other 2-alkylpiperidines. *Tetrahedron*, **63** (46), 11421–11428; (b) Jo, E., Na, Y. and Chang, S. (1999) A highly efficient synthesis of (S)-(+)-N-Boc-coniine using ring-closing olefin metathesis (RCM). *Tetrahedron Lett.*, **40** (30), 5581–5582; (c) Bois, F., Gardette, D. and Gramain, J.-C. (2000) A new asymmetric synthesis of (S)-(+)-pipecoline and (S)-(+)- and (R)-(−)-coniine by reductive photocyclization of dienamides. *Tetrahedron Lett.*, **41** (45), 8769–8772; (d) Pachamuthu, K. and Vankar, Y.D. (2001) Synthesis of (−)-coniine and (−)-pipecoline using ruthenium catalyzed ring closing metathesis. *J. Organomet. Chem.*, **624** (1–2), 359–363.
12. Nicholas, G.M., Blunt, J.W. and Munro, M.H.G. (2001) Cortamidine oxide, a novel disulfide metabolite from the New Zealand Basidiomycete (mushroom) *Cortinarius* species. *J. Nat. Prod.*, **64** (3), 341–344.
13. Nicholas, G.M. and Molinski, T.F. (2000) Structures of cribochalines A and

B, branched-chain methoxylaminoalkyl pyridines from the Micronesian sponge, *Cribochalina* sp. absolute configuration and enantiomeric purity of related O-methyl oximes. *Tetrahedron*, **56** (19), 2921–2927.

14 Teruya, T., Kobayashi, K., Suenaga, K. and Kigoshi, H. (2006) Cyclohaliclonamines A-E: dimeric, trimeric, tetrameric, pentameric, and hexameric 3-alkyl pyridinium alkaloids from a marine sponge *Haliclona* sp. *J. Nat. Prod.*, **69** (1), 135–137.

15 (a) Fusetani, N., Asai, N., Matsunaga, S., Honda, K. and Yasumuro, K. (1994) Cyclostellettamines A-F, pyridine alkaloids which inhibit binding of methyl quinuclidinyl benzilate (QNB) to muscarinic acetylcholine receptors, from the marine sponge, *Stelletta maxima*. *Tetrahedron Lett.*, **35** (26), 3967–3970; (b) de Oliveira, J.H.H.L., Grube, A., Köck, M., Berlinck, R.G.S., Macedo, M.L., Ferreira, A.G. and Hajdu, E. (2004) Ingenamine G and cyclostellettamines G-I, K, and L from the new Brazilian species of marine sponge *Pachychalina* sp. *J. Nat. Prod.*, **67** (10), 1685–1689.

16 (a) Pérez-Balado, C., Nebbioso, A., Rodríguez-Graña, P., Minichiello, A., Miceli, M., Altucci, L. and de Lera, A.R. (2007) Bispyridinium dienes: histone deacetylase inhibitors with selective activities. *J. Med. Chem.*, **50** (10), 2497–2505; (b) Grube, A., Timm, C. and Köck, M. (2006) Synthesis and mass spectrometric analysis of cyclostellettamines H, I, K and L. *Eur. J. Org. Chem.*, **2006** (5), 1285–1295; (c) Baldwin, J.E., Spring, D.R., Atkinson, C.E. and Lee, V. (1998) Efficient synthesis of the sponge alkaloids cyclostellettamines A-F. *Tetrahedron*, **54** (44), 13655–13680.

17 Partheil, A. (1894) Zur Frage der Identität von Cytisin und Ulexin. *Arch. Pharm.*, **232** (6), 486–488.

18 (a) Stead, D. and O'Brien, P. (2007) Total synthesis of the lupin alkaloid cytisine: comparison of synthetic strategies and routes. *Tetrahedron*, **63** (9), 1885–1897; (b) O'Neill, B.T., Yohannes, D., Bundesmann, M.W. and Arnold, E.P. (2000) Total synthesis of (±)-cytisine. *Org. Lett.*, **2** (26), 4201–4204.

19 Spande, T.F., Garraffo, H.M., Edwards, M.W., Yeh, H.J.C., Pannell, L. and Daly, J.W. (1992) Epibatidine: a novel (chloropyridyl)azabicycloheptane with potent analgesic activity from an Ecuadoran poison frog. *J. Am. Chem. Soc.*, **114** (9), 3475–3478.

20 Armstrong, A., Bhonoah, Y. and Shanahan, S.E. (2007) Aza-Prins-pinacol approach to 7-azabicyclo[2.2.1]heptanes: syntheses of (±)-epibatidine and (±)-epiboxidine. *J. Org. Chem.*, **72** (21), 8019–8024.

21 Aida, W., Ohtsuki, T., Li, X. and Ishibashi, M. (2009) Isolation of new carbamate- or pyridine-containing natural products, fuzanins A, B, C, and D from *Kitasatospora* sp. IFM10917. *Tetrahedron*, **65** (1), 369–373.

22 Tsukamoto, S., Takahashi, M., Matsunaga, S., Fusetani, N. and van Soest, R.W.M. (2000) Hachijodines A-G: seven new cytotoxic 3-alkylpyridine alkaloids from two marine sponges of the genera *Xestospongia* and *Amphimedon*. *J. Nat. Prod.*, **63** (5), 682–684.

23 (a) Jain, S.C., Kumar, R., Goswami, R., Pandey, M.K., Khurana, S., Rohatgi, L. and Gyanda, K. (2005) Synthesis of novel non-isoprenoid phenolic acids and 3-alkylpyridines. *Pure Appl. Chem.*, **77** (1), 185–193; (b) Romeril, S.P., Lee, V. and Baldwin, J.E. (2004) Synthesis of marine sponge alkaloid hachijodine B and a comment on the structure of ikimine B and on the absolute configuration of niphatesine D. *Tetrahedron Lett.*, **45** (16), 3273–3277; (c) Goundry, W.R.F., Baldwin, J.E. and Lee, V. (2003) Total synthesis of cytotoxic sponge alkaloids hachijodines F and G. *Tetrahedron*, **59** (10), 1719–1729; (d) Goundry, W.R.F., Lee, V. and Baldwin, J.E. (2002) Total synthesis of cytotoxic sponge alkaloids hachijodines F and G. *Tetrahedron Lett.*, **43** (15), 2745–2747.

24 (a) Fusetani, N., Yasumuro, K., Matsunaga, S. and Hirota, H. (1989) Haliclamines A and B, cytotoxic macrocyclic alkaloids from a sponge of

the genus *Haliclona. Tetrahedron Lett.,* **30** (49), 6891–6894; (b) Volk, C.A., Lippert, H., Lichte, E. and Köck, M. (2004) Two new haliclamines from the arctic sponge *Haliclona viscosa. Eur. J. Org. Chem.,* **2004** (14), 3154–3158; (c) Schmidt, G., Timm, C. and Köck, M. (2009) New haliclamines E and F from the Arctic sponge *Haliclona viscosa. Org. Biomol. Chem.,* **7** (15), 3061–3064.

25 (a) Michelliza, S., Al-Mourabit, A., Gateau-Olesker, A. and Marazano, C. (2002) Synthesis of the cytotoxic sponge metabolite haliclamine A. *J. Org. Chem.,* **67** (18), 6474–6478; (b) Baldwin, J.E., James, D.A. and Lee, V. (2000) Preparation of 3-alkylpyridines. Formal total synthesis of haliclamines A and B. *Tetrahedron Lett.,* **41** (5), 733–736; (c) Morimoto, Y., Yokoe, C., Kurihara, H. and Kinoshita, T. (1998) Total syntheses of macrocyclic marine alkaloids, haliclamines A and B: a convenient and expeditious assembly of 3-substituted pyridine derivatives with different alkyl chains to the bispyridinium macrocycle. *Tetrahedron,* **54** (40), 12197–12214.

26 (a) Charan, R.D., Garson, M.J., Brereton, I.M., Willis, A.C. and Hooper, J.N.A. (1996) Haliclonacyclamines A and B, cytotoxic alkaloids from the tropical marine sponge *Haliclona* sp. *Tetrahedron,* **52** (27), 9111–9120; (b) De Oliveira, J.H.H.L., Nascimento, A.M., Kossuga, M.H., Cavalcanti, B.C., Pessoa, C.O., Moraes, M.O., Macedo, M.L., Ferreira, A.G., Hajdu, E., Pinheiro, U.S. and Berlinck, R.G.S. (2007) Cytotoxic alkylpiperidine alkaloids from the Brazilian marine sponge *Pachychalina alcaloidifera. J. Nat. Prod.,* **70** (4), 538–543; (c) Arai, M., Ishida, S., Setiawan, A. and Kobayashi, M. (2009) Haliclonacyclamines, tetracyclic alkylpiperidine alkaloids, as anti-dormant mycobacterial substances from a marine sponge of *Haliclona* sp. *Chem. Pharm. Bull.,* **57** (10), 1136–1138.

27 Cimino, G., Passeggio, A., Sodano, G., Spinella, A. and Villani, G. (1991) Alarm pheromones from the Mediterranean opisthobranch *Haminoea navicula. Experientia,* **47** (1), 61–63.

28 (a) Solladié, G., Somny, F. and Colobert, F. (1997) Enantioselective synthesis of haminol-1, an alarm pheromone of a Mediterranean mollusk. *Tetrahedron Asymmetry,* **8** (5), 801–810; (b) Matikainen, J., Kaltia, S., Hase, T. and Kuronen, P. (1995) The synthesis of haminols A and B. *J. Nat. Prod.,* **58** (10), 1622–1624.

29 Hayman, A.R. and Gray, D.O. (1989) Hydroxynorcytisine, a quinolizidone alkaloid from *Laburnum anagyroides. Phytochemistry,* **28** (2), 673–675.

30 Yohannes, D., Hansen, C.P., Akireddy, S.R., Hauser, T.A., Kiser, M.N., Gurnon, N.J., Day, C.S., Bhatti, B. and Caldwell, W.S. (2008) First total synthesis of (±)-3-hydroxy-11-norcytisine: structure confirmation and biological characterization. *Org. Lett.,* **10** (23), 5353–5356.

31 Carroll, A.R. and Scheuer, P.J. (1990) Four β-alkylpyridines from a sponge. *Tetrahedron,* **46** (19), 6637–6644.

32 (a) Krauss, J., Wetzel, I. and Bracher, F. (2004) A new approach towards ikimine A analogues. *Nat. Prod. Res.,* **18** (5), 397–401; (b) Wang, Y., Dong, X. and Larock, R.C. (2003) Synthesis of naturally occurring pyridine alkaloids via palladium-catalyzed coupling/migration chemistry. *J. Org. Chem.,* **68** (8), 3090–3098.

33 Surup, F., Wagner, O., von Frieling, J., Schleicher, M., Oess, S., Müller, P. and Grond, S. (2007) The iromycins, a new family of pyridone metabolites from *Streptomyces* sp. I. Structure, NOS inhibitory activity, and biosynthesis. *J. Org. Chem.,* **72** (14), 5085–5090.

34 Shojaei, H., Li-Böhmer, Z. and von Zezschwitz, P. (2007) Iromycins: a new family of pyridone metabolites from *Streptomyces* sp. II. Convergent total synthesis. *J. Org. Chem.,* **72** (14), 5091–5097.

35 Maximo, P. and Lourenco, A. (2000) New quinolizidine alkaloids from *Ulex jussiaei. J. Nat. Prod.,* **63** (2), 201–204.

36 Honda, T., Takahashi, R. and Namiki, H. (2005) Syntheses of (+)-cytisine, (−)-kuraramine, (−)-isokuraramine, and (−)-jussiaeiine A. *J. Org. Chem.,* **70** (2), 499–504.

37 Murakoshi, I., Kidoguchi, E., Haginiwa, J., Ohmiya, S., Higashiyama, K. and Otomasu, H. (1981) (+)-Kuraramine, a possible metabolite of (−)-N-methylcystisine in flowers of *Sophora flavescens*. *Phytochemistry*, **20** (6), 1407–1409.

38 Cheng, X.-F., Meng, Z.-M. and Chen, Z.-L. (1998) A pyridine-type alkaloid from *Mallotus apelta*. *Phytochemistry*, **49** (7), 2193–2194.

39 O'Donnell, G., Poeschl, R., Zimhony, O., Gunaratnam, M., Moreira, J.B.C., Neidle, S., Evangelopoulos, D., Bhakta, S., Malkinson, J.P., Boshoff, H.I., Lenaerts, A. and Gibbons, S. (2009) Bioactive pyridine-N-oxide disulfides from *Allium stipitatum*. *J. Nat. Prod.*, **72** (3), 360–365.

40 Alam, N., Hong, J., Lee, C.O., Im, K.S., Son, B.W., Choi, J.S., Choi, W.C. and Jung, J.H. (2001) Montipyridine, a new pyridinium alkaloid from the stony coral *Montipora* species. *J. Nat. Prod.*, **64** (7), 956–957.

41 Fürstner, A., Leitner, A., Mendez, M. and Krause, H. (2002) Iron-catalyzed cross-coupling reactions. *J. Am. Chem. Soc.*, **124** (46), 13856–13863.

42 Schinz, H., Ruzicka, L., Geyer, U. and Prelog, V. (1946) Muscopyridin, eine Base, $C_{16}H_{25}N$ aus natürlichem Moschus. *Helv. Chim. Acta*, **29** (6), 1524–1528.

43 (a) Suwa, K., Morie, Y., Suzuki, Y., Ikeda, K. and Sato, M. (2008) A highly efficient total synthesis of (R)-(+)-muscopyridine by intramolecular [4 + 2] cycloaddition of bisketene. *Tetrahedron Lett.*, **49** (9), 1510–1513; (b) Fürstner, A. and Leitner, A. (2003) A catalytic approach to (R)-(+)-muscopyridine with integrated 'self-clearance'. *Angew. Chem. Int. Ed.*, **42** (3), 308–311; (c) Hagiwara, H., Katsumi, T., Kamat, V.P., Hoshi, T., Suzuki, T. and Ando, M. (2000) Application of ring closing metathesis to the first total synthesis of (R)-(+)-muscopyridine: determination of absolute stereochemistry. *J. Org. Chem.*, **65** (21), 7231–7234.

44 (a) Kubota, T., Nishi, T., Fukushi, E., Kawabata, J., Fromont, J. and Kobayashi, J. (2007) Nakinadine A, a novel bis-pyridine alkaloid with a β-amino acid moiety from sponge *Amphimedon* sp. *Tetrahedron Lett.*, **48** (29), 4983–4985; (b) Nishi, T., Kubota, T., Fromont, J., Sasaki, T. and Kobayashi, J. (2008) Nakinadines B–F: new pyridine alkaloids with a β-amino acid moiety from sponge *Amphimedon* sp. *Tetrahedron*, **64** (14), 3127–3132.

45 Sleeper, H.L. and Fenical, W. (1977) Navenones A-C: trail-breaking alarm pheromones from the marine opisthobranch *Navanax inermis*. *J. Am. Chem. Soc.*, **99** (7), 2367–2368.

46 Many reports, the most recent: BouzBouz, S., Roche, C. and Cossy, J. (2009) Simple synthesis of conjugated all-(E)-polyenic aldehydes, ketones, and esters using chemoselective cross-metathesis and an iterative sequence of reactions: application to the synthesis of navenone B. *Synlett*, **5**, 803–807.

47 Kem, W.R., Scott, K.N. and Duncan, J.H. (1976) Hoplonemertine worms – a new source of pyridine neurotoxins. *Experientia*, **32** (6), 684–686.

48 (a) Bouillon, A., Voisin, A.S., Robic, A., Lancelot, J.-C., Collot, V. and Rault, S. (2003) An efficient two-step total synthesis of the quaterpyridine nemertelline. *J. Org. Chem.*, **68** (26), 10178–10180; (b) Zoltewicz, J.A. and Cruskie, M.P., Jr (1995) Total synthesis of the incorrectly proposed quaterpyridine isolated from the hoplonemertine sea worm. *Tetrahedron*, **51** (42), 11401–11410.

49 Posselt, W. and Reimann, L. (1828) Chemische Untersuchungen des Tabaks und Darstellung des eigenthumlichen wirksamen Princips dieser Pflanze. *Geigers Mag. Pharm.*, **24**, 138–161.

50 Kobayashi, J., Murayama, T., Kosuge, S., Kanda, F., Ishibashi, M., Kobayashi, H., Ohizumi, Y., Ohta, T., Nozoe, S. and Sasaki, T. (1990) Niphatesines A-D, new antineoplastic pyridine alkaloids from the Okinawan marine sponge *Niphates* sp. *J. Chem. Soc. Perkin Trans. I*, (12), 3301–3303.

51 Krauss, J. and Bracher, F. (2004) New total synthesis of niphatesine C and norniphatesine C based on a

Sonogashira reaction. *Arch. Pharm.*, **337** (7), 371–375.

52 Talpir, R., Rudi, A., Ilan, M. and Kashman, Y. (1992) Niphatoxin A and B; two new ichthyo- and cytotoxic tripyridine alkaloids from a marine sponge. *Tetrahedron Lett.*, **33** (21), 3033–3034.

53 Kaiser, A., Marazano, C. and Maier, M. (1999) First synthesis of marine sponge alkaloid niphatoxin B. *J. Org. Chem.*, **64** (10), 3778–3782.

54 Quinoa, E. and Crews, P. (1987) Niphatynes, methoxylamine pyridines from the marine sponge, *Niphates* sp. *Tetrahedron Lett.*, **28** (22), 2467–2468.

55 Laville, R., Genta-Jouve, G., Urda, C., Fernandez, R., Thomas, O.P., Reyes, F. and Amade, P. (2009) Njaoaminiums A, B, and C: cyclic 3-alkylpyridinium salts from the marine sponge *Reniera* sp. *Molecules*, **14** (11), 4716–4724.

56 Tokuyama, T. and Daly, J.W. (1983) Steroidal alkaloids (batrachotoxins and 4β-hydroxybatrachotoxins), 'indole alkaloids' (calycanthine and chimonanthine) and a piperidinyldipyridine alkaloid (noranabasamine) in skin extracts from the Colombian poison-dart frog *Phyllobates terribilis* (Dendrobatidae). *Tetrahedron*, **39** (1), 41–47.

57 Miao, L., DiMaggio, S.C., Shu, H. and Trudell, M.L. (2009) Enantioselective syntheses of both enantiomers of noranabasamine. *Org. Lett.*, **11** (7), 1579–1582.

58 Huang, X., Ha, C., Yang, R., Jiang, H., Hu, Y. and Zhang, Y. (2007) A new alkaloid from *Lysimachia patungensis*. *Chem. Nat. Compd.*, **43** (2), 170–172.

59 Shin, C., Lim, H., Moon, S., Kim, S., Yong, Y., Kim, B.-J., Lee, C.-H. and Lim, Y. (2006) A novel antiproliferative agent, phenylpyridineylbutenol, isolated from *Streptomyces* sp. *Bioorg. Med. Chem. Lett.*, **16** (21), 5643–5645.

60 Teruya, T., Kobayashi, K., Suenaga, K. and Kigoshi, H. (2005) Phormidinines A and B, novel 2-alkylpyridine alkaloids from the cyanobacterium *Phormidium* sp. *Tetrahedron Lett.*, **46** (23), 4001–4003.

61 Tamura, S., Takahashi, N., Miyamoto, S., Mori, R., Suzuki, S. and Nagatsu, J. (1963) Isolation and physiological activities of piericidin A, a natural insecticide produced by Streptomyces. *Agric. Biol. Chem.*, **27** (8), 576–582.

62 (a) Schnermann, M.J., Romero, F.A., Hwang, I., Nakamaru-Ogiso, E., Yagi, T. and Boger, D.L. (2006) Total synthesis of piericidin A1 and B1 and key analogues. *J. Am. Chem. Soc.*, **128** (36), 11799–11807; (b) Kikuchi, R., Fujii, M. and Akita, H. (2009) Total synthesis of (+)-piericidin A1 and (−)-piericidin B1. *Tetrahedron Asymmetry*, **20** (17), 1975–1983.

63 Tallent, W.H., Stromberg, V.L. and Horning, E.C. (1955) *Pinus* alkaloids. The alkaloids of *Pinus sabiniana* Dougl. and related species. *J. Am. Chem. Soc.*, **77** (23), 6361–6364.

64 (a) Kariya, Y., Kubota, T., Fromont, J. and Kobayashi, J. (2006) Pyrinadine A, a novel pyridine alkaloid with an azoxy moiety from sponge *Cribrochalina* sp. *Tetrahedron Lett.*, **47** (6), 997–998; (b) Kariya, Y., Kubota, T., Fromont, J. and Kobayashi, J. (2006) Pyrinadines B–G, new bis-pyridine alkaloids with an azoxy moiety from sponge *Cribrochalina* sp. *Bioorg. Med. Chem.*, **14** (24), 8415–8419.

65 Anwar, A.M. and Lee, V. (2009) A short biomimetic synthesis of marine sponge alkaloid pyrinadine A. *Tetrahedron*, **65** (29–30), 5834–5837.

66 Coval, S.J. and Scheuer, P.J. (1985) An intriguing C16-alkadienone-substituted 2-pyridine from a marine mollusk. *J. Org. Chem.*, **50** (16), 3024–3025.

67 Matikainen, J., Kaltia, S., Hase, T., Kilpeläinen, I., Drakenberg, T. and Annila, A. (1993) Semipreparative synthesis, carbon-13- and 2D-NMR of pulo'upone. *Tetrahedron*, **49** (36), 8007–8014, and other reports.

68 (a) Tsuda, M., Hirano, K., Kubota, T. and Kobayashi, J. (1999) Pyrinodemin A, a cytotoxic pyridine alkaloid with an isoxazolidine moiety from sponge *Amphimedon* sp. *Tetrahedron Lett.*, **40** (26), 4819–4820; (b) Hirano, K., Kubota, T., Tsuda, M., Mikami, Y. and Kobayashi, J. (2000) Pyrinodemins B–D, potent cytotoxic bis-pyridine alkaloids

from marine sponge *Amphimedon* sp. *Chem. Pharm. Bull.*, **48** (7), 974–977.
69 (a) Pouilhès, A., Amado, A.F., Vidal, A., Langlois, Y. and Kouklovsky, C. (2008) Enantioselective total synthesis of pyrinodemin A. *Org. Biomol. Chem.*, **6** (8), 1502–1510; (b) Ishiyama, H., Tsuda, M., Endo, T. and Kobayashi, J. (2005) Asymmetric synthesis of double bond isomers of the structure proposed for pyrinodemin A and indication of its structural revision. *Molecules*, **10** (1), 312–316; (c) Romeril, S.P., Lee, V., Baldwin, J.E. and Claridge, T.D.W. (2005) On the synthesis of pyrinodemin A. Part 1. The location of the olefin. *Tetrahedron*, **61** (5), 1127–1140; (d) Morimoto, Y., Kitao, S., Okita, T. and Shoji, T. (2003) Total synthesis and assignment of the double bond position and absolute configuration of (-)-pyrinodemin A. *Org. Lett.*, **5** (15), 2611–2614; (e) Romeril, S.P., Lee, V., Claridge, T.D.W. and Baldwin, J.E. (2002) Synthesis of possible structure of pyrinodemin A. *Tetrahedron Lett.*, **43** (2), 327–329; (f) Snider, B.B. and Shi, B. (2001) Synthesis of pyrinodemins A and B. Assignment of the double bond position of pyrinodemin A. *Tetrahedron Lett.*, **42** (9), 1639–1642.
70 Campagnuolo, C., Fattorusso, C., Fattorusso, E., Ianaro, A., Pisano, B. and Taglialatela-Scafati, O. (2003) Simplakidine A, a unique pyridinium alkaloid from the Caribbean sponge *Plakortis simplex*. *Org. Lett.*, **5** (5), 673–676.
71 Fattorusso, E., Romano, A., Scala, F. and Taglialatela-Scafati, O. (2008) Simplexidine, a 4-alkylpyridinium alkaloid from the Caribbean sponge *Plakortis simplex*. *Molecules*, **13** (7), 1465–1471.
72 MacConnell, J.G., Blum, M.S. and Fales, H.M. (1970) Alkaloid from fire ant venom: identification and synthesis. *Science*, **168** (3933), 840–841.
73 (a) Leijondahl, K., Boren, L., Braun, R. and Bäckvall, J.-E. (2009) Enzyme- and ruthenium-catalyzed dynamic kinetic asymmetric transformation of 1,5-diols. Application to the synthesis of (+)-solenopsin A. *J. Org. Chem.*, **74** (5),

1988–1993; (b) Herath, H.M.T.B. and Nanayakkara, N.P.D. (2008) Synthesis of enantiomerically pure fire ant venom alkaloids: solenopsins and isosolenopsins A, B and C. *J. Heterocycl. Chem.*, **45** (1), 129–136; (c) Singh, O.V. and Han, H. (2004) Tandem Overman rearrangement and intramolecular amidomercuration reactions. stereocontrolled synthesis of *cis*- and *trans*-2,6-dialkylpiperidines. *Org. Lett.*, **6** (18), 3067–3070.
74 De Marino, S., Iorizzi, M., Zollo, F., Debitus, C., Menou, J.-L., Ospina, L.F., Alcaraz, M.J. and Payá, M. (2000) New pyridinium alkaloids from a marine sponge of the Genus *Spongia* with a human phospholipase A2 inhibitor profile. *J. Nat. Prod.*, **63** (3), 322–326.
75 Gonzalez, M.A. and Molina-Navarro, S. (2007) Attempted synthesis of spongidines by a radical cascade terminating onto a pyridine ring. *J. Org. Chem.*, **72** (19), 7462–7465.
76 Schildknecht, H., Krauß, D., Connert, J., Essenbreis, H. and Orfanides, N. (1975) Das Spreitungsalkaloid Stenusin aus dem Kurzflügler *Stenus comma* (Coleoptera: Staphylinidae). *Angew. Chem.*, **87** (11), 421–422, DOI: 10.1002/ange.19750871115.
77 Gedig, T., Dettner, K. and Seifert, K. (2007) Short synthesis of stenusine and norstenusine, two spreading alkaloids from *Stenus* beetles (Coleoptera: Staphylinidae). *Tetrahedron*, **63** (12), 2670–2674.
78 Rao, K.V. and Cullen, W.P. (1959–1960) Streptonigrin, an antitumor substance. I. Isolation and characterization. *Antibiot. Annu.*, **7**, 950–953.
79 McElroy, W.T. and DeShong, P. (2006) Synthesis of the CD-ring of the anticancer agent streptonigrin: studies of aryl–aryl coupling methodologies. *Tetrahedron*, **62** (29), 6945–6954.
80 Kobayashi, J., Murayama, T., Ohizumi, Y., Sasaki, T., Ohta, T. and Nozoe, S. (1989) Theonelladins A–D, novel antineoplastic pyridine alkaloids from the Okinawan marine sponge *Theonella swinhoei*. *Tetrahedron Lett.*, **30** (36), 4833–4836.

81 (a) Shorey, B.J., Lee, V. and Baldwin, J.E. (2007) Synthesis of the Arctic sponge alkaloid viscosaline and the marine sponge alkaloid theonelladin C. *Tetrahedron*, **63** (25), 5587–5592; (b) Tsunoda, T., Uemoto, K., Ohtani, T., Kaku, H. and Ito, S. (1999) Arylmethyl phenyl sulfones, a new carbon nucleophile for Mitsunobu-type alkylation. *Tetrahedron Lett.*, **40** (41), 7359–7362, and other reports.

82 Wang, G.Y.S., Kuramoto, M., Uemura, D., Yamada, A., Yamaguchi, K. and Yazawa, K. (1996) Three novel antimicrofouling nitroalkyl pyridine alkaloids from the Okinawan marine sponge *Callyspongia* sp. *Tetrahedron Lett.*, **37** (11), 1813–1816.

83 Volk, C.A. and Köck, M. (2004) Viscosaline: new 3-alkyl pyridinium alkaloid from the Arctic sponge *Haliclona viscosa*. *Org. Biomol. Chem.*, **2** (13), 1827–1830.

84 Volk, C.A. and Köck, M. (2003) Viscosamine: the first naturally occurring trimeric 3-alkyl pyridinium alkaloid. *Org. Lett.*, **5** (20), 3567–3569.

85 Timm, C. and Köck, M. (2006) First total synthesis of the marine natural product viscosamine. *Synthesis*, **15**, 2580–2584.

86 Sakemi, S., Totton, L.E. and Sun, H.H. (1990) Xestamines A, B, and C, three new long-chain methoxylamine pyridines from the sponge *Xestospongia wiedenmayeri*. *J. Nat. Prod.*, **53** (4), 995–999.

87 Corbet, M., de Greef, M. and Zard, S.Z. (2008) A highly conjunctive β-keto phosphonate: application to the synthesis of pyridine alkaloids xestamines C, E, and H. *Org. Lett.*, **10** (2), 253–256.

88 Tilyabaev, Z. and Abduvakhabov, A.A. (1998) Alkaloids of *Anabasis aphylla* and their cholinergic activities. *Chem. Nat. Compd.*, **34** (3), 295–297.

89 Cruskie, M.P. Jr, Zoltewicz, J.A. and Abboud, K.A. (1995) Revised structure and convergent synthesis of nemertelline, the neurotoxic quaterpyridine isolated from the Hoplonemertine sea worm. *J. Org. Chem.*, **60** (23), 7491–7495.

90 Keaton, K.A. and Phillips, A.J. (2006) Titanium(II)-mediated cyclizations of (silyloxy)enynes: a total synthesis of (−)-7-demethylpiericidin A1. *J. Am. Chem. Soc.*, **128** (2), 408–409.

91 Turk, T., Sepčić, K., Mancini, I. and Guella, G. (2008) 3-Alkylpyridinium and 3-alkylpyridine compounds from marine sponges, their synthesis, biological activities and potential use, in *Bioactive natural products (Part O)*, Studies in Natural Products Chemistry, vol. 35 (ed. A. Rahman), Elsevier Science, pp. 355–397.

92 Sepčić, K. (2000) Bioactive alkylpyridinium compounds from marine sponges. *J. Toxicol. Toxin Rev.*, **19** (2), 139–160.

93 Baldwin, J.E., Romeril, S.P., Lee, V. and Claridge, T.D.W. (2001) Studies towards the total synthesis of the cytotoxic sponge alkaloid pyrinodemin A. *Org. Lett.*, **3** (8), 1145–1148.

94 Berlinck, R.G.S. (2007) Polycyclic diamine alkaloids of marine sponges, in *Topics in Heterocyclic Chemistry 10 – Bioactive Heterocycles IV* (eds M.T.H. Khan and R.R. Gupka), Springer-Verlag, Berlin, Heidelberg, Germany, pp. 211–238.

95 Casapullo, A., Pinto, O.C., Marzocco, S., Autore, G. and Riccio, R. (2009) 3-Alkylpyridinium alkaloids from the Pacific sponge *Haliclona* sp. *J. Nat. Prod.*, **72** (2), 301–303.

96 Albrizio, S., Ciminiello, P., Fattorusso, E. and Magno, S. (1995) Amphitoxin, a new high molecular weight antifeedant pyridinium salt from the Caribbean sponge *Amphimedon compressa*. *J. Nat. Prod.*, **58** (5), 647–652.

97 Sepčić, K., Guella, G., Mancini, I., Pietra, F., Dalla Serra, M., Menestrina, G., Tubbs, K., Maček, P. and Turk, T. (1997) Characterization of anticholinesterase-active 3-alkylpyridinium polymers from the marine sponge *Reniera sarai* in aqueous solutions. *J. Nat. Prod.*, **60** (10), 991–996.

98 Mancini, I., Sicurelli, A., Guella, G., Turk, T., Maček, P. and Sepčić, K. (2004) Synthesis and bioactivity of linear

oligomers related to polymeric alkylpyridinium metabolites from the Mediterranean sponge *Reniera sarai*. *Org. Biomol. Chem.*, **2** (9), 1368–1375.

99 Reynolds, T. (2005) Hemlock alkaloids from Socrates to poison aloes. *Phytochemistry*, **66** (12), 1399–1406.

100 Heineman, S.F. and Papke, R.L. (1994) Partial agonist properties of cytisine on neuronal nicotinic receptors containing the $\beta 2$ subunit. *Mol. Pharmacol.*, **45** (1), 142–149.

101 van Tamelen, E.E. and Baran, J.S. (1955) The synthesis of *dl*-cytisine. *J. Am. Chem. Soc.*, **77** (18), 4944–4945.

9
Quinolines and Isoquinolines
Antonio Garrido Montalban

9.1
Introduction

Quinoline **1** and isoquinoline **2** (Figure 9.1) have been known for a long time. Both were originally isolated from coal tar, the former in 1834 and the latter in 1885 [1]. Quinoline is a high-boiling liquid and used as a solvent in organic synthesis. Isoquinoline, on the other hand, is a low melting solid. These isomeric heterocyclic systems are moderately basic (pK_a quinoline = 4.9, pK_a isoquinoline = 5.1) [2] and are structural components of many natural products of which multiple have medicinal value. A compilation of a selection of such natural products and structural types is shown in Table 9.1. The alkaloid quinine **42** (from which the name quinoline is derived), for example, has been known and used for centuries in the treatment of malaria. Several synthetic antimalarial drugs, for example, chloroquine **3**, (Figure 9.1), have been derived thereof. The synthetic tetraethyl homolog **4** of papaverine **36** has, in turn, antispasmodic properties while a related synthetic neuromuscular blocking agent to tubocurarin **52** is mivacurium chloride **5** (Figure 9.1).

In a large number of alkaloids containing the isoquinoline skeleton, the heterocyclic ring system occurs mainly at the 1,2,3,4-tetrahydro-level. Morphine **29** is an example of such an alkaloid. Besides the abundance of these two chemical cores in nature, synthetic analogs that have found applications in fields other than medicine are, for example, the cyanine dyes (e.g., ethyl red **6**, Figure 9.1), which provided the first photographic film sensitizers [2, 3].

The overwhelming majority of quinoline and isoquinoline alkaloids are of flowering plant origin although they have also been isolated from microbial sources and animals [4, 5]. Biologically, quinoline is derived from the amino acid tryptophan **7** whereas the isoquinoline nucleus is derived from tyrosine **8** (Figure 9.2) [6]. The biogenetic link of the former with the indole-containing family of alkaloids was discussed by Woodward as early as 1955 and led to the concept of the Woodward fission [7]. The isoquinoline biosynthetic pathway has recently been reconstructed in an engineered microbial host [6b].

Synthetically, many methods for the construction of both ring systems have been developed [1–3]. It is, however, the complexity of the natural products in

Heterocycles in Natural Product Synthesis, First Edition. Edited by Krishna C. Majumdar and Shital K. Chattopadhyay.
© 2011 Wiley-VCH Verlag GmbH & Co. KGaA. Published 2011 by Wiley-VCH Verlag GmbH & Co. KGaA.

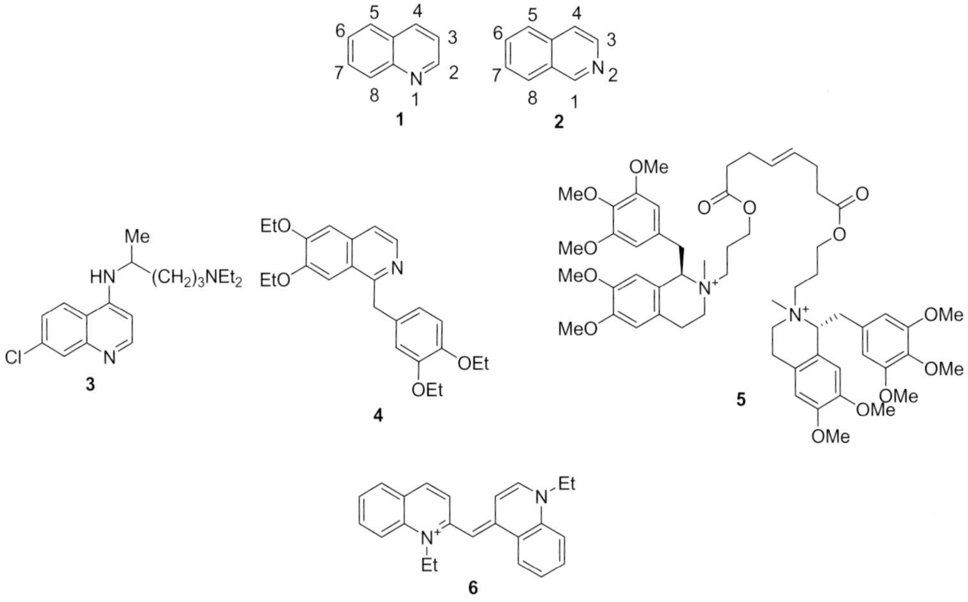

Figure 9.1 Quinoline, isoquinoline and compounds derived thereof.

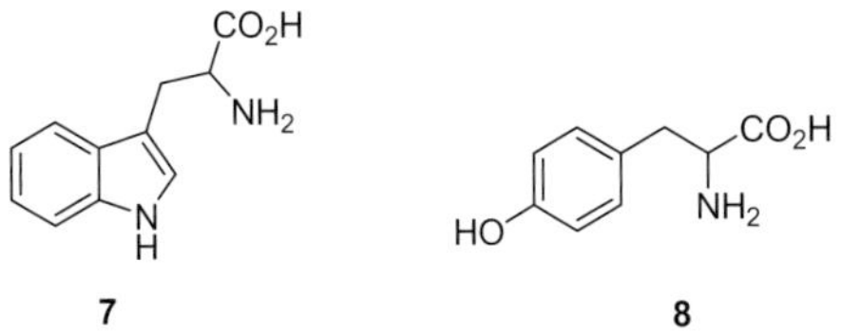

Figure 9.2 Biochemical precursors of quinoline and isoquinoline.

which these substructures appear, that has inspired and fueled the imagination of organic synthetic chemists for decades if not centuries. Quinine **42**, for example, became a Holy Grail of organic synthesis, since William Perkin's failure to prepare it in 1856 [8]. Thus, the application of these two heterocyclic cores in the synthesis of natural products is best illustrated by discussing the synthetic challenges encountered in such endeavors and will be the subject of Section 9.2.

9.2
Application of Quinolines and Isoquinolines in the Synthesis of Natural Products

The use of straightforward quinoline or isoquinoline precursors as starting points for the construction of compounds containing these heterocyclic systems is gener-

9.2 Application of Quinolines and Isoquinolines in the Synthesis of Natural Products | 301

Table 9.1 Natural products containing quinolines and isoquinolines.

Serial No.	Trivial name	Structure	Source	Isolation [Ref]	Biological activity	Synthesis [Ref]
9	Argemonine		*Argemone* Species	[9]		[10]
10	Berberine		Berberidaceae (family of flowering plants)	[11]	Antiprotozoal Antimalarial Antibacterial Antidiarrheal	[12]
11	Berlambine		*Thalictrum foliolosum*	[13]		[14a]
12	Camptothecine		*Camptotheca acuminata*	[15]	Topoisomerase inhibitor (anti-cancer agent)	[16]
13	Canadine		Berberidaceae (family of flowering plants)	[17]		[12, 14, 18]
14	Cinchonine		*Cinchona* (tropical plant species belonging to the Rubiaceae family)	[19]	Antimalarial	

(*Continued*)

Table 9.1 (Continued)

Serial No.	Trivial name	Structure	Source	Isolation [Ref]	Biological activity	Synthesis [Ref]
15	Codeine		Opium		Analgesic Antitussive	[20]
16	Dictamnine		*Dictamnus albus*	[21]		[22]
17	Emetine		Rubiaceae (family of flowering plants)	[23, 24]	Emetic Antiamebic	[25–32]
18	Erysotramidine		*Erythrina genus*			[33]
19	Erythraline		*Erythrina genus*	[34]		[35]
20	Glaucine		*Glaucium genus*	[36]	Antitussive	[37]
21	Haemanthamine		*Narcissus confusus*			[38]

9.2 Application of Quinolines and Isoquinolines in the Synthesis of Natural Products | 303

22	Hydrastine		Hydrastis canadensis	Uterine hemostatic Antiseptic	[39]
23	Hydroquinine		Cinchona (tropical plant species belonging to the Rubiaceae family)	Depigmentor	[40] [41]
24	Jamtine		Cocculus hirsutus	Antihyperglycemic	[42]
25	Lavendamycin		Streptomyces lavendulae	Antitumor Antibiotic	[43] [44–47]
26	Lophocereine		Lophocereus schotti		[48] [49]
27	Lycorine		Lycoris radiata		[50]
28	Maritidine				[51]

(Continued)

Table 9.1 (Continued)

Serial No.	Trivial name	Structure	Source	Isolation [Ref]	Biological activity	Synthesis [Ref]
29	Morphine		Opium	[52]	Analgesic	[53–56]
30	Nandinine		Nandina domestica	[57]		[12, 58]
31	Nantenine		Nandina domestica		$\alpha 1$ adrenergic and 5-HT$_{2A}$ serotonin antagonist	[59]
32	Nitidine		Zanthoxylum nitidum	[60]	Topoisomerase inhibitor (anti-cancer agent)	[61, 62]
33	Noscapine (narcotine)		Opium (Papaver somniferum L. Papaveraceae)	[63]	Antitussive Anticancer	[64]
34	N-norlaudanosine			[65]		[66]

9.2 Application of Quinolines and Isoquinolines in the Synthesis of Natural Products | 305

35	Palmatine		*Jateorhiza palmate* (Calumba Root)	[67]	[68]
36	Papaverine		*Opium*	Vasodilator	[69]
37	Pellotine		*Lophophora williamsii* (pellote cactus)		[70]
38	Protoemetine				[26, 27]
39	Protoemetinol		*Alangium lamarckii*	Emetic Antiamebic Anticancer	[71] [26, 27, 72]
40	Psychotrine		*Uragoga ipecacuanha* (ipecac roots)	HIV-1 transcriptase inhibitor	[73]

(Continued)

Table 9.1 (Continued)

Serial No.	Trivial name	Structure	Source	Isolation [Ref]	Biological activity	Synthesis [Ref]
41	Quinidine		*Cinchona* (tropical plant species belonging to the Rubiaceae family)		Class I antiarrythmic Antimalarial	[74]
42	Quinine		*Cinchona* (tropical plant species belonging to the Rubiaceae family)	[75]	Antimalarial Muscle relaxant	[74, 76–80]
43	d-Quinotoxine					[77, 79, 81]
44	Reticuline		*Anona reticulate* Opium	[82]		[83]
45	Sandramycin		*Norcardioides* sp. (microbes)	[84]	Antitumour Antibiotic	[85]
46	Sanguinarine		*Sanguinaria* (bloodroot)	[86]	Antibacterial	[87]

	Name	Structure	Source	Ref.	Biological Activity	
47	Skimmianine		Skimmia japonica	[88]	Sedative Anticonvulsant	
48	Stepholidine		Stephania intermedia	[89]	D_1 receptor agonist D_2 receptor antagonist	
49	Streptonigrin		Streptomyces flocculus	[90]	Antitumour Antibiotic	[91]
50	Tetrahydro-palmatine		Corydalis family	[94]	Analgesic	[18, 68, 92b, 93]
51	Thalictricavine		Corydalis tuberosa	[94]		[14, 58]
52	Tubocurarin		Chondodendron Tomentosum (tropical plant species)	[95]	Neuromuscular blocking agent	

ally limited to simple natural products [4, 5]. Most of the synthetic methods leading to these simple natural products have been extensively covered in general heterocyclic chemistry textbooks and will, therefore, not be the subject of this chapter [1–3]. On the other hand, in the case of more complex natural products containing these cores, the quinoline or isoquinoline ring system is generally assembled from more advance intermediates later on in the synthetic sequence. Consequently, this section will introduce the reader to some notable syntheses of selected natural products covered in Table 9.1, all of which have fairly complex molecular architectures. When possible, emphasis will be given to the construction of the quinoline and isoquinoline cores.

9.2.1
Quinoline-Containing Natural Products

With no doubt, the most prominent of the quinoline-containing natural products is quinine **42**, to which Section 9.2.1.1 is devoted. The recent controversy [96] revolving around its first reported total synthesis and which lasted until 2008 [76], deserves to be put in historic perspective. This will be followed by referencing the synthesis of sandramycin **45** and detail discussion of elegant syntheses of lavendamycin **25**.

9.2.1.1 Quinine
The word quinine derives from *quina* which is Spanish for the bark of quinine-containing Cinchona species. Indeed, quinine **42** is the primary alkaloid of various species of *Cinchona* and was first isolated in 1820 by Pelletier and Caventau [75]. At the time of World War II, quinine **42** was the only truly effective antimalarial and it has until now outlasted most modern therapeutic agents which have, one by one, fallen to drug resistance. From an organic synthesis viewpoint, quinine **42** came into the spotlight when Woodward and Doering reported its first total synthesis in 1944 [77]. The synthesis by Woodward and Doering, however, would be classified as a "formal" total synthesis in today's jargon. "Formal", because it relied upon a 1918 report by Rabe and Kindler [78], who converted *d*-quinotoxine **43**, a quinoline-containing alkaloid in its own right, into quinine **42** *via* quininone (Scheme 9.1).

Another partial formal synthesis of quinine that relied on the same Rabe and Kindler report, was claimed by Proštenik and Prelog in 1943 [79]. Proštenik and Prelog prepared *d*-quinotoxine **43** through a Claisen-condensation of a protected ethyl-ester derivative **54** of homomeroquinene **53** with ethyl quininate **55** [97], after subsequent hydrolysis and decarboxyation (Scheme 9.2). Proštenik and Prelog obtained homomeroquinene **53** in enantiomerically pure form from degradation of the quinoline-containing natural product cinchonine **14**.

The genius of Woodward, however, was to recognize that isoquinoline has all it atoms arranged as they are in homomeroquinene **53** and that the missing carbon could be introduced by having a hydroxyl group at the 7-position of the

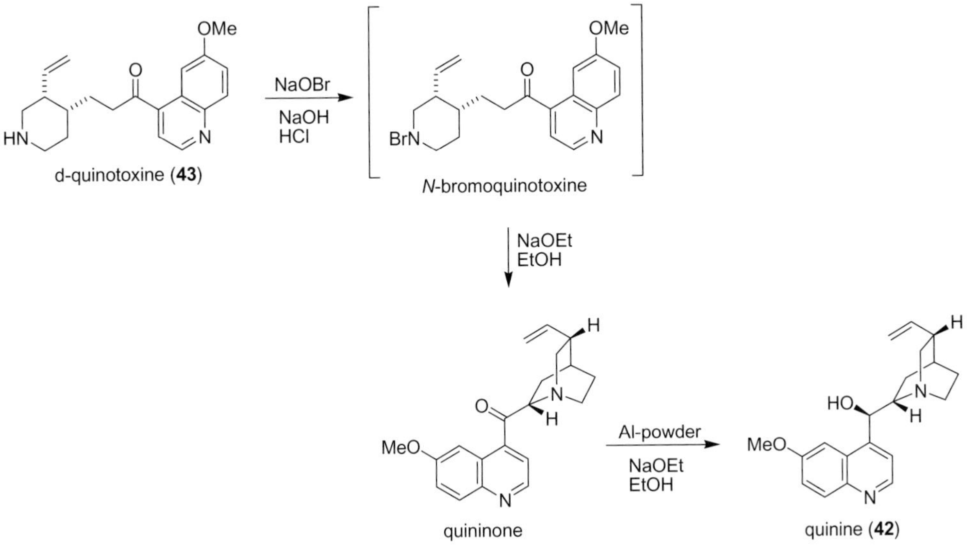

Scheme 9.1

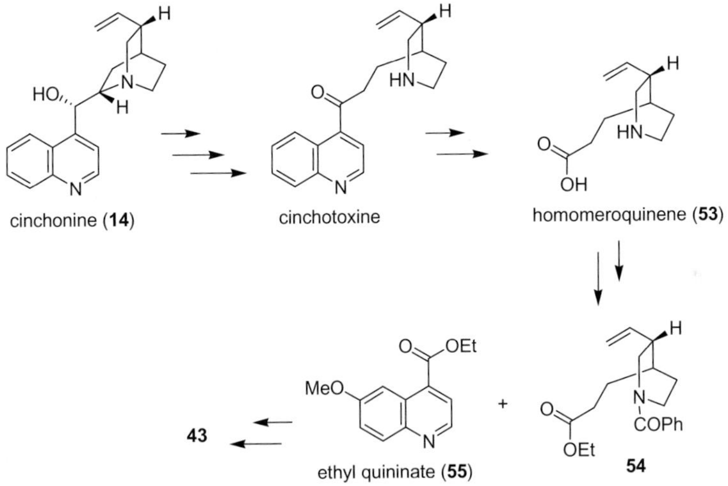

Scheme 9.2

isoquinoline. Thus, 7-hydroxy-isoquinoline **56** [98] was converted through its 8-piperidinomethyl derivative into 7-hydroxy-8-methylisoquinoline **57** (Scheme 9.3) [99]. Hydrogenation over Adam's catalyst, followed by acetylation with acetic anhydride, gave the 1,2,3,4-tetrahydroisoquinoline **58** in almost quantitative yield which was in turn further reduced with Raney nickel and oxidized to a mixture of stereoisomeric ketones with chromic acid. The *cis*-isomer **59** was obtained after

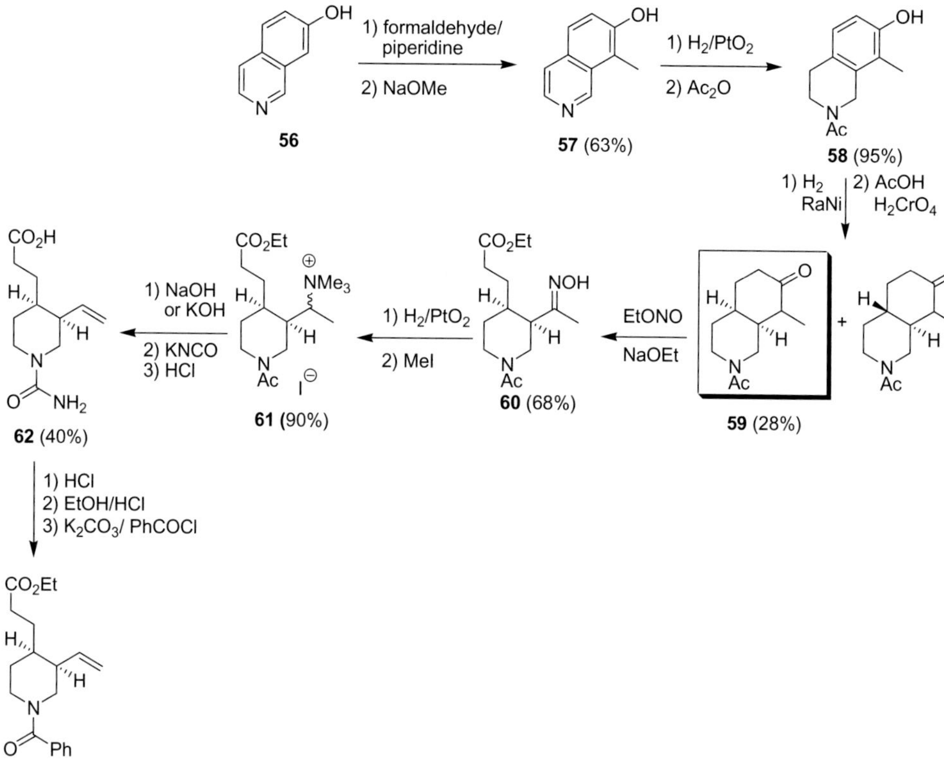

Scheme 9.3

separation from the *trans*-isomer in an average 28% yield. With the pure *cis*-isomer **59** in hand, the carbocycle was cleaved by the action of ethyl nitrite and sodium ethoxide leading to the oxime **60**. Subsequent reduction with Adam's catalyst and quaternization of the primary amine, obtained thereof, with methyl iodide gave the quaternary salt **61** in good yield. Treatment of the quaternary salt **61** with strong alkali resulted in concomitant ester hydrolysis, amide cleavage and Hoffman elimination to give homomeroquinene **53**. Homemeroquinene **53**, however, could not be isolated at this stage and was instead converted *in situ* to the isolatable urea derivative **62**. Finally, cleavage of the urea moiety by the action of diluted hydrochloric acid, followed by esterification of the free acid and benzoylation of the piperidine nitrogen gave the requisite protected racemic homomeroquinene derivative **63** in excellent yield. Further transformation of the *N*-benzoyl derivative **63** into *dl*-quinotoxine was achieved as described earlier by Proštenik and Prelog and the pure alkaloid resolved through its salts with dibenzoyl-*d*-tartrate.

The Woodward–Doering formal total synthesis is remarkable and an early milestone in synthetic organic chemistry [100]. However, this has recently been brought to attention by Gilbert Stork. Since the experimental details of Rabe and Kindler's conversion of d-quinotoxine **43** into quinine **42** were not available, and since neither Proštenik-Prelog nor Woodward–Doering tried to repeat their studies, those experiments required re-establishment. This was finally put to rest when Smith and Williams confirmed the Rabe–Kindler conversion of d-quinotoxine **43** into quinine **42** [76]. Stork et al. [80], nonetheless, accomplished the first asymmetric synthesis of quinine starting from the known (S)-4-vinylbutyrolactone **64** (Scheme 9.4). Thus, 4-vinylbutyrolactone **64** was ring-opened with diethylamine under Lewis-acid catalysis and the resulting primary alcohol protected. The diethylamide derived thereof **65** was alkylated to **66** and the trans-3,4-disubstituted butyrolactone **67** readily obtained by selective removal of the TBS group. Reduction of the butyrolactone **67** to the corresponding lactol followed by Wittig reaction with methoxymethylene triphenylphosphorane gave the enol-ether **68**. Reaction of the latter with diphenylphosphoryl azide to **69** followed by acid-catalyzed hydrolysis of the enol-ether resulted in the requisite intermediate azido aldehyde **70** in ~32% overall yield. Nucleophilic addition of the anion derived from 6-methoxy-4-methylquinoline **71** to the carbonyl group of **70** gave the expected secondary alcohol **72** in 70% yield. Swern oxidation to the intermediate azidoketone **73** followed by an intramolecular Staudinger reaction resulted in the anticipated tetrahydropyridine by refluxing with 1 equiv of triphenylphosphine in tetrahydrofuran. Reduction of the imine proceeded as expected giving the desired stereoisomer which was based on the assumed axial addition of hydride to the imminium intermediate due to the equatorial orientations of the vinyl group and the protected hydroxyethyl chain in the half-chair conformation. Quantitative removal of the silyl protecting group to **74** with aqueous hydrogen fluoride in acetonitrile followed by mesylation-cyclization to the quinuclidine ring system furnished deoxyquinine **75** in moderate yield. Oxidation of deoxyquinine **75** with oxygen in the presence of sodium hydride in anhydrous DMSO completed the quinine **42** synthesis.

The final step was based on the findings and total synthesis of quinine **42** by a Hoffman-La Roche group [74], who prepared deoxyquinine **75** along with deoxyquinidine by reacting the anion derived from 6-methoxy-4-methylquinoline **71** with a derivative of meroquinene **76** [101] after four subsequent synthetic steps (Scheme 9.5).

9.2.1.2 Sandramycin

Sandramycin **45** is an example of a quinoline-containing natural product not isolated from plants. It is a member of a growing class of cyclic decadepsipeptides which possess potent antitumor, antiviral, and antimicrobial activity. The first total synthesis of (−)-sandramycin, and member of this class of natural products, was reported by Boger et al. [85] for which an efficient synthesis of 3-(benzyloxy) quinoline-2-carboxylic acid **80** had to be developed, since a previous report on the indirect synthesis of this compound was only of limited practical value. After some exploration, Boger and Chen found that a modified Friedlander condensation employing the readily accessible O-methyloxime **77** condensed selectively with

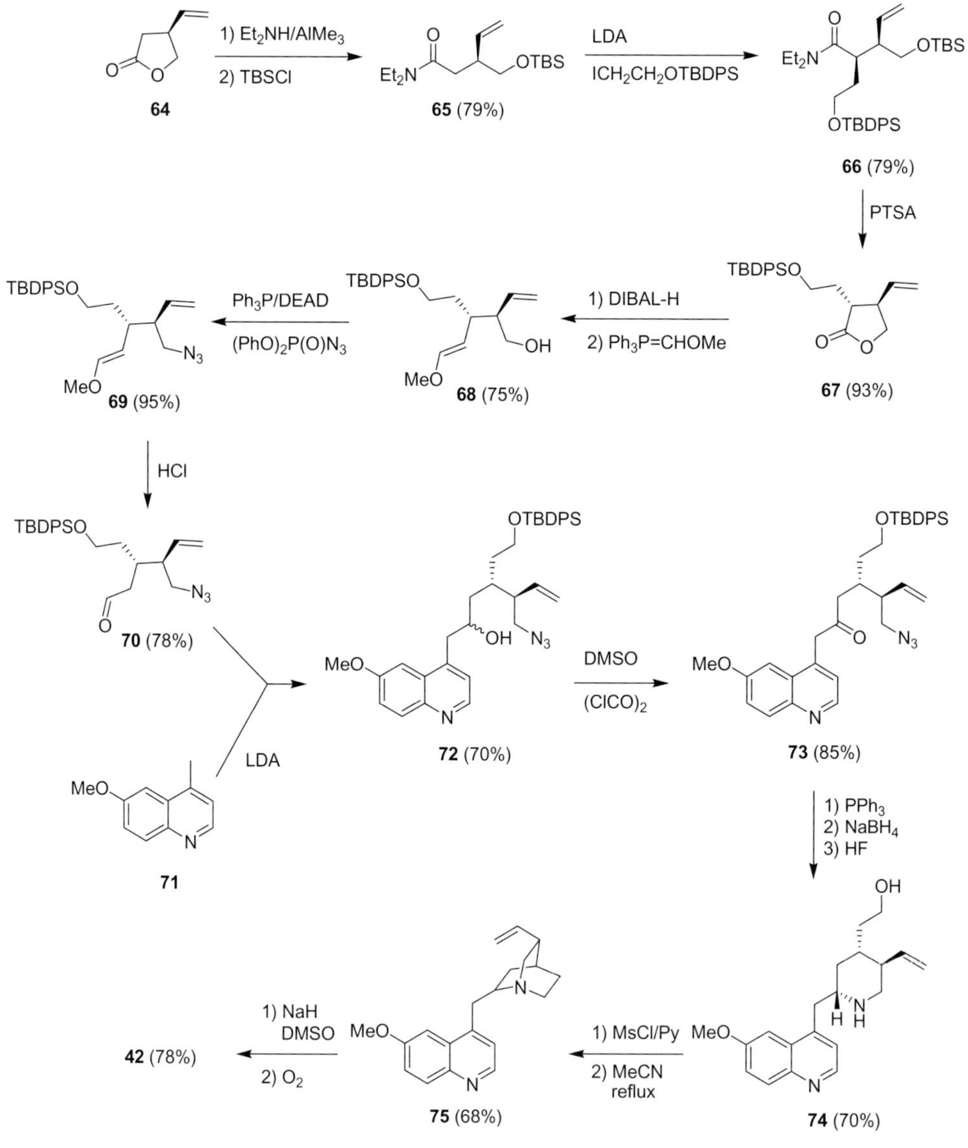

Scheme 9.4

2-aminobenzaldehyde **78** to produce the desired quinoline-2-caboxylic acid **80** [102] (Scheme 9.6). To simplify the purification, the crude acid was converted to its methyl ester derivative **79** and hydrolyzed back with lithium hydroxide. With this protocol, the requisite 3-(benzyloxy)quinoline-2-carboxylic acid **80** could be produced in sufficient quantity for the intended synthesis of (−)-sandramycin.

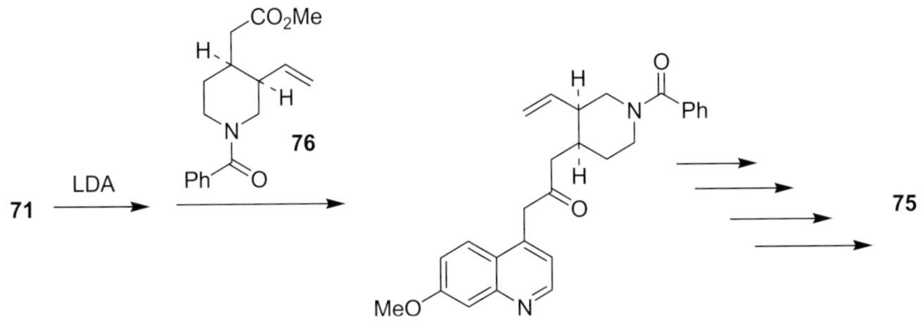

Scheme 9.5

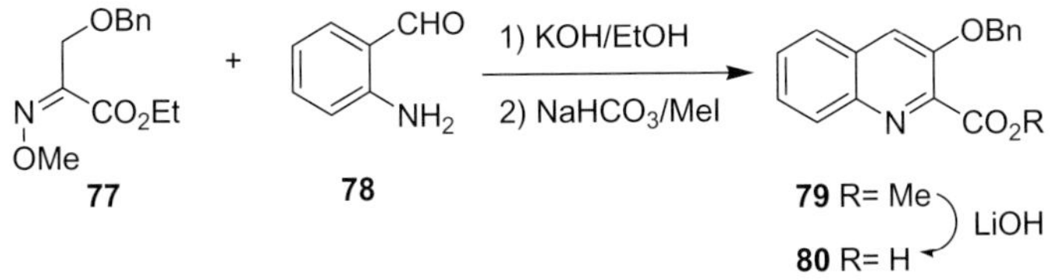

Scheme 9.6

9.2.1.3 Lavendamycin

As for the first total synthesis of sandramycin **45**, Boger et al. also used a Friedlander condensation to build the 7-aminoquinoline-5,8-quinone AB ring system of lavendamycin **25** [44]. Lavendamycin is a highly substituted and functionalized 7-aminoquinoline-5,8-dione, related to streptonigrin **49** [103], which was first isolated and characterized in 1981 by Doyle et al. from the fermentation broths of *Streptomyces lavendulae* [43]. As Sandramycin **45**, this molecule and related natural products have been the focus of much synthetic effort because of their antitumor antibiotic properties [104]. Preparation of the requisite carboline intermediate **88** in Boger's synthesis was based on an inverse electron demand Diels–Alder reaction of the pyrrolidine enamine **81** with 3,5,6-tris(ethoxycarbonyl)-1,2,4-triazine **82** as the first step (Scheme 9.7). The desired pyridine derivative **83**, which resulted from cycloaddition across the C-3/C-6 carbons of the 1,2,4-triazine nucleus, was obtained as the major product along with a small amount of its 3-arylpyridine isomer. Hydrolysis of all three ester groups in **83**, followed by Fischer esterification of the two less hindered acid moieties yielded **84** selectively in 67% yield. A modified Curtius rearrangement with the Yamada–Shioiri reagent of the remaining free acid moiety afforded the primary amine **85** which after acylation was converted to oxazinone **86** in two steps. Treatment of **86** with α-lithiomethyl phenyl sulfoxide and subsequent reductive desulfinylation of the β-keto sulfoxide intermediate afforded **87** after N-deacetylation. Palladium(0)-promoted closure of **87** provided 1-acetyl-3-(methoxycarbonyl)-4-methyl-β-carboline **88**, thus, constituting the CDE ring system of lavendamycin **25**. This newly developed palladium(0)-mediated

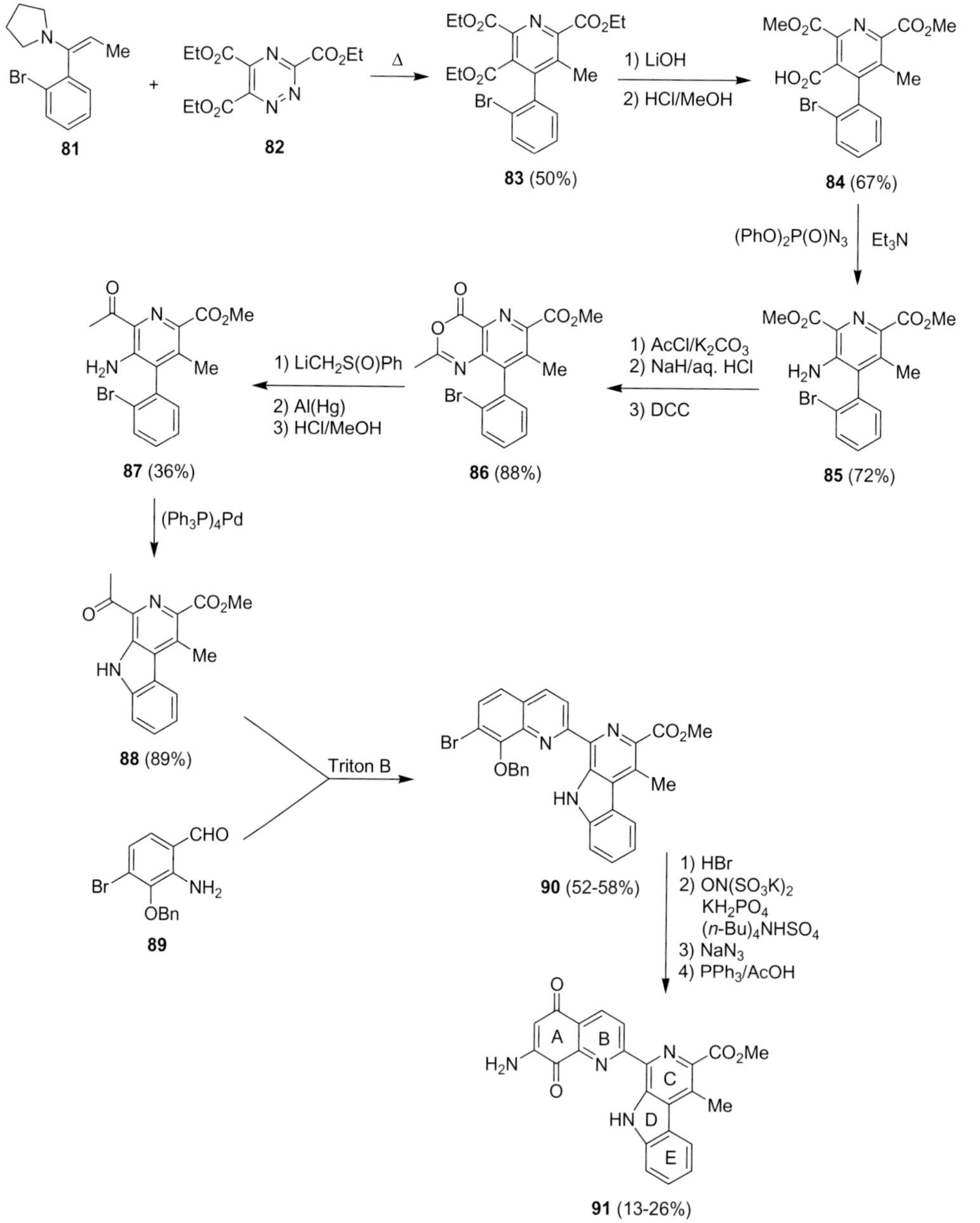

Scheme 9.7

β-carboline synthesis, however, required more than stoichiometric amounts of the palladium catalyst. Triton B (benzyltrimethylammonium hydroxide)-catalyzed Friedlander condensation of **88** with the amino-aldehyde **89** led to the assembly of the carbon skeleton **90** of lavendamycin. Subsequent debenzylation with HBr, oxidation to the *p*-quinone with Fremy's salt under phase transfer conditions, azide

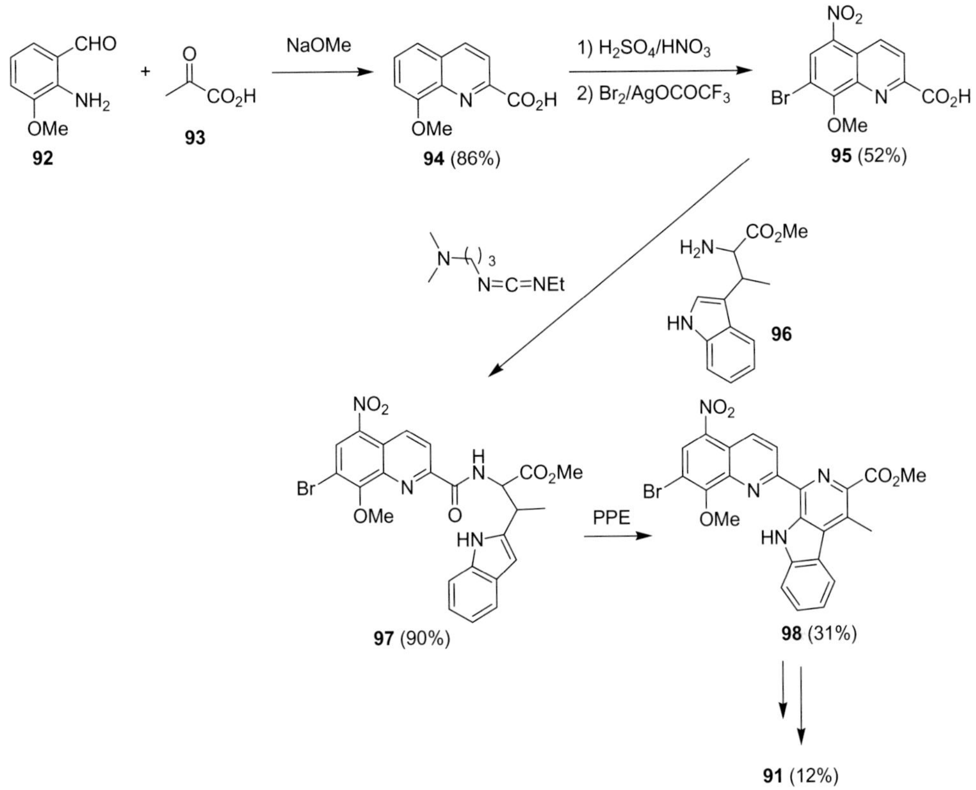

9.2 Application of Quinolines and Isoquinolines in the Synthesis of Natural Products

Scheme 9.8

displacement and reduction to the amine in accordance to the Staudinger reaction led to the methyl ester of lavendamycin **91** without concomitant quinone to hydroquinone conversion.

In the first total synthesis of lavendamycin by Kende and Ebetino in 1984, the Friedlander condensation for the construction of the quinoline core was applied early in the linear synthetic sequence [45]. For this purpose, 2-amino-3-methoxybenzaldehyde **92** was reacted with pyruvic acid **93** to give 8-methoxyquinaldic acid **94** in 86% yield (Scheme 9.8). Nitration followed by bromination of **94** resulted in the formation of the 7-bromo acid **95**. Coupling of this acid with the methyl ester of β-methyltryptophan **96** using a coupling agent gave the corresponding amide **97** in high yield which was in turn converted to the fully aromatic pentacyclic system **98** through a Bischler–Napieralsky condensation/dehydrogenation procedure in polyphosphate esters (PPE). Completion of the total synthesis of the methyl ester of lavendamycin **91** was achieved after reduction of the nitro group, oxidation of the resulting 5-amino derivative to the *p*-quinone with potassium dichromate and azide displacement and reduction to the amine.

Ciufolini and Bishop used the quinoline derivative **99**, which was also obtained through a Friedlander cyclization, as the starting point in their formal total

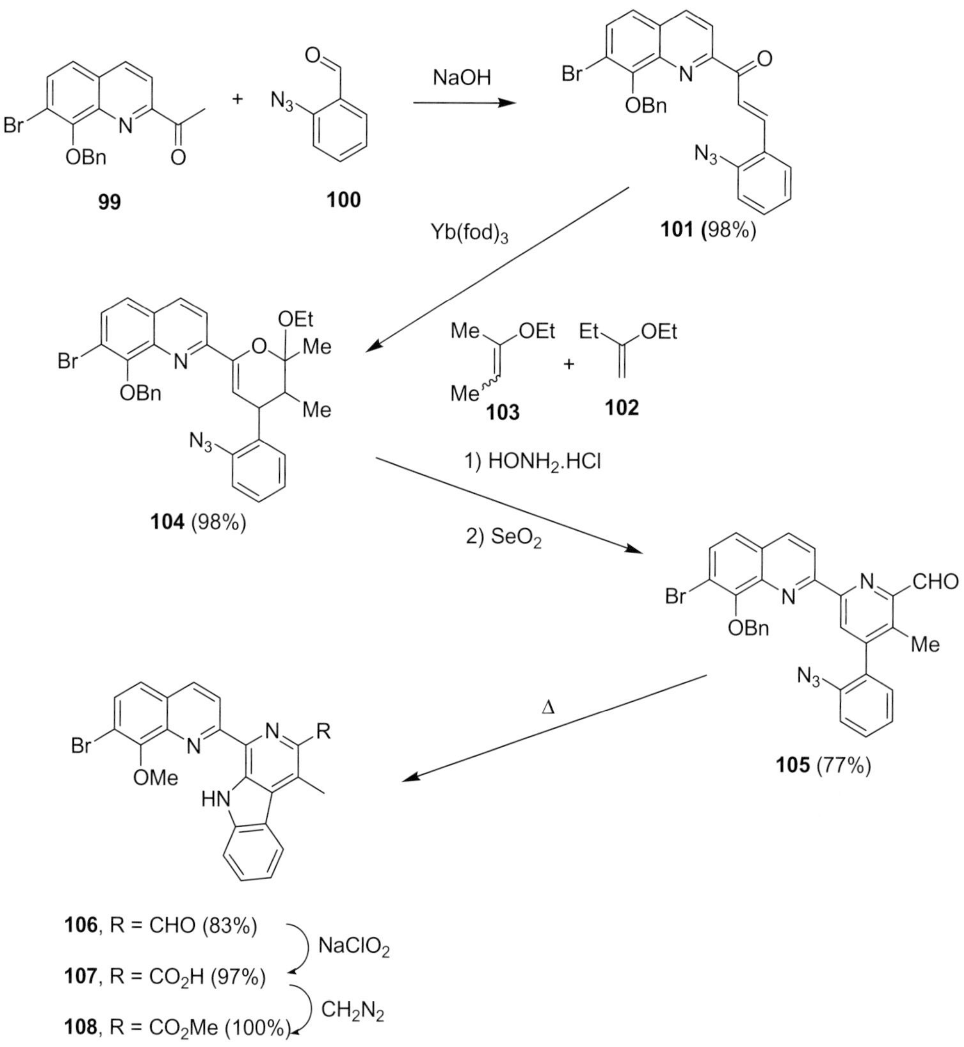

Scheme 9.9

synthesis of lavendamycin methyl ester **91** [46]. Thus, the Knoevenagel-Stobbe condensation of **99** with 2-azidobenzaldehyde **100** provided chalcone **101** in high yield (Scheme 9.9). Selective formal hetero-Diels–Alder reaction of **101** with a mixture of 2-ethoxybut-1-ene **102** and 2-ethoxybut-2-ene **103** in the presence of Yb(fod)$_3$ (where Hfod is 1,1,1,2,2,3,3-heptafluoro-7,7-dimethyloctane-4,6-dione) gave the dihydropyran derivative **104** exclusively. The lanthanide mediated rapid equilibration of the various enol ether isomers appeared to be key for obtaining **104** as a mixture of diastereoisomers in essentially quantitative yield. Reaction of **104** with hydroxylamine hydrochloride, *via* a ring-opening ring-closure mechanism, followed by oxidation with selenium dioxide gave the pyridine derivative **105** in good yield. Thermolysis of **105** gave the requisite pentacyclic lavendamycin **25** core *via* nitrene insertion. This was elaborated further to **108** in high yield through oxidation of the aldehyde **106** to the corresponding acid **107** with sodium chlorite followed by esterification of the acid **107** with diazomethane, thus, resulting in a formal total synthesis of lavendamycin methyl ester **91**.

More recently, Padwa and co-workers reported their efforts towards a total synthesis of lavendamycin through a convergent Stille-coupling strategy [47]. The quinolinyl stannane derivative **113** was prepared in a five-step reaction sequence starting with selective acetylation and subsequent bromination of commercially available 2,8-dihydroxyquinoline **109** to **110** (Scheme 9.10). Exchange of acetyl for methyl was achieved through hydrolysis of the acetyl group followed by reaction of the resulting phenol with methyl iodide to give **111**. Treatment of **111** with HNO_3/H_2SO_4 gave the dinitro derivative **112** in modest yield. Coupling of **112** with $(Bu_3Sn)_2$ and $PdCl_2(PPh_3)_2$ produced the desired 2-stannylquinoline derivative **113** in 61% yield. The second Stille-coupling counterpart **117** was prepared *via* a gold(III)-catalyzed cycloisomerization reaction methodology to access β-carbolines recently reported by the same group [105]. Thus, treatment of N-propargyl-N-tosyl-1-benzylindole-2-carboxamide **114** with a catalytic amount of $AuCl_3$ gave 2-tosyl-β-carbolinone **115** in good yield. N-Detosylation with sodium naphthalenide to **116** followed by treatment of the resulting amide with $POCl_3$ gave the 1-chloro-β-carboline derivative **117** in 76% yield. With both coupling partners in hand, the key Stille-coupling reaction was attempted and found to produce the advance model compound of lavendamycin **118** in reasonable yield after exposure to $PdCl_2(PPh_3)_2$. Conversion of 5,7-dinitro-8-alkoxyquinoline derivatives to the corresponding quinones was demonstrated on simpler derivatives after reduction of the nitro groups and subsequent oxidation under a variety of conditions.

9.2.2
Isoquinoline-Containing Natural Products

This section will cover exemplary syntheses of isoquinoline-containing natural products starting with morphine **29** followed by emetine **17**, the protoberberines tetrahydropalmatine **50**, canadine **13** and stepholidine **48**, and finally nitidine **32**.

9.2.2.1 Morphine
Morphine **29** is the principal of the natural opium alkaloids and indispensable due to its analgesic properties. Morphine is, however, strictly controlled by authorities due to its addictive nature. Nevertheless, its complicated pentacyclic structure containing an octahydroisoquinoline moiety and the desire to develop less addictive morphine-type drugs, has stimulated extensive synthetic efforts since its first total synthesis by Marshall Gates and Gilg Tschudi in 1952 [53]. Morphine's pharmacological activity is dramatically dependent on absolute stereochemistry. Consequently, more recent synthetic efforts have focused on the asymmetric synthesis of this alkaloid. The first asymmetric synthesis of morphine **29** was reported by

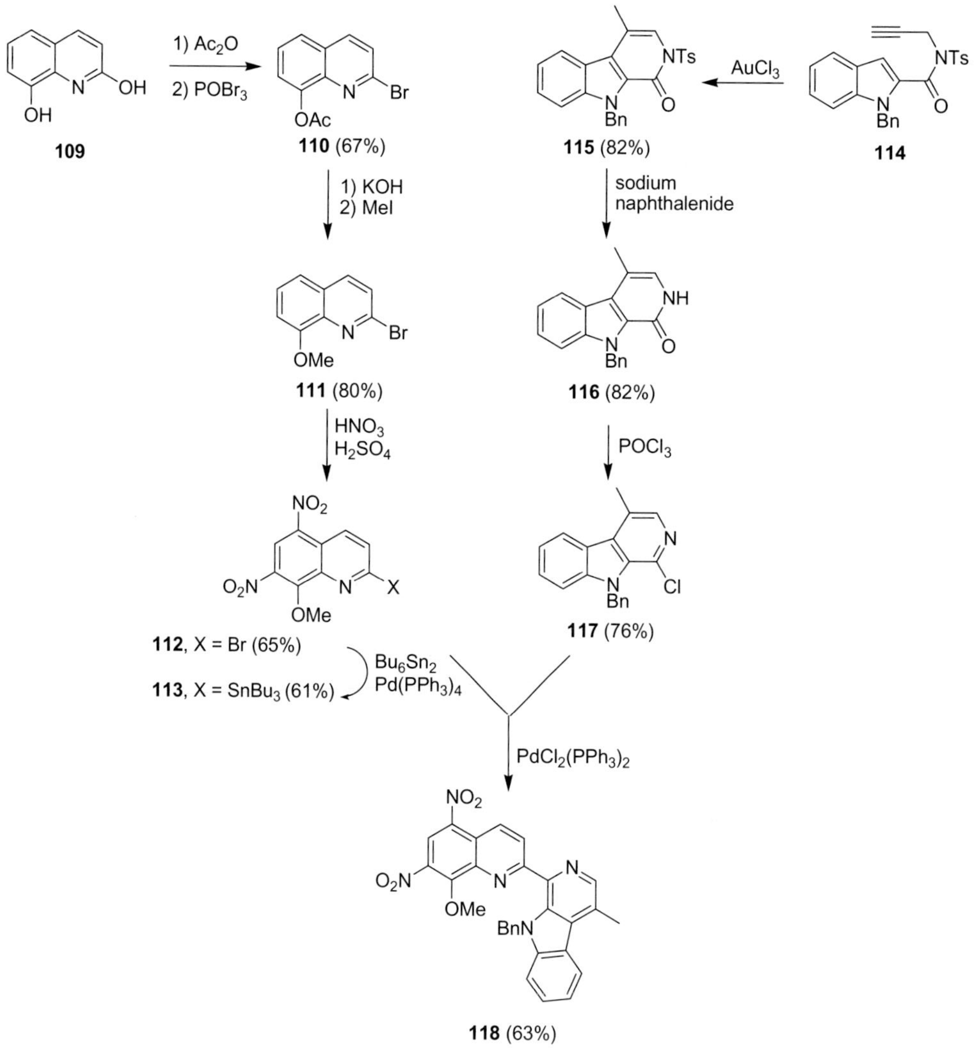

Scheme 9.10

Overman et al. [54]. Their strategy was to first form an enantioenriched octahydroisoquinoline and then employs an intramolecular Heck reaction to forge the critical quaternary center of the morphinan skeleton in a convergent approach. Thus, enantioselective reduction of 2-allylcyclohex-2-en-1-one **119** (Scheme 9.11) with catechol-borane in the presence of a chiral oxazaborolidine catalyst, followed by condensation of the intermediate (S)-cyclohexenol with phenyl isocyanate and selective catalytic dihydroxylation of the terminal double bond and protection provided **120** in 68% overall yield and >96% ee. The allylic carbamate was displaced in an S_N2' fashion by PhMe$_2$SiLi after formation of a copper complex with

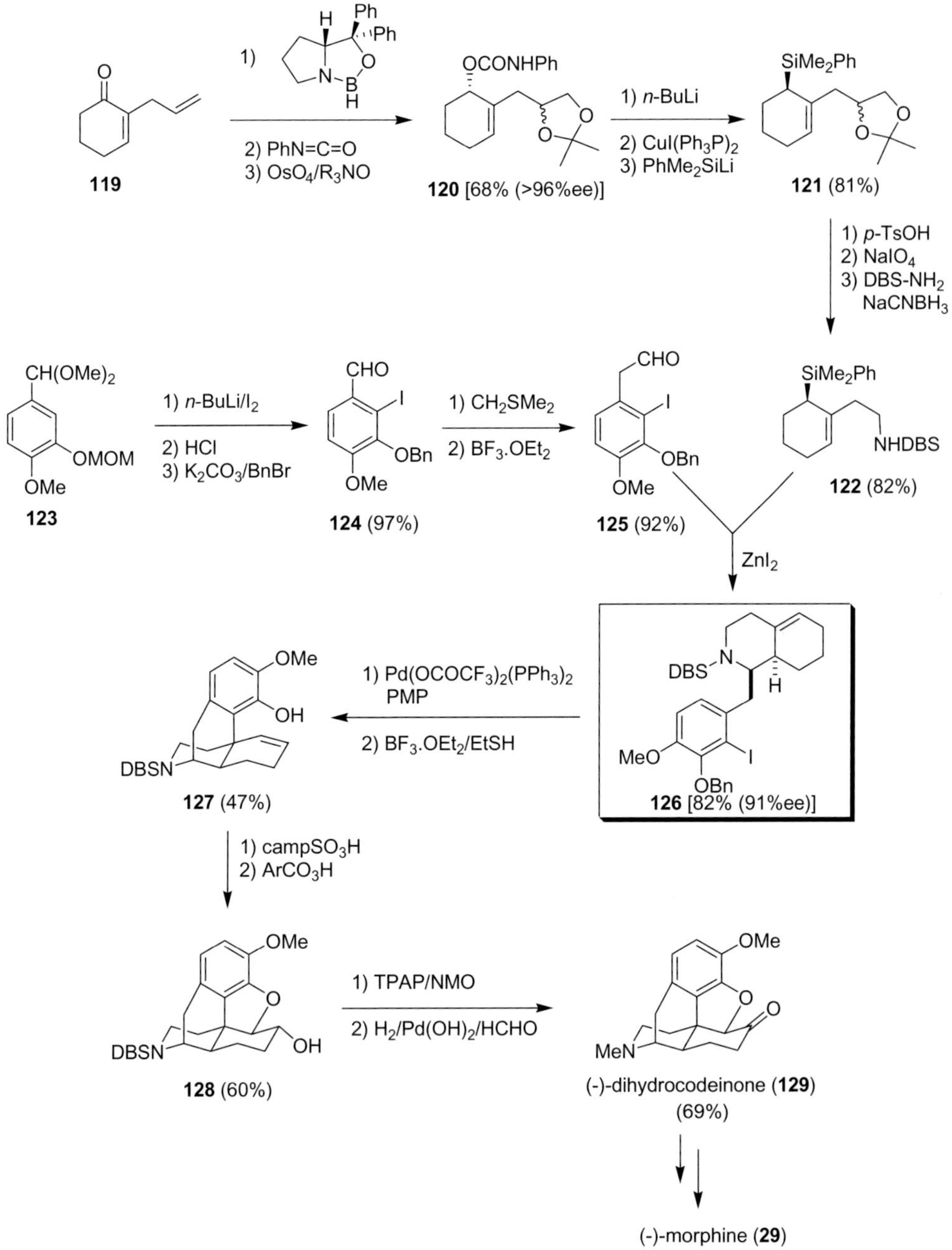

9.2 Application of Quinolines and Isoquinolines in the Synthesis of Natural Products | 319

Scheme 9.11

CuI(Ph$_3$P)$_2$ without significant loss of chiral integrity. Deprotection of the dioxolane moiety of **121** followed by cleavage of the resulting diol and reductive amination with dibenzosuberylamine (DBS-NH$_2$) furnished the homoallylic amine **122** in good yield. The second component for the formation of the requisite enantioenriched octahydroisoquinoline **126** was prepared efficiently from isovanillin. Thus, lithiation-iodination of acetal **123** followed by hydrolysis and exchange of protecting groups gave the benzaldehyde derivative **124** in excellent yield. Homologation of the aldehyde group was achieved by epoxide formation through reaction with a sulfur ylide followed by Lewis acid-catalyzed epoxide rearrangement, thereby completing the synthesis of the required second component **125**. Zn-catalyzed condensation of allylsilane **122** with aryl acetaldehyde **125** proceeded as expected, giving the requisite octahydroisoquinoline **126** in good yield and 91% ee, thus, implying nucleophilic addition of the cyclohexylsilane moiety to the intermediate imminium cation. With **126** in hand, the crucial intramolecular Heck reaction was examined. Fortunately, the expected tetracyclic derivative **127** was obtained in moderate yield after removal of the benzyl protecting group. Formation of the characteristic opioid pentacyclic system was achieved by epoxidation of the double bond and subsequent intramolecular epoxide ring-opening/cyclization to **128**. Oxidation of the resulting alcohol **128**, followed by hydrogenolysis of the DBS group in the presence of formaldehyde gave (–)-dihydrocodeinone **129** in 69% yield and 91%ee after one recrystallization. **129** was transformed into (–)-morphine **29** following a five-step sequence which was developed earlier by Rice [55].

Rice synthesized morphine in 1980 from a key racemic 1,2,3,4-tetrahydroisoquinoline derivative **130**, prepared via a Bischler–Napieralski reaction as shown below (Scheme 9.12)[55].

Several other elegant syntheses of morphine **29** have been described but which, however, do not involved the direct intermediacy of an isoquinoline derivative [56].

9.2.2.2 Emetine

The principal alkaloid found in the root of *Cephaelis ipecacuanha* (Ipecac Root) is emetine **17** [23]. Emetine **17** has been used for centuries as an emetic but of more importance is its antiamebic activity; it is also one of the most synthesized isoquinoline-containing natural products known [24]. Before its total synthesis by Battersby and Turner [25a], Emetine **17** had only been obtained from a synthetic mixture of isomers through resolution. Battersby and Turner, on the other hand,

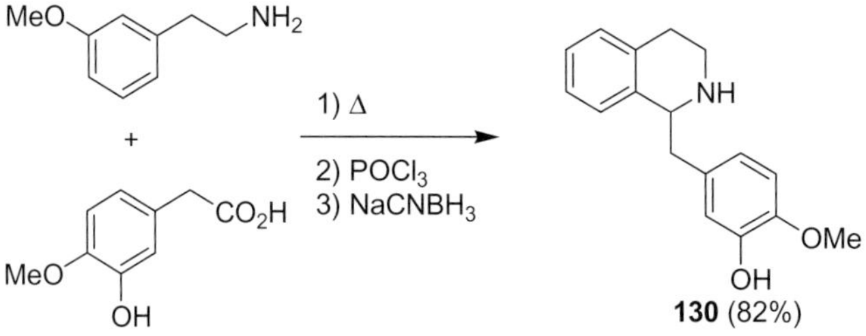

Scheme 9.12

converted the advance intermediate **131** into the enone **132** in moderate overall yield through catalytic hydrogenation or reduction with sodium borohydride followed by acetylation and subsequent β-elimination (Scheme 9.13). Michael addition of the anion derived from diethylmalonate onto dihydropyridone **132** resulted in the exclusive formation of the thermodynamically more stable *trans*-adduct which was hydrolyzed, mono-decarboxylated and re-esterified to the ester **133** in 58% yield. With the relative *trans* stereochemistry in place, the key Bischler–Napieralsky cyclization of intermediate **133** was tried. Indeed, treatment of **133** with phosphorus oxychloride gave the requisite dihydroisoquinoline derivative **134** in good yield. For stability reasons, the product was best isolated as the perchlorate salt. Catalytic hydrogenation using Adam's catalyst produced the tetrahydroisoquinoline derivative **135** selectively and in good yield. Ester hydrolysis followed by formation of the mixed anhydride and subsequent treatment with 3,4-dimethoxyphenethylamine gave the carbon skeleton of emetine as the amide derivative **136** in 91% yield. Final Bischler–Napieralsky cyclization, resolution of the derived racemic O-methyl-psychotrine (an isoquinoline-containing natural product in its own right) as the dibenzoyltartrate salt and catalytic hydrogenation thereof using Adam's catalyst afforded, after neutralization, (–)-emetine **17** in 5.8% overall yield.

Subsequent to Battersby and Turner's work, Takano *et al.* focussed their attention on the synthesis of the tetrahydroisoquinoline derivative **135** for their formal synthesis of emetine [25b]. Two synthetic approaches were developed. In the first approach, condensation of 3-methoxyphenylacetic acid **138** with homoveratrylamine **137** gave the amide derivative **139** in high yield (Scheme 9.14). Subsequent reaction of amide **139** with phosphorus oxychloride followed by reduction with sodium borohydride and acidification resulted in the formation of the requisite 1,2,3,4-tetrahydroisoquinoline derivative **140** as its hydrochloride salt in excellent yield. Mannich condensation of this salt with formaldehyde produced the tetrahydroprotoberberine **141a** (R = H) in good yield. Alternatively, the tetrahydroprotoberberine core was obtained in one step from thermolysis (retro 2 + 2 cycloaddition) of 1-cyano-5-methoxybenzocyclobutane **143** in the presence of 3,4-dihydro-6,7-dimethoxyisoquinoline **142** which resulted in regioselective hetero-Diels–Alder reaction to form tetrahydroprotoberberine **141b** (R = CN) in 50% yield (Scheme 9.14). Both, **141a** and **141b** were subjected separately to Birch reduction conditions using lithium in liquid ammonia. In both cases reduction of the two aromatic rings (A and D) was observed and with **141b** also concomitant decyanation. Selective aromatization of ring A and enol ether cleavage of ring D, however, was achieved using N-chlorosuccinimide. Catalytic hydrogenation of the requisite enone **144** followed by thioacetal formation under Woodward's protocol [106] gave the monothioketal derivative **145**, along with its regioisomer, in moderate yield. Cleavage of **145** with potassium hydroxide, *via* the transient thioacetal anion, afforded the crude thioacetal carboxylic acid, which upon esterification with diazomethane gave rise to the corresponding thioacetal ester **146** in excellent yield. Final desulfurization with Raney nickel completed the formal total synthesis of emetine **17**. The protoemetine derivative **135** was converted to racemic emetine **17** *via* the three step sequence which was developed earlier by Battersby and Turner.

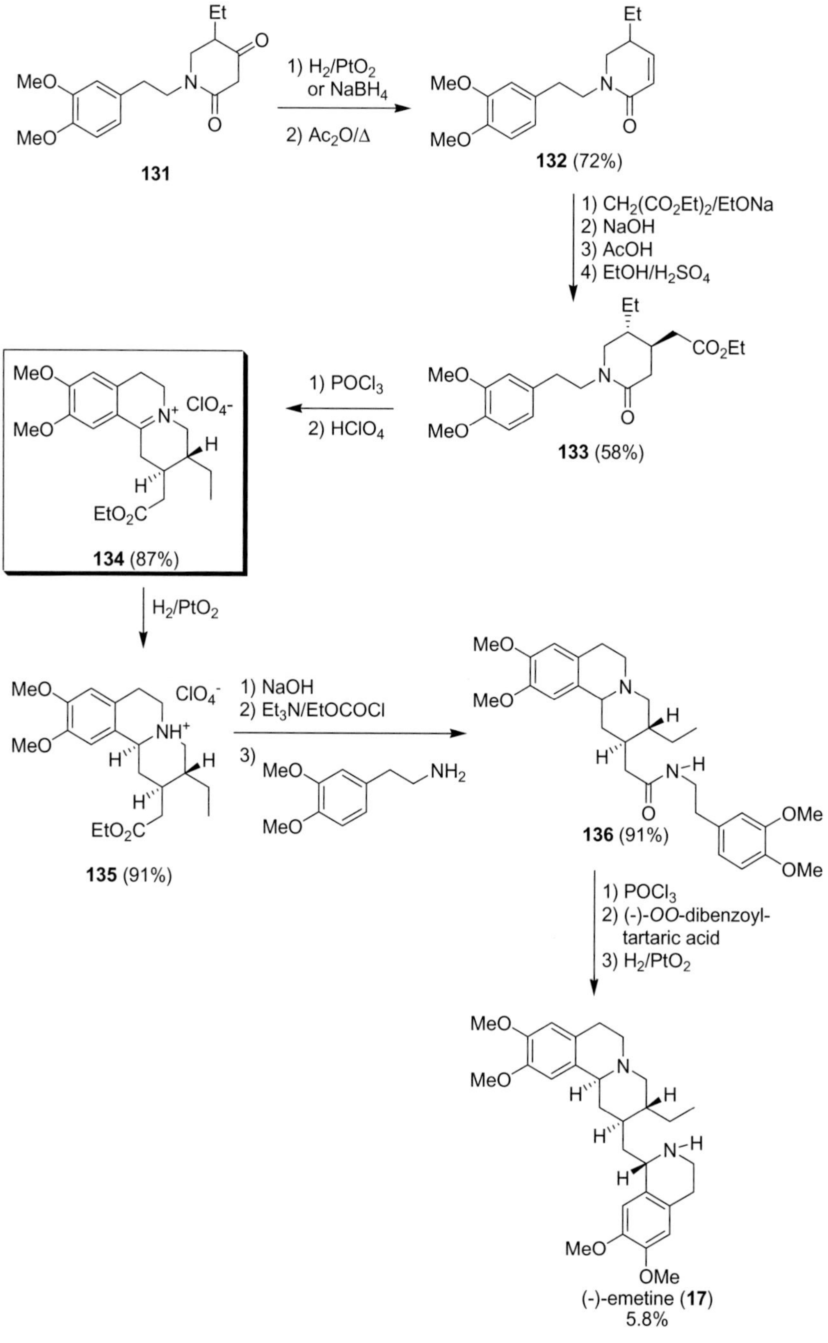

322 | *9 Quinolines and Isoquinolines*

Scheme 9.13

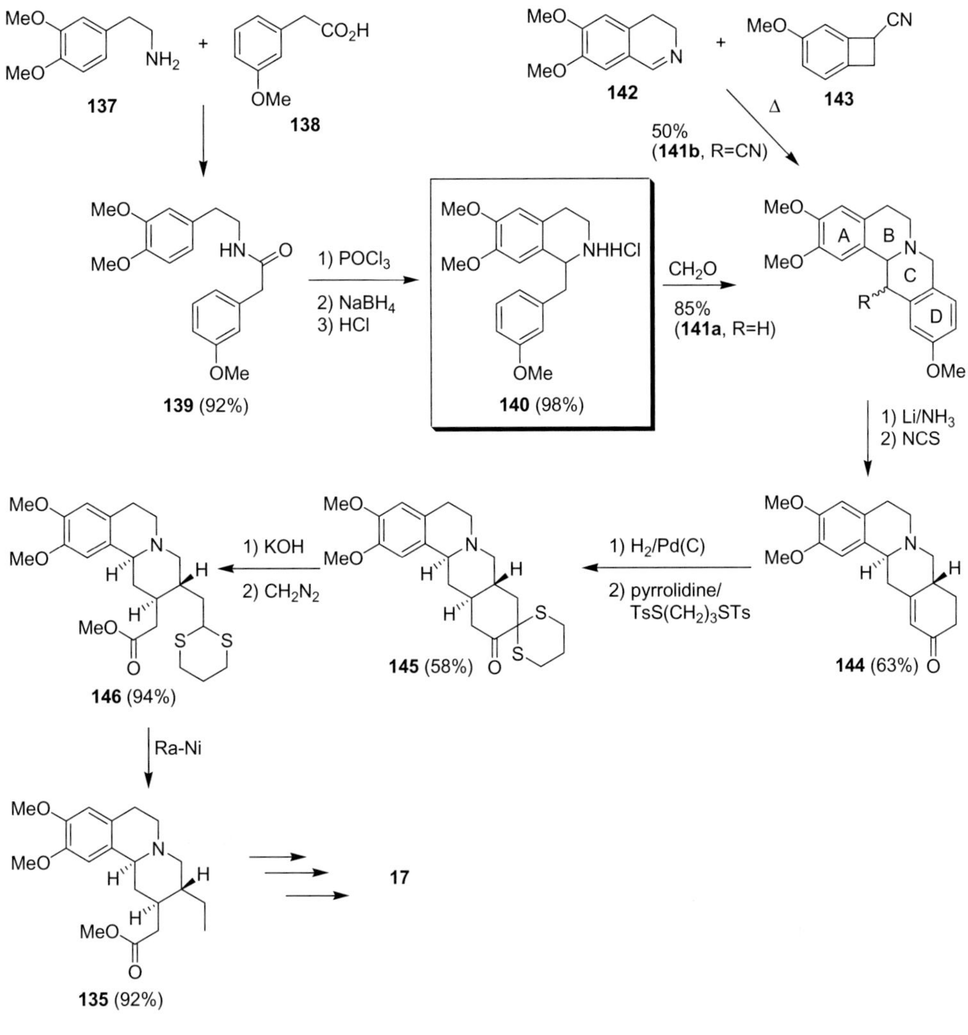

Scheme 9.14

Asymmetric formal total syntheses of emetine have been accomplished by other groups. Fukumoto and co-workers, for example, used the ring opening of a *trans*-substituted valerolactone with 3,4-dimethoxyphenethylamine and subsequent Bischler–Napieralsky cyclization to the dihydroisoquinoline derivative for their entry into the enantiomerically pure emetine core [26]. Thus, the reduction of the (−)-menthyl ester **147** with DIBAL followed by reaction of the primary alcohol with 1,2-dibromoethyl ethyl ether and subsequent deprotection furnished **148** as a mixture of alcohols (Scheme 9.15). Swern oxidation followed by Wadsworth–Emmons olefination, according to Still's procedure [107], gave predominantly the *cis*-esters **149** in 79% overall yield for two steps. Intramolecular radical 1,4-addition

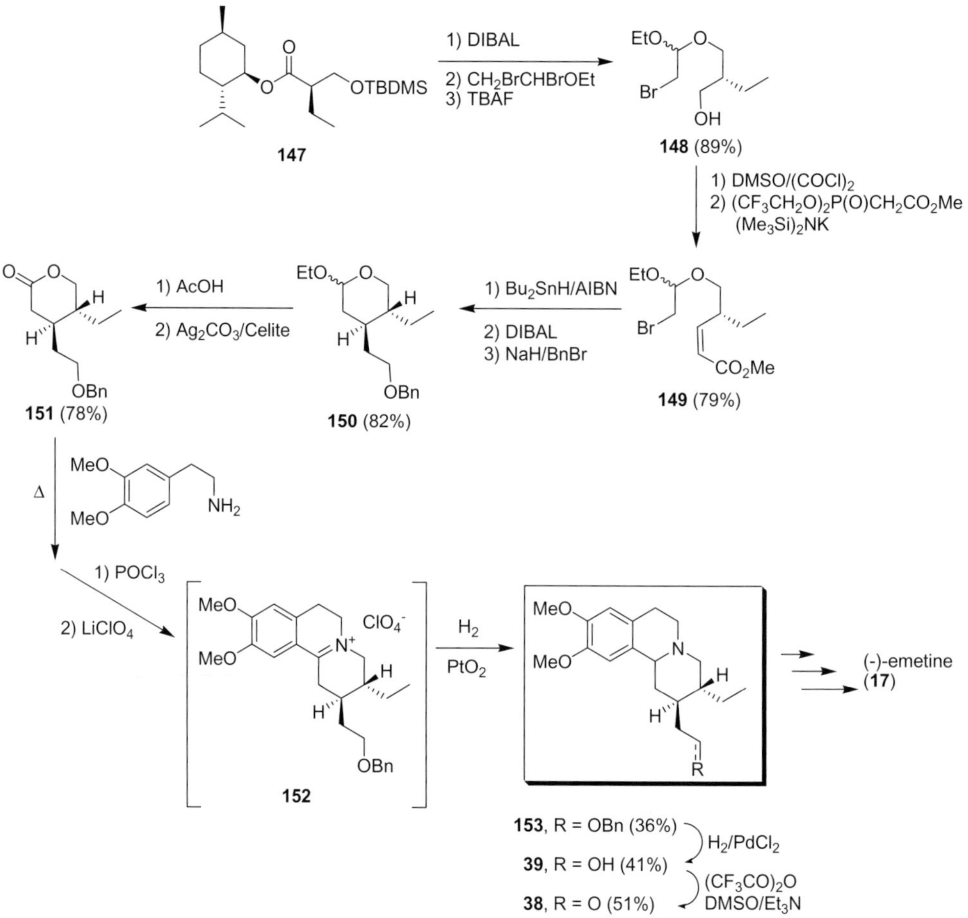

Scheme 9.15

of **149** with tributyltin hydride and AIBN resulted in the desired *trans* relationship between the exocyclic ethyl and ethyl ester groups. Reduction of the ester to the alcohol followed by benzylation, in turn, gave **150** as a mixture of diastereoisomers in good overall yield. The acetal intermediate **150** was first hydrolyzed to the hemiacetal then oxidized to the δ-lactone **151** using Fetizon's reagent. The key cyclization to the tetrahydroisoquinoline derivative **153** was accomplished after ring opening of the lactone **151** with 3,4-dimethoxyphenethylamine, Bischler–Napieralsky condensation of the intermediate amide and *in-situ* reduction of the resulting iminium perchlorate salt **152** using Adam's catalyst. The single diastereoisomer obtained was in turn debenzylated to (−)-protoemetinol **39**. Swern oxidation of **39** resulted in the formation of (−)-protoemetine **38** which was converted into (−)-emetine **17** following previously established procedures [27].

Meyers and Guiles, on the other hand, relied upon the formation of an α-stabilized carbanion activated by a chiral formamidine for the construction of the

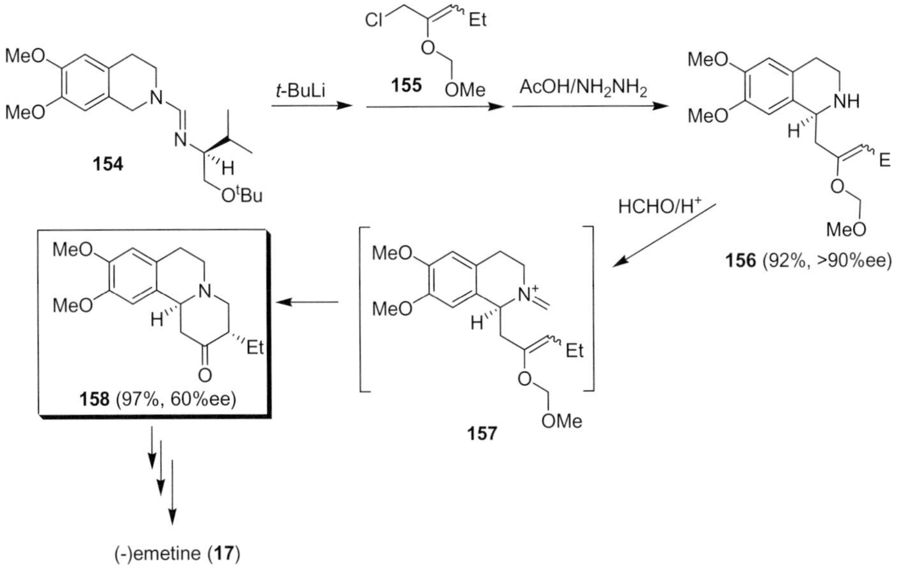

9.2 Application of Quinolines and Isoquinolines in the Synthesis of Natural Products | 325

Scheme 9.16

stereogenic carbon centers in their formal synthesis of (–)-emetine **17** [28]. Thus, trapping of the anion derived from the advance chiral 1,2,3,4-tetrahydroisoquinoline formamidine intermediate **154** with electrophile **155** and subsequent removal of the chiral auxiliary gave the requisite 1-sustituted tetrahydroisoquinoline **156** in good yield and > 90% enantiomeric excess (Scheme 9.16). The final step in the process was based on a variant of the Mannich reaction, namely, the aminomethylation of alkenes. Thus, treatment of the enantiomerically enriched tetrahydroisoquinoline derivative **156** with formaldehyde and catalytic acid produced the intermediate iminium ion **157** which readily underwent cyclization to the quinolizidine **158** in excellent yield. The enantiomeric purity, however, dropped to 60% and was attributed to a competitive rapid aza-Cope rearrangement of **157**, resulting in partial loss of optical purity. Nevertheless, the formal synthesis of enantiomerically enriched (–)-emetine **17** was achieved since several other successful approaches to emetine **17** have used the same ketone **158** as the key intermediate [29].

A recent asymmetric synthesis of (–)-emetine was reported by Itoh et al. in 2006 [30]. Itoh's synthesis started with the copper (I)-catalyzed addition of allyltrimethoxysilane to 3,4-dihydro-6,7-dimethoxyisoquinoline **159** (Scheme 9.17). Best conversions and enantiomeric excess were achieved with tol-BINAP as the chiral ligand. After an additional recrystallization in the presence of (–)-dibenzoyl-L-tartaric acid, the requisite 1-allyl-1,2,3,4-tetrahydro-6,7-dimethoxyisoquinoline **160** was obtained with 97% ee. The allyl group was then further functionalized, after BOC-protection of the isoquinoline nitrogen, to give preferentially the (*E*)-unsaturated ester **161** in good overall yield through an alkene metathesis reaction using the second-generation Grubb's catalyst. Deprotection followed by Michael addition of the free base onto acrolein and subsequent enamine cyclization afforded the

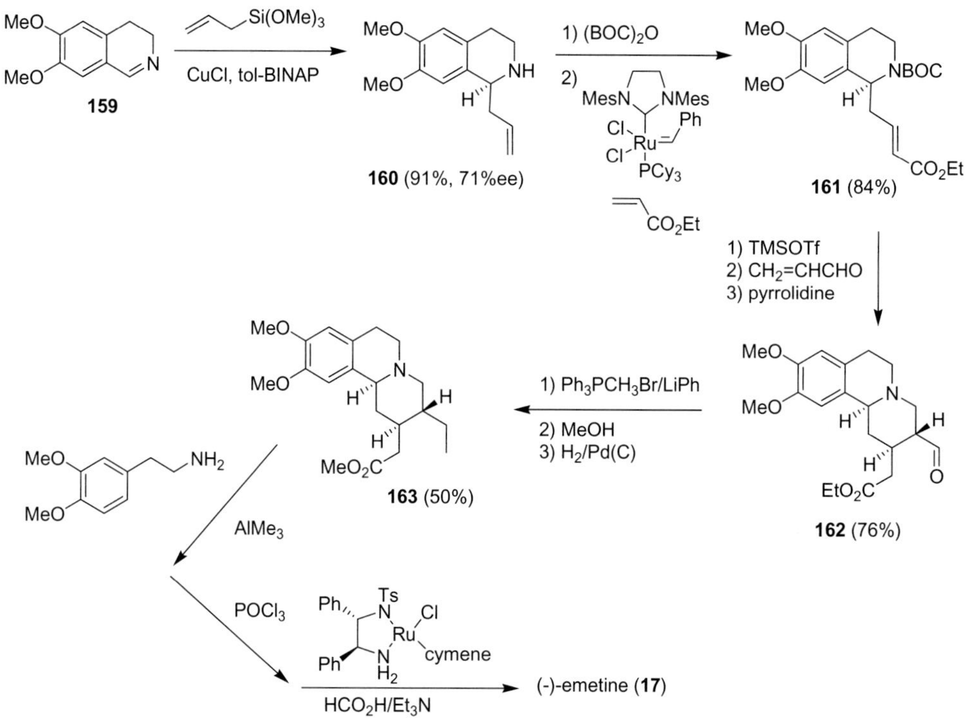

Scheme 9.17

tricyclic isoquinoline **162** in good yield and complete stereoselectivity. The formyl derivative **162** was subjected to the Wittig reaction with concomitant transesterification upon workup with methanol and the resulting alkene was hydrogenated in the presence of palladium to provide **163**. Conversion of **163** into (−)-emethine **17** was accomplished according to Tietze's method [31] involving Lewis acid-catalyzed amide bond formation, Bischler–Napieralsky condensation to form the second dihydroisoquinoline moiety and imine reduction with formic acid and triethylamine in the presence of a chiral ruthenium catalyst.

9.2.2.3 Protoberberines

Protoberberines are a large class of natural products which are characterized by a tetracyclic ring skeleton containing a benzyltetrahydroisoquinoline core. These natural products have been shown to possess anti-inflammatory, antimicrobial, antileukemic as well as antitumor properties. Because of the rich biological activity these molecules display, they have received significant attention from the synthetic community.

Tetrahydropalmatine As for their emetine synthesis, the Meyers group also used a chiral 1,2,3,4-tetrahydroisoquinoline formamidine in their asymmetric total synthesis of (−)-tetrahydropalmatine **50** [92]. Thus, metalation of the chiral formamidine **164** followed by reaction with the advance benzyl chloride intermediate **165** gave after removal of the chiral auxiliary the key benzylsubstituted tetrahydroisoquinoline derivative **166** in moderate yield and enantiomeric excess (Scheme 9.18). Final cyclization to (−)-tetrahydropalmatine **50** was achieved in 65% yield with little epimerization by treating **166** with triphenylphosphine and bromine.

Canadine Another example of a member of the protoberberine family of alkaloids is canadine **13**. Kametani's entry into canadine **13** and other related alkaloids (e.g., nandinine, **30**) was based on a Bischler–Napieralsky isoquinoline synthesis [12]. Thus, condensation of 3,4-methylenedioxyphenethylamine **167** with methyl 5-benzyloxy-2-bromo-4-methoxyphenylacetate **168** afforded the amide **169** which upon treatment with phosphoryl chloride and subsequent reduction with sodium borohydride gave the requisite 1,2,3,4-tetrahydroisoisoquinoline **170** in 36% overall yield (Scheme 9.19). Compound **170** was debenzylated to **171** with concentrated hydrochloric acid and subjected to the Mannich reaction to give the 12-bromonandinine **172** in moderate yield. The tretracyclic derivative **172** was conveniently debrominated with zinc powder in sodium hydroxide to **173** and then elaborated into racemic canadine **13** by methylation with diazomethane. Dehydrogenation of canadine **13** with iodine afforded berberine **10** iodide in 50% yield.

Cushman and Dekow, on the other hand, based their strategy (Scheme 9.20) on the reaction of norhydrastatin **174** with homophthalic anhydride **175** to form the key tetracyclic intermediate **176** in good yield after treatment with acetic acid to equilibrate the initial *cis/trans* mixture to the thermodynamically more stable *cis* diastereoisomer [14a]. This common intermediate could be efficiently

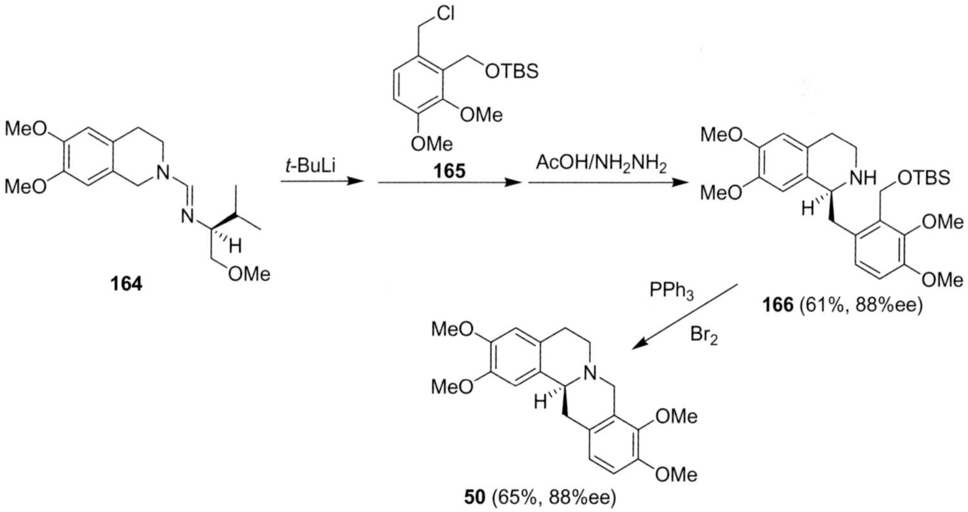

Scheme 9.18

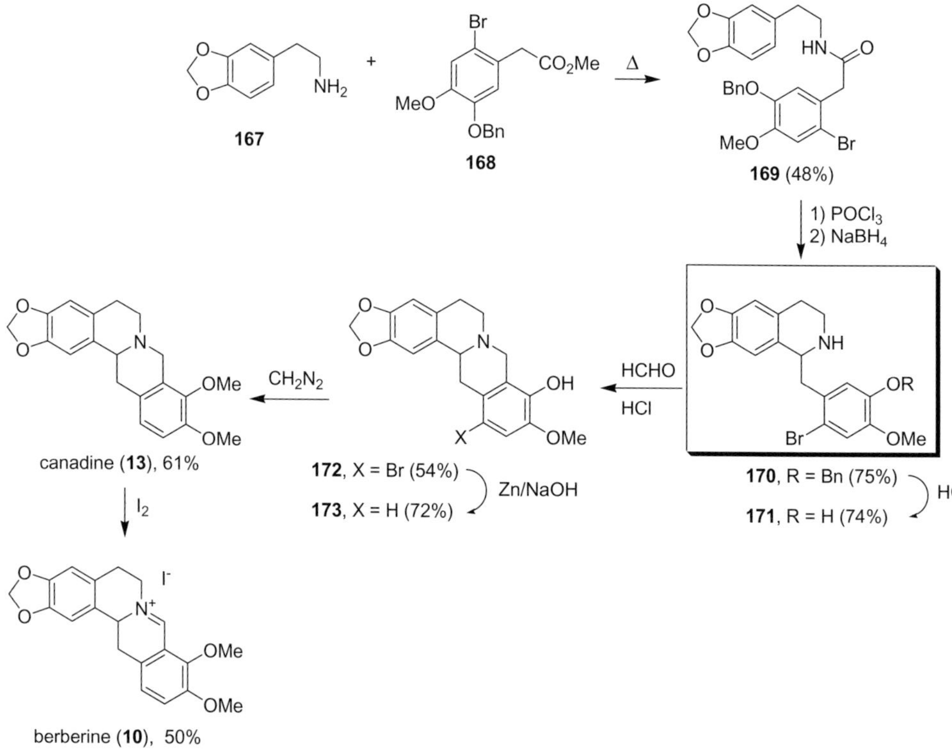

Scheme 9.19

transformed into the racemic protoberberine alkaloids thalictricavine **51**, berlambine **11** and canadine **13**. In the first instance, **176** was methylated with diazomethane. Concomitant reduction of the corresponding methyl ester and amide carbonyl group in **177** with lithium aluminum hydride gave the primary alcohol **178** in good yield. Reduction of the mesylate derivative of **178** with sodium borohydride afforded racemic thalictricavine **51** in over 60% overall yield. Reaction of **176** with lead tetraacetate in the presence of cupric acetate, potassium acetate, acetic acid, and dimethylformamide gave racemic berlambine **11** in moderate yield. The exotic mixture for the last step of the berlambine **11** synthesis, was discovered after classical Hunsdiecker decarboxylative halogenation conditions had failed. Lewis acid-catalyzed aluminum hydride reduction of the amide carbonyl and double bond of berlambine gave racemic canadine **13** in 63% yield.

Similarly, Cushman and co-workers prepared (+)-thalictricavine and (+)-canadine through classical resolution of the common intermediate **176** using (−)-strychnine and following essentially the same subsequent chemical steps [14b].

Stepholidine A more recent synthesis of an alkaloid of the protoberberine family has been reported by Cheng and Yang [89]. Stepholidine **48**, which is extracted

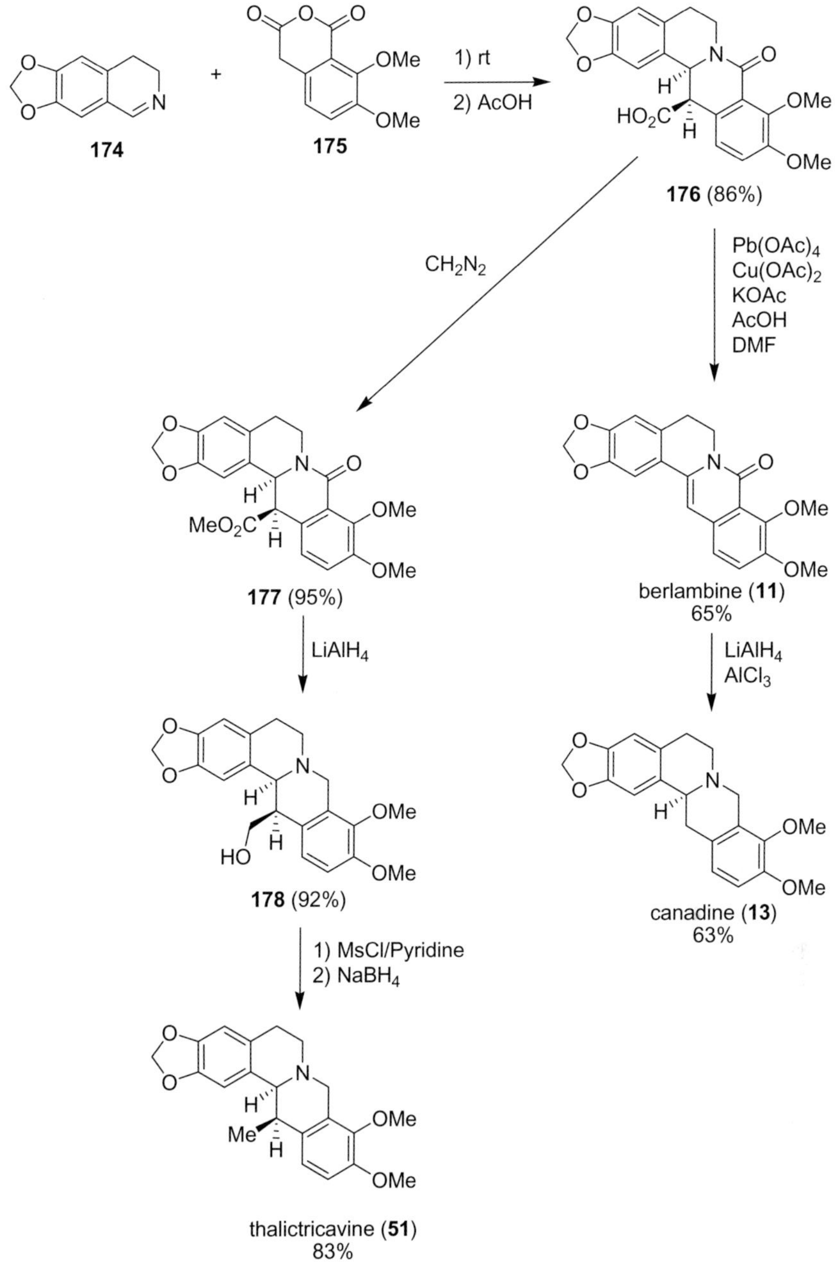

9.2 Application of Quinolines and Isoquinolines in the Synthesis of Natural Products

Scheme 9.20

from *Stephanie intermedi*, has attracted a great deal of attention because of its potential as a drug candidate for the treatment of schizophrenia and/or drug abuse. Cheng and Yang's asymmetric total synthesis of stepholidine **48** was based on the chiral transfer hydrogenation of imines using Noyori's catalyst. The key dihydroisoquinoline **182** intermediate for the asymmetric hydrogenation step was prepared starting with the elaborated lactone **180** (Scheme 9.21). Thus, aminolysis of **180** with phenethylamine **179** afforded after acetylation with acetic anhydride the amide **181** in 66% yield. Bischler–Napieralsky reaction of **181** gave the imine **182** in excellent yield but which, for stability reasons, was used immediately in the next step. Asymmetric transfer hydrogenation in the presence of Noyori's catalyst **183** with formic acid/triethylamine as the hydrogen source gave tetrahydroisoquinoline **184** in good yield. Deacetylation of **184** followed by chlorination with thionyl chloride and base-catalyzed cyclization gave the corresponding tetracyclic derivative **185** in 86% yield and essentially as a single enantiomer. Finally, debenzylation with concentrated hydrochloric acid gave (−)-(S)-stepholidine **48** in 74% isolated yield and 99.6% ee.

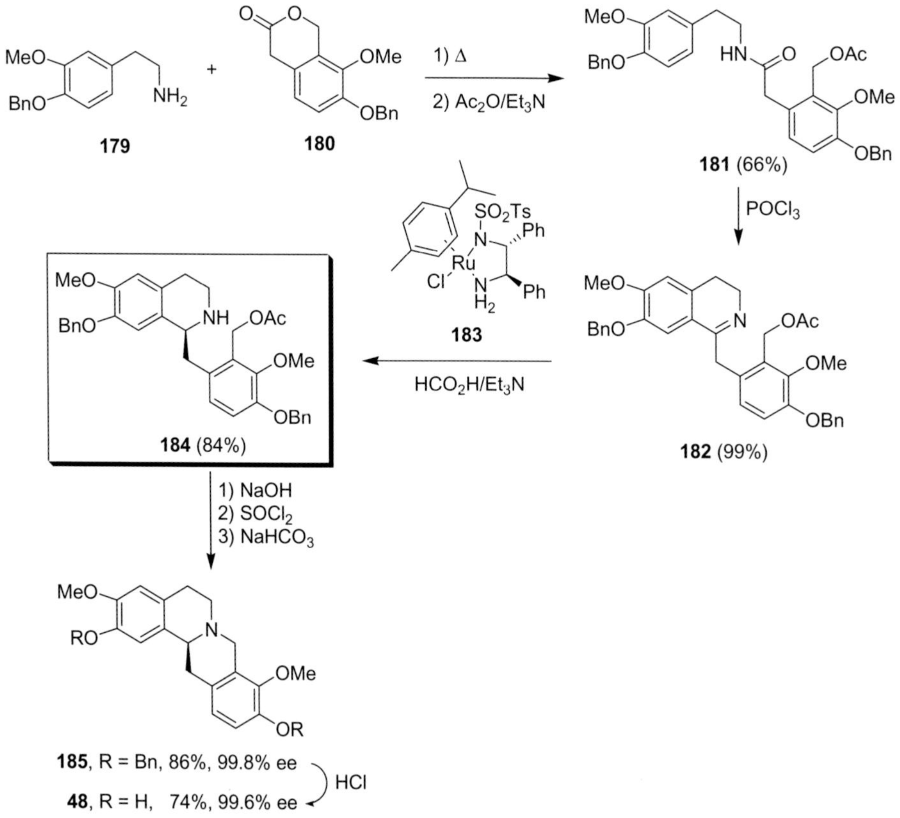

Scheme 9.21

9.2.2.4 Nitidine

Nitidine **32** is an isoquinoline-containing alkaloid belonging to the benzo[c] phenanthridine family. These class of natural products are promising antitumor drug candidates due to their strong antitumor activity by the inhibition of DNA topoisomerase. The recent formal synthesis of nitidine **32** by Takemoto and coworkers is based on a gold(I)-catalyzed tandem cyclization [61]. Their convergent approach involved the preparation of the enol-ether **187** from the aldehyde **186** by conventional Wittig reaction (Scheme 9.22). The second intermediate **189** was

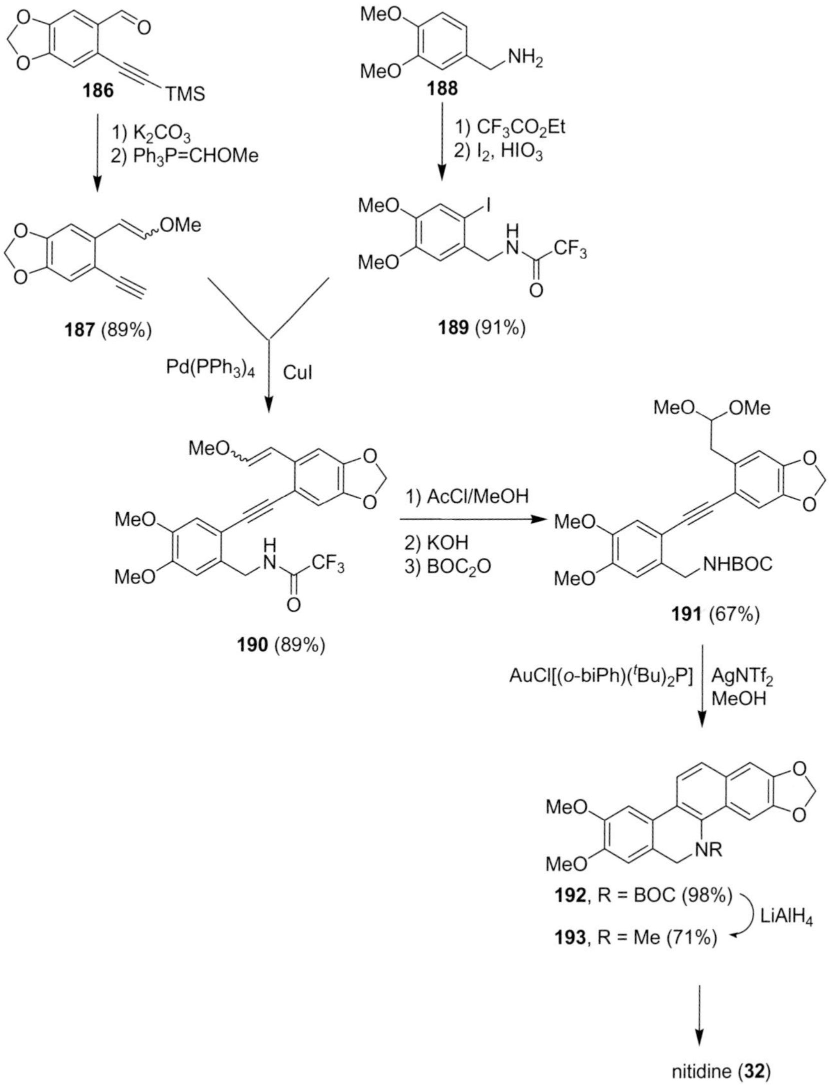

Scheme 9.22

synthesized *via* trifluoroacetylation of the amine **188** followed by iodination with iodine and iodic acid. Pd-catalyzed Sonogashira coupling of **187** and **189** gave the desired alkyne **190** in good yield. Acetalization, hydrolysis and *N*-BOC protection of **190** afforded the acetal **191** in 67% yield. The methanol promoted gold(I)-catalyzed tandem cyclization of **191** gave the desired polycyclic product **192** in excellent yield. Five equivalents of the gold(I)-catalyst were, however, required in order to achieve a high conversion. Final reduction of **192** with lithium aluminum hydride completed the formal synthesis since the conversion of the *N*-methyl derivative **193** to nitidine **32** had previously been demonstrated [108].

9.3
Conclusion

The selected examples of elegant classical and modern total syntheses of natural products in which the quinoline and isoquinoline cores are embedded, presented in this chapter, are as rich and diverse as synthetic organic chemistry can be. For limitations of the chapter, many of these beautiful syntheses are only referenced. Those covered, however, demonstrate the creative and intellectual process which is required to build complex molecular entities and that will lead to further developments in reagents, synthetic methods and ways of assembling these two heterocyclic cores. Fortunately, there will be continued opportunity to do so since, in addition to the thousands of quinoline and isoquinoline-containing natural products known to date, novel intriguing structures belonging to these classes of compounds, displaying rich pharmacological properties, are discovered everyday [109].

References

1 Joule, J.A., Mills, K. and Smith, G.F. (1995) Quinolines and isoquinolines: reactions and synthesis, in *Heterocyclic Chemistry*, 3rd edn, Chapman & Hall, London, UK, pp. 120–145.

2 Davies, D.T. (1993) Quinolines and isoquinolines, in *Aromatic Heterocyclic Chemistry* (ed. S.G. Davies), Oxford University Press, Oxford, UK. pp. 46–52.

3 (a) Gilchrist, T.L. (1992) Quinolines and isoquinolines, *Heterocyclic Chemisty*, 2nd edn, Longman Scientific & Technical, Essex, UK, pp. 152–166; (b) Balasubramanian, M. and Keay, J.G. (1996) *Comprehensive Heterocyclic Chemistry II*, vol. 5 (eds A.R. Katritzky and C.W. Rees), Pergamon Press, London, UK, pp. 245–300.

4 Michael, J.P. (1997) Quinolines, quinazoline and acridone alkaloids. *Nat. Prod. Rep.*, **14**, 605–618.

5 Menachery, M.D., Lavanier, G.L., Wetherly, M.L., Guinaudeau, H. and Shamma, M. (1986) Simple isoquinoline alkaloids. *J. Nat. Prod.*, **49** (5), 745–778.

6 (a) Mann, J. (1994) Alkaloids, in *Natural Products: Their Chemistry and Biological Significance* (eds J. Mann, R.S. Davidson, J.B. Hoffs, D.V. Banthorpe and J.B. Harborne), Longman Group UK Limited, Essex, UK, pp. 389–444.(b) Hawkins, K.M. and Smolke, C.D. (2008) Production of benzylisoquinoline alkaloids in *Saccharomyces cerevisiae*. *Nat. Chem. Biol.*, **4** (9), 564–573.

7 Woddward, R.B. (1956) Neuere Entwicklungen in der Chemie der Naturstoffe. *Angew. Chem.*, **68** (1), 13–20.
8 Bowden, M.E. and Benfey, T. (1992) *Robert Burns Woodward and the Art of Organic Synthesis*, The Beckman Center for the History of Chemistry, Philadelphia, PA, USA.
9 (a) Martell, M.J., Jr., Soine, T.O. and Kier, L.B. (1963) The structure of argemonine; identification as (–) N-methylpavine. *J. Am. Chem. Soc.*, **85** (7), 1022–1023; (b) Stermitz, F.R. and Sieber, J.N. (1966) Alkaloids of the papaveraceae. IV. Argemone hispida Gray and A. munita Dur. & Hilg. subsp. rotunda (Rydb.) G. B. Ownb. *J. Org. Chem.*, **31** (9), 2925–2933.
10 (a) Munchhof, M.J. and Meyers, A.I. (1996) A novel asymmetric route to the 1,3-disubstituted tetrahydroisoquinoline, (–)-argemonine. *J. Org. Chem.*, **61** (14), 4607–4610; (b) Ruchirawat, S. and Namsa-aid, A. (2001) An efficient synthesis of argemonine, a pavine alkaloid. *Tetrahedron Lett.*, **42** (7), 1359–1361.
11 Perkin, W.H., Jr. and Robinson, R. (1910) Strychnine, berberine and allied alkaloids. *J. Chem. Soc.*, (97), 305–323.
12 Kametani, T., Noguchi, I., Saito, K. and Kaneda, S. (1969) Studies on the syntheses of heterocyclic compounds. Part CCCII. Alternative total syntheses of (±)-nandinine, (±)-canadine, and berberine iodide. *J. Chem. Soc. (C)*, 2036–2038.
13 Chattopadhyay, S.K., Ray, A.B., Slatkin, D.J., Knapp, J.E. and Schiff, J.P.L. (1981) The alkaloids of thalictrum foliolosum. *J. Nat. Prod.*, **44** (1), 45–49.
14 (a) Cusman, M. and Dekow, F.W. (1979) Synthesis of (±)-thalictricavine, berlambine, and (±)-canadine from a common intermediate. *J. Org. Chem.*, **44** (3), 407–409; (b) Isawa, K., Gupta, Y.P. and Cushman, M. (1981) Absolute configuration of the cis- and trans-13-methyltetrahydroprotoberberines. Total synthesis of (+)-thalictricavine, (+)-canadine, (±)-, (–)-, and (+)-thalictrifoline, and (±)-, (–)-, and (+)-cavidine. *J. Org. Chem.*, **46** (23), 4744–4750.
15 Wall, M.E., Wani, M.C., Cook, C.E., Palmer, K.H., McPhail, A.T. and Sim, G.A. (1966) Plant antitumor agents. I. The isolation and structure of camptothecin, a novel alkaloidal leukemia and tumor inhibitor from camptotheca acuminata. *J. Am. Chem. Soc.*, **88** (16), 3888–3890.
16 (a) Tang, C. and Rapoport, H. (1972) A total synthesis of dl-camptothecin. *J. Am. Chem. Soc.*, **94** (24), 8615–8616; (b) Comins, D.L., Hong, H. and Jianghua, G. (1994) Asymmetric synthesis of camptothecin alkaloids: a nine-step synthesis of (S)-camptothecin. *Tetrahedron Lett.*, **35** (30), 5331–5334; (c) Curran, D.P., Ko, S.B. and Josien, H. (1996) Cascade radical reactions of isonitriles: a second-generation synthesis of (20S)-camptothecin, topotecan, irinotecan, and Gl-147211C. *Angew. Chem. Int. Ed. Engl.*, **34** (23/24), 2683–2684.
17 Späth, E. and Julian, P.L. (1931) New corydalis alkaloids: d-tetrahydrocoptisine, d-canadine and hydrohydrastinine. *Chem. Ber.*, **64**, 1131–1137.
18 Narasimhan, N.S., Mali, R.S. and Kulkarni, B.K. (1981) Syntheses of 7,8-dimethoxy- and 7,8-methylenedioxyisochroman-3-ones. An efficient synthesis of (±)-tetrahydropalmatine and (±)-canadine. *Tetrahedron Lett.*, **22** (29), 2797–2798.
19 Rabe, P. (1908) Contribution to our knowledge of the cinchona alkaloids. VIII. On the constitution of cinchonine. *Chem. Ber.*, **41**, 62–70.
20 Dauben, W.G., Baskin, C.P. and van Riel, H.C.H.A. (1979) Facile synthesis of codeine from thebaine. *J. Org. Chem.*, **44** (9), 1567–1569.
21 Thoms, H. (1923) Chemical constituents of the rutaceae. *Ber. Deut. Pharm. Ges.*, **33**, 68–83.
22 (a) Tuppy, H. and Böhm, F. (1956) Synthese des dictamnins. *Angew. Chem.*, **68**, 388; (b) Sato, M., Kawakami, K. and Kaneko, C. (1987) Cycloadditions in synthesis. XXXIV. A new method for introducing the 2,2-dichloroethyl group

at the 3-position of the 2-quinoline system and the synthesis of dictamnine. *Chem. Pharm. Bull.*, **35** (3), 1319–1321.

23 (a) Pailer, M. and Porschinski, K. (1949) The constitution of emetine. IV Ipecac alkaloids. *Monatsh. Chem.*, **80**, 94–100; (b) Battersby, A.R. and Openshaw, H.T. (1949) The structure of emetine. IV. Elucidation of the structure of emetine. *J. Chem. Soc.*, 3207–3213.

24 Shamma, M., Blomquist, A.T. and Wasserman, H. (eds) (1972) *The Isoquinoline Alkaloids: Chemistry and Pharmacology*, Academic Press, New York, USA.

25 (a) Battersby, A.R. and Turner, J.C. (1960) Ipecacuanha alkaloids. Part V. Stereospecific synthesis of (+)-O-methylpsychotrine and (−)-emetine. *J. Chem. Soc.*, 717–725; (b) Takano, S., Sasaki, M. and Kanno, H. (1978) New synthesis of (±)-emetine from tetrahydroprotoberberine precursors via an α-diketone monothioketal intermediate. *J. Org. Chem.*, **43** (21), 4169–4172.

26 Ihara, M., Yasui, K., Taniguchi, N. and Fukumoto, K. (1990) Asymmetric total synthesis of (−)-protoemetinol, (−)-protoemetine, (−)-emetine, and (−)-tubulosine by highly stereocontrolled radical cyclisations. *J. Chem. Soc., Perkin Trans. I*, 1469–1476.

27 (a) Battersby, A.R. and Harper, B.J.T. (1959) Ipecacuanha alkaloids. II. Structure of protoemetine and partial synthesis of (−)-emetine. *J. Chem. Soc.*, 1748–1753; (b) Szántay, C., Töke, L. and Kolonits, P. (1966) Synthesis of protoemetine. A new total synthesis of emetine. *J. Org. Chem.*, **31** (5), 1447–1451.

28 Guiles, J.W. and Meyers, A.I. (1991) Asymmetric synthesis of benzoquinolizidines: a formal synthesis of (−)-emetine. *J. Org. Chem.*, **56** (24), 6873–6878.

29 (a) Rubiralta, M., Diez, A., Balet, A. and Bosch, J. (1987) New synthesis of benzo[a]quinolizidine-2-ones via protected 2-aryl-4-piperidones. *Tetrahedron*, **43** (13), 3021–3030; (b) Openshaw, H.T. and Whitaker, N. (1963) The synthesis of emetine and related compounds. IV. A new synthesis of 3-substituted 1,2,3,4,6,7-hexahydro-9,10-dimethoxy-2-oxo-11bH-benzo[α]quinolizines. *J. Chem. Soc.*, (63), 1449–1460; (c) Openshaw, H.T. and Whitaker, N. (1963) A stereochemically favorable synthesis of emetine. *J. Chem. Soc.*, (63), 1461–1471; (d) Fujii, T. and Yoshifuji, S. (1980) Quinolizidines-V: a novel synthetic route to ipecac alkaloids through chemical incorporation of ethyl cincholoiponate derived from the cinchona alkaloid cinchonine. *Tetrahedron*, **36** (11), 1539–1545; (e) Fujii, T., Ohba, M. and Akiyama, S. (1984) A new synthetic approach to benzoquinolizidine alkaloids isolated from alangium lamarckii. *Heterocycles*, **22** (1), 159–164.

30 Itoh, T., Miyazaki, M., Fukuoka, H., Nagata, K. and Ohsawa, A. (2006) Formal total synthesis of (−)-emetine using catalytic asymmetric allylation of cyclic imines as a key step. *Org. Lett.*, **8** (7), 1295–1297.

31 Tietze, L.F., Rackelmann, N. and Sekar, G. (2003) Catalyst-controlled stereoselective combinatorial synthesis. *Angew. Chem. Int. Ed.*, **42** (35), 4254–4257.

32 (a) Kametani, T., Suzuki, Y., Terasawa, H. and Ihara, M. (1979) Studies on the syntheses of heterocyclic compounds. Part 766. A total stereoselective synthesis of emetine and (±)-dihydroprotoemetine. *J. Chem. Soc. Perkin Trans. I*, 1211–1217; (b) van Tamelen, E.E., Placeway, C., Schiemenz, G.P. and Wright, I.G. (1969) Total syntheses of dl-ajmalicine and emetine. *J. Am. Chem. Soc.*, **91** (26), 7359–7371; (c) Burgstahler, A.W. and Bithos, Z.J. (1960) Synthetic applications of hexahydrogallic acid. I. A new route to emetine. *J. Am. Chem. Soc.*, **82** (20), 5466–5474.

33 (a) Lee, H.I., Cassidy, M.P., Rashatasakhon, P. and Padwa, A. (2003) Efficient synthesis of (±)-erysotramidine using an NBS-promoted cyclization reaction of a hexahydroindolinone derivative. *Org. Lett.*, **5** (26), 5067–5070; (b) Gao, S., Tu, Y.Q., Hu, X., Wang, S., Hua, R., Jiang, Y., Zhao, Y., Fan, X. and Zhang, S. (2006) General and efficient

strategy for erythrinan and homoerythrinan alkaloids: syntheses of (±)-3-demethoxyerythratidinone and (±)-erysotramidine. *Org. Lett.*, **8** (11), 2373–2376; (c) Tietze, L.F., Tölle, N., Kratzer, D. and Stalke, D. (2009) Efficient formal total synthesis of the erythrina alkaloid (+)-erysotramidine, using a domino process. *Org. Lett.*, **11** (22), 5230–5233.

34 (a) Folkers, K. and Koniuszy, F. (1939) Erythrina alkaloids. III. Isolation and characterization of a new alkaloid, erythramine. *J. Am. Chem. Soc.*, **61** (5), 1232–1235; (b) Folkers, K. and Koniuszy, F. (1940) Erythrina alkaloids. VII. Isolation and characterization of the new alkaloids, erythraline and erythratine. *J. Am. Chem. Soc.*, **62** (2), 436–441.

35 Banwell, M.G. (2008) New processes for the synthesis of biologically relevant heterocycles. *Pure Appl. Chem.*, **80** (4), 669–679.

36 Fischer, R. (1901) Alkaloids of glaucium luteum. *Arch. Pharm.*, **239**, 426–437.

37 Cava, M.P., Mitchell, M.J., Havlicek, S.C., Lindert, A. and Spangler, R.J. (1970) Photochemical routes to aporphines. New syntheses of nuciferine and glaucine. *J. Org. Chem.*, **35** (1), 175–179.

38 Bohno, M., Sugie, K., Imase, H., Yusof, Y.B., Oishi, T. and Chida, N. (2007) Total synthesis of amaryllidaceae alkaloids, (+)-vittatine and (+)-haemanthamine, starting from d-glucose. *Tetrahedron*, **63** (30), 6977–6989.

39 Falck, J.R. and Manna, S. (1981) An intramolecular Passerini reaction: synthesis of hydrastine. *Tetrahedron Lett.*, **22** (7), 619–620.

40 Rabe, P., Huntenberg, W., Schultze, A. and Volger, G. (1931) Cinchona alkaloids. XXV. Total synthesis of the cinchona alkaloids hydroquinine and hydroquinidine. *Chem. Ber.*, **64B**, 2487–2500.

41 Ahmad, V.U., Rahman, A., Rasheed, T. and Rehman, H. (1987) Jamtine N-oxide. A new isoquinoline alkaloid from cocculus hirsutus. *Heterocycles*, **26** (5), 1251–1255.

42 (a) Padwa, A. and Danca, M.D. (2002) Total synthesis of (±)-jamtine using a thionium/N-acyliminium ion cascade. *Org. Lett.*, **4** (5), 715–717; (b) Simpkins, N.S. and Gill, C.D. (2003) Asymmetric total synthesis of the proposed structure of the medicinal alkaloid jamtine using the chiral base approach. *Org. Lett.*, **5** (4), 535–537; (c) Gill, C.D., Greenhalgh, D.A. and Simpkins, N.S. (2003) Application of the chiral base desymmetrisation of imides to the synthesis of the alkaloid jamtine and the antidepressant paroxetine. *Tetrahedron*, **59** (46), 9213–9230; (d) Pérard-Viret, J., Souquet, F., Manisse, M.-L. and Royer, J. (2010) An expeditious total synthesis of (±)-jamtine using condensation between imine and acid anhydride. *Tetrahedron Lett.*, **51** (1), 96–98.

43 (a) Doyle, T.W., Balitz, D.M., Grulich, R.E., Nettleton, D.E., Gould, S.J., Tann, C.-H. and Moews, A.E. (1981) Structure determination of lavendamycin- a new antitumor antibiotic from streptomyces lanvendulae. *Tetrahedron Lett.*, **22** (46), 4595–4598; (b) Balitz, D.M., Bush, J.A., Bradner, W.T., Doyle, T.W., O'Herron, F.A. and Nettleton, D.E. (1982) *J. Antibiot.*, **35**, 259.

44 (a) Boger, D.L., Duff, S.R., Panek, J.S. and Yasuda, M. (1985) Inverse electron demand Diels–Alder reactions of heterocyclic aza dienes. Studies on the total synthesis of lavendamycin: investigative studies on the preparation of the CDE β-carboline ring system and AB quinoline-5,8-quinone ring system. *J. Org. Chem.*, **50** (26), 5782–5789; (b) Boger, D.L., Duff, S.R., Panek, J.S. and Yasuda, M. (1985) Total synthesis of lavendamycin methyl ester. *J. Org. Chem.*, **50** (26), 5790–5795.

45 (a) Kende, A.S. and Ebetino, F.H. (1984) The regiospecific total synthesis of lavendamycin methyl ester. *Tetrahedron Lett.*, **25** (9), 923–926; (b) Kende, A.S., Ebetino, F.H., Battista, R., Boatman, R.J., Lorah, D.P. and Lodge, E. (1984) New tactics in heterocycles synthesis. *Heterocycles*, **21** (1), 91–106.

46 Ciufolini, M.A. and Bishop, M.J. (1993) Studies towards streptonigrinoids: formal synthesis of lavendamycin

methyl ester. *J. Chem. Soc. Chem. Comm.*, 1463–1464.

47 Verniest, G., Wang, X., De Kimpe, N. and Padwa, A. (2010) Heteroaryl cross-coupling as an entry toward the synthesis of lavendamycin analogues: a model study. *J. Org. Chem.*, **75** (2), 424–433.

48 Djerassi, C., Nakano, T. and Bobbitt, J.M. (1958) Alkaloid studies. XX. Isolation and structure of two new cactus alkaloids, piloceredine and lophocerine. *Tetrahedron*, **2**, 58–63.

49 Bobbitt, J.M. and Chou, T.-T. (1959) Synthesis of isoquinoline alkaloids. I. Lophocerine. *J. Org. Chem.*, **24** (8), 1106–1108.

50 Martin, S.F. and Tu, C.-Y. (1981) General strategies for alkaloid synthesis via intramolecular [4 + 2] cycloadditions of enamides. Application to the formal total synthesis of racemic lycorine. *J. Org. Chem.*, **46** (18), 3763–3764.

51 (a) Bru, C., Thal, C. and Guillou, C. (2003) Concise total synthesis of (±)-maritidine. *Org. Lett.*, **5** (11), 1845–1846; (b) Roe, C. and Stephenson, G.R. (2008) Electrophilic C$_{12}$ building blocks for alkaloids: formal total synthesis of (±)-maritidine. *Org. Lett.*, **10** (2), 189–192.

52 Sertürner, F.W.A. (1805) *Trommsdorffs J. Pharm.*, **13**, 234.

53 (a) Gates, M. and Tschudi, G. (1952) The synthesis of morphine. *J. Am. Chem. Soc.*, **74** (4), 1109–1110; (b) Gates, M. and Tschudi, G. (1956) The synthesis of morphine. *J. Am. Chem. Soc.*, **78** (7), 1380–1393.

54 Hong, C.Y., Kado, N. and Overman, L.E. (1993) Asymmetric synthesis of either enantiomer of opium alkaloids and morphinans. Total synthesis of (−)- and (+)-dihydrocodeinone and (−)- and (+)-morphine. *J. Am. Chem. Soc.*, **115** (23), 11028–11029.

55 (a) Iijima, I., Minamikawa, J.-I., Jacobson, A.E., Brossi, A. and Rice, K.C. (1978) Studies in the (+)-morphinan series. 4. A markedly improved synthesis of (+)-morphine. *J. Org. Chem.*, **43** (7), 1462–1463; (b) Rice, K.C. (1980) Synthetic opium alkaloids and derivatives. A short total synthesis of (±)-dihydrothebainone, (±)-dihydrocodeinone, and (±)-nordihydrocodeinone as an approach to a practical synthesis of morphine, codeine, and congoners. *J. Org. Chem.*, **45** (15), 3135–3137.

56 (a) White, J.D., Hrnciar, P. and Stappenbeck, F. (1997) Asymmetric synthesis of (+)-morphine. The phenanthrene route revisited. *J. Org. Chem.*, **62** (16), 5250–5251; (b) Trost, B.M. and Tang, W. (2002) Enantioselective synthesis of (−)-codeine and (−)-morphine. *J. Am. Chem. Soc.*, **124** (49), 14542–14543; (c) Uchida, K., Yokoshima, S., Kan, T. and Fukuyama, T. (2006) Total synthesis of (±)-morphine. *Org. Lett.*, **8** (23), 5311–5313.

57 Eijkman, J.F. (1884) *Chem. Ber.*, **17**, 441.

58 Kametani, T., Sugai, T., Shoji, Y., Honda, T., Satoh, F. and Fukumoto, K. (1977) Studies on the synthesis of heterocyclic compounds. Part 698. An alternative protoberberine synthesis: total synthesis of (±)-xylopinine, (±)-schefferine, (±)-nandinine, (±)-corydaline, and (±)-thalictricavine. *J. Chem. Soc. Perkin Trans. 1*, 1151–1155.

59 LeGendre, O., Pecic, S., Chaudhary, S., Zimmerman, S.M., Fantegrossi, W.E. and Harding, W.W. (2010) Synthetic studies and pharmacological evaluations of the MDMA ('ecstasy') antagonist nantenine. *Bioorg. Med. Chem. Lett.*, **20** (2), 628–631.

60 (a) Arthur, H.R., Hui, W.H. and Ng, Y.L. (1958) Structures of new benzphenanthridine alkaloids from zanthoxylum nitidum. *Chem. Ind. (Lond.)*, 1514; (b) Arthur, H.R., Hui, W.H. and Ng, Y.L. (1959) Examination of rutaceae of Hong Kong. II. Alkaloids, nitidine and oxynitidine, from zanthoxylum nitidum. *J. Chem. Soc.*, 1840–1845.

61 Enomoto, G., Girard, A.-L., Yasui, Y. and Takemoto, Y. (2009) Gold(I)-catalyzed tandem reactions initiated by hydroamination of alkynyl carbamates: application to the synthesis of nitidine. *J. Org. Chem.*, **74** (23), 9158–9164.

62 Cushman, M. and Cheng, L. (1978) Total synthesis of nitidine chloride. *J. Org. Chem.*, **43** (2), 286–288.

63 Robiquet, P.J. (1817) *Ann. Chim. Phys.*, **5**, 275.

64 (a) Kerekes, P. and Bognar, R. (1971) Synthese des Gnoscopins (dl-Narcotin). *J. Prakt. Chem.*, **313**, 923–928; (b) Soriano, M.D.P.C., Shankaraiah, N. and Santos, L.S. (2010) Short synthesis of noscapine, bicuculline, egenine, and corytensine alkaloids through the addition of 1-siloxy-isobenzofurans to imines. *Tetrahedron Lett.*, **51** (13), 1770–1773.

65 Kakemi, K. (1940) *Yakugaku Zassli*, **60**, 11.

66 Teitel, S., O'Brien, J. and Brossi, A. (1972) Alkaloids in mammalian tissues. 2. Synthesis of (+) and (–)-1-substituted-6,7-dihydroxy-1,2,3,4-tetrahydroisoquinolines. *J. Med. Chem.*, **15** (8), 845–846.

67 Feist, K. and Dschu, G.L. (1925) Alkaloids of columbo root. III. *Arch. Pharm.*, **263**, 294–305.

68 (a) Haworth, R.D., Koepfli, J.B. and Perkin, W.H. (1927) New synthesis of oxyberberine and a synthesis of palmatine. *J. Chem. Soc.*, 548–554; (b) Kiparissides, Z., Fichtner, R.H., Poplawski, J., Nalliah, B.C. and MacLean, D.B. (1980) A regiospecific synthesis of protoberberine alkaloids. *Can. J. Chem.*, **58** (24), 2770–2779.

69 (a) Pictet, A. and Gams, A. (1909) Synthesis of papaverine. *Chem. Ber.*, **42**, 2943–2952; (b) Goldberg, A.A. (1954) Papaverine: its synthesis and pharmacology. *Chem. Prod. Chem. News*, **17**, 371–374.

70 (a) Brossi, A., Schenker, F. and Leimgruber, W. (1964) Syntheses in the isoquinoline series. New syntheses of the cactus alkaloids anhalamine, anhalidine, dl-anhalonidine, and dl-pellotine. *Helv. Chim. Acta*, **47** (7), 2089–2098; (b) Ruchirawat, S., Chaisupakitsin, M., Patranuwatana, N., Cashaw, J.L. and Davis, V.E. (1984) A convenient synthesis of simple tetrahydroisoquinolines. *Synth. Commun.*, **14** (13), 1221–1228.

71 Battersby, A.R., Kapil, R.S., Bhakuni, B.S., Popli, S.P., Merchant, J.R. and Salgar, S.S. (1966) New alkaloids from alamgium lamarckii. *Tetrahedron Lett.*, **7** (41), 4965–4971.

72 Chang, J.-K., Chang, B.-R., Chuang, Y.-H. and Chang, N.-C. (2008) Total synthesis of (±)-protoemetinol. *Tetrahedron*, **64** (41), 9685–9688.

73 (a) Carr, F.H. and Pyman, F.L. (1914) Alkaloids of ipecacuanha. *J. Chem. Soc.*, (105), 1591–1638; (b) Hesse, O. (1914) Beitrag zur Kenntnis der Alkaloide der echten Brechwurzel. *Liebigs Ann. Chem.*, **405** (1), 1–57.

74 Gutzwiller, J. and Uskokovi_ UNDEFINED, M.R. (1978) Total synthesis of cinchona alkaloids. 2. Stereoselective syntheses of quinine and quinidine. *J. Am. Chem. Soc.*, **100** (2), 576–581.

75 Pelletier, P.J. and Caventau, J.B. (1820) *Annal. Chim. Phys.*, **15**, 291.

76 Smith, A.C. and Williamns, R.M. (2008) Rabe rest in peace: confirmation of the Rabe-Kindler conversion of d-quinotoxine into quinine: experimental affirmation of the Woodward-Doering formal total synthesis of quinine. *Angew. Chem. Int. Ed.*, **47**, 1736–1740.

77 Woodward, R.B. and Doering, W.E. (1944) The total synthesis of quinine. *J. Am. Chem. Soc.*, **66** (5), 849.

78 Rabe, P. and Kindler, K. (1918) Über die partielle Synthese des Chinins. Zur Kenntnis der China-Alkaloide XIX. *Chem. Ber.*, **51**, 466–467.

79 Proštenik, M. and Prelog, V. (1943) Synthetische Versuche in der Reihe der China-Alkaloide. (4. Mitteilung). Über Homo-merochinen und über die partielle Synthese des Chinotoxins. *Helv. Chim. Acta*, **26** (6), 1965–1971.

80 Stork, G., Niu, D., Fujimoto, A., Koft, E.R., Balkovec, J.M., Tata, J.R. and Dake, G.R. (2001) The first stereoselective total synthesis of quinine. *J. Am. Chem. Soc.*, **123** (14), 3239–3242.

81 Pasteur, L. (1853) *Comptes rendus*, **37**, 110.

82 (a) Gopinath, K.W., Govindachari, T.R., Pai, B.R. and Viswanathan, N. (1959) The structure of reticuline. *Chem. Ber.*, **92**, 776–779; (b) Brochmann-Hanssen,

E. and Furuya, T. (1964) New opium alkaloid. *J. Pharm. Sci.*, **53** (5), 575.

83 Rice, K.C. and Brossi, A. (1980) Expedient synthesis of racemic and optically active N-norreticuline and N-substituted and 6′-bromo-N-norreticulines. *J. Org. Chem.*, **45** (4), 592–601.

84 Matson, J.A. and Bush, J.A. (1989) Sandramycin, a novel antitumor antibiotic produced by a nocardioides sp. Production, isolation, characterization and biological properties. *J. Antibiot.*, **42** (12), 1763–1767.

85 Boger, D.L., Chen, J.-H. and Saionz, K.W. (1996) (–)-Sandramycin: total synthesis and characterization of DNA binding properties. *J. Am. Chem. Soc.*, **118** (7), 1629–1644.

86 Schmidt, E. et al. (1893) *Arch. Pharm.*, **231**, 145.

87 Dyke, S.F., Moon, B.J. and Sainsbury, M. (1968) The synthesis of sanguinarine. *Tetrahedron Lett.*, **9** (36), 3933–1934

88 Honda, (1904) *Arch. Exp. Pathol. Pharmakol.*, **52**, 83.

89 Cheng, J.-J. and Yang, Y.-S. (2009) Enantioselective total synthesis of (–)-(S)-stepholidine. *J. Org. Chem.*, **74** (23), 9225–9228.

90 Rao, K.V. and Cullen, W.P. (1959–1960) Streptonigrin, an antitumor substance. I. Isolation and characterization. *Antibiot. Ann.*, 950–953.

91 (a) Basha, F.Z., Hibino, S., Kum, D., Pye, W.E., Wu, T.-T. and Weinreb, S.M. (1980) Total synthesis of streptonigrin. *J. Am. Chem. Soc.*, **102** (11), 3962–3964; (b) Kende, A.S., Lorah, D.P. and Boatman, R.J. (1981) A new and efficient total synthesis of streptonigrin. *J. Am. Chem. Soc.*, **103** (5), 1271–1273; (c) Weinreb, S.M., Basha, F.Z., Hibino, S., Khatri, N.A., Kim, D., Pye, W.E. and Wu, T.-T. (1982) Total synthesis of the antitumor antibiotic streptonigrin. *J. Am. Chem. Soc.*, **104** (2), 536–544.

92 Matulenko, M.A. and Meyers, A.I. (1996) Total synthesis of (–)-tetrahydropalmatine via chiral formamidine carbanions: unexpected behavior with certain ortho-substituted electrophiles. *J. Org. Chem.*, **61** (2), 573–580.

93 (a) Bradsher, C.K. and Dutta, N.L. (1961) Aromatic cyclodehydration. XLVI. Synthesis of tetrahydropalmatine and its analogs. *J. Org. Chem.*, **26** (7), 2231–2234; (b) Kametani, T. and Ihara, M. (1967) Synthesis of heterocyclic compounds. CLXXII. An alternative total synthesis of (±)-tetrahydropalmatine. *J. Chem. Soc.*, 530–532.

94 Manske, R.H.F. (1953) The alkaloids of fumariaceous plants. XLIX. Thalictricavine, a new alkaloid from corydalis tuberose DC. *J. Am. Chem. Soc.*, **75** (20), 4928–4929.

95 Dutcher, J.D. (1946) Curare alkaloids from chondodendron tomentosum Ruiz and. Pavon. *J. Am. Chem. Soc.*, **68** (3), 419–424.

96 Seeman, J.I. (2007) The Woodward-Doering/Rabe-Kindler total synthesis of quinine: setting the record straight. *Angew. Chem. Int. Ed.*, **46**, 1378–1413.

97 Campbell, K.N., Tipson, R.S., Elderfield, R.C., Campbell, B.K., Clapp, M.A., Gensler, W.J., Morrison, D. and Moran, W.J. (1946) Synthesis of ethyl quininate. *J. Org. Chem.*, **11** (6), 803–811.

98 Fritsch, P. (1895) Synthesis of isoquinoline derivatives. *Liebigs Ann. Chem.*, **286** (10), 1–26.

99 Woodward, R.B. and Doering, W.E. (1945) The total synthesis of quinine. *J. Am. Chem. Soc.*, **67** (5), 860–874.

100 Nicolaou, K.C. and Snyder, S.A. (2003) *Classics in Total Synthesis II*, Wiley-VCH Verlag GmbH, Weinheim, Germany.

101 (a) Uskoković, M.R., Henderson, T., Reese, C., Lee, H.L., Grethe, G. and Gutzwiller, J. (1978) Total synthesis of cinchona alkaloids. 1. Synthesis of meroquinene. *J. Am. Chem. Soc.*, **100** (2), 571–576; (b) Martinelli, M.J., Peterson, B.C., Khau, V.V., Hutchinson, D.R. and Sullivan, K.A. (1993) A novel, stereoselective synthesis of cis-4a(S),8a(R)-decahydro-6(2H)-isoquinolones from meroquinene esters. *Tetrahedron Lett.*, **34** (34), 5413–5416; (c) Hutchinson, D.R., Khau, V.V., Martinelli, M.J., Nayyar, N.K., Peterson, B.C. and Sullivan, K.A. (1998) Synthesis of cis-4a(S),8a(R)-

perhydro-6(2H)-isoquinolinones from quinine: cis-4a(S),8a(R)-2-bezoyloctahydro-6(2H)-isoquinolinone. *Org. Synth.*, **75**, 223.

102 Boger, D.L. and Chen, J.-H. (1995) A modified Friedlander condensation for the synthesis of 3-hydroxyquinoline-2-carboxylates. *J. Org. Chem.*, **60** (22), 7369–7371.

103 (a) Rao, K.V., Beimann, K. and Woodward, R.B. (1963) The structure of streptonigrin. *J. Am. Chem. Soc.*, **85** (16), 2532–2533; (b) Gould, S.J., Chang, C.C., Darling, D.S., Roberts, J.D. and Squillacote, M. (1980) Streptonigrin biosynthesis. 4. Details of the tryptophan metabolism. *J. Am. Chem. Soc.*, **102** (5), 1707–1712.

104 Boger, D.L., Yasuda, M., Mitscher, L.A., Drake, S.D., Kitos, P.A. and Thompson, S.C. (1987) Streptonigrin and lavendamycin partial structures. Probes for the minimum, potent pharmacophore of streptonigrin, lavendamycin, and synthetic quinoline-5,8-diones. *J. Med. Chem.*, **30** (10), 1918–1928.

105 England, D. and Padwa, A. (2008) Gold-catalyzed cycloisomerization of N-propargylindole-2-carboxamides: application toward the synthesis of lavendamycin analogues. *Org. Lett.*, **10** (16), 3631–3634.

106 Woodward, R.B., Pachter, I.J. and Scheinbaum, M.L. (1971) Dithiotosylates as reagents in organic synthesis. *J. Org. Chem.*, **36** (8), 1137–1139.

107 Still, W.C. and Gennari, C. (1983) Direct synthesis of Z-unsaturated esters. A useful modification of the Horner-Emmons olefination. *Tetrahedron Lett.*, **24** (41), 4405–4408.

108 Hanaoka, M., Yamagishi, H., Marutani, M. and Mukai, C. (1987) Chemical transformation of protoberberines. XIII. A novel and efficient synthesis of antitumor benzo[c]phenanthridine alkaloids, nitidine and fagaronine. *Chem. Pharm. Bull.*, **35** (6), 2348–2354.

109 (a) Ma, Z.-Z., Xu, W., Jensen, N.H., Roth, B.L., Liu-Chen, L.-Y. and Lee, D.Y.W. (2008) Isoquinoline alkaloids isolated from corydalis yanhusuo and their binding affinities at the dopamine D1 receptor. *Molecules*, **13**, 2303–2312; (b) Majak, W., Bai, Y. and Benn, M.H. (2003) Phenolic amides and isoquinoline alkaloids from corydalis sempervirens. *Biochem. Sys. Ecol.*, **31**, 649–651; (c) Pang, S.-Q., Wang, G.-Q., Huang, B.-K., Zhang, Q.-Y. and Qin, L.-P. (2007) Isoquinoline alkaloids from broussonetia papyrifera fruits. *Chem. Nat. Compd.*, **43** (1), 100; (d) Bringmann, G., Gulder, T., Hertlein, B., Hemberger, Y. and Meyer, F. (2010) Total synthesis of the N,C-coupled naphthylisoquinoline alkaloids ancistrocladinium A and B and related analogues. *J. Am. Chem. Soc.*, **132** (3), 1151–1158; (e) Montagnac, A., Litaudon, M. and Pais, M. (1997) Quinine- and quinicine-derived alkaloids from guettarda noumeana. *Phytochem.*, **46** (5), 973–975.

10
Carbazoles and Acridines
Konstanze K. Gruner and Hans-Joachim Knölker

10.1
Introduction to Carbazoles

Carbazoles have been isolated mostly from plants of the genera *Murraya*, *Clausena*, *Glycosmis* and *Micromelum*, all belonging to the family of Rutaceae. Another genus which is a common natural source for carbazoles is *Ekebergia* (family Meliaceae). Nevertheless, *Murraya euchrestifolia* Hayata stands out as the richest source for carbazole alkaloids from terrestrial plants. Several examples of natural carbazoles have been isolated from different *Streptomyces* species. Moreover, carbazoles have been obtained from slime moulds and marine sources. The main representatives are the blue-green alga *Hyella caespitosa*, species of the genera *Aspergillus*, *Actinomadura*, *Didemnum*, *Iotrochota* and the human pathogenic yeast *Malassezia furfur*.

In Table 10.1 we summarize some of the most important carbazole alkaloids, which represent interesting targets for natural product syntheses because of their pharmacological potential. In view of the large number of carbazole alkaloids isolated from nature, this chapter emphasizes only the highlights. For comprehensive reviews, see [1–4].

10.2
Total Synthesis of Carbazole Alkaloids

Due to their biological and pharmacological activities, carbazoles are important targets for natural product synthesis. Therefore, a variety of methodologies for the synthesis of the carbazole framework has been developed [1–4, 174]. The Graebe–Ullmann and the Fischer–Borsche synthesis represent classical routes to carbazoles. However, these methods have limitations regarding the synthesis of highly substituted and functionalized carbazole derivatives.

Today, the most efficient syntheses are based on transition metal-induced coupling reactions of arylamines. The iron-mediated synthesis and palladium-catalyzed procedures emerged as the most versatile methods for the construction of carbazoles [1–4].

Heterocycles in Natural Product Synthesis, First Edition. Edited by Krishna C. Majumdar and Shital K. Chattopadhyay.
© 2011 Wiley-VCH Verlag GmbH & Co. KGaA. Published 2011 by Wiley-VCH Verlag GmbH & Co. KGaA.

Table 10.1 Selected carbazole natural products.

Trivial name	Structure	Source and isolation	Bioactivity	Synthesis
Antiostatin A_1 Antiostatin A_2 Antiostatin A_3 Antiostatin A_4	A_1: R = $(CH_2)_4$Me A_2: R = $(CH_2)_2$CHMeCH$_2$Me A_3: R = $(CH_2)_4$CHMe$_2$ A_4: R = $(CH_2)_6$Me	*Streptomyces cyaneus* 2007-SV$_1$ [5]	Strong inhibition of free radical-induced lipid peroxidation [5]	A_1-A_4 [6], A_1 [7]
Antiostatin B_2 Antiostatin B_3 Antiostatin B_4 Antiostatin B_5	B_2: R = $(CH_2)_5$Me B_3: R = $(CH_2)_4$CHMe$_2$ B_4: R = $(CH_2)_6$Me B_5: R = $(CH_2)_5$CHMe$_2$	*Streptomyces cyaneus* 2007-SV$_1$ [5]	Strong inhibition of free radical-induced lipid peroxidation [5]	B_2-B_5 [6]
Arcyriaflavin A	Arcyriaflavin A	A, B, C: *Arcyria nutans, Arcyria denudata* [8, 9] B, C: *Metatrichia vesparium* [10]	Antibiotic and antifungal [8]	A: [3, 11–13], B: [14] B-D: [15]
Arcyriaflavin B				
Arcyriaflavin C		D: *Dictydiaethalium plumbeum* [8], see also [3]		
Arcyriaflavin D	Arcyriaflavin B: R, R' = H Arcyriaflavin C: R = H, R' = OH Arcyriaflavin D: R = OH, R' = H			

Calothrixin A		Calothrix cyanobacteria [16]	Antiplasmodial (antimalarial) and anticancer [16]	[17, 19]
Calothrixin B				[17–26]
Carazostatin		Streptomyces chromofuscus [27]	Radical scavenger [27, 28]	[29–32]
Carbalexin C		Glycosmis parviflora [33]	Antifungal [33]	[34]
Carbazomadurin A		Actinomadura madurae 2808-SV1 [35] abs. config. [36]	Antioxidative (neuronal cell protector) [35]	A: [37], B: [36]
Carbazomadurin B				

(Continued)

Table 10.1 (Continued)

Trivial name	Structure	Source and isolation	Bioactivity	Synthesis
Carbazomycin A Carbazomycin B	A: R = Me B: R = H	A–H: *Streptoverticillium ehimense* H 1051-MY 10 [38–41]	A, B: antifungal, antibacterial [38, 40] B: radical scavenger [28]	A: [44] A, B: [45–49] B: [50–52] C, D: [53] E: [45, 54] G, H: [55–57] G: [58]
Carbazomycin C Carbazomycin D	C: R = H D: R = Me		B, C: inhibition of 5-lipoxygenase [43]	
Carbazomycin E Carbazomycin F	E: R = H F: R = OMe	E, F: *Streptoverticillium* KCC U-0166 [42]	C, D, E, F: antimicrobial [40, 42]	
Carbazomycin G Carbazomycin H	G: R = H H: R = OMe		G, H: antifungal [38]	
Carquinostatin A		*Streptomyces exfoliatus* 2419-SVT2 [59]	Neuronal protector (radical scavenger) [59]	[60, 61]
Chlorohyellazole		*Hyella caespitosa* [62]		[63–66]
Clausenol		*Clausena anisata* [67]	Antibiotic [67]	[67–69]

10.2 Total Synthesis of Carbazole Alkaloids | 345

Clausine I		*Clausena excavata* [70, 71]	Inhibition of platelet aggregation [70]	[69]
Clausine K (Clauszoline-J)		*Clausena excavata* [70, 72]	Inhibition of platelet aggregation [70] Anti-HIV-1 [73] Anti-TB [74, 75]	[76]
Clausine L		*Clausena excavata* [77]		[78, 79]
Clausine Z		*Clausena excavata* [80]	Inhibition of CDK5, neuronal cell protector [80]	[69]
trans-Dihydroxy-girinimbine		*Murraya euchrestifolia* [81]		[81b, 82]
Ellipticine		*Ochrosia elliptica* Labill., *O. sandwicensis* A.DC. [83]; see also [3]	Anticancer [84]	[1, 85–90]
Epocarbazolin A Epocarbazolin B		*Streptomyces anulatus* T688-8 [91]	Inhibition of 5-lipoxygenase and antibacterial [91]	[92]
Euchrestifoline		*Murraya euchrestifolia* [93]		[94]

(*Continued*)

Table 10.1 (Continued)

Trivial name	Structure	Source and isolation	Bioactivity	Synthesis
Eustifoline-A (Glycomaurin) Eustifoline-B	A: R = H; B: R = prenyl; R = geranyl	A: *Glycosmis mauritiana* [95] A, B, C, D: *Murraya euchrestifolia* [96]		A, B, C, D: [97] D: [98]
Eustifoline-C				
Eustifoline-D				
Furoclausine A		*Clausena excavata* [99]		[100]
Furostifoline		*Murraya euchrestifolia* [96]		[101–107]
Girinimbine		*Murraya koenigii* [108], *Clausena heptaphylla* [109]	Antitumor [110] Apoptosis inducer [111]	[82, 94]
Glycomaurrol		*Glycosmis mauritiana* [95]		[97, 98]

Hyellazole		Hyella caespitosa [62]	[30, 31, 46, 63–66, 112–119]
Lavanduquinocin		Streptomyces viridochromogenes 2942-SVS3 [120]	Neuronal cell protector [120] [121, 122]
Micromeline		Micromelum hirsutum [123]	Anti-TB [123] [98]
Mukonidine		Murraya koenigii [124] Clausena excavata [77]	[79, 125, 126]
Murrayacine		Murraya koenigii [127, 128] Clausena heptaphylla [129]	[81b, 82]
Murrayanine		Murraya koenigii [130], Clausena heptaphylla [131], Glycosmis stenocarpa [132]	Antimicrobial [130a] [133–135]
Neocarazostatin B		Streptomyces species strain GP 38 [136], abs. config. [137]	Strong inhibition of free radical-induced lipid peroxidation [136] [137, 138]

(Continued)

Table 10.1 (Continued)

Trivial name	Structure	Source and isolation	Bioactivity	Synthesis
Olivacine		*Peschiera buchtieni* [139], see also [3]		[1, 140–143]
Pityriazole		*Malassezia furfur* [144]		[78]
Rebeccamycin		*Saccharothrix aerocolonigenes* C38383-RK-2 [145]	Anticancer [146]	[147–149]
Siamenol		*Murraya siamensis* [150]	Anti-HIV [150]	[151, 152]
(+)-Staurosporine (AM-2282)		*Streptomyces staurosporeus* [153], abs. config. [154]	Antimicrobial, hypotensive, cytotoxic, inhibition of PKC [155]	[156–159]

Staurosporinone (K-252c)		*Nocardiopsis* sp. K-290 [160, 161], *Eudistoma* sp. [162]	Anticancer, inhibition of PKC isoenzymes [160, 162]	[3, 12, 163–169]
Tijipanazole B		*Tolypothrix tjipanasensis* [170]	Antifungal [170]	D, E: [170] B, D, E, I: [171] F1: [172] F2: [173]
Tjipanazole D	D: R = Cl I: R = H			
Tjipanazole E				
Tjipanazole F1	B: R', R''' = Cl, R'''' = H E: R', R''' = Cl, R'''' = CH$_2$OH F1: R' = Cl, R'', R'''' = H			
Tjipanazole F2	Tjipanazole F2			
Tjipanazole I				

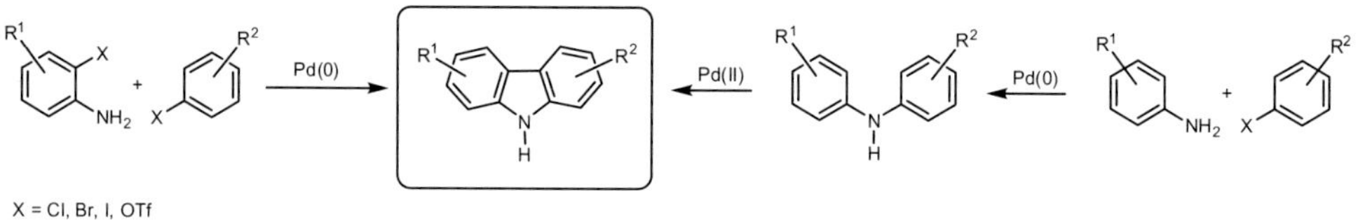

Figure 10.1 Synthetic approaches by Sakamoto/Bedford (left) and by Knölker (right).

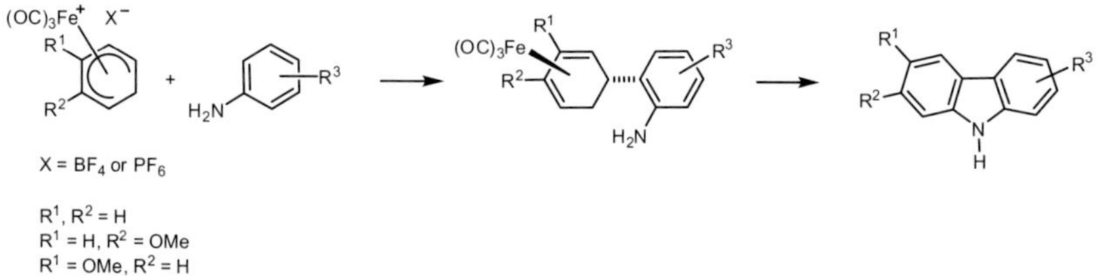

Figure 10.2 General route for the iron-mediated carbazole synthesis.

A two-step process of Buchwald–Hartwig amination and subsequent palladium(II)-catalyzed oxidative cyclization of the *N,N*-diarylamine by double C–H bond activation was applied to the synthesis of a wide range of carbazole alkaloids [2, 4] (Figure 10.1). Sakamoto and Bedford have developed an alternative one-pot palladium(0)-catalyzed route to carbazoles consisting of Buchwald–Hartwig amination using *ortho*-haloanilines and in situ ring closure of the *N,N*-diarylamine via C–H bond activation [175, 176].

The iron-mediated carbazole synthesis involves two steps: the electrophilic aromatic substitution of an arylamine with a tricarbonyl(η^5-cyclohexadienylium)iron salt followed by oxidative cyclization (Figure 10.2). There are three common tricarbonyliron complex salts used for synthesizing carbazole alkaloids [6, 53, 100]. In general these tricarbonyliron complex salts are prepared by complexation of the corresponding cyclohexadiene with pentacarbonyliron [177].

10.2.1
Palladium-Catalyzed Synthesis of Carbazoles

10.2.1.1 Total Synthesis of Pityriazole

Because of the 3-indolyl substituent at the 1-position, pityriazole is a carbazole alkaloid with a unique structure. Three years after Steglich *et al.* had reported the isolation of this natural product [144], we described the first total synthesis of pityriazole using a highly efficient palladium-catalyzed route [78]. Pityriazole is structurally related to mukonidine and clausine L. Therefore, these alkaloids were considered as potential key intermediates for the total synthesis of pityriazole. A sequence of Buchwald–Hartwig amination, palladium(II)-catalyzed oxidative cyclization and Suzuki–Miyaura coupling was envisaged for framework construction (Scheme 10.1).

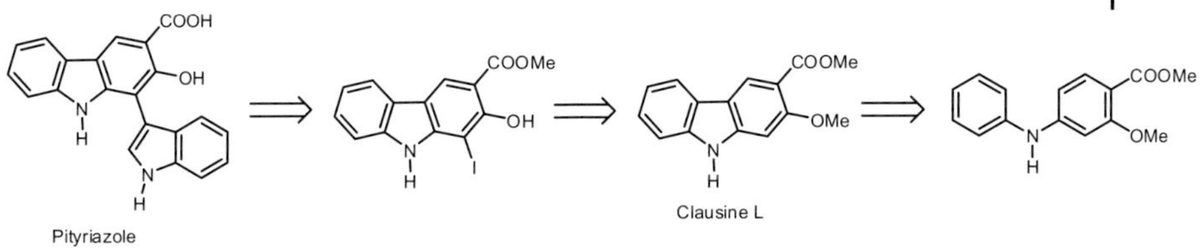

Scheme 10.1 Retrosynthetic analysis of pityriazole.

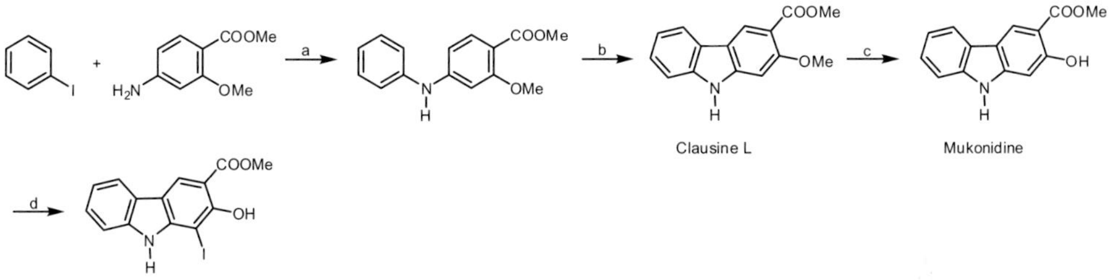

Scheme 10.2 Synthesis of 1-iodomukonidine. Reaction conditions: (a) cat. Pd(OAc)$_2$, cat. BINAP, Cs$_2$CO$_3$, toluene, reflux, 2 d, 100%; (b) cat. Pd(OAc)$_2$, cat. Mn(OAc)$_3$·2 H$_2$O, pivalic acid, air, 100 °C, 3 d, 62%; (c) BBr$_3$, CH$_2$Cl$_2$, −78 to −4 °C, 1 h, 95%; (d) I$_2$, Cu(OAc)$_2$, 80 °C (microwave), 2 h, 85%.

Buchwald–Hartwig amination of iodobenzene with commercial methyl 4-amino-2-methoxybenzoate led to the N,N-diarylamine which on palladium(II)-catalyzed oxidative cyclization afforded clausine L. Ether cleavage of clausine L provided mukonidine, the structure of which had been under discussion for some time [79, 125, 126]. The hydroxy group of mukonidine directs the subsequent electrophilic iodination to the 1-position of the carbazole framework (Scheme 10.2).

Suzuki–Miyaura coupling of 1-iodomukonidine with 1-(phenylsulfonyl)indol-3-yl-boronic acid provided the 1-(indol-3-yl)carbazole. Finally, cleavage of the methyl ester and the phenylsulfonyl group led to pityriazole which was obtained in six steps and 35% overall yield (Scheme 10.3). Our approach to pityriazole also includes the first total synthesis of clausine L.

10.2.1.2 Total Synthesis of Euchrestifoline and Girinimbine

Euchrestifoline and girinimbine belong to the pyrano[3,2-a]carbazole alkaloid family which is characterized by a cyclic terpenoid unit. Biogenetically, pyranocarbazoles and furocarbazoles are presumably formed by cyclization of a corresponding prenylated precursor. Euchrestifoline was obtained by Chakraborty and Islam as an intermediate in their synthesis of girinimbine [178]; 25 years later, this

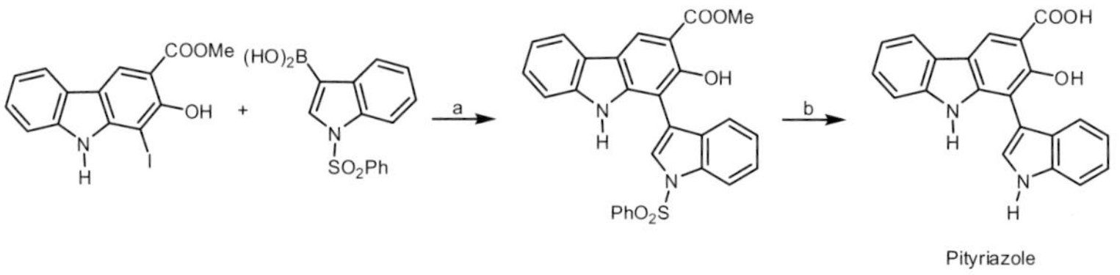

Scheme 10.3 Synthesis of pityriazole. Reaction conditions: (a) cat. Pd(OAc)$_2$, cat. S-Phos, moist K$_3$PO$_4$, dioxane, 90 °C (microwave), 3 h, 82%; (b) KOH, EtOH, 40 °C, 2 d, 86%.

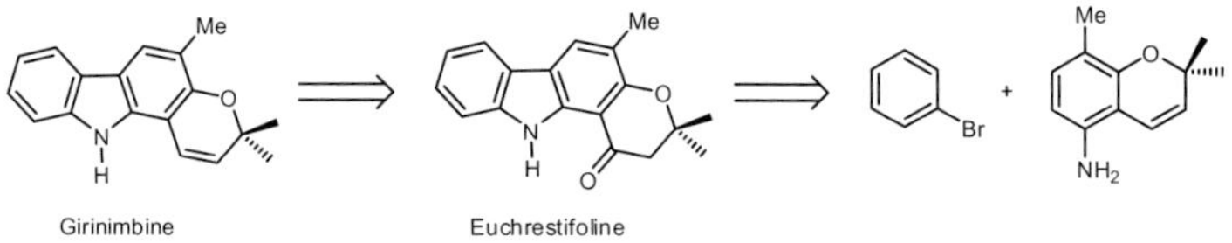

Scheme 10.4 Retrosynthetic analysis of girinimbine and euchrestifoline.

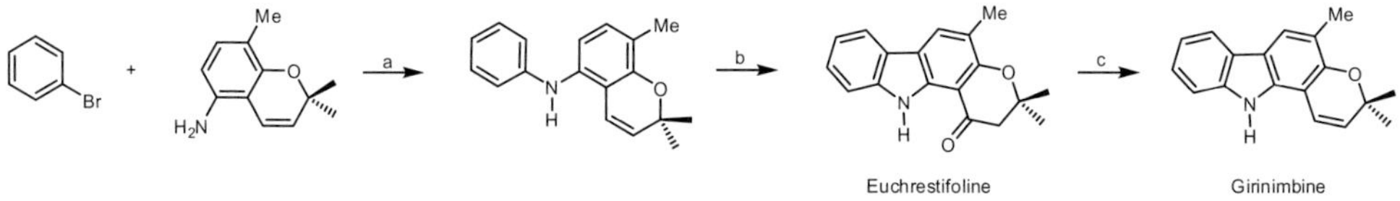

Scheme 10.5 Synthesis of euchrestifoline and girinimbine. Reaction conditions: (a) cat. Pd(OAc)$_2$, cat. BINAP, Cs$_2$CO$_3$, toluene, reflux, 36 h, 93%; (b) cat. Pd(OAc)$_2$, cat. Cu(OAc)$_2$, HOAc-H$_2$O (10:1), 90 °C, 48 h, 40%; (c) 1. LiAlH$_4$, THF, 0 °C to rt, 17 h, 2. HCl, 60 °C, 1 h, 70%.

compound was isolated from natural sources [93]. We have developed a palladium(II)-catalyzed synthesis of euchrestifoline by application of a one-pot triple C–H bond activation. This remarkable reaction sequence consists of a Wacker oxidation to generate the arylketone followed by an aryl–aryl bond formation to construct the carbazole framework (Scheme 10.4) [94].

Buchwald–Hartwig amination of bromobenzene with 5-amino-2,2,8-trimethyl-2H-chromene afforded an N,N-diarylamine [94]. Subsequent heating in a solvent mixture of acetic acid and water in the presence of catalytic amounts of palladium(II) acetate and copper(II) acetate for 48 hours in air directly provided euchrestifoline. Activation of the vinyl C–H bond takes place first and subsequent double aryl C–H bond activation forms the biaryl axis. This sequence was confirmed by isolation of the initial Wacker product and its subsequent conversion to euchrestifoline. Our approach provides euchrestifoline in two steps and 37% overall yield based on bromobenzene. Moreover, the transformation of euchrestifoline to girinimbine has been improved (Scheme 10.5).

10.2.2
Iron-Mediated Synthesis of Carbazoles

10.2.2.1 Total Syntheses of the Antiostatins

The antiostatins belong to the 3-oxygenated carbazole alkaloids, which are known for their strong inhibitory activity of free radical induced lipid peroxidation. In 1990, Seto et al. isolated the antiostatins A_1 to A_4 and B_2 to B_5 [5]. The antiostatins are unique carbazole alkaloids due to the nitrogen substituent at the 4-position. For the antiostatins of the A series, this nitrogen is acetylated and in the B series the nitrogen is part of a 5-isobutylbiuret substituent. Our synthetic strategy was based on an iron-mediated carbazole synthesis. Introduction of the nitrogen substituent at C4 was envisaged after formation of the carbazole framework (Scheme 10.6).

As an example for the synthesis of the whole antiostatin family, our approach to antiostatin B_2 with an isobutylbiuret group at C4 and a hexyl side chain at C1 is discussed in detail. The starting material 2,6-dimethoxytoluene was first nitrated followed by selective ether cleavage. Conversion of the phenol to the triflate paved the way for the palladium(0)-catalyzed Sonogashira–Hagihara coupling with 1-hexyne. Finally, simultaneous hydrogenation of the triple bond and the nitro group led to the desired arylamine, which was obtained in five steps and 69% overall yield (Scheme 10.7).

Electrophilic substitution of this arylamine by reaction with tricarbonyl (η^5-cyclohexadienylium)iron tetrafluoroborate afforded the corresponding 5-aryl-substituted cyclohexa-1,3-diene complex. Subsequent oxidative cyclization by treatment with ferrocenium hexafluorophosphate in presence of sodium carbonate

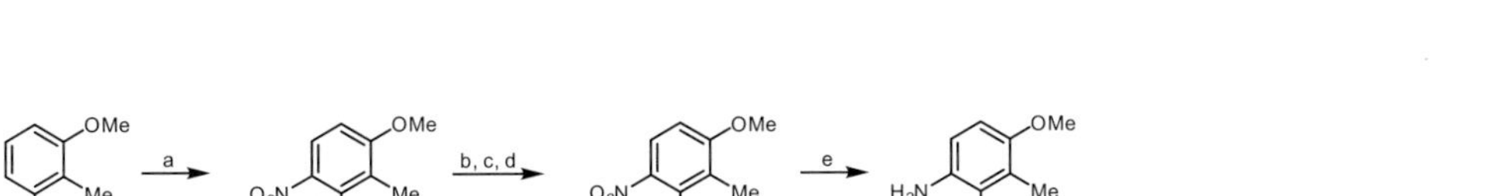

R = alkyl
R' = Ac or CONHCONH*i*-Bu

Scheme 10.6 Retrosynthetic analysis of the antiostatins.

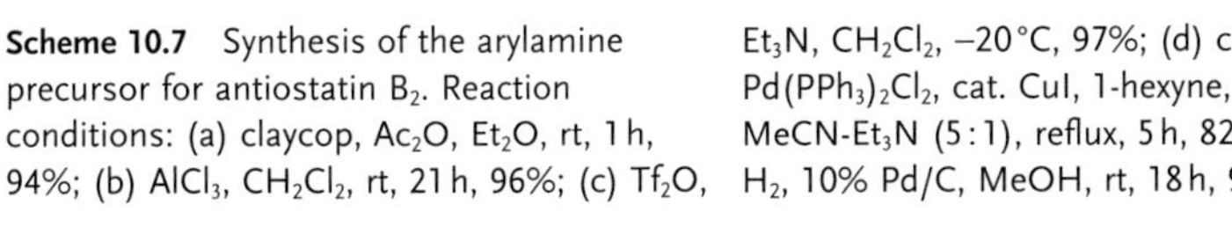

Scheme 10.7 Synthesis of the arylamine precursor for antiostatin B_2. Reaction conditions: (a) claycop, Ac_2O, Et_2O, rt, 1 h, 94%; (b) $AlCl_3$, CH_2Cl_2, rt, 21 h, 96%; (c) Tf_2O, Et_3N, CH_2Cl_2, −20 °C, 97%; (d) cat. $Pd(PPh_3)_2Cl_2$, cat. CuI, 1-hexyne, Bu_4NI, MeCN-Et_3N (5:1), reflux, 5 h, 82%; (e) 1 bar H_2, 10% Pd/C, MeOH, rt, 18 h, 96%.

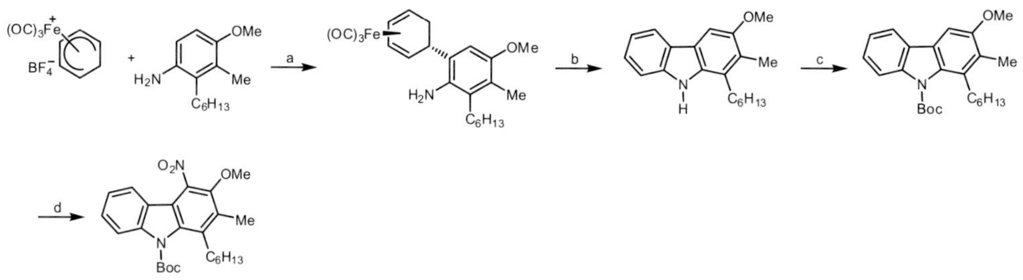

Scheme 10.8 Synthesis of the 9-Boc-4-nitrocarbazole precursor for antiostatin B_2. Reaction conditions: (a) MeCN, reflux, 1 h, 96%; (b) Cp_2FePF_6, Na_2CO_3, CH_2Cl_2, rt, 22 h, 81%; (c) Boc_2O, DMAP, MeCN, rt, 20 h, 99%; (d) claycop, Ac_2O, Et_2O, rt, 3 h, 90%.

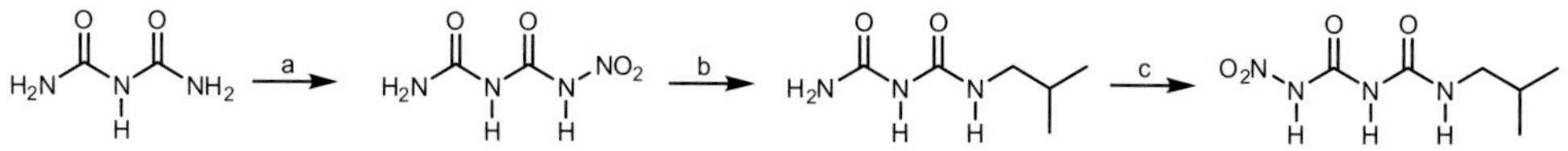

Scheme 10.9 Synthesis of 5-isobutyl-1-nitrobiuret. Reaction conditions: (a) conc. H_2SO_4, conc. HNO_3, 0 °C to rt, 2 h, 51%; (b) *i*-$BuNH_2$, EtOH, rt, 2 h, then 150 °C until melt, 99%; (c) conc. H_2SO_4, conc. HNO_3, 0 °C, 2 h, 24%.

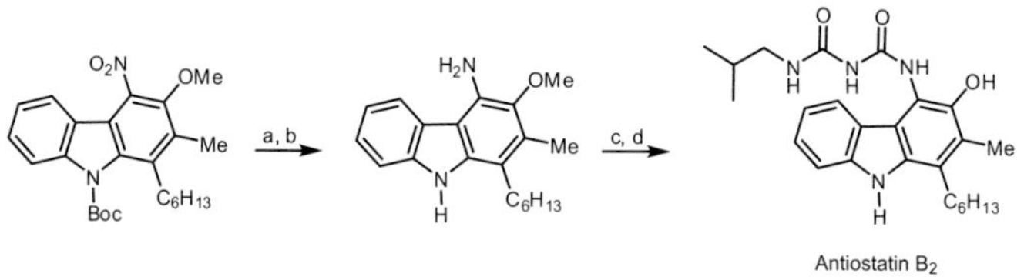

Antiostatin B_2

Scheme 10.10 Synthesis of antiostatin B_2. Reaction conditions: (a) 180 °C, 45 min, 100%; (b) 1 bar H_2, 30% Pd/C, EtOAc, rt, 1 d, 82%; (c) 5-isobutyl-1-nitrobiuret, MeCN, reflux, 4.5 h, 87%; (d) BBr_3, CH_2Cl_2, −78 to 0 °C, 4.5 h, 79%.

furnished the carbazole. In order to direct the subsequent nitration into the 4-position, the carbazole-nitrogen was protected by the Boc group (Scheme 10.8).

Antiostatins of the B series have an isobutylbiuret substituent at C4. This structural feature required the development of an appropriate reagent for its efficient introduction. For this purpose, we have achieved nitration of biuret according to a classical procedure from 1898 followed by reaction of the resulting 1-nitrobiuret with isobutylamine and further nitration to afford 5-isobutyl-1-nitrobiuret (Scheme 10.9).

Deprotection of 9-Boc-4-nitrocarbazole and reduction of the nitro group afforded a 4-aminocarbazole. Subsequent reaction with 5-isobutyl-1-nitrobiuret and cleavage of the methyl ether completed the first total synthesis of antiostatin B_2. This highly efficient synthesis provides antiostatin B_2 in eight steps and 39% overall yield based on the tricarbonyl(η^5-cyclohexadienylium)iron salt (Scheme 10.10).

10.2.2.2 Total Synthesis of R-(–)-Neocarazostatin B and Carquinostatin A

The free radical scavenger neocarazostatin B represents a 3,4-dioxygenated carbazole alkaloid with a stereogenic center in the alkyl side chain at C1. The absolute configuration of neocarazostatin B had not been assigned by Kato *et al.* during isolation and structural elucidation [136]. However, neocarazostatin B is structurally related to carquinostatin A which has an R configuration at the stereogenic center as determined by transformation into the corresponding Mosher ester [59]. Thus, because of the structural similarity, we assumed that neocarazostatin B also has an R configuration and designed our enantioselective synthesis accordingly [137]. Key steps are the iron-mediated synthesis of the carbazole nucleus and a nickel-mediated prenylation (Scheme 10.11).

The chiral arylamine was prepared in eight steps and 65% overall yield starting from guaiacol. The stereogenic center in the side chain resulted from R-(+)-propene oxide. Reaction of the arylamine with the iron complex salt in air provided the corresponding tricarbonyl(η^4-4a,9a-dihydrocarbazole)iron complex (Scheme 10.12). For aromatization and demetallation, this compound was treated with NBS in basic medium. Subsequent electrophilic substitution by reaction with NBS in the presence of a catalytic amount of HBr afforded the 6-bromocarbazole. An X-ray

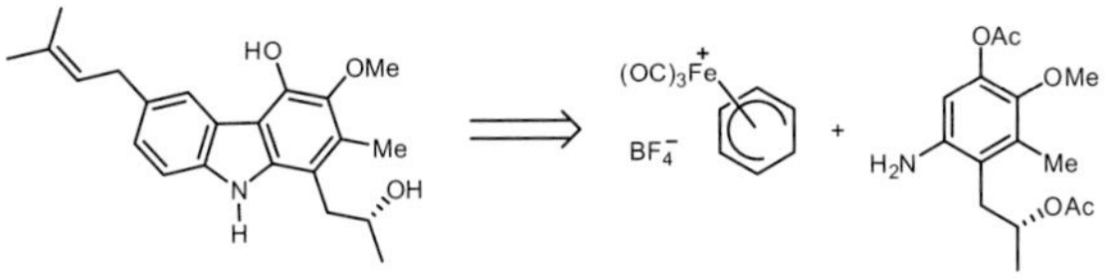

Scheme 10.11 Retrosynthetic analysis of R-(–)-neocarazostatin B.

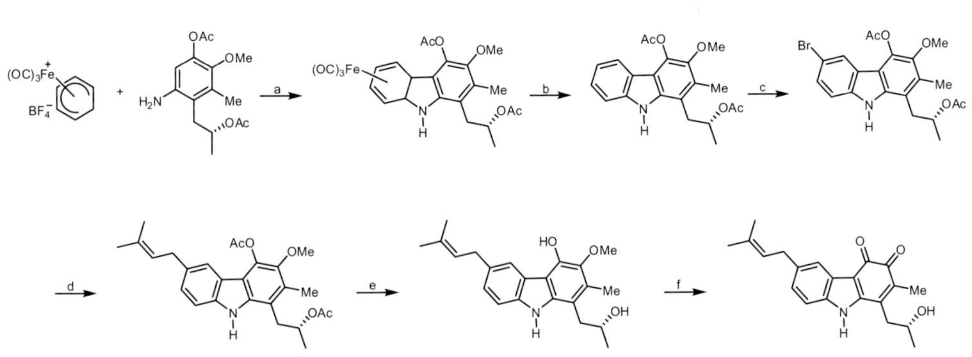

Scheme 10.12 Synthesis of R-(–)-neocarazostatin B and carquinostatin A. Reaction conditions: (a) MeCN, air, rt, 7 d, 68%; (b) NBS, Na$_2$CO$_3$, MeCN, rt, 100%; (c) NBS, cat. HBr, MeCN, rt, 88%; (d) 1. prenyl bromide, Ni(CO)$_4$, benzene, 60 °C, 2. bromocarbazole, DMF, 65 °C, 16 h, 66%; (e) LiAlH$_4$, Et$_2$O, rt, 45 min, 92%; (f) CAN, MeCN-H$_2$O, 0 °C, 2 h, 92%.

crystal structure determination of this compound confirmed the R configuration of the stereogenic center in the alkyl side chain. Prenylation of the 6-bromocarbazole was achieved using a dimeric π-prenylnickel bromide complex generated in situ from tetracarbonylnickel and prenyl bromide following Wilke's procedure. Finally, cleavage of both acetyl groups led to R-(–)-neocarazostatin B. Comparison of the specific rotation of our synthetic neocarazostatin B with the value reported for the natural product confirmed the absolute configuration for the natural product. Moreover, R-(–)-neocarazostatin B was converted to carquinostatin A and subsequently into the corresponding (R,R)-Mosher ester. Comparison of the spectroscopic data of this derivative with those of the (R,R)-Mosher ester of natural carquinostatin A described by Seto provided another proof for the R-configuration of both natural products [59]. The present route affords enantiopure R-(–)-neocarazostatin B in five steps and 36% overall yield based on the tricarbonyl(η^5-cyclohexadienylium)iron salt [137].

10.2.3
Total Syntheses of Ellipticine and Staurosporinone

Due to their useful pharmacological activities, ellipticine and staurosporinone are among the most important carbazole alkaloids. Based on these two parent compounds, a range of synthetic pyrido[4,3-b]carbazole and indolo[2,3-a]carbazole derivatives have been prepared and some compounds are in clinical use. Continuous efforts towards the development of improved synthetic routes to these frameworks are still being made. Two representative recent approaches are discussed below.

10.2.3.1 Synthesis of Ellipticine
Ellipticine belongs to the family of the pyrido[4,3-b]carbazole alkaloids and was first isolated in 1959 [83]. This alkaloid was found to be a promising antitumor drug, which probably derives from the ability to intercalate into DNA based on the planarity of ellipticine and inhibition of topoisomerase II. Therefore, a large number of ellipticine syntheses have been reported over the past decades. In 2006, Ho and Hsieh described an elegant total synthesis of ellipticine [89] (Scheme 10.13).

The synthesis started from 4,7-dimethylindan-2-one which was obtained by oxidation of the corresponding indene [179]. Reduction of the ketone was followed by iodination. Acetylation of the hydroxy group and subsequent palladium(0)-catalyzed Suzuki–Miyaura coupling afforded a 2-nitrobiaryl derivative. Cadogan cyclization furnished the carbazole framework. Subsequent oxidation with DDQ afforded a keto acetate which on reduction with lithium aluminum hydride led to a mixture of the corresponding cis- and trans-diols. For the oxidative cleavage of the C–C bond, this mixture was treated with sodium periodate. Work-up of the resulting crude product with ammonium acetate provided ellipticine in seven steps and 27% overall yield based on 4,7-dimethylindan-2-one.

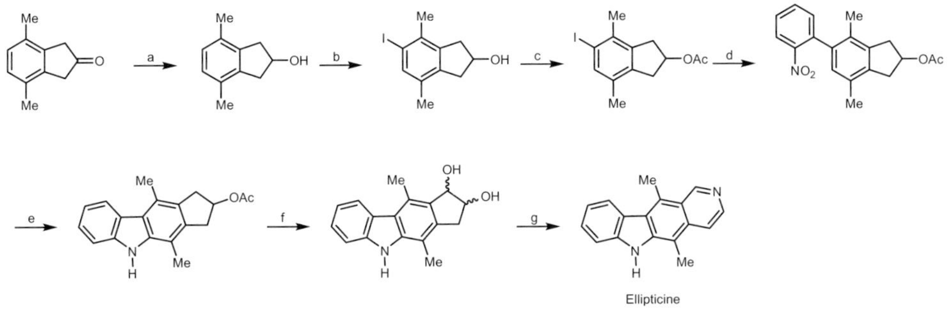

Scheme 10.13 Synthesis of ellipticine. Reaction conditions: (a) NaBH$_4$, CH$_2$Cl$_2$, MeOH, 99%; (b) I$_2$, CH$_2$Cl$_2$, *silfen* (prepared by grinding of Fe(NO$_3$)$_3$·9 H$_2$O with SiO$_2$), 82%; (c) Ac$_2$O, pyridine, DMAP, rt, 99%; (d) 2-nitrophenylboronic acid, Pd(PPh$_3$)$_2$Cl$_2$, NaHCO$_3$, DME, H$_2$O, 80 °C, 79%; (e) (EtO)$_3$P, 150 °C, 73%; (f) 1. DDQ, THF, H$_2$O; 2. aq. K$_2$CO$_3$; 3. LiAlH$_4$, THF, 68%; (g) 1. NaIO$_4$, *t*-BuOH, H$_2$O; 2. aq. NH$_4$OAc, 87%.

10.2.3.2 Synthesis of Staurosporinone

Because of its pharmacological profile (including antitumor activity), a variety of synthetic approaches to staurosporinone has been developed. The shortest access to staurosporinone was described by Faul *et al.* [167] (Scheme 10.14). The key reaction is a double substitution reaction of dichloromaleimide with two equivalents of indolylmagnesium bromide which provides the natural product arcyriarubin A in 72% yield. The use of dibromomaleimide for this transformation, as previously reported by Hill *et al.*, provided arcyriarubin A only in moderate yield [12]. Oxidative cyclization of arcyriarubin A to arcyriaflavin A was achieved by using one equivalent of palladium(II) acetate. Reduction of the imide to the hydroxylactam followed by hydrogenolysis afforded staurosporinone [12]. The combined approach reported by Faul and Hill *et al.* leads to staurosporinone in four steps and 34% overall yield based on dichloromaleimide. Moreover, the natural products arcyriarubin A and arcyriaflavin A have been obtained as synthetic intermediates of this approach which thus, may be considered a biomimetic synthesis.

In 2003, Uang *et al.* published an improved synthesis which requires only two steps for the transformation of arcyriarubin A into staurosporinone [180]. This is

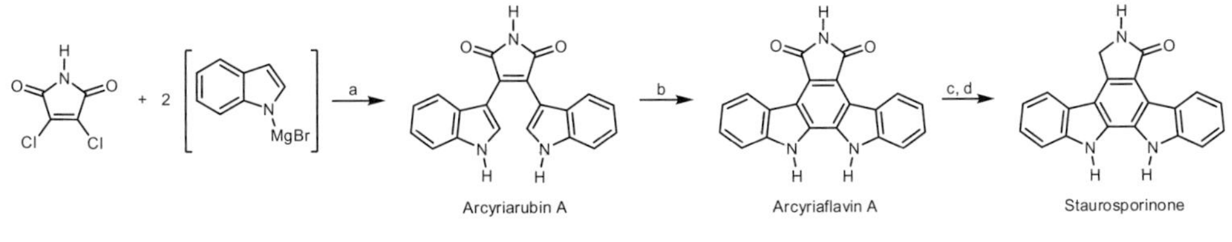

Scheme 10.14 Synthesis of staurosporinone. Reaction conditions: (a) toulene-Et$_2$O-THF, rt, 24 h, 72%; (b) Pd(OAc)$_2$, HOAc, 110 °C, 75%; (c) LiAlH$_4$, Et$_2$O, rt; d) 10% Pd/C, H$_2$, EtOH, rt, 63% over 2 steps.

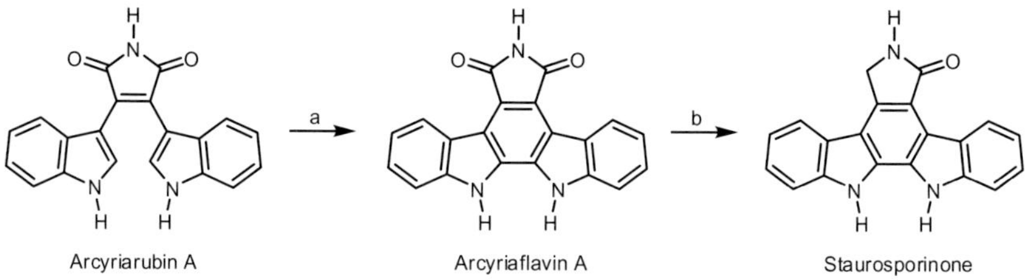

Scheme 10.15 Improved transformation of arcyriarubin A into staurosporinone. Reaction conditions: (a) hv, cat. I$_2$, THF-MeCN, rt, 12 h, 85%; (b) Zn/Hg, THF, 6 M HCl, reflux, 1.5 h, 68%.

achieved by oxidative photocyclization of arcyriarubin A in the presence of catalytic amounts of iodine which provided arcyriaflavin A in 85% yield (Scheme 10.15). Application of a Clemmensen reduction for the final step was shown to afford directly staurosporinone in 68% yield. Overall, this approach provides staurosporinone in three steps and 42% yield based on dichloromaleimide.

10.3
Introduction to Acridines

Acridines are heterocyclic compounds with a dibenzo[b,e]pyridine framework. They were first isolated from charcoal and initially used as dyes and pigments [181]. In 1948, Hughes and co-workers isolated the first acridone alkaloids from Australian plants of the family Rutaceae [182]. Since then, acridones became important targets for natural product synthesis and served as lead structures for the development of potential new drugs. Acridones are related to acridines by their tautomeric form, the 9-hydroxyacridines (Figure 10.3). Section 10.3 gives a brief overview of natural acridone alkaloids and synthetic acridines with useful pharmacological activities (Table 10.2). More comprehensive reviews provide detailed information on recent developments in acridone chemistry [183, 184] and biosynthesis, structure elucidation, occurrence, synthesis and biological properties of acridone alkaloids [185]. For recent progress of acridines in anticancer research, see [186–188].

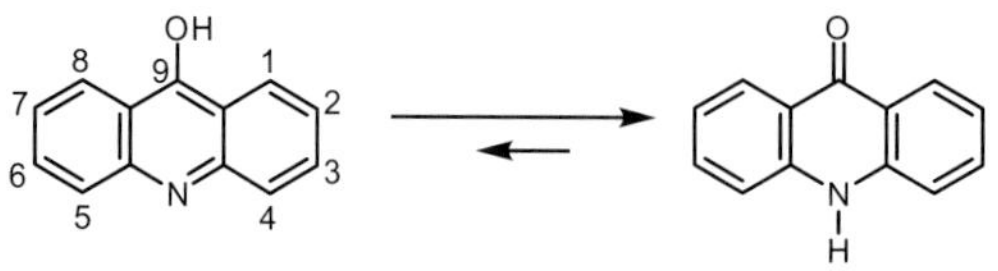

Figure 10.3 Tautomerism of 9-hydroxyacridine and acridone.

Table 10.2 Selected acridines and acridones.

Trivial name	Structure	Source and isolation	Bioactivity	Synthesis
Acronycine		*Acronychia baueri* Schott [182], see also [189, 190]	Antitumor [191]	[192–194]
Aminacrine (9-Aminoacridine)		Synthetic drug	Antiprotozoal, antibacterial [187] Inhibition of fibroblast mitosis [195]	[196]
Amphimedine		*Amphimedon* sp. [197]	Antitumor [198]	[199, 200a]
Amsacrine (*m*-AMSA)		Synthetic drug	Anti-lymphoblastic leukemia [201]	[202]
Arborinine		*Glycosmis pentaphylla* [203], see also [204, 205]	Antifeedant [206] Anticancer [207]	
Atalaphylline		*Atalantia monophylla* Correa [208]		[209]
Atalaphyllidine		*Severinia buxifolia* [210]	Anticancer [211, 212]	[213]

(Continued)

Table 10.2 (Continued)

Trivial name	Structure	Source and isolation	Bioactivity	Synthesis
Glyfoline		*Glycosmis citrifolia* [214]	Anticancer [211]	[215, 216]
Mepacrine (Quinacrine, Atebrine)		Synthetic drug	Antimalarial [217, 218]	[217]
Proflavine		Synthetic drug	Antiseptic [219]	[220]
(−)-Rutacridone		*Ruta graveolens* [221], abs. config. [222]	Anticancer [207]	[223]
Tacrine (Cognex)		Synthetic drug	Acetylcholinesterase inhibition [224–226]	[227, 228]
Tecleanthine		*Teclea natalensis* [229] *Vepris sclerophylla* [230]		[231]
Xanthoxoline		*Evodia xanthoxyloides* [232]	Antifeedant [206]	[233]

10.4
Synthesis of Acridines and Acridones

10.4.1
Total Synthesis of Acronycine

Acronycine, a pyrano[2,3-c]acridone, is one of the first isolated acridones [182] and also one of the most prominent. Acronycine shows an inhibitory activity against many different murine tumor models, but only low activity against leukemia models [191]. It was tested in clinical phase I-II at patients with multiple myeloma, but has the disadvantage of low water-solubility. Therefore, acronycine represents a lead compound and many derivatives are still being synthesized and investigated.

Diverse synthetic approaches to acronycine have been developed [185]. The first total synthesis was reported by Beck et al. in 1967 [192]. In 1996, Anand et al. described a short and efficient route to acronycine (Scheme 10.16) [194].

Prenylation of commercial 3,5-dimethoxyacetanilide with 3-methylbut-1-en-3-ol in the presence of catalytic amounts of boron trifluoride etherate afforded the 2-prenylated acetanilide. Alkaline hydrolysis led to the free aniline, which directly reacts with diphenyliodonium-2-carboxylate in an Ullmann–Jourdan coupling to the diarylamine-2-carboxylic acid. Formation of the acridone framework was accomplished by an intramolecular Friedel–Crafts acylation using polyphosphate ester under strictly anhydrous conditions. Cleavage of both methyl ethers and subsequent oxidative cyclization by using DDQ led to bis(desmethyl)acronycine. Finally, methylation by using an excess of methyl iodide provided acronycine which was obtained in seven steps and 60% overall yield.

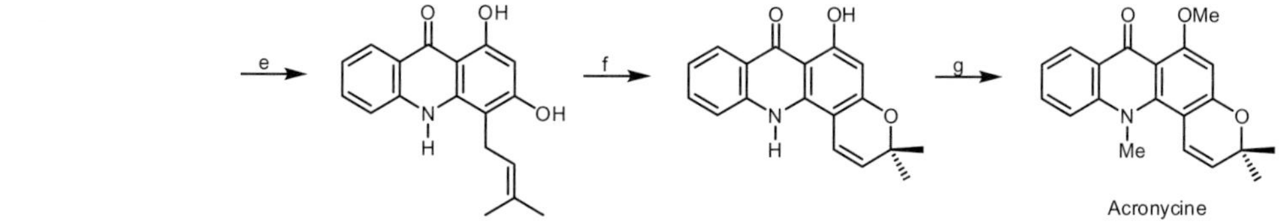

Scheme 10.16 Synthesis of acronycine. Reaction conditions: (a) 3-methylbut-1-en-3-ol, cat. BF$_3$·OEt$_2$, dioxane, reflux, 89–92%; (b) alkaline hydrolysis; (c) diphenyliodonium-2-carboxylate, Cu(OAc)$_2$, i-PrOH, 92–94% over two steps; (d) anhydrous polyphosphate ester, 88–91%; (e) EtSNa-DMF; (f) DDQ, o-dichlorobenzene; (g) MeI, KOH, Me$_2$CO.

10.4.2
Synthesis of Amsacrine

Amsacrine (m-AMSA) was designed especially for elucidation of the interaction of acridine derivatives with DNA and topoisomerase II. Amsacrine is a powerful chemotherapeutic agent for the treatment of acute leukemia, and on the basis of its structure many new derivatives have been prepared [188]. It has been recognized that amsacrine forms a ternary complex in which the acridine nucleus intercalates into DNA and the side chain interacts with topoisomerase II [234]. Although much work has been done in recent years, not all details of this mechanism are clear yet.

Amsacrine was synthesized in 1975 along with many other so-called AMSAs [202, 235]. Ullmann–Jourdan coupling of 2-chlorobenzoic acid with aniline afforded N-phenylanthranilic acid (Scheme 10.17). Subsequent ring closure with phosphoryl chloride led directly to 9-chloroacridine. Finally, reaction with N-(4-amino-3-methoxyphenyl)methanesulfonamide methanesulfonamide provided amsacrine.

10.4.3
Total Syntheses of Amphimedine

Amphimedine is an important member of the pyridoacridine class of natural products. It was isolated first in 1983 from the marine sponge *Amphimedon* sp. [197]. Only five years later, the first total synthesis was reported by Stille and Echavarren [199]. The key steps of their approach are a palladium(0)-catalyzed Stille coupling and a hetero-Diels–Alder reaction (Scheme 10.18).

5,8-Dimethoxyquinolin-4-yl triflate was prepared in three steps and 65–67% yield from 2,5-dimethoxyaniline. The palladium(0)-catalyzed coupling of this triflate with N-Boc-2-(trimethylstannyl)aniline afforded a 4-arylquinoline derivative. Conversion of the *t*-butyl carbamate to the trifluoromethanesulfonamide followed by oxidation with ceric ammonium nitrate (CAN) led to the corresponding 4-arylquinoline-5,8-dione. Hetero-Diels–Alder cycloaddition of the 4-arylquinoline-5,8-dione with 1,3-bis(*t*-butyldimethylsilyloxy)-2-aza-1,3-butadiene and subsequent protodesilylation with pyridine hydrofluoride provided via elimination the

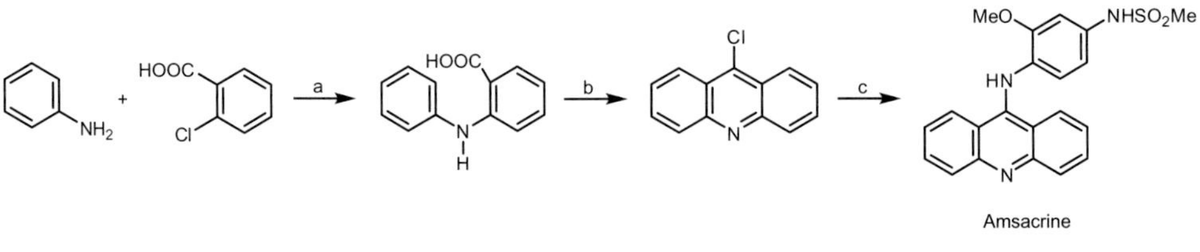

Scheme 10.17 Synthesis of amsacrine. Reaction conditions: (a) K₂CO₃, Cu, 2-ethoxyethanol; (b) POCl₃, reflux; (c) N-(4-amino-3-methoxyphenyl)methanesulfonamide, 12 N HCl, EtOH-H₂O (2:1), reflux.

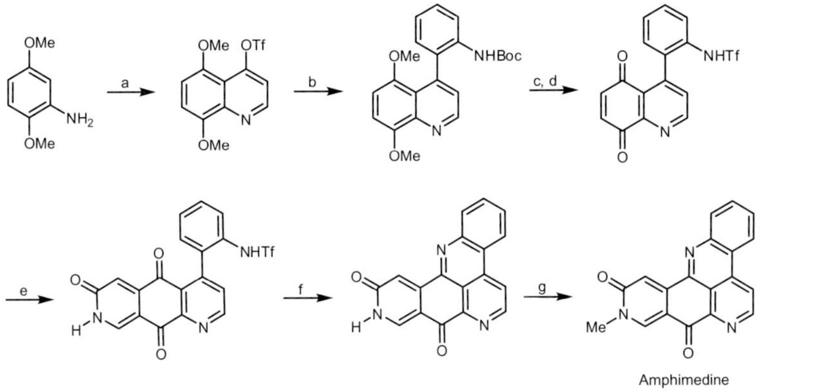

10.4 Synthesis of Acridines and Acridones

Scheme 10.18 First synthesis of amphimedine. Reaction conditions: (a) 1. Meldrum's acid, trimethyl orthoformate, reflux, 4–7 h; 2. Ph₂O, reflux, 15 min; 3. Tf₂O, 2,6-lutidine, DMAP, CH₂Cl₂, 0–23 °C, 3.5 h, 65–67%; (b) N-Boc-2-(trimethylstannyl)aniline, Pd(PPh₃)₄, LiCl, dioxane, 100 °C, 5–7 h, 87%; (c) 1. TFA, 23 °C, 1 h; 2. Tf₂O, i-Pr₂EtN, THF, 0 °C, 30 min., 94-100% over 2 steps; (d) CAN, MeCN-H₂O (2:1), 23 °C, 15 min, 85%; (e) 1,3-bis(t-butyldimethylsilyloxy)-2-aza-1,3-butadiene, THF, 23 °C, 6 h; then Py·HF, 48%; (f) 6M HCl-THF (1:1), 70-80 °C, 3 h, 86%; (g) Me₂SO₄, K₂CO₃, DMF, 23 °C, 3 h, 96%.

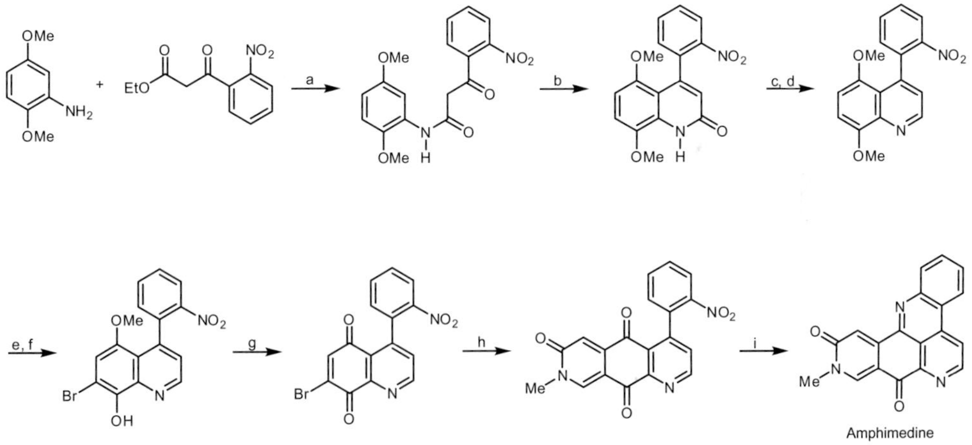

Scheme 10.19 Improved synthesis of amphimedine. Reaction conditions: (a) cat. pyridine, toluene, 140 °C, 6 h, 100%; (b) 80% H₂SO₄, 75 °C, 30 min, 53%; (c) Tf₂O, Et₃N, CH₂Cl₂, −20 °C, 30 min, 93%; (d) cat. Pd(OAc)₂, Et₃N, PPh₃, formic acid, DMF, 0 to 60 °C, 2 h, 87%; (e) BBr₃, CH₂Cl₂, reflux, 15 h, 95%; (f) Br₂, K₂CO₃, CHCl₃, 25 °C, 1 h, 83%; (g) CAN, MeCN-H₂O, 0 °C, 15 min, 70%; (h) 1,3-bis(t-butyldimethylsilyloxy)-2-aza-1,3-butadiene, CHCl₃, 25 °C, 1 h, then K₂CO₃, MeI, TDA-1, DMF, 25 °C, 1 h, 27%; (i) 10% Pd/C, H₂, MeOH, 1.5 h, 41%.

1,7-diazaanthraquinone. Hydrolysis of the trifluoromethanesulfonamide followed by intramolecular condensation led to desmethylamphimedine. Finally, methylation with dimethyl sulfate provided amphimedine which was obtained in ten steps and 19–20% overall yield based on 2,5-dimethoxyaniline.

In 1996, Kubo *et al.* described another synthesis of amphimedine (Scheme 10.19) [200]. This route is also relying on a hetero-Diels–Alder reaction of a 2-aza-1,3-butadiene as key step. The required dienophile was obtained in seven steps starting from 2,5-dimethoxyaniline. Heating of 2,5-dimethoxyaniline with ethyl (2-nitrobenzoyl)acetate in the presence of small amounts of pyridine afforded a (2-nitrobenzoyl)acetanilide which on Knorr cyclization provided the 2-quinolinone. Removal of the carbonyl functionality was achieved via transformation into the triflate and subsequent palladium-catalyzed reaction with formic acid to give 5,8-dimethoxy-4-(2-nitrophenyl)quinoline. Selective monodemethylation was followed by regioselective bromination at C-7 of the quinoline framework. Subsequent oxidation with CAN afforded 7-bromo-4-(2-nitrophenyl)quinoline-5,8-dione. Hetero-Diels–Alder reaction with 1,3-bis(*t*-butyldimethylsilyloxy)-2-aza-1,3-butadiene followed by methylation led to the 1,7-diazaanthraquinone framework. Catalytic hydrogenation of the nitro group and subsequent intramolecular condensation provided amphimedine in nine steps and 26% overall yield based on 2,5-dimethoxyaniline.

References

1 Knölker, H.-J. and Reddy, K.R. (2002) Isolation and synthesis of biologically active carbazole alkaloids. *Chem. Rev.*, **102** (11), 4303–4428.

2 (a) Knölker, H.-J. (2004) Synthesis of biologically active carbazole alkaloids using organometallic chemistry. *Curr. Org. Synth.*, **1** (4), 309–331; (b) Knölker, H.-J. (2005) Occurrence, biological activity, and convergent organometallic synthesis of carbazole alkaloids. *Top. Curr. Chem.*, **244**, 115–148.

3 Knölker, H.-J. and Reddy, K.R. (2008) Chemistry and biology of carbazole alkaloids, in *The Alkaloids*, vol. 65, (ed. G.A. Cordell), Academic Press, Amsterdam, The Netherlands, pp. 1–430.

4 Knölker, H.-J. (2009) Synthesis of biologically active carbazole alkaloids using selective transition metal-catalyzed coupling reactions. *Chem. Lett.*, **38** (1), 8–13.

5 Mo, C.-J., Shin-ya, K., Furihata, K., Furihata, K., Shimazu, A., Hayakawa, Y. and Seto, H. (1990) Isolation and structural elucidation of antioxidative agents, antiostatins A_1 to A_4 and B_2 to B_5. *J. Antibiot.*, **43** (10), 1337–1340.

6 Knott, K.E., Auschill, S., Jäger, A. and Knölker, H.-J. (2009) First total synthesis of the whole series of the antiostatins A and B. *Chem. Commun.*, (12), 1467–1469.

7 Alayrac, C., Schollmeyer, D. and Witulski, B. (2009) First total synthesis of antiostatin A1, a potent carbazole-based naturally occurring antioxidant. *Chem. Commun.*, (12), 1464–1466.

8 Steglich, W. (1989) Slime moulds (*Myxomycetes*) as a source of new biologically active metabolites. *Pure Appl. Chem.*, **61** (3), 281–288.

9 Gill, M. and Steglich, W. (1987) Pigments of fungi (macromycetes), in *Progress in the Chemistry of Organic Natural Products*, vol. 51, (eds W. Herz, H. Grisebach, G.W. Kirby and C. Tamm), Springer-Verlag, Wien, Austria, pp. 216–286.

10 Kopanski, L., Li, G.-R., Besl, H. and Steglich, W. (1982) Naphthochinon-Farbstoffe aus den Schleimpilzen *Trichia florimis* und *Metatrichia vesparium* (Myxomycetes). *Liebigs Ann. Chem.*, (9), 1722–1729.

11 Gribble, G.W. and Berthel, S.J. (1992) Synthetic approaches to indolo[2,3-*a*] carbazole alkaloids. Syntheses of arcyriaflavin A and AT2433-B aglycone. *Tetrahedron*, **48** (41), 8869–8880.

12 Harris, W., Hill, C.H., Keech, E. and Malsher, P. (1993) Oxidative cyclizations with palladium acetate. A short synthesis of staurosporine aglycone. *Tetrahedron Lett.*, **34** (51), 8361–8364.

13 Bergman, J., Koch, E. and Pelcman, B. (2000) Coupling reactions of indole-3-acetic acid derivatives. Synthesis of arcyriaflavin A. *J. Chem. Soc. Perkin Trans. 1*, (16), 2609–2614.

14 Hughes, I. and Raphael, R.A. (1983) Synthesis of arcyriaflavin B. *Tetrahedron Lett.*, **24** (13), 1441–1444.

15 Ohkubo, M., Nishimura, T., Jona, H., Honma, T. and Morishima, H. (1996) Practical synthesis of indolopyrrolocarbazoles. *Tetrahedron*, **52** (24), 8099–8112.

16 Rickards, R.W., Rothschild, J.M., Willis, A.C., de Chazal, N.M., Kirk, J., Kirk, K., Saliba, K.J. and Smith, G.D. (1999) Calothrixins A and B, novel pentacyclic metabolites from *Calothrix* cyanobacteria with potent activity against malaria parasites and human cancer cells. *Tetrahedron*, **55** (47), 13513–13520.

17 Kelly, T.R., Zhao, Y., Cavero, M. and Torneiro, M. (2000) Synthesis of the potent antimalarials calothrixin A and B. *Org. Lett.*, **2** (23), 3735–3737.

18 Bernardo, P.H., Chai, C.L.L. and Elix, J.A. (2002) A simple and concise route to calothrixin B. *Tetrahedron Lett.*, **43** (16), 2939–2940.

19 Bernardo, P.H. and Chai, C.L.L. (2003) Friedel–Crafts acylation and metalation strategies in the synthesis of calothrixins A and B. *J. Org. Chem.*, **68** (23), 8906–8909.

20 Sissouma, D., Collet, S.C. and Guingant, A.Y. (2004) A synthesis of calothrixin B. *Synlett*, (14), 2612–2614.

21 Bernardo, P.H., Chai, C.L.L., Heath, G.A., Mahon, P.J., Smith, G.D., Waring, P. and Wilkes, B.A. (2004) Synthesis, electrochemistry, and bioactivity of the cyanobacterial calothrixins and related quinones. *J. Med. Chem.*, **47** (20), 4958–4963.

22 Tohyama, S., Choshi, T., Matsumoto, K., Yamabuki, A., Ikegata, K., Nobuhiro, J. and Hibino, S. (2005) A new total synthesis of an indolo[3,2-*j*] phenanthridine alkaloid calothrixin B. *Tetrahedron Lett.*, **46** (32), 5263–5264.

23 Bennasar, M.L., Roca, T. and Ferrando, F. (2006) A new radical-based route to calothrixin B. *Org. Lett.*, **8** (4), 561–564.

24 Yamabuki, A., Fujinawa, H., Choshi, T., Tohyama, S., Matsumoto, K., Ohmura, K., Nobuhiro, J. and Hibino, S. (2006) A biomimetic synthesis of the indolo[3,2-*j*] phenanthridine alkaloid, calothrixin B. *Tetrahedron Lett.*, **47** (33), 5859–5861.

25 Sissouma, D., Maingot, L., Collet, S. and Guingant, A. (2006) Concise and efficient synthesis of calothrixin B. *J. Org. Chem.*, **71** (22), 8384–8389.

26 Sperry, J., McErlean, C.S.P., Slawin, A.M.Z. and Moody, C.J. (2007) A biomimetic synthesis of calothrixin B. *Tetrahedron Lett.*, **48** (2), 231–234.

27 Kato, S., Kawai, H., Kawasaki, T., Toda, Y., Urata, T. and Hayakawa, Y. (1989) Studies on free radical scavenging substances from microorganisms I. Carazostatin, a new free radical scavenger produced by *Streptomyces*. *J. Antibiot.*, **42** (12), 1879–1881.

28 Kato, S., Kawasaki, T., Urata, T. and Mochizuki, J. (1993) *In vitro* and *ex vivo* free radical scavenging activities of carazostatin, carbazomycin B and their derivatives. *J. Antibiot.*, **46** (12), 1859–1865.

29 Knölker, H.-J. and Hopfmann, T. (1995) Total synthesis of the naturally occurring free radical scavenger carazostatin. *Synlett*, (9), 981–983.

30 Choshi, T., Sada, T., Fujimoto, H., Nagayama, C., Sugino, E. and Hibino, S. (1996) Total syntheses of carazostatin and hyellazole by allene-mediated electrocyclic reaction. *Tetrahedron Lett.*, **37** (15), 2593–2596.

31 Choshi, T., Sada, T., Fujimoto, H., Nagayama, C., Sugino, E. and Hibino, S. (1997) Total syntheses of carazostatin, hyellazole, and carbazoquinocins B-F. *J. Org. Chem.*, **62** (8), 2535–2543.

32 Knölker, H.-J. and Hopfmann, T. (2002) Iron-mediated synthesis of carazostatin, a free radical scavenger from *Streptomyces chromofuscus*, and O-methylcarazostatin. *Tetrahedron*, **58** (44), 8937–8945.

33 Pacher, T., Bacher, M., Hofer, O. and Greger, H. (2001) Stress induced carbazole phytoalexins in *Glycosmis* species. *Phytochemistry*, **58** (1), 129–135.

34 Schmidt, M. and Knölker, H.-J. (2009) Palladium-catalyzed approach to 2,6-dioxygenated carbazole alkaloids – first total synthesis of the phytoalexin carbalexin C. *Synlett*, (15), 2421–2424.

35 Kotoda, N., Shin-ya, K., Furihata, K., Hayakawa, Y. and Seto, H. (1997) Isolation and structure elucidation of novel neuronal cell protecting substances, carbazomadurins A and B produced by *Actinomadura madurae*. *J. Antibiot.*, **50** (9), 770–772.

36 Knöll, J. and Knölker, H.-J. (2006) First total synthesis and assignment of the absolute configuration of the neuronal cell protecting alkaloid carbazomadurin B. *Synlett*, (4), 651–653.

37 (a) Knölker, H.-J. and Knöll, J. (2003) First total synthesis of the neuronal cell protecting carbazole alkaloid carbazomadurin A by sequential transition metal-catalyzed reactions. *Chem. Commun.*, (10), 1170–1171; (b) Hieda, Y., Choshi, T., Kishida, S., Fujioka, H. and Hibino, S. (2010) A novel total synthesis of the bioactive poly-substituted carbazole alkaloid carbazomadurin A. *Tetrahedron Lett.*, **51** (27), 3593–3596.

38 Sakano, K.-I., Ishimaru, K. and Nakamura, S. (1980) New antibiotics, carbazomycins A and B I. Fermentation, extraction, purification and physico-chemical and biological properties. *J. Antibiot.*, **33** (7), 683–689.

39 (a) Sakano, K.-I. and Nakamura, S. (1980) New antibiotics, carbazomycins A and B II. Structural elucidation. *J. Antibiot.*, **33** (9), 961–966; (b) Kaneda, M., Sakano, K.-I., Nakamura, S., Kushi, Y. and Iitaka, Y. (1981) The structure of carbazomycin B. *Heterocycles*, **15** (2), 993–998.

40 Naid, T., Kitahara, T., Kaneda, M. and Nakamura, S. (1987) Carbazomycins C, D, E and F., minor components of the carbazomycin complex. *J. Antibiot.*, **40** (2), 157–164.

41 Kaneda, M., Naid, T., Kitahara, T., Nakamura, S., Hirata, T. and Suga, T. (1988) Carbazomycins G and H, novel carbazomycin congeners containing a quinol moiety. *J. Antibiot.*, **41** (5), 602–608.

42 Kondo, S., Katayama, M. and Marumo, S. (1986) Carbazomycinal and 6-methoxycarbazomycinal as aerial mycelium formation inhibitory substances of *Streptoverticillium* species. *J. Antibiot.*, **39** (5), 727–730.

43 Hook, D.J., Yacobucci, J.J., O'Connor, S., Lee, M., Kerns, E., Krishnan, B., Matson, J. and Hesler, G. (1990) Identification of the inhibitory activity of carbazomycins B and C against 5-lipoxygenase, a new activity for these compounds. *J. Antibiot.*, **43** (10), 1347–1348.

44 Knölker, H.-J., Bauermeister, M., Bläser, D., Boese, R. and Pannek, J.-B. (1989) Highly selective oxidations of Fe(CO)$_3$-cyclohexadiene complexes: synthesis of 4b,8a-dihydrocarbazol-3-ones and the first total synthesis of carbazomycin A. *Angew. Chem.*, **101** (2), 225–227. *Angew. Chem. Int. Ed. Engl.* **28** (2), 223–225.

45 Knölker, H.-J. (1992) Iron-mediated synthesis of heterocyclic ring systems and applications in alkaloid chemistry. *Synlett*, (5), 371–387.

46 Moody, C.J. and Shah, P. (1989) Synthesis of the carbazole alkaloids carbazomycin A and B and hyellazole. *J. Chem. Soc. Perkin Trans. 1*, (2), 376–377.

47 Knölker, H.-J. and Bauermeister, M. (1989) Transition metal-diene complexes in organic synthesis, Part 2. The total synthesis of the carbazole antibiotic carbazomycin B and an improved route to carbazomycin A. *J. Chem. Soc. Chem. Commun.*, (19), 1468–1470.

48 Knölker, H.-J. and Bauermeister, M. (1993) Transition metal-diene complexes in organic synthesis. Part 15. Iron-mediated total synthesis of carbazomycin A and B. *Helv. Chim. Acta*, **76** (7), 2500–2514.

49 Knölker, H.-J. and Fröhner, W. (1999) Improved total syntheses of the antibiotic alkaloids carbazomycin A and B. *Tetrahedron Lett.*, **40** (38), 6915–6918.

50 Clive, D.L.J., Etkin, N., Joseph, T. and Lown, J.W. (1993) Synthesis of carbazomycin B. *J. Org. Chem.*, **58** (9), 2442–2445.

51 Beccalli, E.M., Marchesini, A. and Pilati, T. (1996) Diels-Alder reactions of (Z)-ethyl 3-[(1-ethoxycarbonyloxy-2-methoxy)ethenyl]-2-(ethoxycarbonyloxy)-indole-1-carboxylate. Synthesis of the carbazole alkaloid carbazomycin B. *Tetrahedron*, **52** (8), 3029–3036.

52 Crich, D. and Rumthao, S. (2004) Synthesis of carbazomycin B by radical arylation of benzene. *Tetrahedron*, **60** (7), 1513–1516.

53 Knölker, H.-J. and Schlechtingen, G. (1997) First total synthesis of carbazomycin C and D. *J. Chem. Soc. Perkin Trans. 1*, (4), 349–350.

54 Knölker, H.-J. and Bauermeister, M. (1991) First total synthesis of carbazomycinal. *Heterocycles*, **32** (12), 2443–2450.

55 Knölker, H.-J. and Fröhner, W. (1997) First total synthesis of carbazomycin G and H. *Tetrahedron Lett.*, **38** (23), 4051–4054.

56 Knölker, H.-J. and Fröhner, W. (1998) Palladium-catalyzed total synthesis of the antibiotic carbazole alkaloids carbazomycin G and H. *J. Chem. Soc. Perkin Trans. 1*, (2), 173–175.

57 Knölker, H.-J., Fröhner, W. and Reddy, K.R. (2003) Iron-mediated synthesis of carbazomycin G and carbazomycin H, the first carbazole-1,4-quinol alkaloids from Streptoverticillium ehimense. *Eur. J. Org. Chem.*, (4), 740–746.

58 Hagiwara, H., Choshi, T., Fujimoto, H., Sugino, E. and Hibino, S. (2000) A novel total synthesis of antibiotic carbazole alkaloid carbazomycin G. *Tetrahedron*, **56** (32), 5807–5811.

59 Shin-ya, K., Tanaka, M., Furihata, K., Hayakawa, Y. and Seto, H. (1993) Structure of carquinostatin A, a new neuronal cell protecting substance produced by Streptomyces exfoliatus. *Tetrahedron Lett.*, **34** (31), 4943–4944.

60 Knölker, H.-J. and Fröhner, W. (1997) First total synthesis of the neuronal cell protecting substance (±)-carquinostatin A via iron- and nickel-mediated coupling reactions. *Synlett*, (9), 1108–1110.

61 Knölker, H.-J., Baum, E. and Reddy, K.R. (2000) First enantioselective total synthesis of the potent neuronal cell protecting substance carquinostatin A from (R)-propene oxide. *Tetrahedron Lett.*, **41** (8), 1171–1174.

62 Cardellina, J.H., Kirkup, M.P., Moore, R.E., Mynderse, J.S., Seff, K. and Simmons, C.J. (1979) Hyellazole and chlorohyellazole, two novel carbazoles from the blue-green alga Hyella caespitosa born. et flah. *Tetrahedron Lett.*, **20** (51), 4915–4916.

63 Kano, S., Sugino, E., Shibuya, S. and Hibino, S. (1981) Synthesis of carbazole alkaloids hyellazole and 6-chlorohyellazole. *J. Org. Chem.*, **46** (19), 3856–3859.

64 Beccalli, E.M., Marchesini, A. and Pilati, T. (1994) Synthesis of the carbazole alkaloids hyellazole and 6-chlorohyellazole and related derivatives. *J. Chem. Soc. Perkin Trans. 1* (5), 579–586.

65 Duval, E. and Cuny, G.D. (2004) Synthesis of substituted carbazoles and β-carbolines by cyclization of diketoindole derivatives. *Tetrahedron Lett.*, **45** (28), 5411–5413.

66 Knölker, H.-J., Fröhner, W. and Heinrich, R. (2004) Total synthesis of the marine alkaloid 6-chlorohyellazole. *Synlett*, (15), 2705–2708.

67 Chakraborty, A., Chowdhury, B.K. and Bhattacharyya, P. (1995) Clausenol and clausenine—two carbazole alkaloids from Clausena anisata. *Phytochemistry*, **40** (1), 295–298.

68 Lin, G. and Zhang, A. (1999) The first synthesis of optically pure biscarbazoles and determination of their absolute configurations. *Tetrahedron Lett.*, **40** (2), 341–344.

69 Börger, C. and Knölker, H.-J. (2008) A general approach to 1,6-dioxygenated carbazole alkaloids – first total synthesis of clausine G, clausine I, and clausine Z. *Synlett*, (11), 1698–1702.

70 Wu, T.-S., Huang, S.-C., Wu, P.-L. and Teng, C.-M. (1996) Carbazole alkaloids from *Clausena excavata* and their biological activity. *Phytochemistry*, **43** (1), 133–140.

71 Wu, T.-S., Huang, S.-C. and Wu, P.-L. (1996) Carbazole alkaloids from stem bark of *Clausena excavata*. *Phytochemistry*, **43** (6), 1427–1429.

72 Ito, C., Katsuno, S., Ohta, H., Omura, M., Kajiura, I. and Furukawa, H. (1997) Constituents of *Clausena excavata*. Isolation and structural elucidation of new carbazole alkaloids. *Chem. Pharm. Bull.*, **45** (1), 48–52.

73 Kongkathip, B., Kongkathip, N., Sunthitikawinsakul, A., Napaswat, C. and Yoosook, C. (2005) Anti-HIV-1 constitutes from *Clausena excavata*: Part II. Carbazoles and a pyranocoumarin. *Phytother. Res.*, **19** (8), 728–731.

74 Sunthitikawinsakul, A., Kongkathip, N., Kongkathip, B., Phonnakhu, S., Daly, J.W., Spande, T.F., Nimit, Y. and Rochanaruangrai, S. (2003) Coumarins and carbazoles from *Clausena excavata* exhibited antimycobacterial and antifungal activities. *Planta Med.*, **69** (2), 155–157.

75 (a) Choi, T.A., Czerwonka, R., Fröhner, W., Krahl, M.P., Reddy, K.R., Franzblau, S.G. and Knölker, H.-J. (2006) Synthesis and activity of carbazole derivatives against *Mycobacterium tuberculosis*. *ChemMedChem*, **1** (8), 812–815; (b) Choi, T.A., Czerwonka, R., Forke, R., Jäger, A., Knöll, J., Krahl, M.P., Krause, T., Reddy, K.R., Franzblau, S.G. and Knölker, H.-J. (2008) Synthesis and pharmacological potential of carbazoles. *Med. Chem. Res.*, **17** (2–7), 374–385.

76 Kataeva, O., Krahl, M.P. and Knölker, H.-J. (2005) First total synthesis of the biologically active 2,7-dioxygenated tricyclic carbazole alkaloids 7-methoxy-O-methylmukonal, clausine H (clauszoline-C), clausine K (clauszoline-J) and clausine O. *Org. Biomol. Chem.*, **3** (17), 3099–3101.

77 Wu, T.-S., Huang, S.-C., Lai, J.-S., Teng, C.-M., Ko, F.-N. and Kuoh, C.-S. (1993) Chemical and antiplatelet aggregative investigation of the leaves of *Clausena excavata*. *Phytochemistry*, **32** (2), 449–451.

78 Forke, R., Jäger, A. and Knölker, H.-J. (2008) First total synthesis of clausine L and pityriazole, a metabolite of the human pathogenic yeast *Malassezia furfur*. *Org. Biomol. Chem.*, **6** (14), 2481–2483.

79 Forke, R., Krahl, M.P., Däbritz, F., Jäger, A. and Knölker, H.-J. (2008) An efficient palladium-catalyzed route to 2-oxygenated and 2,7-dioxygenated carbazole alkaloids – total synthesis of 2-methoxy-3-methylcarbazole, glycosinine, clausine L, mukonidine, and clausine V. *Synlett*, (12), 1870–1876.

80 Potterat, O., Puder, C., Bolek, W., Wagner, K., Ke, C., Ye, Y. and Gillardon, F. (2005) Clausine Z, a new carbazole alkaloid from *Clausena excavata* with inhibitory activity on CDK5. *Pharmazie*, **60** (8), 637–639.

81 (a) Furukawa, H., Wu, T.-S. and Kuoh, C.-S. (1985) Dihydroxygirinimbine, a new carbazole alkaloid from *Murraya euchrestifolia*. *Heterocycles*, **23** (6), 1391–1393; (b) Gruner, K.K., Hopfmann, T., Matsumoto, K., Jäger, A., Katsuki, T. and Knölker, H.-J. (2011) Efficient iron-mediated approach to pyrano[3,2-a]carbazole alkaloids – first total syntheses of O-methylmurrayamine A and 7-methoxymurrayacine, first asymmetric synthesis and assignment of the absolute configuration of (−)-trans-dihydroxygirinimbine. *Org. Biomol. Chem.*, **9** (7), 2057–2061.

82 Knölker, H.-J. and Hofmann, C. (1996) Molybdenum-mediated total synthesis of girinimbine, murrayacine, and dihydroxygirinimbine. *Tetrahedron Lett.*, **37** (44), 7947–7950.

83 Goodwin, S., Smith, A.F. and Horning, E.C. (1959) Alkaloids of *Ochrosia elliptica* Labill. *J. Am. Chem. Soc.*, **81** (8), 1903–1908.

84 Stiborová, M., Bieler, C.A., Wiessler, M. and Frei, E. (2001) The anticancer agent ellipticine on activation by cytochrome P450 forms covalent DNA adducts. *Biochem. Pharmacol.*, **62** (12), 1675–1684.

85 Sainsbury, M. (1977) The synthesis of 6H-pyrido[4,3-b]carbazoles. *Synthesis*, (7), 437–448.
86 Suffness, M. and Cordell, G.A. (1985) Antitumor alkaloids; XI. Ellipticine, in *The Alkaloids* (ed. A. Brossi), Academic Press, New York, USA, p. 89.
87 Suffness, M. and Cordell, G.A. (1985) Antitumor alkaloids; XI. Ellipticine, in *The Alkaloids* (ed. A. Brossi), Academic Press, New York, USA, p. 304.
88 Hibino, S. and Sugino, E. (1990) Synthesis of ellipticine and olivacine by the thermal electrocyclic reaction *via* pyridine 3,4-quinodimethane intermediates. *J. Heterocycl. Chem.*, **27** (6), 1751–1755.
89 Ho, T.-L. and Hsieh, S.-Y. (2006) Regioselective synthesis of ellipticine. *Helv. Chim. Acta*, **89** (1), 111–116.
90 Konakahara, T., Kiran, Y.B., Okuno, Y., Ikeda, R. and Sakai, N. (2010) An expedient synthesis of ellipticine *via* Suzuki–Miyaura coupling. *Tetrahedron Lett.*, **51** (17), 2335–2338.
91 Nihei, Y., Yamamoto, H., Hasegawa, M., Hanada, M., Fukagawa, Y. and Oki, T. (1993) Epocarbazolines A and B, novel 5-lipoxygenase inhibitors taxonomy, fermentation, isolation, structures, and biological activities. *J. Antibiot.*, **46** (1), 25–33.
92 Knöll, J. and Knölker, H.-J. (2006) First total synthesis of (±)-epocarbazolin A and epocarbazolin B, and asymmetric synthesis of (–)-epocarbazolin A *via* Shi epoxidation. *Tetrahedron Lett.*, **47** (34), 6079–6082.
93 Wu, T.-S., Wang, M.-L. and Wu, P.-L. (1996) Seasonal variations of carbazole alkaloids in *Murraya euchrestifolia*. *Phytochemistry*, **43** (4), 785–789.
94 Gruner, K.K. and Knölker, H.-J. (2008) Palladium-catalyzed total synthesis of euchrestifoline using a one-pot Wacker oxidation and double aromatic C–H bond activation. *Org. Biomol. Chem.*, **6** (21), 3902–3904.
95 Kumar, V., Reisch, J. and Wickramasinghe, A. (1989) Glycomaurin and glycomaurrol, new carbazole alkaloids from *Glycosmis mauritiana* (Rutaceae). *Bark. Aust. J. Chem.*, **42** (8), 1375–1379.
96 Ito, C. and Furukawa, H. (1990) New carbazole alkaloids from *Murraya euchrestifolia* Hayata. *Chem. Pharm. Bull.*, **38** (6), 1548–1550.
97 Lebold, T.P. and Kerr, M.A. (2007) Total synthesis of eustifolines A-D and glycomaurrol *via* a divergent Diels–Alder strategy. *Org. Lett.*, **9** (10), 1883–1886.
98 Forke, R., Krahl, M.P., Krause, T., Schlechtingen, G. and Knölker, H.-J. (2007) First total synthesis of methyl 6-methoxycarbazole-3-carboxylate, glycomaurrol, the anti-TB active micromeline, and the furo[2,3-c]carbazole alkaloid eustifoline-D. *Synlett*, (2), 268–272.
99 Wu, T.-S., Huang, S.-C. and Wu, P.-L. (1997) Pyrano- and furocarbazole alkaloids from the root bark of *Clausena excavata*. *Heterocycles*, **45** (5), 969–973.
100 Knölker, H.-J. and Krahl, M.P. (2004) First total synthesis of furoclausine-A. *Synlett*, (3), 528–530.
101 Knölker, H.-J. and Fröhner, W. (1996) First total synthesis of furostifoline. *Tetrahedron Lett.*, **37** (51), 9183–9186.
102 (a) Hagiwara, H., Choshi, T., Fujimoto, H., Sugino, E. and Hibino, S. (1998) New syntheses of murrayaquinone A and furostifoline. *Chem. Pharm. Bull.*, **46** (12), 1948–1949; (b) Hagiwara, H., Choshi, T., Nobuhiro, J., Fujimoto, H., Sugino, E. and Hibino, S. (2001) Novel syntheses of murrayaquinone A and furostifoline through 4-oxygenated carbazoles by allene-mediated electrocyclic reactions starting from 2-chloroindole-3-carbaldehyde. *Chem. Pharm. Bull.*, **49** (7), 881–886.
103 Beccalli, E.M., Clerici, F. and Marchesini, A. (1998) A new synthesis of furostifoline. *Tetrahedron*, **54** (38), 11675–11682.
104 Soós, T., Timári, G. and Hajós, G. (1999) A concise synthesis of furostifoline. *Tetrahedron Lett.*, **40** (49), 8607–8609.
105 Knölker, H.-J. and Fröhner, W. (2000) Convergent iron-mediated syntheses of the furo[3,2-a]carbazole alkaloid furostifoline. *Synthesis*, (14), 2131–2136.
106 Yasuhara, A., Suzuki, N. and Sakamoto, T. (2002) A concise synthesis of furostifoline by tetrabutylammonium

fluoride-promoted indole ring formation. *Chem. Pharm. Bull.*, **50** (1), 143–145.

107 Fröhner, W., Krahl, M.P., Reddy, K.R. and Knölker, H.-J. (2004) Synthetic routes to naturally occurring furocarbazoles. *Heterocycles*, **63** (10), 2393–2407.

108 Chakraborty, D.P., Barman, B.K. and Bose, P.K. (1964) On the structure of girinimbine, a pyrano-carbazole derivative, isolated from *Murraya koenigii* Spreng. *Sci. Cult.*, **30** (9), 445.

109 Joshi, B.S., Kamat, V.N. and Gawad, D.H. (1970) On the structures of girinimbine, mahanimbine, isomahanimbine, koenimbidine and murrayacine. *Tetrahedron*, **26** (5), 1475–1482.

110 Chakraborty, D.P. and Roy, S. (2003) Carbazole alkaloids IV., in *Progress in the Chemistry of Organic Natural Products* (eds W. Herz, H. Grisebach, G.W. Kirby, W. Steglich and C. Tamm), Springer-Verlag, Wien, Austria, pp. 125–230.

111 Cui, C.-B., Yan, S.-Y., Cai, B. and Yao, X.-S. (2002) Carbazole alkaloids as new cell cycle inhibitor and apoptosis inducers from *Clausena Dunniana* Levl. *J. Asian Nat. Prod. Res.*, **4** (4), 233–241.

112 Kano, S., Sugino, E. and Hibino, S. (1980) Synthesis of the carbazole alkaloid hyellazole. *J. Chem. Soc. Chem. Commun.*, (24), 1241–1242.

113 Takano, S., Suzuki, Y. and Ogasawara, K. (1981) A simple synthesis of the blue-green alga alkaloid, hyellazole. *Heterocycles*, **16** (9), 1479–1480.

114 Kawasaki, T., Nonaka, Y. and Sakamoto, M. (1989) A new efficient synthesis of the 3-methoxycarbazole alkaloid hyellazole. *J. Chem. Soc. Chem. Commun.*, (1), 43–44.

115 Danheiser, R.L., Brisbois, R.G., Kowalczyk, J.J. and Miller, R.F. (1990) An annulation method for the synthesis of highly substituted polycyclic aromatic and heteroaromatic compounds. *J. Am. Chem. Soc.*, **112** (8), 3093–3100.

116 Kawasaki, K., Nonaka, Y., Akahane, M., Maeda, N. and Sakamoto, M. (1993) New approach to 3-oxygenated carbazoles. Synthesis of hyellazole and 4-deoxycarbazomycin B. *J. Chem. Soc. Perkin Trans. 1*, (15), 1777–1781.

117 Knölker, H.-J., Baum, E. and Hopfmann, T. (1995) Total synthesis of the marine alkaloid hyellazole. *Tetrahedron Lett.*, **36** (30), 5339–5342.

118 Knölker, H.-J., Baum, E. and Hopfmann, T. (1999) Iron-mediated synthesis of hyellazole and isohyellazole. *Tetrahedron*, **55** (34), 10391–10412.

119 Witulski, B. and Alayrac, C. (2002) A highly efficient and flexible synthesis of substituted carbazoles by rhodium-catalyzed inter- and intramolecular alkyne cyclotrimerizations. *Angew. Chem.*, **114** (17), 3415–3418. *Angew. Chem. Int. Ed.* **41** (17), 3281–3284.

120 Shin-Ya, K., Shimizu, S., Kunigami, T., Furihata, K., Furihata, K. and Seto, H. (1995) A new neuronal cell protecting substance, lavanduquinocin, produced by *Streptomyces viridochromogenes*. *J. Antibiot.*, **48** (7), 574–578.

121 Knölker, H.-J. and Fröhner, W. (1998) First total synthesis of the potent neuronal cell protecting substance (±)-lavanduquinocin *via* iron- and nickel-mediated coupling reactions. *Tetrahedron Lett.*, **39** (17), 2537–2540.

122 Knölker, H.-J., Baum, E. and Reddy, K.R. (2000) First enantioselective total synthesis of lavanduquinocin, a potent neuronal cell protecting substance from *Streptomyces viridochromogenes*. *Chirality*, **12** (5–6), 526–528.

123 Ma, C., Case, R.J., Wang, Y., Zhang, H.-J., Tan, G.T., Hung, N.V., Cuong, N.M., Franzblau, S.G., Soejarto, D.D., Fong, H.H.S. and Pauli, G.F. (2005) Anti-tuberculosis constituents from the stem bark of *Micromelum hirsutum*. *Planta Med.*, **71** (3), 261–267.

124 Chakraborty, D.P., Roy, S. and Guha, R. (1978) Structure of mukonidine. *J. Indian Chem. Soc.*, **55** (11), 1114–1115.

125 Knölker, H.-J. and Wolpert, M. (1997) Cyclization of tricarbonyliron complexes by oxygen to 4a,9a-dihydro-9*H*-carbazoles: application to the synthesis of mukonine, mukonidine, and pyrido[3,2,1-*jk*]carbazoles. *Tetrahedron Lett.*, **38** (4), 533–536.

126 Knölker, H.-J. and Wolpert, M. (2003) Iron-mediated total synthesis of mukonine and mukonidine by oxidative

127 Chakraborty, D.P. and Das, K.C. (1968) Structure of murrayacine. *Chem. Commun.*, (16), 967–969.

128 Chakraborty, D.P., Das, K.C. and Chowdhury, B.K. (1971) Chemical taxonomy. XXVI. Structure of murrayacine. *J. Org. Chem.*, **36** (5), 725–727.

129 Ray, S. and Chakraborty, D.P. (1976) Murrayacine from *Clausena heptaphylla*. *Phytochemistry*, **15** (2), 356–356.

130 (a) Das, K.C., Chakraborty, D.P. and Bose, P.K. (1965) Antifungal activity of some constituents of *Murraya koenigii* Spreng. *Experientia*, **21** (6), 340; (b) Chakraborty, D.P., Barman, B.K. and Bose, P.K. (1965) On the constitution of murrayanine, a carbazole derivative isolated from *Murraya koenigii* Spreng. *Tetrahedron*, **21** (2), 681–685.

131 Bhattacharyya, P. and Chakraborty, D.P. (1973) Murrayanine and dentatin from *Clausena heptaphylla*. *Phytochemistry*, **12** (7), 1831–1832.

132 Cuong, N.M., Hung, T.Q., Sung, T.V. and Taylor, W.C. (2004) A new dimeric carbazole alkaloid from *Glycosmis stenocarpa* roots. *Chem. Pharm. Bull.*, **52** (10), 1175–1178.

133 Knölker, H.-J. and Bauermeister, M. (1990) Iron-mediated total synthesis of the cytotoxic carbazole koenoline and related alkaloids. *J. Chem. Soc. Chem. Commun.*, (9), 664–665.

134 Knölker, H.-J. and Bauermeister, M. (1993) Iron-mediated total synthesis of 1-oxygenated carbazole alkaloids. *Tetrahedron*, **49** (48), 11221–11236.

135 Benavides, A., Peralta, J., Delgado, F. and Tamariz, J. (2004) Total synthesis of the natural carbazoles murrayanine and murrayafoline A, based on the regioselective Diels–Alder addition of *exo*-2-oxazolidinone dienes. *Synthesis*, (15), 2499–2504.

136 Kato, S., Shindo, K., Kataoka, Y., Yamagishi, Y. and Mochizuki, J. (1991) Studies on free radical scavenging substances from microorganisms II. Neocarazostatins A, B and C, novel free radical scavengers. *J. Antibiot.*, **44** (8), 903–907.

137 Czerwonka, R., Reddy, K.R., Baum, E. and Knölker, H.-J. (2006) First enantioselective total synthesis of neocarazostatin B, determination of its absolute configuration and transformation into carquinostatin A. *Chem. Commun.*, (7), 711–713.

138 Knölker, H.-J., Fröhner, W. and Wagner, A. (1998) First total synthesis of the free radical scavenger (±)-neocarazostatin B via iron- and nickel-mediated coupling reactions. *Tetrahedron Lett.*, **39** (19), 2947–2950.

139 Azoug, M., Loukaci, A., Richard, B., Nuzillard, J.-M., Moreti, C., Zèches-Hanrot, M. and Le Men-Olivier, L. (1995) Alkaloids from stem bark and leaves of *Peschiera buchtieni*. *Phytochemistry*, **39** (5), 1223–1228.

140 Wittwer, H. and Schmutz, J. (1960) Die Synthese von Olivacin, Dihydro-olivacin, Tetrahydro-olivacin, *N*-Methyl-tetrahydro-olivacin, und die Konstitution von u-Alkaloid D. *Helv. Chim. Acta*, **43** (3), 793–799.

141 Yokoyama, Y., Okuyama, N., Iwadate, S., Momoi, T. and Murakami, Y. (1990) Synthetic studies of indoles and related compounds, Part 22. The Vilsmeier–Haak reaction of *N*-benzyl-1,2,3,4-tetrahydrocarbazoles and its synthetic application to olivacine and ellipticine. *J. Chem. Soc. Perkin Trans. 1*, (11), 1319–1329.

142 Gribble, G.W., Saulnier, M.G., Obaza-Nutaitis, J.A. and Ketcha, D.M. (1992) A versatile and efficient construction of the 6*H*-pyrido[4,3-*b*] carbazole ring system. Syntheses of the antitumor alkaloids ellipticine, 9-methoxyellipticine, and olivacine, and their analogs. *J. Org. Chem.*, **57** (22), 5891–5899.

143 Miki, Y., Tsuzaki, Y., Hibino, H. and Aoki, Y. (2004) Synthesis of 3-methoxyolivacine and olivacine by Friedel-Crafts reaction of indole-2,3-dicarboxylic anhydride with 2,4,6-trimethoxypyridine. *Synlett*, (12), 2206–2208.

144 Irlinger, B., Bartsch, A., Krämer, H.-J., Mayser, P. and Steglich, W. (2005) New tryptophan metabolites from cultures of the lipophilic yeast

Malassezia furfur. Helv. Chim. Acta, **88** (6), 1472–1484.

145 Nettleton, D.E., Doyle, T.W., Krishnan, B., Matsumoto, G.K. and Clardy, J. (1985) Isolation and structure of rebeccamycin – a new antitumor antibiotic from *Nocardia aerocoligenes*. *Tetrahedron Lett.*, **26** (34), 4011–4014.

146 Bush, J.A., Long, B.H., Catino, J.J., Bradner, W.T. and Tomita, K. (1987) Production and biological activity of rebeccamycin, a novel antitumor agent. *J. Antibiot.*, **40** (5), 668–678.

147 Kaneko, T., Wong, H., Okamoto, K.T. and Clardy, J. (1985) Two synthetic approaches to rebeccamycin. *Tetrahedron Lett.*, **26** (34), 4015–4018.

148 (a) Gallant, M., Link, J.T. and Danishefsky, S.J. (1993) A stereoselective synthesis of indole-β-N-glycosides: an application to the synthesis of rebeccamycin. *J. Org. Chem.*, **58** (2), 343–349; (b) Faul, M.M., Winneroski, L.L. and Krumrich, C.A. (1999) Synthesis of rebeccamycin and 11-dechlororebeccamycin. *J. Org. Chem.*, **64** (7), 2465–2470.

149 Wang, J., Rosingana, M., Watson, D.J., Dowdy, E.D., Discordia, R.P., Soundarajan, N. and Li, W.-S. (2001) Practical synthesis of the rebeccamycin aglycone and related analogs by oxidative cyclization of bisindolylmaleimides with a Wacker-type catalytic system. *Tetrahedron Lett.*, **42** (51), 8935–8937.

150 Meragelman, K.M., McKee, T.C. and Boyd, M.R. (2000) Siamenol, a new carbazole alkaloid from *Murraya siamensis*. *J. Nat. Prod.*, **63** (3), 427–428.

151 Krahl, M.P., Jäger, A., Krause, T. and Knölker, H.-J. (2006) First total synthesis of the 7-oxygenated carbazole alkaloids clauszoline-K, 3-formyl-7-hydroxycarbazole, clausine M, clausine N and the anti-HIV active siamenol using a highly efficient palladium-catalyzed approach. *Org. Biomol. Chem.*, **4** (17), 3215–3219.

152 Naffziger, M.R., Ashburn, B.O., Perkins, J.R. and Carter, R.G. (2007) Diels–Alder approach for the construction of halogenated, o-nitro biaryl templates and application to the total synthesis of the anti-HIV agent siamenol. *J. Org. Chem.*, **72** (26), 9857–9865.

153 Omura, S., Iwai, Y., Hirano, A., Nakagawa, A., Awaya, J., Tsuchiya, H., Takahashi, Y. and Masuma, R. (1977) A new alkaloid AM-2282 of *Streptomyces* origin. Taxonomy, fermentation, isolation and preliminary characterization. *J. Antibiot.*, **30** (4), 275–282.

154 Funato, N., Takayanagi, H., Konda, Y., Toda, Y., Harigaya, Y., Iwai, Y. and Omura, S. (1994) Absolute configuration of staurosporine by X-ray analysis. *Tetrahedron Lett.*, **35** (8), 1251–1254.

155 Omura, S., Sasaki, Y., Iwai, Y. and Takeshima, H. (1995) Staurosporine, a potential important gift from a microorganism. *J. Antibiot.*, **48** (7), 535–548.

156 Link, J.T., Raghavan, S. and Danishefsky, S.J. (1995) First total synthesis of staurosporine and ent-staurosporine. *J. Am. Chem. Soc.*, **117** (1), 552–553.

157 Link, J.T., Raghavan, S., Gallant, M., Danishefsky, S.J., Chou, T.C. and Ballas, L.M. (1996) Staurosporine and ent-staurosporine: the first total syntheses, prospects for a regioselective approach, and activity profiles. *J. Am. Chem. Soc.*, **118** (12), 2825–2842.

158 Wood, J.L., Stoltz, B.M. and Goodman, S.N. (1996) Total synthesis of (+)-RK-286c, (+)-MLR-52, (+)-staurosporine, and (+)-K252a. *J. Am. Chem. Soc.*, **118** (43), 10656–10657.

159 Wood, J.L., Stoltz, B.M., Goodman, S.N. and Onwueme, K. (1997) Design and implementation of an efficient synthetic approach to pyranosylated indolocarbazoles: total synthesis of (+)-RK286c, (+)-MLR-52, (+)-staurosporine, and (−)-TAN-1030a. *J. Am. Chem. Soc.*, **119** (41), 9652–9661.

160 Nakanishi, S., Matsuda, Y., Iwahashi, K. and Kase, H. (1986) K-252b, c and d, potent inhibitors of protein kinase C from microbial origin. *J. Antibiot.*, **39** (8), 1066–1071.

161 Yasuzawa, T., Iida, T., Yoshida, M., Hirayama, N., Takahashi, M., Shirahata, K. and Sano, H. (1986) The structures of the novel protein kinase C inhibitors

K-252a, b, c and d. *J. Antibiot.*, **39** (8), 1072–1078.

162 Horton, P.A., Longley, R.E., McConnell, O.J. and Ballas, L.M. (1994) Staurosporine aglycon (K252-c) and arcyriaflavin A from marine ascidian *Eudistoma* sp. *Experientia*, **50** (9), 843–845.

163 Sarstedt, B. and Winterfeldt, E. (1983) Reactions with indole derivatives, XLVIII. A simple synthesis of the staurosporine aglycon. *Heterocycles*, **20** (3), 469–476.

164 Moody, C.J. and Rahimtoola, K.F. (1990) Synthesis of the staurosporine aglycone. *J. Chem. Soc. Chem. Commun.*, (23), 1667–1668.

165 Hughes, I., Nolan, W.P. and Raphael, R.A. (1990) Synthesis of the indolo[2,3-*a*] carbazole natural products staurosporinone and arcyriaflavin B. *J. Chem. Soc. Perkin Trans. 1*, (9), 2475–2480.

166 Moody, C.J., Rahimtoola, K.F., Porter, B. and Ross, B.C. (1992) Synthesis of the staurosporine aglycon. *J. Org. Chem.*, **57** (7), 2105–2114.

167 Faul, M.M., Sullivan, K.A. and Winneroski, L.L. (1995) A general approach to the synthesis of bisindolylmaleimides: synthesis of staurosporine aglycone. *Synthesis*, (12), 1511–1516.

168 Mahboobi, S., Eibler, E., Koller, M., Kumar, K.C.S., Popp, A. and Schollmeyer, D. (1999) Synthesis of pyrrolidin-2-ones and of staurosporine aglycon (K-252c) by intermolecular Michael reaction. *J. Org. Chem.*, **64** (13), 4697–4704.

169 Gaudêncio, S.P., Santos, M.M.M., Lobo, A.M. and Prabhakar, S. (2003) A short synthesis of staurosporinone (K-252c). *Tetrahedron Lett.*, **44** (12), 2577–2578.

170 Bonjouklian, R., Smitka, T.A., Doolin, L.E., Molloy, R.M., Debono, M., Shaffer, S.A., Moore, R.E., Stewart, J.B. and Patterson, G.M.L. (1991) Tjipanazoles, new antifungal agents from the blue-green alga *Tolypothrix tjipanasensis*. *Tetrahedron*, **47** (37), 7739–7750.

171 Kuethe, J.T., Wong, A. and Davies, I.W. (2003) Effective strategy for the preparation of indolocarbazole aglycons and glycosides: total synthesis of tjipanazoles B, D, E, and I. *Org. Lett.*, **5** (20), 3721–3723.

172 Gilbert, E.J., Ziller, J.W. and Van Vranken, D.L. (1997) Cyclizations of unsymmetrical bis-1,2-(3-indolyl) ethanes: synthesis of (–)-tjipanazole F1. *Tetrahedron*, **53** (48), 16553–16564.

173 Gilbert, E.J. and Van Vranken, D.L. (1996) Control of dissymmetry in the synthesis of (+)-tjipanazole F2. *J. Am. Chem. Soc.*, **118** (23), 5500–5501.

174 Chakraborty, D.P. (1977) Carbazole alkaloids, in *Progress in the Chemistry of Organic Natural Products*, vol. 34, (eds W. Herz, H. Griesebach and G.W. Kirby), Springer, Wien, Austria, pp. 299–371.

175 Iwaki, T., Yasuhara, A. and Sakamoto, T. (1999) Novel synthetic strategy of carbolines via palladium-catalyzed amination and arylation reaction. *J. Chem. Soc. Perkin Trans. 1*, (11), 1505–1510.

176 (a) Bedford, R.B. and Cazin, C.S.J. (2002) A novel catalytic one-pot synthesis of carbazoles via consecutive amination and C-H activation. *Chem. Commun.*, (20), 2310–2311; (b) Ackermann, L. and Althammer, A. (2007) Domino N–H/C–H bond activation: palladium-catalyzed synthesis of annulated heterocycles using dichloro(hetero)arenes. *Angew. Chem.*, **119** (10), 1652–1654. *Angew. Chem. Int. Ed.* **46** (10), 1627–1629.

177 Knölker, H.-J. (2000) Efficient synthesis of tricarbonyliron-diene complexes – development of an asymmetric catalytic complexation. *Chem. Rev.*, **100** (8), 2941–2962.

178 Chakraborty, D.P. and Islam, A. (1971) Synthesis of girinimbine. *J. Indian Chem. Soc.*, **48** (1), 91–92.

179 Leino, R., Luttikhedde, H.J.G., Lehtonen, A., Sillanpää, R., Penninkangas, A., Strandén, J., Mattinen, J. and Näsman, J.H. (1998) Bis[2-(*tert*-butyldimethylsiloxy)-4,7-dimethylindenyl]zirconium dichloride: synthesis, torsional isomerism and olefin polymerization catalysis. *J. Organomet. Chem.*, **558** (1-2), 171–179.

180 Reddy, G.M., Chen, S.-Y. and Uang, B.-J. (2003) A facile synthesis of indolo[2,3-a]pyrrolo[3,4-c]carbazoles via oxidative photocyclization of bisindolylmaleimides. *Synthesis*, (4), 497–500.

181 Graebe, C. and Caro, H. (1870) Ueber Acridin. *Ber. Dtsch. Chem. Ges.*, **3** (2), 746–747.

182 Hughes, G.K., Lahey, F.N., Price, G.R. and Webb, L.J. (1948) Alkaloids of the Australian rutaceae. *Nature*, **162** (4110), 223–224.

183 Michael, J.P. (2008) Quinoline, quinazoline and acridone alkaloids. *Nat. Prod. Rep.*, **25** (1), 166–187.

184 Michael, J.P. (2007) Quinoline, quinazoline and acridone alkaloids. *Nat. Prod. Rep.*, **24** (1), 223–246.

185 Skaltsounis, A.L., Mitaku, S. and Tillequin, F. (2000) Acridone alkaloids, in *The Alkaloids*, vol. 54, (ed. G.A. Cordell), Academic Press, Amsterdam, The Netherlands, pp. 259–377.

186 Demeunynck, M., Charmantray, F. and Martelli, A. (2001) Interest of acridine derivatives in the anticancer chemotherapy. *Curr. Pharm. Des.*, **7** (17), 1703.

187 Denny, W.A. (2004) Chemotherapeutic effects of acridine derivatives. *Med. Chem. Rev*, **1** (3), 257–266.

188 Belmont, P., Bosson, J., Godet, T. and Tiano, M. (2007) Acridine and acridone derivatives, anticancer properties and synthetic methods: where are we now? *Anticancer Agents Med. Chem.*, **7** (2), 139–169.

189 Michael, J.P. (2001) Quinoline, quinazoline and acridone alkaloids. *Nat. Prod. Rep.*, **18** (5), 543–559.

190 Michael, J.P. (2005) Quinoline, quinazoline and acridone alkaloids. *Nat. Prod. Rep.*, **22** (5), 627–646.

191 Tillequin, F. (2007) Rutaceous alkaloids as models for the design of novel antitumor drugs. *Phytochem. Rev.*, **6** (1), 65–79.

192 Beck, J.R., Booher, R.N., Brown, A.C., Kwok, R. and Pohland, A. (1967) Synthesis of acronycine. *J. Am. Chem. Soc.*, **89** (15), 3934–3935.

193 Loughhead, D.G. (1990) Synthesis of des-N-methylacronycine and acronycine. *J. Org. Chem.*, **55** (7), 2245–2246.

194 Anand, R.C. and Selvapalam, N. (1996) A practical regiospecific approach towards acronycine and related alkaloids. *Chem. Commun.*, (2), 199–200.

195 Lasnitzki, T. and Wilkinson, J.H. (1948) The effect of acridine derivatives on the growth and mitosis of cells in vitro. *Br. J. Cancer*, **2** (4), 369–375.

196 Sebestík, J., Hlaváček, J. and Stibor, I. (2007) A role of the 9-aminoacridines and their conjugates in a life science. *Curr. Protein Pept. Sci.*, **8** (5), 471–483.

197 Schmitz, F.J., Agarwal, S.K., Gunasekera, S.P., Schmidt, P.G. and Shoolery, J.N. (1983) Amphimedine, new aromatic alkaloid from a pacific sponge, *Amphimedon* sp. Carbon connectivity determination from natural abundance $^{13}C-^{13}C$ coupling constants. *J. Am. Chem. Soc.*, **105** (14), 4835–4836.

198 Marshall, K.M. and Barrows, L.R. (2004) Biological activities of pyridoacridines. *Nat. Prod. Rep.*, **21** (6), 731–751.

199 Echavarren, A.M. and Stille, J.K. (1988) Total synthesis of amphimedine. *J. Am. Chem. Soc.*, **110** (12), 4051–4053.

200 (a) Nakahara, S., Tanaka, Y. and Kubo, A. (1996) Total synthesis of amphimedine. *Heterocycles*, **43** (10), 2113–2123; (b) Nakahara, S., Tanaka, Y. and Kubo, A. (1993) Total synthesis of eilatin. *Heterocycles*, **36** (5), 1139–1144.

201 Horstmann, M.A., Hassenpflug, W.A., zur Stadt, U., Escherich, G., Janka, G. and Kabisch, H. (2005) Amsacrine combined with etoposide and high-dose methylprednisolone as salvage therapy in acute lymphoblastic leukemia in children. *Haematologica*, **90** (12), 1701–1703.

202 Cain, B.F., Atwell, G.J. and Denny, W.A. (1975) Potential antitumor agents. 16. 4'-(Acridin-9-ylamino)methanesulfonanilides. *J. Med. Chem.*, **18** (11), 1110–1117.

203 Ito, C., Kondo, Y., Rao, K.S., Tokuda, H., Nishino, H. and Furukawa, H. (1999) Chemical constituents of *Glycosmis pentaphylla*. Isolation of a novel naphtoquinone and a new acridone alkaloid. *Chem. Pharm. Bull.*, **47** (11), 1579–1581.

204 Reisch, J., Szendrei, K., Novak, I. and Minker, E. (1972) Vorkommen und

biologische Wirkung von Acridon-Derivaten. *Scientia Pharmaceutica*, **40** (3), 161–178.
205 Michael, J.P. (2000) Quinoline, quinazoline and acridone alkaloids. *Nat. Prod. Rep.*, **17** (6), 603–620.
206 Tringali, C., Spatafora, C., Calì, V. and Simmonds, M.S.J. (2001) Antifeedant constituents from *Fagara macrophylla*. *Fitoterapia*, **72** (5), 538–543.
207 Réthy, B., Zupkó, I., Minorics, R., Hohmann, J., Ocsovszki, I. and Falkay, G. (2007) Investigation of cytotoxic activity on human cancer cell lines of arborinine and furanoacridones isolated from *Ruta graveolens*. *Planta Med.*, **73** (1), 41–48.
208 Govindachari, T.R., Viswanathan, N., Pai, B.R., Ramachandran, V.N. and Subramaniam, P.S. (1970) Alkaloids of *Atalantia monophylla* Correa. *Tetrahedron*, **26** (12), 2905–2910.
209 Bahar, M.H., Shringarpure, J.D., Kulkarni, G.H. and Sabata, B.K. (1982) Structure and synthesis of atalaphylline and related alkaloids. *Phytochemistry*, **21** (11), 2729–2731.
210 Wu, T.-S. and Chen, C.-M. (2000) Acridone Alkaloids from the root bark of *Severinia buxifolia* in Hainan. *Chem. Pharm. Bull.*, **48** (1), 85–90.
211 Chou, T.-C., Tzeng, C.-C., Wu, T.-S., Watanabe, K.A. and Su, T.-L. (1989) Inhibition of cell growth and macromolecule biosynthesis of human promyelocytic leukemic cells by acridone alkaloids. *Phytother. Res.*, **3** (6), 237–242.
212 Kawaii, S., Tomono, Y., Katase, E., Ogawa, K., Yano, M., Takemura, Y., Ju-ichi, M., Ito, C. and Furukawa, H. (1999) Acridones as inducers of HL-60 cell differentiation. *Leuk. Res.*, **23** (3), 263–269.
213 Ramesh, K. and Kapil, R.S. (1986) Synthesis of acridone alkaloids: glycocitrine-I, N-methylatalaphylline, atalaphyllidine, 11-hydroxyacronycine and 11-hydroxynoracronycine. *Indian J. Chem. Sect. B*, **25** (7), 684–687.
214 Wu, T.-S., Furukawa, H. and Kuoh, C.-S. (1982) Acridone alkaloids IV. Structures of four new acridone alkaloids from *Glycosmis citrifolia* (Willd.) Lindl. *Heterocycles*, **19** (6), 1047–1051.
215 Su, T.-L., Dziewiszek, K. and Wu, T.-S. (1991) Synthesis of glyfoline, a constituent of *glycosmis citrifolia* (Willd.) Lindl. and a potential anticancer agent. *Tetrahedron Lett.*, **32** (12), 1541–1544.
216 Su, T.L., Kohler, B., Chou, T.C., Chun, M.W. and Watanabe, K.A. (1992) Synthesis of the acridone alkaloids, glyfoline and congeners. Structure-activity relationship studies of cytotoxic acridones. *J. Med. Chem.*, **35** (14), 2703–2710.
217 Mietzsch, F. and Mauss, H. (1934) Gegen Malaria wirksame Acridinverbindungen. *Angew. Chem.*, **47** (37), 633–636.
218 Coatney, G.R. and Greenberg, J. (1952) The use of antiobiotics in the treatment of malaria. *Ann. N. Y. Acad. Sci.*, **55** (6), 1075–1081.
219 Colledge, L., Drummond, H., Worthington, R.T., McNee, J.W., Sladden, A.F. and McCartney, J.E. (1917) A report on the treatment of a series of recently inflicted war wounds with 'proflavine'. *Lancet*, **190** (4914), 676–677.
220 Benda, L. (1912) Über das 3.6-Diamino-acridin. *Ber. Dtsch. Chem. Ges.*, **45** (2), 1787–1799.
221 Reisch, J., Szendrei, K., Minker, E. and Novak, I. (1967) Chemistry of natural substances. XVI. Rutacridone from *Ruta graveolens*. *Acta Pharm. Suec.*, **4** (4), 265–266.
222 Meepagala, K.M., Schrader, K.K., Wedge, D.E. and Duke, S.O. (2005) Algicidal and antifungal compounds from the roots of *Ruta graveolens* and synthesis of their analogs. *Phytochemistry*, **66** (22), 2689–2695.
223 Mester, I., Reisch, J., Rózsa, S. and Szendrei, K. (1981) Synthesis of (±)-rutacridone. *Heterocycles*, **16** (1), 77–79.
224 McKenna, M.T., Proctor, G.R., Young, L.C. and Harvey, A.L. (1997) Novel tacrine analogues for potential use against Alzheimer's disease: potent and selective acetylcholinesterase inhibitors and 5-HT uptake inhibitors. *J. Med. Chem.*, **40** (22), 3516–3523.
225 Carlier, P.R., Chow, E.S.H., Han, Y., Liu, J., Yazal, J.E. and Pang, Y.-P. (1999) Heterodimeric tacrine-based acetylcholinesterase inhibitors:

investigating ligand–peripheral site interactions. *J. Med. Chem.*, **42** (20), 4225–4231.
226 Lahiri, D.K., Farlow, M.R. and Sambamurti, K. (1998) The secretion of amyloid β-peptides is inhibited in the tacrine-treated human neuroblastoma cells. *Mol. Brain Res.*, **62** (2), 131–140.
227 Albert, A. and Gledhill, W. (1945) Improved syntheses of aminoacridines. Part IV. Substituted 5-aminoacridines. *J. Soc. Chem. Ind.*, **64**, 169–172.
228 Moore, J.A. and Kornreich, L.D. (1963) A direct synthesis of 4-aminoquinolines. *Tetrahedron Lett.*, **4** (20), 1277–1281.
229 Pegel, K.H. and Wright, W.G. (1969) South african plant extractives Part II. Alkaloids of *Teclea natalensis*. *J. Chem. Soc.(C)*, (18), 2327–2329.
230 Rasoanaivo, P., Federici, E., Palazzino, G. and Galeffi, C. (1999) Acridones of *Vepris sclerophylla*: their ^{13}C-NMR data. *Fitoterapia*, **70** (6), 625–627.
231 Ramachandran, V.N., Pai, B.R. and Santhanam, R. (1972) Synthesis of tecleanthine. *Indian J. Chem.*, **10** (1), 14–15.
232 Hughes, G.K., Neill, K.G. and Ritchie, E. (1952) Alkaloids of the Australian rutaceae: *Evodia xanthoxyloides* F. Muell. II. Isolation of the alkaloids from the leaves. *Aust. J. Sci. Res.*, **5** (2), 401–405.
233 Cannon, J.R., Hughes, G.K., Neill, K.G. and Ritchie, E. (1952) Alkaloids of the Australian Rutaceae: *Evodia xanthoxyloides* F. Muell. III. The structures of the coloured alkaloids, Evoxanthidine, Xanthevodine and Xanthoxoline. *Aust. J. Sci. Res.*, **5** (2), 406–411.
234 Chourpa, I., Morjani, H., Riou, J.-F. and Manfait, M. (1996) Intracellular molecular interactions of antitumor drug amsacrine (m-AMSA) as revealed by surface-enhanced Raman spectroscopy. *FEBS Lett.*, **397** (1), 61–64.
235 Cain, B.F., Seelye, R.N. and Atwell, G.J. (1974) Potential antitumor agents. 14. Acridylmethanesulfonanilides. *J. Med. Chem.*, **17** (9), 922–930.

11
Thiophene and Other Sulfur Heterocycles

Krishna C. Majumdar and Shovan Mondal

11.1
Introduction

Thiophenes are sulfur-containing heterocyclic compounds present in the roots of many *Asteraceous* plants such as *Tagetes minuta, T. patula, T. erecta, T. laxa, T. terniflora* HBK, *T. campanulata, T. mendocina* and *T. argentina* [1]. Thiophenes show a variety of biological activities against nematelmints, bacteria, fungi, virus, insects and algae [2, 3]. The five main thiophenes present in *Tagetes* tissue are 2-T (2-terthienyl), BBT (butenenylbithiophene) and BBTOH (hydroxybutenenylbithiophene), BBTOAc (acetoxybutenenylbithiophene) and BBT(OAc)$_2$ (diacetoxybutenenylbithiophene). Two of these thiophenes (BBT and BBTOAC) account for over 80% of total thiophene content [1]. The "thiophene" was first discovered as a contaminant in benzene in 1883 [4]. It was observed that isatin forms a blue dye if it is mixed with sulfuric acid and crude benzene. The formation of the blue indophenin was long believed to be a reaction with benzene. Victor Meyer was able to isolate the substance responsible for this reaction from commercial benzene and also achieved its synthesis from acetylene and sulfur. This new heterocyclic compound was thiophene [5]. Thiophenes are classically prepared by the reaction of 1,4-diketones, diesters, or dicarboxylates with reagents such as P_4S_{10}, Lawesson's reagent or *via* the Gewald reaction [6], which involves the condensation of two esters in the presence of elemental sulfur.

Theoretical calculations suggest that the degree of aromaticity of thiophene is less than that of benzene. The "electron pairs" on sulfur are significantly delocalized in the pi electron system. As a consequence of its aromaticity, thiophene does not exhibit the properties observed for conventional thioethers. For example, the sulfur atom resists alkylation and oxidation. Although the sulfur atom is relatively unreactive, the flanking carbon centers, the 2- and 5-positions, are highly susceptible to attack by electrophiles and this has been exploited for the synthesis of thiophene-containing natural products from thiophene-based substrates. Therefore, two different strategies are available for the synthesis of natural products containing sulfur heterocycles: (i) by the application of the reactivity of the heterocycle; and (ii) by constructing the heterocyclic moieties from appropriate

Heterocycles in Natural Product Synthesis, First Edition. Edited by Krishna C. Majumdar
and Shital K. Chattopadhyay.
© 2011 Wiley-VCH Verlag GmbH & Co. KGaA. Published 2011 by Wiley-VCH Verlag GmbH & Co. KGaA.

substrates. In this chapter we cover recent developments in this area including citations from earlier literature [7]. We have excluded thiazole-containing natural products which are the subject of Chapter 13 in this book.

The representative members of natural products containing thiophene and other sulfur heterocycles (other than thiazole) are listed in Table 11.1. Besides, there are also reports of biologically active natural artifacts containing disulfide linkage arising out of dimerization of cysteine [8, 9].

11.2
Synthesis of Natural Products Containing Thiophene

The thiophene ring system is ubiquitous in nature and many examples are easily found in foods and plants [53]. Thiophene-containing natural products can be prepared by two different routes: (i) from thiophene-containing substrates; (ii) by the formation of the thiophene nucleus using appropriate substrates.

11.2.1
Synthesis of Natural Products from Thiophene-Based Substrates

Functionalized thiophenes have found extensive use as precursors for functional materials [54], natural products [53], and pharmaceuticals [55]. Such functionalized thiophenes are commonly generated from various halothiophenes, which can provide access to the desired product through the application of catalytic cross-coupling chemistry. Only a few reports are available in literature for the synthesis of thiophene-containing natural products by the application of the reactivity of the thiophene nucleus.

Many naturally occurring and synthetic acetylenic thiophenes have been found to possess UV-mediated antimicrobial activity. Acetylenic thiophenes have shown remarkable antimicrobial activity against *Candida albicans* and *Staphylococcus albus* only when irradiated with UV-A light [39]. Thiophenic compounds having 1-alkynyl groups linked to a thiophene in the 2-position, or to a 2,2'-bithienyl system in the 5-position have exhibited the most potent UV-mediated antibiotic activity [56]. These compounds also possess photoactive antiviral and cytotoxic activities [57]. The compounds possessing a conjugated linear configuration of two or three thiophene rings and a substituent with an acetylenic linkage were found to be most active. A naturally occurring terthiophene, α-terthiophene, has shown anti-HIV activity on irradiation with UV-A light [58].

Rossi and co-workers reported the synthesis of 5-(3-buten-1-ynyl)-2,2'-bithienyl **6** and 2-phenyl-5-(3-buten-1-ynyl) thiophene **11** *via* two different routes [37]. The first involves the palladium-catalyzed cross-coupling of vinyl bromide with the Grignard reagents derived from 5-ethynyl-2,2'-bithienyl **5** and 2-ethynyl-5-phenylthiophen **10**. The second method uses the coupling reaction of vinyl bromide with **5** and **10**, respectively, in the presence of a catalytic amount of $(PPh_3)_4Pd$ and CuI (Scheme 11.1).

11.2 Synthesis of Natural Products Containing Thiophene | 379

Table 11.1 Natural products containing thiophene and other sulfur heterocycles.

Serial No.	Trivial name	Structure	Source	Isolation [Ref]	Biological activity	Synthesis [Ref]
1	Foetithiophene A One similar compound is known		*Ferula foetida*	[10]	NR	NR
2	Xanthopappin A Four similar compounds are known		*Xanthopappus subcaulis*	[11]	Potent photoactivated insecticidal activity against the fourth-instar larvae of the Asian tiger mosquito, *Aedes albopictus*.	NR
3	Tonghaosu analog		*Chrysanthemum coronarium l* and *Athemdeae*	[12]	NR	[13]
4	Tonghaosu analog		*Chrysanthemum coronarium l* and *Athemdeae*	[12]	NR	[13]
5	Banminth		Laboratory assays using *Nematospiroides dubius* in mice and *Nippostrongylus muris* in rats.	[14]	Most promising member of anthelmintics	NR
6	Makaluvamine F		Fijian sponge *Zyzzya cf. marsailis*	[15]	cytotoxicity towards the human colon tumor cell-line HCT-116 and inhibition of topoisomerase II	[16]

(Continued)

Table 11.1 (Continued)

Serial No.	Trivial name	Structure	Source	Isolation [Ref]	Biological activity	Synthesis [Ref]
7	Echinoynethiophene A		*Echinops grijisii* and *Pluchea indica*	[17, 18]	Anti-amebic activity	NR
8	2-(Penta-1,3-diynyl)-5-(3,4-dihydroxybut-1-ynyl)thiophene		*Echinops grijsii* Hance	[19]	Potent quinone oxidoreductase1 inducing Agent	NR
9	Ineupatoriol		*Inula eupatorioides*	[20]	Potent fish poison ichthyothereol	NR
10	Echinothiophene		Roots of *Echinops grijissii*	[21]	NR	NR
11	Bryoanthrathiophene		Bryozoan, *Watersipora subtorquata*	[22]	Potent antiangiogenic activity on bovine aorta endothelial cell proliferation	NR
12	5,7-Dihydroxy-1-methoxycarbonyl-6-oxo-6H-anthra[1,9-bc]thiophene		Bryozoan, *Dakaira subovoidea*	[23]	Antiangiogenic activity	[24]

13	Urothione		Human urine	[25]	Molybdenum cofactor	[25]
14	Discorhabdin A		New Zealand sponges of the genus Latrunculia	[26, 27]	Potent antitumor activity	[28]
15	Discorhabdin B		New Zealand sponges of the genus Latrunculia	[27]	Potent antitumor activity	NR
15	Discorhabdin D		New Zealand sponges of the genus Latrunculia	[29]	Potent antitumor activity	NR
17	Breynin A		Taiwanese woody shrub *Breynia officinalis* Hemsl	[30]	Orally active, lowering rat serum cholesterol by 30–60% after 10 daily doses of 10-20 mg/kg	NR
18	(+)-Biotin		Egg yolk, liver and some vegetables	[31]	B-complex vitamin that aids in body growth	[32]

(Continued)

Table 11.1 (Continued)

Serial No.	Trivial name	Structure	Source	Isolation [Ref]	Biological activity	Synthesis [Ref]
19	Epicoccin I Few more similar compounds are known.		Endophytic fungus *Epicoccum nigrum*	[33]	Fungal metabolites	NR
20	Onionin A		*Allium cepa*	[34]	Inhibits macrophage activation	[34]
21	Epibreynin B Ten more members are known.		*Breynia fruticosa*	[35]	NR	NR
22	Sulfolane		*Batzella sp./ Lissoclinum sp.*	[36]	NR	NR
23	5-(3-Buten-1-ynyl)-2,2′-bithienyl		*Tagetes* root	[37]	Nematicidal and photo-induced fungicidal activity	[37]
24	7-Chloroarctinone-b		roots of *Rhaponticum uniflorum*	[38]	Efficiently antagonizes both hormone and rosiglitazone induced adipocyte differentiation in cell culture.	NR
25	5′-methyl-5-(4-[3-methyl-l-oxobutoxy]-1-butynyl)-2,2′-bithiophen		*Blumea obliqua*	[39]	Antifungal activity against *Epidermopbyton floccosum*	NR

26	5'-hydroxymethyl-5-[butyl-3-en-1-yn]-2,2'-bithiophene iso valeroxy ester		Blumea obliqua.	[39]	Antifungal activity against Pleurotus ostreatus	NR
27	Xanthopappin C		Xanthopappus subacaulis	[11]	Potent photoactivated insecticidal activity against the fourth-instar larvae of the Asian tiger mosquito, Aedes albopictus.	NR
28	Gerrardine		Cassipourea guianensis	[40]	Antimicrobial activity	NR
29	Spiruchostatin A		Pseudomonas sp.	[41]	Potent inhibitory activity against histone deacetylases	[42]
30	Aspirochlorine		Aspergillus tamari	[43]	Antifungal activity	[44]
31	Aplidinone A Two more similar compounds are known.		Aplidium conicum	[45]	NR	NR

(Continued)

Table 11.1 (Continued)

Serial No.	Trivial name	Structure	Source	Isolation [Ref]	Biological activity	Synthesis [Ref]
32	Thiaplidiaquinone A		Ascidian *Aplidium conicum*	[46]	Disruption of mitochondrial potential and cell death by apoptosis	NR
33	Thiaplidiaquinone B		Ascidian *Aplidium conicum*	[46]	Disruption of mitochondrial potential and cell death by apoptosis	NR
34	Shermilamine B		*Diplosoma* sp.	[47]	Antibacterial activity	NR
35	3-Ketoadociaquinone B		*Xestospongia* sp.	[48]	Inhibitors of Cdc25B phosphatase	NR

36	Lissoclibadin 3		Ascidian *Lissoclinum* cf. *badium*	[49]	Cytotoxic against HL-60	NR
37	Gliocladine A Five more members are reported.		*Gliocladium roseum*	[50]	Antinematodal activity against *Caenorhabditis elegans* and *Panagrellus redivivus*	NR
38	Lissoclibadin 1		Ascidian *Lissoclinum* cf. *badium*	[49]	Inhibited the growth of the marine bacterium *Ruegeria atlantica* and also cytotoxic against HL-60	NR
39	Tetrathioaspirochlorine		*Aspergillus flavus*	[51]	Azole-resistant *Candida albicans*	NR
40	Lissoclintoxin A		Tunicate *Lissoclinum perforatum*	[52]	NR	NR

NR denotes not reported.

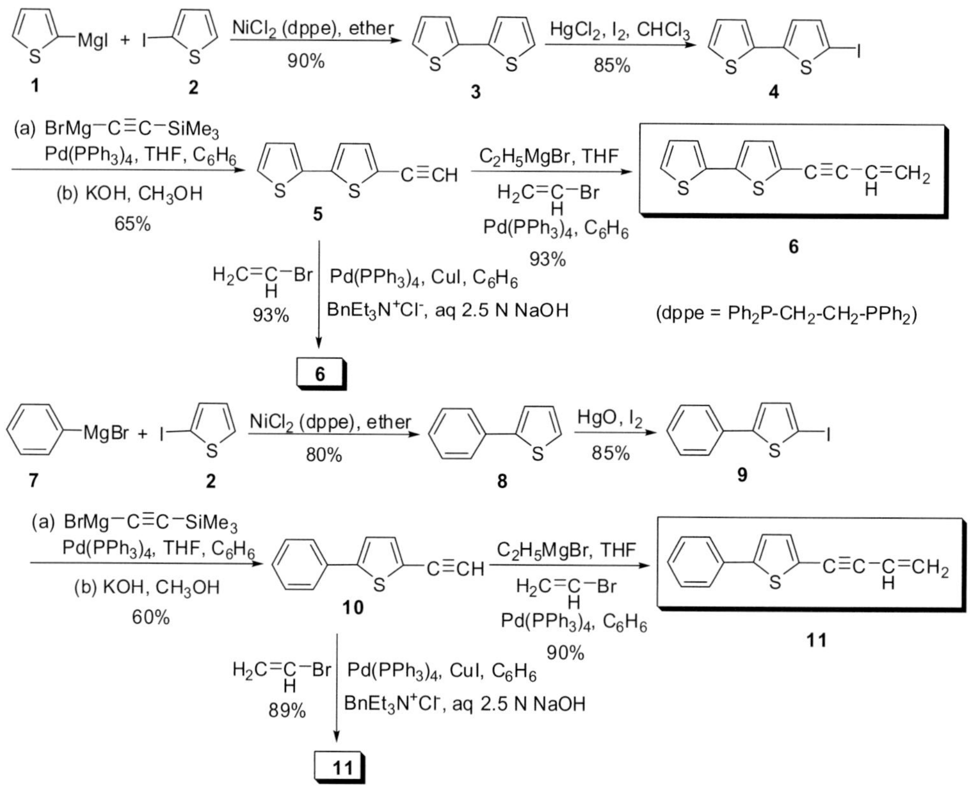

Scheme 11.1 Rossi synthesis of naturally occurring acetylenic thiophenes.

The spiroketal unit is present in a number of natural compounds isolated from many sources such as plants, insects, marine organisms, and microorganisms. Extensive efforts have been made to their structure elucidation and synthesis [59]. Wu and co-workers synthesized [13] naturally occurring spiroketal enol ethers containing thiophene **14a** and **14b** starting from 3-(2′-furyl)-propan-l-ol (**12a**) and 4-(2′-furyl)-butan-l-ol (**12b**). Treatment of **12** with butyl lithium and 2-thiophenecarboxaldehyde gave the furandiols **13a** and **13b** in 77% and 71% yields. Several acidic conditions have been tested for the last dehydration-spiroketalization step, and it was found that the reaction of **13** in toluene in the presence of 1 equivalent of $CuSO_4 \cdot H_2O$ afforded smoothly the desired thiophene attached spiroketal enol ethers **14a** and **14b** in almost quantitative yields (Scheme 11.2).

11.2.2
Synthesis of Natural Products by Construction of the Thiophene Nucleus

A thiophene nucleus may be constructed by one of the established routes which include: (i) Paal–Knorr synthesis; (ii) Fiesselmann synthesis; (iii) Gewald synthe-

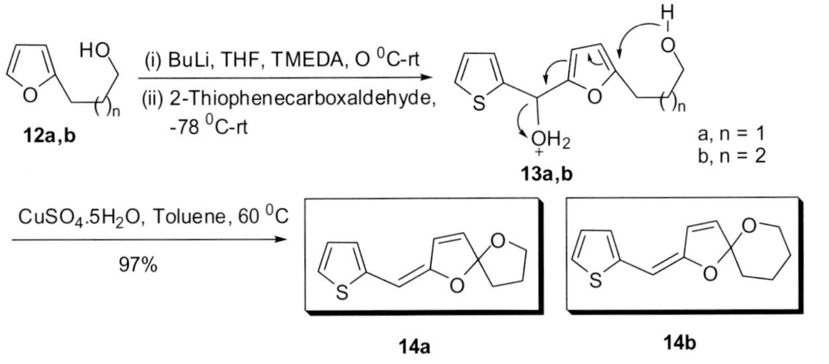

Scheme 11.2 Wu synthesis of thiophene substituted spiroketal enol ethers.

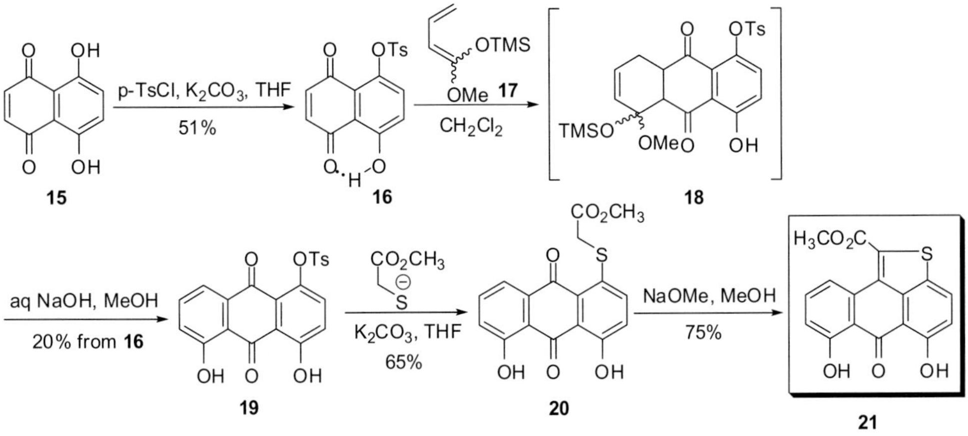

Scheme 11.3 Kelly anthrathiophene synthesis.

sis; (iv) Victor Meyer synthesis and many others as found in the standard text books [60].

In 2000, Kelly et al. reported [24] the synthesis of anthrathiophene **21** by base-catalyzed Knoevenagel-type cyclization of **20**. Tosylate **19** was regioselectively prepared in two operations from naphthazarin **15**. Monotosylation of **15** gave **16**, which underwent a regiospecific Diels–Alder reaction with the diene **17** to give **19** via **18**. Treatment of the quinone tosylate **19** with methyl mercaptoacetate and potassium carbonate in THF resulted in formation of the expected compound **20** by nucleophilic aromatic substitution. The compound **20** was then cyclized with methanolic methoxide to afford **21** in good yield (Scheme 11.3).

Taylor and Reiter reported the total synthesis of urothione [25], the urinary metabolite of the molybdenum cofactor, in racemic form in ten steps from pyrazine **27** in 19% overall yield. Thus, a thiophene ring was fused to pyrazine **27** to

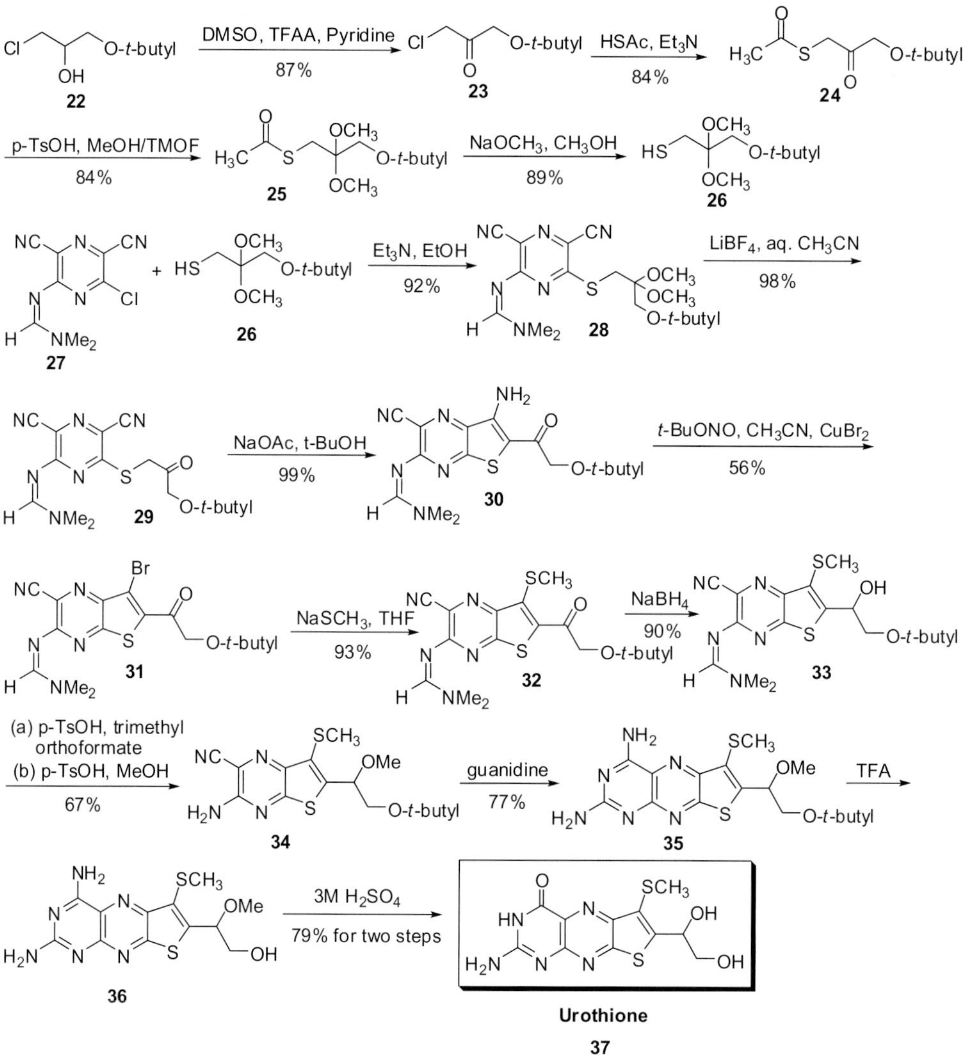

Scheme 11.4 Taylor and Reiter synthesis of urothione.

give a thieno[2,3-b]pyrazine **30** which contained latent functionality suitable for the introduction of other substituents in the natural product. The 7-amino group of **30** was converted to a methylthio substituent in two steps, and this transformation was followed by reduction of the ketocarbonyl of **32** to give **33**, which possesses the requisite 1,2-diol oxidation state on the ethyl side chain. Subsequent adjustment of the protecting groups was followed by pyrimidine ring fusion, giving the 2,4-diaminopteridine **35**. Subsequent acid treatments of **35** afforded the recemic urothione **37** (Scheme 11.4).

Makaluvamine F **49** exhibits potent biological activity (e.g., cytotoxicity towards the human colon tumor cell-line HCT-116 [$IC_{50} = 0.17$ mM] and inhibition of topoisomerase II) [15] and has an α-aminodihydrobenzothiophene skeleton. The synthesis of this natural product is rare perhaps due to the difficulty in construction of the labile and highly strained N,S-acetal skeleton. Kita *et al.* has reported [16] the total synthesis of (±)-makaluvamine F **49** via the facile construction of the labile N,S-acetal skeleton by a combination of hypervalent iodine oxidation reactions. The key intermediate 2-azido-5-bromo-6-hydroxy-dihydrobenzo-thiophene **47** was prepared by the azidation of compound **45** using a combination of PhINO and Me$_3$SiN$_3$ followed by the hydrolytic deprotection of the 6-acetoxy group of compound **46**. Notably the direct azidation of **44** gave only a trace of the expected α-azido compound. This is because there appears to be a large number of reactive sites on phenol ether **44** towards the hypervalent iodine-induced azidation. Finally, the catalytic hydrogenation of **47** using 10% Pd-C in the presence of 4 equiv. of TFA resulted in complete reduction to give **48** as a TFA salt in quantitative yield devoid any side reaction. The final coupling reaction between the synthetic precursors, **48** (TFA salt) and **40**, proceeded in MeOH to give the TFA salt of **49** in 86% yield (Scheme 11.5).

In a subsequent report [28], Kita *et al.* also disclosed another unique sulfur-containing pyrroloiminoquinone alkaloid, discorhabdin A **60**, possessing potent antitumor activity. The key step in the stereocontrolled total synthesis of **60** involves both a diastereoselective oxidative spirocyclization of **54** using a hypervalent iodine (III) reagent and efficient construction of the labile and highly strained N,S-acetal skeleton. The synthetic route is depicted in Scheme 11.6.

(+)-Biotin **72** has received considerable attention due to its significant biological activities for human nutrition and animal health. The compound **72** possesses highly functionalized sulfur-containing bicyclic ureido core with a 4-carboxybutyl chain at C-4 and all-*cis* configuration. Among a number of synthetic approaches to **72** [32], the Goldberg and Sternbach approach [61], has been the most efficient method. They developed the first total synthesis of **72** using readily accessible fumaric acid **61** as the starting material. Bromination of **61** followed by treatment with benzylamine (BnNH$_2$) and cyclization afforded *meso*-2-imidazolidinone dicarboxylic acid **63** stereoselectively. Dehydration of **63** and reduction with zinc dust in the presence of acetic anhydride (Ac$_2$O) gave bicyclic acetoxylactone *rac*-**65**, which, upon treatment with hydrogen sulfide and subsequent reduction with zinc dust (Zn), provided thiolactone *rac*-**67**. Introduction of 4-carboxybutyl chain to *rac*-**67** was initiated by the Grignard reaction with methoxypropyl magnesium chloride [ClMg(CH$_2$)$_3$OMe]. Vinyl sulfide *rac*-**68**, formed through acid-catalyzed dehydration, was subjected to hydrogenation to afford the required all *cis*-configuration. The chain elongation, introduction of carboxyl group and subsequent cleavage of protective groups were achieved through a reaction sequence involving bromination of *rac*-**69**, treatment with malonate followed by decarboxylation and removal of benzyl protective groups with hydrobromic acid (HBr). Optical resolution of *rac*-**70** was achieved by salt

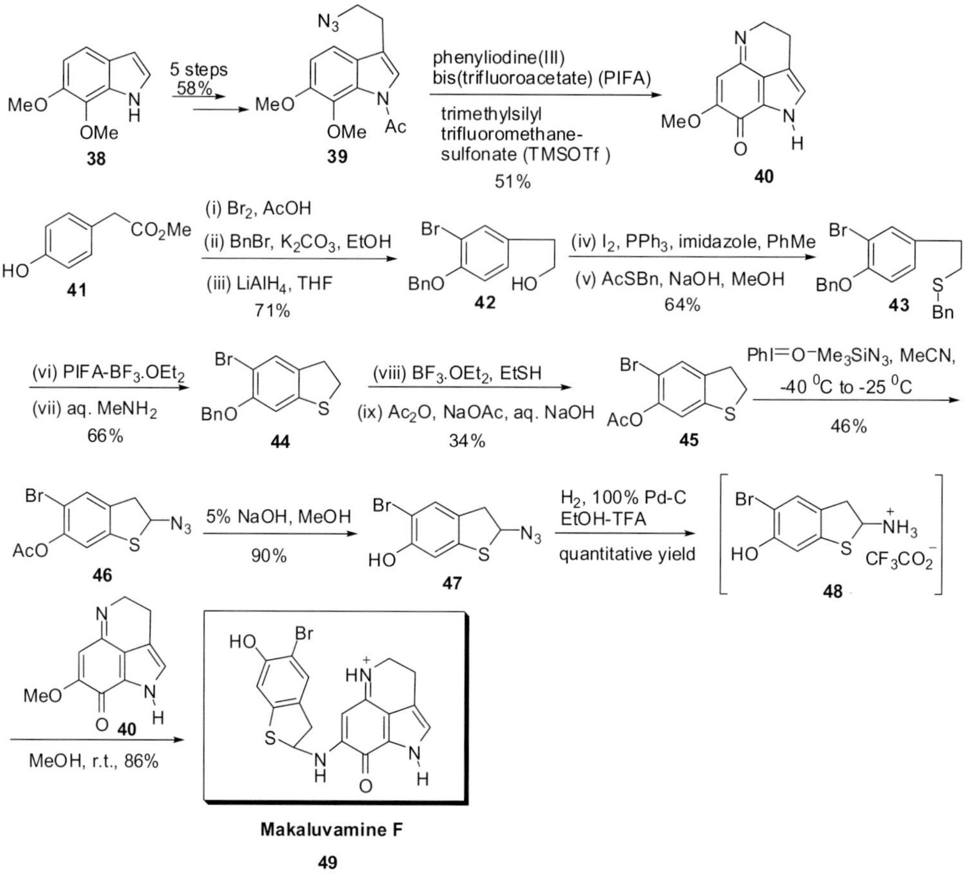

Scheme 11.5 Kita synthesis of makaluvamine F.

formation with d-camphorsulfonic acid. The salt obtained, chiral sulfonium d-camphorsulfonate **71**, gave **72** by treatment of dimethyl malonate followed by HBr (Scheme 11.7).

Baggiolini et al. also synthesized d-biotin from L-cystine via intramolecular [3 + 2] cycloaddition [62]. First L-cystine dimethyl ester **73** was acylated at the nitrogen with 5-hexynoyl chloride which was then treated with zinc dust in acetic acid to give **74**. Compound **74** was then reduced with diisobutylaluminum hydride followed by treatment with benzylhydroxylamine hydrochloride to give nitrone **75**. On refluxing in toluene, **75** underwent cycloaddition in the anticipated fashion with the exclusive formation of the tricyclic intermediate **76**. Cleavage of the isoxazolidine ring and acylation of the free amine gave the bicyclic intermediate **77**, which on treatment with barium hydroxide in refluxing aqueous dioxane underwent hydrolysis of the lactam moiety and concomitant cyclization to the imidazolidinone **78**. Thionyl chloride treatment in ether and subsequent quenching with

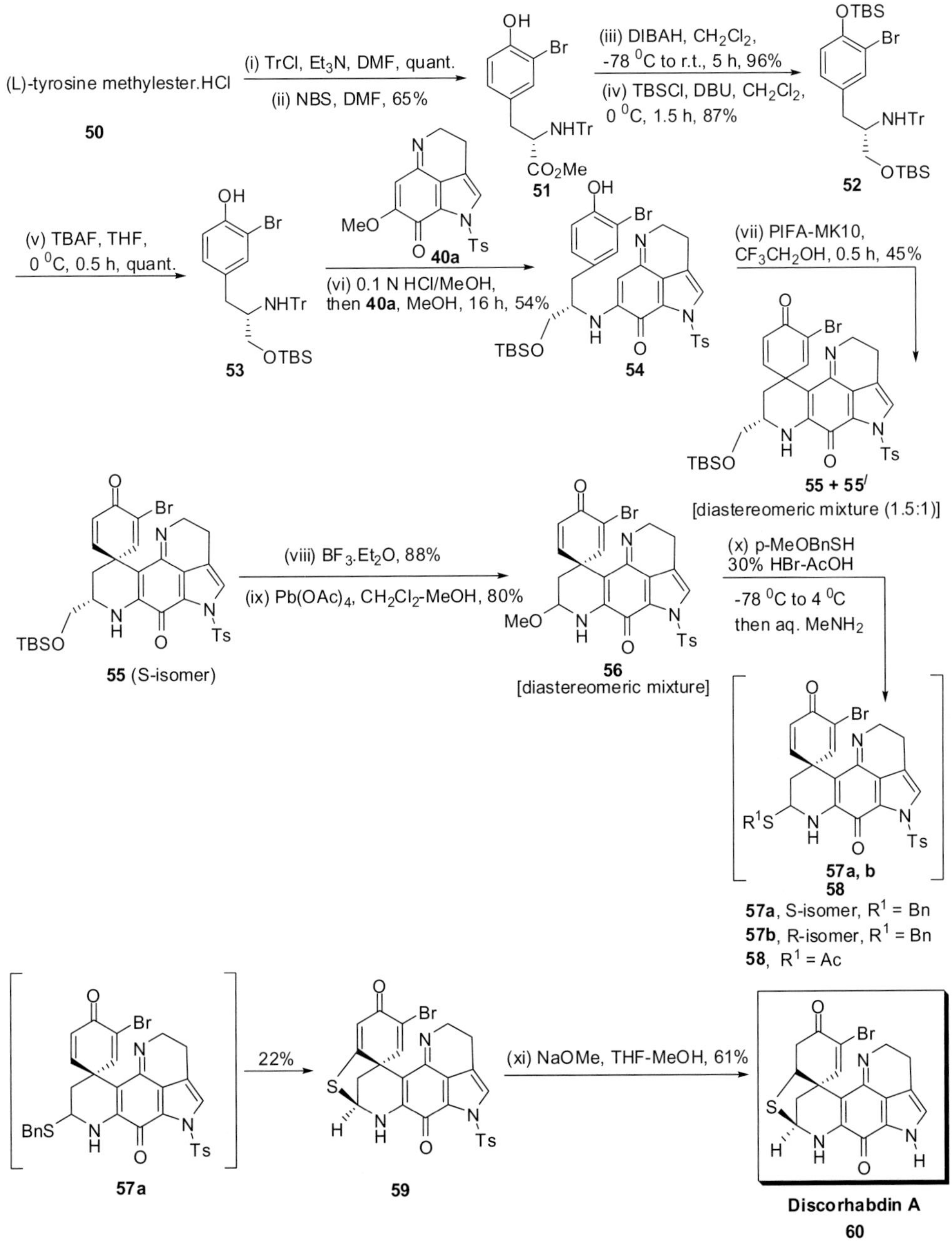

Scheme 11.6 Kita synthesis of discorhabdin A.

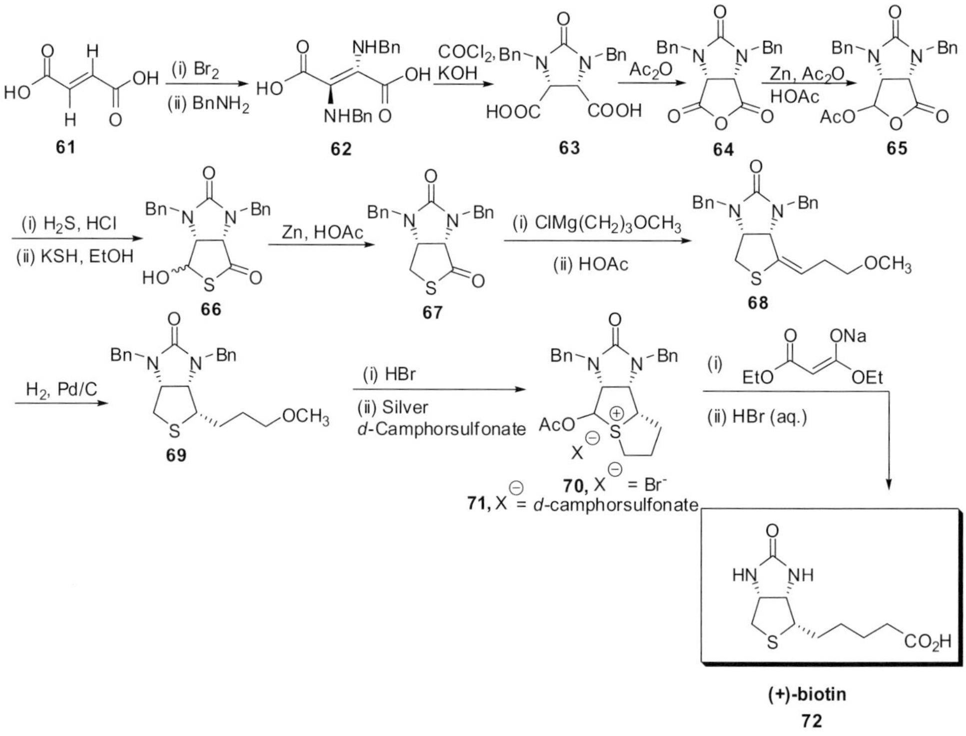

Scheme 11.7 Goldberg and Sternbach synthesis of biotin.

methanol gave the chloro ester **79**. Dechlorination to **80** was effected with excess of sodium borohydride in dimethylformamide. Finally, treatment of **80** with aqueous hydrobromic acid gave d-biotin **72** (Scheme 11.8).

Breynins A and B, novel thiophene-containing glycosides possess significant oral hypocholesterolemic activity [63]. The total syntheses of breynins have not yet been reported. Exhaustive hydrolysis of breynin A afforded breynolide **100**, along with d-glucose, l-rhamnose, and p-hydroxybenzoic acid. The first total synthesis of breynolide **100** was reported by Williams et al. in 1990 [64]. In a consecutive report [65], Smith et al. described another alternate, stereochemically linear approach for the synthesis of racemic breynolide. Highlights of the Smith breynolide synthesis are: (i) anomerically driven spiroketalization-equilibration of enedione **94**; (ii) a chemoselective elimination effecting regiocontrolled epoxide ring opening (**83** to **84**); (iii) thiophene ring formation of **85** by the addition of thiolactic acid followed by the treatment with NaOMe in MeOH (actually three steps occurred here, first 1,4-addition initially occurred anti to the hydroxyl group; then base treatment of the resultant adduct induced acetyl migration to the vicinal hydroxyl, generating a thiolate anion which cyclized via displacement of chloride

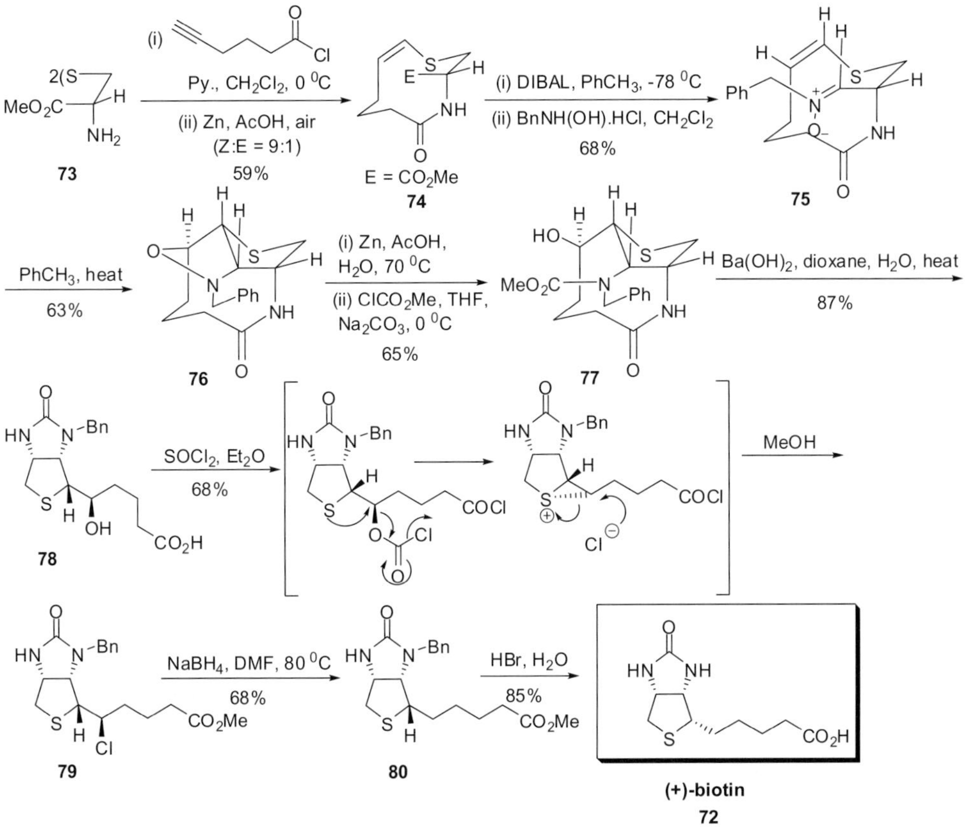

Scheme 11.8 Baggiolini synthesis of (+)-biotin.

and last of all equlilibration of the bicyclic system provided the more stable *cis*-fused isomer); (iv) an end game exploiting the strain in enone **96** to permit Michael addition of allyl alcohol, followed by the generation and hydroxylation of an enolate bearing two β-alkoxy groups (Scheme 11.9).

11.3
Synthesis of Natural Products Containing Other Sulfur Heterocycles

Aromatic alkaloids possessing polysulfide structures have been isolated from ascidians of the genera *Lissoclinum* [66, 67], *Eudistoma* [68], and *Polycitor* [69]. More than ten monomeric cyclic polysulfides [66–69] and four dimeric polysulfides [67] have been reported. These compounds exhibit various biological activities, for example, antifungal activity, antibacterial activity, cytotoxicity, antimalarial

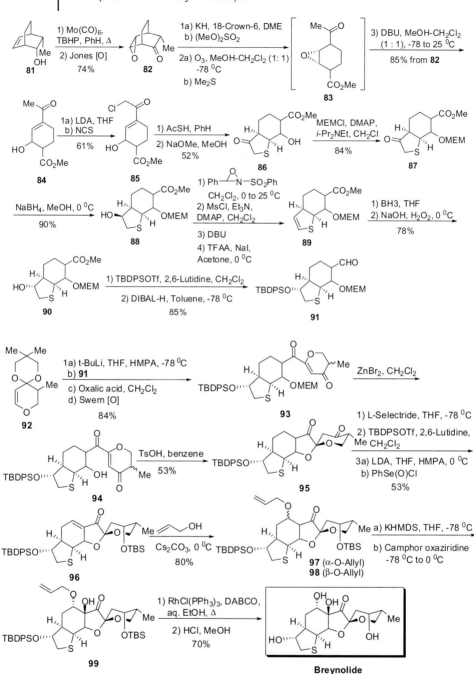

Scheme 11.9 Smith synthesis of breynolide.

(*Plasmodium falciparum*) activity, inhibition of protein kinase C, and inhibition of IL-8 Ra and Rb receptors [49]. To our knowledge there are only two recent reports of the synthesis of natural products containing heterocycles with di/polysulfide linkage(s).

Aspirochlorine **114** is a novel seven-membered epidithiapiperazine-2,5-dione isolated from *Aspergillus tamari* in 1976 [43]. Aspirochlorine is the only known natural product containing epipolythiapiperazine-2,5-dione, derived from glycine, which places a free amide (i.e., NH) adjacent to the S–S bridge. In addition, it is a rare example of an amino acid metabolite to incorporate the unusual N-methoxyl moiety in a diketopiperazine ring. Aspirochlorine was first synthesized by Williams *et al.* [44] in a diastereoselective fashion from commercially available 5-chlororesorcinol **101** in 13 steps. The synthesis involves an efficient stereoselective cycloaddition reaction of a hydroxamic ester **107** to form the parent spiro[benzofuran-2(3H),2′-piperazine] ring system **108**. In addition the synthesis employs a 2-nitrobenzyl group as an amide protecting group which is easily removed under photolytic conditions. The total synthesis of aspirochlorine is dipicted in Scheme 11.10.

Spiruchostatin A **126** isolated from *Pseudomonas* sp., exhibits potent inhibitory activity against histone deacetylases (HDACs) [41]. Spiruchostatin A is acyclic depsipeptide consisting of (3S,4R)-statine, D-cysteine, D-alanine, and (E)-3-hydroxy-7-thio-4-heptenoic acid. The two thiol-groups are connected as a disulfide forming a bicyclic depsipeptide. Potent activity was observed for **126** in models for anticancer activity and cardiac hypertrophy. Spiruchostatin A inhibits the growth of breast cancer cells with an IC_{50} of approximately 10 nM, compared with 100 nM for the HDAC inhibitor trichostatin A. Therefore, the total synthesis of spiruchostatin A attracted the synthetic community. The first total synthesis of spiruchostatin A was reported by Ganesan *et al.* in 2004 [70]. Doi *et al.* have demonstrated [42] an asymmetric aldol reaction using the Zr-enolate of acetyl N-oxazolidin-2-one derivative **115** that proceeds with high diastereoselectivity. The synthesis of the statine involves a malonate condensation under mild conditions. The use of an allyl ester protecting group was found to be compatible with highly functionalized intermediates. Finally, the macrolactonization of **125** was efficiently achieved with the Shiina reagent. The total synthesis of spiruchostatin A according to Doi's approach is shown in Scheme 11.11.

11.4
Conclusion

In this chapter we have highlighted (i) the synthesis of natural products containing thiophene and other sulfur heterocycles and (ii) the use of thiophene building blocks in the synthesis of allied natural products. Thiazole-containing natural products have been excluded as they are included Chapter 13. The progress so far made shows the state of the art in contemporary organic synthesis. However, more remains to be done to achieve efficient and cost-effective routes to synthesize

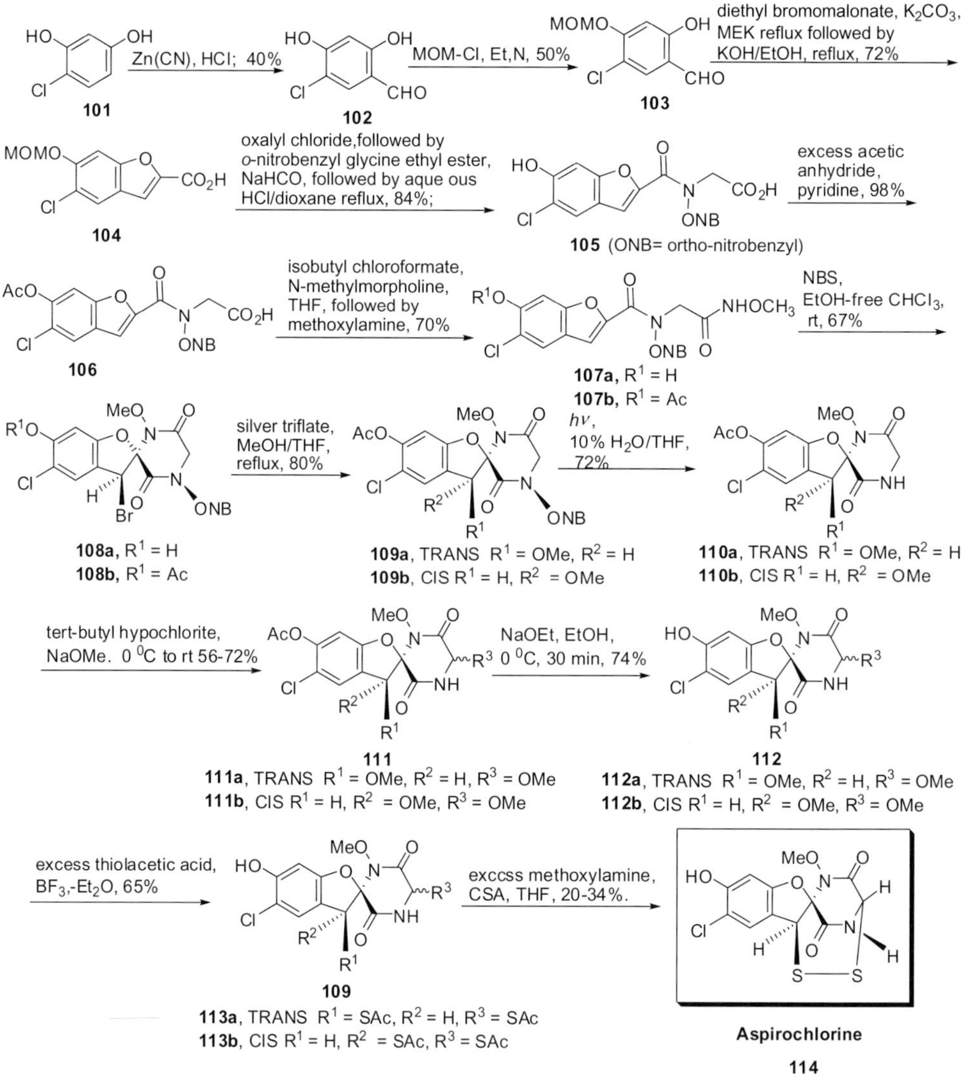

Scheme 11.10 Williams synthesis of a spirochlorine.

biogically active complex natural products containing heterocyclic moieties and their analogs for bio-assay and pre-clinical studies.

Acknowledgments

The authors are thankful to the DST (New Delhi) and the CSIR (New Delhi) for financial assistance and the University of Kalyani for facilities.

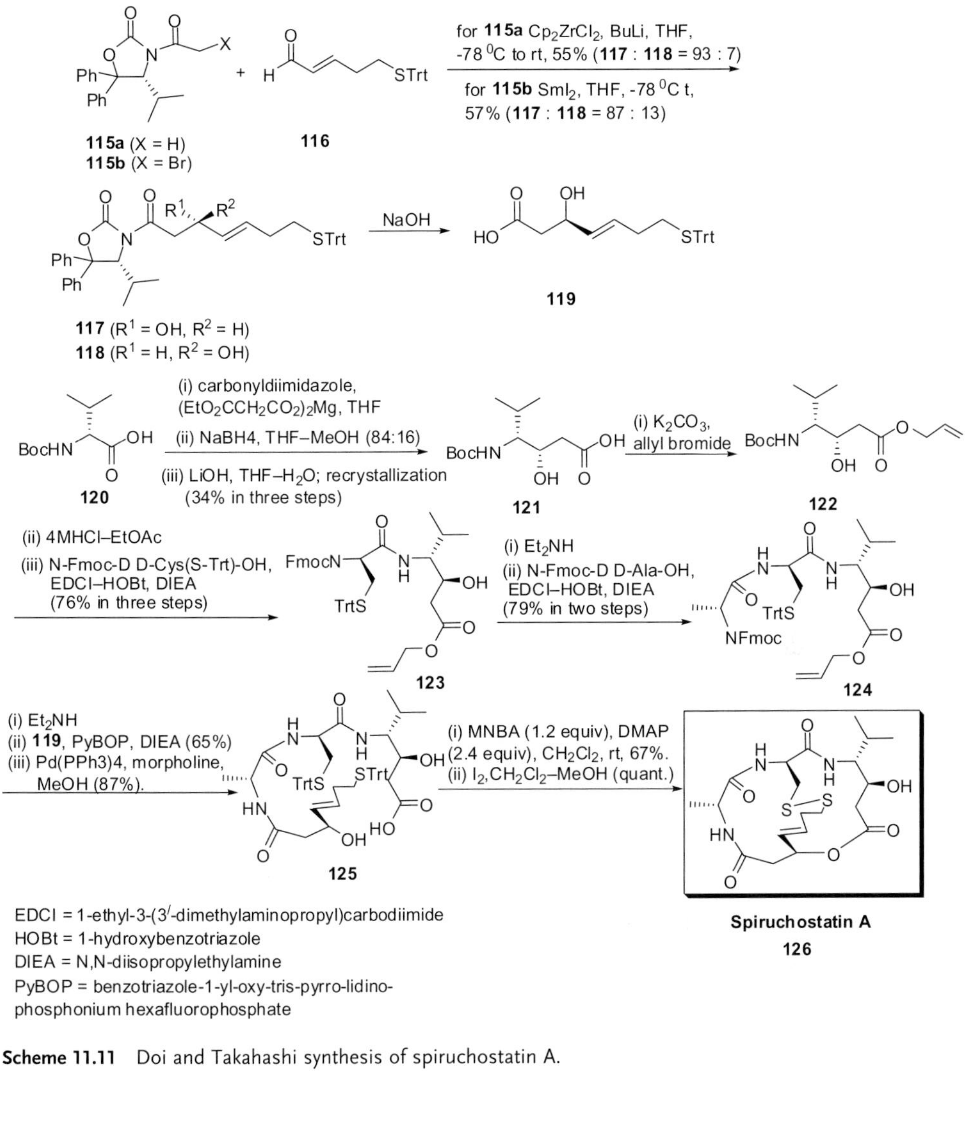

EDCI = 1-ethyl-3-(3′-dimethylaminopropyl)carbodiimide
HOBt = 1-hydroxybenzotriazole
DIEA = N,N-diisopropylethylamine
PyBOP = benzotriazole-1-yl-oxy-tris-pyrro-lidino-phosphonium hexafluorophosphate

Scheme 11.11 Doi and Takahashi synthesis of spiruchostatin A.

References

1 Massera, P.E., Rodriquez Talou, J. and Giulietti, A.M. (1998) Thiophene production in transformed roots cultures of *Tagetes filifolia*. *Biotechnol. Lett.*, **20** (6), 573–577.

2 Chan, G.F.Q., Neil Towers, G.H. and Mitchell, J.C. (1975) Ultraviolet-mediated antibiotic activity of thiophene compounds of *Tagetes*. *Phytochemistry*, **14** (10), 2295–2296.

3 Hudson, J.B., Graham, E.A., Micki, N., Hudson, L. and Towers, G.H.N. (1986) Antiviral activity of the photoactive thiophene α-terthienyl. *Photochem. Photobiol.*, **44** (4), 477–482.

4 Meyer, V. (1883) Ueber den Begleiter des Benzols im Steinkohlenteer. *Berichte der Deutschen chemischen Gesellschaft*, **16** (1), 1465–1478.

5 Sumpter, W.C. (1944) The chemistry of isatin. *Chem. Rev.*, **34** (3), 393–434.

6 Barnes, D.M., Haight, A.R., Hameury, T., McLaughlin, M.A., Mei, J., Tedrowy, J.S. and Toma, J.D.R. (2006) New conditions for the synthesis of thiophenes via the Knoevenagel/Gewald reaction sequence. Application to the synthesis of a multitargeted kinase inhibitor. *Tetrahedron*, **62** (49), 11311–11319.

7 Hepworth, J.D. and Heron, B.M. (1999) Six-membered ring systems: with O and/or S atoms. *Prog. Heterocycl. Chem.*, **11**, 299–318.

8 Kotoku, N., Cao, L., Aoki, S. and Kobayashi, M. (2005) Absolute stereo-structure of kendarimide A, a novel MDR modulator, from a marine sponge. *Heterocycles*, **65** (3), 563–578.

9 Williams, D.E., Austin, P., Diaz-Marrero, A.R., Soest, R.V., Matainaho, T., Roskelley, C.D., Roberge, M. and Andersen, R.J. (2005) Neopetrosiamides, peptides from the marine sponge neopetrosia sp. That inhibit amoeboid invasion by human tumor cells. *Org. Lett.*, **7** (19), 4173–4176.

10 Duan, H., Takaishi, Y., Tori, M., Takaoka, S., Honda, G., Ito, M., Takeda, Y., Kodzhimatov, O.K., Kodzhimatov, K. and Ashurmetov, O. (2002) Polysulfide derivatives from *Ferula foetida*. *J. Nat. Prod.*, **65** (11), 1667–1669.

11 Tian, Y., Wei, X. and Xu, H. (2006) Photoactivated insecticidal thiophene derivatives from *Xanthopappus subacaulis*. *J. Nat. Prod.*, **69** (8), 1241–1244.

12 Gao, Y., Wu, W.L., Wu, Y.L., Ye, B. and Zhou, R. (1998) A straightforward synthetic approach to the spiroketal-enol ethers synthesis of natural antifeeding compound tonghaosu and its analogs. *Tetrahedron*, **54** (41), 12523–12538.

13 Gao, Y., Wu, W.L., Ye, B., Zhou, R. and Wu, Y.L. (1996) Convenient syntheses of tonghaosu and two thiophene substituted spiroketal enol ether natural products. *Tetrahedron Lett.*, **37** (6), 893–896.

14 Aubry, M.L., Cowell, P., Davey, M.J. and Shevde, S. (1970) Aspects of the pharmacology of a new anthelmintic: pyrantel. *Br. J. Pharmac.*, **38**, 332–344.

15 Radisky, D.C., Radisky, E.S., Barrows, L.R., Copp, B.R., Kramer, R.A. and Ireland, C.M. (1993) Novel cytotoxic topoisomerase ii inhibiting pyrroloiminoquinones from Fijian sponges of the genus *Zyzzya*. *J. Am. Chem. Soc.*, **115** (5), 1632–1638.

16 Kita, Y., Egi, M. and Tohma, H. (1999) Total synthesis of sulfur-containing pyrroloiminoquinone marine product, (±)-makaluvamine F using hypervalent iodine(III)-induced reactions. *Chem. Commun.*, (2), 143–144.

17 Liu, Y., Ye, M., Guo, H.Z., Zhao, Y.Y. and Guo, D.A. (2002) New thiophenes from *Echinops grijisii*. *J. Asian Nat. Prod. Res.*, **4** (3), 175–178.

18 Biswas, R., Dutta, P.K., Achari, B., Bandyopadhyay, D., Mishra, M., Pramanik, K.C. and Chatterjee, T.K. (2007) Isolation of pure compound R/J/3 from *Pluchea indica* (L.) Less. and its anti-amoebic activities against *Entamoeba histolytica*. *Phytomedicine*, **14** (7-8), 534–537.

19 Shi, J., Zhang, X. and Jiang, H. (2010) 2-(Penta-1,3-diynyl)-5-(3,4-dihydroxybut-1-ynyl)thiophene, a novel NQO1 inducing agent from *Echinops grijsii* Hance. *Molecules*, **15** (8), 5273–5281.

20 Baruah, R.N., Sharma, R.P., Baruah, J.N., Mondeshka, D., Hertz, W. and Watanabe, K. (1982) Ineupatoriol, a thiophene analogue of ichthyothereol, from *Inula eupatorioides*. *Phytochemistry*, **21** (3), 665–667.

21 Koike, K., Jia, Z., Nikaido, T., Liu, Y., Zhao, Y. and Guo, D. (1999) Echinothiophene, a novel benzothiophene glycoside from the roots of *Echinops grijissii*. *Org. Lett.*, **1** (2), 197–198.

22 Jeong, S.J., Higuchi, R., Miyamoto, T., Ono, M., Kuwano, M. and Mawatari, S.F. (2002) Bryoanthrathiophene, a new antiangiogenic constituent from the bryozoan *Watersipora subtorquata* (d'Orbigny, 1852). *J. Nat. Prod.*, **65** (9), 1344–1345.

23 Shindo, T., Sato, A., Kasanuki, N., Hasegawa, K., Sato, S., Iwata, T. and

Hata, T. (1993) 6H-anthra[1,9-*bc*] thiophene derivatives from a bryozoan, *Dakaira subovoidea*. *Cell. Mol. Life Sci.*, **49** (2), 177–178.

24 Kelly, T.R., Fu, Y., Sieglen, Jr., J.T. and Silva, H.D. (2000) Synthesis of an orange anthrathiophene pigment isolated from a Japanese bryozoan. *Org. Lett.*, **2** (15), 2351–2352.

25 Taylor, E.C. and Reiter, L.A. (1989) Studies on the molybdenum cofactor. An unequivocal total synthesis of (±) Urothione. *J. Am. Chem. Soc.*, **111** (1), 285–291.

26 Kobayashi, J., Cheng, J.-F., Ishibashi, M., Nakamura, H., Ohizumi, Y., Hirata, Y., Sasaki, T., Lu, H. and Clardy, J. (1987) Prianosin A, a novel antileukemic alkaloid from the Okinawan marine sponge *Prianos melanos*. *Tetrahedron Lett.*, **28** (42), 4939–4942.

27 Perry, N.B., Blunt, J.W. and Munro, M.H.G. (1988) Cytotoxic pigments from New Zealand sponges of the genus latrunculia: discorhabdins A, B and C. *Tetrahedron*, **44** (6), 1727–1734.

28 Tohma, H., Harayama, Y., Hashizume, M., Iwata, M., Kiyono, Y., Egi, M. and Kita, Y. (2003) The first total synthesis of discorhabdin A. *J. Am. Chem. Soc.*, **125** (37), 11235–11240.

29 Perry, N.B., Blunt, J.W., Munro, M.H.G., Higa, T. and Sasaki, R. (1988) Discorhabdin D, an antitumor alkaloid from the sponges *Latrunculia brevis* and *Prianos* sp. *J. Org. Chem.*, **53** (17), 4127–4128.

30 Smith, A.B., III, Keenan, T.P., Gallagher, R.T., Furst, G.T. and Dormer, P.G. (1992) Structures of breynins A and B, architecturally complex, orally active hypocholesterolemic spiroketal glycosides. *J. Org. Chem.*, **57** (19), 5115–5120.

31 Hofmann, K., Melville, D.B. and Vigneaud, V.D. (1941) Characterization of the functional groups of biotin. *J. Biol. Chem.*, **141** (1), 207–214.

32 Seki, M. (2008) (+)-Biotin: a challenge for industrially viable total synthesis of natural products. *Stud. Nat. Prod. Chem.*, **34**, 265–307.

33 Wang, J.M., Ding, G.Z., Fang, L., Dai, J.G., Yu, S.S., Wang, Y.H., Chen, X.G.,

Ma, S.G., Qu, J., Xu, S. and Du, D. (2010) Thiodiketopiperazines produced by the endophytic fungus *Epicoccum nigrum*. *J. Nat. Prod.*, **73** (7), 1240–1249.

34 El-Aasr, M., Fujiwara, Y., Takeya, M., Ikeda, T., Tsukamoto, S., Ono, M., Nakano, D., Okawa, M., Kinjo, J., Yoshimitsu, H. and Nohara, T. (2010) Onionin A from *Allium cepa* inhibits macrophage activation. *J. Nat. Prod.*, **73** (7), 1306–1308.

35 Meng, D., Chen, W. and Zhao, W. (2007) Sulfur-containing spiroketal glycosides from *Breynia fruticosa*. *J. Nat. Prod.*, **70** (5), 824–829.

36 Barrow, R.A. and Capon, R.J. (1992) Sulfolane as a natural product from the sponge/tunicate composite, *Batzella* sp./*Lissoclinum* sp. *J. Nat. Prod.*, **55** (9), 1330–1331.

37 Rossi, R., Carpita, A. and Lezzi, A. (1984) Palladium-catalyzed synthesis of naturally-occurring acetylenic thiophenes and related compounds. *Tetrahedron*, **40** (14), 2773–2779.

38 Li, Y.T., Li, L., Chen, J., Hu, T.C., Huang, J., Guo, Y.W., Jiang, H.L. and Shen, X. (2009) 7-Chloroarctinone-B as a new selective PPARγ antagonist potently blocks adipocyte differentiation. *Acta Pharmacol. Sin.*, **30** (9), 1351–1358.

39 Ahmad, V.U. and Alam, N. (1995) New antifungal bithienylacetylenes from *Blumea obliqua*. *J. Nat. Prod.*, **58** (9), 1426–1429.

40 Kato, A., Okada, M. and Hashimot, Y. (1984) Occurrence of gerrardine in *Casszpourea guzanenszs*. *J. Nat. Prod.*, **47** (4), 706–707.

41 Masuoka, Y., Nagai, A., Shin-ya, K., Furihata, K., Nagai, K., Suzuki, K., Hayakawa, Y. and Seto, H. (2001) Spiruchostatins A and B, novel gene expression-enhancing substances produced by *Pseudomonas* sp. *Tetrahedron Lett.*, **42** (1), 41–44.

42 Doi, T., Iijima, Y., Shin-ya, K., Ganesand, A. and Takahashi, T. (2006) A total synthesis of spiruchostatin A. *Tetrahedron Lett.*, **47** (7), 1177–1180.

43 Berg, D.H., Massing, P., Hoehn, M.M., Boeck, L.D. and Hamill, R.L. (1976) A30641, A new epidithiodiketopiperazine

with antifungal activity. *J. Antibiot.*, **29** (4), 394–397.
44. Miknis, G.F. and Williams, R.M. (1993) Total synthesis of (±)-aspirochlorine. *J. Am. Chem. Soc.*, **115** (2), 536–547.
45. Aiello, A., Fattorusso, E., Luciano, P., Mangoni, A. and Menna, M. (2005) Isolation and structure determination of aplidinones A–C from the Mediterranean ascidian *Aplidium conicum*: a successful regiochemistry assignment by quantum mechanical ^{13}C NMR chemical shift calculations. *Eur. J. Org. Chem.*, **23**, 5024–5030.
46. Aiello, A., Fattorusso, E., Luciano, P., Macho, A., Menna, M. and Muñoz, E. (2005) Antitumor effects of two novel naturally occurring terpene quinones isolated from the mediterranean ascidian *Aplidium conicum*. *J. Med. Chem.*, **48** (9), 3410–3416.
47. Ciufolini, M.A., Shen, Y.C. and Bishap, M.J. (1995) A unified strategy for the synthesis of sulfur-containing pyridoacridine alkaloids: antitumor agents of marine origin. *J. Am. Chem. Soc.*, **117** (50), 12460–12469.
48. Cao, S., Foster, C., Brisson, M., Lazob, J.S. and Kingstona, D.G.I. (2005) Halenaquinone and xestoquinone derivatives, inhibitors of Cdc25B phosphatase from a *Xestospongia* sp. *Bioorg. Med. Chem.*, **13** (4), 999–1003.
49. Liu, H., Fujiwara, T., Nishikawa, T., Mishima, Y., Nagai, H., Shida, T., Tachibana, K., Kobayashi, H., Mangindaane, R.E.P. and Namikoshi, M. (2005) Lissoclibadins 1–3, three new polysulfur alkaloids, from the ascidian *Lissoclinum cf. badium*. *Tetrahedron*, **61** (36), 8611–8615.
50. Dong, J.Y., He, H.P., Shen, Y.M. and Zhang, K.Q. (2005) Nematicidal epipolysulfanyldioxopiperazines from *Gliocladium roseum*. *J. Nat. Prod.*, **68** (10), 1510–1513.
51. Klausmeyer, P., McCloud, T.G., Tucker, K.D., Cardellina, II, J.H. and Shoemaker, R.H. (2005) Aspirochlorine class compounds from *Aspergillus flavus* inhibit azole-resistant *Candida albicans*. *J. Nat. Prod.*, **68** (8), 1300–1302.
52. Aebisher, D., Brzostowska, E.M., Sawwan, N., Ovalle, R. and Greer, A. (2007) Implications for the existence of a heptasulfur linkage in natural o-benzopolysulfanes. *J. Nat. Prod.*, **70** (9), 1492–1494.
53. Bohlman, F. and Zdero, C. (1986) *Thiophene and Its Derivatives*, (ed. S. Gronowitz), John Wiley & Sons, Ltd, Chichester, UK, Part 3, pp. 261–323.
54. (a) Skotheim, T.A. and Reynolds, J.R. (eds) (2007) *Handbook of Conducting Polymers*, 3rd edn, CRC Press, Boca Raton, FL, USA; (b) Fichou, D. (ed.) (1999) *Handbook of Oligo- and Polythiophenes*, Wiley-VCH Verlag GmbH, Weinheim, Germany.
55. Press, J.B. (1991) Thiophene and its Derivatives, in *The Chemistry of Heterocyclic Compounds*, vol. 44 (Part 4) (ed. S. Gronowitz), John Wiley & Sons, Inc., New York, USA, pp. 397–502.
56. Marles, R.J., Hudson, J.B., Graham, E.A., Soucy-Breau, C., Morand, P., Compadre, R.L., Compadre, C.M., Towers, G.H.N. and Arnason, J.T. (1992) Structure-activity studies of photoactivated antiviral and cytotoxic tricyclic thiophenes. *Photochem. Photobiol.*, **56** (4), 479–487.
57. Hudson, J.B., Graham, E.A., Miki, N., Towers, G.H.N., Hudson, L.L., Rossi, R., Carpita, A. and Neri, D. (1989) Photoactive antiviral and cytotoxic activities of synthetic thiophenes and their acetylenic derivatives. *Chemosphere*, **19** (8–9), 1329–1343.
58. Hudson, J.B., Harris, L., Marles, R.J. and Arnason, J.T. (1993) The anti-HIV activities of photoactive terthiophenes. *Photochem. Photobiol.*, **58** (2), 246–250.
59. Perron, F. and Albizati, K.F. (1989) Chemistry of spiroketals. *Chem. Rev.*, **89** (7), 1617–1661.
60. (a) Joule, J.A. and Mills, K. (2010) *Heterocyclic Chemistry*, John Wiley & Sons, Ltd, Chichester, UK.(b) Taylor, E.C. and Jones, R.A. (1990) *Pyrroles*, John Wiley & Sons, Inc., New York, USA.(c) Gilchrist, T.L. (1997) *Heterocyclic Chemistry*, 3rd edn, Addison Wesley Longman, Essex, UK.
61. Goldberg, M.W. and Sternbach, L.H. (1949) US Pat 2489232, Nov 22 1949, *Chem. Abst.* 1951, 45, 184b.
62. Baggiolini, E.G., Lee, H.L., Pizzolato, G. and Uskoković, M.R. (1982) Synthesis

of *d*-Biotin from L-cystine via intramolecular [3 + 2] cycloaddition. *J. Am. Chem. Soc.*, **104** (23), 6460–6462.

63 Sasaki, K. and Hirata, Y. (1973) Structure of breynolide. *Tetrahedron Lett.*, **14** (27), 2439–2442.

64 Williams, D.R., Jass, P.A., Tse, H.-L.A. and Gaston, R.D. (1990) Total synthesis of (+)-breynolide. *J. Am. Chem. Soc.*, **112** (11), 4552–4554.

65 Smith, A.B., III, Empfield, J.R., Rivero, R.A. and Vaccaro, H.A. (1991) Total synthesis of (±)-breynolide: an aglycon derivative of a potent, orally active hypocholesterolemic agent. *J. Am. Chem. Soc.*, **113** (10), 4037–4038.

66 Davidson, B.S., Molinski, T.F., Barrows, L.R. and Ireland, C.M. (1991) Varacin: a novel benzopentathiepin from lissoclinum vareau that is cytotoxic toward a human colon tumor. *J. Am. Chem. Soc.*, **113** (12), 4709–4710.

67 Searle, P.A. and Molinski, T.F. (1994) Five new alkaloids from the tropical ascidian, *Lissoclinum* sp. lissoclinotoxin A is chiral. *J. Org. Chem.*, **59** (22), 6600–6605.

68 Compagnone, R.S., Faulkner, D.J., Carte, B.K., Chan, G., Hemling, M.A., Hofmann, G.A. and Mattern, M.R. (1994) Pentathiepins and trithianes from two *Lissoclinum* species and a *Eudistoma* sp.: inhibitors of protein kinase C. *Tetrahedron*, **50** (45), 12785–12792.

69 Makawva, T.N., Stonik, V.A., Dmitrenok, A.S., Grebne, B.B., Isakov, V.V. and Reeiachyk, N.M. (1995) Varacin and three new marine antimicrobial ascidian *Polycztor* sp. polysulfides from the far-eastern. *J. Nat. Prod.*, **58** (2), 254–258.

70 Yurek-George, A., Habens, F., Brimmell, M., Packham, G. and Ganesan, A. (2004) Total synthesis of spiruchostatin A, a potent histone deacetylase inhibitor. *J. Am. Chem. Soc.*, **126** (4), 1030–1031.

12
Oxazole and Its Derivatives

David W. Knight

12.1
Introduction

The single theme of this chapter is to summarize the various ways in which naturally-occurring oxazoles have been synthesized. Most examples of this class of natural product have been isolated only during the past 20 years or so and only in the 1990s were methods of reasonably general applicability and often improved efficacy developed for their synthesis. That these methods are relatively efficient is essential, given the often exceptional complexity of such oxazoles-containing targets. Of course, the classical methods of oxazole synthesis, which still enjoy applications to these modern targets, are featured in all standard texts concerning heteroaromatic synthesis. In addition, both more obscure and more contemporary methodologies are amply described in the three relevant chapters in *Comprehensive Heterocyclic Synthesis* [1]. Three reviews have also summarized synthetic efforts towards natural products in this and related heteroaromatic areas up to around 2005 [2]. In view of the volume of work in the oxazole area alone, this present review is focused entirely on oxazole synthesis; neither discussion of the elaboration of reduced derivatives (oxazolines, oxazolidines) nor of the eventual completion of a particular total synthesis is included. While completeness of coverage is certainly not claimed, an attempt to include all important methods reported up to 2009, together with a few 2010 references, has been made.

Given that, biosynthetically, an oxazole can be derived by an apparently straightforward cyclodehydration of a serine residue, usually contained in a peptide chain, followed by oxidation of the resulting oxazoline, it is perhaps surprising that such natural compounds have only become prominent during the past 20 years or so and, further, that their efficient synthesis has necessitated the development of some rather sophisticated reagent cocktails. Overall, this area contains some of the most complex and synthetically challenging compounds yet isolated from nature. This would have provided quite enough stimuli for subsequent synthetic efforts, but these have been significantly amplified by the spectacular bioactivities displayed by members of this group, especially in the anticancer and antibiotic areas.

Heterocycles in Natural Product Synthesis, First Edition. Edited by Krishna C. Majumdar and Shital K. Chattopadhyay.
© 2011 Wiley-VCH Verlag GmbH & Co. KGaA. Published 2011 by Wiley-VCH Verlag GmbH & Co. KGaA.

This review is organized sequentially into mono-, bis-, tris- and poly-oxazole structural types. It will soon become clear that some of the more popular methodologies have been applied repeatedly to these targets. There will be considerable variations in the depth of discussion as well, as in some examples, the oxazole is almost a peripheral bystander group but in others it is a crucial and central structural feature.

12.2
Mono-Oxazoles

12.2.1
Pimprinin

Serial No.	Trivial name	Structure	Source	Isolation	Biological activity	Synthesis
1	Pimprinin		*Streptomyces pimprina*	[3]	Active against hyperkinetic diseases; anticholinergic, antihistaminic, MAO-inhibiting	[4–6]

Pimprinin **1** is one of the simplest and earliest known mono-oxazole derivatives isolated from nature [3] but which nevertheless displays some valuable bioactivities. An early synthesis features cyclodehydration, induced by polyphosphate ester in hot chloroform, of the keto amide **2**, readily obtained from 3-acetylindole [4]. The now-classical van Leusen oxazole synthesis has also been used to synthesize pimprinin **1**, by reaction between indole-3-carboxaldehyde **3** and the TOSMIC-based carbanion **4** [5]. 3-Acetylindole has been used in a different approach to pimprinin **1** by conversion into the diazo derivative **5** followed by a mechanistically appealing if not always very efficient cycloaddition to acetonitrile, initiated by carbenoid generation [6]. (See also Refs [7, 8].)

Perhaps surprisingly, more modern methods involving palladium-catalyzed coupling reactions do not seem to have been applied to this target; this is certainly not true of the following group.

12.2.2
Texamine and Relatives

Serial No.	Trivial name	Structure	Source	Isolation	Biological activity	Synthesis
6	Texamine	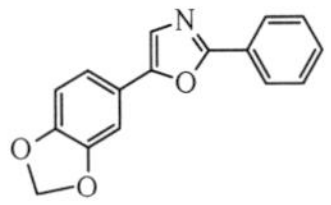	Amyris sp.	[9]	Activity against Mycobacteria claimed, incl. M. tuberculosis	[10–14]

Texamine **6** belongs to a small group of 2,5-diaryl oxazoles **6–9**, which have somewhat confusing names and over whose bioactivity there has been some disagreement [10]. However, there does seem to be potential in this area for identifying therapies for tuberculosis.

The 3-pyridyl derivative, texaline **8**, has been synthesized in a classical manner by cyclodehydration of the keto amide **10** upon exposure to hot acid in moderate yield [10] (Scheme 12.1). Useful though this is, much more significant are some applications of palladium-catalyzed coupling methodologies to these targets. An earlier study showed that 2-chlorooxazole-4-carboxylates and some 4-bromooxazoles undergo smooth arylations of many kinds including Suzuki, Stille and Negishi couplings although, somewhat unusually, Sonogashira reactions sometimes fail [11]. More significantly, it has been found subsequently that direct couplings with unfunctionalized oxazoles are readily achieved. Under one set of conditions applicable to this group of targets, ligands are also not required (Scheme 12.2). The initial aryloxazoles **11** were obtained from the corresponding aryl aldehydes and TOSMIC using the van Leusen method (see above). The same type of diaryloxazole

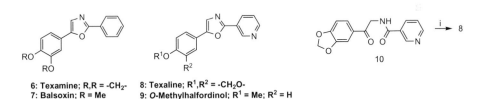

6: Texamine; R,R = -CH$_2$-
7: Balsoxin; R = Me
8: Texaline; R^1,R^2 = -CH$_2$O-
9: O-Methylhalfordinol; R^1 = Me; R^2 = H

Scheme 12.1 *Reagents and conditions:* i. c.H$_2$SO$_4$, Ac$_2$O, 90°C, (46%).

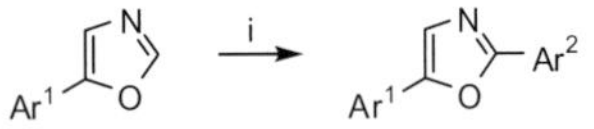

Ar^1—[**11**]—$\xrightarrow{i}$—Ar^1—[**12**]—Ar^2

Scheme 12.2 *Reagents and conditions:* i. Ar^{2}Br, Pd(OAc)$_2$, CuI, K$_2$CO$_3$, DMF, mW, 150°C, 4–15 min; 63–81%.

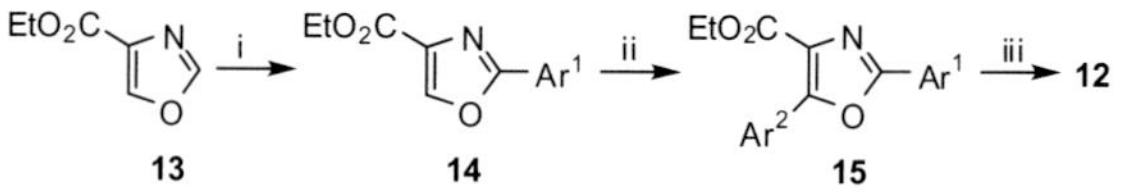

Scheme 12.3 Reagents and conditions: Ar1X (X = Cl, Br, I), 5% Pd (OAc)2, Cs2CO3, 10% ligand, 110 °C, 4 h; ii. as i; iii. ester hydrolysis then CuO, DMF, 160 °C, 70 h.

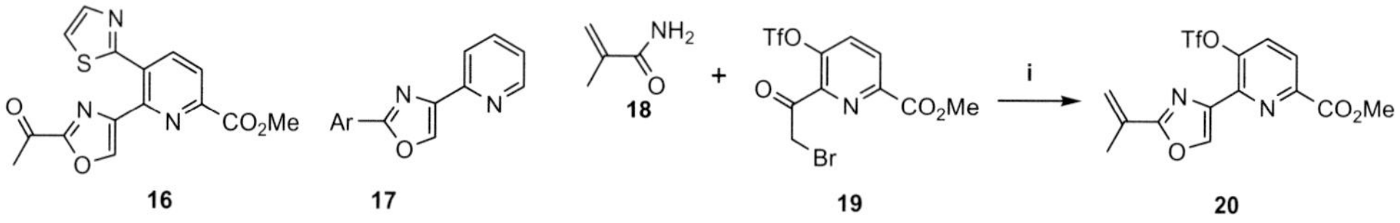

Scheme 12.4 Reagents and conditions: i. THF, 100 °C, sealed tube, 3 days (63%).

12 can also be synthesized in a related stepwise manner starting from an oxazole-4-carboxylate 13 using various types of ligands and adding a final decarboxylation step [13] (Scheme 12.3). Clearly, this methodology should also find many applications in general oxazole synthesis of aryl- 14 and diaryloxazole-4-carboxylates 15; when applied to syntheses of balsoxin 7 and texaline 8, overall yields of 54 and 61% were obtained, despite the brutality of the last step. Direct couplings of 5-aryloxazoles to aryl iodides to give the diaryloxazoles 12 can also be achieved very efficiently using a combination of silver carbonate and Pd(dppf)Cl₂ with triphenylphosphine in wet dichloromethane [14]. 2-Aryloxazoles can also be obtained directly from oxazole and a Negishi reagent [ArZnCl] in the presence of Pd(Ph₃P)₄.

12.2.3
Synthesis of Sulfomycin Fragments

Various degradation products derived from the enormous oxazole-containing macrolide sulfomycin have provided some useful illustrations of methodology in this area. Dimethyl sulfomycinamate 16 has been synthesized by an approach which was initially designed to make extensive use of palladium-catalyzed coupling reactions [15]. However, while 2-aryloxazoles 17 were successfully obtained from Stille couplings between a 2-pyridylstannane and the corresponding 2-aryl-oxazoyl triflates, extensions to more delicate targets (e.g., 20) were not productive. Hence, the necessary oxazole function was prepared using a more classical Hantzsch procedure (Scheme 12.4) by extended thermolysis of methyl methacrylamide 18 and the bromoketone 19; despite the brutal conditions, a very respectable 63% yield of the pyridyl oxazole 20 was realised.

The Bagley group have reported an alternative strategy to this target in which a main aim was to exploit their highly efficient version of the Bohlmann–Rahtz pyridine synthesis (Scheme 12.5). A similar strategy, but one featuring much milder modified Hantzsch conditions, delivered the initial oxazole 22 from reaction between acrylamide 18 and the bromo keto-ester 21. Subsequent homologa-

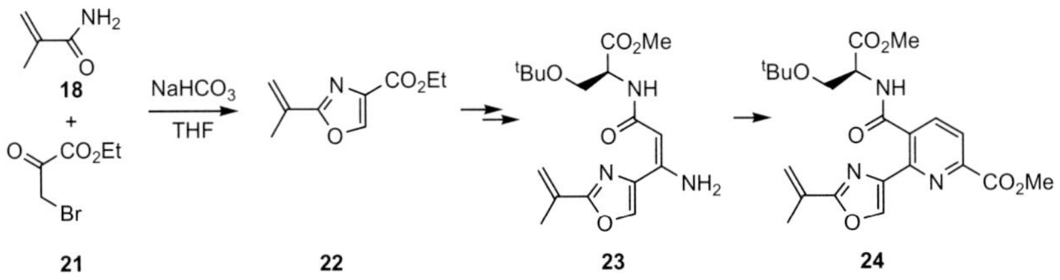

Scheme 12.5

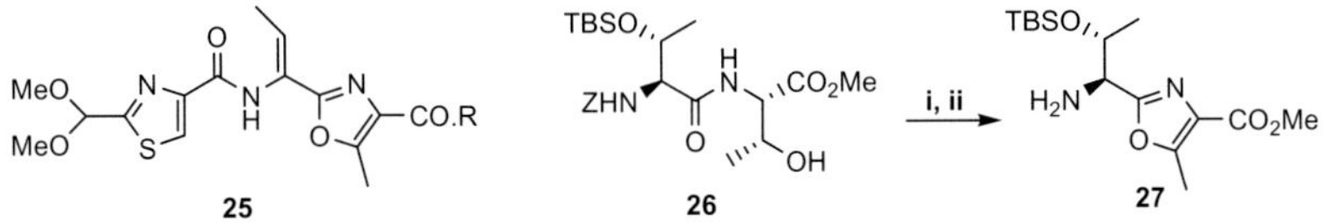

Scheme 12.6 Reagents and conditions: i. Deoxo-Fluor; ii. CBrCl$_3$, DBU then H$_2$, Pd-C, MeOH (~50%).

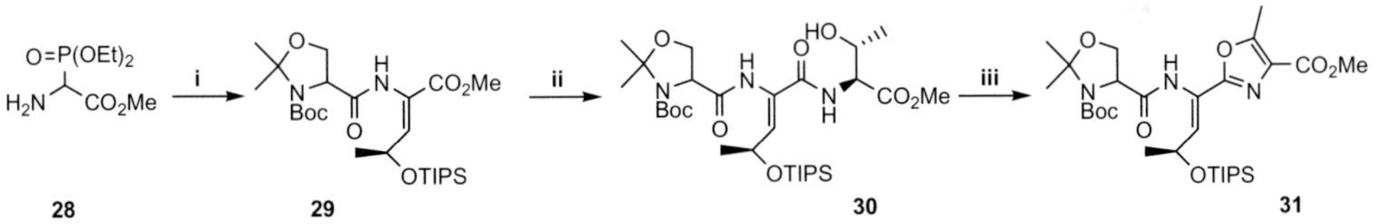

Scheme 12.7 Reagents and conditions: i. Garner's acid coupling, then TIPSOCH(Me)CHO, DBU; ii. saponify, coupling; iii. 2.67M Jones reagent then I$_2$, Ph$_3$P, Et$_3$N, CHCl$_3$.

tion led to the vinylogous amide **23** containing suitable functionality for the later construction of the thiazole ring, which was then converted, using the Bohlmann–Rahtz method, into the pyridine **24** by reaction with the alkyne TMSCCC(O).CO$_2$Me [16].

The oxazole function in a second degradation product of sulfomycin, methyl sulfomycinate **25**, has also been synthesized by the Bagley group but using more modern methodology to obtain the oxazole group **27** from the corresponding masked dipeptide **26** [17] (Scheme 12.6). This combination of reagents and relatives constitutes one of the more popular methods for the elaboration of this type of oxazole substitution pattern and is discussed in greater detail below [18]. In this particular example, a reasonable overall yield was obtained following an additional hydrogenolysis step to release the amine group ready for introduction of the thiazole moiety.

The protected derivative **31** of a central portion of sulfomycin **1** has been prepared using yet another common strategy for formation of oxazoles of this type [19] (Scheme 12.7). Starting with the phosphonate **28**, sequential coupling with

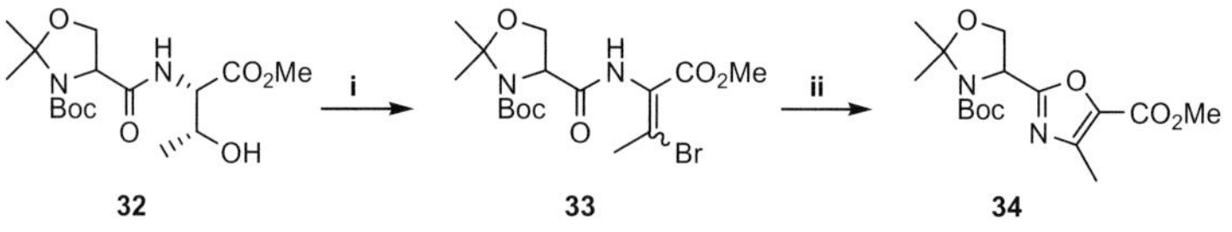

Scheme 12.8 Reagents and conditions: i. MsCl, DBU then NBS; ii. Cs$_2$CO$_3$, dioxane.

Garner's acid and Wadsworth–Emmons homologation provided the acylated enamine **29** which was further homologated to the tripeptide **30**. This was then converted into the oxazole **31** by the commonly successful Robinson–Gabriel procedure which consists of an initial oxidation of the secondary alcohol group, here using Jones reagent, followed by cyclodehydration. Many reagent combinations have been used to achieve this transformation with varying degrees of success; in this example, the commonly used pairing of iodine and triphenylphosphine delivered a decent yield of the target **31**.

A somewhat different approach was used by the same authors to prepare the simpler sulfomycin fragment **34** [19] (Scheme 12.8). Thus, an initial dipeptide **32** was converted into the vinylic bromides **33** by dehydration followed by enamine bromination [20]; cyclization was then induced under basic conditions.

12.2.4
Ajudazol A and B

Serial No.	Trivial name	Structure	Source	Isolation	Biological activity	Synthesis
35	Ajudazol A	See below	*Chondromyces crocatus*.	[21]	Antimicrobial	[22–24]

Syntheses directed towards the highly unsaturated antimicrobial Ajudazol A **35** has been the subject of three reports, each of which highlights a different method for the elaboration of the central oxazole. Krebs and Taylor have established suitable conditions (PdCl$_2$(PPh$_3$)$_2$, DMF, 50 °C) for attaching the model oxazole stannane **36** to the polyene chain, specifically as the vinyl iodide **37**, using a Stille coupling; the related bromide was insufficiently reactive [22].

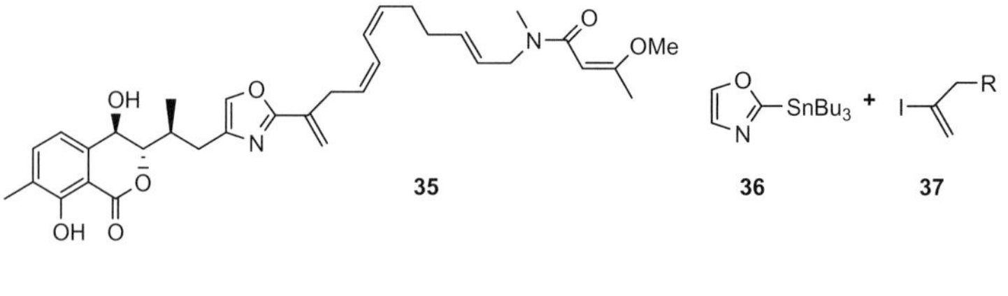

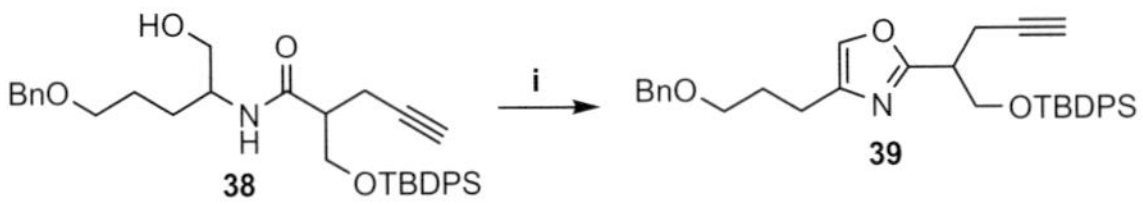

Scheme 12.9 Reagents and conditions: i. Dess–Martin oxidation then Ph$_3$P, (BrCl$_2$C)$_2$, 2,6-di-t-butyl-4-methylpyridine then DBU (90%).

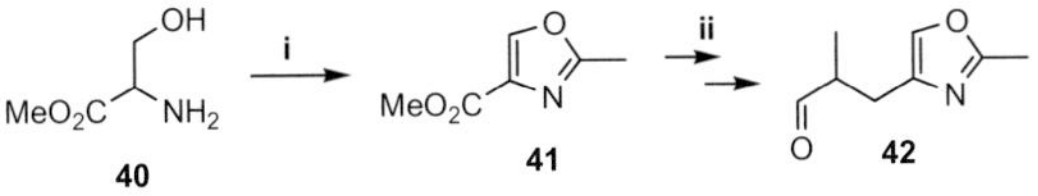

Scheme 12.10 Reagents and conditions: i. Ethyl acetimidate, Et$_3$N then CBrCl$_3$, DBU; ii. Dibal, Swern, Wittig (MeC(=PPh$_3$).CO$_2$Et, Dibal, H$_2$/Pd-C, Swern.

In a completely different approach, the central oxazole has been synthesized [23] using what has turned out to be one of the most popular of the newer methodologies for such ring synthesis, a Wipf cyclodehydration method [25] (Scheme 12.9). Essentially a modified Robinson–Gabriel method, this involves the oxidation of a suitable amide, here compound **38**, followed by exposure to a bromonium ion source and finally base-induced elimination of hydrogen bromide from the resulting 5-bromo-oxazoline to give the oxazole **39**. The efficiency can be excellent: the present example gives oxazole **39** in 90% yield for the three steps.

A related but lengthier approach to the oxazole **42** has been defined during a project which was more focused on ways to obtain the isobenzofuran unit in ajudazol A [24]. Firstly, serine **40** was condensed with ethyl acetimidate and the resulting oxazolidine oxidized to the oxazole **41** by sequential bromination and base-induced elimination. A rather repetitive sequence was then used to homologate the side chain to the key aldehyde **42** (Scheme 12.10).

12.2.5
Rhizoxin

Serial No.	Trivial name	Structure	Source	Isolation	Biological activity	Synthesis
43	Rhizoxin	See below	Rhizopus chinensis	[26]	Antitumor	[27–32]

The peripheral oxazole ring in rhizoxin **43** [27] is perhaps, synthetically, one of the least important components of this small group of macrolides. Its incorporation has featured some relatively routine chemistry, both in terms of formation of the trienyloxazole unit itself and of the various precursors, which include the Stille dienylstannane **44** [28] and the unsaturated aldehyde **45** [29], which participates well in Wittig couplings. The related phosphonium salt **46a** [28a] is useful for Wittig-type couplings in the reverse sense, as are the HWE phosphonate **46b** [30] and the phosphine oxide **46c** [31], along with the sulfone **47** [32] which participates well in Julia–Lythgoe olefinations. Doubtless, this group of intermediates will find applications in other areas of oxazole target synthesis.

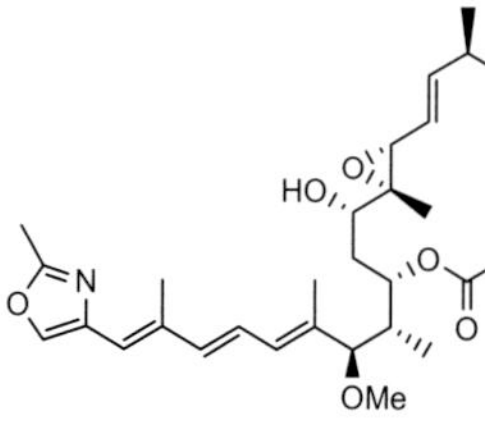

a) X = PPh$_3^+$Br$^-$
b) X = P(O)OMe$_2$
c) X = P(O)Ph$_2$

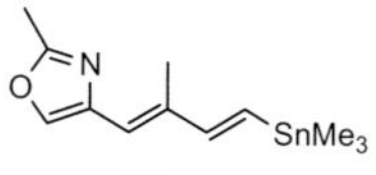

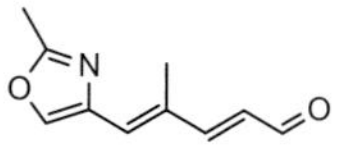

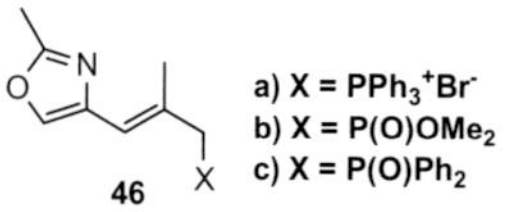

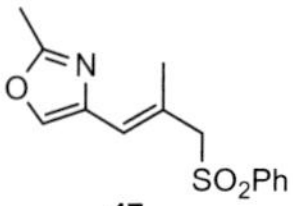

12.2.6
The Calyculins

Serial No.	Trivial name	Structure	Source	Isolation	Biological activity	Synthesis
48	Calyculin A–H and relatives	See below	*Discodermia calyx*	[33]	Cytotoxicity to leukemia cells; protein phosphatase inhibition (~1nM)	[34–41]

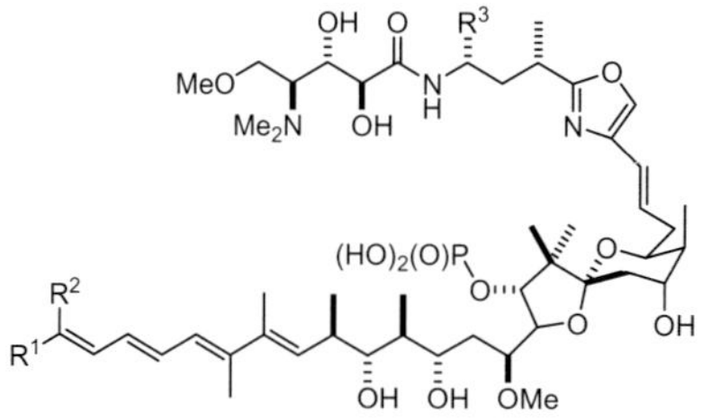

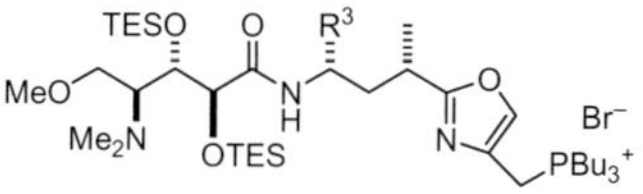

The calyculins **48** vary by having either R^1 or R^2 as a nitrile group, R^3 as a proton or methyl and occasionally, a (Z)-alkene at C$_6$; in some cases they display spectacular levels of cytotoxicity [33]. Synthetic efforts thus far have tended to follow the strategy of the first total synthesis, which was completed in 1992 by Evans, Gage and Leighton [34] and which used a Wittig coupling of the phosphonium salt **49** to establish the central alkene linkage.

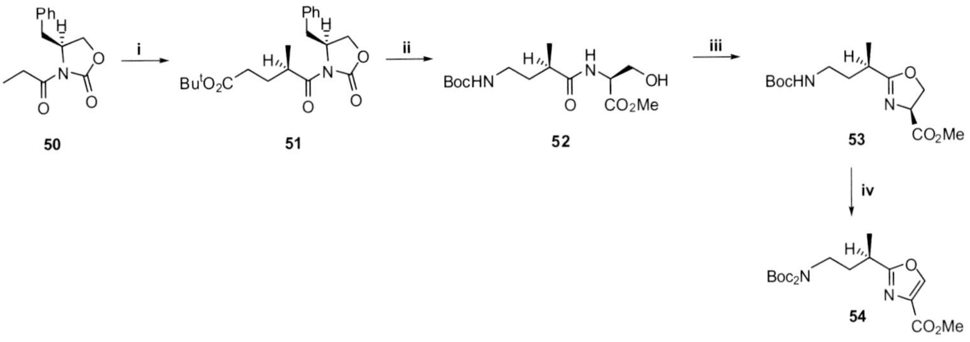

Scheme 12.11 *Reagents and conditions:* i. CH$_2$=CHCO$_2$Bu^t, Ti(OPr^i)Cl$_3$, ^{i}Pr$_2$NEt, CH$_2$Cl$_2$, 88%; ii. four steps, 60% overall; iii. SOCl$_2$, py, 78%; iv. a) Boc$_2$O, DMAP, MeCN, b) KHMDS, PhSeCl, c) H$_2$O$_2$, 50% overall.

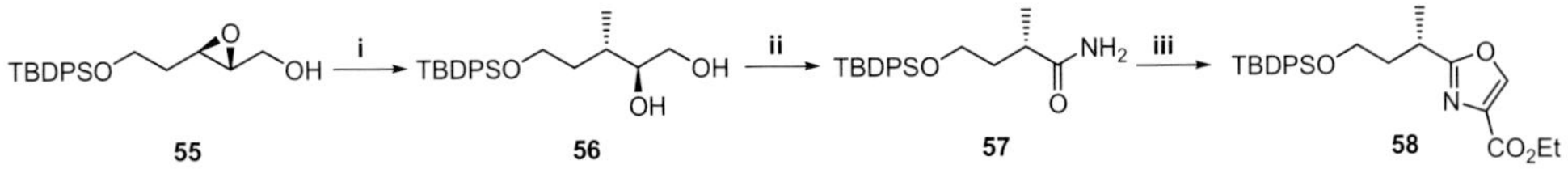

Scheme 12.12 *Reagents and conditions:* i. Me$_3$Al, pentane, 91%; ii. four steps, 93%; iii. BrCH$_2$C(O)CO$_2$Et, excess epoxycyclopentane, 63%.

Specifically (Scheme 12.11), Evans' enolate chemistry starting with the oxazolidinone chemistry was used to establish the methyl-substituted stereogenic center by conversion of precursor **50** into the homolog **51** by a Lewis acid-catalyzed Michael addition and manipulation through to the serine derivative **52**. Cyclodehydration was efficiently achieved using thionyl chloride but subsequent attempts at oxidation of the oxazoline **53** using the classic method of heating with nickel peroxide proved very poor. Hence, an alternative if lengthier strategy was employed which involved sequential double protection of the nitrogen, enolization and selenation and finally oxidative elimination; the return of oxazole **54** of around 50% was compensated for by the reliability of this lengthier protocol. The subsequent Wittig coupling delivered around 65% yields of the expected alkene.

An alternative strategy was to use a Julia olefination for this key step; on the way to the required aldehyde, the oxazole **58** was a key intermediate [35] (Scheme 12.12). Starting with the Sharpless epoxidation product **55**, nucleophilic ring opening established the stereogenic center; subsequent cleavage and further manipulations of the resulting diol **56** led efficiently to the amide **57**, which was converted into the oxazole **58** using a classical Hantzsch two-step method. The presence of excess epoxide was necessary to neutralize the hydrogen bromide formed and hence to prevent epimerization.

The Wittig strategy (*cf.* **49**) was also used by the Smith group, although a very different approach was used to access the necessary oxazole. (*D*)-Erythronolactone was used as the source of the stereogenic center in the azido-acid **59** [36]

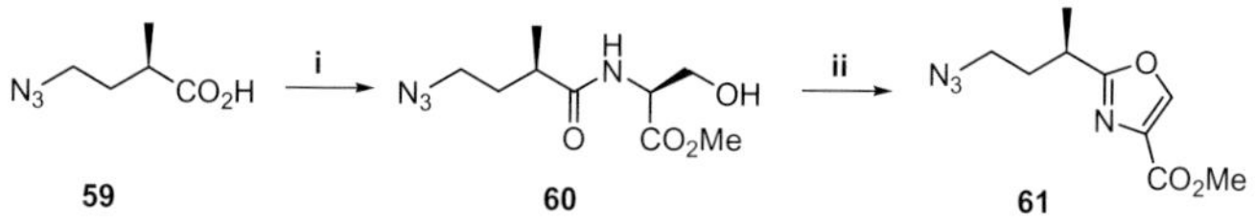

Scheme 12.13 Reagents and conditions: i. steps; ii. a) Burgess reagent [MeO$_2$CN$^-$.SO$_2$NEt$_3^+$], THF, b) CuBr$_2$, HMTA, DBU, CH$_2$Cl$_2$.

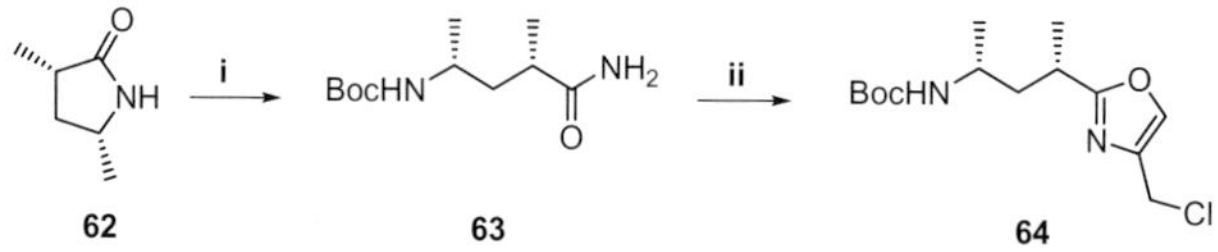

Scheme 12.14 Reagents and conditions: i. a) Boc$_2$O, b) Me$_3$Al, NH$_3$; ii. ClCH$_2$C(O)CH$_2$Cl, K$_2$CO$_3$, CHCl$_3$, 100 °C, 61%.

(Scheme 12.13). Routine chemistry was then employed to reach the serine derivative **60**, cyclodehydration and oxidation of which was achieved using another of Wipf's contributions to this area [37]: the combination of the Burgess reagent [38] for the first step and copper(II) bromide in the presence of DBU and HMPA for the overall oxidation step, known as the Barrish–Singh procedure [39]. In general, this method is suitable for the synthesis of oxazole-4-carboxylates.

This has become one of the most popular methods for oxazole synthesis, as the subsequent examples will illustrate. Its mildness is indicated by the lack of epimerization, in this case during the formation of oxazole **61**; cyclodehydration using thionyl chloride in pyridine or acid-catalyzed methods suffered from either poor yields or such epimerization. Also notable in this strategy is the prolonged survival of the azide group.

A much more classical approach has been highlighted by Ogawa and Armstrong [40] (Scheme 12.14).

(S)-Pyroglutamic acid was the source of the dimethyl pyrrolidinone **62**, opening of which gave the amide **63**, condensation of which with 1,3-dichloroacetone under warm basic conditions led to a decent yield of the chloromethyl oxazole **64**. Epimerization only occurred at higher temperatures. Clearly, both the rapidity and the elaboration of the useful chloromethyl group are significant features of this approach.

Old and new methods of cyclodehydration to the intermediate oxazoline **66** have been compared in a second approach to the calyculins, starting with the same pyrrolidinone **62**, which was converted into the serine amide **65** [41] (Scheme 12.15); as shown, the newer Burgess method won, but only just, in contrast to the foregoing results obtained during the Smith synthesis. Subsequent overall oxidation to the required oxazole **67** by either iodination and elimination or the Barrish–Singh method ended in a draw.

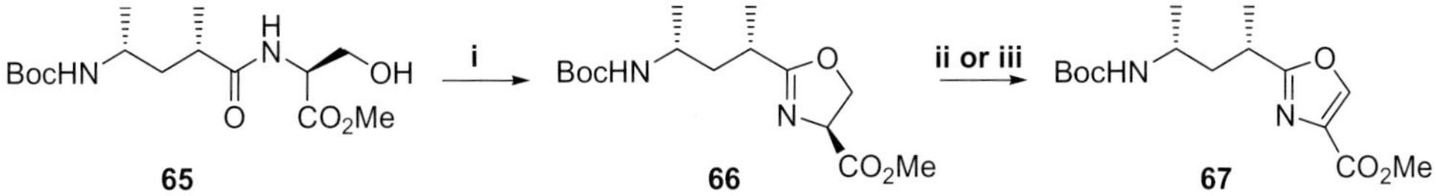

Scheme 12.15 *Reagents and conditions:* i. Burgess reagent, THF, Δ (82%) or SOCl$_2$, py (75%); ii. a) LHMDS, TMSCl, b) KHMDS, I$_2$, 42%; iii. CuBr$_2$, HMTA, DBU, CH$_2$Cl$_2$, 42%.

12.2.7
Leucascandrolide A, B and Neopeltolide

Serial No.	Trivial name	Structure	Source	Isolation	Biological activity	Synthesis
68	Leucascandrolide A, B and neopeltolide	See below	*Leucascandra caveolata*	[42]	Cytotoxicity to KB asnd P388 cells; strong inhibition of the yeast *Candida albicans*	[7, 8, 18, 43–48]

Leucascandrolide A **68** was isolated from dead sponge material and hence its exact origin is uncertain; subsequent attempts to re-isolate it have been unsuccessful. Interestingly, its cytotoxicity appears to derive from the macrolide portion whereas its antifungal properties are associated with the oxazole-containing side chain. Leucascandrolide B does not contain an oxazole ring while the closely related oxazole macrolide neopeltolide (not shown) differs from leucascandrolide A by having a less elaborate macrolide ring. Clearly, the unsaturated acid **69** is a viable precursor, as exemplified in an early approach by Wipf and Graham [43]. The easily prepared amide **70** was converted into the acid **69** by sequential modified Robinson–Gabriel cyclization, as detailed in Scheme 12.9 [25], Lindlar reduction of the alkyne and, finally, conversion into the corresponding aldehyde and a Still–Gennari homologation.

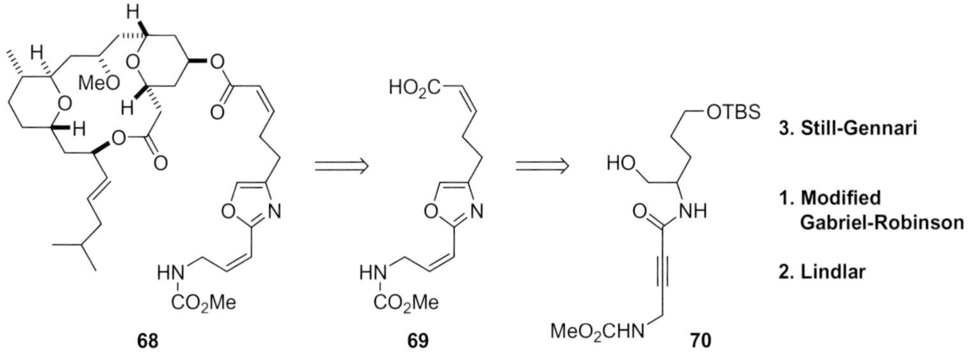

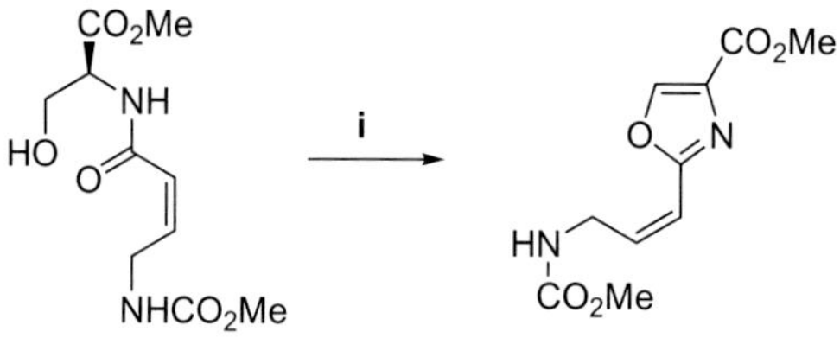

414 | *12 Oxazole and Its Derivatives*

Scheme 12.16 Reagents and conditions: i. a) DAST [Et$_2$NSF$_3$], CH$_2$Cl$_2$, −20 °C; b) BrCCl$_3$, DBU, 0 °C.

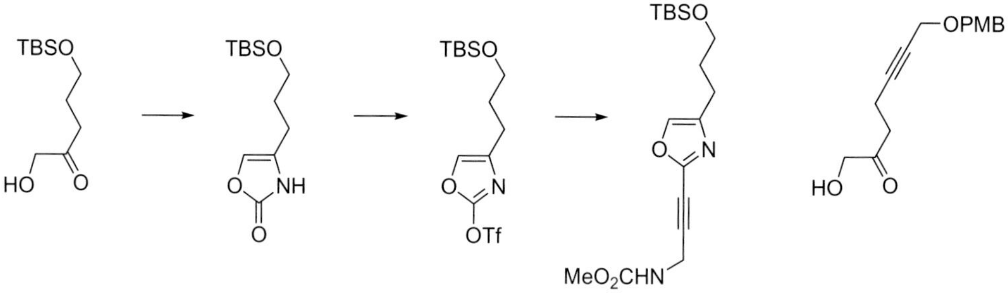

Scheme 12.17

The presence of two *cis*-alkenyl linkages makes this a somewhat more demanding target; clearly, the corresponding alkynes and attendant coupling reactions often play a key role. The sensitivity of a (*Z*)-alkene during oxazole construction has resulted in the deployment of yet another novel tactic for cyclodehydration, now known as the Wipf–Williams method [18], illustrated by the conversion of the serine amide **71**, a familiar intermediate type, into the vinyloxazole **72** [44] by sequential reaction with the electrophilic fluorine source DAST [Et$_2$NSF$_3$] followed by BrCCl$_3$ and DBU (Scheme 12.16). Overall yields are usually in excess of 60%. The Barrish–Singh protocol [39] can also be used in the second step of this mechanistically somewhat obscure reagent combination (the authors describe the mechanistic details as "scarce"!). The delightfully named Deoxo-Fluor™ [(MeOCH$_2$CH$_2$)$_2$SF$_3$] [18] can be substituted for DAST, as illustrated above [17].

Perhaps inevitably, a combination of a Sonogashira coupling and a Lindlar reduction has been used extensively to construct the (*Z*)-vinyloxazole fragment, a tactic summarized in Scheme 12.17 wherein an initial hydroxyl-ketone **73** is converted into the oxazolinone **74** (e.g., Cl$_2$CO, NH$_4$OH, H$_2$SO$_4$) and thence into the triflate (e.g., Tf$_2$O, 2,6-lutidine) and finally coupled with a protected propargylamine to give the target **76** [45].

Again, the final (*Z*)-enoate was usually added using a Still–Gennari coupling, although Paterson and Tudge employed the slightly shorter strategy of starting

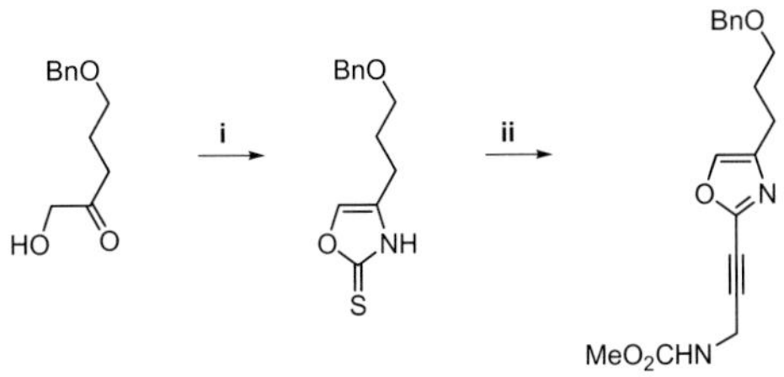

Scheme 12.18 *Reagents and conditions:* i. KSCN, HCl, aq. EtOH, Δ, (92%) ii. MeO$_2$CNHCH$_2$CCH, 5 mol% Pd(PPh$_3$)$_4$, 60 mol% CuTC (Cu(I) thiophene-2-carboxylate), 50 mol% CuI, DMF, μw, 100 °C, (52%).

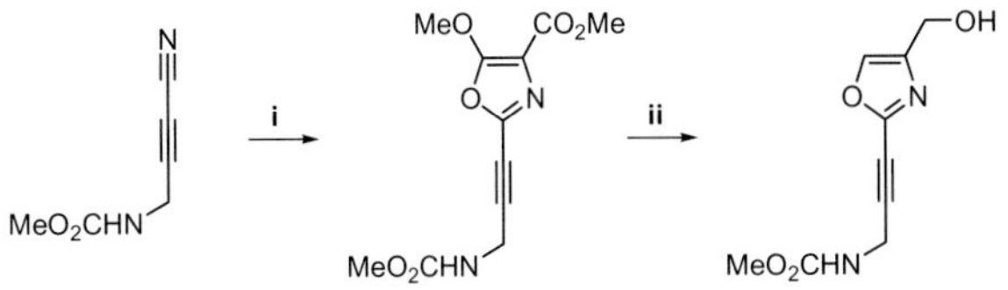

Scheme 12.19 *Reagents and conditions:* i. (MeO$_2$C)$_2$CN$_2$, 5 mol% Rh(I) (60%); ii. a) Lindlar, b) LiEt$_3$BH.

with the alkynyl hydroxyl-ketone **77** and using a later double Lindlar reduction to establish both (Z)-alkenes in a single step [45b, c].

A slightly different and briefer approach [46] features as the key step a desulfurative cross coupling [47]: a similar starting hydroxyl-ketone **78** is converted into the thio-oxazoline **79** in one efficient step, which is then directly coupled with the protected propargylamine to give a respectable 52% yield of the oxazole **80** (Scheme 12.18).

In a very different approach, the idea of adding a carbenoid to a nitrile to directly form an oxazole ([7]; see also [6]) has been used in a quite efficient synthesis of the oxazole-4-methanol **83** from the alkynyl nitrile **81** [8] (Scheme 12.19). The direct oxazole formation is followed by Lindlar reduction of the alkynyl-oxazole **82** and simultaneous ester reduction and removal of the methoxy group using Super-Hydride®.

Alternative tactics including alkene synthesis using a Still–Gennari coupling in the reverse sense have also made useful contributions to this area of synthesis [48].

12.2.8
Chivosazole

Serial No.	Trivial name	Structure	Source	Isolation	Biological activity	Synthesis
84	Chivosazole	See below	*Streptomyces cellulosum* sp.	[49]	Antifungal, highly toxic to some mammalian cell cultures.	[50]

Chivosazole **84** is a relatively recent addition to this large group of oxazole metabolites; the complex structure of this 31-membered macrolide will provide many synthetic challenges. As a useful contribution, the serine amide **85**, derived using a Masamune *anti*-aldol which in this case worked better than the related Evans and Paterson protocols, has been converted into the oxazole **86** in 70% yield using the Wipf–Williams procedure based on DAST [50].

12.2.9
Madumycin II

Serial No.	Trivial name	Structure	Source	Isolation	Biological activity	Synthesis
87	Madumycin II	See below	*Streptomyces cellulosum* sp.	[51]	Antibiotic; synergistic with vancomycin.	[52–56]

(–)-Madumycin II **87** [51] is a member of the Group A streptogramin antibiotics, which includes virginiamycin and griseoviridin among others. All these compounds are 23-membered macrolides containing 2,4-disubstituted oxazoles together with an (*E,E*)-dienylamine and poly-β-ketide dioxygenation patterns, which exhibit powerful synergistic effects against, for example, vancomycin-resistant strains of Gram positive bacteria. Much of the synthetic work in this area was reported some time ago [52] and had been reviewed previously [2]. A later contribution from the Meyers' group [53] features the now common tactic of cyclodehydration of a serine derivative, in this case the dioxolane **88**, derived in ten steps from (*S*)-malic acid, followed by oxidation of the resulting oxazoline to

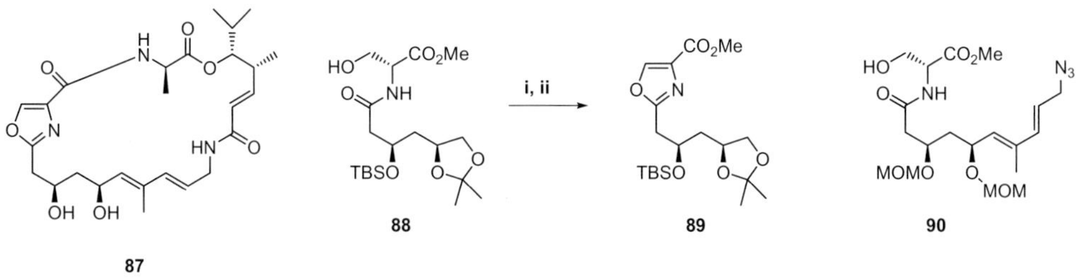

Scheme 12.20 Reagents and conditions: i. Burgess reagent; ii. tBuOOCOPh, CuBr, Cu(OAc)$_2$, C$_6$H$_6$, (57%).

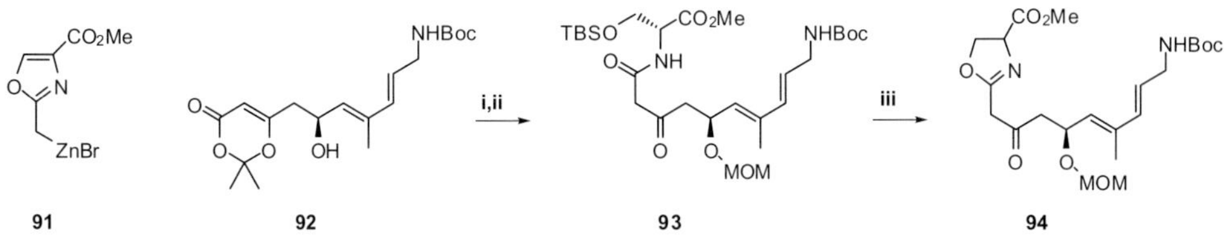

Scheme 12.21 Reagents and conditions: i. MOM-Cl, CH$_2$Cl$_2$, 0 °C to 20 °C, 48 h; ii. H-Ser(OTBS)-OMe, toluene, 110 °C, 1 h (68%); iii. HF-py, THF, 0 °C to 20 °C, 4 h then DAST, K$_2$CO$_3$, CH$_2$Cl$_2$, −78 °C.

obtain the key intermediate oxazole **89** (Scheme 12.20). In this case, this was achieved using a protocol related to the now popular Barrish–Singh method [39] (cf. Scheme 12.23 below).

A subsequent total synthesis of (−)-madumycin II **87** [54] relied on the intermediacy of a similar serine derivative **90**, which was converted into the corresponding oxazole using the now standard combination of Burgess reagent dehydration and Barrish–Singh oxidation introduced by Wipf [37, 39]. A notable feature of this approach is the early inclusion and subsequent survival of the pendant azide group (see also Scheme 12.13).

Metallation chemistry, while a seemingly viable pathway for incorporation of an oxazole group, has not quite matched its initial promise but one notable contribution was the definition of methodology to obtain the zinc derivative **91** [55].

A later contribution [56] established the key asymmetric center in precursor **92** using a Carreira enantioselective vinylogous Mukaiyama-aldol condensation (80%; 81% ee) between the corresponding dienyl aldehyde and the silyl enol ether of 2,2,6-trimethyl-4H-1,3-dioxin-4-one catalyzed by CuF·(R)-tolBinap (Scheme 12.21). Homologation to a familiar serine derivative **93** was followed by cyclodehydration to the corresponding oxazoline **94** using the DAST method [18]. However, the final oxidation step was reported to fail when using the Wipf–Williams [18] or the Barrish–Singh [39] methods and hence recourse was made the older nickel peroxide protocol; as is usually the case, while successful, this delivered a poor yield of 34%.

A key step in a notable total synthesis of the closely related metabolite, virginiamycin M$_2$ **95**, relied on a successful Reformatsky condensation of the foregoing zinc reagent **91** [57]; a full recipe for the precursor bromide has now been published [58], which relies on the Cornforth oxazole synthesis; the tactic has been suggested as a method for completing an alternative approach to this target [59].

12.2.10
14,15-Anhydropristinamycin II$_B$

Serial No.	Trivial name	Structure	Source	Isolation	Biological activity	Synthesis
96	14,15-Anhydropristinamycin II$_B$	See above		[60]		[61]

The closely related and rather sensitive 14,15-anhydropristinamycin IIB **96** has been synthesized using an alternative strategy in which the dianion **97** derived from the corresponding alcohol (BuLi, −78 °C) was condensed with the dienal **98** to give the expected diol in ~45% yield; macrocyclization was then achieved not by a ubiquitous amide or macrolactonization or macrolactamization but rather by an intramolecular Stille coupling [61].

12.2.11
Griseoviridin

Serial No.	Trivial name	Structure	Source	Isolation	Biological activity	Synthesis
99	Griseoviridin	See below	*Streptomyces* sp.	[62]	Antibiotic	[63]

Another member of this family, griseoviridin **99**, posses a increased synthetic challenge by reason of its unusual and more complex additional thialactone ring. After many studies, the Meyers' group achieved a thus far unique distinction of a completed enantioselective total synthesis [63]. The key oxazole component **100** was again derived from (S)-malic acid.

12.2.12
Thiangazole

Serial No.	Trivial name	Structure	Source	Isolation	Biological activity	Synthesis
101	Thiangazole	See below	*Polyangium* sp. of gliding bacteria.	[64]	HIV-1 (not confirmed)	[65–70]

The most remarkable feature of the Heathcock synthesis of thiangazole **101** [65] is the triple cascade reaction used to stitch up the central tris-thiazoline from the corresponding α-methylcysteine-derived tripeptide. In contrast, the oxazole unit was elaborated as a last step using relatively conventional modified Robinson–Gabriel methodology (Scheme 12.22) by initial oxidation of the threonine derivative **102** [25] followed by acid-catalyzed cyclodehydration, probably only feasible because all of the stereogenic centers are quaternary. Possibly because of the presence of the sensitive ene-sulfides, many alternatives failed, including oxazoline oxidation (derived using the Burgess method) along with other dehydration methods such as iodine-triphenylphosphine and thionyl chloride and relatives.

The Wipf synthesis of thiangazole **101** by contrast begins with a synthesis of the oxazole **104** again by sequential oxidation and cyclodehydration (Scheme 12.23), for which step the iodine–triphenylphosphine–Et$_3$N was now suitable in the absence of the sulfur functions [66]. The subsequent steps featured a similar triple zipper cyclization but of a tripeptide and subsequent exchange of oxygen and sulfur by ring opening (AcSH) and reclosure (NH$_3$) to generate the tris-thiazoline.

The Pattenden approach to thiangazole [67] proceeded through a similar oxazole derivative **105**, which was manipulated rather inefficiently to the corresponding oxazole using sequential Burgess cyclodehydration and oxidation using the Meyers' method [68] (*t*BuO$_3$Ph, CuBr; *cf.* Scheme 12.20). Less sophisticated methods were used in an earlier synthesis of the closely related tantazole, an oxazolyl-*tetra*-thazoline, in which a suitable oxazole was prepared by the Robinson–Gabriel

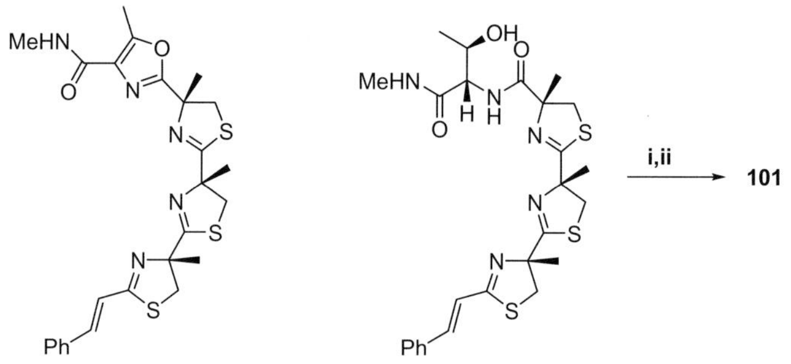

Scheme 12.22 Reagents and conditions: i. Dess–Martin (71%); ii. p-TsOH, 4A mol. sieves, C_6H_6, 80 °C (84%).

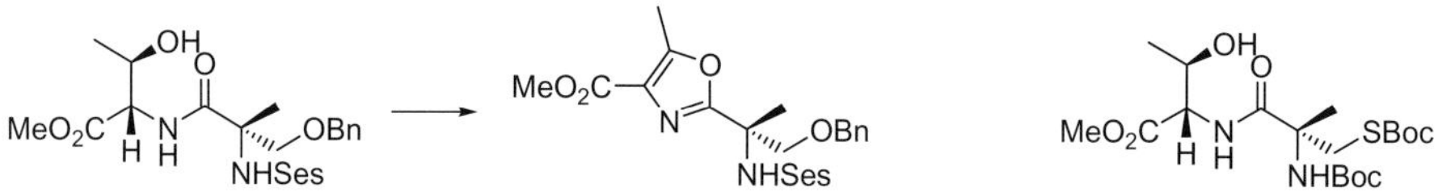

Scheme 12.23 Reagents and conditions: i. Dess–Martin; ii. Ph_3P, I_2, Et_3N.

strategy using Jones reagent for the oxidation step and thionyl chloride for the cyclodehydration. Clearly, the exact nature of the desired product is highly significant in guiding such reagent choices.

In direct contrast, a very recent synthesis of oxazole-containing analogs of the oxazoline-based metabolite Lissoclinum has employed the Wipf–Williams method [18] to dehydrate threonine derivatives, specifically using DAST followed by $BrCCl_3$-DBU (rather than the Barrish–Singh method) [70].

12.3
Unconnected Bis- and Tris-Oxazoles

12.3.1
Disorazole C_1

Serial No.	Trivial name	Structure	Source	Isolation	Biological activity	Synthesis
106	Disorazole C_1	See below	*Sorangium cellulosum*	[71]	Anticancer (antitubulin)	[72–76]

The disorazoles possess significant antitumor properties, some of which are greater than those displayed by epothiolin B [71]. Despite its symmetrical dimeric bis-lactone structure, disorazole C₁ **106** represents a notable but popular synthetic challenge, one addressed by Wipf and Graham in their total synthesis which featured a typical oxazole synthesis from a precursor serine derivative **107** using the Wipf–Williams method [18] (DAST then BrCCl₃-DBU). A somewhat unexpected but not unreasonable outcome was the generation of approximately equal amounts of the free alkyne **108a** and the corresponding bromide **108b**; conversion of the former into the latter (NBS, AgNO₃) allowed continuation of the sequence [72].

An approach by the Hoffmann group [73] employed the Panek-modified Hantzsch procedure [74] to prepare the useful oxazole aldehyde **109** and subsequently the corresponding alkyne **110** using the Ohira–Bestmann reagent. Alternatives due both to the Meyers and Hoffmann groups, sadly thwarted by later steps, returned to the Wipf–Williams protocol [18] for the efficient conversion in 79% yield of the serine derivative **111**, ultimately derived from (L)-malic acid, into the oxazole **112** [75]. A closely related differentially protected oxazole was used subsequently [76].

12.3.2
Phorboxazoles

Serial No.	Trivial name	Structure	Source	Isolation	Biological activity	Synthesis
113	Phorboxazole A and B	See below	*Phorbas* sp. (Ocean sponge)	[77]	General high level anticancer; antifungal	[78–93]

The remarkably complex and active phorboxazoles **113** have understandably been the subject of many partial and total syntheses. An early effort is distinguished by a very late formation of the "upper" oxazole using the standard Wipf protocol [25] based on a Robinson–Gabriel oxidation-ring closure of the peptide **114**, which delivered a notable 77% yield [78]. The second "lower" oxazole was formed from another serine residue **115** during the penultimate step using the same Wipf method followed by global desilylation to provide the target in 33% yield for the two steps. No doubt, the success of these cyclizations attests to the mild and specific nature of this methodology.

113 R^1, R^2 = H, OH.   **114**   **115**

Although formed at earlier stages, a subsequent approach to the phorboxazoles used the same Wipf procedure to obtain the key intermediate **116** [79], while a second synthesis used optimized conditions, especially in the Dess–Martin oxidation step [80].

116   **117**   **118**   **119**

In the first Smith synthesis [81], the upper vinyloxazole function was accessed using a Wittig condensation, with a phosphonium salt derived from the monooxazole derivative **119**. The oxazole unit was readily incorporated starting with the aldehyde **117** and forming the enol acetal **118**. This then was induced to undergo a remarkable Petasis–Tebbe rearrangement followed by a Petasis–Ferrier reaction upon exposure to dimethylaluminum chloride [82]. The transformation proceeds by sequential ring opening to a oxonium enolate and 6-*endo*-trig reclosure and is probably assisted by prior complexation between the oxazole nitrogen and the incoming aluminum reagent as shown.

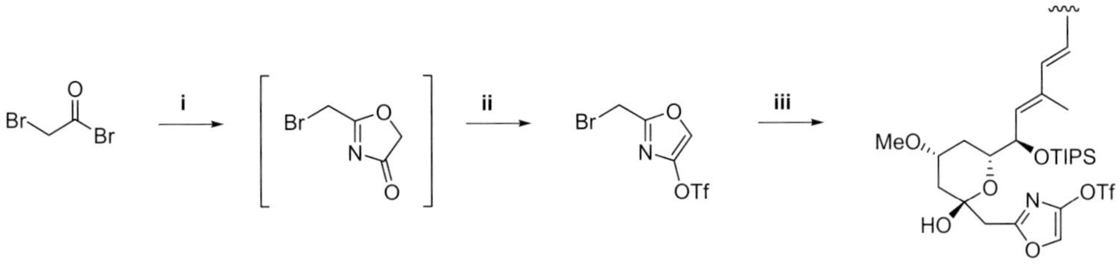

Scheme 12.24 *Reagents and conditions:* i. AgOCN, CH$_2$N$_2$; ii. Tf$_2$O, Et$_3$N (48%); iii. ^{i}PrMgCl, add lactone (72%).

The "lower" oxazole was obtained using more conventional but uncommon methodology, as shown in Scheme 12.24. Starting with bromoacetyl bromide **120**, the intermediate oxazolone **121** is not isolated but rather immediately converted into the triflate **122**, isolated in 48% yield. Subsequent Grignard reagent exchange and condensation with a protected lactone provided the advanced intermediate **123** in 76% yield, ready for further elaboration towards the target using a Stille coupling.

In an early total synthesis achieved by the Evan's group, the enantiopure oxazole derivative **124** was obtained by an asymmetric aldol condensation using the corresponding oxazole-4-carboxazdehyde [74] while the lower oxazole was incorporated employing the carbanion **125** [83]. This illustrates an important principle of regiochemistry with respect to oxazole metallation: such carbanions are usually in rapid equilibrium with the corresponding 5-lithio species, but if formed in the presence of an unhindered amine such as diethylamine, then the 2-methyl carbanion is favored. This, along with formation of the "lower" 4-vinyl-oxazole moiety by the HWE reagent **126** [84, 85] followed by such a metallation [86, 87] represents an efficient strategy for the elaboration of the lower half of the phorboxazoles. In the latter more recent synthesis, once again the Wipf procedure [25] was used to obtain the "upper" oxazole.

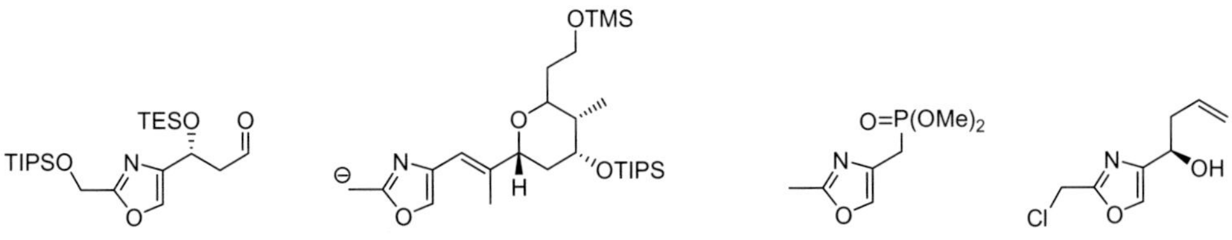

A variation of forming the "upper" combination of an oxazole and a tetrahydropyran is to form the latter using a hetero-Diels–Alder cycloaddition with a

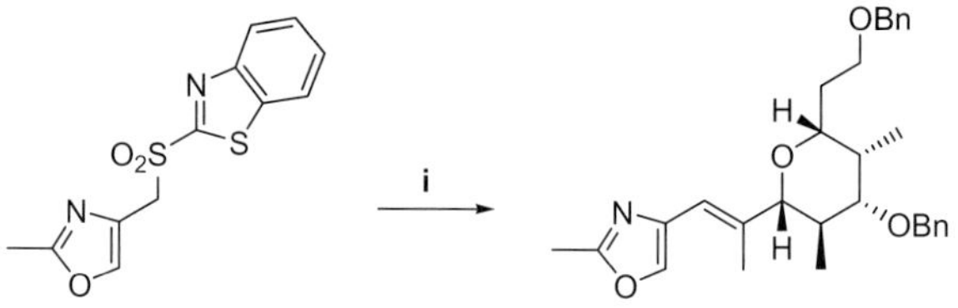

Scheme 12.25 Reagents and conditions: i. NaHMDS, THF, −78 °C, add aldehyde (70%, based on recovered starting material; E/Z:9:1).

2-substituted oxazole-4-carboxaldehyde as the dienophile [88]; alternatively, this functionality can also be established by multiple aldol condensations, while the lower oxazole is again formed using the combined HWE-metallation strategy [89].

An alternative use of such oxazole aldehydes is in the formation of enantiopure secondary alcohols such as **127** by asymmetric addition of allyltin species catalyzed by (R)-BINOL, ready for pyran formation using Prins cyclizations [90].

More recent progress in this area has seen the development of a second generation synthesis by the Smith group, but which uses much the same strategies with respect to oxazole incorporation [91], the definition of the sulfone **128** as a useful alternative for synthesis of the lower vinyloxazole array **129** of phorboxazole [92] (Scheme 12.25) and the development of strategies to prepare phorboxazole analogs [93].

12.3.3
Leucamide A

Serial No.	Trivial name	Structure	Source	Isolation	Biological activity	Synthesis
130	Leucamide A	See below	Leucetta microraphis (Marine sponge)	[94]	Cytotoxic	[95]

By the standards of many of the foregoing metabolites, leucamide A is a relative simple target, the synthesis of which relies mainly upon efficient peptide coupling methods along with yet another successful application of the Wipf–Williams method (DAST then BrCCl$_3$, DBU [18]); while not particularly efficient, delivering a 51% yield of the corresponding oxazole from the thiazolyl peptide **131**, no epimerization of the stereogenic center was evident [95].

130 **131** **132**

12.3.4
Promothiocin A

Serial No.	Trivial name	Structure	Source	Isolation	Biological activity	Synthesis
132	Promothiocin A	See below		[96]		[97]

The synthesis of the required oxazoles in the somewhat more complex metabolite promothiocin A **132** has a rather different emphasis in that the required precursors (e.g., **134**) were prepared by carbenoid insertion into the N–H bond of a starting amino-amide (e.g., **133**) [97]. Subsequent oxazole formation was again achieved using iodine, triphenylphosphine and triethylamine to complete the heteroaromatic ring formation (Scheme 12.26).

12.3.5
Berninamycin A

Serial No.	Trivial name	Structure	Source	Isolation	Biological activity	Synthesis
136	Berninamycin A	See below		[98]		[99]

In synthetic terms, berninamycin A appears to be as much of a test of stamina as synthetic design, with its multiple repeated peptide linkages and dehydroserine residues. In two remarkably compressed publications, the required methyl oxazoles **138** were prepared in two ways, firstly a familiar oxidation and aromatization of a threonine derivative **139** but the other a less familiar Michael addition-elimination sequence from a vinyl bromide **137** (Scheme 12.27) [99]. This latter method was first reported in 1992 and can also be used for 5-bromo oxazoles by

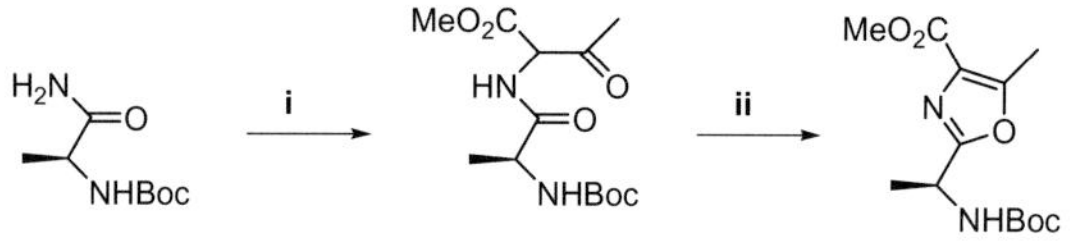

426 | 12 Oxazole and Its Derivatives

Scheme 12.26 *Reagents and conditions:* i. MeC(O)C(N$_2$)CO$_2$Me, Rh(I); ii. I$_2$, PPh$_3$, Et$_3$N, CH$_2$Cl$_2$ (70%).

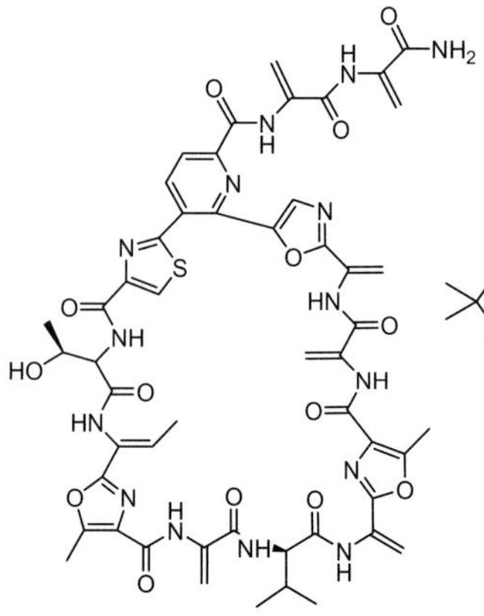

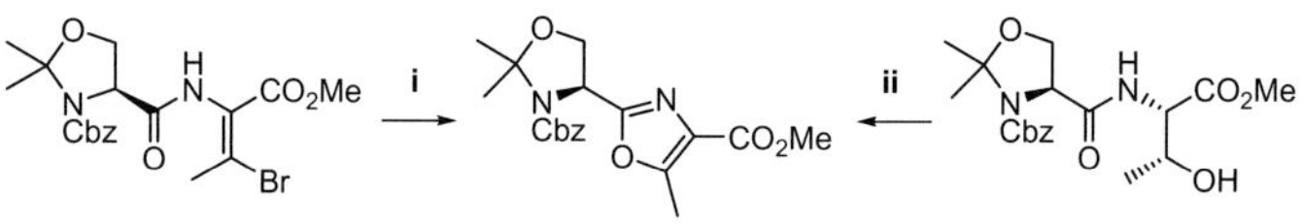

Scheme 12.27 *Reagents and conditions:* i. Cs$_2$CO$_3$, dioxane, 50 °C, 24 h (87%); ii. Jones oxidation then I$_2$, PPh$_3$, Et$_3$N (86% for 2 steps).

using either copper(II) bromide and DBU or methanolic hydroxide to effect the cyclization [100].

12.4
Cyclic Polyheterocyclic Metabolites Containing Single Oxazole Residues

12.4.1
Dendroamide A

Serial No.	Trivial name	Structure	Source	Isolation	Biological activity	Synthesis
140	Dendroamide A	See below	*Stigonema Dendroideum fremy*	[101]	Reversal of MDR.; MRP1 antagonist	[102–104]

12.4 Cyclic Polyheterocyclic Metabolites Containing Single Oxazole Residues

The small group of cyclic peptides in which some or all (see later) serine, threonine or cysteine residues have undergone cyclodehydration and oxidation to the corresponding oxazoles or thiazoles, form a most fascinating group of natural products; to this reviewer, it seems that they must show some useful properties!

A formal total synthesis of dendroamide A **140** highlights common problems in this area, both associated with the presence of stereogenic centers in these metabolites, those of structural assignment and synthesis without racemization. Oxazole synthesis was carried out using the Wipf approach [25] based on the Robinson–Gabriel method, but with the crucial modification that the isolated but very sensitive Dess–Martin oxidation product **141** was cyclodehydrated using bis(triphenyl)oxodiphosphonium triflate, readily obtained from triphenylphosphine oxide and triflic anhydride, to give the oxazole **142** in 84% yield and with 98% enantiomeric enrichment [102]. Other methods such as the popular iodine–triphenylphosphine require the presence of a base and hence carry an inherent risk of epimerization. Alternatively, cyclodehydration of substrates at the lower oxidation level (e.g., **143**) using the Burgess reagent provide the corresponding oxazolines in excellent yields; the problem comes with the subsequent oxidation step to reach the oxazole. While the Wipf–Williams method of $BrCCl_3$-DBU can give good yields (~70%), the presence of the base can induce epimerization and older methods using oxides such as NiO_2 or MnO_2 deliver much poorer returns [103].

140 **141** **142** **143**

Cyclizations of substrates very similar to peptide **143** using the Wipf–Williams procedure [18] proved relatively inefficient (<25% yield), whereas the combination of a Dess–Martin oxidation and cyclization using iodine–PPh_3–Et_3N gave better returns (46–65%); epimerization levels seemed to be low [104].

12.4.2
Nostocyclamide

Serial No.	Trivial name	Structure	Source	Isolation	Biological activity	Synthesis
144	Nostocyclamide	See below		[105]		[103c, 106]

The related nostocyclamide **144** has been the subject of a single total synthesis by Bagley and Moody who used the method shown in Scheme 12.26 to obtain the necessary oxazole derivative [[106]; see also [103c]].

12.4.3
Bistratamides

Serial No.	Trivial name	Structure	Source	Isolation	Biological activity	Synthesis
145	Bistratamides	See below	*Lissoclinium bistratum*	[107]	Moderate cytotoxicity; antimicrobial	[108–111]

The bistratamides (e.g., bistratamide H **145**) are a series of valine-derived metabolites which vary both in oxidation state (oxazole, oxazoline, open chain) and by having one or two oxazole functions.

144   145   146

The total syntheses of many members of the series by You and Kelly used the same methodology to prepare optically pure oxazole **146** and relatives as described above in a synthesis of dendroamide A *via* oxazole **142** [108]. An optimized sequence using the Wipf–Williams procedure [18], with Deoxo-Fluor has also proven especially effective for such oxazole synthesis [109] as did the Meyers' method (CuBr, Cu(OAc)$_2$, *t*BuOOCOPh) in earlier syntheses [110]. A more recent synthesis of bistratamide H **145** features the use of a new fluorous amine protecting group, tris(perfluorodecyl)silyl-ethoxycarbonyl (FTeoc), which may also contribute to the avoidance of epimerization during deprotection [111]. The necessary oxazoles were once again prepared using the Wipf–Williams method in good overall yield [18]. An additional notable feature was the use of immobilized Burgess reagent.

12.4.4
Tenuecyclamides A–D

Serial No.	Trivial name	Structure	Source	Isolation	Biological activity	Synthesis
147	Tenuecyclamides A–D	See below	*Nostoc spongiaeforme var. tenue*	[112]	Inhibitors of cell division	[113]

The four known tenuecyclamides (e.g., **147**) vary by the nature of the amino-acid residues (ala, gly or met) as well as by stereochemistry. A single synthesis [113] relies upon solid state couplings, a novel departure in this area, to assemble the key fragments, one of which, the oxazole acid **148**, was obtained by the Kelly modification for cyclodehydration (cf. **141** → **142**) using a bisphosphonium salt in order to minimize any epimerization.

147 **148** **149** **150**

12.4.5
Dolastatin I

Serial No.	Trivial name	Structure	Source	Isolation	Biological activity	Synthesis
149	Dolastatin I	See below	*Dolabella auricularia*	[114]	Cytotoxic	[115, 116]

The two dolastatins, dolastatin I **149** and dolaststin E vary by the presence of an oxazoline or a thiazoline ring respectively. During a total synthesis of the former, the oxazole acid **150** was obtained from the related theonine-based dipeptide by sequential cyclodehydration using the Burgess reagent followed by Meyers radical oxidation (CuBr, tBuOOCOPh) in decent yields [115]. By contrast, the synthesis of some analogs relied upon the use of $BrCCl_3$-DBU for the second step; epimerization did not seem to be a problem despite the presence of a strong base [116].

12.5
Conjugated Bis-Oxazoles

12.5.1
(−)-Hennoxazole A

Serial No.	Trivial name	Structure	Source	Isolation	Biological activity	Synthesis
151	Hennoxazole A	See below	*Polyfibrospongia* sp.	[117]	Active against herpes simplex virus (0.6 µg mL^{-1})	[118–124]

Wipf and Lim succeeded in a total synthesis of (+)-hennoxazole A, thereby establishing the structure of the natural material as (−)-**151** [118]. Serine-based peptides were the precursors to the two oxazoles. The right-hand ring **152** was prepared from the corresponding peptide by sequential cyclodehydration using the Burgess reagent followed by oxidation (CuBr, DBU). The second oxazole was synthesized as the final step, a very brave move! The key serine-like peptide **153** was successfully oxidized by the Dess–Martin method [25] but subsequent attempted oxidation using this Wipf method [(BrCl$_2$C)$_2$; hindered base] stopped at the 10-bromo-oxazoline; it was only when this was exposed to DBU in acetonitrile that oxazole generation was achieved, in acceptable yields.

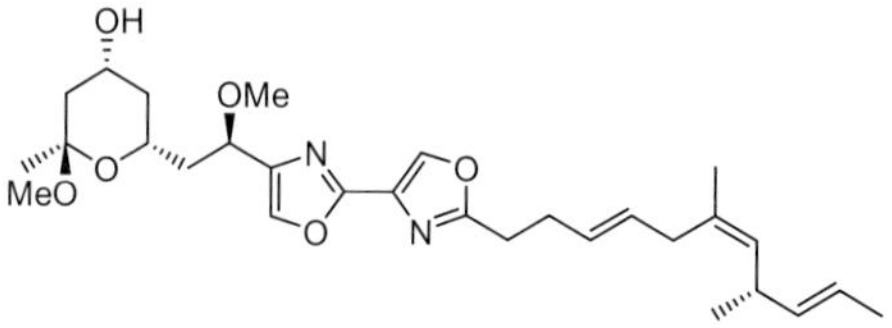

151 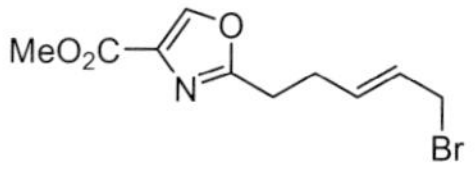 **152** 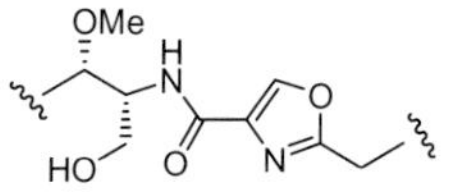 **153**

A later synthesis based on much the same strategy employed the Wipf–Williams method and Deoxo-Fluor [18], which gave improved yields [119]. An alternative approach to the central bis-oxazole unit **156** in hennoxazole A features the iterative use of the Wipf–Williams procedure, starting with the serine derivative **154** and proceeding via the oxazole-4-carboxylate **155** (Scheme 12.28) [120].

Curiously, another group had great difficulties in using this procedure to form the bis-oxazole unit and hence used a rather different set of recipes (Scheme 12.29) [121]. Starting from the serine derivative **157** as usual, conversion into the oxazole-4-carboxylate **158** was relatively routine although a slightly different set of reagents was used. Subsequently, all attempts to cyclodehydrate the serine derivative formed from the corresponding oxazole-4-acid (see Scheme 12.28) were unsuccessful.

Recourse was therefore made to an older protocol due to Shapiro which proceeded via the malonate **159** and which delivered the desired bis-oxazole **160** in decent yield. Careful reduction and oxidation then led to the desired 1,3-dithiane **161**, ready for further coupling.

A perhaps obvious general approach to the bis-oxazole unit is to use a palladium-catalyzed coupling. However, a model study showed that these can require rather

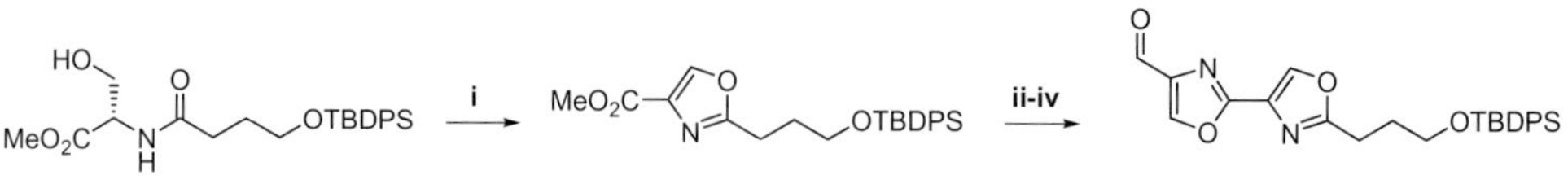

154 **155** **156**

Scheme 12.28 Reagents and conditions: i. DAST, K$_2$CO$_3$, CH$_2$Cl$_2$ then BrCCl$_3$, DBU (61%); ii. LiOH then ^{i}BuOC(O)Cl, serine.OMe; iii. as i.; iv. Dibal-H (51%).

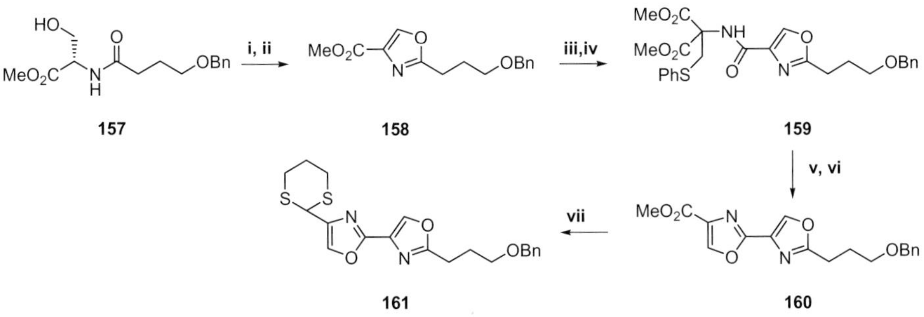

Scheme 12.29 *Reagents and conditions:* i. Ph₃P, CCl₄, ⁱPr₂NEt, 4:1 MeCN-CH₂Cl₂, 0°C–20°C (70%); ii. CuBr, DBU, HMTA, CH₂Cl₂ (90%); iii. K₂CO₃, aq. MeOH (89%); iv. couple; v. NCS, DMF, 0°C then ⁱPr₂NEt, DMF, 20°C; vi. K₂CO₃, MeOH, 60°C (72% for v and vi); vii. LiBH₄, MeOH (64%) then Dess–Martin and HS(CH₂)₃SH.

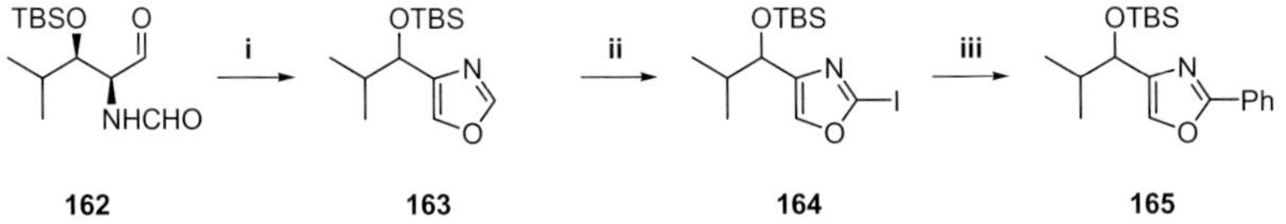

Scheme 12.30 *Reagents and conditions:* i. I₂, Ph₃P, Et₃N, CH₂Cl₂ (44%); ii. BuLi, THF, −78°C then I₂; iii. PhSnMe₃, PdCl₂(PPh₃)₂, DMPU, 100°C, 24h (50%).

brutal conditions and are not necessarily efficient. Thus, the aldehyde **162** was converted into the oxazole **163** using the Robinson–Gabriel method and then metallated specifically at the 2-position to provide a route to the useful iodide **164** (Scheme 12.30) [122]. A subsequent Stille coupling eventually provided a reasonable yield of the 2-phenyloxazole **165**. Attempts to form a tin derivative corresponding to the iodide **164** were unsuccessful.

A more recent synthesis of (−)-hennoxazole A **151** relied on a similar iterative approach to that outlined in Scheme 12.28 to form a 2-methyl bis-oxazole unit, which was followed by an extensive series of metallation studies to establish conditions suitable for introduction of the remaining sections of the target using this methodology [123]. As was found by Evans [83], the presence of small amines (here, Et₂NH from LiNEt₂) tends to favor reaction at the 2-methyl group rather than at the 5-position of the oxazole.

A novel method (*cf.* Scheme 12.28) for the cyclodehydration of serine derivatives (e.g., **166**) employs molybdenum (IV) and (VI) oxides under a Dean and Stark water separator and can give very high yields and is sufficiently mild that a second pre-formed oxazole survives; hence its use in the synthesis of the hennoxazole bis-oxazole **167** (Scheme 12.31) [124].

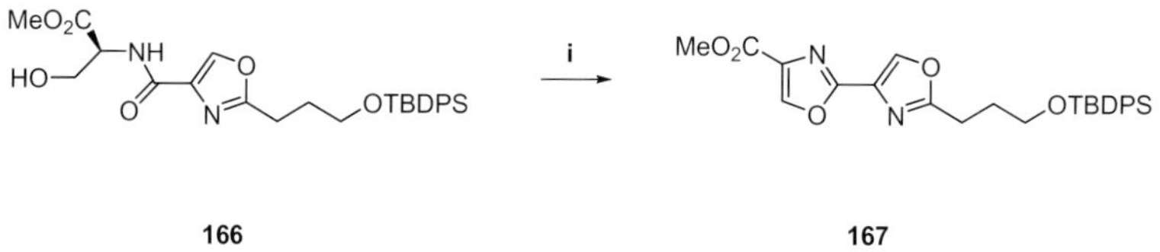

Scheme 12.31 Reagents and conditions: i. MoO$_2$.MoO$_3$, 10% (NH$_4$)$_2$MoO$_4$, toluene, 110 °C, (Dean and Stark water separator) (80–90%) then BrCCl$_3$, DBU.

12.5.2
Muscoride A

Serial No.	Trivial name	Structure	Source	Isolation	Biological activity	Synthesis
168	Muscoride A	See below	*Nostoc muscorum*	[125]	Weak antibiotic	[126–129]

Two early syntheses of muscoride A relied once again on an iterative construction of the bis-oxazole feature. The first synthesis by Wipf and Venkatraman used a modified Robinson–Gabriel method starting from the dipeptide Pro-Thr.OMe and a reagent combination of a Dess–Martin oxidation followed by cyclodehydration using iodine–Ph$_3$P-Et$_3$N to arrive at the bis-oxazole **169** [126]. Overall yields for each oxazole formation were around 65%. The Pattenden approach started from the other end. After a first oxazole synthesis from a suitably protected Thr-Thr.OMe dipeptide using Burgess reagent and CCl$_4$-DBU-pyridine, a slightly unconventional recipe introduced by Jung [127], the key intermediate **170** was elaborated using standard deprotection and coupling reactions. While the original Burgess reagent method was extremely capricious, exposure of this intermediate to DAST at low temperature and, once again, the Jung procedure delivered the desired bis-oxazole reproducibly in around 50% yield [128].

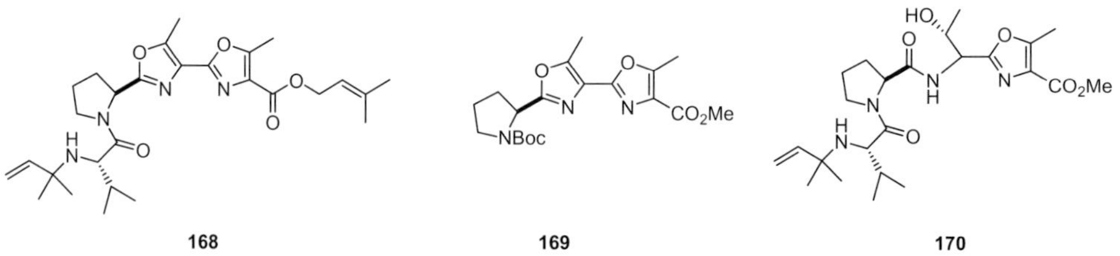

A completely different iterative procedure has been used to obtain the bis-oxazole unit in a later synthesis which features a 5-*exo*-dig cyclization followed by

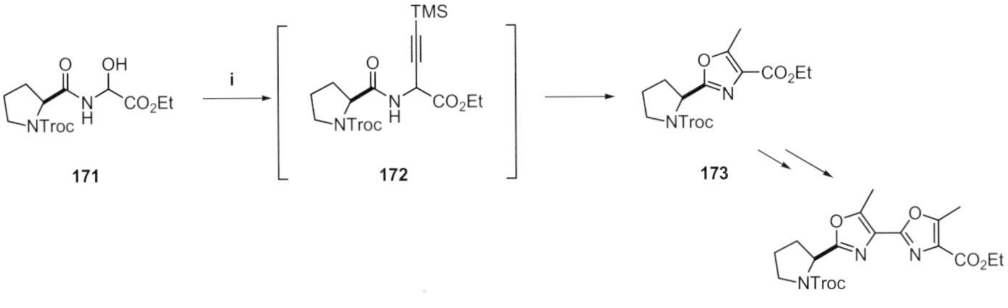

Scheme 12.32 Reagents and conditions: i. SOCl$_2$ then Me$_2$AlCCTMS.

alkene migration to establish the oxazole ring (Scheme 12.32) [129]. Thus, the proline amide-glyoxalate adduct **171** is chlorinated and the resulting chloride displaced by an alkynyl aluminum reagent to give the alkyne **172** which then undergoes spontaneous cyclization and isomerization to give the prolinyl-oxazole-4-carboxylate **173** in 72% yield. Conversion of the ester function into the corresponding amide and repetition of the sequence then delivers the bis-oxazole **174** in 45% yield.

12.5.3
Diazonamide A

Serial No.	Trivial name	Structure	Source	Isolation	Biological activity	Synthesis
175	Diazonamide A	See below	Diazona chinensis	[130]	Antitumor at the nanomolar level	[131–141]

The highly complex diazonamide A **175** has attracted considerable synthetic interest [130b], not the least because the initially proposed structure was incorrect, a reflection of its complexity rather than any crass error on the part of the researchers [130a]. However, such errors were not in the vicinity of the bis-oxazole feature. Hence, the following is a summary of the various ways in which this array has been assembled, with little reference to the remaining multiple(!) steps, and includes the initial synthesis of the erroneous structure by the Harran group, as such oxazole syntheses are clearly of relevance to syntheses of the correct structure. Most syntheses of the central bis-oxazole unit tend to be iterative but these two preparations are usually interspersed by many other steps needed to construct the remainder of the target.

In an initial synthesis of the bis-oxazole unit towards the incorrect structure [131], the left hand side ring was first formed using Freeman's oxazole synthesis

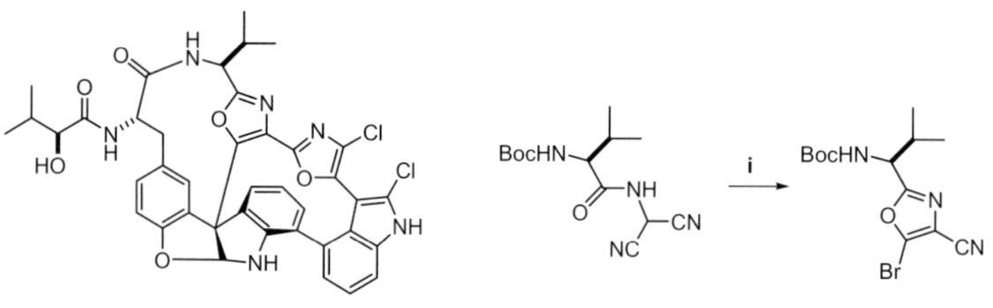

Scheme 12.33 Reagents and conditions: i. tBuONO, CuBr$_2$, MeCN (35%).

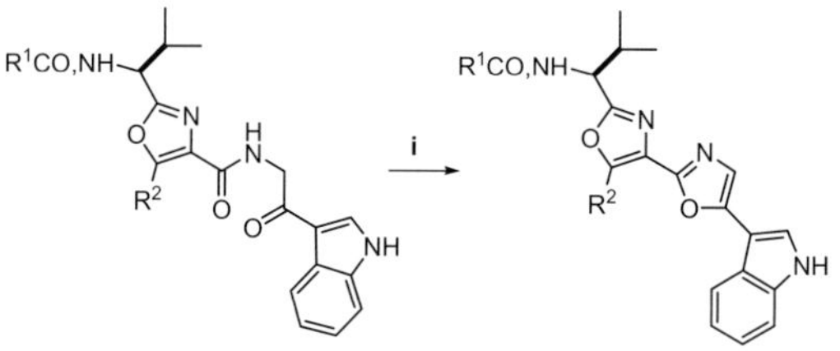

Scheme 12.34 Reagents and conditions: i. (Cl$_3$C)$_2$, Ph$_3$P, Et$_3$N, THF (68%).

[132] wherein N-Boc-Val was coupled with aminomalononitrile and the resulting dinitrile **176** cyclized under oxidative conditions similar to the Meyers' procedure (Scheme 12.33). The resulting oxazole **177**, formed in 35% yield, was well set up for subsequent elaboration, through multiple steps, until the synthesis was nearly complete.

At this stage, the keto amide **178** was set up (see below) and cyclized under typical Wipf conditions [25] to give the indolyl-bis-oxazole **179** in 68% yield (Scheme 12.34). The two chlorine atoms were added using N-chlorosuccinimide.

A neat trick, which has been used a number of times in diazonamide A syntheses, exploits the presence of two indole fragments at various stages, one of which ends up as a semi-saturated ring, while the second is retained.

This is illustrated in Harran's synthesis of the true structure of the natural product and consists of using the indole ring to induce DDQ to carry out a benzylic oxidation, thereby setting up a keto amide perfect for a Robinson–Gabriel cyclodehydration to give an oxazole. In this case, the peptide **180** was thus oxidized and the resulting keto amide simply exposed to acid to deliver the oxazole **181** (Scheme 12.35) [133]. The second oxazole was again formed much later, but using the same type of DDQ oxidation but of the advanced intermediate **182**; subsequent cyclization was induced using the Wipf procedure on this much more sensitive substrate.

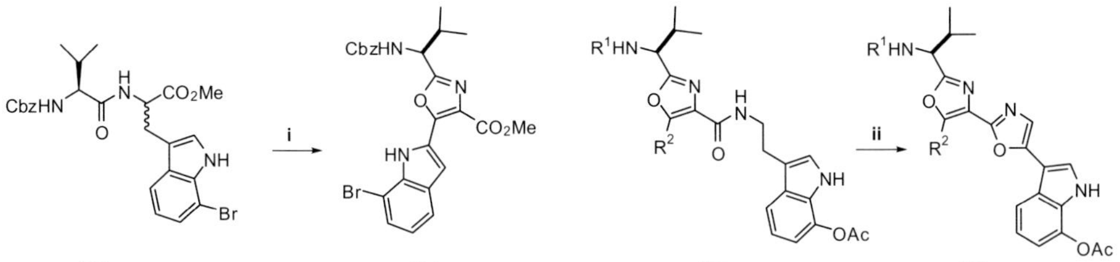

Scheme 12.35 Reagents and conditions: i. DDQ, aq THF then HBr, HOAc; ii. DDQ then $(Cl_3C)_2$, Ph_3P, Et_3N, CH_2Cl_2 (47%).

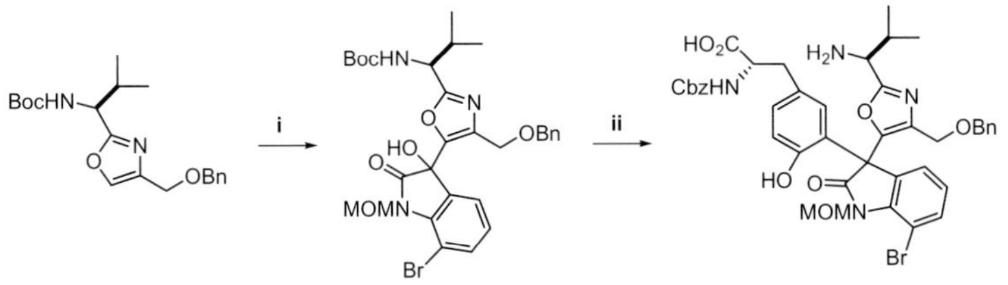

Scheme 12.36 Reagents and conditions: i. 2.BuLi then N-MOM-(7-bromo)isatin (73%); ii. N-Cbz-tyrosine-OMe, TsOH.

In the first approach to diazonamide A by the Nicolaou group [134], new methodology was used to build up the key intermediate **186** from the oxazole **184**, previously used by Meyers in a synthesis of a bistratamide (see Section 12.3.3). Exposure to two equivalents of butyl lithium generated a dianion which condensed smoothly with an N-MOM isatin to give the intermediate **185**, which was further homologated by an acid-catalyzed alkylation of Cbz-tyrosine.OMe (Scheme 12.36). The second oxazole was elaborated in much the same way as outlined in Scheme 12.35 except that the keto amide cyclization was effected using $POCl_3$ in pyridine.

In a second approach, the oxazole synthesis was in reverse order, with an initial indolyl oxazole being prepared by a relatively routine Wipf modification [25] of the Robinson–Gabriel method; a similar intermediate to the foregoing was used much later to form the second oxazole, but again using $POCl_3$-py as the cyclodehydrating agent [135].

Efforts towards the original erroneous structure proposed for diazonamide A **175** featured the synthesis of an oxazole **188** from an isomeric keto amide **187** using a standard Wipf procedure. When an intended RCM reaction failed, an intramolecular olefination featuring derivatives of the phosphonate **189** was used [136].

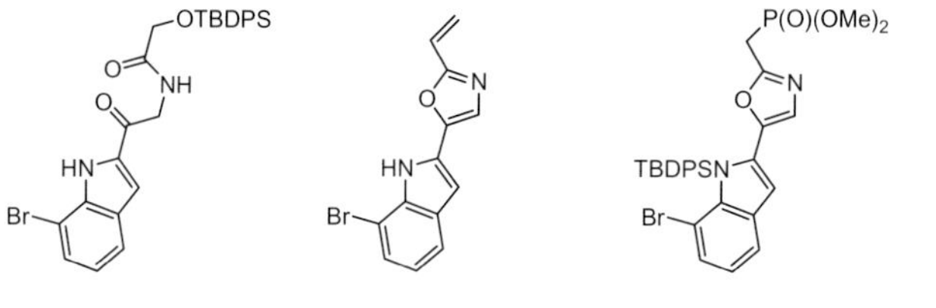

Scheme 12.37 Reagents and conditions: i. tBuLi, ClCO₂Me then LiCH₂NC, −40 °C–0 °C; ii. TsOH (71%).

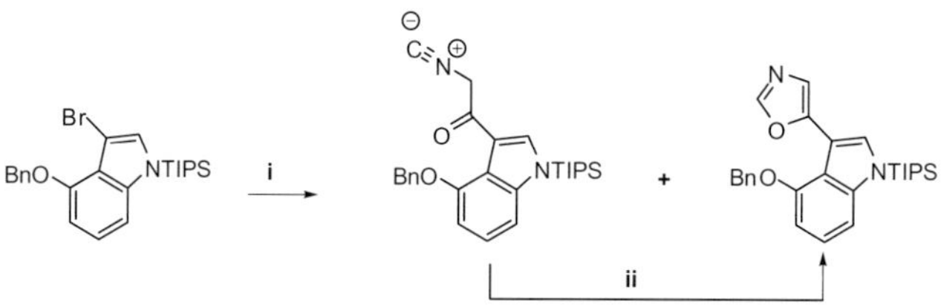

In related studies, a number of the foregoing protocols for oxazole formation failed for various reasons but, once again, keto amide closure using POCl$_3$-py was successful [137]. Of course, it would be asking too much to extrapolate these results too extensively in view of the extreme complexity of such intermediates.

Zajac and Vedejs have contributed a rather different approach to the indolyl oxazole unit in diazonamide, consisting of 3-methoxycarbonylation of the 3-bromoindole **190** followed by the addition of lithiated methyl isocyanide, which gives a 2:1 mixture of the isonitrile **190.1** and the desired oxazole **190.2** (Scheme 12.37) [138]. Conversion of the former into the latter is readily achieved by exposure to acid. These studies also provide methods for the 4-chlorination of the oxazole and various metallation and coupling methods for further homologation. In common with the Nicolaou approaches, the second oxazole is constructed from a keto amide using POCl$_3$-pyridine, but in only moderate yield. In contrast, during a formal total synthesis, recourse was made to the Burgess/BrCCl$_3$-DBU methods for oxazole construction [139].

Construction of the necessary keto amides for oxazole formation using a modified Robinson–Gabriel cyclization can also be carried out by carbenoid insertion into an N–H bond of an amide function (Scheme 12.26). This method has been exploited by the Moody group in a neat if not especially efficient synthesis of the mono-oxazole **191** [140].

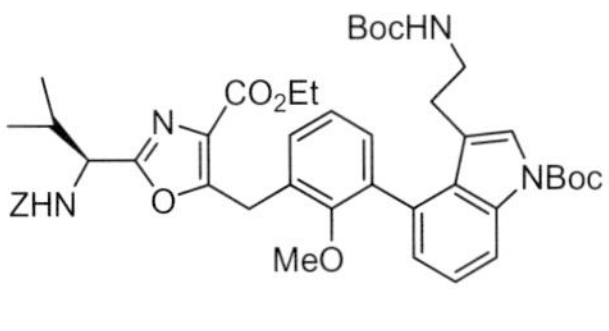

Scheme 12.38 *Reagents and conditions:* i. DDQ, THF, reflux (17%).

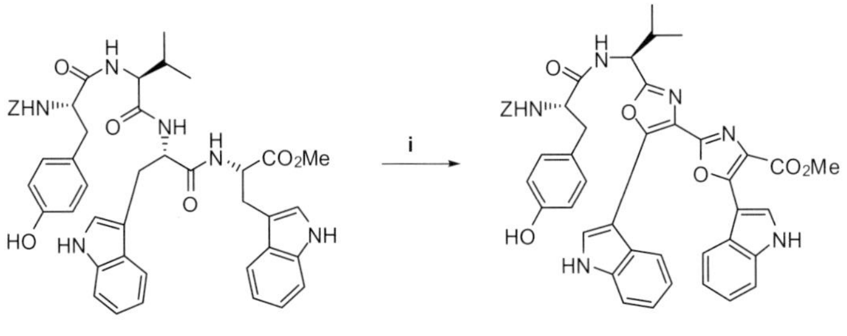

In a remarkable contribution, the idea of setting up a keto amide precursor to an oxazole by benzylic oxidation adjacent to an indole residue (Scheme 12.35) has been taken a step further with the double oxidation and *in situ* cyclization of the tetrapeptide **192** to give a 17 % yield of the bis-oxazole **193**, a productivity of course ameliorated by the fact that, effectively, four steps have been carried out in one (Scheme 12.38) [141].

Serial No.	Trivial name	Structure	Source	Isolation	Biological activity	Synthesis
194	Bengazole A	See below	*Jaspis* sp. (sponge)	[142]	Powerful antifungal; anthelmintic activity.	[143–147]

12.5.4
Bengazole A

This far, some 22 members of the bengazole family have been identified [142], but synthetic efforts have been focused largely on bengazole A **194**. Despite not being a conjugated bis-oxazole, it is included in this section because each oxazole ring can strongly influence the chemistry of the second, especially during the construction of intermediates.

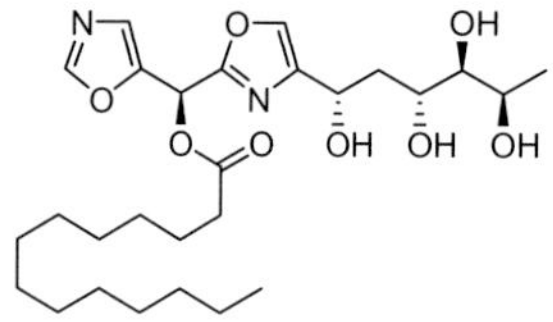

194

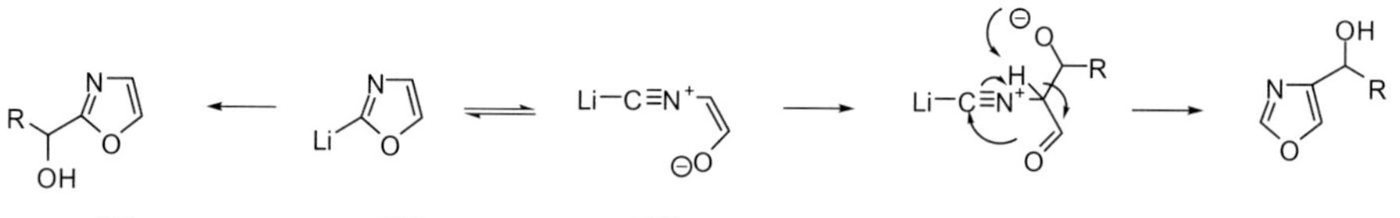

196 195 197 198 199

Scheme 12.39

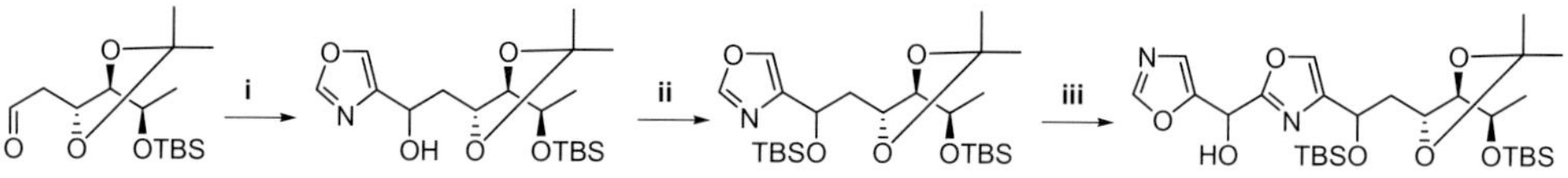

200 201 202 203

Scheme 12.40 *Reagents and conditions:* i. 10 equiv. 2-lithio-oxazole, 20:1 THF-hexanes, −78 °C (57%); ii. TBSOTf, 2,6-lutidine; iii. BH_3. THF, *t*BuLi THF, −78 °C, then oxazole-2-carboxaldehyde (40%).

The first total synthesis of one of these unprecedented structures featured some clever metallation chemistry, summarized in Scheme 12.39. The 2-lithio oxazole species **195** can undergo a predictable condensation with an aldehyde, RCHO, to give the expected 2-substituted oxazoles **196**. However, this required special conditions (see below); if these are not applied, then ring opening occurs to give the isocyanide enolate **197**, which then condenses with the aldehyde to generate the alkoxide **198**. This then undergoes proton transfer and reclosure to give the 4-substituted oxazole **199**.

Both condition sets have been used in the first synthesis of bengazole A (Scheme 12.40) [143]. Initial condensation between the aldehyde **200**, derived in nine known steps from D-galactose, and a large excess of 2-lithio-oxazole delivered a workable yield of the 4-substituted oxazole **201** *via* the ring opening-reclosure mechanism (Scheme 12.39). Subsequent alcohol protection as the silyl ether **202** was followed by a "simple" condensation at the 2-position with oxazole-2-carboxaldehyde to give the final product **203**, the trick here being to stabilize the metallated oxazole with respect to ring opening by complexation with borane.

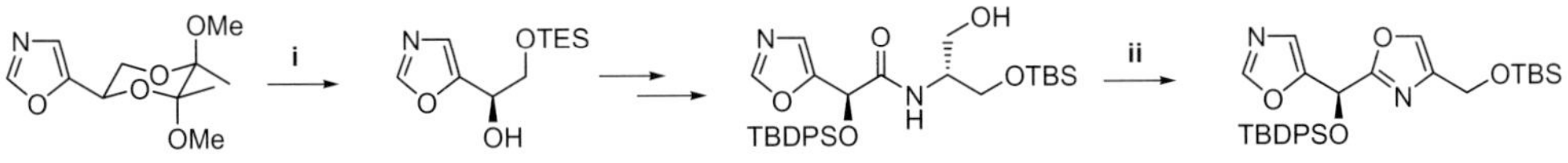

Scheme 12.41 Reagents and conditions: i. aq. TFA, 20 °C then TESCl, ^{i}Pr$_2$NEt, CH$_2$Cl$_2$, −78 °C (94%); ii. Dess–Martin periodinane, NaHCO$_3$, CH$_2$Cl$_2$, 0 °C then (BrCCl$_2$)$_2$, Ph$_3$P, CH$_2$Cl$_2$, 2,6-di-t-butylpyridine, 0 °C then Et$_3$N, MeCN, 0–20 °C (89%).

Sadly, the synthesis was not notable for its stereochemical features: in the first step the wrong isomer was the major product (1:7) while there was no control in the relatively inefficient second step.

A later synthesis started with the butane-2,3-diacetal of D-glyceraldehyde, the aldehyde group of which was reacted under Schollkopf conditions to provide the oxazole **204**, hydrolysis and protection of which led to the alcohol **205**, thereby establishing the initial and very sensitive stereogenic center (Scheme 12.41) [144]. Subsequent "routine" manipulations required great care to preserve this center unmolested and to arrive at the serine amide **206**, which was then successfully converted into the bis-oxazole **207** using a version of the Wipf procedure [25]. Notably, highly basic DBU was replaced in the last step by milder triethylamine to prevent epimerization. The other stereogenic center adjacent to the oxazole was introduced using [1, 3]-dipolar cycloaddition methodology followed by selective reduction.

A later publication details these tribulations, along with many other seemingly sensible approaches, many of which failed due to the extreme sensitivity of the doubly benzylic alcohol group, which resulted in either decomposition or epimerization [145].

A very recent synthesis of bengazole A proceeds *via* the same alcohol **205**, but which was generated using an AD-mix bis-hydroxylation of the corresponding 5-vinyloxazole [146]. The remainder of the bis-oxazole synthesis was essentially the same as that shown in Scheme 12.41, except that the Dess–Martin oxidation step was achieved using a TEMPO-based method. The side chain hydroxyls were protected as their MOM derivatives throughout; removal of these using TiCl$_4$ proceeded without incident. Various analogs of this highly active antifungal compound have also been synthesized recently [147].

12.5.5
Siphonazole

Serial No.	Trivial name	Structure	Source	Isolation	Biological activity	Synthesis
207	Siphonazole	See below	*Herpetosiphon* sp.	[148]	–	[149, 150]

The first synthesis of siphonazole by the Moody group [149] featured the assembly of two oxazoles **208** and **209** using their favored strategy of carbenoid insertion into an amide N–H followed by Robinson–Gabriel cyclodehydration of the resulting keto amide (cf. Scheme 12.26). The key linking step was a Reformatsky condensation between these two components; many other related reactions failed.

207; R = H, Me **208** **209**

A subsequent synthesis of siphonazole was developed following an extensive investigation of methyloxazole metallation chemistry and featured reactions of the sulfone-stabilized carbanion **210**, which will doubtless find many applications in the future [150].

210

12.6
Tris- and Poly-Oxazoles

12.6.1
Ulapualide A

Serial No.	Trivial name	Structure	Source	Isolation	Biological activity	Synthesis
211	Ulapualide A	See below		[151]	Antifungal, ichthyotoxic, active against L1210 leukemia cells	[152–158]

That oxazoles were virtually unknown in nature some 20 years ago makes the identification of a series of tris-oxazoles all the more extraordinary. The best known of these marine metabolites is ulapualide A **211**, which very much typifies this group which, of course, pose a diversity of synthetic challenges.

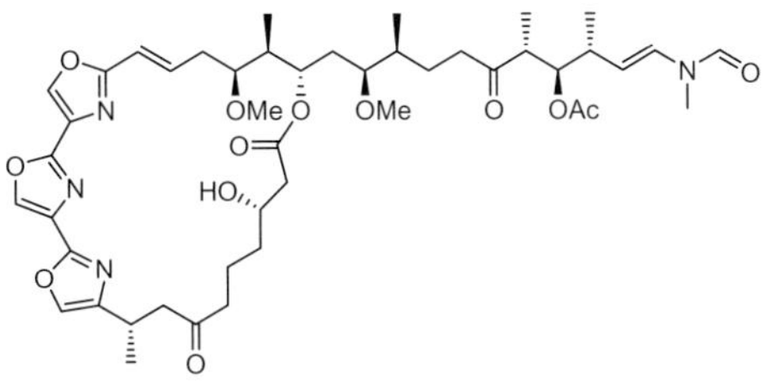

211

Early synthetic studies focused on the oxazole unit were not especially appealing, especially as the stepwise assembly of the oxazoles one at a time rendered these vulnerable to repeated inefficient step which, in early studies, was often the oxidation step [152]. The introduction of many of the newer methods exemplified in many places above certainly saved the day!

Panek and Beresis [153] also employed an iterative procedure to build up the tris-oxazole, which benefitted considerably from the availability of newer methodology as well as optimization of older procedures; regrettably though, this is still a rather lengthy and labor-intensive endeavor (Scheme 12.42). Starting with amide **212**, a modified and optimized Hantzsch condensation using trifluoroacetic anhydride to complete the aromatization provided a very reasonable yield of the oxazole **213**. Selective alkene cleavage (ozone reacts with oxazoles) gave the aldehyde **214**

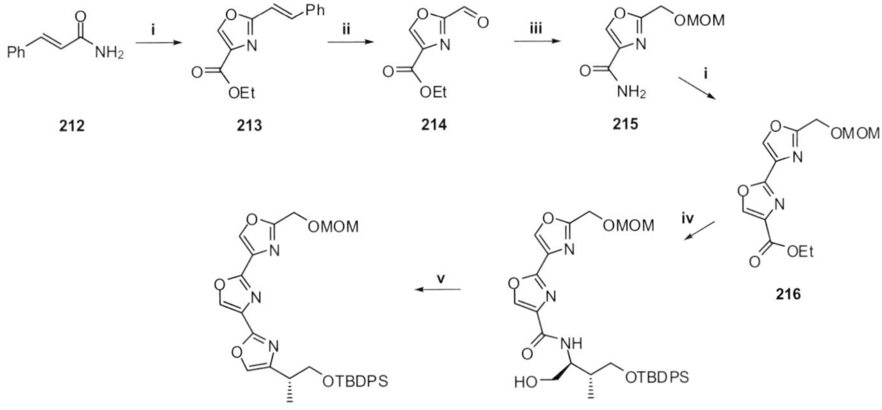

Scheme 12.42 Reagents and conditions: i. BrCH$_2$C(O)CO2Et (85%) then Pb(OAc)$_4$, K$_2$CO$_3$, MeOH, 0 °C (84%); iii. NaBH$_4$, EtOH (87%) then (MeO)$_2$CH$_2$, P$_2$O$_5$, CHCl$_3$ (75%) then aq. NH$_4$OH (99%); iv. LiOH, aq THF, (98%) then BnOCH$_2$CH(NH$_2$)CH(Me)CH$_2$OTBDPS, DCC, HOBT, DMF (68%) then H$_2$, Pd/C (87%); v. Dess.Martin oxidation then (ClBr$_2$C)$_2$, Ph$_3$P, 2,6-di-*t*-bytylpyridine then DBU, MeCN (92%).

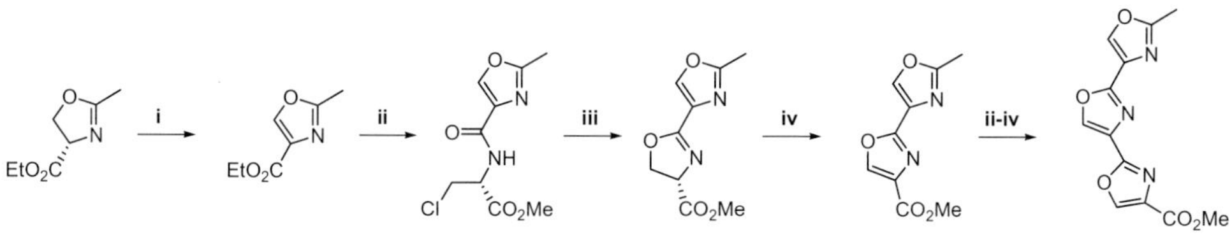

Scheme 12.43 *Reagents and conditions:* i. NiO$_2$ [155]; ii. three steps: saponify, couple then OH to Cl (SOCl$_2$); iii. AgOTf; iv. LDA, −78 °C, PhSeSePh, then H$_2$O$_2$.

and thence the protected amide **215** following reduction, MOM ether formation and amidation. A repetition of the optimized Hantzsch procedure then gave a 60% yield of the bis-oxazole **216**. Hydrolysis and coupling then led to the peptide **217** and finally to the tris-oxazole **218** using the now popular Wipf procedure [25].

The magnificent achievement of an ulapualide A total synthesis was realised in the first instance by Chattopadhyay and Pattenden. Once again, the approach used to access the tris-oxazole unit was distinctly iterative but employed some rather different tactics to the Panek approach (Scheme 12.43) [154]. The initial oxazoline **219** was obtained by condensation of serine ethyl ester with ethyl acetimidate in the presence of triethylamine. Subsequent oxidation using the Meyers' nickel peroxide method [155] proved sufficiently efficient to deliver the oxazole ester **220**, which was converted by three standard steps into the chloride **221** which was then cyclized by exposure to silver(I) triflate, a somewhat unusual method, to give the oxazolyl-oxazoline **222**. Subsequent attempts to again use nickel peroxide was this time very unproductive as was photo-induced bromination; recourse to another standard method (LDA, selenenation, oxidation and [2.3]-sigmatropic elimination) eventually delivered the desired bis-oxazole **223**. Repetition of this latter sequence then led to the key tris-oxazole **224**. Homologation at the methyl group was achieved by bromination (NBS, AIBN), phosphonium salt formation and Wittig condensation.

In a second generation synthesis of ulapualide A, the tris-oxazole unit was prepared very late on by formation of the central ring after macrolactamization, thereby combining at least two key steps [156]. These and alternative approaches are fully detailed in these two publications; probably the best approach consisted of preparing the ester **225** and selectively deprotecting this to give the amino-alcohol **226** (Scheme 12.44). This was selectively cyclized through the more nucleophilic nitrogen and the resulting macrolactam **227** dehydrated by mesylation in the presence of base to give the enamide **228**. In yet another departure from the normal, guided by previous work by Shin *et al.* [157] and their own successful model studies, this was successfully cyclized by bromo-methoxylation of the new alkene and base-induced cyclization, to give the methoxy oxazoline **229**, acidification of which completed the synthesis of the aromatic system **230**.

The second paper of these two [156b] contains a particularly well-drawn summary of all of the currently known naturally occurring tris-oxazoles. The tris-oxazole unit in one of these, mycalolide, structurally very similar to ulapualide A, has been synthesized by Panek's group, in much the same manner to that outlined in

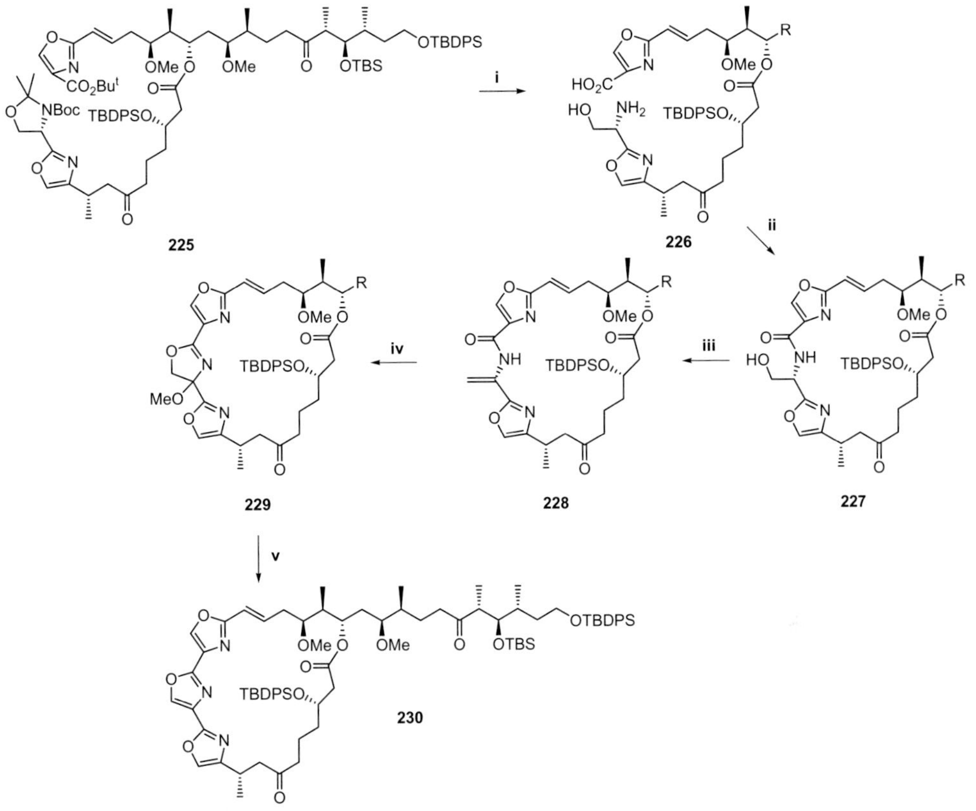

Scheme 12.44 *Reagents and conditions:* i. TMSOTf, Et₃N, CH₂Cl₂, 0 °C; ii. HATU, Et₃N, CH₂Cl₂, 0 °C (67%); iii. MsCl, *i*Pr₂NEt, CH₂Cl₂, DBU, 0 °C (75%); iv. NBS, MeOH, CH₂Cl₂, (92%) then Cs₂CO₃, dioxane, 60 °C (92%); v. CSA, benzene, 5A mol. sieves, 80 °C (58%).

Scheme 12.42 [158]. In a later approach, a Japanese group have used the same tactic as the Pattenden group (Scheme 12.44) in assembling the central oxazole ring. However, in this case, a serine-derived peptide (cf. **227**) was cyclized by sequential exposure to DAST [18] followed by oxidation with nickel peroxide [159]. The two yields were 82% and 51%, respectively, showing that the often maligned NiO₂ method *can* be successful.

12.6.2
(R)-Telomestatin

Serial No.	Trivial name	Structure	Source	Isolation	Biological activity	Synthesis
231	Telomestatin	See below	*Streptomyces anulatus* 3533-SV4	[160]	Potent and specific telomerase inhibitor IC$_{50}$ = 5 nM)	[161–164]

The quite extraordinary *septa*-oxazole-thiazoline unit found in telomestatin **231** [160] is, presumably in reality, nothing more than a fully cyclodehydrated octapeptide. Its exceptional level of bioactivity renders this an important find, especially as it can be regarded as filling a currently blank area of molecular space. The one fully detailed total synthesis by the Takahashi group is fully described by Mohd Nor, Xu and Ye in Chapter 16 of this volume (Figure 16.7 and Scheme 16.10), which will not be repeated here. One only notes that, once again, the last oxazole ring to be synthesized in a virtually complete target proved difficult until, in common with the Pattenden second generation approach to ulapualide A (Scheme 12.44), recourse was made to the Shin method [157]; see also [19, 20].

Various analogs of telomestatin **231** have very recently been reported, starting from the mono-oxazole **232**, by a most convoluted nitrile hydrolysis protocol previously developed by the Nicolaou group (LiOH caused extensive decomposition) and subsequent couplings and Burgess dehydrations *etc*. [162]. The key tris-oxazole **233** undergoes smooth coupling with, for example, phenylboronic acid but under quite forcing conditions (Pd(OAc)$_2$, KF, 110 °C). Other model studies have resulted in a synthesis of telomestatin analog **235**, from the tris-oxazole **234**, related to that fund in ulapualide A, essentially by an overall dimerization. [163]. A related but extended *penta*-oxazole has also been synthesized along quite similar lines [164].

12.6.3
IB-01211

Serial No.	Trivial name	Structure	Source	Isolation	Biological activity	Synthesis
236	IB-01211; Mechercharmycin A	See below	Strain ES7-008 close to a *Thermoactinomyces* sp.	[165]	Cytotoxic to lung and leukemia cancer cells	[166]

IB-01211, also called mechercharmycin A, **236** has been the subject of a single total synthesis [166]; again, this is described in detail in Chapter 16 (Schemes 16.8 and 16.9).

236

The same group have reported a number of analog syntheses [167]; these studies contain a wealth of information which will be of use to anyone wishing to carry out synthetic work in this area. Very recently, Chattopadhyay and [168] came very close to completing a synthesis, based on previous studies [169], which foundered due to the insolubility of some advanced intermediates, despite precedent for formation of the final oxazole ring.

12.6.4
YM-216391

Serial No.	Trivial name	Structure	Source	Isolation	Biological activity	Synthesis
237	YM-216391	See below	*Streptomyces nobilis*	[170]	Cytotoxic	[171]

The related bacterial metabolite YM-216391 **237** [170] has been the subject of one total synthesis, which has resulted in a structural revision at the valine stereogenic center [171]. The synthesis featured the preparation of a tris-oxazole **239** using the Wipf–Williams method applied to precursor **238** itself derived from the corresponding oxazole 4-carboxylic acid and the oxazole-2-amino-alcohol. Further couplings then provided the extended peptide **240**, which was converted into the target **240** by sequential macrolactamization, thiazoline formation (DAST, −78 °C) and finally oxidation (MnO$_2$).

237   **238**   **239**   **240**

In summary, it is clear that the Wipf [25], the combined Burgess dehydration-oxidation method [37, 39] and the Wipf–Williams [18] protocols have made major positive impacts in this synthetic area. However, these are not always successful and recourse to alternative, less popular recipes has often proven beneficial.

References

1 Katritzky, A.R., Ramsden, C.A., Scriven, E.F.V. and Taylor, R.J.K. (eds) (2008) *Comprehensive Heterocyclic Chemistry III, (1995–2007)*, Elsevier, St. Louis, USA and Oxford, UK, and previous volumes I and II.

2 (a) Yeh, V.S.C. (2004) Recent advances in the total synthesis of oxazole-containing natural products. *Tetrahedron*, **60**, 11995–12042; (b) Jin, Z. (2005) Muscarine, imidazole, oxazole and thiazole alkaloids. *Nat. Prod. Rpts.*, **22**, 196–229; (c) Jin, Z. (2006) Imidazole, oxazole and thiazole alkaloids. *Nat. Prod. Rpts.*, **23**, 464–496; (d) Riego, E. Hernandez, D. Albericio, F. and Alvarez, M. (2005) Directly linked polyazoles: important moieties in natural products. *Synthesis*, 1907–1922; (e) Jin, Z. (2009) Muscarine, imidazole, oxazole and thiazole alkaloids. *Nat. Prod. Rpts.*, **26**, 382–445. For an earlier summary, see Wipf, P. (1995) Synthetic studies of biologically active marine cyclopeptides. *Chem. Rev.*, **95**, 2115–2134.

3 Bhate, D.S., Hulyalkar, R.K. and Menon, S.K. (1960) Isolation of iso-butyropyrrothine along with thiolutin and aureothricin from a *Streptomyces* sp. *Experientia*, **16**, 504–505.

4 Somei, M., Sato, H., Komura, N. and Kaneko, C. (1985) The chemistry of indoles .25. Synthetic study for 1-methoxyindoles and 1-methoxy-2-oxindoles. *Heterocycles*, **23**, 1101–1106; for a much earlier synthesis also starting from 3-acetylindole, see Joshi, B.S., Taylor, W.I., Bhate, D.S. and Karmarker,

S.S. (1963) The structure and synthesis of pimprinine. *Tetrahedron*, **19**, 1437–1439; for the synthesis of 2-amino derivatives using an aza-Wittig approach, see Molina, P., Fresneda, P.M. and Almendros, P. (1993) Iminophosphorane-mediated synthesis of 3-(oxazol-5-yl)indoles–application to the preparation of pimprinine type alkaloids. *Synthesis*, 54–56.

5 Houwing, H.A., Wildeman, J. and van Leusen, A.M. (1981) Synthesis of 2,5-disubstituted oxazoles from aldehydes and N-(tosylmethyl)imino synthons–application to the synthesis of pimprinine analogs. *J. Het. Chem.*, **16**, 1133–1139.

6 Doyle, K.J. and Moody, C.J. (1994) Synthesis of oxazolylindolyl alkaloids via rhodium carbenoids. *Synthesis*, 1021–1022.

7 Connell, R.D., Tebbe, M., Helquist, P. and Akermark, B. (1991) Direct preparation of 4-carboethoxy-1,3-oxazoles. *Tetrahedron Lett.*, **32**, 17–20; Connell, R.D., Tebbe, M., Gangloff, A.R. and Helquist, P. (1993) Rhodium-catalyzed heterocycloaddition route to 1,3-oxazoles as building blocks in natural products synthesis. *Tetrahedron*, **49**, 5445–5459.

8 Wang, Y., Janjic, J. and Kozmin, S.A. (2002) Synthesis of leucascandrolide A via a spontaneous macrolactolization. *J. Am. Chem. Soc.*, **124**, 13670–13671; Kozmin, S.A. (2001) Efficient stereochemical relay en route to leucascandrolide A. *Org Lett.*, **3**, 755–758; Wang, Y., Janjic, J. and Kozmin, S.A. (2005) Synthesis of leucascandrolide A. *Pure Appl. Chem.*, **77**, 1161–1169; Fuwa, H., Naito, S., Goto, T. and Sasaki, M. (2008) Total synthesis of (+)-neopeltolide. *Angew. Chem. Int. Edn.*, **47**, 4737–4739; Fuwa, H., Saito, A. and Sasaki, M. (2010) A concise total synthesis of (+)-neopeltolide. *Angew. Chem. Int. Edn.*, **49**, 3041–3044.

9 Dominguez, X.A., Fuente, G., Gonzalez, A.G., Reina, M. and Timon, I. (1988) Two new oxazoles from *Amyris texana* P. Wilson. *Heterocycles*, **27**, 35–38.

10 Giddens, A.C., Boshoff, H.I.M., Franzblau, S.G., Barry, C.E., III and Copp, B.R. (2005) Amtimycobacterial natural products: synthesis and preliminary biological evaluation of the oxazole-containing alkaloid texaline. *Tetrahedron Lett.*, **46**, 7355–7357; for a previous synthesis of balsoxin 7, see Burke, B., Parkins, H. and Talbot, A.M. (1979) Oxazole and its precursor in *Amyris balsamifera*. *Heterocycles*, **12**, 349–351.

11 Hodgetts, K.J. and Kershaw, M.T. (2002) Ethyl 2-Chlorooxazole-4-carboxylate: a versatile intermediate for the synthesis of substituted oxazoles. *Org. Lett.*, **4**, 2905–2907.

12 Besselièvre, F., Mahuteau-Betzer, F., Grierson, D.S. and Piguel, S. (2008) Ligandless microwave-assisted Pd/Cu-catalyzed direct arylation of oxazoles. *J. Org. Chem.*, **73**, 3278–3280.

13 Verrier, C., Martin, T., Hoarau, C. and Marsais, F. (2008) Palladium-catalyzed direct (hetero)arylation of ethyl oxazole-4-carboxylate: an efficient access to (hetero)aryloxazoles. *J. Org. Chem.*, **73**, 7383–7386.

14 Ohnmacht, S.A., Mamone, P., Culshaw, A.J. and Greaney, M.F. (2008) Direct arylations on water: synthesis of 2,5-disubstituted oxazoles balsoxin and texaline. *Chem. Commun.*, 1241–1243.

15 Kelly, T.R. and Lang, F. (1996) Total synthesis of dimethyl sulfomycinamate. *J. Org. Chem.*, **61**, 4623–4633.

16 Bagley, M.C., Chapaneri, K., Dale, J.W., Xiong, X. and Bower, J. (2005) One-pot multistep Bohlmann–Rahtz heteroannulation reactions: synthesis of dimethyl sulfomycinamate. *J. Org. Chem.*, **70**, 1389–1399.

17 Bagley, M.C. and Glover, C. (2006) Synthesis of methyl sulfomycinate, sulfomycinic amide and sulfomycinine, degradation products of the sulfomycin thiopeptide antibiotics. *Tetrahedron*, **62**, 66–72.

18 Phillips, A.J., Uto, Y., Wipf, P., Reno, M.J. and Williams, D.R. (2000) Synthesis of functionalized oxazolines and oxazoles with DAST and deoxofluor. *Org. Lett.*, **2**, 1165–1168.

19. Kayano, T., Yonezawa, Y. and Shin, C. (2004) Convenient synthesis of the main tridehydropentapeptide skeleton for a macrocyclic antibiotic, sulfomycin I. Chem. Lett., 33, 72–73.
20. Coleman, R.S. and Carpenter, A.J. (1993) Stereoselective bromination of dehydroamino acids with controllable retention or inversion of olefin configuration. J. Org. Chem., 58, 4452–4461.
21. Jansen, R., Kunze, B., Reichenbach, H. and Höfle, G. (2002) The ajudazols A and B, novel isochromanones from Chondromyces crocatus (Myxobacteria): isolation and structure elucidation. Eur. J. Org. Chem., 917–921.
22. Krebs, O. and Taylor, R.J.K. (2005) Synthesis of the eastern portion of ajudazol a based on Stille coupling and double acetylene carbocupration. Org. Lett., 7, 1063–1066.
23. Ganame, D., Quach, T., Poole, C. and Rizzacasa, M.A. (2007) Synthesis of the C9–C29 fragments of ajudazols A and B. Tetrahedron Lett., 48, 5841–5843.
24. Hobson, S.J., Parkin, A. and Marquez, R. (2008) Oxidative rearrangements of isobenzofurans: studies toward the synthesis of the ajudazols. Org. Lett., 10, 2813–2816.
25. Wipf, P. and Miller, C.P. (1993) A new synthesis of highly functionalized oxazoles. J. Org. Chem., 58, 3604–3606; See also Wipf, P. and Lim, S. (1995) Total synthesis of the enantiomer of the antiviral marine natural product hennoxazole A. J. Am. Chem. Soc., 117, 558–559. Wipf, P. and Lim, S. (1996) Total synthesis and structural studies of the antiviral marine natural product hennoxazole A. Chimica, 50, 157–167.
26. Iwasaki, S., Kobayashi, H., Furukawa, J., Namikoshi, M., Okuda, S., Sato, Z., Matsuda, I. and Noda, T. (1984) Studies on macrocyclic lactone antibiotics .7. Structure of a phytotoxin rhizoxin produced by Rhizopus chinensis. J. Antibiot., 37, 354–362.
27. For a review of the chemistry and biology of the rhizoxins, see Hong, J. and White, J.D. (2004) The chemistry and biology of rhizoxins, novel antitumor macrolides from Rhizopus chinensis. Tetrahedron, 60, 5653–5681.
28. (a) Kende, A.S., Blasse, B.E. and Henry, J.R. (1995) Enantioselective total synthesis of didesepoxyrhizoxin. Tetrahedron Lett., 36, 4741–4744; (b) White, J.D., Blackmore, P.R., Green, N.J., Hanser, E.B., Holoboski, M.A., Keown, K.E., Nyland Kolz, C.S. and Phillips, B.W. (2002) Total synthesis of rhizoxin D, a potent antimitotic agent from the fungus Rhizopus chinensis. J. Org. Chem., 67, 7750–7760.
29. Lafontaine, J.A., Provencal, D.P., Gardelli, C. and Leahy, J.W. (2003) Enantioselective total synthesis of the antitumor macrolide rhizoxin D. J. Org. Chem., 68, 4215–4234.
30. Jiang, Y., Hong, J. and Burke, S.D. (2004) Stereoselective total synthesis of antitumor macrolide (+)-rhizoxin D. Org. Lett., 6, 1445–1448; N'Zoutani, M.-A., Lensen, N., Pancrazi, A., Ardisson, J. (2005) Total synthesis of rhizoxin D. Synlett 491–495.
31. Mitchell, I.S., Pattenden, G. and Stonehouse, J. (2005) A total synthesis of the antitumour macrolide rhizoxin D. Org. Biomol. Chem., 3, 4412–4431.
32. Keck, G.E., Wager, C.A., Wager, T., Savin, K.A., Covel, J.A., McLaws, M.D., Krishnamurthy, D. and Cee, V.J. (2001) Asymmetric total synthesis of rhizoxin D. Angew. Chem. Int. Edn., 40, 231–234.
33. Kato, Y., Fusetani, N., Matsunaga, S., Hashimoto, K., Fujita, S. and Furuya, T. (1986) Bioactive marine metabolites. Part 16. Calyculin A. A novel antitumor metabolite from the marine sponge Discodermia calyx. J. Am. Chem. Soc., 108, 2780–2781; Matsunaga, S. and Fusetani, N. (1991) Absolute Stereochemistry of the calyculins, potent inhibitors of protein phosphatases 1 and 2A. Tetrahedron Lett., 32, 5605–5606; Kato, Y., Fusetani, N., Matsunaga, S., Hashimoto, K. and Koseki, K. (1988) Bioactive marine metabolites. 24. Isolation and structure elucidation of calyculins B, C, and D, novel antitumor metabolites, from the marine sponge Discodermia calyx. J. Org.

Chem., **53**, 3930–3930; Matsunaga, S., Wakimoto, T., Fusetani, N. and Suganuma, M. (1997) Isolation of dephosphonocalyculin a from the marine sponge, *Discodermia calyx*. Tetrahedron Lett., **38**, 3763–3764; Dumdei, E.J., Blunt, J.W., Munro, M.H.G. and Pannell, L.K. (1997) Isolation of calyculins, calyculinamides, and swinholide h from the new zealand deep-water marine sponge *Lamellomorpha strongylata*. J. Org. Chem., **62**, 2636–2639; Matsunaga, S., Wakimoto, T. and Fusetani, N. (1997) Isolation of four new calyculins from the marine sponge *Discodermia calyx*. J. Org. Chem., **62**, 2640–2642; Matsunaga, S., Wakimoto, T. and Fusetani, N. (1997) Isolation of four new calyculins from the marine sponge *Discodermia calyx*. J. Org. Chem., **62**, 9388. (erratum).

34 Evans, D.A., Gage, J.R. and Leighton, J.L. (1992) Total synthesis of (+)-calyculin A. J. Am. Chem. Soc., **114**, 9434–9453.

35 Tanimoto, N., Gerritz, S.W., Sawabe, A., Noda, T., Filla, S.A. and Masamune, S. (1994) The synthesis of naturally occurring (−)-calyculin A. Angew. Chem. Int. Edn., **33**, 673–675.

36 Smith, A.B., III, Friestad, G.K., Barbosa, J., Bertounesque, E., Hull, K.G., Iwashima, M., Qui, Y., Salvatore, B.A., Spoors, P.G. and Duan, J.J.-W. (1999) Total synthesis of (+)-calyculin A and (−)-calyculin B: asymmetric synthesis of the C(9–25) spiroketal dipropionate subunit. J. Am. Chem. Soc., **121**, 10468–10477; Smith, A.B., III, Friestad, G.K., Barbosa, J., Bertounesque, E., Duan, J.J.-W., Hull, K.G., Iwashima, M., Qui, Y., Spoors, P.G. and Salvatore, B.A. (1999) Total synthesis of (+)-calyculin A and (−)-calyculin B: cyanotetraene construction, asymmetric synthesis of the C(26–37) oxazole, fragment assembly, and final elaboration. J. Am. Chem. Soc., **121**, 10478–10486.

37 Wipf, P. and Miller, C.P. (1992) A short, stereospecific synthesis of dihydrooxazoles from serine and threonine derivatives. Tetrahedron Lett., **33**, 907–910; Wipf, P. and Miller, C.P. (1992) Total synthesis of westiellamide. J. Am. Chem. Soc., **114**, 10975–10977;Wipf, P. and Miller, C.P. (1993) Stereospecific synthesis of peptide analogs with allo-threonine and D-allo-threonine residues. J. Org. Chem., **58**, 1575–1578.

38 Burgess, E.M., Penton, H.R., Jr., Taylor, E.A. and Williams, W.M. (1988) Conversion of primary alcohols to urethanes *via* the inner salt of methyl (carboxysulfamoyl)triethylammonium hydroxide: methyl *n*-hexylcarbamate. Org. Synth., **6**, 788–791.

39 Barrish, J.C., Singh, J., Spergel, S.H., Han, W.-C., Kissick, T.P., Kronenthal, D.R. and Mueller, R.H. (1993) Cupric bromide mediated oxidation of 4-carboxyoxazolines to the corresponding oxazoles. J. Org. Chem., **58**, 4494–4496.

40 Ogawa, A.K. and Armstrong, R.W. (1998) Total synthesis of calyculin C. J. Am. Chem. Soc., **120**, 12435–12442.

41 Pihko, P.M. and Koskinen, A.P.M. (1998) Synthesis of the C26–C32 oxazole fragment of calyculin C: a test case for oxazole syntheses. J. Org. Chem., **63**, 92–98; See also Habrant, D., Stewart, A.J.W. and Koskinen, A.M.P. (2009) Towards the total synthesis of Calyculin C: preparation of the C13–C25 spirocyclic core. Tetrahedron, **65**, 7927–7934. for synthesis of other fragments.

42 D'Ambrosio, M., Guerriero, A., Debitus, C. and Pietra, F. (1996) Leucascandrolide A, a new type of macrolide: the first powerfully bioactive metabolite of calcareous sponges (*Leucascandra caveolata*, a new genus from the Coral Sea). Helv. Chim. Acta, **79**, 51–60.

43 Wipf, P. and Graham, T.H. (2001) Synthesis of the C1-C11 Segment of leucascandrolide A. J. Org. Chem., **66**, 3242–3245.

44 Hornberger, K.R., Hamblett, C.L. and Leighton, J.L. (2000) Total synthesis of leucascandrolide A. J. Am. Chem. Soc., **122**, 12894–12895; Custar, D.W., Zabawa, T.P. and Scheidt, K.A. (2008) Total synthesis and structural revision of the marine macrolide neopeltolid. J. Am. Chem. Soc., **130**, 804–805; Custar, D.W.,

Zabawa, T.P., Hines, J., Crews, C.M. and Scheidt, K.A. (2009) Total synthesis and structure-activity investigation of the marine natural product neopeltolide. *J. Am. Chem. Soc.*, **131**, 12406–12414.

45 (a) Dakin, L.A., Langille, N.F. and Panek, J.S. (2002) Synthesis of the C1–C11 oxazole-containing side chain of leucascandrolide A. application of a sonogashira cross-coupling. *J. Org. Chem.*, **67**, 6812–6815; (b) Paterson, I. and Tudge, M. (2003) Stereocontrolled total synthesis of (+)-leucascandrolide A. *Angew. Chem. Int. Edn.*, **42**, 343–347; (c) Paterson, I. and Tudge, M. (2003) A fully stereocontrolled total synthesis of (+)-leucascandrolide A. *Tetrahedron*, **59**, 6833–6849; (d) Su, Q. and Panek, J.S. (2005) Total synthesis of (+)-leucascandrolide A. *Angew. Chem. Int. Edn.*, **44**, 1223–1225; (e) Su, Q., Dakin, L.A. and Panek, J.S. (2007) Application to the total synthesis of leucascandrolide A. *J. Org. Chem.*, **72**, 2–24; (f) Woo, S.K., Kwon, M.S. and Lee, E. (2008) Total synthesis of (+)-neopeltolide by a Prins macrocyclization. *Angew. Chem. Int. Edn.*, **47**, 3242–3244; (g) Vintonyak, V.V., Kunze, B., Sasse, F. and Maier, M.E. (2008) Total synthesis and biological activity of neopeltolide and analogues. *Chem. Eur. J.*, **14**, 11132–11140.

46 Guinchard, X. and Roulland, E. (2009) Total synthesis of the antiproliferative macrolide (+)-neopeltolide. *Org. Lett.*, **11**, 4700–4703.

47 Silva, S., Sylla, B., Suzenet, F., Tatibouët, A., Rauter, A.P. and Rollin, P. (2008) Oxazolinethiones and oxazolidinethiones for the first copper-catalyzed desulfurative cross-coupling reaction and first sonogashira applications. *Org. Lett.*, **10**, 853–856.

48 Jung, H.H., Seiders, J.R., II and Floreancig, P.E. (2007) Oxidative cleavage in the construction of complex molecules: synthesis of the leucascandrolide a macrolactone. *Angew. Chem. Int. Edn.*, **46**, 8464–8467; Youngsaye, W., Lowe, J.T., Pohlki, F., Ralifo, P. and Panek, J.S. (2007) Total synthesis and stereochemical reassignment of (+)-neopeltolide. *Angew. Chem. Int. Edn.*, **46**, 9211–9214; See also Van Orden, L.J., Patterson, B.D. and Rychnovsky, S.D. (2007) Total synthesis of leucascandrolide A: a new application of the Mukaiyama aldol–Prins reaction. *J. Org. Chem.*, **72**, 5784–5793; Paterson, I. and Miller, N.A. (2008) Total synthesis of the marine macrolide (+)-neopeltolide. *Chem. Commun.*, 4708–4710; Fuwa, H., Saito, A., Naito, S., Konoki, K., Yotsu-Yamashita, M. and Sasaki, M. (2009) Total synthesis and biological evaluation of (+)-neopeltolide and its analogues. *Chem. Eur. J.*, **15**, 12807–12818.

49 Janssen, D., Albert, D., Jansen, R., Muller, R. and Kalesse, M. (2007) Chivosazole A – Elucidation of the absolute and relative configuration. *Angew. Chem. Int. Edn.*, **46**, 4898–4901.

50 Janssen, D. and Kalesse, M. (2007) Synthesis of the C15-C35 segment of chivosazole A. *Synlett*, 2667–2670.

51 Brazhnikova, M.G., Kudinova, M.K., Potapova, N.P., Filippova, T.M., Borowski, E., Zelinski, Y. and Golik, J. (1976) Structure of antibiotic madumycin. *Bioorgan. Khim.*, **2**, 149–157; Chamberlin, J.W. and Chen, S. (1977) A2315, New antibiotics produced by *Actinoplanes philippinensis* .2. structure of A2315A. *J. Antibiot.*, **30**, 197–201.

52 Meyers, A.I., Lawson, J.P., Walter, D.G. and Linderman, R.J. (1986) Synthetic studies on the streptogramin antibiotics. Enantioselective synthesis of the oxazole dienyl amine moiety. *J. Org. Chem.*, **51**, 5111–5123.

53 Tavares, F., Lawson, J.P. and Meyers, A.I. (1996) Total synthesis of streptogramin antibiotics. (−)-Madumycin II. *J. Am. Chem. Soc.*, **118**, 3303–3304.

54 Ghosh, A.K. and Liu, W. (1997) Convergent, enantioselective total synthesis of streptogramin antibiotic (−)-Madumycin II. *J. Org. Chem.*, **62**, 7908–7909.

55 Gangloff, A.R., Akermark, B. and Helquist, P. (1992) Generation and use of a zinc derivative of a functionalised 1,3-oxazole. Solution of the virginiamycin/madumycin oxazole

problem. *J. Org. Chem.*, **57**, 4797–4799.

56 Brennan, C.J. and Campagne, J.-M. (2001) Synthetic studies towards the group A streptogramin antibiotics. Synthesis of the C9-C23 fragment. *Tetrahedron Lett.*, **42**, 5195–5197.

57 Schlessinger, R.H. and Li, Y.-J. (1996) Total synthesis of (−)-virginiamycin M2 using second-generation vinylogous urethane chemistry. *J. Am. Chem. Soc.*, **118**, 3301–3302.

58 White, J.D., Kranemann, C.L. and Kuniyong, P. (2003) 4-Methoxycarbonyl-2-methyl-1,3-oxazole. *Org. Synth.*, **79**, 244–250.

59 Mortensen, M.S., Osbourn, J.M. and O'Doherty, G.A. (2007) De novo formal synthesis of (−)-virginiamycin M_2 via the asymmetric hydration of dienoates. *Org. Lett.*, **9**, 3105–3108.

60 Entwistle, D.A., Jordan, S.I., Montgomery, J. and Pattenden, G. (1996) Total synthesis of oxazole-based virginiamycin antibiotics: 14,15-anhydropristinamycin IIB. *J. Chem. Soc. Perkin Trans. 1*, 1315–1317.

61 Entwistle, D.A., Jordan, S.I., Montgomery, J. and Pattenden, G. (1998) Total synthesis of oxazole-based virginiamycin antibiotics: 14,15-anhydropristinamycin IIB. *Synthesis*, 603–612.

62 Birnbaum, G.I. and Hall, S.R. (1976) Structure of antibiotic griseoviridin. *J. Am. Chem. Soc.*, **98**, 1926–1931; Bycroft, B.W. and King, T.J. (1976) Revised constitution, absolute-configuration and conformation of griseoviridin, a modified cyclic peptide antibiotic. *J. Chem. Soc. Perkin Trans. 1*, 1996–2004.

63 Dvorak, C.A., Schmitz, W.D., Poon, D.J., Pryde, D.C., Lawson, J.P., Amos, R.A. and Meyers, A.I. (2000) The synthesis of streptogramin antibiotics: (−)-griseoviridin and its C-8 epimer. *Angew. Chem. Int. Edn.*, **39**, 1664–1666.

64 Jensen, R., Kunze, B., Reichenbach, H., Jurkiewicz, E., Hunsmann, G. and Höfle, G. (1992) Antibiotics from gliding bacteria .47. thiangazole–a novel inhibitor of HIV-1 from polyangium-spec. *Liebigs Ann. Chem.*, 357–359; Jansen, R., Schomburg, D. and Höfle, G. (1993) Antibiotics from gliding bacteria .52. thiangazole, a new tris(thiazoline) derivative from polyangium-spec–absolute-configuration. *Liebigs Ann. Chem.*, 701–704.

65 Parsons, R.L., Jr. and Heathcock, C.H. (1994) Total synthesis of (−)-thiangazole, a naturally-occurring HIV-1 inhibitor. *J. Org. Chem.*, **59**, 4733–4734; Akaji, K. and Kiso, Y. (1999) Total synthesis of thiangazole. *Tetrahedron*, **55**, 10685–10694.

66 Wipf, P. and Venkatraman, S. (1995) Total synthesis of (−)-thiangazole and structurally related polyazoles. *J. Org. Chem.*, **60**, 7224–7229.

67 Boyce, R.J., Mulqueen, G.C. and Pattenden, G. (1995) Total synthesis of thiangazole, a novel naturally-occurring HIV-1 inhibitor from polyangium sp. *Tetrahedron*, **51**, 7321–7330.

68 Meyers, A.I. and Tavares, F. (1996) Oxidation of oxazolines and thiazolines to oxazoles and thiazoles. Application of the Kharasch-Sosnovsky reaction. *J. Org. Chem.*, **61**, 8207–8215.

69 Fukuyama, T. and Xu, L. (1993) Total synthesis of (−)-tantazole B. *J. Am. Chem. Soc.*, **115**, 8449–8450.

70 Black, R.J.G., Dungan, V.J., Li, R.Y.T., Young, P.G. and Jolliffe, K.A. (2010) Cyclooligomerisation approach to backbone-modified cyclic peptides bearing guanidinium arms. *Synlett.*, 551–554.

71 Hopkins, C.D. and Wipf, P. (2009) Isolation, biology and chemistry of the disorazoles: new anti-cancer macrodiolides. *Nat. Prod. Rpts.*, **26**, 585–601.

72 Wipf, P. and Graham, T.H. (2004) Total synthesis of (−)-disorazole C_1. *J. Am. Chem. Soc.*, **126**, 15346–15347.

73 Haustedt, L.O., Panicker, S.B., Kleinert, M., Hartung, I.V., Eggert, U., Niess, B. and Hoffmann, H.M.R. (2003) Synthetic studies toward the disorazoles: synthesis of a masked northern half of disorazole D-1 and a cyclopropane analog of the masked northern half of disorazole A(1). *Tetrahedron*, **59**, 6967–6977.

12 Oxazole and Its Derivatives

74 Hartung, I.V., Eggert, U., Haustedt, L.O., Niess, B., Schäfer, P.M. and Hoffmann, H.M.R. (2003) Toward the total synthesis of disorazole A(1): asymmetric synthesis of the masked northern half. *Synthesis*, 1844–1850. See also reference 154.

75 Hillier, M.C., Price, A.T. and Meyers, A.I. (2001) Studies on the total synthesis of disorazole C1. An advanced macrocycle intermediate. *J. Org. Chem.*, **66**, 6037–6045; Hillier, M.C., Park, D.H., Price, A.T., Ng, R. and Meyers, A.I. (2000) The synthesis of the monomeric moiety of disorazole C-1. *Tetrahedron Lett.*, **41**, 2821–2824; Hartung, I.V., Niess, B., Haustedt, L.O. and Hoffmann, H.M.R. (2002) Toward the total synthesis of disorazole A(1) and C-1: asymmetric synthesis of a masked southern segment. *Org. Lett.*, **4**, 3239–3242.

76 Niess, B., Hartung, I.V., Haustedt, L.O. and Hoffmann, H.M.R. (2006) Synthesis of 2 tetradehydro-disorazole C-1. *Eur. J. Org. Chem.*, 1132–1143.

77 Searle, P.A. and Molinski, T.F. (1996) Phorboxazoles A and B: potent cytostatic macrolides from marine sponge *Phorbas* species. *J. Am. Chem. Soc.*, **117**, 8126–8131.

78 Forsyth, C.J., Ahmed, F., Cink, R.D. and Lee, C.S. (1998) Total synthesis of phorboxazole A. *J. Am. Chem. Soc.*, **120**, 5597–5598.

79 Lucas, B.S., Luther, L.M. and Burke, S.D. (2004) Synthesis of the C1–C17 segment of phorboxazole B. *Org. Lett.*, **6**, 2965–2968.

80 Wang, B. and Forsyth, C.J. (2006) Stereoselective synthesis of the phorboxazole A macrolide by ring-closing metathesis. *Org. Lett.*, **8**, 5223–5226.

81 Smith, A.B., III, Minbiole, K.P., Verhoest, P.R. and Schelhaas, M. (2001) Total synthesis of (+)-phorboxazole A exploiting the Petasis–Ferrier rearrangement. *J. Am. Chem. Soc.*, **123**, 10942–10953.

82 For a review, see Smith, A.B., III, Fox, R.J. and Razler, T.M. (2008) Evolution of the Petasis–Ferrier union/rearrangement tactic: construction of architecturally complex natural products possessing the ubiquitous cis-2,6-substituted tetrahydropyran structural element. *Acc. Chem. Res.*, **41**, 675–687.

83 Evans, D.A., Fitch, D.M., Smith, T.E. and Cee, V.J. (2000) Application of complex aldol reactions to the total synthesis of phorboxazole B. *J. Am. Chem. Soc.*, **122**, 10033–10046.

84 Li, D.R., Sun, C.Y., Su, C., Lin, G.-Q. and Zhou, W.-S. (2004) Toward the total synthesis of phorboxazole B: an efficient synthesis of the C20–C46 segment. *Org. Lett.*, **6**, 4261–4264.

85 Pattenden, G., Gonzalez, M.A., Little, P.B., Millan, D.S., Plowright, A.T., Tornos, J.A. and Ye, T. (2003) Total synthesis of (+)-phorboxazole A, a potent cytostatic agent from the sponge *Phorbas* sp. *Org. Biol. Chem.*, **1**, 4173–4208.

86 White, J.D., Kuntiyong, P. and Lee, T.H. (2006) Total synthesis of phorboxazole A. 1. Preparation of four subunits. *Org. Lett.*, **6**, 6039–6042; White, J.D., Lee, T.H. and Kuntiyong, P. (2006) Total synthesis of phorboxazole A. 2. assembly of subunits and completion of the synthesis. *Org. Lett.*, **6**, 6043–6046.

87 Lucas, B.S., Gopalsamuthiram, V. and Burke, S.D. (2007) Total synthesis of phorboxazole B. *Angew. Chem. Int. Ed.*, **46**, pp. 769–772.

88 Paterson, I., Steven, A. and Luckhurst, C.A. (2004) Phorboxazole B synthetic studies: construction of C(1-32) and C(33-46) subtargets. *Org. Biol. Chem.*, **2**, 3026–3038.

89 Li, D.-R., Zhang, D.-H., Sun, C.-Y., Zhang, J.-W., Yang, L., Chen, J., Liu, B., Su, C., Zhou, W.-S. and Lin, G.-Q. (2006) Total synthesis of phorboxazole B. *Chem. Eur. J.*, **12**, 1185–1204; Lucas, B.S., Luther, L.M. and Burke, S.D. (2005) A catalytic enantioselective hetero Diels-Alder approach to the C20-C32 segment of the phorboxazoles. *J. Org. Chem.*, **70**, 3757–3760.

90 Vitale, J.P., Wolckenhauer, S.A., Do, N.M. and Rychnovsky, S.D. (2005) Synthesis of the C3-C19 segment of phorboxazole B. *Org. Lett.*, **7**, 3255–3258.

91 Smith, A.B., III, Razler, T.M., Ciavarri, J.P., Hirose, T., Ishikawa, T. and Meis,

R.M. (2008) A second-generation total synthesis of (+)-phorboxazole A. *J. Org. Chem.*, **73**, 1192–1200.
92 Yadav, J.S., Satyanarayana, M., Srinivasulu, G. and Kunwar, A.C. (2007) A stereoselective synthesis of the C20-C32 fragment of the phorboxazoles. *Synlett.*, 1577–1580.
93 Forsyth, C.J., Ying, L., Chen, J. and LaClair, J.J. (2006) Phorboxazole analogues induce association of cdk4 with extranuclear cytokeratin intermediate filaments. *J. Am. Chem. Soc.*, **128**, 3858–3859; Smith, A.B., III, Razler, R.M. and Pettit, G.R. (2006) Design and synthesis of a potent phorboxazole C(11-15) acetal analogue. *Org. Lett.*, **8**, 797–799; See also Smith, A.B., III, Razler, T.M., Meis, R.M. and Pettitt, G.R. (2008) Synthesis and biological evaluation of phorboxazole congeners leading to the discovery and preparative-scale synthesis of (+)-chlorophorboxazole A possessing picomolar human solid tumor cell growth inhibitory activity. *J. Org. Chem.*, **73**, 1201–1208.
94 Kehraus, S., Konig, G.M. and Wright, A.D. (2002) Leucamide A: a new cytotoxic heptapeptide from the Australian sponge *Leucetta microraphis*. *J. Org. Chem.*, **67**, 4989–4992.
95 Wang, W.-L. and Nan, F.-J. (2003) First total synthesis of Leucamide A. *J. Org. Chem.*, **68**, 1636–1639; Wang, W.-L., Yao, D.-Y., Gu, M., Fan, M.-Z., Li, J.-Y., Xing, Y.-C., Nan, F.-J. (2005) Synthesis and biological evaluation of novel bisheterocycle-containing compounds as potential anti-influenza virus agents. *Bioorg. Med. Chem. Lett.*, **15**, 5284–5287.
96 Yun, B.-S., Hidaka, T., Furihata, K. and Seto, H. (1994) Promothiocin-A and promothiocin-B, novel thiopeptides with a tipa promoter inducing activity produced by *Streptomyces sp* SF2741. *J. Antibiot.*, **47**, 510–514.
97 Bagley, M.C., Bashford, K.E., Hesketh, C.L. and Moody, C.J. (2000) Total synthesis of the thiopeptide promothiocin A. *J. Am. Chem. Soc.*, **122**, 3301–3313; Bagley, M.C., Buck, R.T., Hind, S.L. and Moody, C.J. (1998) Synthesis of functionalised oxazoles and bis-oxazoles. *J. Chem. Soc., Perkin Trans. 1*, 591–600.
98 Liesch, J.M. and Rinehart, K.L. (1977) Berninamycin .3. Total structure of berninamycin-A. *J. Am. Chem. Soc.*, **99**, 1645–1646; Lau, R.C.M. and Rinehart, K.L. (1994) Berninamycin-B, berninamycin-C and berninamycin-D, minor metabolites from *Streptomyces bernensis*. *J. Antibiot.*, **47**, 1466–1472. and references therein.
99 Saito, H., Yamada, T., Okumura, K., Yonezawa, Y. and Shin, C.G. (2002) Convenient synthesis of the main dehydrohexapeptide skeleton constituting a macrocyclic antibiotic, berninamycin A. *Chem. Lett.*, 1098–1099; Yamada, T., Okumura, K., Yonezawa, Y., Shin, C.G. (2001) Useful synthesis of the main dehydrohexapeptide segment of a macrocyclic antibiotic, berninamycin B. *Chem. Lett.*, 102–103.
100 Das, J., Reid, J.A., Kronenthal, D.R., Singh, J., Pansegrau, P.D. and Mueller, R.H. (1992) Novel methods for syntheses of substituted oxazoles by cyclization of vinyl bromides. *Tetrahedron Lett.*, **33**, 7835–7838.
101 Ogino, J., Moore, R.E., Patterson, G.M.L. and Smith, C.D. (1996) Dendroamides, new cyclic hexapeptides from a blue-green alga. Multidrug-resistance reversing activity of dendroamide A. *J. Nat. Prod.*, **59**, 581–586.
102 You, S.-L. and Kelly, J.W. (2003) Total synthesis of dendroamide A: oxazole and thiazole construction using an oxodiphosphonium salt. *J. Org. Chem.*, **68**, 9506–9509.
103 (a) Xia, Z. and Smith, C.D. (2001) Total synthesis of dendroamide A, a novel cyclic peptide that reverses multiple drug resistance. *J. Org. Chem.*, **66**, 3459–3466; (b) Bertram, A. and Pattenden, G. (2000) Self-assembly of amino acid-based thiazoles and oxazoles. Total synthesis of dendroamide A, a cyclic hexapeptide from the cyanobacterium *Stigonema dedroideum*. *Synlett.*, 1519–1521; (c) Bertram, A. and Pattenden, G. (2002) Dendroamide A, nostocyclamide and related cyclopeptides from cyanobacteria. Total synthesis,

together with organised and metal-templated assembly from oxazole and thiazole-based amino acids. *Heterocycles*, **58**, 521–561. See also; (d) Yonezawa, Y. and Tani, N. (2005) New total synthesis of dendroamide a from dehydrodi- and tripeptides. *Heterocycles*, **65**, 95–105; (e) Matsumoto, T., Morishita, E. and Shioiri, T. (2007) Investigation of macrocyclization sites for the synthesis of dendroamide A–an approach from a conformational search. *Tetrahedron*, **63**, 8571–8575.

104 Somogyi, L., Haberhauer, G. and Rebek, J., Jr. (2001) Improved synthesis of functionalized molecular platforms related to marine cyclopeptides. *Tetrahedron*, **57**, 1699–1708.

105 Todorova, A.K. and Jüttner, F. (1995) Nostocyclamide–a new macrocyclic, thiazole-containing allelochemical from nostoc-sp-31 (cyanobacteria). *J. Org. Chem.*, **60**, 7891–7895.

106 Moody, C.J. and Bagley, M.C. (1998) Total synthesis of (+)-nostocyclamide. *J. Chem. Soc. Perkin Trans. 1*, 601–607. See also reference 103c.

107 Perez, L.J. and Faulkner, D.J. (2003) Bistratamides E-J, modified cyclic hexapeptides from the Philippines ascidian *Lissoclinum bistratum*. *J. Nat. Prod.*, **66**, 247–250.

108 You, S.-L. and Kelly, J.W. (2005) The total synthesis of bistratamides F-I. *Tetrahedron*, **61**, 241–249; You, S.L. and Kelly, J. W. (2004) Highly efficient biomimetic total synthesis and structural verification of bistratamides E and J from *Lissoclinum bistratum*. *Chem. Eur. J.*, **10**, 71–75.

109 Bertram, A., Maulucci, N., New, O.M., Nor, S.M.N. and Pattenden, G. (2007) Synthesis of libraries of thiazole, oxazole and imidazole-based cyclic peptides from azole-based amino acids. A new synthetic approach to bistratamides and didmolamides. *Org. Biol. Chem.*, **5**, 1541–1553.

110 Downing, S.V., Aguilar, E. and Meyers, A.I. (1999) Total synthesis of bistratamide D. *J. Org. Chem.*, **64**, 826–831.

111 Nakamura, Y., Okumura, K., Kojima, M. and Takeuchi, S. (2006) An expeditious synthesis of bistratamide H using a new fluorous protecting group. *Tetrahedron Lett.*, **47**, 239–243.

112 Banker, R. and Carmeli, S. (1998) Tenuecyclamides A-D, cyclic hexapeptides from the cyanobacterium *Nostoc spongiaeforme* var. tenue. *J. Nat. Prod.*, **61**, 1248–1251.

113 You, S.-L., Deechongkit, S. and Kelly, J.W. (2004) Solid-phase synthesis and stereochemical assignments of tenuecyclamides A-D employing heterocyclic amino acids derived from commercially available Fmoc α-amino acids. *Org. Lett.*, **6**, 2627–2630.

114 Ojika, M., Nemoto, T., Nakamura, M. and Yamada, K. (1995) Dolastatin-E, a new cyclic hexapeptide isolated from the sea hare *Dolabella-auricularia*. *Tetrahedron Lett.*, **36**, 5057–5058; Sone, H., Kigoshi, H. and Yamada, K. (1997) Isolation and stereostructure of dolastatin I, a cytotoxic cyclic hexapeptide from the Japanese sea hare *Dolabella auricularia*. *Tetrahedron*, **53**, 8149–8154; Pettit, G.R. (1997) The dolastatins. *Fortschr. Chem. Org. Naturst.*, **70**, 1–79.

115 Kigoshi, H. and Yamada, S. (1999) Synthesis of dolastatin I, a cytotoxic cyclic hexapeptide from the sea hare *Dolabella auricularia*. *Tetrahedron*, **55**, 12301–12308.

116 Mink, D., Mecozzi, S. and Rebek, J., Jr. (1998) Natural products analogs as scaffolds for supramolecular and combinatorial chemistry. *Tetrahedron Lett.*, **39**, 5709–5712.

117 Ichiba, T., Yoshida, W.Y., Scheuer, P.J., Higa, T. and Gravalos, D.G. (1991) Hennoxazoles, bioactive bisoxazoles from a marine sponge. *J. Am. Chem. Soc.*, **113**, 3173–3174.

118 Wipf, P. and Lim, S. (1995) Total synthesis of the enantiomer of the antiviral marine natural product hennoxazole A. *J. Am. Chem. Soc.*, **117**, 558–559.

119 Yokokawa, F., Asano, T. and Shioiri, T. (2001) Total synthesis of (–)-hennoxazole A. *Tetrahedron*, **57**, 6311–6327.

120 Williams, D.R., Brooks, D.A. and Berliner, M.A. (1999) Total synthesis of (–)-hennoxazole A. *J. Am. Chem. Soc.*,

121 Zylstra, E.J., She, M.W.-L., Salamant, W.A. and Leahy, J.W. (2007) A flexible synthetic approach to the hennoxazoles. *Synlett*, 623–627.

122 Barrett, A.G.M. and Kohrt, J.T. (1995) Model studies on the synthesis of hennoxazole-A. *Synlett*, 415–416.

123 Smith, T.E., Kuo, W.-H., Balskus, E.P., Bock, V.D., Roizen, J.L., Theberge, A.B., Carroll, K.A., Kurihara, T. and Wessler, J.D. (2008) Total synthesis of (−)-hennoxazole A. *J. Org. Chem.*, **73**, 142–150.

124 Sakakura, A., Kondo, R. and Ishihara, K. (2005) Molybdenum oxides as highly effective dehydrative cyclization catalysts for the synthesis of oxazolines and thiazolines. *Org. Lett.*, **7**, 1971–1974; Sakakura, A., Kondo, R., Ishihara, K., Umemura, S. (2009) Dehydrative cyclization of serine, threonine, and cysteine residues catalyzed by molybdenum(VI) oxo compounds. *Tetrahedron*, **65**, 2102–2109.

125 Nagatsu, A., Kajitani, H. and Sakakibara, J. (1995) Muscoride-A – a new oxazole peptide alkaloid from fresh-water cyanobacterium nostoc-muscorum. *Tetrahedron Lett.*, **36**, 4097–4100.

126 Wipf, P. and Venkatraman, S. (1996) Total synthesis of (−)-muscoride A. *J. Org. Chem.*, **61**, 6517–6522.

127 Videnov, G., Kaiser, D., Kempfer, C. and Jung, G. (1996) Synthesis of naturally occurring, conformationally restricted oxazole- and thiazole-containing di- and tripeptide mimetics. *Angew. Chem. Int. Edn.*, **35**, 1503–1506.

128 Muir, J.C., Pattenden, G. and Thomas, R.M. (1998) Total synthesis of (−)-muscoride A: a novel bis-oxazole based alkaloid from the cyanobacterium *Nostoc muscorum*. *Synthesis*, 613–618.

129 Coqueron, P.-Y., Didier, C. and Ciufolini, M.A. (2003) Iterative oxazole assembly via α-chloroglycinates: total synthesis of (−)-muscoride A. *Angew. Chem. Int Edn.*, **42**, pp. 1411–1414.

130 (a) Lindquist, N., Fenical, W., Duyne, G.D.V. and Clardy, J. (1991) Isolation and structure determination of diazonamides A and B, unusual cytotoxic metabolites from the marine ascidian Diazona chinensis. *J. Am. Chem. Soc.*, **113**, 2303–2304. (b) For a review, see Lachia, M. and Moody, C.J. (2008) The synthetic challenge of diazonamide A, a macrocyclic indole bis-oxazole marine natural product. *Nat. Prod. Rpts.*, **25**, 227–253.

131 Li, J., Jeong, S., Esser, L. and Harran, P.G. (2001) Total synthesis of nominal diazonamides – Part 1: convergent preparation of the structure proposed for (−)-diazonamide A. *Angew. Chem. Int Edn.*, **40**, 4765–4769; Li, J., Burgett, A.W.G., Esser, L., Amezcua, C. and Harran, P.G. (2001) Total synthesis of nominal diazonamides – Part 2: on the true structure and origin of natural isolates. *Angew. Chem. Int. Edn.*, **40**, 4770–4773; Li, J., Chen, X., Burgett, A.W.G. and Harran, P.G. (2001) Synthetic seco forms of (−)-diazonamide A. *Amgew. Chem. Int. Edn.*, **40**, 2682–2685; Chen, X., Esser, L. and Harran, P.G. (2000) Stereocontrol in pinacol ring-contraction of cyclopeptidyl glycols: the diazonamide C-10 problem. *Angew. Chem. Int. Edn.*, **39**, 937–940; Jeong, J., Chen, X. and Harran, P.G. (1998) Macrocyclic triarylethylenes via Heck endocyclization: a system relevant to diazonamide synthesis. *J. Org. Chem.*, **63**, 8640–8641.

132 Freeman, F., Chen, T. and vander Linden, J.B. (1997) Synthesis of highly functionalized 1,3-oxazoles. *Synthesis*, 861–862.

133 Burgett, A.W.G., Li, Q., Wei, Q. and Harran, P.G. (2003) A concise and flexible total synthesis of (−)-diazonamide A. *Angew. Chem. Int. Edn.*, **42**, 4961–4966.

134 Nicolaou, K.C., Bella, M., Chen, D.Y.-K., Huang, X., Ling, T. and Snyder, S.A. (2002) Total synthesis of diazonamide A. *Angew. Chem. Int. Edn.*, **41**, 3495–3499; Nicolaou, K.C., Chen, D.Y.-K., Huang, X., Ling, T., Bella, M. and Snyder, S.A. (2004) Chemistry and biology of diazonamide A: first total synthesis and confirmation of the true structure. *J. Am. Chem. Soc.*, **126**, 12888–12896.

135 Nicolaou, K.C., Rao, P.B., Hao, J., Reddy, M.V., Rassias, G., Huang, X., Chen, D.Y.-K. and Snyder, S.A. (2003)

The second total synthesis of diazonamide A. *Angew. Chem. Int. Edn.*, **42**, 1753–1758; Nicolaou, K.C., Hao, J., Reddy, M.V., Rao, P.B., Rassias, G., Snyder, S.A., Huang, X., Chen, D.Y.-K., Brenzovich, W.E., Giuseppone, N., Giannakakou, P. and O'Brate, A. (2004) Chemistry and biology of diazonamide A: second total synthesis and biological investigations. *J. Am. Chem. Soc.*, **126**, 12897–12906.

136 Nicolaou, K.C., Snyder, S.A., Huang, X., Simonsen, K.B., Koumbis, A.E. and Bigot, A. (2004) Studies toward diazonamide a: initial synthetic forays directed toward the originally proposed structure. *J. Am. Chem. Soc.*, **126**, 10162–10173.

137 Nicolaou, K.C., Snyder, S.A., Giuseppone, N., Huang, X., Bella, M., Reddy, M.V., Rao, P.B., Koumbis, A.E., Giannakakou, P. and O'Brate, A. (2004) Studies toward diazonamide A: development of a hetero-pinacol macrocyclization cascade for the construction of the bis-macrocyclic framework of the originally proposed structure. *J. Am. Chem. Soc.*, **126**, 10174–10182.

138 Zajac, M.A. and Vedejs, E. (2004) A synthesis of the diazonamide heteroaromatic biaryl macrocycle/hemiaminal core. *Org. Lett.*, **6**, 237–240.

139 Cheung, C.-M., Goldberg, F.W., Magnus, P., Russell, C.J., Turnbull, R. and Lynch, V. (1999) An expedient formal total synthesis of (−)-diazonamide A via a powerful, stereoselective O-aryl to C-aryl migration to form the C10 quaternary center. *J. Am. Chem. Soc.*, **129**, 12320–12327; See also Booker, J.E.M., Boto, A., Churchill, G.H., Green, C.P., Ling, M., Meek, G., Prabhakaran, J., Sinclair, D., Blake, A.J. and Pattenden, G. (2006) Approaches to the quaternary stereocentre and to the heterocyclic core in diazonamide A using the Heck reaction and related coupling reactions. *Org. Biomol. Chem.*, **4**, 4193–4205.

140 Palmer, F.N., Lach, F., Poriel, C., Pepper, A.G., Bagley, M.C., Slawin, A.M.Z. and Moody, C.J. (2005) The diazo route to diazonamide A: studies on the tyrosine-derived fragment. *Org. Biomol. Chem.*, **3**, 3805–3811.

141 Sperry, J. and Moody, C.J. (2006) Biomimetic approaches to diazonamide A. Direct synthesis of the indole bis-oxazole fragment by oxidation of a TyrValTrpTrp tetrapeptide. *Chem. Commun.*, 2397–2399.

142 Adamczeski, M., Quinoa, E. and Crews, P. (1988) Unusual anthelminthic oxazoles from a marine sponge. *J. Am. Chem. Soc.*, **110**, 1598–1602; Searle, P. A., Richter, R. K. and Molinski, T. F. (1996) Bengazoles C-G from the sponge *Jaspis* sp synthesis of the side chain and determination of absolute configuration. *J. Org. Chem.*, **61**, 4073–4079; Fernandez, R., Dherbomez, M., Letourneux, Y., Nabil, M., Verbist, J.F. and Biard, J.F. (1999) Antifungal metabolites from the marine sponge *Pachastrissa* sp.: new bengamide and bengazole derivatives. *J. Nat. Prod.*, **62**, 678–680. and references therein.

143 Mulder, R.J., Shafer, C.M. and Molinski, T.F. (1999) First total synthesis of bengazole A. *J. Org. Chem.*, **64**, 4995–4998. [See also Erratum (2000). *J. Org. Chem.*, **65**, 8126].

144 Bull, J.A., Balskus, E.P., Horan, R.A.J., Langner, M. and Ley, S.V. (2006) Stereocontrolled total synthesis of bengazole A: a marine bisoxazole natural product displaying potent antifungal properties. *Angew. Chem. Int. Edn.*, **45**, 6714–6718.

145 Bull, J.A., Balskus, E.P., Horan, R.A.J., Langner, M. and Ley, S.V. (2007) Total synthesis of potent antifungal marine bisoxazole natural products bengazoles A and B. *Chem. Eur. J.*, **13**, 5515–5538.

146 Chandrasekhar, S. and Sudhakar, A. (2010) Total synthsis of bengazole A. *Org. Lett.*, **12**, 236–238.

147 Enriquez-Garcia, A. and Ley, S.V. (2009) Total synthesis of the potent antifungal agents bengazole C and E. *Coll. Czech. Chem. Commun.*, **74**, 887–900; Mulder, R.J., Shafer, C.M., Dalisay, D.S. and Molinski, T.F. (2009) Synthesis and structure-activity relationships of bengazole A analogs. *Bioorg. Med. Chem. Lett.*, **19**, 2928–2930.

148 Nett, M., Erol, O., Kehraus, S., Kock, M., Krick, A., Eguereva, E., Neu, E. and Konig, G.M. (2006) Siphonazole, an unusual metabolite from *Herpetosiphon* sp. *Angew. Chem. Int. Edn.*, **45**, 3863–3867.

149 Linder, J., Blake, A.J. and Moody, C.J. (2008) Total synthesis of siphonazole and its O-methyl derivative, structurally unusual bis-oxazole natural products. *Org. Biomol. Chem.*, **6**, 3908–3916.

150 Zhang, J., Polishchuk, E.A., Chen, J. and Ciufolini, M.A. (2009) Development of an oxazole conjunctive reagent and application to the total synthesis of siphonazoles. *J. Org. Chem.*, **74**, 9140–9151; Zhang, J. and Ciufolini, M.A. (2009) Total synthesis of siphonazoles by the use of a conjunctive oxazole building block. *Org. Lett.*, **11**, 2389–2392.

151 Isolation: for the X-ray structural determination (of a Ulapualide-actin complex), see Allington, J.S., Tanaka, J., Marroitt, G. and Rayment, I. (2004) Absolute stereochemistry of ulapualide A. *Org. Lett.*, **6**, 597–599.

152 Knight, D.W., Pattenden, G. and Rippon, D.E. (1990) Synthesis of the tris-oxazole ring-system of ulapualides. *Synlett.*, 36–37; Yoo, S.K. (1992) Synthesis of poly-oxazole systems found in marine metabolites. *Tetrahedron Lett.*, **33**, 2159–2162; Shapiro, R. (1993) Dimethyl amino[(phenylthio)methyl] malonate – a useful C-3 unit in a mild, direct synthesis of oxazole-4-carboxylates. *J. Org. Chem.*, **58**, 5759–5764.

153 Panek, J.S. and Beresis, R.T. (1996) Studies directed toward the synthesis of ulapualide A. Symmetric synthesis of the C8-C25 tris-oxazole fragment. *J. Org. Chem.*, **61**, 6496–6497; See alsoLiu, P., Celatka, C.A. and Panek, J.S. (1997) Synthesis of the fully functionalized tris-oxazole fragment found in metabolites derived from marine organisms. *Tetrahedron Lett.*, **38**, 5445–5448. for a very similar approach to a tris-oxazole unit.

154 Chattopadhyay, S.K. and Pattenden, G. (2000) A total synthesis of the unique tris-oxazole macrolide ulapualide A produced by the marine nudibranch *Hexabranchus sanguineus*. *J. Chem. Soc. Perkin Trans. 1*, 2429–2454; Chattopadhyay, S.K., Kempson, J., McNeil, A., Pattenden, G., Reader, M., Rippon, D.E. and Waite, D. (2000) Towards a total synthesis of ulapualide A. Concise synthetic routes to the tris-oxazole ring system and tris-oxazole macrolide core in ulapualides, kabiramides, halichondramides, mycalolides and halishigamides. *J. Chem. Soc., Perkin Trans. 1*, 2415–2428.

155 Evans, D.L., Minster, D.K., Jordis, U., Hecht, S.M., Mazzu, A.L. and Meyers, A.I. (1979) Nickel peroxide dehydrogenation of oxygen-containing, sulfur-containing, and nitrogen-containing heterocycles. *J. Org. Chem.*, **44**, 497–501.

156 (a) Pattenden, G., Ashweek, N.J., Baker-Glenn, C.A.G., Walker, G.M. and Yee, J.G.K. (2007) Total synthesis of (-)-ulapualide A: the danger of overdependence on NMR spectroscopy in assignment of stereochemistry. *Angew. Chem. Int. Ed.*, **46**, 4359–4363.: (b) Pattenden, G., Ashweek, N.J., Baker-Glenn, C.A.G., Kempson, J., Walker, G.M. and Yee, J.G.K. (2008) Total synthesis of (-)-ulapualide A, a novel tris-oxazole macrolide from marine nudibranchs, based on some biosynthesis speculation. *Org. Biomol. Chem.*, **6**, 1478–1497.

157 Endoh, N., Tsuboi, K., Kim, R., Yonezawa, Y. and Shin, C. (2003) Useful synthesis of the longer array oxazole rings for telomestatin. *Heterocycles*, **60**, 1567–1572.

158 Panek, J.S. and Liu, P. (2000) Total synthesis of the actin-depolymerizing agent (–)-mycalolide A: application of chiral silane-based bond construction methodology. *J. Am. Chem. Soc.*, **122**, 11090–11097; Liu, P. and Panek, J.S. (2000) Total synthesis of (–)-mycalolide A. *J. Am. Chem. Soc.*, **122**, 1235–1236.

159 Kimura, T., Kuribayashi, S., Sengoku, T., Matsui, K., Ueda, S., Hayakawa, I., Suenaga, K. and Kigoshi, H. (2007) Synthetic studies on mycalolide B:

synthesis of the C7–C35 fragment. *Chem. Lett.*, 1490–1491.

160 Shin-Ya, K., Wierzba, K., Matsuo, K., Ohtani, T., Yamada, Y., Furihata, K., Hayakawa, Y. and Seto, H. (2001) Telomestatin, a novel telomerase inhibitor from *Streptomyces anulatus*. *J. Am. Chem. Soc.*, **123**, 1262–1263.

161 Doi, T., Yoshida, M., Shin-ya, K. and Takahashi, T. (2006) Total synthesis of (*R*)-telomestatin. *Org. Lett.*, **8**, 4165–4167; See also Yamada, S., Shigeno, K., Kitagawa, K., Okajima, S. and Asao, T. (2002) Process for preparing substance GM-95. WO 200248153.

162 Shibata, K., Yoshida, M., Doi, T. and Takahashi, T. (2010) Derivatization of a tris-oxazole using Pd-catalyzed coupling reactions of a 5-bromooxazole moiety. *Tetrahedron Lett.*, **51**, 1674–1677.

163 Chattopadhyay, S.K. and Biswas, S. (2006) Convergent synthesis of a 24-membered macrocyclic hexaoxazole derivative related to the novel telomerase inhibitor telomestatin. *Tetrahedron Lett.*, **47**, 7897–7900; Chattopadhyay, S.K., Biswas, S. and Pal, B.K. (2006) Efficient construction of a doubly functionalized trisoxazole derivative relevant to the synthesis of the novel telomerase inhibitor telomestatin and its analogues. *Synthesis*, 1289–1294.

164 Marson, C.M. and Saadi, M. (2006) Synthesis of the penta-oxazole core of telomestatin in a convergent approach to poly-oxazole macrocycles. *Org. Biomol. Chem.*, **4**, 3892–3893.

165 Romero, P., Malet, L., Cañedo, M.L., Cuevas, C. and Reyes, J. (2005) New cytotoxic depsipeptide IB-01211 isolated from actinomycete strain ES7-008 with antitumor activity, derivatives, synthesis, and use as antitumoral agents. WO 2005000880 A2.; Kanoh, K., Matsuo, Y., Adachi, K., Imagawa, H., Nishizawa, M. and Shizuri, Y. (2005) Mechercharmycins A and B, cytotoxic substances from marine-derived *Thermoactinomyces* sp. YM3-251. *J. Antibiot.*, **58**, 289–292.

166 Hernández, D., Vilar, G., Riego, E., Cañedo, L.M., Cuevas, C., Albericio, F. and Alvarez, M. (2007) Synthesis of IB-01211, a cyclic peptide containing 2,4-concatenated thia- and oxazoles, via Hantzsch macrocyclization. *Org. Lett.*, **9**, 809–811.

167 Hernández, D., Altuna, M., Cuevas, C., Aligué, R., Albericio, F. and Álvarez, M. (2008) Synthesis and antitumor activity of mechercharmycin A analogues. *J. Med. Chem.*, **51**, 5722–5730; Hernández, D., Riego, E., Albericio, F. and Álvarez, M. (2008) Synthesis of natural product derivatives containing 2,4-concatenated oxazoles. *Eur. J. Org. Chem.*, 3389–3396; Hernández, D., Riego, E., Francesch, A., Cuevas, C., Albericio, F. and Álvarez, M. (2007) Preparation of penta-azole containing cyclopeptides: challenges in macrocyclisation. *Tetrahedron*, **63**, 9862–9870.

168 Chattopadhyay, S.K. and Singha, S.K. (2010) Efficient construction of the carbon skeleton of the novel polyoxazole-based cyclopeptide IB-01211 via a biomimetic macrocyclisation. *Synlett*, 555–558.

169 Chattopadhyay, S.K., Biswas, S. and Ghosh, S.K. (2008) Convergent synthesis of contiguous tetra- and pentaoxazole ring systems related to some bioactive natural products. *Synthesis*, 1029–1032.

170 Sanada, K., Takebayashi, Y., Nagai, K. and Hiramoto, M. (1999) New antitumor compound. JP 11180997-A.

171 Pattenden, G. and Deeley, J. (2005) Synthesis and establishment of stereochemistry of the unusual polyoxazole-thiazole based cyclopeptide YM-216391 isolated from *Streptomyces nobilis*. *Chem. Commun.*, 797–799; Deeley, J., Bertram, A. and Pattenden, G. (2008) Novel polyoxazole-based cyclopeptides from *Streptomyces* sp. Total synthesis of the cyclopeptide YM-216391 and synthetic studies towards telomestatin. *Org. Biomol. Chem.*, **6**, 1994–2010.

13
Thiazoline and Thiazole and Their Derivatives
Zhengshuang Xu and Tao Ye

13.1
Introduction

Thiazoles and thiazolines are important heterocycles that contain sulfur and nitrogen atoms within a five-membered ring, with thiazoline as the lower oxidative state of thiazole. Both thiazole- and thiazoline-containing chemicals have distinctive bioactivities. One interesting thiazoline-containing natural product is the firefly luciferin **1**, which is responsible for the characteristic yellow light emission from fireflies when oxidized under the catalytic effects of luciferase and ATP. Another example of thiazole-containing natural product is vitamin B_1 (VB_1) **2**, also known as thiamine or thiamin, whose phosphate derivatives are involved in many cellular processes. VB_1 derivatives and VB_1-dependent enzymes play important roles in all the cells of the human body. VB_1 deficiency would adversely affect all organ systems (Figure 13.1).

Thiazoles and thiazolines are present in many natural and synthetic products with different pharmacological activities, such as antimicrobial, antiviral, anticancer, and anti-inflammatory activities. Besides the above-mentioned VB_1, many pharmaceuticals also contain these heterocycles. The anticonvulsant riluzole **3**, the antiparkinsonian talipexole **4**, the antischistosomal miridazole **5**, the antiviral ritonavir **6**, and the antibacterial sulfathiazole **7** shown in Figure 13.2 are some of the representative examples [1].

More and more thiazole- and/or thiazoline-containing natural products have been discovered. With a wide variety of bioactivities, these compounds provide humans with promising lead compounds for drug discovery. This chapter will discuss important thiazole/thiazoline-containing natural products (Section 13.3), and the methods developed for the synthesis of these heterocycles.

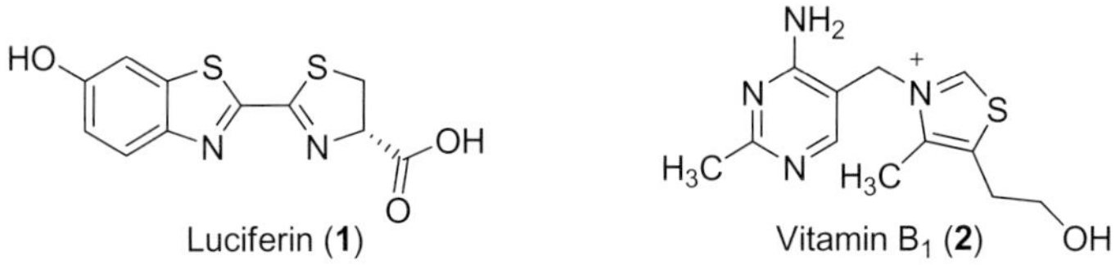

13 Thiazoline and Thiazole and Their Derivatives

Figure 13.1 Structures of luciferin and vitamin B$_1$.

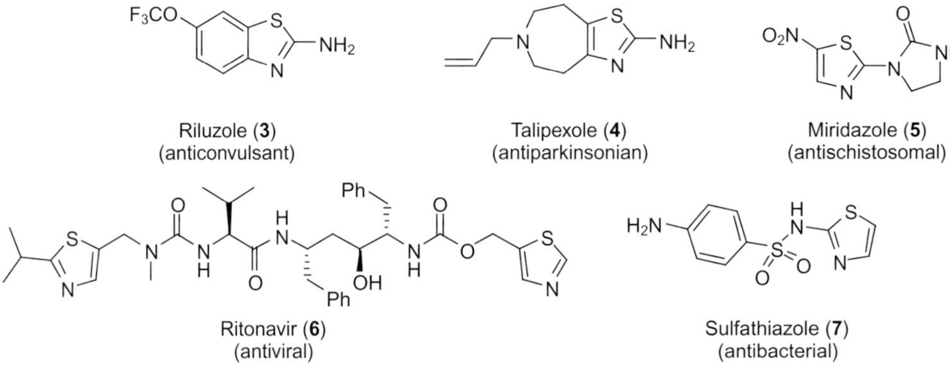

Figure 13.2 Thiazole-containing pharmaceuticals.

13.2
General Methods for the Synthesis of Thiazoline and Thiazole Derivatives

There are several excellent reviews on the chemistry of thiazole and thiazoline [2]. In this chapter, the general methods for the construction of these heterocycles will be presented before the introduction of thiazole/thiazoline-containing natural products and the efforts toward their total synthesis. Thiazoles are the dehydrogenation products of thiazolines. Therefore, oxidation of thiazoline to the 3,4-didehydro form has been employed in the practical synthesis of thiazoles. The general methods for thiazole synthesis also depend on approaches for the generation of thiazolines. Many natural products contain both thiazole and thiazoline moieties, hence the methods for the stereocontrolled preparation of thiazolines will be discussed first.

13.2.1
Methods for the Preparation of Thiazolines

From the biosynthetic origination of thiazoline synthesis, it is possibly derived from a cysteine-containing dipeptide. From a synthetic point of view, it is well known that the stereogenic centers at C-4 and C-2 exomethine of the thiazoline ring **8** are easy to racemize/epimerize under either acidic or basic conditions (Scheme 13.1) [3]. This remained as a major challenge for the stereocontrolled synthesis of 2,4-disubstituted thiazolines.

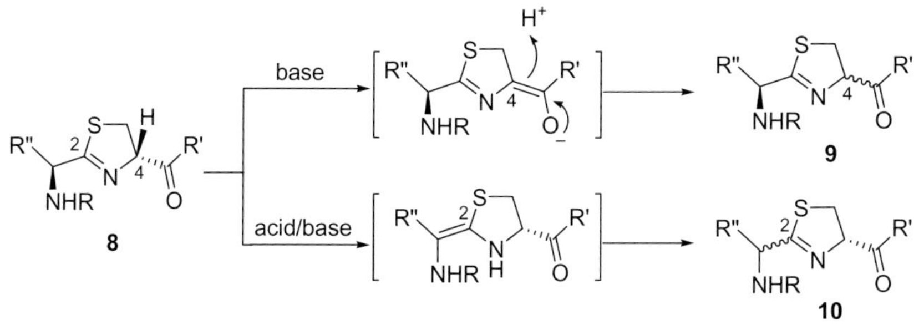

Scheme 13.1 Epimerization of stereogenic centers on and adjacent to thiazoline ring.

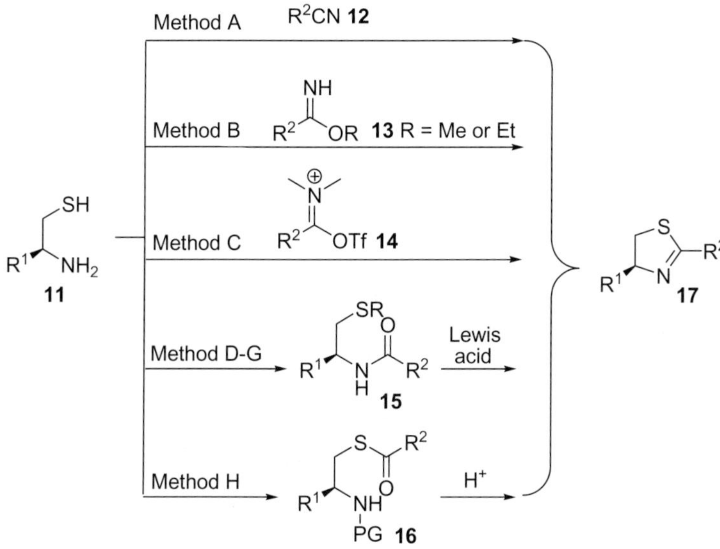

Scheme 13.2 Preparation of thiazoline ring from vicinal amino thiol **11**.

Thiazolines can be synthesized from either vicinal amino thiols or amino alcohols via condensation with a suitable partner, that could be cyanides, iminoethers, iminium triflates and carboxylic acids or esters, followed by a cyclodehydration reaction. While using vicinal amino alcohols as the starting material, this approach usually needs one or more steps to introduce a sulfur functional group before the condensation or cyclodehydration; thionation and thioacylation have been employed for this purpose.

13.2.1.1 Using Vicinal Amino Thiols as Starting Materials

Vicinal amino thiols **11** are readily condensed with cyanides **12**, imino ethers **13** or iminium triflates **13** to give thiazoline derivatives **17** directly (Scheme 13.2, Methods A–C). Either amides **15** or thioesters **16** were formed followed by a cyclodehydration process (Scheme 13.2, Methods D–H).

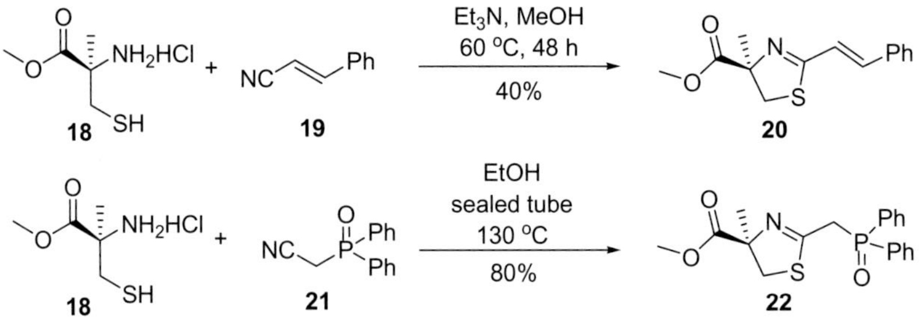

Scheme 13.3 Condensation of a vicinal amino thiol **18** with nitriles.

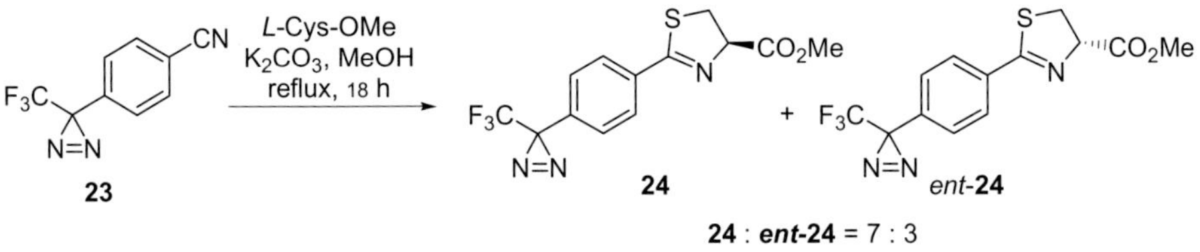

Scheme 13.4 Epimerization of the C-4 stereogenic center.

Method A: Condensation of a Vicinal Amino Thiol with a Nitrile Nitriles are electrophilic and thus can be attacked by amine or thiol groups. This reaction is usually carried out in hot methanol or ethanol with or without a base [4]. During the total synthesis of thiangazole, Ehrler and co-workers employed nitriles **19**, **21** as coupling partners to condense with the α-methyl cysteine methyl ester **18** to produce thiazolines **20**, **22** [5]. Similarly, condensation reaction was also employed in the total synthesis of thiangazole by Pattenden (Scheme 13.3) [6].

The reaction may involve direct nucleophilic attack on the nitrile group by amino or thiol group, but the stepwise reaction sequence, in which the nitrile is first converted into the imino ether cannot be ruled out. When L-cysteine methyl ester (L-Cys-OMe) is used as the coupling partner, the C-4 stereogenic center racemizes to some extent. For example, when nitrile **23** was condensed with L-Cys-OMe, this afforded **24** and **ent-24** in a ratio of 7:3 (Scheme 13.4) [7]. When chiral nitrile was employed as a coupling partner, the stereogenic center on C-2 exomethine also suffers from high risk of racemization. These drawbacks have hampered the application of this method. It has only been used for the construction of thiazolines that are not racemizable or have no adjacent stereogenic centers, for example, the use of the α-methylated cysteine as starting material.

Condensation between nitrile and amino thiol can also be catalyzed by Lewis acids. Kunieda and co-workers employed zinc chloride as the catalyst for the condensation of amino thiol with a nitrile to produce thiazoline **26** (Scheme 13.5) [8].

Method B: Condensation of a Vicinal Amino Thiol with an Iminoether As discussed in Method A, a nitrile might go through an iminoether intermediate before its

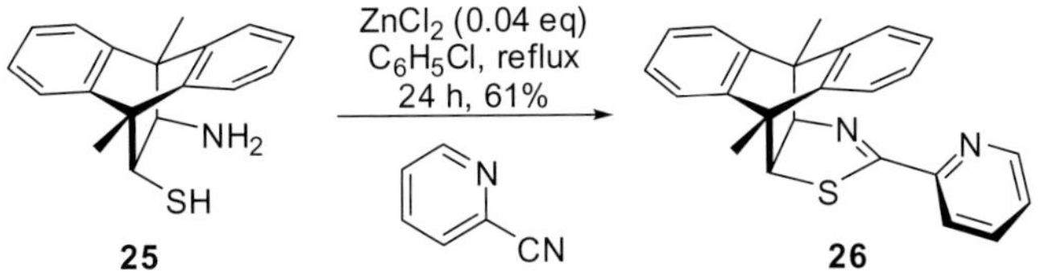

Scheme 13.5 Lewis acid-catalyzed thiazoline formation.

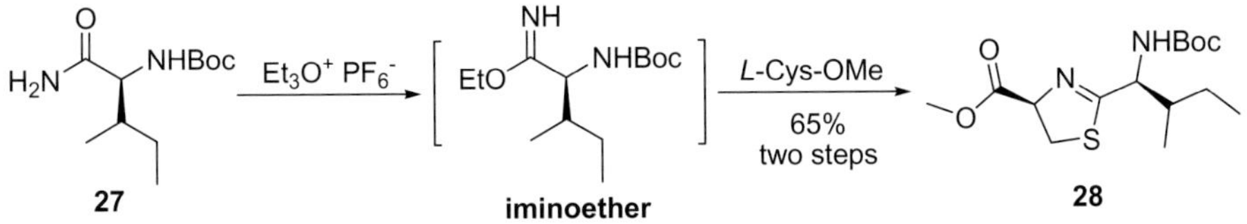

Scheme 13.6 Condensation of a vicinal amino thiol with an iminoether.

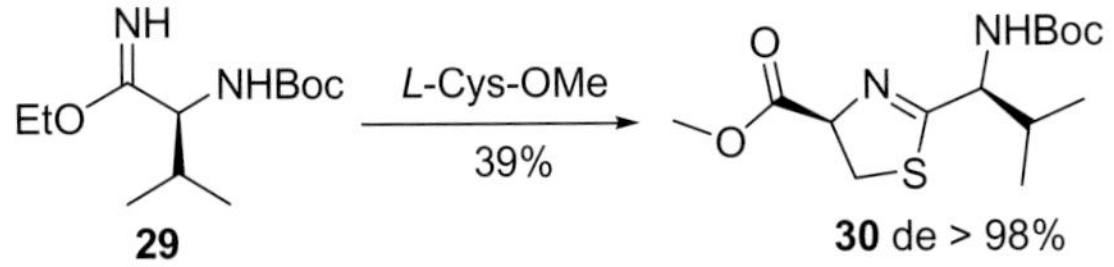

Scheme 13.7 Condensation of a vicinal amino thiol with an iminoether.

condensation with an amino thiol partner. Iminoethers are more active than the nitrile toward the condensation reaction with vicinal amino thiols. They are usually prepared from nitriles or primary amides; and primary amides are more frequently employed in the total synthesis of natural products. This is because amides are easily prepared from the corresponding carboxylic acids. For example, this method had been applied in the total synthesis of (−)-didehydromirabazole A (Scheme 13.6) [9]. Treatment of the primary amide **27** with triethyloxonium hexafluorophosphate afforded the corresponding **iminoether** that was subjected to condensation with L-Cys-OMe to produce the thiazoline **28**.

Racemization of the chiral center on thiazoline ring (C-4) could be minimized by working up the reaction immediately after the iminoether was consumed (Scheme 13.7) [10]. Reaction of iminoether **29** with L-Cys-OMe produced thiazoline **30** in 39% yield with >98% de.

Method C: Condensation of a Vicinal Amino Thiol with an Iminium Triflate Iminium triflate is more reactive towards the condensation reaction with an amino thiol. It can be easily prepared by treating primary or secondary amide with triflic anhydride (Tf$_2$O) in the presence of pyridine [11]. Charette and Chua have screened a wide variety of amides with functionalities, such as silyl ether **31e**, ester **31f**,

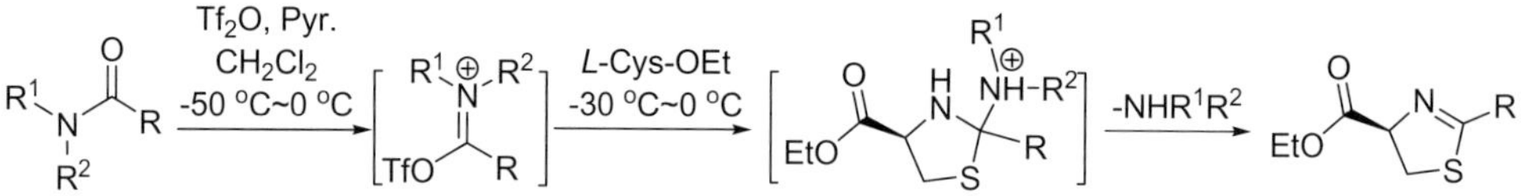

464 | *13 Thiazoline and Thiazole and Their Derivatives*

Scheme 13.8 Iminium triflate-mediated thiazoline formation.

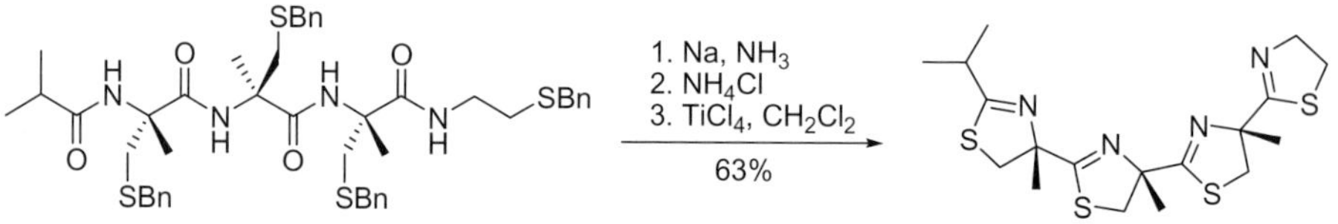

Compd. No.	R	R^1	R^2	Yield
31a	PhCH$_2$CH$_2$	Et	Et	90%
31b	PhCH$_2$CH$_2$	H	Bn	91%
31c	PhCH$_2$CHMe	Me	Me	55%
31d	2-Naphthyl	Et	Et	65%
31e	TBDPSOCH$_2$CH$_2$CH$_2$	H	Me	73%
31f	4-Me-C$_6$H$_4$CO$_2$(CH$_2$)$_3$	H	Me	76%
31g	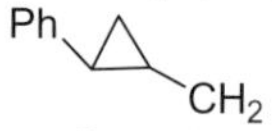 Ph-cyclopropyl-CH$_2$	H	Me	77%
31h	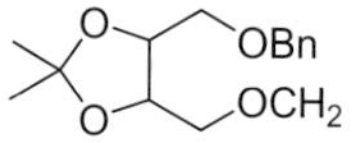 acetonide-OBn/OCH$_2$	H	Bn	80%

Scheme 13.9 Titanium (IV)-mediated cyclization of amide thiol **33**.

cyclopropane **31g**, and acetonide **31h**. As shown in Scheme 13.8, the desired thiazolines were obtained in moderate to very good yields.

Although this method is tolerable to most of the functional groups, the use of one equivalent of triflic anhydride would expose the intermediate to strong acid, and thus impose the risk of racemization to the stereogenic centers, especially to the one attached to the C-2 exomethine position of the thiazoline ring. This method has been applied in the total synthesis of (+)-cystothiazole A [12].

Method D: Titanium (IV)-Mediated Cyclization of Amide Thiol (Heathcock) During the total synthesis of (−)-mirabazole C, Heathcock and co-workers reported a thiazoline-formation approach based on the cyclodehydration of amide thiol in the presence of TiCl$_4$. The efficiency of this method was also demonstrated in their total synthesis of (−)-thiangazole [13]. As illustrated in Scheme 13.9, treatment of a tetrathiol derived from **33** with TiCl$_4$ afforded four consecutive thiazoline rings **34** simultaneously in 63% yield.

13.2 General Methods for the Synthesis of Thiazoline and Thiazole Derivatives

As indicated in Scheme 13.9, the titanium-mediated cyclodehydration process has only been employed to precursors that were derived from 2-methylcysteine since these quaternary stereogenic centers on the thiazolines would not undergo racemization.

Method E: Titanium (IV)-Mediated Cyclization of Amide Thiol (Kelly's Modification) The TiCl$_4$-mediated thiazoline formation approach was further modified by Kelly and co-workers [14]. Treatment of S-trityl-protected cysteine amides with TiCl$_4$ initiated a tandem deprotection-cyclodehydration process to afford thiazolines in good yields. The selection of trityl to replace the benzyl for thiol protection made this modified method more practical. However the reaction condition might not be able to prevent potential epimerization/racemization of the stereogenic centers at the C-4 and/or C-2 exomethine of the thiazoline ring. The C-4 stereogenic center is less prone to racemization than the one attached to C-2, but when an electron withdrawing group such as p-NO$_2$-benzyl group attached to C-2 carbon, the cyclization process afforded the corresponding thiazoline with extensive racemization at the C-4 stereogenic center (Scheme 13.10).

When this method was applied to Phe-Cys(Tr) dipeptide **37**, complete racemization at the C-2 exomethine stereogenic center occurred (Scheme 13.11). This method has been applied in the total synthesis of lyngbyabellins A and B [15].

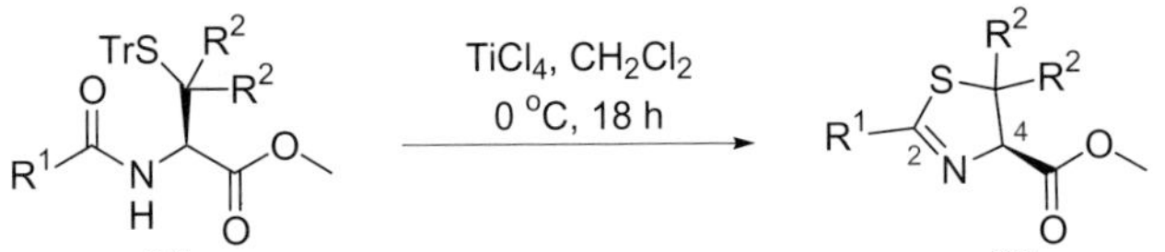

Compd. No.	R^1	R^2	Yield	ee
35a	Ph	Me	96%	99%
35b	4-NO$_2$C$_6$H$_4$	H	77%	22%
35c	4-MeOC$_6$H$_4$	H	51%	93%
35d	PhCH$_2$CH$_2$	H	61%	91%

Scheme 13.10 Titanium (IV)-mediated cyclization of amide thiol **35**.

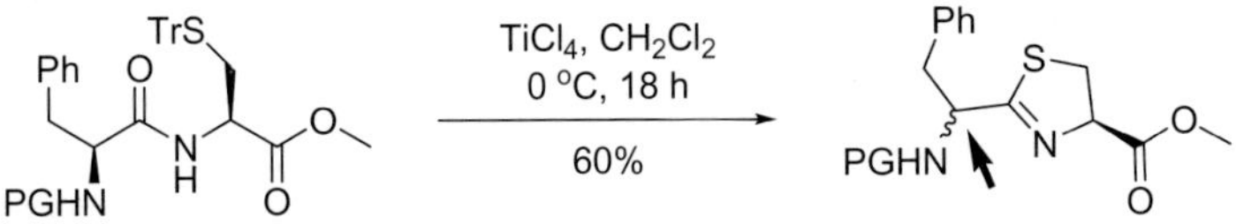

Scheme 13.11 Titanium (IV)-mediated cyclization of amide thiol **37**.

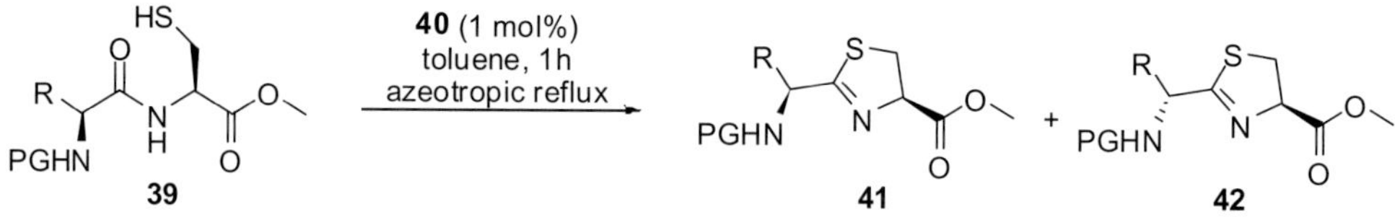

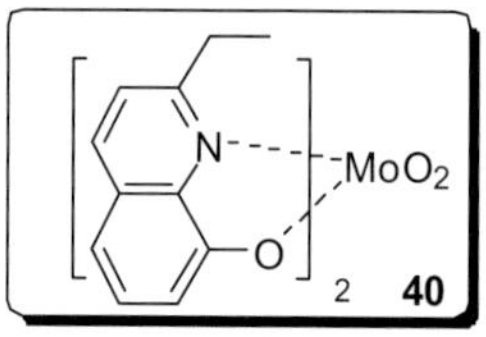

Scheme 13.12 Molybdenum (IV)-mediated cyclization of amide thiol **39**.

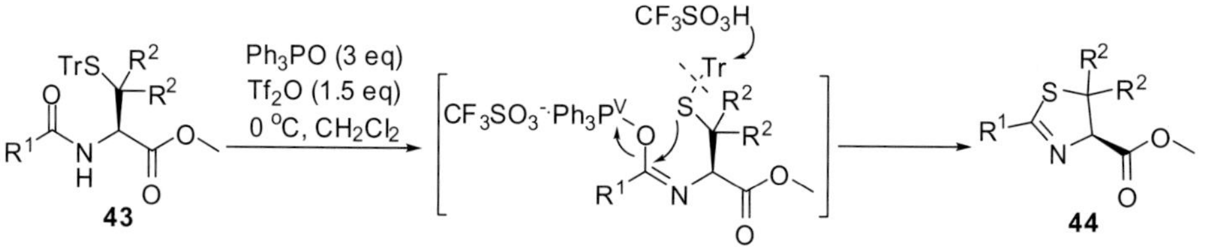

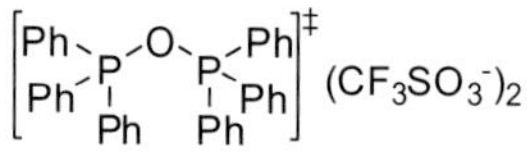

Compd. No.	PG	R	Yield	Ratio (**41** : **42**)
39a	Cbz	Me	83%	99 : 1
39b	Cbz	Bn	85%	98 : 2
39c	Boc	Me	82%	94 : 6
39d	Fmoc	Me	91%	94 : 6

Compd. No.	R^1	R^2	Yield	ee
43a	Ph	Me	98%	99%
43b	Ph	H	98%	>99.5%
43c	4-$NO_2C_6H_4$	H	86%	>99.5%
43d	4-MeOC_6H_4	H	92%	>99.5%
43e	PhCH$_2$CH$_2$	H	84%	>99.5%

Scheme 13.13 Hexaphenyloxodiphosphonium trifluoromethanesulfonate-mediated cyclization of amide thiol **43**.

Method F: Molybdenum (IV)-Mediated Cyclization of Amide Thiol   Ishihara and coworker employed molybdenum (IV) complex **40** for the cyclodehydration of the amide thiol substrates **39** [16] which led to the corresponding thiazoline with less than 6% of epimerization at the C-2 exomethine position (Scheme 13.12).

Method G: Hexaphenyloxodiphosphonium Trifluoromethanesulfonate-Mediated Cyclization of Amide Thiol   In 2003, Kelly reported a biomimetic procedure that employed hexaphenyloxodiphosphonium trifluoromethanesulfonate for the synthesis of thiazolines. The O-bridged bistriphenylphosphonium salt (active species) not only promoted the deprotection of Tr group but also facilitated the dehydrative cyclization step [17]. A few fully protected cysteine N-amide derivatives **43** were subjected to the deprotection and dehydrocyclization conditions to afford the corresponding thiazolines **44** in excellent chemical and optical yields (Scheme 13.13).

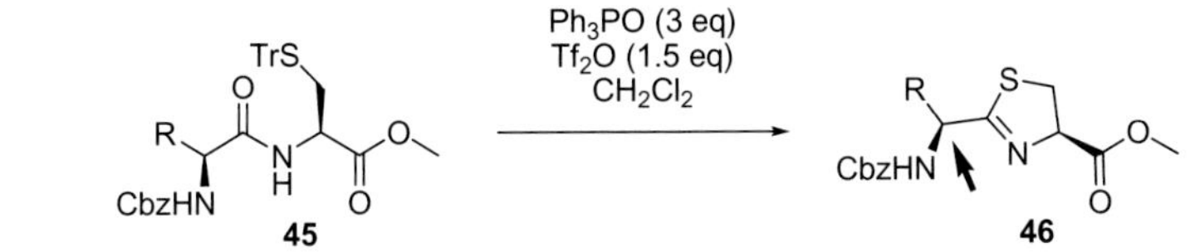

13.2 General Methods for the Synthesis of Thiazoline and Thiazole Derivatives

Scheme 13.14 Preservation of stereochemistry.

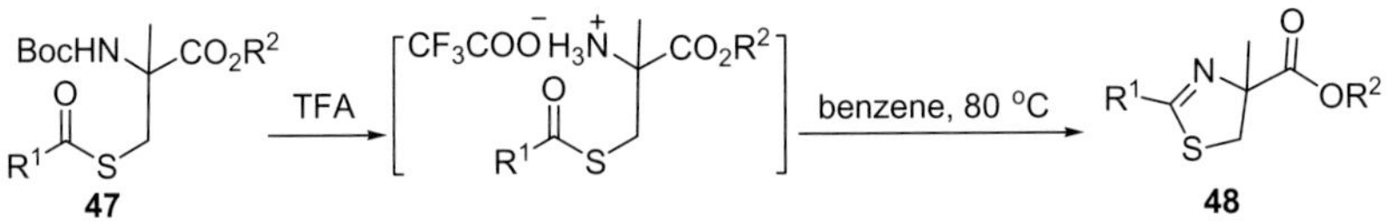

Compd. No.	R	T [°C]	t (min)	Starting material (d.r.)	Yield	Product (d.r.)	ee
45a	Bn	0	10	>99:1	84%	93:7	>99.5%
45b	Bn	-20	120	>99:1	98%	97:3	>99.5%
45c	Me	-20	120	99:1	84%	91:9	>99.5%
45d	^{i}Pr	-20	120	98:2	86%	97:3	>99.5%
45e	^{s}Bu	-20	120	96:4	95%	96:4	>99.5%

Scheme 13.15 Cyclization of thioesters stimulated by TFA.

When this method was applied to dipeptide **45**, the stereochemical integrity was well transformed to the product **46** (Scheme 13.14). This method has been successfully applied in the total synthesis of a number of natural products including apratoxin A [18], halipeptin A [19], largazole [20], didmolamides [21], dendroamide A [22], bistratamides [22, 23], tenuecyclamides [24] and telomestatin [25].

Method H: Cyclization of Thioesters Fukuyama and co-workers developed a method for the synthesis of thiazoline based on the TFA-mediated cyclization of thioesters in refluxing benzene [26]. As shown in Scheme 13.15, deprotection of the N-Boc group of **47** with trifluoroacetic acid followed by heating at 80 °C in benzene furnished the desired thiazoline **48**. In general, the N-protecting group was selected to be an acid sensitive one, which could facilitate the deprotection and cyclodehydration in one step. It was observed that acyl migration from sulfur to nitrogen occurred during the cyclodehydration process.

Due to the acidic reaction media, the stereogenic centers are at high risk of racemization, therefore, this method is only applicable to those substrates bearing quaternary stereogenic centers. This method was employed in the total synthesis of tantazole B (Scheme 13.16) [26]. Treatment of thioester **49** with TFA, followed by heating in benzene for 2 h, produced the thiazoline **50**. Further coupling reaction between carboxylic acid **50** and methyl 3-mercaptopropanoate in the presence of BOPCl provided **51** in 77% yield, which was ready for the next step of reaction.

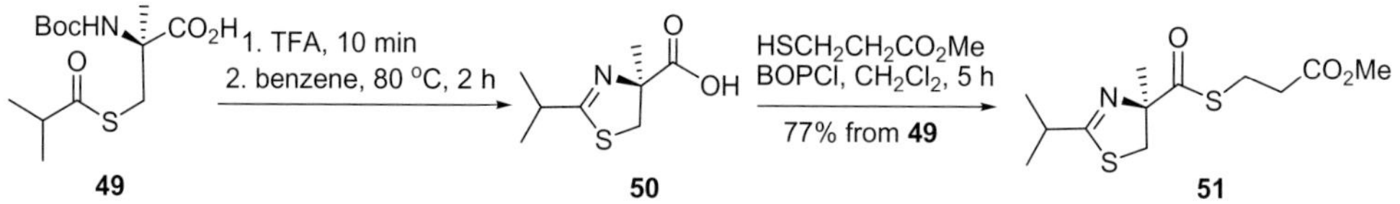

Scheme 13.16 Application in total synthesis of tantazole B.

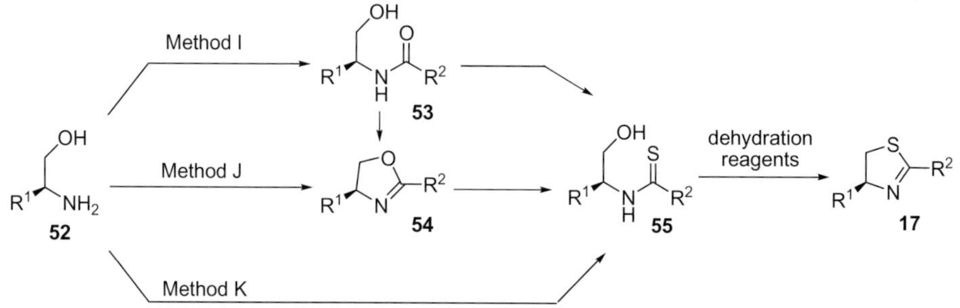

Scheme 13.17 Synthesis of thiazoline from vicinal amino alcohol.

Additional examples of this method include the construction of the thiazoline moieties presented in thiangazole [9] and mirabazole [13a].

13.2.1.2 From Vicinal Amino Alcohol

Thioamide **55** served as the key intermediate for thiazoline synthesis when vicinal amino alcohol **52** was employed as the starting material. There are three main methods for the preparation of the key thioamide intermediate (Scheme 13.17). During the cyclodehydration process, the hydroxyl group will be converted into a leaving group, while the sulfur atom will act as a nucleophile to complete the cyclization. The choice of an appropriate dehydration reagent was crucial for this kind of transformation.

The afore-mentioned three practical methods for the preparation of thioamide **55** [27] are: (i) using thionation reagents to convert the amide precursors **53** to thioamides **55** (Method I); (ii) using the thiolysis of oxazoline intermediate **54** (Method J); and (iii) using thioacylating reagents to couple with amino alcohols **52** to produce the thioamide **55** directly (Method K) (Scheme 13.17).

The cyclodehydration process can be facilitated by Mitsunobu reaction or other reagents that can activate the hydroxyl group, such as diethylaminosulfur trifluoride (DAST), Burgess reagent, tosyl chloride, and so on [2c]. These reaction conditions will be discussed in detail in this section.

Method I: Thionation of Amide It has been well documented that amide carbonyl groups are more readily thionated than other carbonyl groups such as ketones and esters. The most frequently employed thionating reagents are phosphorus pentasulfide [28] and Lawesson's reagent [29]. (Scheme 13.18).

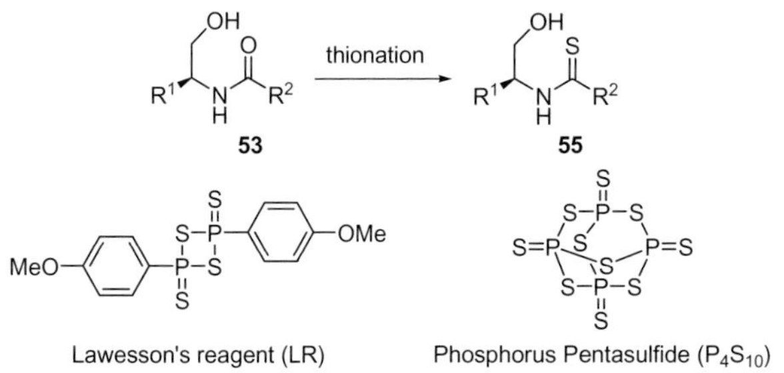

Scheme 13.18 Thionation of amide with the corresponding reagents.

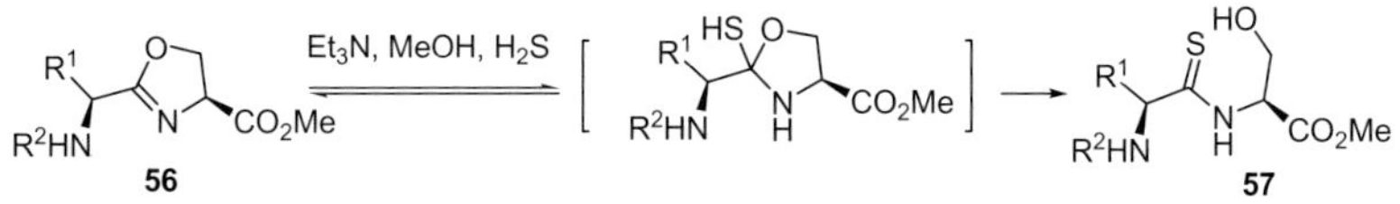

Scheme 13.19 Thiolysis of oxazoline.

When phosphorus pentasulfide is employed for thionation reaction, preactivation of P_4S_{10} with Na_2CO_3 is necessary. This limits the application of this method to only simple substrates. Lawesson's reagent converts amides into corresponding thioamides under mild reaction conditions. There is a general conclusion that Lawesson's reagent is superior to P_4S_{10}, particularly in terms of yields. Recent improvement on the use of phosphorus pentasulfide indicated that the use of hexamethyldisiloxane (HMDO) together with P_4S_{10} [30] gives superior or comparable yields to those obtained with Lawesson's reagent. However, it should be noted that thionation of precursors with multi-amide bonds with these thionating reagents usually provide products with no regio-selectivity.

Method J: Thiolysis of Oxazoline (Wipf's Method) Generally, oxazolines are synthetically more accessible than the corresponding thiazolines, so the conversion of oxazoline to thiazoline is a desirable method for thiazoline synthesis. Although direct treatment of oxazoline with P_2S_5 can produce thiazoline in moderate yield [31], the procedure developed by Wipf and co-workers proved to be more practical to bring this kind of transformation [32]. Thus, treatment of oxazoline **56** with saturated methanolic H_2S furnished the corresponding thioamide **57** smoothly (Scheme 13.19).

The scope of this transformation is quite general and chemoselectivity can also be achieved since thiolysis of threonine-derived oxazolines requires longer exposure to H_2S than thiolysis of C-5 unsubstituted serine-derived oxazolines (1–3 days vs. 1–3 hours). This method was elaborated for the total synthesis of lissoclinamide

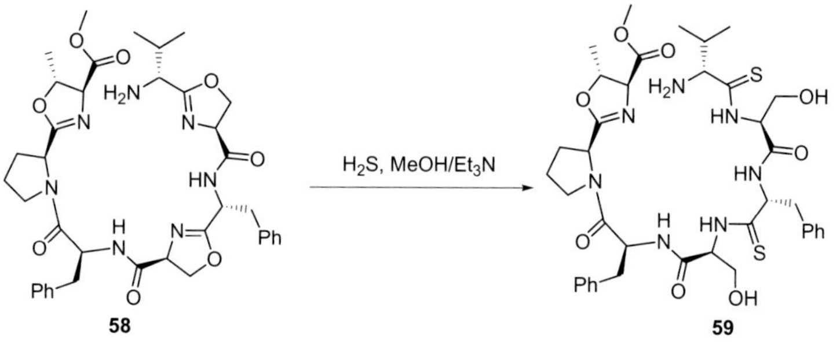

Scheme 13.20 Thiolysis of oxazoline **58**.

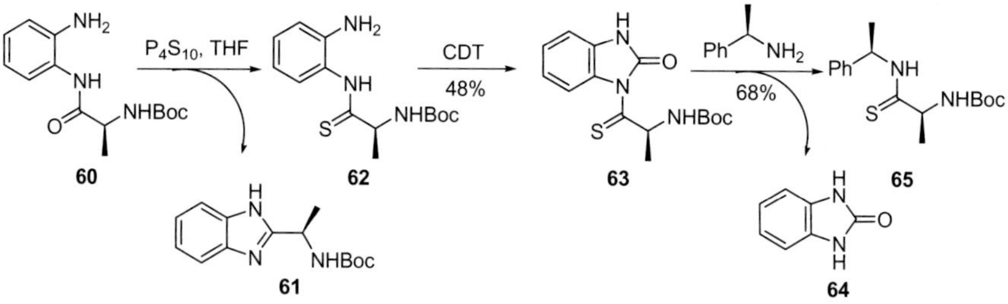

CDT = 1,1'-carbonyldi(1,2,4-triazole)

Scheme 13.21 Using the thioacyl-*N*-benzimidazolinone as thioacylating reagent.

7 [33]. Thus, treatment of trioxazoline **58** with H_2S selectively gave the bis-thioamide **59** with the 5-substituted oxazoline and other amide carbonyl groups untouched (Scheme 13.20).

The advantage of this method is that the thiocarbonyl group can be selectively introduced at a specific site for the polyamide-containing substrates. Although this method can be applied in the total synthesis of a number of thiazoline- or thiazole-bearing natural products such as curacin A [34], kalkitoxin [35], tubulysins [36], sclerltodermin A [37], lyngbyabellin B [15] and halipeptin A [38], it requires an additional cyclodehydration reaction to pre-form the oxazoline as the precursor for thiolysis. In addition, H_2S is not a commonly available reagent in most laboratories.

Method K: Using the Thioacylating Reagents This method was first developed for the site-specific incorporation of thioamide linkages into a growing peptide under mild conditions. The pioneering work was done by Zacharie and co-workers [39], who employed thioacyl-*N*-benzimidazolinones as the activated thioacylating agent. The original protocol was hampered by side reactions (e.g., resulted in the formation of **61**) and low reactivity toward aminolysis (Scheme 13.21).

13.2 General Methods for the Synthesis of Thiazoline and Thiazole Derivatives

Ko and co-workers also reported that thioacyl-*N*-phthalimides can be useful thioacylating agents, but their preparation requires more steps [40].

The most practical method for thioacylation was developed by Rapoport and co-workers [41]. This method includes the condensation of carboxylic acid with 1,2-phenylenediamine derivatives to produce the corresponding amides, followed by thionation of the amide with P_4S_{10} or Lawesson's reagent to give thioanilide **67**. Treatment of the thioanilide **67** with sodium nitrite produces, in the presence of a weak acid, the corresponding thiobenzotriazole species that is ready to react with amines to afford thioamides in high yield. (Scheme 13.22).

4-Nitro-1,2-phenylenediamine can also serve as the starting material for the generation of the corresponding thioacylating reagent, which is more reactive than that derived from 1,2-phenylenediamine. As shown in Scheme 13.22, thioacylating reagent **67**, derived from 4-nitro-1,2-phenylenediamine, can be obtained in high yield. This also indicated that the side reaction leading to side products (similar to the structure of **61**) was suppressed.

The thioacylation protocol is a very practical method to incorporate thioamide regioselectively. Pattenden and co-workers employed thioacyl-*N*-benzotriazole **70** in their total synthesis of curacin A [42]. The highly sensitive *cis*-cyclopropane-containing moiety **72** was obtained in 87% yield. (Scheme 13.23).

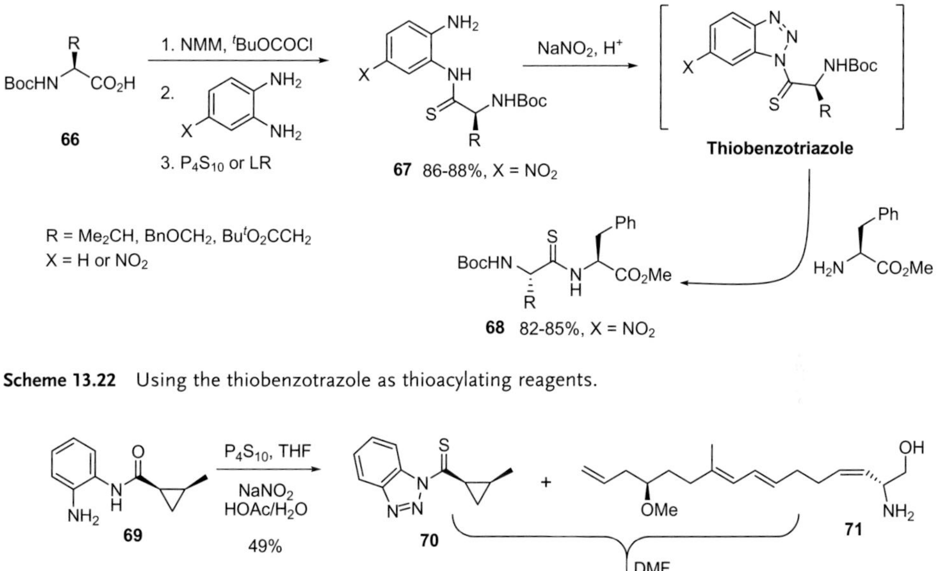

Scheme 13.22 Using the thiobenzotrazole as thioacylating reagents.

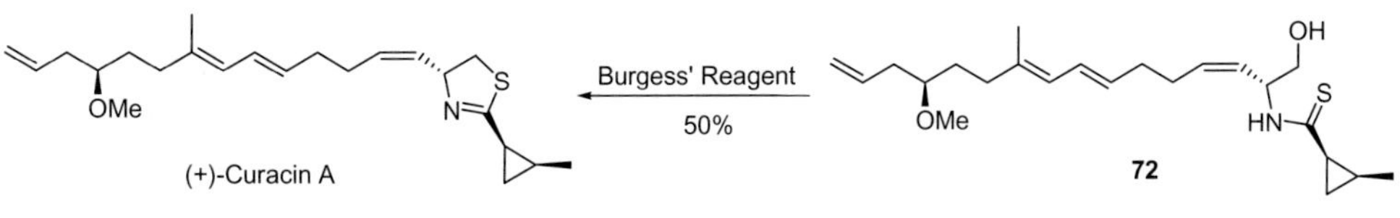

Scheme 13.23 Application of 1,2-phenylenediamine-derived reagent in total synthesis.

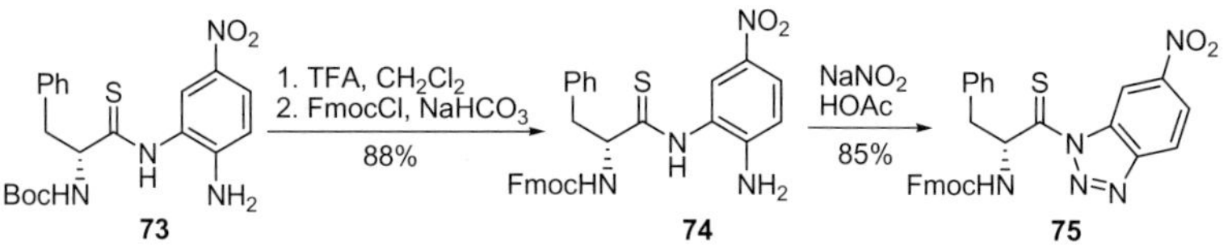

Scheme 13.24 Synthesis of building block for SPPS.

Pattenden also applied this protocol in the total synthesis of mollamide [43]. Furthermore, Ma and co-workers employed this protocol for the construction of a thiazoline ring in their total synthesis of halipeptin A [44]. Recently, this strategy served as one of the key steps in the total synthesis of grassypeptolide. The bisthioamide precursor, prepared using the thioacylation protocol, underwent a DAST-mediated cyclodehydration to give rise to the tandem thiazoline heterocycles [45].

Thioacylating reagent **75** was prepared from its precursor **74**, which was in turn prepared by manipulating the *N*-protecting group in thioamide **73**. This Fmoc-protected thiobenzotriazole reagent is quite stable and has been incorporated in solid-phase peptide synthesis of trunkamide A [46]. (Scheme 13.24).

Reagents for Cyclodehydration Conversion of β-hydroxy thioamide into the corresponding thiazoline involves intramolecular dehydration. Several reagents have been developed for this key cyclodehydration process. Thionyl chloride ($SOCl_2$), MsCl and TsCl were first employed for this transformation, but the functional group compatibility, extensive epimerization at the C-2 exomethine position and side reactions induced by β-elimination have restricted their application in the total synthesis of complex natural products.

The Mitsunobu reaction also effectively promotes the cyclodehydration of β-hydroxy thioamides [47]. However, further research conducted by Wipf and Fritch [3a] indicated that extensive epimerization at the C-2 exomethine position occurred under the Mitsunobu conditions. Difficulties for this Mitsunobu process also arise from the workup and purification procedures. It is known that triphenylphosphine oxide (Ph_3PO) and hydrazine derivative ($EtO_2CNHNHCO_2Et$) generated from the Mitsunobu conditions, are not easily removed from the reaction mixture.

Wipf and co-workers systematically investigated a number of reagents, including thionyl chloride, tosyl chloride/triethylamine, Mistunobu reaction and Burgess reagent for the cyclodehydration of β-hydroxy thioamides. Among these reaction conditions explored, Burgess reagent gave the best results [48]. Treatment of thioamide **76** with Burgess reagent afforded the corresponding thiazoline in excellent yield with high preservation of the stereochemical integrity at the C-2 exomethine position (ratio **77/78** > 97:3) (Scheme 13.25) [3a, 49].

Wipf and co-workers also developed a polyethylene glycol-linked Burgess reagent. This PEG-linked Burgess reagent was found to be equal to or more efficient than Burgess reagent for some cyclodehydration process with less than 2%

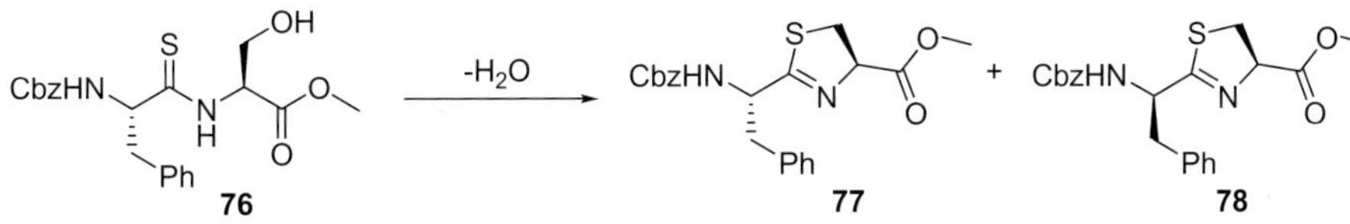

13.2 General Methods for the Synthesis of Thiazoline and Thiazole Derivatives

Method	Yield	Ratio (77 : 78)		
TsCl, Et$_3$N, CH$_2$Cl$_2$, 42 °C, 1 h	40	1	:	1
1. SOCl$_2$, 0 °C, 2 h; 2. Pyr., THF, 0 °C, 15 min	49	1	:	1
Ph$_3$P, DIAD, CH$_2$Cl$_2$, −78 ~ 22 °C, 30 min	80	78	:	22
Burgess Reagent, THF, 65 °C, 10 min	96	97	:	3

Scheme 13.25 Cyclodehydration of β-hydroxy thioamide **76**.

Burgess reagent: Et$_3$N$^{\oplus}$–S(=O)$_2$–N$^{\ominus}$–CO$_2$Me

PEG Burgess reagent: Et$_3$N$^{\oplus}$–S(=O)$_2$–N$^{\ominus}$–CO$_2$(CH$_2$)$_2$O(CH$_2$CH$_2$O)$_n$(CH$_2$)$_2$OMe

DAST: SF$_3$NEt$_2$ **Deoxo Fluor:** (MeOCH$_2$CH$_2$)$_2$NSF$_3$

Figure 13.3 Dehydration reagents.

epimerization. PEG-linked Burgess reagent has the advantage of easy removal from the reaction system, simply by filtration through a pad of silica gel [50].

Other cyclodehydration reagents include DAST [51] and its modified version, [bis(2-methoxyethyl)amino]-sulfur trifluoride (Deoxo-Fluor™) [52]. DAST-mediated cyclodehydration of β-hydroxy thioamides is best carried out at low temperature. DAST is more air- and moisture-sensitive than Burgess reagent. Deoxo-Fluor, which is thermally more stable than DAST, proved to be more efficient for the cyclodehydration process. Structures of Burgess reagent, PEG-conjugated Burgess reagent, DAST and Deoxo-Fluor are illustrated in Figure 13.3.

In Scheme 13.26 the results of cyclodehydration of β-hydroxy thioamide **79** with Deoxo-Fluor and PEG-Burgess reagents [52]are summarized. For cyclodehydration of precursor **79b**, PEG-conjugated Burgess reagent produced the desired product and the major side-product (40%) was derived from a β-elimination of the acetoxy group.

13.2.1.3 Miscellaneous

Method L: Nondehydrative Thiazoline Formation via Staudinger Reaction/Intramolecular aza-Wittig (S-AW) Process When the thiazoline ring was directly connected to an α,β-unsaturated system, methods based on cyclodehydration of β-hydroxy

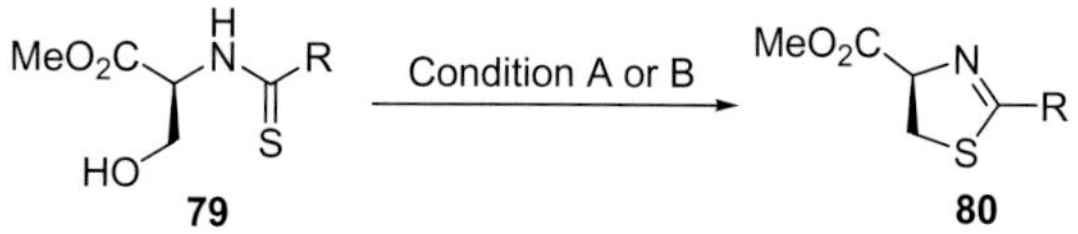

Condition A: Deoxo-Fluor, CH_2Cl_2, -20 °C
Condition B: PEG-Burgess, THF/dioxane 1/1, 85 °C

Substrates	Products	Yield (%) A/B	Reaction time (h) A/B
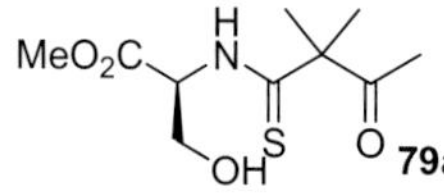 79a	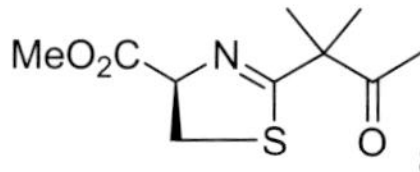 80a	80/64	1/3
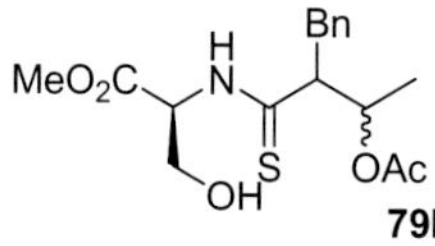 79b	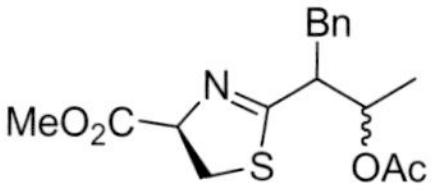 80b	90/0	0.5/8
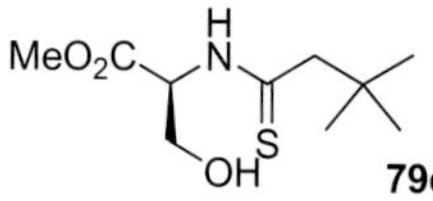 79c	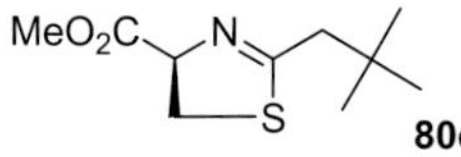 80c	85/71	2/2

Scheme 13.26 Cyclodehydration of *β*-hydroxy thioamide **79** with Deoxo-Fluor and PEG-Burgess reagents.

thioamide would normally not work since thioamides would readily undergo a Michael addition reaction prior to cyclodehydration [53]. Forsyth and co-workers employed the Staudinger / aza-Wittig process [54] for the construction of the thiazoline moiety. This nondehydrative process involves initial Staudinger reduction of **81** to form the phosphinimine that participates in the aza-Wittig reaction [55] with the neighboring thioester carbonyl to produce the desired thiazoline **82**. More examples are shown in Scheme 13.27. This method has been employed in the total synthesis of apratoxin A [56].

13.2.2 Methods for Preparation of Thiazoles

Thiazole is an important heterocycle that has generated many synthetic endeavors. Despite the fact that thiazole can be obtained from dehydrogenation of the corresponding thiazoline, methods leading to the direct formation of thiazoles have also been developed. Section 13.2.2 will introduce some practical methods for the construction of thiazoles.

13.2.2.1 Dehydrogenation of Thiazolines or Thiazolidines

Thiazole is the aromatic form of thiazoline and thiazolidine, thus the oxidation (or dehydrogenation) of these two precursors can be employed for the preparation

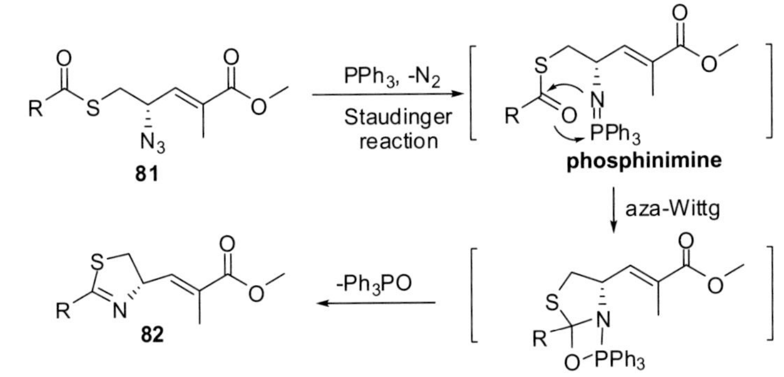

13.2 General Methods for the Synthesis of Thiazoline and Thiazole Derivatives | 475

Compd. No.	R	Yield	Compd. No.	R	Yield
81a	Me	88%	81c	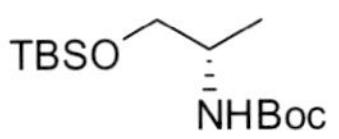 TBSO-CH(CH$_3$)-NHBoc	82%
81b	 Ph	87%	81d	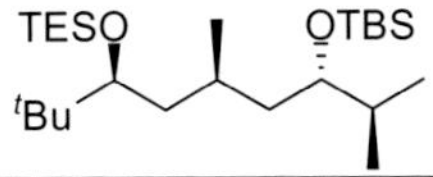 ^{t}Bu-CH(TESO)-CH$_2$-CH(OTBS)-CH$_3$	67%

Scheme 13.27 Construction of thiazolines with Staudinger aza-Wittig reaction.

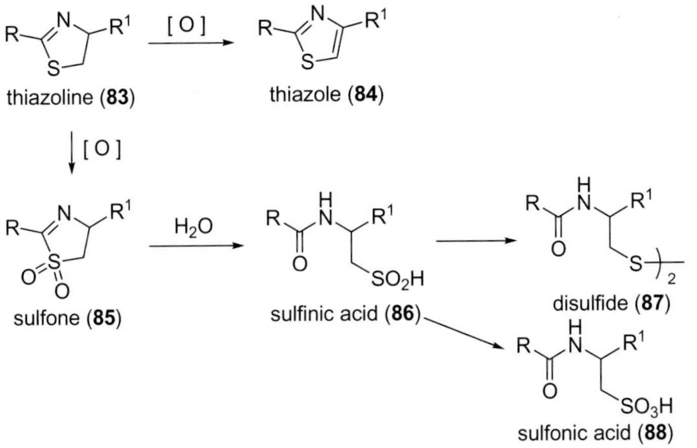

Scheme 13.28 Oxidation of thiazoline.

of thiazole derivatives. This has been extensively employed in the total synthesis of natural products. However, thiazoline and thiazolidine have proven to be unstable toward strong oxidants, such as oxone, mCPBA, KMnO$_4$, H$_2$O$_2$, peracids, and oxaziridines. These oxidants always give sulfone **85** or other undesired products because the sulfur atom could be oxidized to sulfinic acid **86**, disulfide **87**, or sulfonic acid **88** (Scheme 13.28) [31].

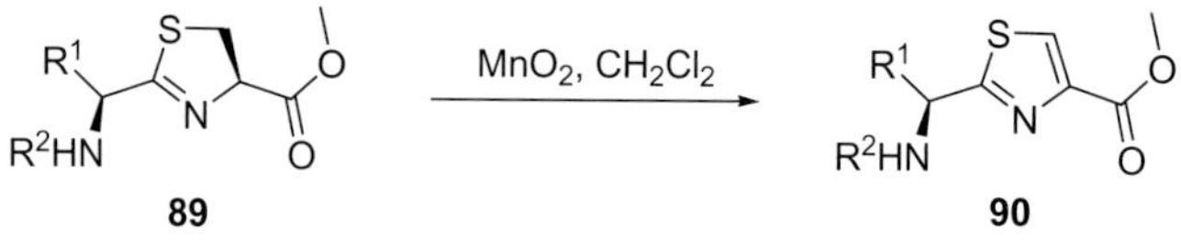

Scheme 13.29 Dehydrogenation with activated manganese dioxide.

It can be seen from Scheme 13.28 that careful selection of oxidants and reaction conditions for the dehydrogenation of thiazoline is crucial. Only a few methods for the transformation of thiazoline into the corresponding thiazoles have proven useful in the total synthesis of natural products. These methods will be introduced in the following sub-sections.

Method M: Oxidation of Thiazoline with Activated Manganese Dioxide Manganese dioxide (MnO_2) is the most frequently used oxidant for the transformation of thiazolines to the corresponding thiazoles. This method has been applied in the synthesis of a number of natural products (Scheme 13.29). The activated MnO_2 refers to the chemical prepared in the laboratory [57], and usually refers to its gamma-form [58].

The oxidation process is usually carried out at room temperature using dichloromethane as solvent. The quality of MnO_2 is crucial for the success of the reaction [22, 59]. The extent of racemization of the stereogenic center on C-2 exomethine is usually low. This method, in combination with methods for thiazoline formation, was applied in the total synthesis of cis,cis-ceratospongamide [60], didmolamides [21], dendroamide A [22], bistratamides [23], tenuecyclamides [24], lissoclinamide [61], cyclodidemnamide [62] and YM-216391 [63].

Method N: Oxidation of Thiazolidine with Chemical Manganese Dioxide (CMD) Chemical manganese dioxide (CMD) is commercially available. It is a by-product of dry battery manufacture. It is a useful oxidant for benzylic and allylic alcohols [64], and for the transformation of thiazoline and thiazolidine to the corresponding thiazoles [65]. The pioneering work was done by Shioiri and co-workers for the synthesis of optically pure thiazole-containing amino acids. Condensation of L-cystenine ester with α-amino aldehydes **91** afforded the corresponding thiazolidines **92**, which were transformed into thiazoles **90** by CMD oxidation (Scheme 13.30).

This method features the versatile condensation of aldehyde with vicinal amino thiol, followed by a direct oxidation of the resulting thiazolidine. The reaction yield is usually good to excellent for N-protected amino acid derivatives and the optical purities of the thiazole amino acid derivatives are retained. It has been used in the total synthesis of mycothiazole [66], tubulysins [67], lyngbyabellins [15], ceratospongamides [68] and amythiamicin D [69].

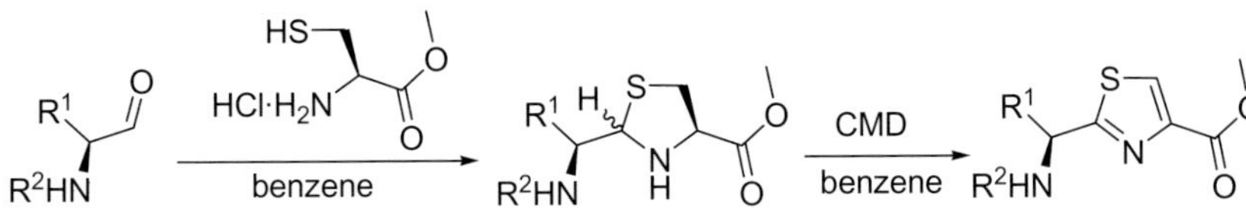

13.2 General Methods for the Synthesis of Thiazoline and Thiazole Derivatives

[Scheme showing conversion: 91 → 92 → 90, with reagents HS-containing amino acid·HCl/benzene, then CMD/benzene]

Compd. No.	R^1	Configuration	R^2	Yield (%)*	ee
91a	CH$_3$	D-	Boc	84/69	>99%
91b	(CH$_3$)$_2$CH	D-	Boc	87/61	>98%
91c	(CH$_3$)$_2$CH	L-	Boc	88/59	>97%
91d	(CH$_3$)$_2$CHCH$_2$	D-	Boc	93/69	>98%
91e	CH$_3$CH$_2$CH(CH$_3$)	L-	Boc	92/58	98.2%
91f	PhCH$_2$	D-	Boc	84/63	98.8%
91g	^{t}BuOCOCH$_2$CH$_2$	D-	Cbz	70/50	97.3%
91h	^{t}BuOCOCH$_2$CH$_2$	L-	Cbz	87/50	98.6%
91i	H	-	Cbz	74/29	-

*yield for step 1/step 2.

Scheme 13.30 Dehydrogenation with chemical manganese dioxide (CMD).

Method O: Cu(I/II)-Mediated Radical Oxidation Process In 1996, Meyers and coworkers investigated the Kharasch–Sosnovsky reaction for the oxidation of oxazoline and/or thiazoline to oxazole/thiazole in the presence of copper salts and peroxidants (Scheme 13.31). It was found that the use of the CuBr-Cu(OAc)$_2$-^{t}BuOOCOPh combination was superior to the NBS-AIBN process for the conversion of oxazoline/thiazoline **93** to oxazole/thiazole **94** [70]. The method tolerates a variety of functional groups without compromising the stereochemical integrity adjacent to the oxazoline/thiazoline rings.

One carboalkoxy group (C-4 substitution) is required for the effective transformation. This method has been applied in the synthesis of a series of natural products [71].

Method P: Dehydrogenation of Thiazoline with BrCCl$_3$ and DBU (Williams' Method) This method was developed by Williams and co-workers [72]. It has proven to be more efficient than the methods mentioned above and has been applied in the total synthesis of many natural products. Treatment of thiazolines **95** with bromotrichloromethane and 1,5-diazabicyclo-[5.4.0]-undecane (DBU) at 0 °C in methylene chloride afforded the corresponding thiazoles **96** in high yield (Scheme 13.32). Again, similar to Method O, an electron-withdrawing group (e.g., ester) attached to the C-4 position is a prerequisite to this process.

This method is also more efficient in terms of avoiding side reactions. For example, it was found to be difficult to oxidize the thiazoline **97** into thiazole **99** by the use of conventional methods, such as the activated manganese dioxide, which produced only the thiazole-furan derivate **98**. When the Williams' method was employed for this transformation, the desired thiazole **99** can be obtained in 61% yield (Scheme 13.33) [73].

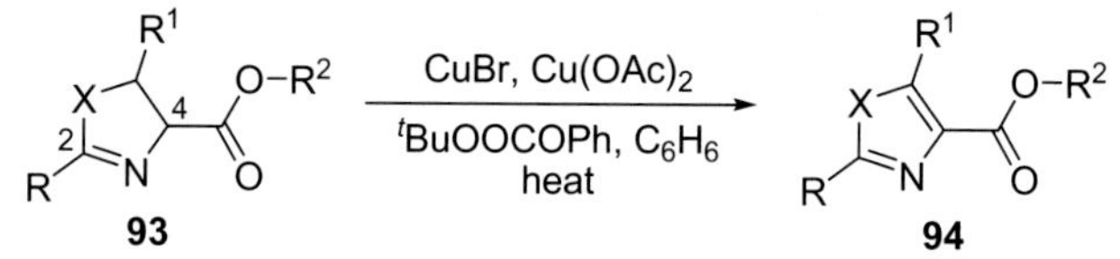

13 Thiazoline and Thiazole and Their Derivatives

Compd. No.	R	R¹	R²	X	Time (h)	Yield
93a	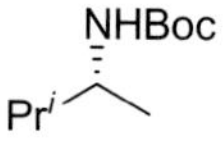 Pr^i–CH(NHBoc)–	Me	Et	S	4.5h	85%
93b	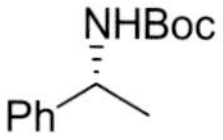 Ph–CH(NHBoc)–	H	Et	S	4.5h	76%
93c	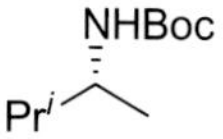 Pr^i–CH(NHBoc)–	Me	Me	O	8.5h	56%
93d	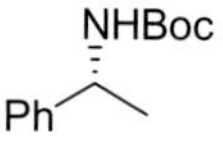 Ph–CH(NHBoc)–	H	Me	O	8.0h	54%

Scheme 13.31 Cu(I/II)-mediated radical oxidation process for thiazole formation.

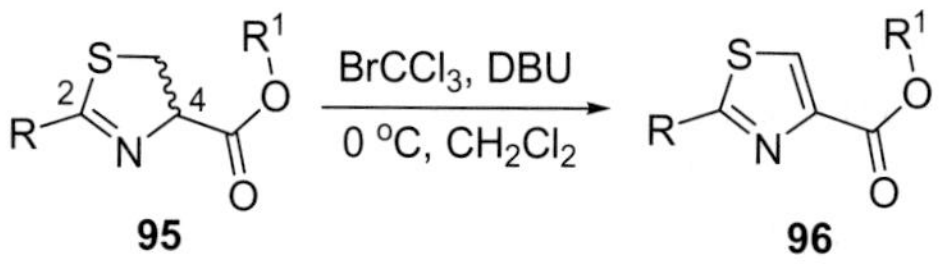

Substrates	Products	Yield (%)
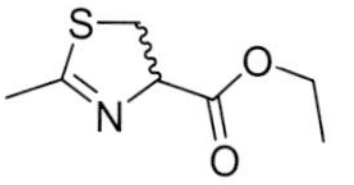 95a	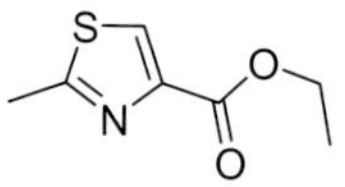 96a	95
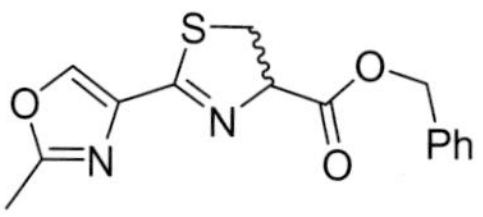 95b	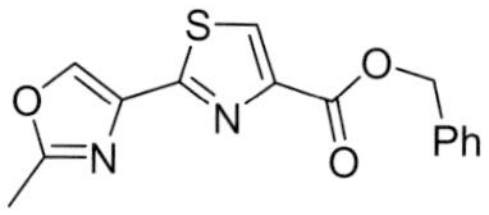 96b	92
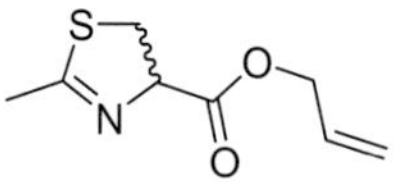 95c	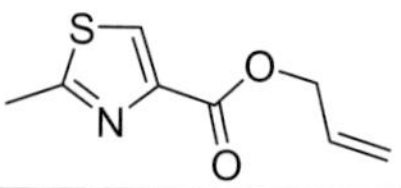 96c	95

Scheme 13.32 Dehydrogenation with BrCCl$_3$ and DBU.

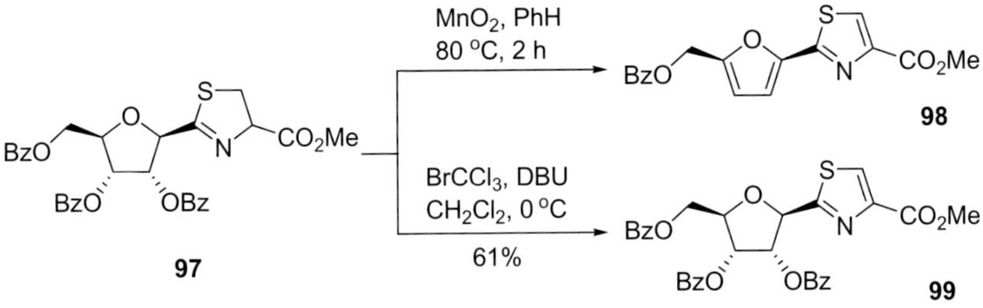

Scheme 13.33 Thiazole formation with the combination of BrCCl$_3$ and DBU.

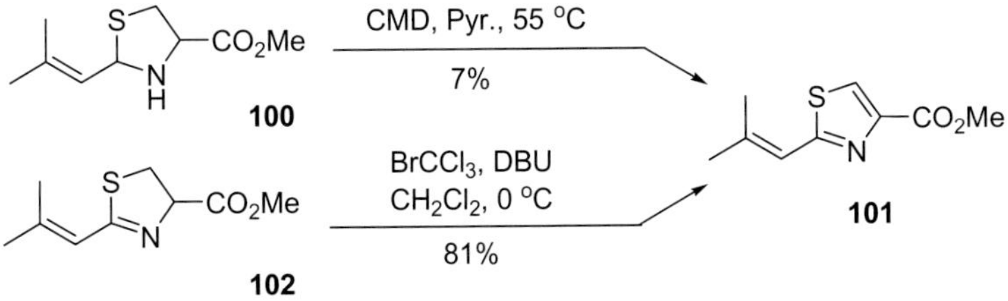

Scheme 13.34 Thiazole formation with the combination of BrCCl$_3$ and DBU.

Another example came from Shioiri's total synthesis of lyngbyabellins A and B. Treatment of thiazolidine **100** with CMD in pyridine at 55 °C only produced the corresponding thiazole **101** in 7% yield. On the other hand, dehydrogenation of thiazoline **102** with the Williams' method proved successful and afforded thiazole **101** in 81% yield (Scheme 13.34) [15].

Further application of this method in the total synthesis includes the formation of thiazole moieties of cystothiazoles [12, 74], tubulysins [36], sclerotidermin A [37], largazole [75], and GE2270s [76].

13.2.2.2 The Hantzsch Method and Its Modifications

Method Q: The Hantzsch Method The Hantzsch method is the most cited method for the preparation of thiazole-containing compounds [77]. Hantzsch's original procedure employed ethyl bromopyruvate **103** to condense with primary thioamide **104**, followed by acid-catalyzed dehydration of intermediate **105** to form the thiazole ring **106**. Although this method gives a high chemical yield of the desired product, it also leads to complete racemization of the α-stereogenic center of the thioamide partner (the C-2 exomethine of thiazole). The loss of stereochemistry was first proposed to be caused by a late stage deprotection of the α-amino group, but later it was found that it happened prior to the formation of the thiazole system due to an imine-enamine type equilibration (Scheme 13.35) [78].

The starting primary thioamides **104** are usually prepared by the thionation of the corresponding amides using P$_4$S$_{10}$ or Lawesson's reagent. Although the loss of stereochemical integrity has restricted this method to simple substrates, it had

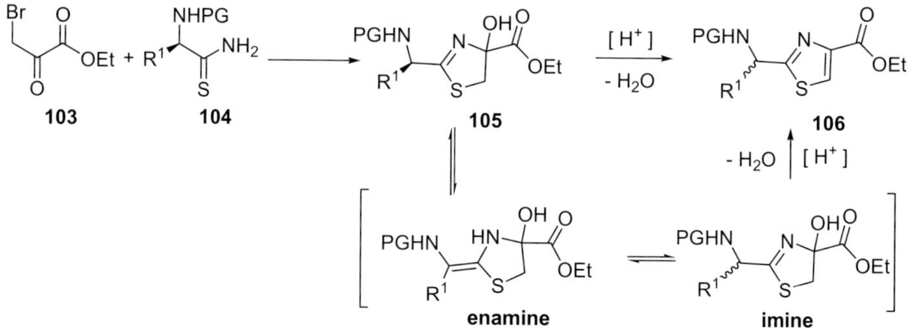

Scheme 13.35 Mechanism of racemization which occurs in the Hantzsch thiazole formation.

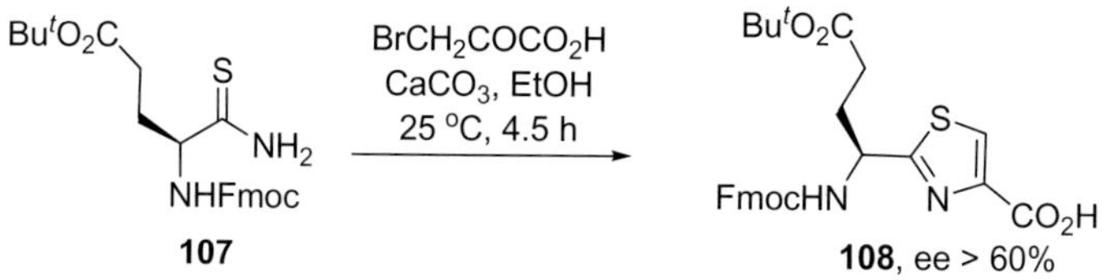

Scheme 13.36 Kelly-modified Hantzsch thiazole formation.

been employed in the total synthesis of WS75624s [79], myxothiazoles [80], cystothiazoles [81], epothilones [82], archazolids [83], tubulysins [84], tallysomycins [85], bleomycin A2 [86], didehydromirabazole [9], largazole, [20a, 87], and micrococcin P1 [88]. In order to reduce racemization and improve the feasibility of chiral thiazole synthesis, some modified protocols were developed.

Method R: Kelly's Modification In 1986, Kelly and co-workers employed $CaCO_3$ as an additive to buffer the reaction, and the ee value of the corresponding thiazole **108** was found to be higher than 60% (Scheme 13.36) [89].

Method S: Schmidt's Modification In order to suppress the racemization caused by HBr during the Hantzsch condensation, Schmidt and co-workers adopted a two-step reaction sequence by using ethyloxirane (1,2-epoxybutane) **110**, as an additive to trap the hydrobromic acid and stop the reaction at the stage of dihydrothiazole **111**. Further dehydration was then facilitated by trifluoroacetic anhydride. The ee value was improved to higher than 80%. Protective groups on the α-nitrogen and reaction solvents were crucial for the optical purity of the products. Schmidt and co-workers also found that ethanol was the best solvent for substrates **109a–109e** since only completely racemized products were obtained when dioxane or toluene was employed. While for the amino acid substrates **109f–109h**, acetone was also a suitable solvent for this reaction (Scheme 13.37) [90].

Method T: Holzapfel's Modification In 1990, Holzapfel and co-workers further optimized the Hantzsch reaction into a one-pot procedure [91]. Treatment of

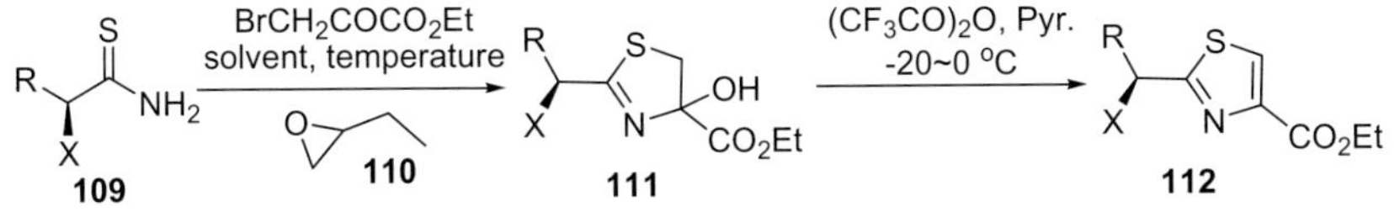

13.2 General Methods for the Synthesis of Thiazoline and Thiazole Derivatives

Scheme 13.37 Schmidt-modified Hantzsch thiazole formation.

Compd. No.	R	X	Solvent	Temp. (°C)	Yield (%)	ee (%)
109a	CH_3	OAc	EtOH	60	73	94
109b	$^iC_4H_9$	OAc	EtOH	60	91	98
109c	$^sC_4H_9$	OAc	EtOH	60	79	90*
109d	$^iC_3H_7$	OAc	EtOH	60	91	90
109e	$PhCH_2$	OAc	EtOH	60	84	>95
109f	CH_3	CbzNH	CH_3COCH_3	-10 ~ 0	64	76
109g	$^iC_4H_9$	BocNH	CH_3COCH_3	-10 ~ 0	83	>96
109h	$^sC_4H_9$	CbzNH	CH_3COCH_3	-10 ~ 0	97	>94*

* d.r. value

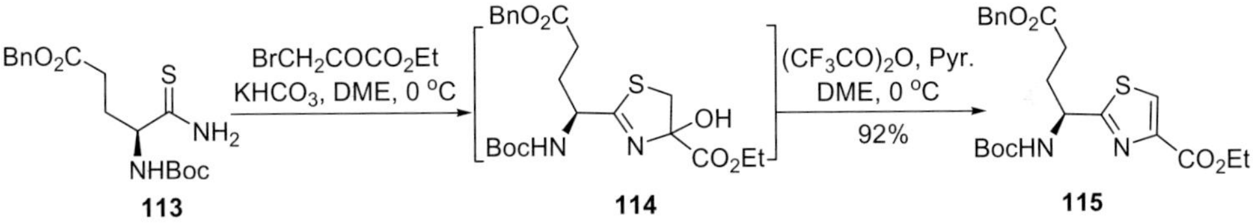

Scheme 13.38 Holzapfel-modified Hantzsch thiazole formation.

primary thioamide **113**, ethyl bromopyruvate and $KHCO_3$ in DME, the resulted intermediate was then treated with trifluoroacetic anhydride in the presence of pyridine to produce the corresponding thiazole **115** in good yield (Scheme 13.38). Holzapfel and co-workers claimed that this reaction resulted in retention of stereochemistry of the stereogenic center on C-2 exomethine.

This modified thiazole-formation procedure is highly practical and hence is frequently applied in the total synthesis of thiazole-containing natural products, such as myxothiazoles [80], epothilones [92], pateamine [93], argyrin B [94], largazole [95], ceratospongamides [96], dendroamide A [97], bistratamide D [98], mechercharmycin A [99], micrococcin P1 [100], thiostrepton [101], and amythiamicin D [69].

Method U: Meyers' Modification The modified Hantzsch thiazole synthesis was reinvestigated by Meyers and co-workers in 1994 [102]. They found that when thioamide derived from glutamic acid **116a**, valine **116b**, and phenylalanine **116c** were used as starting materials, Holzapfel's protocol produced the desired thiazoles with moderate yield and excellent enantiocontrol. However, when Holzapfel's method was applied to thioamide derived from phenylglycine **116d** or alanine **116e**, the ee value of the thiazole dropped dramatically; this was especially so for

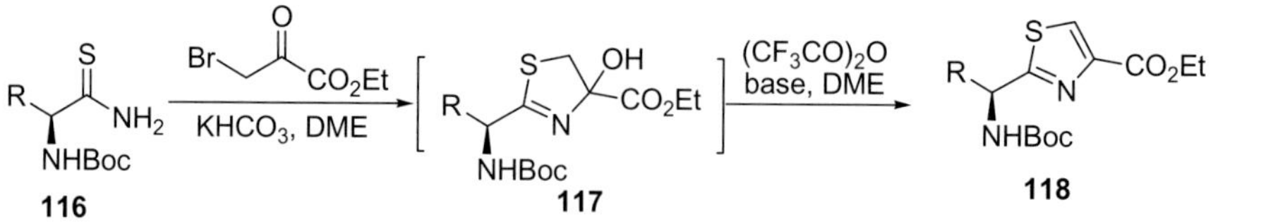

Scheme 13.39 Meyers-modified Hantzsch thiazole formation.

Compd. No.	R	Method	Base	Temp. (°C)	Yield (%)	ee (%)
116a	$BnO_2CCH_2CH_2$	Holzapfel	Pyridine	0	53	98
116b	$i\text{-}C_3H_7$	Holzapfel	Pyridine	0	84	94
116c	$PhCH_2$	Holzapfel	Pyridine	0	69	>98
116d	Ph	Holzapfel	Pyridine	0	87	2
116e	CH_3	Holzapfel	Pyridine	0	69	48
116f	CH_3	Meyers	Lutidine	-15	96	98

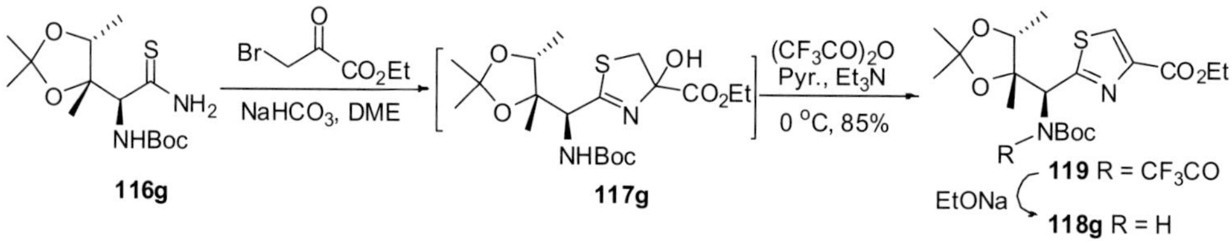

Scheme 13.40 Nicoloau-modified Hantzsch thiazole formation.

the phenylglycine derivative **116d**, as complete racemization occurred. In order to fulfill the requirement of the stereocontrolled construction of alanine-derived thiazole, Meyers and co-workers further modified Holzapfel's protocol by simply changing the base from pyridine to lutidine and lowering the reaction temperature to −15 °C. The isolation yield for the desired product **116f** was improved to 96%, and the ee value was higher than 98% (Scheme 13.39).

Method V: Nicolaou's Modification As indicated by Pattenden [3b], and Nicolaou [103] and their respective co-workers, neither the Schmidt- nor Holzapfel-modified procedures were reliable for large scale preparation of some sensitive substrates. Nicolaou and co-workers reported that pyridine-triethylamine could promote the dehydration step and concomitantly introduce the trifluoroacetyl to the protected NH-group in the substrate **119** at 0 °C. This procedure proved scalable, except that it needed one more step to remove the N-trifluoroacetyl group (Scheme 13.40).

Merritt and Bagley compared the Holzapfel-, Meyers- and Nicolaou-modified conditions for Hantzsch thiazole synthesis [104]. They noticed that the racemization in the Hantzsch thiazole synthesis occurs in the elimination step of the procedure. Modifications to the temperature and reagents employed could play critical roles in the Hantzsch thiazole synthesis and enable the preparation of chiral thiazole building blocks with complete stereocontrol and in excellent yield.

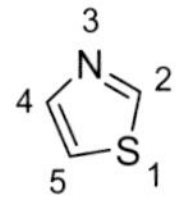

Atom number	1	2	3	4	5
calculated pi-electron density	1.97	0.87	1.19	0.96	1.01

Figure 13.4 Calculated π-electron density of thiazole ring.

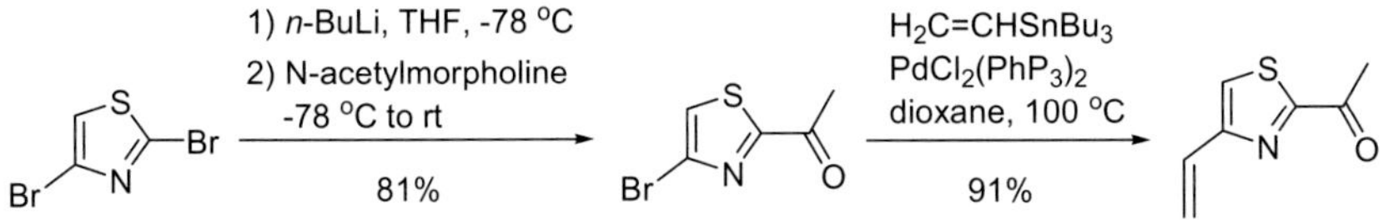

Scheme 13.41 Carbon chain elongation of 2/4-halo thiazole.

13.2.2.3 Alkylation of Thiazole or Thiazole Derivatives

Thiazole, which possesses a planar aromatic ring, has greater aromaticity than the oxazole heterocycle. The π-electron densities on its three carbons are different from each other (Figure 13.4); C2-H is the most acidic proton, C4-H is the least acidic, while C-2 is the primary site for nucleophilic substitution and C-5 is the site for electrophilic substitution.

Method W: Carbon Chain Elongation of 2/4-Halo-Thiazole and Its Derivatives Direct lithiation of the thiazole ring is possible, but most commonly lithium-halogen exchange protocol is chosen to generate the active species for further carbon-carbon bond formation starting from halide-substituted thiazoles. At the same time, thiazole halide can also be used for direct palladium-catalyzed cross-coupling reactions, such as the Stille and Negishi reactions. The most useful substrate is 2,4-dibromothiazole **120**. The different electron densities make it very easy to differentiate between C-2 and C-4 sites. These carbon-carbon bond formation reactions have been proven to be useful in natural product synthesis (Scheme 13.41) [105].

Addition of active species, derived from lithium-halogen exchange protocol, to carbonyl group was also extensively studied [106]. Due to the differences in acidity at different positions of thiazole, the halogen dance reaction might happen after lithiation [107].

13.2.2.4 Miscellaneous

Method X: Multicomponent Reaction to Assemble 2-Acyloxymethyl Thiazoles
Domling and co-workers developed a new one-pot multicomponent reaction for the synthesis of substituted 2-acyloxymethyl thiazoles. This three-component reaction involves the treatment of methyl 3-(N,N-dimethylamino)-2-isocyanoacrylate

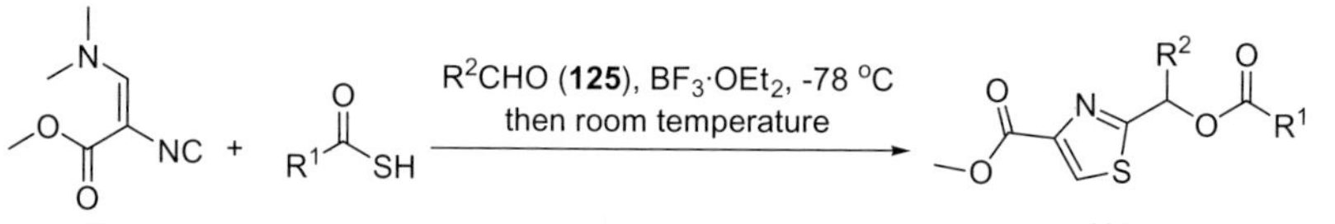

Scheme 13.42 Multicomponent reaction for thiazole synthesis.

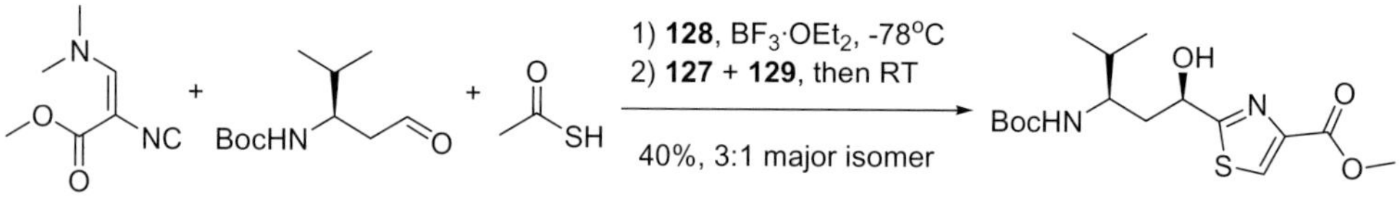

Entries	R¹	R²	Yield (%)	Entries	R¹	R²	Yield (%)
1	Me	p-MeC$_6$H$_4$	12	7	Me	c-C$_3$H$_5$	35
2	Me	PhCH$_2$CH$_2$	32	8	Me	CH$_2$CHCH$_2$(CH$_3$)$_2$C	15
3	Me	Me$_3$C	29	9	CF$_3$	Me$_2$CHCH$_2$	11
4	Me	c-C$_6$H$_{11}$	31	10	Me	p-MeOC$_6$H$_4$	10
5	Me	n-C$_5$H$_{11}$	28	11	Ph	Me$_2$CH	19
6	Me	Me$_2$CHCH$_2$	31	12	Me	MeSCH$_2$CH$_2$	23

Scheme 13.43 Synthesis of Tuv fragment of tubulysines.

123, thiocarboxylic acids 124 and aldehydes 125 in the presence of suitable Lewis acid catalyst (Scheme 13.42) [108]. The reaction is performed under mild conditions and is compatible with a wide range of functionalized starting materials.

BF$_3$·OEt$_2$ was found to be the most effective catalyst in a survey of Lewis acids, and the reaction is usually performed at −78 °C. Despite the relatively low yield, the process is still quite practical compared with the multistep synthesis of thiazole derivatives. In addition, this method has been successfully employed for the construction of the Tuv fragment 130 of tubulysins U, V and B (Scheme 13.43) [109].

Method Y: Method for the Synthesis of 5-Substituted Thiazoles 5-Substituted thiazoles or thiazolines are unique structural moieties in natural products. The routine cyclodehydration of threonine thioamide followed by dehydrogenation and the selective alkylation of the C-5 position of thiazoles were carried out for the construction of 5-methyl thiazole. One-pot thionation and cyclodehydration of the β-ketone amide 131 in the presence of Lawesson's reagent proved to be effective for the synthesis of various 5-substituted thiazoles (Scheme 13.44) [110]. This method has been employed in the construction of thiazole moieties of GE2270A [111] and amythiamicin [69].

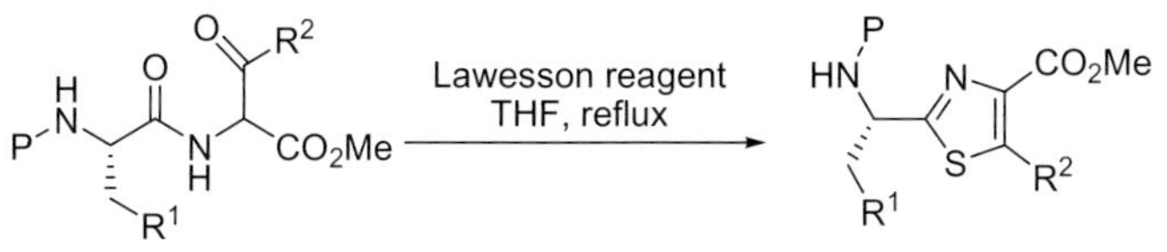

Compd.	P	R^1	R^2	Yield (%)
131a	Cbz	3-indolyl	phenyl	66
131b	Boc	phenyl	2-naphthyl	65
131c	Boc	phenyl	3-pyridyl	60
131d	Boc	3-indolyl	isopropyl	68
131e	Boc	phenyl	methyl	76

Scheme 13.44 Synthesis of 5-substituted thiazoles.

13.3
Thiazole and Thiazoline-Containing Natural Products

Natural products containing thiazoline and/or thiazole rings are listed in Table 13.1. The isolation and bioactivities of the natural products are discussed, as are the synthetic methods employed for the construction of thiazoline/thiazole moieties. Section 13.3 is primarily concerned with acyclic thiazoline/thiazole-containing natural products. Thiazoline/thiazole-containing macrocyclic natural products, including cyclopeptides, cyclodepsipeptides and macrolides are discussed in Chapter 16.

13.3.1
Thiazoline and Thiazole Embedded in Polyketides

The first natural product with a 2,4-disubstituted thiazole embedded between two acyclic polyketide chains is the mycothiazole **133**, which was first isolated from the marine sponge *Spongia mycofijiensis* collected from Vanuatu by Crews and co-workers [112] in 1988 and later by Cutignano and co-workers in the extracts of the marine sponge of the genus *Dactylospongia* [113]. Mycothiazole exhibits anthelmintic activity (*in vitro*) and high toxicity for mice. Further bioactivity study indicated that it has selective toxicity toward lung cancer cells. Its total synthesis was completed by Shioiri and co-workers in 2003 and the thiazole moiety was constructed by the use of the unique CMD oxidation of thiazoline (Method N) [66]. Shioiri and co-workers also indicated that the thiazole moiety could not be constructed *via* the standard Hantzsch method. Later, a racemic synthesis of mycothiazole was accomplished by Cossy and co-workers [105b]. The thiazole moiety was prepared by carbon chain elongation of 2,4-dibromothiazole (Method W).

Table 13.1 Thiazoline- and thiazole-containing natural products.

Items	Name	Resources	Bioactivity	Method for heterocycle [Code]	Total synthesis [Refs]
1	Mycothiazole (**133**)	*Spongia mycofijiensis* of *Dactylospongia*	Anthelmintic and cytotoxicity	N, W	[66, 105]
2	WS75624 A and B (**134**)	*Saccharothrix* sp. No. 75624	ECE inhibitors and antihypertensive activity	Q, W	[79, 106, 107]
3	Curacin A (**135**)	*Lyngbya majuscula*	Antiproliferative activity	I, J, K, B	[42b, 116]
4	Kalkitoxin (**136**)	*Lyngbya majuscula*	Ichthyotoxic and neurotoxicity	D, J	[35, 118]
5	Myxothiazoles (**137**)	*Myxobacteria*	Fungicide	Q, T	[80]
6	Melithiazoles (**138**)	*Melihangium*	Antifungal, cytotoxic	W	[105a]
7	Cystothiazoles (**139**)	*Cystobacter fuscus*	Antifungal	C-P, B-P, Q, W	[12, 74, 81, 105]
8	Epothilones (**140**)	*Myxobacterium Sorangium cellulosum*	Antitumor	T, Q	[82, 92]
9	Archazolids (**144**)	*Archangium gephyra* and *Cystobacter violaceus*	Antiproliferative activity	Q (MS)	[83]
10	Bacitracin A (**145**)	*Bacillus subtilis* and *B. licheniformis*	Antibacterial ointment	B	[10, 129]
11	Tubulysins (**146**)	*Angiococcus disciformis* and *Archangium gephyra*	Antitumor	Q, X, N, J-P, W, S	[67, 84, 110, 133]
12	Tallysomycins (**147**)	*Streptoalloteichus hindustanus*	Antitumor, antibiotic	Q	[85]
13	Bleomycin A2 (**148**)	*Streptomyces verticellus*	Antitumor, antibiotic	Q	[85, 86]
14	Tantazoles (**149**)	*Scytonema mirabile* and *Polyangium* sp.	Antitumor and anti-HIV-1 *in vitro*	B, Q, H, D	[26]
15	Thiangazole (**150**)	*Scytonema mirabile* and *Polyangium* sp.	Antitumor and anti-HIV-1 *in vitro*	A, B, Q, H, D	[13a]
16	Mirabazoles (**151**)	*Scytonema mirabile* and *Polyangium* sp.	Antitumor and anti-HIV-1 *in vitro*	B, Q, H, D	[9]
17	Pateamine	*Mycale* sp.	Antifungal and immunosuppressant	T	[93]
18	Trunkamide A	Ascidians	Antitumor	I-DAST, K (SPPS)	[46]
19	Mollamide	Ascidians	Cytotoxicity	K	[43]
20	Apratoxins	*Lyngbya* sp.	Cytotoxicity	L, G	[18, 56]
21	Halipeptins	*Haliclona* sp.	Anticancer or antimicrobial	L, G	[19, 38, 44]
22	Grassypeptolide	*Lyngbya confervoides*	Anticancer	K-DAST	[45]

Table 13.1 (Continued)

Items	Name	Resources	Bioactivity	Method for heterocycle [Code]	Total synthesis [Refs]
23	Scleritodermin A	*Scleritoderma nodosum* Thiele 1900	Cytotoxicity and antitumor	J-DAST-P	[37]
24	Argyrin B	*Archangium gephyra*	Antitumor	T	[94]
25	Hectochlorin	*Lyngbya majuscula*	Antifungal and antiproliferative activity	W	[105d]
26	Lyngbyabellins	*Lyngbya majuscula* and Dry Tortugas	Cytotoxicity	N, E-P, J-DAST, T, N	[15]
27	Largazole	*Symploca* sp.	Antiproliferative activity as HDACi	Q, T, C, G	[20, 75, 87, 95]
28	Ceratospongamides	*Sigmadocia symbiotica*	Anticancer	T, N, I-Burgess-M	[60, 68, 96]
29	Didmolamides A and B	Ascidian *Didemnum molle*	Cytotoxicity	G, M	[21]
30	Dolastatin E	Sea hare *Dolabella auricularia*	Cytotoxicity	S, I-Mitsunobu	[78, 89]
31	Dendroamide A	*Stigonema dendroideum fremy*	Antitumor	T, G-M	[22, 97]
32	Bistratamides A–J	*Lissoclinum bistratum*	Antimicrobial and antitumor	T, G-M	[23, 98]
33	Tenuecyclamides A–D	*Nostoc spongiaeforme* var. *tenue*	Antiproliferative activity	G-M	[24]
34	lissoclinamides	*Lissoclinum patella*	Cytotoxicity	B-M, K-Burgess	[33, 61]
35	Cyclodidemnamide	Sea squirt *Didemnum molle*	Cytotoxicity	B-P, K-Burgess	[62]
36	Telomestatin	*Streptomyces anulatus* 3533-SV4	Telomerase inhibitor	G (*t*-Bu ether)	[25, 63]
37	Mechercharmycin A	*Thermoactinomyces* sp.	Cytotoxicity	T	[99]
38	YM-216391	*Streptomyces nobilis*	Anticancer	I-DAST-M	[63]
39	Micrococcin P1	*Streptomyces* sp.	Antibiotic	T, Q	[88, 100]
40	Thiostrepton	*Streptomyces azureus* and *Streptomyces laurentii*	Antibiotic	T	[101]
41	Amythiamicin D	*Amycolatopsis* sp.	Antibiotic	T, N, U	[69]
42	GE2270s	*Planobispora rosea*	Antibiotic	I-DAST-P, Q, W, Y	[76, 111]

WS75624 A and B, **134a**, **134b**, were isolated from the fermentation broth of *Saccharothrix* sp. No. 75624 [114]. These compounds are potent endothelin converting enzyme (ECE) inhibitors and are potential antihypertensive agents. The thiazole ring in WS75624s is attached to a polyketide chain and a multiple substituted pyridine ring. The total synthesis of WS75624 B **134b** was completed by three research groups. Patt and co-workers employed the Hantzsch method to construct the thiazole ring [79] (Method Q). Both Gordon [106a] and Sammakia [107] and their respective co-workers applied carbon chain elongation of 2/4-bromothiazole (Method W).

Mycothiazole (**133**)

R = Me, n = 0, WS75624 A (**134a**)
R = H, n = 1, WS75624 B (**134b**)

Curacin A (**135**)

Kalkitoxin (**136**)

Curacin A and kalkitoxin are two thiazoline-containing polyketides with extraordinary bioactivities. Curacin A **135** was isolated from the cyanobacterium *Lyngbya majuscula* collected off the coast of Curaçao [115]. Curacin A exhibits mammalian cell antiproliferative activity (IC_{50} 6.8 ng mL^{-1}) and mechanism studies revealed that its bioactivity is related to its capacity to inhibit tubulin polymerization at the colchicine site. Due to its outstanding bioactivity and unique structure with a thiazoline attached to a chiral cyclopropane moiety and a Z-alkene, curacin A has generated intensive total synthetic studies. Almost all of the synthetic strategies differ in their methods of installing the chiral thiazoline moiety in the molecule [42b, 116]. The cyclodehydration of an amino-thioester was used by White *et al.* and Iwasaki *et al.*, while Aubé *et al.* and Falck *et al.* employed hydroxy thioamide as the cyclodehydration precursor, which was derived by thionation (Method I), whereas Wipf *et al.* applied the thiolysis of oxazoline followed by cyclization of the corresponding thioamide (Method J). Pattenden and co-workers developed a new strategy and featured the facile and selective thioacylation of amino alcohol with the benzotriazole-derived thioamide as the key step (Method K). Condensation of the imino ether with vicinal amino thiol substrates was employed by Kobayashi *et al.* to access the thiazoline ring in curacin A (Method B).

Kalkitoxin **136** was isolated from cyanobacterium *Lyngbya majuscula* in 2000 [35a]. It was strongly ichthyotoxic to the common goldfish with (LC_{50} 700 nM for

Carassius auratus) and toxic to brine shrimp (LC$_{50}$ 170 nM for *Artemia salina*), inhibited cell division in a fertilized sea urchin embryo assay (IC$_{50}$ 25 nM) and displayed neurotoxicity (LC$_{50}$ 3.86 nM) toward a primary cell culture of rat neurons [117]. White and co-workers reported their total synthesis of kalkitoxin [118], where the thiazoline moiety was constructed by the use of the titanium-mediated cyclodehydration of cystenine *N*-amide (Method D). Shioiri and co-workers constructed all possible isomers of kalkitoxin and established its absolute stereochemistry [35]. In this study, an oxazoline-thioamide-thiazoline protocol was employed for the construction of the heterocycle (Method J).

Myxothiazoles **137**, melithiazoles **138**, and cystothiazoles **139** all belong to a large family of secondary metabolic antibiotics that contains the β-methoxyl acrylates (β-MOAs) moiety [119]. Myxothiazoles **137** were isolated from different strains of myxobacteria containing either a bisthiazole or thiazoline-thiazole motif and exhibiting potency as agrochemical fungicides [120]. For the total synthesis of myxothiazoles A **137a** and Z **137b**, Hantzsch's condensation was applied to the left thiazole heterocycle (Method Q), and the Holzapfel method was employed for the right one to minimize racemization of the C-2 exomethine stereogenic center (Method T) [80].

Melithiazoles **138** were isolated from the cultures of *Melittangium lichenicola*, *Archangium gephyra* and *Myxococcus stipitatus* [121]. Melithiazoles A **138a** and B **138b** exhibit antifungal and cytotoxic activities, and inhibit NADH oxidation. Melithiazole C **138c** contains only one thiazole ring in the molecule and its bioactivity is much less potent than that of the other natural congeners. The total synthesis of melithiazole C **138c** was completed by an approach that involved acetylation followed by Stille alkylation of 2,4-dibromothiazole (Method W) [105a].

Cystothiazoles **139** were isolated from the myxobacterium culture broth of *Cystobacter fuscus*. They showed potent antifungal activity against the phytopathogenic fungus *Phytophthora capsici* with no effect on bacterial growth [122]. Several research groups have achieved the total synthesis of the cystothiazole family of natural products [12, 74, 81, 105]. In the total synthesis of cystothiazole A **139a** carried out by Williams and co-workers, the right-hand thiazole was prepared by the condensation of cysteine ethyl ester with the corresponding iminoether to form the thiazoline followed by dehydrogenation with BrCCl$_3$ and DBU (Method B and Method P); and the left-hand thiazole was constructed with the Hantzsch method (Method Q). Charette and co-workers constructed both thiazole moieties of cystothiazole A **139a** by condensing of iminium triflates with the cystenine ethyl esters, followed by dehydrogenation with BrCCl$_3$ and DBU (Method C and Method P). Panek and co-workers accomplished the total synthesis of cystothiazoles A **139a** and B **139b**. The bisthiazole fragment was united with the side chain through a Stille cross-coupling of a terminal (*E*)-vinylstannane with a 4-trifloylsubstituted thiazole (Method W). Akita and co-workers employed the Hantzsch thiazole synthesis for the preparation of both thiazoles in cystothiazoles A **139a** and B **139b** (Method Q).

Myxothiazole A (**137a**) R = NH$_2$
Myxothiazole Z (**137b**) R = OMe

Melithiazole C (**138c**)

Cystothiazole A (**139a**) R^1 = CH$_3$, R^2 = H
Cystothiazole B (**139b**) R^1 = CH$_3$, R^2 = OH
Cystothiazole C (**139c**) R^1 = R^2 = H

Melithiazole A (**138a**) R^1 = R^2 = CH$_2$, R^3 = CH$_3$, R^4 = H
Melithiazole B (**138b**) R^1 = R^2 = CH$_2$, R^3 = CH$_3$, Thiazole

The epothilones **140** are a family of macrolides. Epothilones A **140a** and B **140b** were the first two members isolated from culture extracts of the myxobacterium *Sorangium cellulosum* (Myxococcales) and identified as potent antitumor agents [123]. The epothilones have the same mechanism of action as taxol, but with the advantages of better solubility, obtainable in multigram quantities via total synthesis and more potent even toward multidrug-resistant cancer cell lines. In 2007, ixabepilone (Ixempra®) **141**, a semisynthetic analog of epothilone B, was approved by the FDA for the treatment of certain advanced breast cancers[1]. Extensive total synthesis and structure activity relationship studies have made the epothiolones the most promising natural product family for new anticancer drugs [124]. Only one thiazole heterocycle ring is located outside the macrocycle, and this heterocycle has been proven to be very important in terms of the potency and several analogs were in clinical trials **142** or pre-clinical trials **143** [125]. In most of the efforts toward the total synthesis of epothilones, the thiazole moiety was retrosynthetically cleaved from the macrocycle and can be constructed conveniently from

	R^1	R^2	R^3
Epothilone A (**140a**)	Me	O	H
Epothilone B (**140b**)	Me	O	Me
Epothilone C (**140c**)	Me	alkene	H
Epothilone D (**140d**)	Me	alkene	Me
Epothilone E (**140e**)	CH$_2$OH	O	H
Epothilone F (**140f**)	CH$_2$OH	O	Me

Ixabepilone (Ixempra, **141**)

1) http://www.medicalnewstoday.com/articles/85726.php (accessed 6 January 2011).

2-methyl-4-formyl thiazole derivatives. The thiazole intermediate can be easily prepared by either the Holzapfel-modified Hantzsch thiazole synthesis [92] (Method T) or the Hantzsch method [82] (Method Q).

Archazolids **144**, isolated from myxobacterial fermentation broths of *Archangium gephyra* and *Cystobacter violaceus* [126] are a group of macrolides with polyene moieties on the macrocycle and a thiazole moiety on the side-chain. Archazolid A **144a** exhibited high potency of antiproliferative activity against various cancer cell lines in subnanomolar concentrations. Its total synthesis was completed by Menche and co-workers [83], with the thiazole heterocycle constructed by the Hantzsch method (Method Q).

Archazolid A (**144a**) R^1 = Me R^2 = R^3 = H
Archazolid B (**144b**) R^1 = R^2 = R^3 = H
Archazolid C (**144c**) R^1 = Me R^2 = H R^3 =
Archazolid D (**144d**) R^1 = Me R^2 = OH R^3 =

Bacitracin A (**145**)

13.3.2
Thiazoline and Thiazole Embedded in Peptides

Bacitracin A **145**, produced nonribosomally by *Bacillus subtilis* and *B. licheniformis* [127], is widely used as an ingredient in some topical antibacterial ointments. It is active in cell wall biosynthesis. A solid-phase total synthesis of bacitracin A was completed by Griffin and co-workers [10]. The enantiomer of bacitracin A was also synthesized [128] and the *ent*-bacitracin A was found to be as potent as the natural product. This provided some proof that bacitracin exerts its antibacterial effects through interacting with bactoprenyl pyrophosphate. During the total synthesis of bacitracin A **145**, the thiazoline moiety was prepared by condensation of the imino ether with Cys-OMe (Method B), and then incorporated into the solid-phase synthesis.

Tubulysins **146** are highly active antimitotic tetrapeptides isolated from fermentation of the myxobacteria *Angiococcus disciformis* and *Archangium gephyra* [109a, 129]. Tubulysins A **146a**, D **146d** and U **146e** are representatives of the basic structures of the family. They inhibit tubulin polymerization and induce apoptosis at the cell level. Their cell growth inhibitory activity exceeds that of taxol or vinblastine by 10 times to more than 100 times [130]. Structurally, tubulysins contain four amino acids: *D-N*-methyl pipecolic acid (Mep), *L*-isoleucine (Ile), the thiazole-containing fragment tubuvaline (Tuv), and tubutyrosine (Tut) or tubuphenylalanine (Tup). The main differences among different tubulysins lie in three substitutes, the R^1-R^3 groups. From the known SAR studies, the R^1 and R^2 groups have large impact on the potency of inhibitory activity; R^3 plays a less important role, while the variation of the ring size of the Mep moiety could alter the bioactivity dramatically [131]. Due to the striking structure features and the potent bioactivities, tubulysins have attracted attention as targets for synthesis and further biological evaluation [132]. Among those elegant synthetic endeavors, various strategies have been employed for the construction of the thiazole moiety.

Ellman and co-workers applied the Hantzsch method to construct the thiazole ring, and a 3Å molecular sieve was used for the condensation of diethoxythioacetamide and ethyl bromopyruvate during the total synthesis of tubulysin D **146d** [84] (Method Q). For the total synthesis of tubulysins U **146e** and V **146f**, Domling and co-workers employed the MCR reaction to prepare the Tuv fragment. The major diastereoisomer, derived from the substrate-controlled reaction, contains the desired configuration [110] (Method X). Zanda and co-workers employed a two-step sequence for the synthesis of the thiazole-containing Tuv moiety; it involved the condensation of pyruvaldehyde with L-Cys-OMe to form the corresponding thiazolidine, followed by oxidization with activated MnO_2 [67a] (Method N). Chandrasekhar and co-workers applied a similar protocol, but used a different aldehyde derived from protected valine for the synthesis of the thiazole-containing Tuv [67b] (Method N). Wipf and co-workers employed various approaches for the construction of the Tuv-Tup fragment of tubulysins. A five-step sequence involving thiolysis of oxazoline, DAST-mediated thiazoline formation and dehydrogenation with DBU and $BrCCl_3$ (Method J and Method P) produced the targeted fragment in 24% yield. With the same starting material, a three-step approach based on the Schmidt-modified Hantzsch method (Method S) proved to be more straightforward and the overall yield was improved 41% [36]. In addition, Wipf, and Fecik and their respective co-workers [133] carried out another approach that involves the activation of 2-bromo-4-hydroxymethyl thiazole at the C-2 position with the Grignard reagent or n-BuLi, followed by addition of the metallated thiazole to aldehyde or Weinreb amide to afford the precursor for the Tuv fragment (Method W).

Tubulysin A (**146a**) R^1 = OH, R^2 = Ac, R^3 = $CH_2OC(O)CH_2CHMe_2$
Tubulysin B (**146b**) R^1 = OH, R^2 = Ac, R^3 = $CH_2OC(O)CH_2CH_2Me$
Tubulysin C (**146c**) R^1 = OH, R^2 = Ac, R^3 = $CH_2OC(O)CH_2Me$
Tubulysin D (**146d**) R^1 = H, R^2 = Ac, R^3 = $CH_2OC(O)CH_2CHMe_2$
Tubulysin U (**146e**) R^1 = H, R^2 = Ac, R^3 = H
Tubulysin V (**146f**) R^1 = R^2 = R^3 = H

Tallysomycins A, B **147** and bleomycin A_2 **148** are glycopeptides derived antitumor antibiotics. Tallysomycins were isolated from fermentation broths of *Streptoalloteichus hindustanus* [134], while bleomycin A2 was isolated from *Streptomyces verticellus*. The structure of bleomycin A2 was revised after its isolation [135] and its absolute stereochemistry was confirmed by total synthesis [86]. Bleomycin A_2

148 is now a clinical anticancer drug under the name as Blenoxane for the treatment of Hodgkin's lymphoma, carcinomas of the skin, head and neck, and testicular cancers. Bleomycin A$_2$ **148** exerts its biological effects through DNA binding and degradation. The synthesis of tallysomycins A, B **147** and bleomycin A$_2$ **148** had been intensively studied by Hecht [85], and Boger [86], and their respective co-workers. Both thiazole motifs of these glycopeptides were constructed by the use of the Hantzsch method (Method Q).

Tallysomycin A (**147a**) R$_1$ = H, R$_2$ = [sugar] OH NH$_2$ R$_3$ = OH, R$_4$ = NH(CH$_2$)$_3$CH(NH$_2$)CH$_2$CONH(CH$_2$)$_3$NH(CH$_2$)$_4$NH$_2$

Tallysomycin B (**147b**) R$_1$ = H, R$_2$ = [sugar] R$_3$ = OH, R$_4$ = NH(CH$_2$)$_3$NH(CH$_2$)$_4$NH$_2$

Bleomycin A$_2$ (**148**)   R$_1$ = CH$_3$, R$_2$ = R$_3$ = H, R$_4$ =NH(CH$_2$)$_3$S$^+$(CH$_3$)$_2$X$^-$

Tantazoles **149**, thiangazole **150**, and mirabazoles **151** are a class of linear fused thiazole/thiazoline-containing natural products. The consecutive thiazole/thiazoline moieties in these natural products are thought to be derived from the non-natural 2-methylcysteine. Both tantazoles **149** and mirabazoles **151** were isolated from the terrestrial cyanophyte *Scytonema mirabile* with pronounced solid tumor selective toxicity [136]. Thiangazole **150** was isolated from a metabolyte of *Polyangium* sp. strain P13007, and exhibited unusually high inhibitory activity against HIV-1 *in vitro* [137]. Since most of these natural products are derived from 2-methyl cysteine, the stereogenic centers on thiazoline will not epimerize during the total synthesis. For the total synthesis of didehydromirabazole **151d**, Pattenden and co-workers employed imino ether-based condensation approach for the synthesis of thiazoline moieties and the Hantzsch method for thiazole ring formation (Methods B and Q) [9]. For the total synthesis of tantazole B **149b**, Fukuyama and co-workers applied the acid-mediated tandem deprotection and cyclodehydration of the corresponding thioesters to afford the tandem-thiazoline ring system (Method H) [26]. In addition, during the total synthesis of (−)-mirabazole C **150c**, Heathcock and co-workers developed an approach based on titanium-mediated cyclodehydration of 2-methylcysteine amide for the construction of tandem-thiazolines [13a] (Method D).

Tantazole A (**149a**) (R^1 = H, R^2 = Me)
Tantazole B (**149b**) (R^1 = R^2 = Me)
Tantazole F (**149c**) (R^1 = Me, R^2 = H)

Thiangazole (**150**)

Mirabazole A (**151a**) (R^1 = R^2 = H, R^3 = Me)
Mirabazole B (**151b**) (R^1 = Me, R^2 = H, R^3 = Me)
Mirabazole C (**151c**) (R^1 = Me, R^2 = R^3 = H)
Didehydromirabazole A (**151d**) (R^1, R^2 = alkene, R^3 = Me)

13.4
Conclusions

In this chapter we have given an updated perspective on naturally occurring thiazoline/thiazole derivatives isolated from various natural sources. Most of these compounds display interesting pharmacological and biological properties, such as antimicrobial, antiviral, anticancer, and anti-inflammatory activities. They have been arranged according to similarities in structural characteristics and their biological significance, where applicable, has been adumbrated. In the past decades a wide range of synthetic methods have been reported for the synthesis of thiazoline/thiazole-containing natural products. Synthetic routes to both thiazoline and thiazole moieties have been examined and the advantages, scope, and the limitations of each method in terms of expediency, flexibility, and stereoselectivity are described. Particular emphasis has been placed on the strategies developed for the application of thiazoline- and thiazole-formation approaches to total synthesis. It is expected that the results collected here will be useful in spurring on new improvements and developments in this active and attractive area of the organic synthesis of natural products.

References

1 (a) Fontecave, M., Ollagnier-de-Choudens, S. and Mulliez, E. (2003) Biological radical sulfur insertion reactions. *Chem. Rev.*, **103**, 2149–2166; (b) Kleemann, A. and Engel, J. (2001) *Pharmaceutical Substances*, 4th edn. Thieme Medical Publishers, Stuttgart, Germany.

2 (a) Nora De Souza, M.V. (2005) Synthesis and biological activity of natural thiazoles: an important class of heterocyclic compounds. *J. Sulfur Chem.*, **26**, 429–449; (b) Dondoni, A. and Marra, A. (2004) Thiazole-mediated synthetic methodology. *Chem. Rev.*, **104**, 2557–2599; (c) Gaumont, A.C., Gulea, M. and Levillain, J. (2009) Overview of

the chemistry of 2-thiazolines. *Chem. Rev.*, **109**, 1371–1401.
3 (a) Wipf, P. and Fritch, P.C. (1994) Synthesis of peptide thiazolines from β-hydroxythiamides, an investigation of racemization in cyclodehydration protocols. *Tetrahedron Lett.*, **35**, 5397–5400; (b) Boden, C.D.J., Pattenden, G. and Ye, T. (1995) The synthesis of optically active thiazoline and thiazole derived peptides from N-protected α-amino acids. *Synlett*, 417–419.
4 Baganz, H. and Domaschke, L. (1962) Notiz über 2-Alkoxymethyl-2-thiazoline. *Chem. Ber.*, **95**, 1842–1843.
5 Ehrler, J. and Farooq, S. (1994) Total synthesls of thiangazole. *Synlett*, 702–704.
6 Boyce, R.J., Mulqueen, G.C. and Pattenden, G. (1994) Total synthesis of thiangazole, a novel inhibitor of HIV-1 from *Polyangium* sp. *Tetrahedron Lett.*, **35**, 5705–5708.
7 Kwiatkowski, S., Cracker, P.J., Chavan, A.J., Imai, N., Haley, B.E. and Watt, D.S. (1990) Total synthesis of thiangazole, a novel inhibitor of HIV-1 from *Polyangium* sp. *Tetrahedron Lett.*, **31**, 2093–2096.
8 Yamakuchi, M., Matsunaga, H., Tokuda, R., Ishizuka, T., Nakajima, M. and Kunieda, T. (2005) Sterically congested 'roofed' 2-thiazolines as new chiral ligands for copper(II)-catalyzed asymmetric Diels–Alder reactions. *Tetrahedron Lett.*, **46**, 4019–4022.
9 Pattenden, G. and Thorn, S.M. (1993) Naturally occurring linear fused thiazoline-thiazole containing metabolites: total synthesis of (−)-didehydromirabazole A, a cytotoxic alkaloid from blue-green algae. *J. Chem. Soc. Perkin Trans. I*, 1629–1636.
10 Lee, J. and Griffin, J.H. (1996) Solid-phase total synthesis of bacitracin A. *J. Org. Chem.*, **61**, 3983–3986.
11 Charette, A.B. and Chua, P. (1998) Mild method for the synthesis of thiazolines from secondary and tertiary amides. *J. Org. Chem.*, **63**, 908–909.
12 DeRoy, P.L. and Charette, A.B. (2003) Total synthesis of (+)-cystothiazole A. *Org. Lett.*, **5**, 4163–4165.
13 (a) Walker, M.A. and Heathcock, C.H. (1992) Total synthesis of (−)-mirabazole C. *J. Org. Chem.*, **57**, 5566–5568; (b) Parsons, R.L. and Heathcock, C.H. (1994) Total synthesis of (−)-thiangazole, a naturally-occurring HIV-1 inhibitor. *J. Org. Chem.*, **59**, 4733–4734; (c) Akaji, K. and Kiso, Y. (1999) Total synthesis of thiangazole. *Tetrahedron*, **55**, 10685–10694.
14 Raman, P., Razavi, H. and Kelly, J.W. (2000) Titanium (IV)-mediated tandem deprotection- cyclodehydration of protected cysteine N-amides: biomimetic syntheses of thiazoline and thiazole-containing heterocycles. *Org. Lett.*, **2**, 3289–3292.
15 Yokokawa, F., Sameshima, H., Katagiri, D., Aoyama, T. and Shioiri, T. (2002) Total syntheses of lyngbyabellins A and B, potent cytotoxic lipopeptides from the marine cyanobacterium *Lyngbya majuscula*. *Tetrahedron*, **58**, 9445–9458.
16 Sakakura, A., Kondo, R., Umemura, S. and Ishihara, K. (2007) Catalytic synthesis of peptide-derived thiazolines and oxazolines using bis(quinolinolato) dioxomolybdenum(VI) complexes. *Adv. Synth. Catal.*, **349**, 1641–1646.
17 You, S.L., Razavi, H. and Kelly, J.W. (2003) A biomimetic synthesis of thiazolines using hexaphenyl-oxodiphosphonium trifluoromethanesulfonate. *Angew. Chem. Int. Ed.*, **42**, 83–85.
18 (a) Ma, D., Zou, B., Cai, G., Hu, X. and Liu, J.O. (2006) Total synthesis of the cyclodepsipeptide apratoxin A and its analogues and assessment of their biological activities. *Chem. Eur. J.*, **12**, 7615–7626; (b) Doi, T., Numajiri, Y., Munakata, A. and Takahashi, T. (2006) Total synthesis of apratoxin A. *Org. Lett.*, **8**, 531–534; (c) Numajiri, T., Takahashi, T. and Doi, T. (2009) Total synthesis of (−)-apratoxin A, 34-epimer, and its oxazoline analogue. *Chem. Asian J.*, **4**, 111–125.
19 Hara, S., Makino, K. and Hamada, Y. (2006) Total synthesis of halipeptin A, a potent anti-inflammatory cyclodepsipeptide from a marine sponge. *Tetrahedron Lett.*, **47**, 1081–1085.

20 (a) Ghosh, A.K. and Kulkarni, S. (2008) Enantioselective total synthesis of (+)-largazole, a potent inhibitor of histone deacetylase. *Org. Lett.*, **10**, 3907–3909; (b) Numajiri, Y., Takahashi, T., Takagi, M., Shin-ya, K. and Doi, T. (2008) Total synthesis of largazole and its biological evaluaion. *Synlett*, 2483–2486. .

21 You, S.L. and Kelly, J.W. (2005) Total synthesis of didmolamides A and B. *Tetrahedron Lett.*, **46**, 2567–2570.

22 You, S.L. and Kelly, J.W. (2003) Total synthesis of dendroamide A: oxazole and thiazole construction using an oxodiphosphonium salt. *J. Org. Chem.*, **68**, 9506–9509.

23 (a) You, S.L. and Kelly, J.W. (2004) Highly efficient biomimetic total synthesis and structural verification of bistratamides e and j from lissoclinum bistratum. *Chem. Eur. J.*, **10**, 71–75; (b) You, S.L. and Kelly, J.W. (2005) The total synthesis of bistratamides F–I. *Tetrahedron*, **61**, 241–249.

24 You, S.L., Deechongkit, S. and Kelly, J.W. (2004) Solid-phase synthesis and stereochemical assignments of tenuecyclamides A–D employing heterocyclic amino acids derived from commercially available fmoc α-amino acids. *Org. Lett.*, **6**, 2627–2630.

25 Doi, T., Yoshida, M., Shin-ya, K. and Takahashi, T. (2006) Total synthesis of (R)-telomestatin. *Org. Lett.*, **8**, 4165–4167.

26 Fukuyama, T. and Xu, L. (1993) Total synthesis of (−)-tantazole B. *J. Am. Chem. Soc.*, **115**, 8449–8450.

27 (a) Hurd, R.N. and Delamater, G. (1961) The preparation and chemical properties of thionamides. *Chem. Rev.*, **61**, 45–86; (b) Schaumann, E. (1991) *Comprehensive Organic Synthesis*, vol. 6, (eds B.M. Trost and I. Fleming), Pergamon Press, Oxford, UK, pp. 419–434; (c) Jagodzinski, T.S. (2003) Thioamides as useful synthons in the synthesis of heterocycles. *Chem. Rev.*, **103**, 197–227.

28 (a) Wenker, H. (1935) The synthesis of Δ^2-oxazolines and Δ^2-thiazolines from N-acyl-2-aminoethanols. *J. Am. Chem. Soc.*, **57**, 1079–1080; (b) Bach, G. and Zahn, M. (1959) Darstellung von 2-Thiazolinen und 2-Dihydro-1,3-thiazinen. *J. Prakt. Chem.*, **8**, 68–72; (c) Handrick, G.R., Atkinson, E.R., Granchelli, F.E. and Bruni, R.J. (1965) Potential antiradiation drugs. II. 2-amino-1-alkanethiols, 1-amino-2-alkanethiols, 2-thiazolines, and 2-thiazoline-2-thiols. *J. Med. Chem.*, **8**, 762–766.

29 (a) Ozturk, T., Ertas, E. and Mert, O. (2007) Use of Lawesson's reagent in organic syntheses. *Chem. Rev.*, **107**, 5210–5278; (b) Scheibye, S., Pedersen, B.S. and Lawesson, S.O. (1978) The dimer of *p*-methoxyphenylthionophosphine sulfide as thiation reagent. a new route to thiocarboxamides. *Bull. Soc. Chim. Belg.*, **87**, 229–238; (c) Yde, B., Yousif, N.M., Pedersen, U., Thomsen, I. and Lawesson, S.O. (1984) Preparation of thiated synthons of amides, lactams and imides by use of some new phosphorus- and sulfur-containing reagents. *Tetrahedron*, **40**, 2047–2052; (d) Jensen, O.E., Lawesson, S.O., Bardi, R., Piazzesi, A.M. and Toniolo, C. (1985) Synthesis and crystal structure of two monothiated analogs of Boc-Gly-S-Ala-Aib-OMe. *Tetrahedron*, **41**, 5595–5606; (e) Unverzagt, C., Geyer, A. and Kessler, H. (1992) Chain elongation of thiodipeptides with proteases. *Angew. Chem. Int. Ed.*, vol. **31**, pp. 1229–1230. .

30 (a) Curphey, T.J. (2000) A superior procedure for the conversion of 3-oxo esters to 3H-1,2-dithiole-3-thiones. *Tetrahedron Lett.*, **41**, 9963–9966; (b) Curphey, T.J. (2002) Thionation with the reagent combination of phosphorus pentasulfide and hexamethyldisiloxane. *J. Org. Chem.*, **67**, 6461–6473; (c) Curphey, T.J. (2002) Thionation of esters and lactones with the reagent combination of phosphorus pentasulfide and hexamethyldisiloxane. *Tetrahedron Lett.*, **43**, 371–373; (d) Szostak, M. and Aube, J. (2009) Studies on the deamination of the ethyl ester of 5-amino-3-methylisoxazole-4-carboxylic acid. *Chem. Commun.*, 7122–7124.

31 Aitken, R.A., Armstrong, D.P., Galt, R.H.B. and Mesher, S.T.E. (1997) Synthesis and oxidation of chiral

2-thiazolines (4,5-dihydro-1,3-thiazoles). *J. Chem. Soc. Perkin Trans. 1*, 935–943.

32 Wipf, P., Miller, C.P., Venkatraman, S. and Fritch, P.C. (1995) Thiolysis of oxazolines: a new, selective method for the direct conversion of peptide oxazolines into thiazolines. *Tetrahedron Lett.*, **36**, 6395–6398.

33 Wipf, P. and Fritch, P.C. (1996) Total synthesis and assignment of configuration of lissoclinamide 7. *J. Am. Chem. Soc.*, **118**, 12358–12367.

34 Wipf, P. and Xu, W. (1996) Total synthesis of the antimitotic marine natural product (+)-curacin A. *J. Org. Chem.*, **61**, 6556–6562.

35 (a) Wu, M., Okino, T., ogle, N.L.M., Marquez, B.L., Williamson, R.T., Sitachitta, N., Berman, F.W., Murray, T.F., McGough, K., Jacobs, R., Colsen, K., Asano, T., Yokokawa, F., Shioiri, T. and Gerwick, W.H. (2000) Structure, synthesis, and biological properties of kalkitoxin, a novel neurotoxin from the marine cyanobacterium *Lyngbya majuscule*. *J. Am. Chem. Soc.*, **122**, 12041–12042; (b) Yokokawa, F., Asano, T., Okino, T., Gerwick, W.H. and Shioiri, T. (2004) An expeditious total synthesis of kalkitoxins: determination of the absolute stereostructure of natural kalkitoxin. *Tetrahedron*, **60**, 6859–6880.

36 Wipf, P., Takada, T. and Rishel, M.J. (2004) Synthesis of the tubuvaline-tubuphenylalanine (Tuv-Tup) fragment of tubulysin. *Org. Lett.*, **6**, 4057–4060.

37 (a) Liu, S., Cui, Y.M. and Nan, F.J. (2008) Total synthesis of the originally proposed and revised structures of scleritodermin A. *Org. Lett.*, **10**, 3765–3768; (b) Sellanes, D., Manta, E. and Serra, G. (2007) Toward the total synthesis of scleritodermin A: preparation of the C1–N15 fragment. *Tetrahedron Lett.*, **48**, 1827–1830.

38 (a) Nicolaou, K.C., Kim, D.W., Schlawe, D., Lizos, D.E., de Noronha, R.G. and Longbottom, D.A. (2005) Total synthesis of halipeptins A and D and analogues. *Angew. Chem. Int. Ed.*, **44**, 4925–4929; (b) Nicolaou, K.C., Lizos, D.E., Kim, D.W., Schlawe, D., de Noronha, R.G., Longbottom, D.A., Rodriquez, M., Bucci, M. and Cirino, G. (2006) Total synthesis and biological evaluation of halipeptins A and D and analogues. *J. Am. Chem. Soc.*, **128**, 4460–4470.

39 Zacharie, B., Sauve, G. and Penney, C. (1993) Thioacylating agents: Use of thiobenzimidazolone derivatives for the preparation of thiotuftsin analogs. *Tetrahedron*, **49**, 10489–10500.

40 Brain, C.T., Hallett, A. and Ko, S.Y. (1997) Thioamide synthesis: thioacyl-*N*-phthalimides as thioacylating agents. *J. Org. Chem.*, **62**, 3808–3809.

41 Shalaby, M.A., Grote, C.W. and Rapoport, H. (1996) Thiopeptide synthesis. α-amino thionoacid derivatives of nitrobenzotriazole as thioacylating agents. *J. Org. Chem.*, **61**, 9045–9048.

42 (a) Muir, J.C., Pattenden, G. and Ye, T. (2002) Total synthesis of (+)-curacin A, a novel antimitotic metabolite from a cyanobacterium. *J. Chem. Soc. Perkin Trans. 1*, 2243–2250; (b) Muir, J.C., Pattenden, G. and Ye, T. (1998) A concise total synthesis of (+)-curacin A, a novel cyclopropyl-substituted thiazoline from the cyanobacterium *Lyngbya majuscula*. *Tetrahedron Lett.*, **39**, 2861–2864.

43 McKeever, B. and Pattenden, G. (2003) Total synthesis of the cytotoxic cyclopeptide mollamide, isolated from the sea squirt *Didemnum molle*. *Tetrahedron*, **59**, 2701–2712.

44 Yu, S., Pan, X., Lin, X. and Ma, D. (2005) Total synthesis of halipeptin A: a potent antiinflammatory cyclic depsipeptide. *Angew. Chem. Int. Ed.*, **44**, 135–138.

45 Liu, H., Liu, Y., Kwong, S.Q., Xing, X.Y., Zhang, H., Xu, Z. and Ye, T. (2010) Total synthesis of grassypeptolide. *Chem. Commun.*, **46**, 7486–7488.

46 Caba, J.M., Rodriguez, I.M., Manzanares, I., Giralt, E. and Albericio, F. (2001) Solid-phase total synthesis of trunkamide A. *J. Org. Chem.*, **66**, 7568–7574.

47 (a) Galeotti, N., Montagne, C., Poncet, J. and Jouin, P. (1992) Formation of oxazolines and thiazolines in peptides by the Mitsunobu reaction. *Tetrahedron Lett.*, **33**, 2807–2810; (b) Wipf, P. and Miller, C. (1992) An investigation of the

Mitsunobu reaction in the preparation of peptide oxazolines, thiazolines, and aziridines. *Tetrahedron Lett.*, **33**, 6267–6270.

48 (a) Atkins, G.M. and Burgess, E.M. Jr. (1968) The reactions of an N-sulfonylamine inner salt. *J. Am. Chem. Soc.*, **90**, 4744–4745; (b) Burgess, E.M., Penton, H.R., Taylor, E.A. Jr. and Williams, W.M. (1977) Conversion of primary alcohols to urethanes via the inner salt of methyl (carboxysulfamoyl) triethylammonium hydroxide: methyl n-hexylcarbamate. *Org. Synth.*, **56**, 40–43.

49 Wipf, P., Miller, C.P. and Short, A. (1992) Stereospecific synthesis of dihydrooxazoles from serine and threonine derivatives. *Tetrahedron Lett.*, **33**, 907–910.

50 Wipf, P. and Venkatraman, S. (1996) An improved protocol for azole synthesis with PEG-supported Burgess reagent. *Tetrahedron Lett.*, **37**, 4659–4662.

51 Lafargue, P., Guenot, P. and Lellouche, J.P. (1995) Preparation of 2-thiazolines from (1,2)-thioamido-alcohols; DAST as a useful reagent. *Synlett*, **2**, 171–172.

52 Mahler, S.M., Serra, G.L., Antonow, D. and Manta, E. (2001) Deoxo-Fluor-mediated cyclodehydration of β-hydroxy thioamides to the corresponding thiazolines. *Tetrahedron Lett.*, **42**, 8143–8146.

53 Xu, Z. and Ye, T. (2005) Synthesis of 2,4,5-trisubstituted thiazoline via a novel stereoselective intramolecular conjugate addition. *Tetrahedron Asymmetry*, **16**, 1905–1912.

54 Chen, J.H. and Forsyth, C.J. (2003) Synthesis of the apratoxin 2,4-disubstituted thiazoline via an intramolecular aza-Wittig reaction. *Org. Lett.*, **5**, 1281–1283.

55 (a) Molina, P. and Vilaplana, M.J. (1994) Iminophosphoranes: useful building blocks for the preparation of nitrogen-containing heterocycles. *Synthesis*, 1197–1218; (b) Eguchi, E., Matsushita, Y. and Yamashita, K. (1992) The aza-Wittig reaction in heterocyclic synthesis. a review. *Org. Prep. Proced. Int.*, **24**, 209–243.

56 (a) Chen, J. and Forsyth, C.J. (2003) Total synthesis of apratoxin A. *J. Am. Chem. Soc.*, **125**, 8734–8735; (b) Chen, J. and Forsyth, C.J. (2004) Total synthesis of the marine cyanobacterial cyclodepsipeptide apratoxin A. *Proc. Natl. Acad. Sci. U.S.A.*, **101**, 12067–12072.

57 Goldman, I.M. (1969) Activation of manganese dioxide by azeotropic removal of water. *J. Org. Chem.*, **34**, 1979–1981.

58 (a) Fatiadi, A.J. (1976) Active manganese dioxide oxidation in organic chemistry–Part I. *Synthesis*, **65**, 65–104; (b) Fatiadi, A.J. (1976) Active manganese dioxide oxidation in organic chemistry–Part II. *Synthesis*, **65**, 133–167.

59 North, M. and Pattenden, G. (1990) Synthetic studies towards cyclic peptides. concise synthesis of thiazoline and thiazole containing amino acids. *Tetrahedron*, **46**, 8267–8290.

60 Chen, Z.Y., Deng, J.G. and Ye, T. (2003) Total synthesis of cis, cis-ceratospongamide. *ARKIVOC*, 268–285.

61 (a) Boden, C.D.J. and Pattenden, G. (1995) Total syntheses and re-assignment of configurations of the cyclopeptides lissoclinamide 4 and lissoclinamide 5 from *Lissoclinum patella*. *Tetrahedron Lett.*, **36**, 6153–6156; (b) Boden, C.D.J. and Pattenden, G. (2000) Total syntheses and re-assignment of configurations of the cyclopeptides lissoclinamide 4 and lissoclinamide 5 from *Lissoclinum patella*.*J. Chem. Soc. Perkin Trans. 1*, 875–882.

62 Boden, C.D.J., Norley, M. and Pattenden, G. (2000) Total synthesis and assignment of configuration of the thiazoline based cyclopeptide cyclodidemnamide isolated from the sea squirt *Didemnum molle*. *J. Chem. Soc. Perkin Trans. 1*, 883–888.

63 (a) Deeley, J., Bertram, A. and Pattenden, G. (2008) Novel polyoxazole-based cyclopeptides from *Streptomyces* sp. total synthesis of the cyclopeptide YM-216391 and synthetic studies towards telomestatin. *Org. Biomol. Chem.*, **6**, 1994–2010; (b) Deeley, J. and

Pattenden, G. (2005) Synthesis and establishment of stereochemistry of the unusual polyoxazole-thiazole based cyclopeptide YM-216391 isolated from *Streptomyces nobilis*. *Chem. Commun.*, 797–799.

64 Aoyama, T., Sonoda, N., Yamauchi, M., Toriyama, K., Anzai, M., Ando, A., and Shioiri, T. (1998) Chemical manganese dioxide (CMD), an efficient activated manganese dioxide. application to oxidation of benzylic and allylic alcohols. *Synlett*, 35–36.

65 Hamada, Y., Shibata, M., Sugiura, T., Kato, S. and Shioiri, T. (1987) New methods and reagents in organic synthesis. 67. A general synthesis of derivatives of optically pure 2-(l-aminoalkyl) thiazole-4-carboxylic acids. *J. Org. Chem.*, 52, 1252–1255.

66 (a) Sugiyama, H., Yokokawa, F. and Shioiri, T. (2000) Asymmetric total synthesis of (−)-mycothiazole. *Org. Lett.*, 2, 2149–2152; (b) Sugiyama, H., Yokokawa, F. and Shioiri, T. (2003) Total synthesis of mycothiazole, a polyketide heterocycle from marine sponges. *Tetrahedron*, 59, 6579–6593.

67 (a) Sani, M., Fossati, G., Huguenot, F. and Zanda, M. (2007) Total synthesis of tubulysins U and V. *Angew. Chem. Int. Ed.*, 46, 3526–3529; (b) Chandrasekhar, S., Mahipal, B. and Kavitha, M. (2009) Toward tubulysin: gram-scale synthesis of tubuvaline-tubuphenylalanine fragment. *J. Org. Chem.*, 74, 9531–9534.

68 Yokokawa, F., Sameshima, H. and Shioiri, T. (2001) Total synthesis of *cis,cis*-ceratospongamide, a bioactive thiazole-containing cyclic peptide from marine origin. *Synlett*, (SI), 986–988.

69 Hughes, R.A., Thompson, S.P., Alcaraz, L. and Moody, C.J. (2005) Total synthesis of the thiopeptide antibiotic amythiamicin D. *J. Am. Chem. Soc.*, 127, 15644–15651.

70 Meyers, A.I. and Tavares, F.X. (1996) Oxidation of oxazolines and thiazolines to oxazoles and thiazoles. Application of the Kharasch-Sosnovsky reaction. *J. Org. Chem.*, 61, 8207–8215.

71 (a) Meyers, A.I. and Tavares, F. (1994) The oxidation of 2-oxazolines to 1,3-oxazoles. *Tetrahedron Lett.*, 35, 2481–2484; (b) Tavares, F. and Meyers, A.I. (1994) Further studies on oxazoline and thiazoline oxidations. A reliable route to chiral oxazoles and thiazoles. *Tetrahedron Lett.*, 35, 6803–6806.

72 Williams, D.R., Lowder, P.D., Yu, Y.G. and Brooks, D.A. (1997) Studies of mild dehydrogenations in heterocyclic systems. *Tetrahedron Lett.*, 38, 331–334.

73 Brown, R.S., Dowden, J., Moreau, C. and Potter, B.V.L. (2002) A concise route to tiazofurin. *Tetrahedron Lett.*, 43, 6561–6562.

74 Williams, D., Patnaik, S. and Clark, M. (2001) Total synthesis of cystothiazoles A and C. *J. Org. Chem.*, 66, 8463–8469.

75 Seiser, T., Kamena, D.F. and Cramer, N. (2008) Synthesis and biological activity of largazole and derivatives. *Angew. Chem. Int. Ed.*, 47, 6483–6485.

76 Nicolaou, K.C., Zou, B., Dethe, D.H., Li, D.B., Chen, D. and Chen, Y.K. (2006) Total synthesis of antibiotics GE2270A and GE2270T. *Angew. Chem. Int. Ed.*, 45, 7786–7792.

77 (a) Meyer, R. (1979) Thiazole carozylic acids, thiazole carboxyaldehydes, and thiazole ketones, in *Thiazole and Its Derivatives, Pt. 1* (ed. J.V. Metzger), John Wiley & Sons, Inc., New York, USA, p. 519; (b) Kelly, T.R., Echavarren, A., Chandrakumar, N.S. and Kobaal, Y. (1984) [2.2](4,4′) benzophenono-(2,6)-naphthalenophane: synthesis, structure, and spectroscopic study. *Tetrahedron Lett.*, 25, 2187–2190; (c) Riordan, J.M. and Sakai, T.T. (1981) Bleomycin analogs. Synthesis and proton NMR spectral assignments of thiazole amides related to bleomycin A2. *J. Heterocycl. Chem.*, 18, 1213–1221.

78 Holzapfel, C.W. and Pettit, G.R. (1985) Synthesis of the dolastatin thiazole amino acid component (g1n)thz. *J. Org. Chem.*, 50, 2323–2327.

79 (a) Patt, W.C. and Massa, M.A. (1997) The total synthesis of the natural product endothelin converting enzyme (ECE) inhibitor, WS75624 B. *Tetrahedron Lett.*, 38, 1297–1300; (b) Massa, M.A., Patt, W.C., Ahn, K., Sisneros, A.M., Herman, S.B. and Doherty, A. (1998)

Synthesis of novel substituted pyridines as inhibitors of endothelin converting enzyme-1 (ECE-1). *Bioorg. Med. Chem. Lett.*, **8**, 2117–2122.

80 Clough, J.M., Dube, H., Martin, B.J., Pattenden, G., Reddy, K.S. and Waldron, I.R. (2006) Total synthesis of myxothiazols, novel bis-thiazole β-methoxyacrylate-based anti-fungal compounds from myxobacteria. *Org. Biomol. Chem.*, **4**, 2906–2911.

81 (a) Kato, K., Sasaki, T., Takayama, H. and Akita, H. (2003) New total synthesis of (+)-cystothiazole a based on palladium-catalyzed cyclization-methoxycarbonylation. *Tetrahedron*, **59**, 2679–2685; (b) Sasaki, T., Kato, K. and Akita, H. (2004) Determination of the absolute structure of (+)-cystothiazole B. *Chem. Pharm. Bull.*, **52**, 770–771.

82 Bode, J.W. and Carreira, E.M. (2001) Stereoselective syntheses of epothilones A and B via nitrile oxide cycloadditions and related studies. *J. Org. Chem.*, **66**, 6410–6424.

83 Menche, D., Hassfeld, J., Mayer, K., Li, J. and Rudolph, S. (2009) Modular total synthesis of archazolid A and B. *J. Org. Chem.*, **74**, 7220–7229.

84 (a) Peltier, H.M., McMahon, J.P., Patterson, A.W. and Ellman, J.A. (2006) The total synthesis of tubulysin D. *J. Am. Chem. Soc.*, **128**, 16018–16019; (b) Inami, K. and Shiba, T. (1985) Total synthesis of antibiotic althiomycin. *Bull. Chem. Soc. Jpn.*, **58**, 352–360.

85 Sznaidman, M.L. and Hecht, S.M. (2001) Studies on the total synthesis of tallysomycin. synthesis of the threonylbithiazole moiety containing a structurally unique glycosylcarbinolamide. *Org. Lett.*, **3**, 2811–2814.

86 (a) Zou, Y., Fahmi, N.E., Vialas, C., Miller, G.M. and Hecht, S.M. (2002) Total synthesis of deamido bleomycin A2, the major catabolite of the antitumor agent bleomycin. *J. Am. Chem. Soc.*, **124**, 9476–9488; (b) Boger, D.L. and Cai, H. (1999) Bleomycin: synthetic and mechanistic studies. *Angew. Chem. Int. Ed.*, **38**, 448–476.

87 Nasveschuk, C.J., Ungermannova, D., Liu, X. and Phillips, A.J. (2008) A concise total synthesis of largazole, solution structure, and some preliminary structure activity relationships. *Org. Lett.*, **10**, 3595–3598.

88 Ciufolini, M.A. and Shen, Y.C. (1997) Studies toward thiostrepton antibiotics: assembly of the central pyridine-thiazolecluster of micrococcins. *J. Org. Chem.*, **62**, 3804–3805.

89 Kelly, R.C., Gebhard, I. and Wicnienski, N. (1986) Synthesis of (R)-and (S)-(g1u) Thz and the corresponding bisthiazole dipeptide of dolastatin 3. *J. Org. Chem.*, **51**, 4590–4594.

90 Schmidt, U., Gleich, P., Griesser, H. and Utz, R. (1986) Synthesis of optically active 2-(1-hydroxyalkyl)-thiazole-4-carboxylic acids and 2-(1-aminoalky)-thiazole-4-carboxylic acids. *Synthesis*, 992–998.

91 Bredenkamp, M.W., Holzapfel, C.W. and Vanzyl, W.J. (1990) The chiral synthesis of thiazole amino acid enantiomers. *Synth. Commun.*, **20**, 2235–2249.

92 (a) Meng, D., Sorenson, E.J., Bertinato, P. and Danishefsky, S.J. (1996) Studies toward a synthesis of epothilone A: use of hydropyran templates for the management of acyclic stereochemical relationships. *J. Org. Chem.*, **61**, 7998–7999; (b) Bertinato, P., Sorenson, E.J., Meng, D. and Danishefsky, S.J. (1996) Studies toward a synthesis of epothilone a: stereocontrolled assembly of the acyl region and models for macrocyclization. *J. Org. Chem.*, **61**, 8000–8001; (c) Nicolaou, K.C., He, Y., Vourloumis, D., Vallberg, H. and Yang, Z. (1996) An approach to epothilones based on olefin metathesis. *Angew. Chem. Int. Ed.*, **35**, 2399–2401; (d) Yang, Z., Vourloumis, D., Y.H., Vallberg, H., Nicolaou and K.C. (1997) Total synthesis of epothilone A: the olefin metathesis approach. *Angew. Chem. Int. Ed.*, **36**, 166–168.

93 (a) Romo, D., Rzasa, R.M., Shea, H.A., Park, K., Langenhan, J. M., Sun, L., Akhiezer, A. and Liu, J.O. (1998) Total synthesis and immunosuppressive activity of (−)-pateamine A and related compounds: implementation of a β-lactam-based macrocyclization. *J. Am.*

Chem. Soc., **120**, 12237–12254; (b) Pattenden, G., Critcher, D.J. and Remuiñán, M. (2004) Total synthesis of (−)-pateamine a, a novel immunosuppressive agent from *Mycale* sp. *Can. J. Chem.*, **82**, 353–365.

94 Ley, S.V., Priour, A. and Heusser, C. (2002) Total synthesis of the cyclic heptapeptide argyrin B: a new potent inhibitor of T-cell independent antibody formation. *Org. Lett.*, **4**, 711–714.

95 (a) Ying, Y., Taori, K., Kim, H., Hong, J. and Luesch, H. (2008) Total synthesis and molecular target of largazole, a histone deacetylase inhibitor. *J. Am. Chem. Soc.*, **130**, 8455–8459; (b) Bowers, A., West, N., Taunton, J., Schreiber, S.L., Bradner, J.E. and Williams, R.M. (2008) Total synthesis and biological mode of action of largazole: a potent class I histone deacetylase inhibitor. *J. Am. Chem. Soc.*, **130**, 11219–11222; (c) Ren, Q., Dai, L., Zhang, H., Tan, W., Xu, Z. and Ye, T. (2008) Total synthesis of largazole. *Synlett*, 2379–2383.

96 Deng, S. and Taunton, J. (2002) Kinetic control of proline amide rotamers: total synthesis of *trans,trans*- and *cis,cis*-ceratospongamide. *J. Am. Chem. Soc.*, **124**, 916–917.

97 Xia, Z. and Smith, C.D. (2001) Total synthesis of dendroamide a, a novel cyclic peptide that reverses multiple drug resistance. *J. Org. Chem.*, **66**, 3459–3466.

98 Downing, S.V., Aguilar, E. and Meyers, A.I. (1999) Total synthesis of bistratamide D. *J. Org. Chem.*, **64**, 826–831.

99 (a) Hernandez, D., Vilar, G., Riego, E., Canedo, L.M., Cuevas, C., Albericio, F. and Alvarez, M. (2007) Synthesis of IB-01211, a cyclic peptide containing 2,4-concatenated thia- and oxazoles, via Hantzsch macrocyclization. *Org. Lett.*, **9**, 809–811; (b) Hernandez, D., Riego, E., Francesch, A., Cuevas, C., Albericio, F. and Alvarez, M. (2007) Preparation of pentaazole containing cyclopeptides: challenges in macrocyclization. *Tetrahedron*, **63**, 9862–9870.

100 Ciufolini, M.A. and Shen, Y.C. (1999) Synthesis of the bycroft-gowland structure of micrococcin P1. *Org. Lett.*, **1**, 1843–1846.

101 (b) Nicolaou, K.C., Safina, B.S., Zak, M., Estrada, A.A. and Lee, S.H. (2004) Total synthesis of thiostrepton, Part 1: construction of the dehydropiperidine/thiazoline-containing macrocycle. *Angew. Chem. Int. Ed.*, **43**, 5087–5092; (c) Nicolaou, K.C., Zak, M., Safina, B.S., Lee, S.H. and Estrada, A.A. (2004) Total synthesis of thiostrepton, part 2: construction of the quinaldic acid macrocycle and final stages of the synthesis. *Angew. Chem. Int. Ed.*, **43**, 5092–5097.

102 Aguilar, E. and Meyers, A.I. (1994) Reinvestigation of a modified Hantzsch thiazole synthesis. *Tetrahedron Lett.*, **35**, 2473–2476.

103 (a) Nicolaou, K.C., Safina, B.S., Zak, M., Lee, S.H., Nevalainen, M., Bella, M., Estrada, A.A., Funke, C., Zécri, F.J. and Bulat, S. (2005) Total synthesis of thiostrepton. Retrosynthetic analysis and construction of key building blocks. *J. Am. Chem. Soc.*, **127**, 11159–11175; (b) Nicolaou, K.C., Zak, M., Safina, B.S., Estrada, A.A., Lee, S.H. and Nevalainen, M. (2005) Total synthesis of thiostrepton. Assembly of key building blocks and completion of the synthesis. *J. Am. Chem. Soc.*, **127**, 11176–11183.

104 Merritt, E.A. and Bagley, M.C. (2007) Holzapfel-Meyers-Nicolaou modification of the Hantzsch thiazole synthesis. *Synthesis*, 3535–3541.

105 (a) Gebauer, J., Arseniyadis, S. and Cossy, J. (2007) A concise total synthesis of melithiazole C. *Org. Lett.*, **9**, 3425–3427; (b) Flohic, A.L., Meyer, C. and Cossy, J. (2005) Total synthesis of (±)-mycothiazole and formal enantioselective approach. *Org. Lett.*, **7**, 339–342; (c) Shao, J. and Panek, J.S. (2004) Total synthesis of cystothiazoles A and B. *Org. Lett.*, **6**, 3083–3085; (d) Cetusic, J.R.P., Green, F.R. III, Graupner, P.R. and Oliver, M.P. (2002) Total synthesis of hectochlorin. *Org. Lett.*, **4**, 1307–1310.

106 (a) Huang, S.T. and Gordon, D.M. (1998) Total synthesis of endothelin-converting enzyme antagonist WS75624 B. *Tetrahedron Lett.*, **39**, 9335–9338; (b) O'Dowd, H., Ploypradith, P., Xie, S.,

Shapiro, T.A. and Posner, G.H. (1999) Antimalarial artemisinin analogs. Synthesis via chemoselective C-C bond formation and preliminary biological evaluation. *Tetrahedron*, **55**, 3625–3636; (c) Ung, A.T. and Pyne, S.G. (1998) Asymmetric synthesis of (1R,2S,3R)-2-acetyl-4- (1,2,3,4-tetrahydroxybutyl) thiazole. *Tetrahedron Asymmetry*, **9**, 1395–1407.

107 Stangeland, E.L. and Sammakia, T. (2004) Use of thiazoles in the halogen dance reaction: application to the total synthesis of WS75624 B. *J. Org. Chem.*, **69**, 2381–2385.

108 Henkel, B., Beck, B., Westner, B., Mejat, B. and Domling, A. (2003) Convergent multicomponent assembly of 2-acyloxymethyl thiazoles. *Tetrahedron Lett.*, **44**, 8947–8950.

109 (a) Domling, A., Beck, B., Eichelberger, U., Sakamuri, S., Menon, S., Chen, Q.Z., Lu, Y. and Wessjohann, L.A. (2006) Total synthesis of tubulysin U and V. *Angew. Chem. Int. Ed.*, **45**, 7235–7239; (b) Pando, O., Dorner, S., Preusentanz, R., Denkert, A., Porzel, A., Richter, W. and Wessjohann, L. (2009) First total synthesis of tubulysin B. *Org. Lett.*, **11**, 5567–5569.

110 Gordon, T.D., Singh, J., Hansen, P.E. and Morgan, B.A. (1993) Synthetic approaches to the 'azole' peptide mimetics. *Tetrahedron Lett.*, **34**, 1901–1904.

111 Muller, H.M., Delgado, O. and Bach, T. (2007) Total synthesis of the thiazolyl peptide GE2270 A. *Angew. Chem. Int. Ed.*, **46**, 4771–4774.

112 Crews, P., Kakou, Y. and Quinoa, E. (1988) Mycothiazole, a polyketide heterocycle from a marine sponge. *J. Am. Chem. Soc.*, **110**, 4365–4368.

113 Cutignano, A., Bruno, I., Bifulco, G., Casapullo, A., Debitus, C., Gomez-Paloma, L. and Riccio, R. (2001) Dactylolide, a new cytotoxic macrolide from the vanuatu sponge *Dactylospongia* sp. *Eur. J. Org. Chem.*, 775–778.

114 (a) Tsurumi, Y., Ueda, H., Hayashi, K., Takase, S., Nishikawa, M., Kiyoto, S. and Okuhara, M. (1995) WS75624 A and B, new endothelin converting enzyme inhibitors isolated from *Saccharothrix* sp. No. 75624. I. taxonomy, fermentation, isolation, physico-chemical properties and biological activities. *J. Antibiot.*, **48**, 1066–1072; (b) Yoshimura, S., Tsuruni, T., Takase, S. and Okuhara, M. (1995) WS75624 A and B, new endothelin converting enzyme inhibitors isolated from *Saccharothrix* sp. No. 75624. *J. Antibiot.*, **48**, 1073–1075.

115 (a) Gerwick, W.H., Proteau, P.J., Nagle, D.G., Hamel, E., Blokhin, A. and Slate, D.L. (1994) Structure of curacin A, a novel antimitotic, antiproliferative and brine shrimp toxic natural product from the marine cyanobacterium *Lyngbya majuscula*. *J. Org. Chem.*, **59**, 1243–1245; (b) Yoo, H.D. and Gerwick, W.H. (1995) Curacins B and C, new antimitotic natural products from the marine cyanobacterium *Lyngbya majuscule*. *J. Nat. Prod.*, **58**, 1961–1965.

116 See also ref. 34 and 42a. (a) White, J.D., Kim, T.-S. and Nambu, M. (1995) Synthesis of curacin A: a powerful antimitotic /PCHO from the cyanobacterium *Lyngbya majuscule*. *J. Am. Chem. Soc.*, **117**, 5612–5613; (b) White, J.D., Kim, T.-S. and Nambu, M. (1997) Absolute configuration and total synthesis of (+)-curacin A, an antiproliferative agent from the cyanobacterium, *Lyngbya majuscule*. *J. Am. Chem. Soc.*, **119**, 103–111; (c) Hoemann, M.Z., Agrios, K.A. and Aubé, J. (1996) Total synthesis of curacin A. *Tetrahedron Lett.*, **37**, 953–956; (d) Hoemann, M.Z., Agrios, K.A. and Aubé, J. (1997) Total synthesis of (+)-curacin A, a marine cytotoxic agent. *Tetrahedron*, **53**, 11087–11098; (e) Ito, H., Imai, N., Tanikawa, S. and Kobayashi, S. (1996) Enantioselective synthesis of curacin A. 1. construction of C1–C7, C8–C17, and C18–C22 segments. *Tetrahedron Lett.*, **37**, 1795–1798; (f) Ito, H., Imai, N., Takao, K.I. and Kobayashi, S. (1996) Enantioselective synthesis of curacin A. 2. Total synthesis of curacin A by condensation of C1–C7, C8–C17, and C18–C22 segments. *Tetrahedron Lett.*, **37**, 1799–1780; (g) Onoda, T., Shirai, R., Koiso, Y. and Iwasaki, S. (1996) Asymmetric total synthesis of curacin A.

Tetrahedron Lett., **37**, 4397–4400; (h) Lai, J.-Y., Yu, J., Mekkonnen, B. and Falck, J.R. (1996) Synthesis of curacin A, an antimitotic cyclopropane-thiazoline from the marine cyanobacterium *Lyngbya majuscule*. *Tetrahedron Lett.*, **37**, 7167–7170.

117 (a) Berman, F.W., Gerwick, W.H. and Murray, T.F. (1999) Antillatoxin and kalkitoxin, ichthyotoxins from the tropical cyanobacterium lyngbya majuscula, induce distinct temporal patterns of NMDA receptor-mediated neurotoxicity. *Toxicon*, **37**, 1645–1648; (b) Manger, R.L., Leja, L.S., Lee, S.Y., Hungerford, J.M., Hokama, Y., Dickey, R.W., Granade, H.R., Lewis, R., Yasumoto, T. and Wekell, M.M. (1995) Detection of sodium channel toxins: directed cytotoxicity assays of purified ciguatoxins, brevetoxins, saxitoxin, and seafood extracts. *J. AOAC Int.*, **78**, 521–527.

118 (a) White, J.D., Lee, C.S. and Xu, Q. (2003) Total synthesis of (+)-kalkitoxin. *Chem. Commun.*, 2012–2013; (b) White, J.D., Xu, Q., Lee, C.S. and Valeriote, F.A. (2004) Total synthesis and biological evaluation of (+)-kalkitoxin, a cytotoxic metabolite of the cyanobacterium *Lyngbya majuscule*. *Org. Biomol. Chem.*, **2**, 2092–2102.

119 Sauter, H., Steglich, W. and Anke, T. (1999) Strobilurins-new fungicides for crop protection. *Angew. Chem. Int. Ed.*, **38**, 1328–1349.

120 Myxothiazole A: (a) Gerth, K., Irschik, H., Reichenbach, H. and Trowitzsch, W. (1980) Myxothiazol, an antibiotic from *Myxococcus fulvus* (myxobacterales). I. Cultivation, isolation. *J. Antibiot.*, **33**, 1474–1479; (b) Trowitzsch, W., Reifenstahl, G., Wray, V. and Gerth, K. (1980) Myxothiazol, an antibiotic from *Myxococcus fulvus* (myxobacterales). II. Structure elucidation. *J. Antibiot.*, **33**, 1480–1490; (c) Trowitzsch, W., Hofle, G. and Sheldrick, W.S. (1981) The stereochemistry of myxothiazol. *Tetrahedron Lett.*, **22**, 3829–3232. Myxothiazole Z: (a) Ahn, J.W., Woo, S.H., Lee, C.O., Cho, K.Y. and Kim, B.S. (1999) KR025, a new cytotoxic compound from *Myxococcus fulvus*. *J. Nat. Prod.*, **62**, 495–496; (b) Steinmetz, H., Forche, E., Reichenbach, H. and Hofle, G. (2000) Biosynthesis of myxothiazol Z, the ester-analog of myxothiazol A in *Myxococcus fulvus*. *Tetrahedron*, **56**, 1681–1684.

121 Bohlendorf, B., Herrmann, M., Hecht, H.J., Sasse, F., orche, F.E., Kunze, B., Reichenbach, H. and Hofle, G. (1999) Antibiotics from gliding bacteria, 85 melithiazols A–N: new antifungal β-methoxyacrylates from myxobacteria. *Eur. J. Org. Chem.*, **10**, 2601–2608.

122 (a) Ojika, M., Suzuki, Y., Tsukamoto, A., Sakagami, Y., Fudou, R., Yoshimura, T. and Yamanaka, S. (1998) Ystothiazoles A and B, new bithiazole-type antibiotics from the myxobacterium *Cystobacter fuscus*. *J. Antibiot.*, **51**, 275–281; (b) Suzuki, Y., Ojika, M., Sakagami, Y., Fudou, R. and Yamanaka, S. (1998) Cystothiazoles C–F, new bithiazole-type antibiotics from myxobacterium *Cystobacter fuscus*. *Tetrahedron*, **54**, 11399–11404.

123 (a) Bollag, D.M., McQueney, P.A., Zhu, J., Hensens, O., Koupal, L., Leisch, J., Geotz, M., Lazarides, E. and Woods, C.M. (1995) Epothilones, a new class of microtubule-stabilizing agents with a taxol-like mechanism of action. *Cancer Res.*, **55**, 2325–2333; (b) Hofle, G., Bedorf, N., Steinmetz, H., Schomburg, D., Gerth, H. and Reichenbach, H. (1996) Epothilone A and B-novel 16-membered macrolides with cytotoxic activity: isolation, crystal structure, and conformation in solution. *Angew. Chem. Int. Ed.*, **35**, 1567–1569; (c) Gerth, K., Bedorf, N., Hofle, G., Irschik, H. and Reichenback, H. (1996) Epothilons A and B : antifungal and cytotoxic compounds from *Sorangium cellulosum* (myxobacteria) production, physico-chemical and biological properties. *J. Antibiot.*, **49**, 560–563.

124 For reviews, see: (a) Mulzer, J. (2000) Epothilone B and its derivatives as novel antitumor drugs: total and partial synthesis and biological evaluation. *Monatsh. Chem.*, **131**, 205–238; (b) Nicolaou, K.C., Roschanger, F. and Vourloumis, D. (1998) Chemical biology

of epothilones. *Angew. Chem. Int. Ed.*, **37**, 2014–2045; (c) Altmann, K.H., Bold, G., Caravatti, G., End, N., Florsheimer, A., Guagnano, V., O'Reilly, T. and Wartmann, M. (2000) Epothilones and their analogs – potential new weapons in the fight against cancer. *Chimia*, **54**, 612–621; (d) Altmann, K.H. (2004) The merger of natural product synthesis and medicinal chemistry: on the chemistry and chemical biology of epothilones. *Org. Biomol. Chem.*, **2**, 2137–2152; (e) Nicolaou, K.C., Ritzen, A. and Namoto, K. (2001) Recent developments in the chemistry, biology and medicine of the epothilones. *Chem. Commun.*, 1523–1535; (f) Watkins, E.B., Chittiboyina, A.G. and Avery, M.A. (2006) Recent developments in the syntheses of the epothilones and related analogues. *Eur. J. Org. Chem.*, 4071–4084; (g) Altmann, K.-H., Pfeiffer, B., Arseniyadis, S., Pratt, B.S. and Nicolaou, K.C. (2007) The chemistry and biology of epothilones – the wheel keeps turning. *Chem. Med. Chem.*, **2**, 396–423.

125 Nicolaou, K.C., Chen, J.S. and Dalby, S.M. (2009) From nature to the laboratory and into the clinic. *Bioorg. Med. Chem.*, **17**, 2290–2303.

126 Sasse, F., Steinmetz, H., Hofle, G. and Reichenbach, H. (2003) Archazolids, new cytotoxic macrolactones from *Archangium gephyra* (Myxobacteria). *J. Antibiot.*, **56**, 520–525.

127 Ressler, C. and Kashelikar, D.V. (1966) Identification of asparaginyl and glutaminyl residues in *endo* position in peptides by dehydration-reduction. *J. Am. Chem. Soc.*, **88**, 2025–2035.

128 McDougaly, P.G. and Griffin, J.H. (2003) Enantiotracin. *Bioorg. Med. Chem. Lett.*, **13**, 2239–2240.

129 (a) Sasse, F., Steinmetz, H., Heil, J., Hofle, G. and Reichenbach, H. (2000) Tubulysins, new cytostatic peptides from myxobacteria acting on microtubuli. production, isolation, physico-chemical and biological properties. *J. Antibiot.*, **53**, 879–885; (b) Steinmetz, H., Glaser, N., Herdtweck, E., Sasse, F., Reichenbach, H. and Hofle, G. (2004) Crystal and solution structure determination, and biosynthesis of tubulysins-powerful inhibitors of tubulin polymerization from myxobacteria. *Angew. Chem., Int. Ed.*, **43**, 4888–4892.

130 Khalil, M.W., Sasse, F., Lunsdorf, H., Elnakady, Y.A. and Reichenbach, H. (2006) Mechanism of action of tubulysin, an antimitotic peptide from myxobacteria. *Chem. Bio. Chem.*, **7**, 678–683.

131 Wang, Z.Y., McPherson, P.A., Raccor, B.S., Balachandran, R., Zhu, G.Y., Day, B.W., Vogt, A. and Wipf, P. (2007) Structure-activity and high-content imaging analyses of novel tubulysins. *Chem. Biol. Drug Des.*, **70**, 75–86.

132 Neri, D., Fossati, G. and Zanda, M. (2006) Efforts toward the total synthesis of tubulysins: new hopes for a more effective targeted drug delivery to tumors. *Chem. Med. Chem.*, **1**, 175–180.

133 (a) Wipf, P. and Wang, Z.Y. (2007) Total synthesis of N^{14}-desacetoxytubulysin H. *Org. Lett.*, **9**, 1605–1607; (b) Raghavan, B., Balasubramanian, R., Stelle, J.C., Sackett, D.L. and Fecik, R.A. (2008) Cytotoxic simplified tubulysin analogues. *J. Med. Chem.*, **51**, 1530–1533.

134 (a) Kawaguchi, H., Tsukiura, H., Tomita, K., Konishi, M., Saito, K.I., Kobaru, S., Numata, K.I., Fujisawa, K.I., Miyaki, T., Hatori, M. and Koshiyama, H. (1977) Tallysomycin, a new antitumor antibiotic complex related to bleomycin I. Production, isolation, and properties. *J. Antibiot.*, **30**, 779–788; (b) Konishi, M., Saito, K.I., Numata, K.I., Tsuono, T., Asama, K., sukiura, T.H., Naito, T. and Kawaguchi, H. (1977) Tallysomycin, a new antitumor antibiotic complex related to bleomycin II structure determination of tallysomycis A and B. *J. Antibiot.*, **30**, 789–805.

135 (a) Umezawa, H., Maeda, K., Takeuchi, T. and Okami, Y. (1966) New antibiotics, bleomycin A and B. *J. Antibiot.*, **19**, 200–209; (b) Takita, T., Muraoka, Y., Nakatani, T., Fujii, A., Umezawa, Y., Naganawa, H. and Umezawa, H. (1978) Chemistry of bleomycin. XIX revised structure of bleomycin and phleomycin. *J. Antibiot.*, **31**, 801–804.

136 Tantazoles: Carmeli, S., Moore, R.E., Patterson, G.M.L., Corbett, T.H., Valeriote, F.J. (1990) Tantazoles: unusual cytotoxic alkaloids from the blue-green alga *Scytonema mirabile*. *J. Am. Chem. Soc.*, **112**, 8195–8197; Mirabazoles: Carmeli, S., Moore, R.E., Patterson, G.M.L. (1991) Mirabazoles, minor tantazole-related cytotoxins from the terrestrial blue-green alga *Scytonema mirabile*. *Tetrahedron Lett.*, **32**, 2593–2596.

137 Jansen, R., Kunze, B., Reichenbach, H., Jurkiewicz, E., Hunsmann, G. and Hofle, G. (1992) Thiangazol, a novel inhibitory HIV-1 infection in cell cultures from *Polyangium* sp. *Liebigs Ann. Chem.*, 357–359.

14
Pyrimidine and Imidazole
Vipan Kumar and Mohinder P. Mahajan

14.1
General Introduction

As the cradle of mankind, nature not only supports people's needs, but also provides many novelties. During the past two decades, natural product research in the field of pharmaceutical development has witnessed an enormous expansion, and numerous biologically active compounds currently in clinical application are either natural products or their derivatives. These naturally-occurring secondary metabolites not only serve as drugs or leads for pharmaceutical development, but in many instances lead to the discovery of novel chemical synthetic methodologies and biological interaction pathways. Even with the ready availability of advanced analytical and spectroscopic techniques, the structural revisions or assignments are made on the basis of synthetic undertakings. In addition, the isolation of new natural products continues to provide leads for pharmaceutical chemists and new methodological challenges for the synthetic chemists. The present chapter provides an insight into the natural products based on pyrimidine and imidazole nuclei because of their well-established role as biologically imperative scaffolds.

14.2
Pyrimidine-Based Natural Products

14.2.1
Introduction

The pyrimidine system, over the years has proven to be an important precursor for the synthesis of a variety of natural products of medicinal potential. It is also one of the essential pharmacophores in a number of drugs like HIV drug zidovudine, ultrashort acting barbiturates pentothal, and antimalarials pyrimethamine and trimethoprim (Figure 14.1). Because of the biological latency of pyrimidine-based heterocycles, the synthesis of this heterocyclic ring system has been exhaustively reviewed. Section 14.2 provides a detailed insight into the naturally-occurring

Heterocycles in Natural Product Synthesis, First Edition. Edited by Krishna C. Majumdar and Shital K. Chattopadhyay.
© 2011 Wiley-VCH Verlag GmbH & Co. KGaA. Published 2011 by Wiley-VCH Verlag GmbH & Co. KGaA.

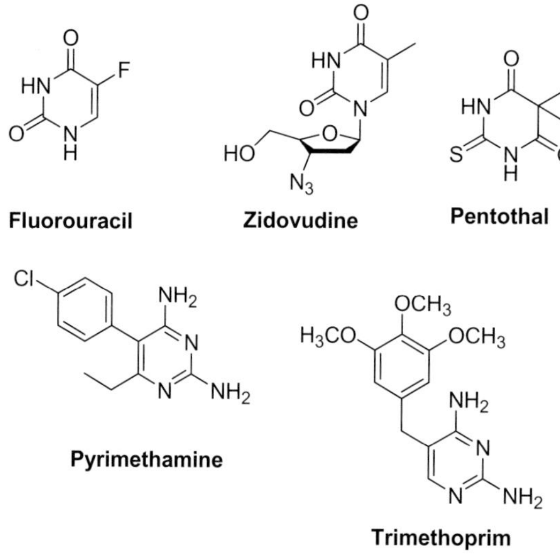

Fluorouracil

Zidovudine

Pentothal

Pyrimethamine

Trimethoprim

Figure 14.1 Pyrimidine-based natural products.

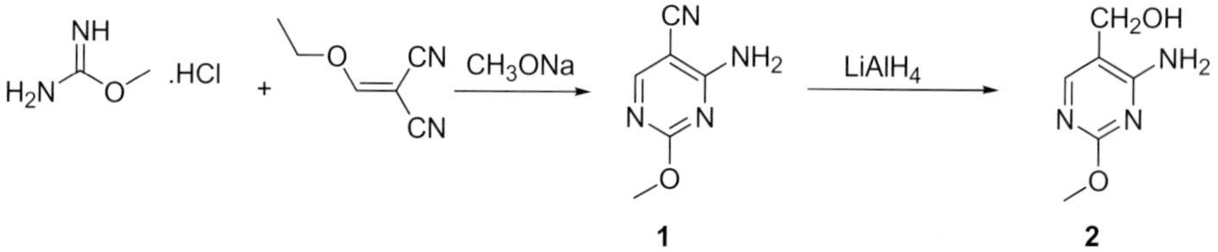

Scheme 14.1

pyrimidine derivatives, brief description of their biological activities and synthetic strategies developed over the years for the construction of pyrimidine analogs of naturally-occurring scaffolds (Table 14.1).

14.2.2
Synthesis of Pyrimidine-Based Natural Products

The simplest pyrimidine antibiotic bacimethrin **2** is a naturally-occurring thiamine antimetabolite which is active against several yeasts and bacteria *in vitro* as well as against staphylococcal infections *in vivo*. It has been prepared by the condensation of 2-methylisourea with 2-(ethoxymethylidene)malononitrile and the 4-amino-2-methoxypyrimidine-5-carbonitrile **1**, so formed was reduced easily to bacimethrin (Scheme 14.1) [9].

Crambescin, a group of cytotoxic compounds bearing 2-aminopyrimidine moieties was isolated from a Mediterranean sponge. Murphy *et al.* have recently reported the total synthesis of crambescidin 359 **9** *via* a double Michael addition

14.2 Pyrimidine-Based Natural Products

Table 14.1 Pyrimidine-based natural products.

Serial No.	Name	Structure	Source	Isolation [Ref]	Biological activity	Synthesis [Ref]
1	Amicetin		*Streptomyces vinaceusdrappus* and *S. fascicularis*	[1]	Bacterial peptidyl transferase inhibitor [2], protection against *Hepes simplex* virus Type-I [3]	[4]
2	Bacimethrin		*Bacillus megatherium*	[5, 6]	Inhibits the growth of *Salmonella enterica* serovar Typhimurium (Typhoid fever) [7] and anticancer agent [8]	[9]
3	Blastcidin-S		*Streptomyces griseochromogenes*	[10]	Has a potent curative effect on rice blast disease [11]	[12]
4	Bleomycin		*Streptomyces verticillus.*	[13]	Treatment of squamous cell cancers, melanoma, sarcoma, testicular cancer, Hodgkin's and non-Hodgkin's lymphoma [14]	[15]

(Continued)

Table 14.1 (Continued)

Serial No.	Name	Structure	Source	Isolation [Ref]	Biological activity	Synthesis [Ref]
5	Crambescin		*Crambe crambe* (Mediterranean sponge)	[16]	Inhibits HIV-1 envelope-mediated fusion [17], alters the morphology of neuronal cell lines [18] general cytotoxicity toward mammalian cells [19] potent and reversible blockage of Ca^{2+} channels [20]	[21]
6	Cylindrospermopsin		*Cylindrospermopsis raciborskii*	[22, 23]	Responsible for severe outbreak of hepatoenteritis in tropical Australia, Cyanotoxin toxic to liver and kidney tissues [24]	[25]
7	Ezomycin		*Nocardia mesenterica*	[26]	Antifungal antibiotic [27]	[28]
8	Febrifugin		*Dichroa febrifuga* and *Hydrangea umbellate*	[29]	Antimalarial [30]	[31]
9	Isofebrifugin		*Dichroa febrifuga* and *Hydrangea umbellate*	[32]	Antimalarial [33]	[31]

14.2 Pyrimidine-Based Natural Products

10	Manzacidin	*Hymeniacidon* and *Astrosclera willeyana*	[34, 35]	α-adrenoreceptor blockers, serotonin antagonists and actomyosin ATPase activators [36]	[37]
11	Meridianin	*Aplidium meridianum*	[38]	Anticancer [39]	[40]
12	Polyoxin	*Streptomyces cacoi var. asoenis*	[41]	Antifungal [42]	[43]
13	Sparsomycin	*Streptomyces sparsogenes* or *Streptomyces cuspidosporus*	[44]	Antitumor antibiotic [45]	[46]
14	Toxopyrimidine	*Toxoplasma gondii*	[5, 6]	Anti-cancer agent [8]	[47]
15	Variolin	*Kirkpatrickia variolosa*	[48]	Anticancer [49]	[50]

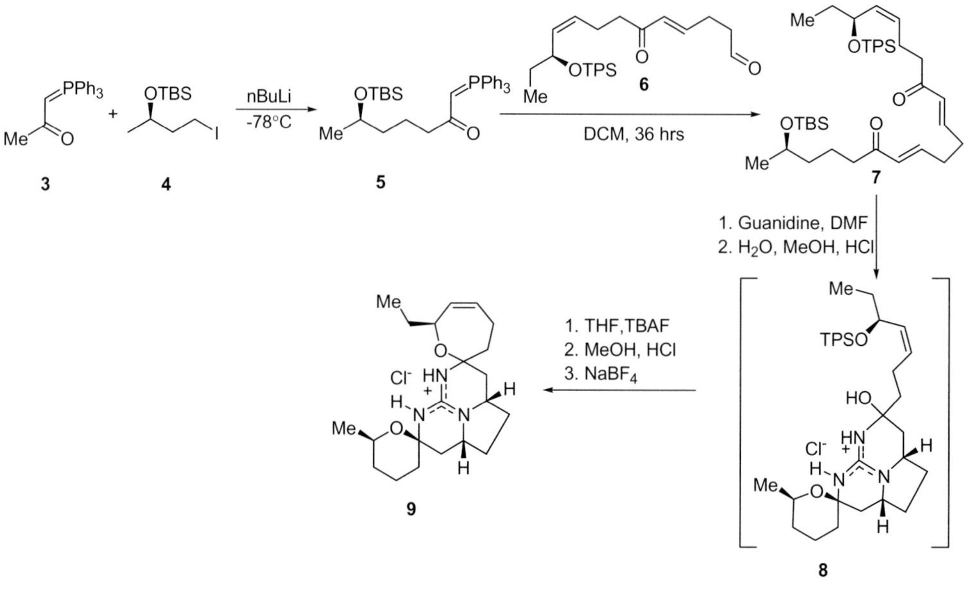

Scheme 14.2

of guanidine to a suitably functionalized bis-enone **7** obtained by the Wittig reaction of ylide **5** with aldehyde **6** [21]. The methodology involved the preparation of ylide **5** by reaction of iodide **4** with lithiated acetylmethylene triphenylphosphorane **3**. The reaction of guanidine with **7** at 0 °C followed by addition of a solution of methanolic HCl gave **8** which upon treatment with 5 equiv of TBAF in THF and subsequent addition of methanolic HCl and sodium fluoroborate led to **9** (Scheme 14.2).

A number of alkaloids containing an ester linked to a pyrimidine moiety have been isolated from marine sponges. The bromopyrrole derivative manzacidin C **16** was isolated from the Okinawan sponge *Hymeniacidon* sp. The methodology employed for the synthesis of manzacidine C is shown in Scheme 14.3 and involved a single step synthesis of oxathiazinanes **11** from ethyl glyoxalate **10** followed by its conversion to carbamate *via* protection with Boc$_2$O. The carbamate on treatment with sodium azide gave the ring opened product **12** which was sequentially reduced and acylated to generate formamide **13**. The treatment of **13** with POCl$_3$ and 2,6-di(*t*-butyl)-4-methylpyridine resulted in the formation of tetrahydropyrimidine **14** without attendant epimerization at C-4. The deprotection of **14** to **15** followed by acylation with pyrrole derivative led to the formation of the desired manzacidin C **16** (Scheme 14.3) [37].

Another approach employed for the synthesis of manzacidin used an isothiourea-iodocyclization strategy. The methodology involved the formation of glycine ester imine **18**, by the condensation reaction of **17** with benzophenone, followed by its alkylation in the presence of NaH to obtain **19**. Selective hydrolysis of imine **19** to amine **20** followed by introduction of thiourea moiety resulted in the formation of

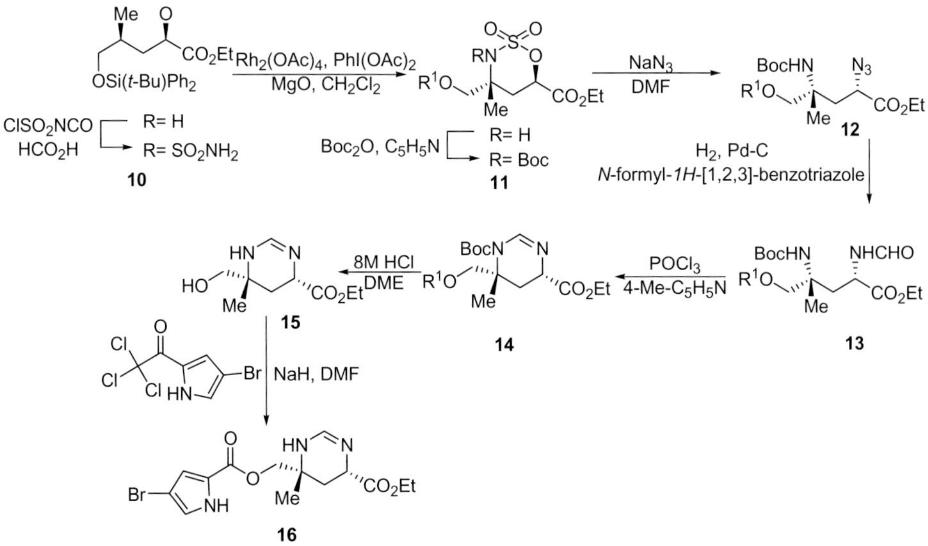

14.2 Pyrimidine-Based Natural Products

Scheme 14.3

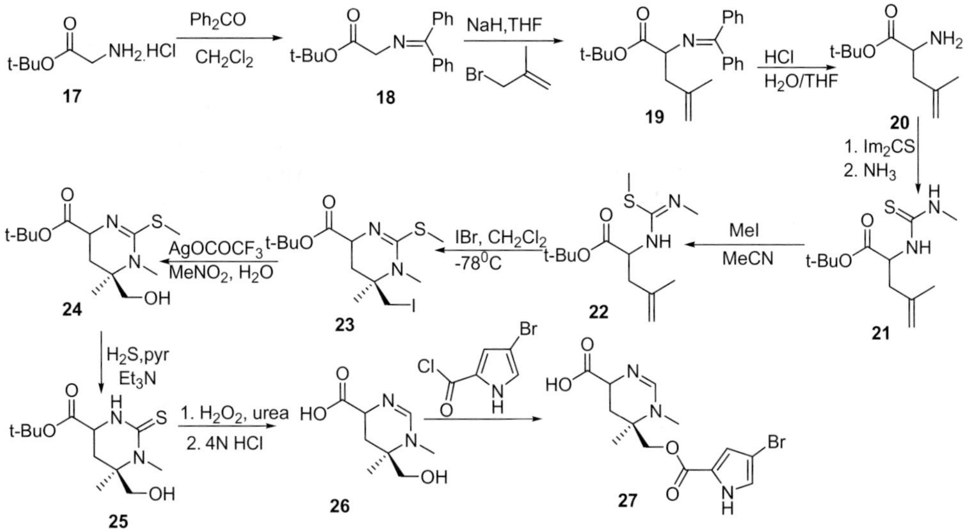

Scheme 14.4

21. Its methylation to isothiourea **22** and iodocylization with IBr led to the diasteroselective synthesis of tetrahydropyrimidine **23**. Ag-salt promoted solvolysis of **23** afforded alcohol **24** which through a sequence of reactions *via* **25** led to **26**, a substrate suitable for condensation with the pyrrole derivatives to result in manzacidin D **27** (Scheme 14.4) [37].

Variolins constitute non-traditional guanidine-based alkaloids which possess a broad spectrum of bioactivities and contain the guanidine moiety in the guise of 2-aminopyrimidine rings. Variolins are the first examples of either terrestrial or marine natural products having a pyrido[3',2',4,5]pyrrolo[1,2-c]-pyrimidine system. Recently, variolin B **35** has attracted considerable interest as a synthetic target because of its potent antitumor activity, and also because it is no longer possible to collect the original sponge source. A variety of strategies has been developed to synthesize the pyrido[3',2':4,5]pyrrolo[1,2-c]pyrimidine core of the variolins, the majority of which employ conventional heterocyclic chemistry. One of the recently reported synthetic strategies involved an initial aza-Wittig reaction of iminophosphorane **28** with α-methyl-benzylisocyanate to yield tricyclic pyrimidopyrrolopyridine **29**, the central core of variolin. The bromination of the pyrrole ring resulted in **30** which on acylation using 2-(ethoxyvinyl)trimethyltin in presence of $PdCl_2(PPh_3)_2$ afforded **31**. The reaction of **31** with dimethylformamide di(t-butyl) acetal in DMF provided the corresponding enaminone **32** which upon condensation with guanidine hydrochloride resulted in the formation of 2-aminopyrimidine ring **33** with concomitant ester hydrolysis. Heating of **33** in Ph_2O at 260 °C not only led to the decarboxylation but also O-methyl deprotection to yield **34** which upon N-deprotection with triflic acid finally yielded variolin-B **35** (Scheme 14.5) [50].

Another route for the synthesis of variolin B was explicated by Morris et al. that involved the reaction of 4-lithiated-2-methylthio pyrimidine **36** with diethyl carbonate **37** at −90 °C to result in the formation of symmetrical ketone **38** [50]. Addition of **38** to lithiated pyridine **39** resulted in the formation of corresponding triarylalcohol **40**. Deoxygenation of the triaryl alcohol in the presence of triethylsilane and trifluoroacetic acid led to the formation of **41**. Oxidation of **41** with m-chloroperbenzoic acid gave the disulfoxide **42**, which upon heating in p-methoxybenzylamine at 85 °C followed by treatment with an excess of sodium ethanethiolate in dry DMF yielded bis-protected pyridinol **43**. The p-methoxybenzyl groups on the amines were quantitatively removed with triflic acid at room temperature leading to the isolation of variolin B **35** (Scheme 14.6).

Meridianins, closely related to variolins are isolated from the tunicate *Aplidium meridianum*, an ascidian collected in South Atlantic. Molina et al. [40] have recently reported a facile synthesis of meridianins involving an initial acylation of 4-benzyloxy-7-bromoindole **44** to obtain 3-acetylindole **45** which on N-protection resulted in N-tosyl-3-acetyl indole **46**. The Bredereck protocol was employed for the construction of 2-amino pyrimidine ring **48** by reacting **46** with DMF-DMA to result in enaminone **47** followed by its condensation with guanidine hydrochloride. The hydrogenation of **48** with Pd over 10% charcoal led to the formation of hydroxylated meridianin A **49** and selective O-deprotection led to the synthesis of meridianin E **50** (Scheme 14.7).

The alkaloid mixture febrifugine **56** and isofebrifugine **57** originally isolated from the root of *Dichroa febrifuga* and extracted from the leaves and buds of *Hydrangea macrophylla var. Otaksa*, has been used in Chinese traditional medicine to treat malaria for over 4000 years. Febrifugine is believed to block the prolifera-

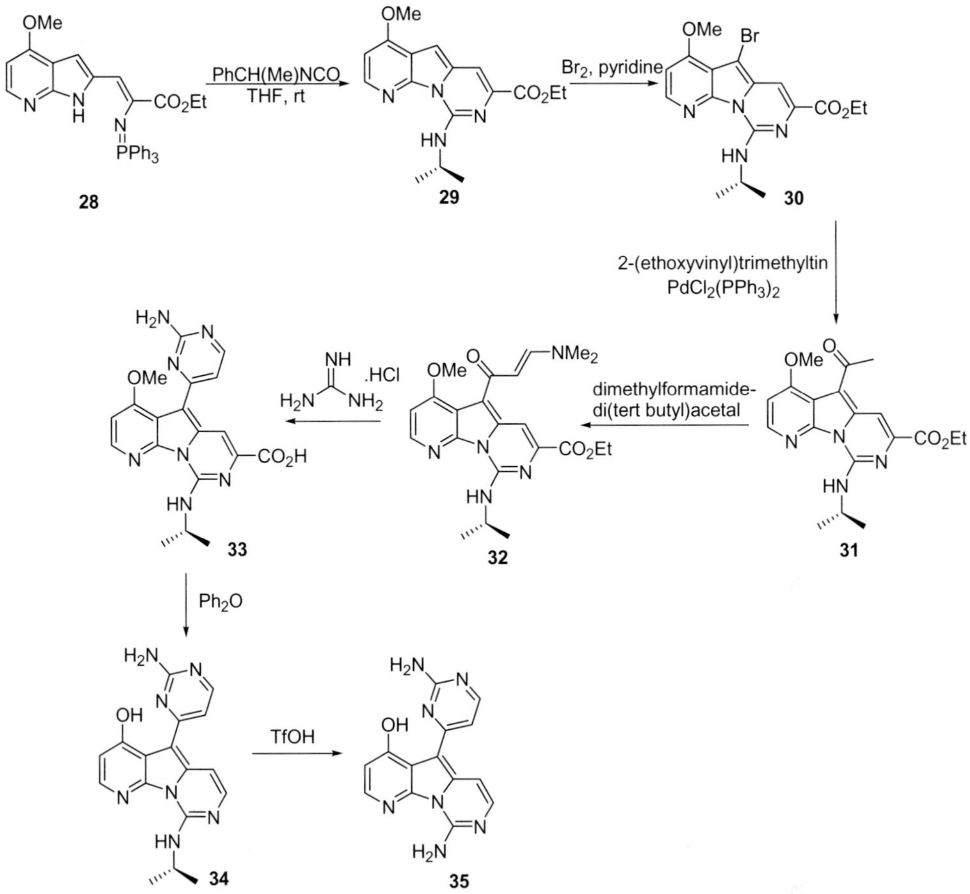

14.2 Pyrimidine-Based Natural Products

Scheme 14.5

tion of malarial parasites and acts by impairing haemazoin formation required for maturation of the parasite at the trophozoite stage. However, strong liver toxicity has precluded the development of febrifugine as a potential clinical drug candidate.

Takeuchi et al. reported the asymmetric synthesis of febrifugine **56** and isofebrifugine **57** from chiral piperidin-3-ol (+)-**52**, which was prepared by the reductive dynamic optical resolution of the 3-piperidone derivatives (±)-**51** using Baker's yeast [31]. The intramolecular bromoetherification of (+)-**52** using N-bromosuccinimide afforded octahydrofuro[3,2-b]-pyridine **53**. The 2-methoxy intermediate **54** was prepared from **53** as a diastereomeric mixture by dehydrobromination using potassium t-butoxide and bromoetherification using N-bromosuccinimide and methanol. Deacetalization of **54** followed by a coupling reaction with 4(3H)–quinazolinone resulted in the formation of **55**. The

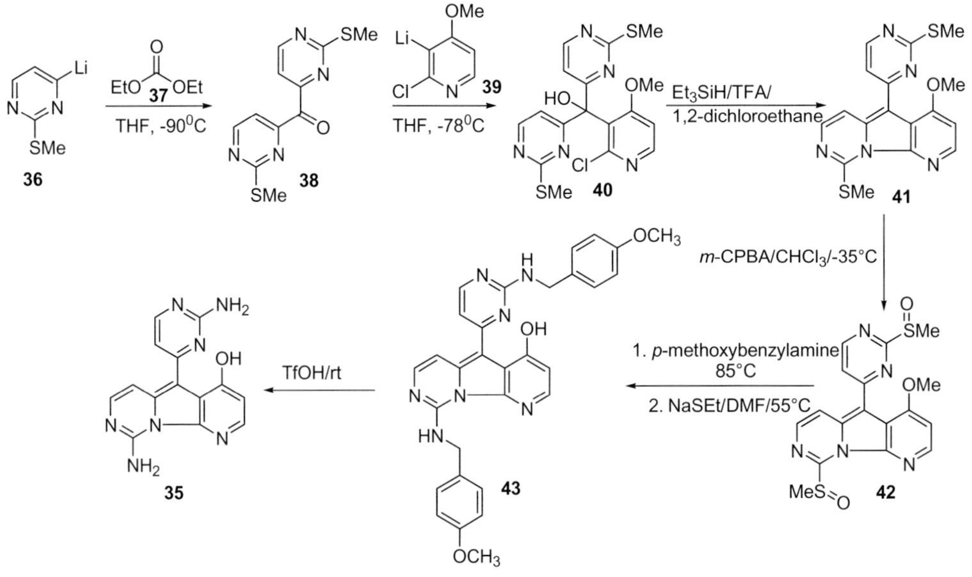

Scheme 14.6

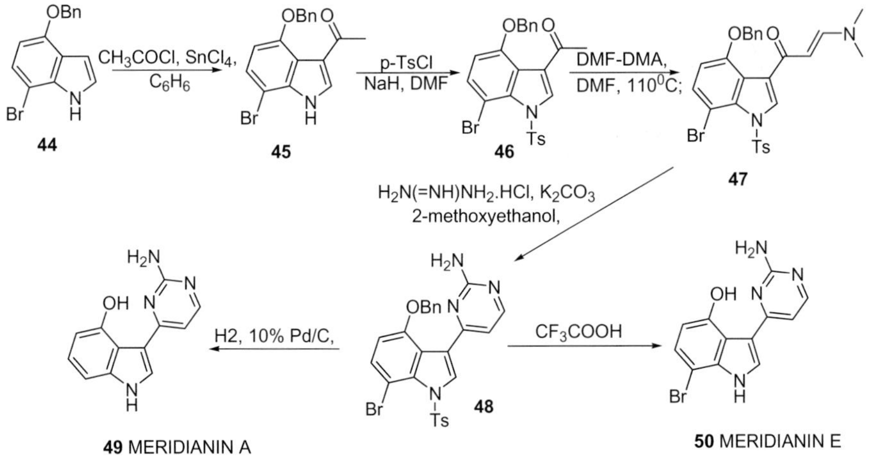

Scheme 14.7

hydrogenolysis of this compound gave isofebrifugine **56**, which upon heating at 80 °C in water resulted in the largest ratio (2 : 1) of (+)-**56** to (+)-**57**. (Scheme 14.8).

In another approach, Kobayashi et al. [31] proposed the catalytic asymmetric synthesis of febrifugine and isofebrifugine using tin (II)-catalyzed asymmetric aldol and lanthanide-catalyzed aqueous three-component reactions. In the presence of a chiral tin (II) Lewis acid (20 mol %), 3-*t*-butyldimethylsiloxypropanal **58**

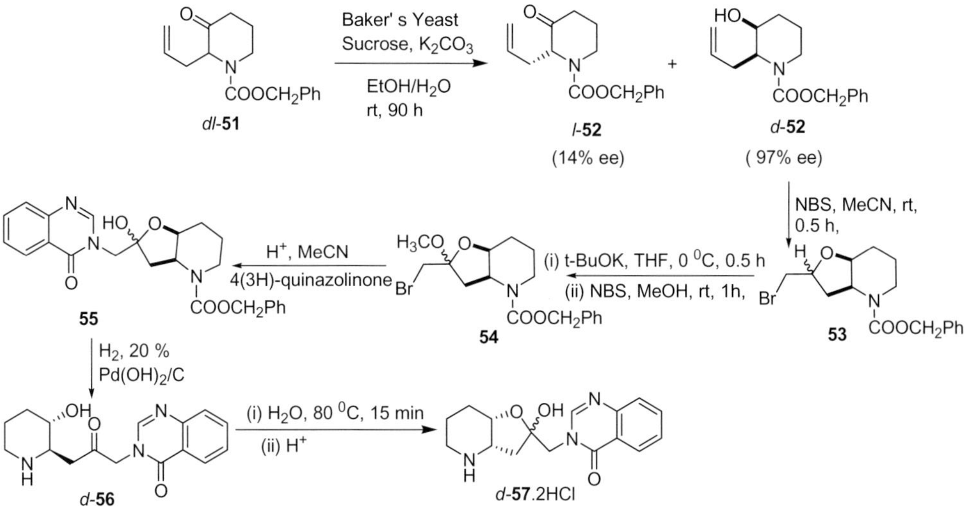

14.2 Pyrimidine-Based Natural Products

Scheme 14.8

was reacted with 2-benzyloxy-1-trimethylsiloxy-1-phenoxyethene **59** in propionitrile at −78 °C to afford the corresponding aldol-type adduct **60** with excellent diastero- and enantioselectivities. The hydroxyl group at position 3 was removed via a 2-step sequence, and the phenyl ester **61** so obtained was converted to the aldehyde **62** under the Swern oxidation conditions (Scheme 14.9). The three component reaction of aldehyde **62**, 2-methoxyaniline, and 2-methoxypropene was performed in the presence of 10 mol% of ytterbium triflate (Yb(OTf)$_3$) to afford the desired Mannich type adduct **63** as a mixture of syn- and anti- adducts. These syn- and anti-diastereomeric adducts were used in the synthesis of isofebrifugine and febrifugine. The anti-adduct was treated with HF to remove the TBS protecting group followed by bromination to give a spontaneously cyclized adduct whose N-protected (2-methoxyphenyl) group was removed using ceric ammonium nitrate (CAN) to afford piperidine **64**. It was then protected as its N-Boc and treated with lithiumhexamethyldisilazide (LHMDS) followed by trimethylsilyl chloride (TMSCl). The resulting silyl enol ether was oxidized and brominated to give bromoacetone **65**. The coupling reaction of this compound with 4-hydroxyquinazoline was carried out using potassium hydroxide to afford the penultimate **66**, whose protecting groups were successfully removed using 6N HCl to afford febrifugine **56** (Scheme 14.9).

Sparsomycin, a metabolite of *Streptomyces sparsogenes* is a peptidyl transferase drug and interferes with peptide bond formation, a central process in protein biosynthesis in bacteria. The stereoselective synthesis of sparsomycin has been described by Nakajima et al. [46]. The key step of this synthesis involves stereoselective oxidation of the sulfide **67** to the chiral sulfoxide **68** under asymmetric conditions. The deprotection of the amino group followed by sulfenylation gave the dithioacetal mono-oxide **69** which was coupled with β-(6-methyluracil)acrylic

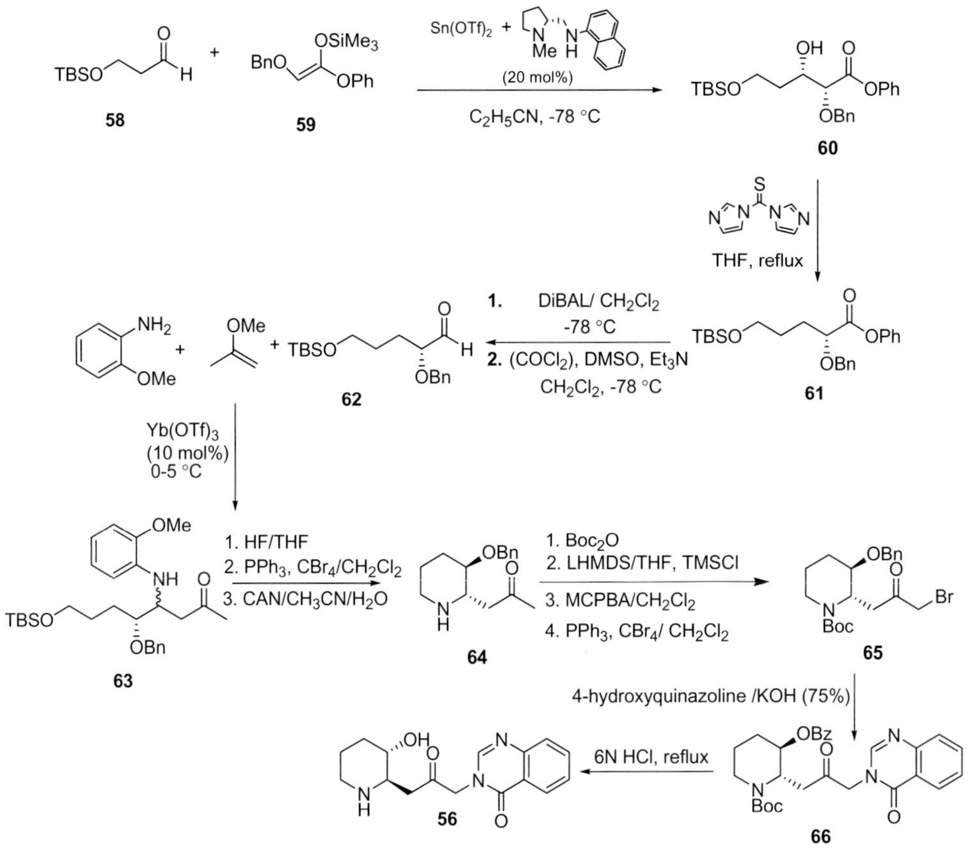

Scheme 14.9

acid in the presence of DCC and 1-hydroxy-7-azabanzotriazole (HOAt) to obtain **70**. The deprotection of the MOM group in **71** led to sparsomycin **72a** whereas oxidation of the sulfide yielded a mixture of MOM-protected sparoxomycins **71a/b** which after deprotection afforded a mixture of **72b/c**, separated by HPLC (Scheme 14.10).

14.3
Imidazole-Based Natural Products

14.3.1
Introduction

The chemistry of nitrogen heterocyclic compounds, especially imidazoles, has attracted more attention during recent years due to their reactivity and novel bio-

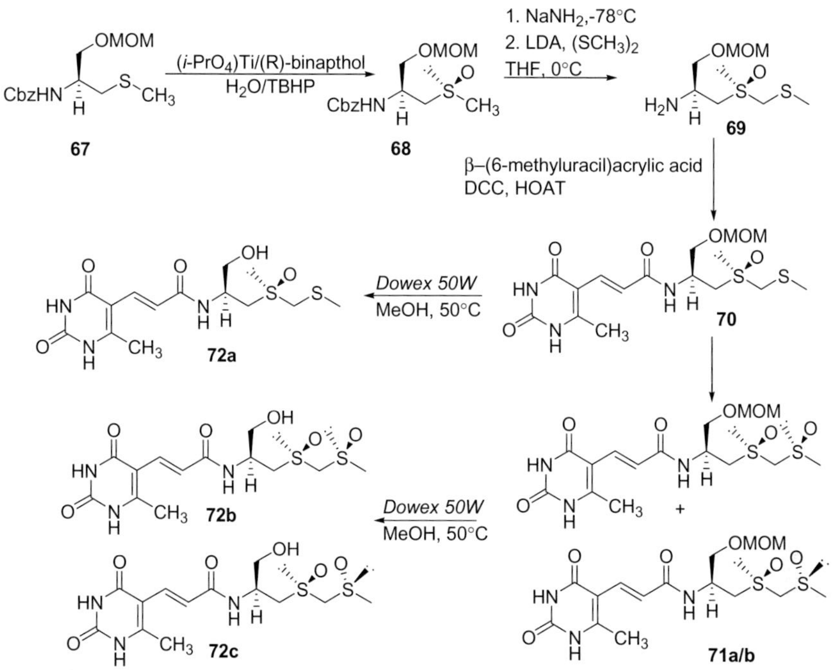

14.3 Imidazole-Based Natural Products

Scheme 14.10

Figure 14.2 Imidazole-based natural products.

azomycin   metronidazole   misonidazole   clotrimazole

logical activities. Compounds bearing the imidazole nucleus are known to show unique anti-edema and anti-inflammatory activities. Differently substituted imidazoles have also been found to be anthelmintic, analgesic, antibacterial, antifungal, antiviral, antitubercular, anticancer and COX-2/LOX inhibitors. 2-Nitroimidazole (azomycin) and 1-(2-hydroxyethyl)-2-methyl-5-nitroimidazole (metronidazole) are good antimicrobial agents with particular applications as trichomonacides along with metronidazole and clotrimazole which are important anticancer drugs (Figure 14.2).

Marine sponges produce a plethora of fascinating, structurally diverse secondary metabolites usually containing imidazole moieties. Since the discovery in 1971 of the first alkaloid of the imidazole family, oroidin, many hundreds of such compounds have been isolated, ranging from relatively simple compounds containing intact imidazole systems such as hymenidine, parazoanthoxanthin A, cribrostatin 6, and girolline to considerably more complex metabolites. Compounds such as palau'amine, styloguanidine and axinellamine A are architecturally complex molecules containing significant stereogenicity. In recent years, these unique and challenging heterocyclic structures have attracted the attention of synthetic chemists. Section 14.3 describes some examples of recent work in the area of imidazole alkaloid total synthesis. Target molecules have been chosen which exemplify approaches to intact imidazole systems as well as to metabolites where the imidazole ring has been modified. A detailed account of the imidazole-based natural products, the source of their origin along with their structures, bioactivities and total synthesis is provided in Table 14.2.

14.3.2
Synthesis of Imidazole-Based Natural Products

One of the simplest members of this series, spongotines A, has been isolated from the marine sponge *Spongosorites* sp. collected off the coast of Jeju Island, Korea. Fujioka and co-workers have recently reported a facile route towards the synthesis of spongotine A [77]. The methodology involved the initial preparation of vinyl indole **73** from 6-bromoindole by formylation, tosylation, and the Wittig olefination reaction. The asymmetric dihydroxylation of vinyl indole **73** resulted in chiral diol **74**, with (R)- configuration, 98% ee. The diol **74** was then converted into diazide **75** by the Mitsunobu reaction with diphenylphosphoryl azide (DPPA) which on reduction with triphenylphosphine in refluxing toluene produced indole diamine **76**. The condensation of diamine **76** with keto aldehyde followed by NCS oxidation in CH_3CN proceeded smoothly and compound **77**, which has the imidazoline/ketone moiety, was obtained that was directly deprotected by NaOH in refluxing MeOH to give spongotine A **77** (Scheme 14.11).

A novel Pd-catalyzed asymmetric annulation between 5-bromopyrrole-2-carboxylate esters and vinyl aziridines has been developed by Trost *et al.* to efficiently construct pyrrolopiperazinones, which can serve as key intermediates in the enantioselective syntheses of imidazole-based alkaloid natural product agesamide [55]. Thus, the reaction between methyl 5-bromopyrrole-2-carboxylate **78** and vinyl aziridine **79** in the presence of $[Pd(C_3H_5)Cl]_2$ and (R,R)-L resulted in the annulation product **80** with 95% ee. Hydroboration of **80** with 9-BBN followed by oxidation with sodium perborate gave primary alcohol **81**. Under optimized conditions, the DMB group was cleaved by treating **81** with five equiv of tetrahydrothiophene in TFA/DCM (1:1) to obtain **82**. Treatment of **82** with Dess–Martin periodane (DMP) under buffered conditions gave the desired aldehyde **83** which was subsequently subjected to the Bucherer–Bergs hydantoin formation protocol to yield agesamide (Scheme 14.12).

Cribrostatin, a constituent of the Republic of Maldives' marine sponge *Cribrochalina* sp., was found to be a cancer cell growth inhibitor (P388 ED50 0.3 µg/mL).

Table 14.2 Imidazole-based natural products.

S.No.	Name	Structure	Source	Isolation [Ref]	Biological activity	Synthesis [Ref]
16	Ageliferin		*Agelas conifera*	[51]	Antibiotic and antiviral activity and is a useful agent for the study of actin-myosin contractile systems [52]	[53]
17	Agesamide		Okinawan marine sponge *Agelas* sp.	[54]	No Bioactivity reported till date	[55]
18	Axinellamine A		Marine sponge *Axinella* sp.	[56]	Bactericidal activity against *Helicobacter pylori*, a bacterium implicated in pepticular and gastric cancer [57]	[58]
19	Cribrostatin		*Cribrochalina* sp.	[59]	Cancer cell growth inhibitor, inhibit the growth of a number of pathogenic bacteria and fungi [60]	[61]
20	Dragmacidin		Mediterranean sponge *Halicortex* sp.	[62]	Anticancerous, antifungal, antiviral, and anti-inflammatory [63]	[64]
21	Hymenidine		*Agelas longissima*	[65]	Antiserotonergic activity [65]	[66]

(Continued)

14.3 Imidazole-Based Natural Products | 521

Table 14.2 (Continued)

S.No.	Name	Structure	Source	Isolation [Ref]	Biological activity	Synthesis [Ref]
22	Nagelamide		Okinawan marine sponge *Agelas sp.*	[67]	Antimicrobial activity against *Staphylococcus aureus* and *Cryptococcus neoformans* [67]	[68]
23	Palauamine		*Stylotella aurantium*	[69]	Antibacterial [69]	[70]
24	Sarcodictyin		Soft Coral *Bellonella albiflora*	[71]	Stabilize microtubules by competing with the paclitaxel binding site on microtubule polymers [72]	[73]
25	Securamine		Marine Bryozoan *Securiflustra securifrons*	[74]	No Bioactivity reported till date	[75]
26	Spongotine		*Spongosorites sp.*	[76]	Cytotoxic [76]	[77]

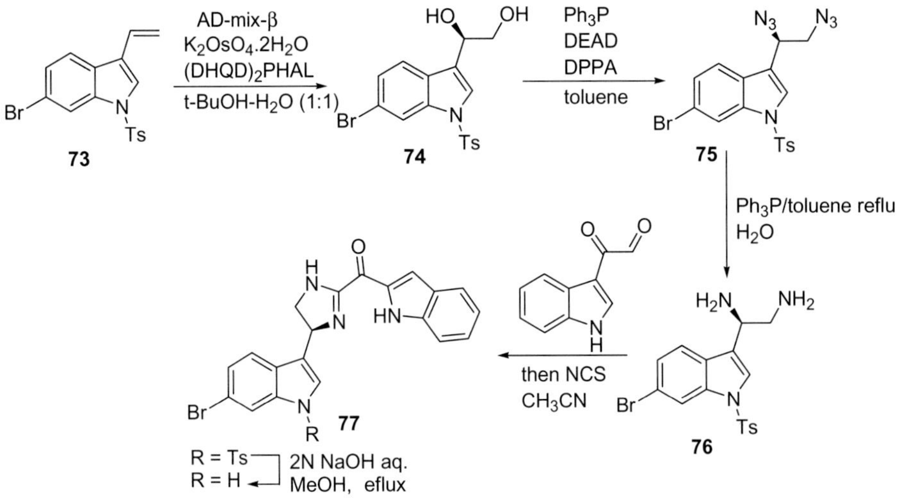

14.3 Imidazole-Based Natural Products | 523

Scheme 14.11

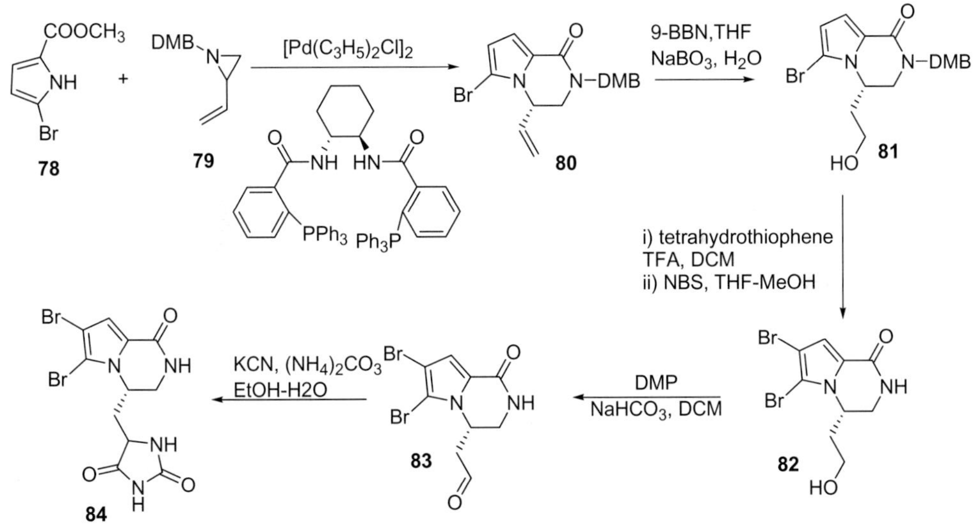

Scheme 14.12

Cribrostatin **101** is also the first known naturally-occurring example of the imidazo[5,1-*a*]isoquinoline ring system. It was synthesized from the known protected imidazole **86**, available in a single step by the reaction of 2-methylimidazole **85** and 2-(trimethylsilyl)ethoxymethyl chloride (SEMCl). The directed *ortho*-metallation of **86** with *n*-butyllithium and subsequent quenching with tributyltin chloride gave the desired embodiment **87**. The precursor **92** was obtained from the ether analog 89 of commercially available 2-methylresorcinol 88. Hydroxylation of 89 with hydrogen peroxide in acetic acid gave phenol 90 which on bromination gave 91 and subsequent protection of the hydroxyl group as its triisopropylsilyl (TIPS) ether afforded bromide 92. Palladium-catalyzed cross-coupling between bromide 92 and stannane 87 forged the biaryl bond to give intermediate 93. Introduction of the final two carbons began with a regioselective bromination of intermediate 93 with bromine-1,4-dioxane in a 1:1 (v:v) solvent mixture of diethyl ether (Et_2O) and trifluoroacetic acid (TFA) to give bromide 92. The side chain was introduced with palladium-catalyzed cross-coupling reaction between 94 and allyltributyltin to give allyl biaryl 95. A two-step deprotection/reprotection of 95 was necessary to set the stage for the intramolecular cyclization. Thus, 95 was first deprotected with TFA to get 96 and then regioisomerically reprotected with di-*t*-butyl dicarbonate (Boc_2O) to afford carbamate 97. Biaryl 97 underwent intramolecular cyclization upon exposure to catalytic osmium tetroxide and excess sodium periodate to generate aminal 98. Dehydration of 98 was performed with methanesulfonyl chloride (MsCl) and triethylamine to give imidazo[5,1-*a*]isoquinoline 99. Finally, removal of the TIPS group from 99 with tetrabutylammonium fluoride (TBAF) followed by the known oxidation of phenol 100 with HNO_3 delivered cribrostatin 101 (Scheme 14.13) [61].

The dragmacidins represent a small but growing family of marine alkaloids that possess a variety of interesting structural and biological features. This structurally interesting natural product possesses a variety of synthetic challenges, namely, the differentially substituted pyrazinone, the bridged [3.3.1] bicyclic ring system, which is fused to both the trisubstituted pyrrole and aminoimidazole heterocycles along with the installation and maintenance of the 6-bromoindole fragment.

The enantiospecific total synthesis of (+)-dragmacidin F [64], features a palladium-mediated intramolecular oxidative carbocyclization, a halogen-selective Suzuki cross-coupling reaction, and a late-stage Neber rearrangement. Synthesis commenced with bicyclic lactone 102, a compound available by lactonization and selective silylation of (−)-quinic acid. Oxidation of 102 followed by Wittig olefination of the resultant ketone produced *exo*-methylene lactone 103 (Scheme 14.14). A homogeneous Pd-catalyzed π-allyl hydride addition to allylic lactone 103, led quantitative reductive isomerization to the desired unsaturated carboxylic acid 104. Oxidative cyclization of 104 *via* Weinreb amide formation followed by the addition of lithiopyrrole led to the formation of 105. Exposure of 105 to Pd(OAc)$_2$ under a variety of conditions led to carbocyclization, and under optimized conditions, produced the desired pyrrole-fused bicycle 106 as a single stereo- and regioisomer. With the formation of [3.3.1] bicyclic framework, the final stereocenter present in the natural product was installed *via* catalytic hydrogenation of olefin 106 followed by methylation to produce 3° ether 107. Conversion of 107 to 108 was realized by regioselective bromination of the pyrrole followed by metallation and coupling to boronic ester 108. In the critical halogen-selective Suzuki fragment-coupling reaction, pyrroloboronic ester 108 and dibromide 109 were reacted under Pd(0) catalysis. An exquisitely selective bond formation constructed the dragmacidin F framework by fusion of the pyrrole and alkoxypyrazine subunits while leaving the

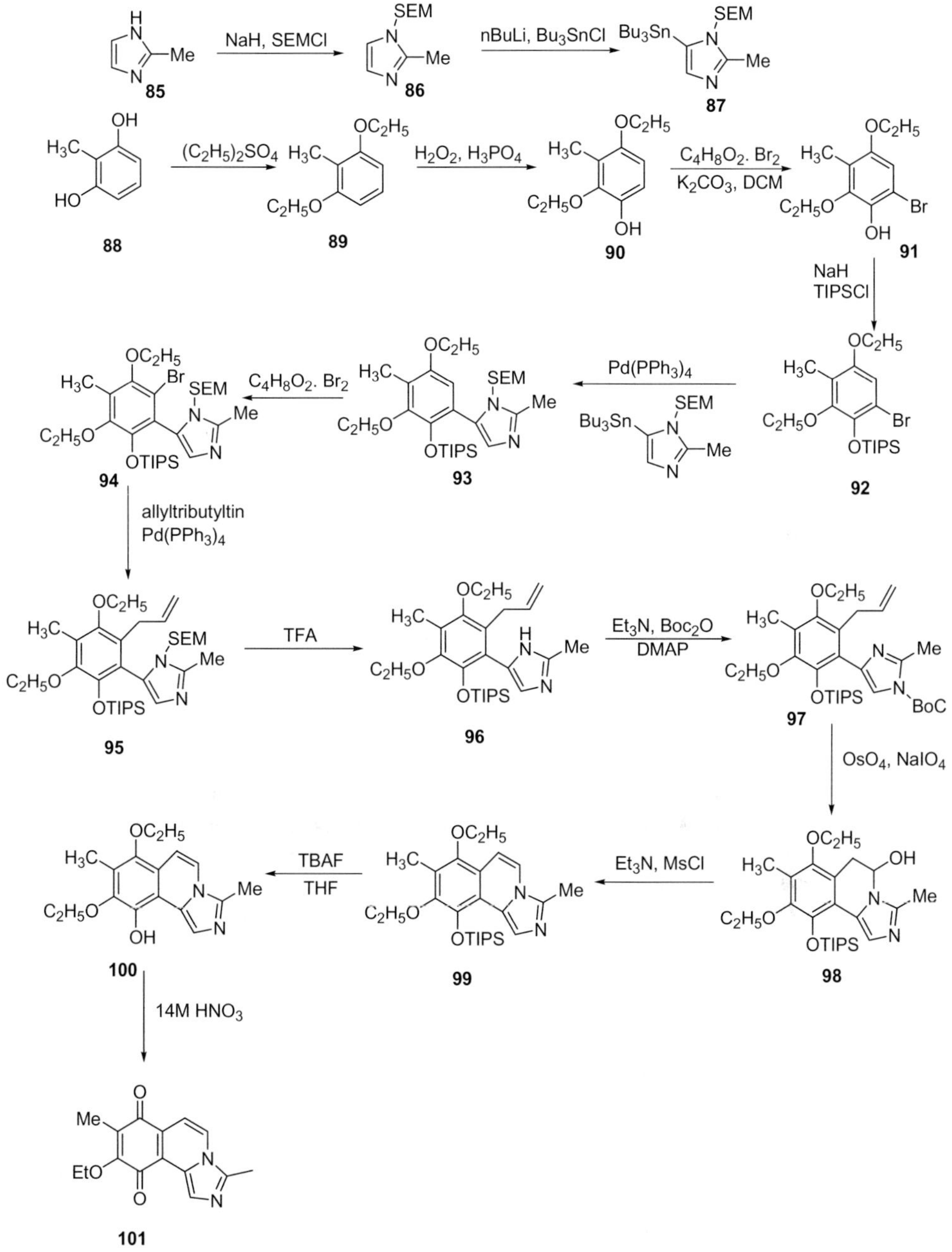

14.3 *Imidazole-Based Natural Products* | 525

Scheme 14.13

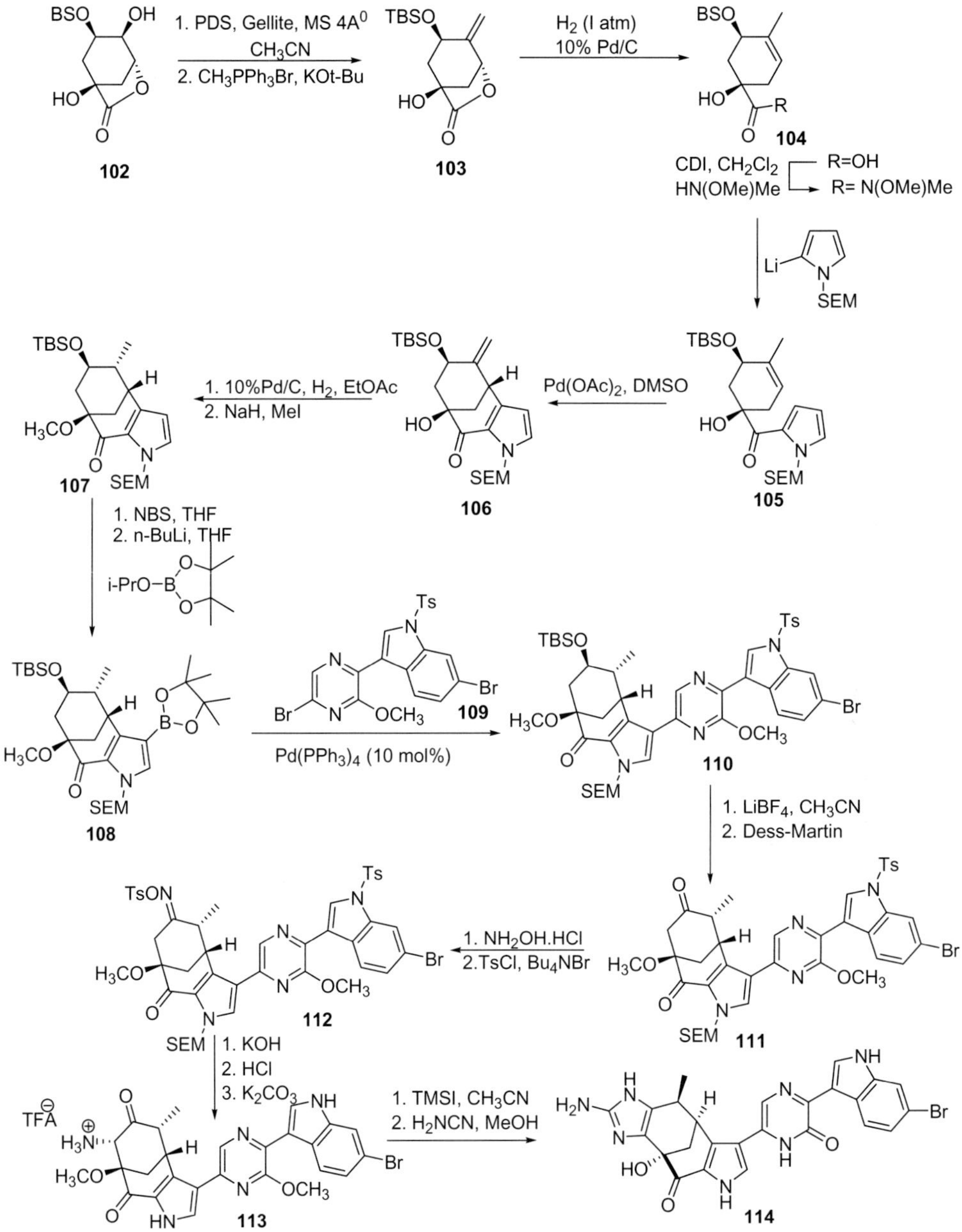
BSO
OH
HO
O
O
102
1. PDS, Gellite, MS 4A0
CH3CN
2. CH3PPh3Br, KOt-Bu
TBSO
HO
O
O
103
H2 (I atm)
10% Pd/C
BSO
HO
R
O
104
CDI, CH2Cl2
HN(OMe)Me
R=OH
R= N(OMe)Me
Li
N
SEM
TBSO
HO
O
N
SEM
105
Pd(OAc)2, DMSO
TBSO
H
HO
O
N
SEM
106
1. 10%Pd/C, H2, EtOAc
2. NaH, MeI
TBSO
H
H3CO
O
N
107
SEM
1. NBS, THF
2. n-BuLi, THF
i-PrO–B
TBSO
H
H3CO
O
B–O
N
SEM
108
Ts
N
N
Br
N
OCH3
Br
109
Pd(PPh3)4 (10 mol%)
TBSO
H
H3CO
O
N
SEM
N
N
OCH3
Ts
N
Br
110
1. LiBF4, CH3CN
2. Dess-Martin
O
H
H3CO
O
N
SEM
111
N
N
OCH3
Ts
N
Br
1. NH2OH.HCl
2.TsCl, Bu4NBr
TsON
H
H3CO
O
N
SEM
N
N
OCH3
Ts
N
Br
112
1. KOH
2. HCl
3. K2CO3
TFA
H3N
O
H
H3CO
O
HN
N
N
OCH3
NH
Br
113
1. TMSI, CH3CN
2. H2NCN, MeOH
H2N
H
N
N
H
HO
O
N
H
N
N
H
O
NH
Br
114

Scheme 14.14

indolyl bromide of **109** and, in turn, **110** intact. Selective deprotection of silyl ether **109** and oxidation with DMP produced ketone **111** which was subjected to Neber rearrangement to yield amino ketone **113** as a single regio- and stereo-chemical isomer in excellent yield. Both the tosyl and SEM protecting groups were quantitatively removed from the corresponding heterocycles **112**. Finally, liberation of the 3° hydroxyl and pyrazinone functionalities by exposure of bis-ether **113** to TMSI, followed by treatment of the penultimate amino ketone with cyanamide and aqueous NaOH produced (+)-dragmacidin F **114** (Scheme 14.14).

Nagelamide is a novel dimeric alkaloid isolated from an Okinawan marine sponge *Agelas* sp. with antimicrobial activity. Nagelamide J is the first bromopyrrole alkaloid possessing a cyclopentane ring fused to an amino imidazole ring.

The (*E*)-vinylstannane **118** required for the desired synthesis was prepared *via* the hydrostannylation of TBS protected propargyl alcohol **117**. The imidazole substituted propargyl alcohol **117** in turn was prepared *via* a Sonogashira reaction between 4-iodoimidazole **116** and TBS-protected propargyl alcohol. The imidazolyl iodide component **119** was assembled in rapid order from the dimethylaminosulfonyl (DMAS)-protected 4,5-diiodoimidazole **115**, which was initially formylated by metallation at C5 with EtMgBr, treated with *N*-(2-pyridyl)-*N*-methylformamide followed by Horner–Wadsworth–Emmons reaction. With the synthesis of requisite fragments **118** and **119**, Stille cross-coupling conditions provided the bis-vinylimidazole **120**. The removal of the silyl group provided **121** which on catalytic hydrogenation led to saturation of both the double bonds to give the diol **122** and then protected as the bis-silyl ether **123** (Scheme 14.15). The deprotonation of the imidazole C2 positions with *n*-BuLi and trapping with TsN$_3$ provided the bis-azide **124**. Deprotection of the silyl ethers with TBAF led to the formation of the expected diol **125** which was subjected to a double Mitsunobu reaction with the dibromopyrrolehydantoin leading to **126**. In order to prevent competitive formation of the iminophosphorane by reaction of the triphenylphosphine with the azide moieties, triphenylphosphine was reacted with DIAD prior to introduction of the diol. Exposure of **126** to aqueous sodium hydroxide led to the hydrolysis of the ureas, providing the pyrrolecarboxamide which upon deprotection with methanolic HCl provided **127**. The catalytic hydrogenation of **127** provided the desired nagelamide D **128** (Scheme 14.15) [68].

14.4
Conclusion

Pyrimidine- and imidazole-based novel molecular scaffolds continue to be an integral part of the rich source of natural products. The large biodiversity of tropical marine organisms as well as plants provide a huge resource for the extension of the chemodiversity of these two molecular frameworks. The exhaustive chemodiversity and established bioactivity profiles of pyrimidines and imidazoles continue to provide stimulus to synthetic organic chemists for the development of flexible novel synthetic protocols for the synthesis of libraries of their natural and synthetic analogs for structure–activity relationship studies.

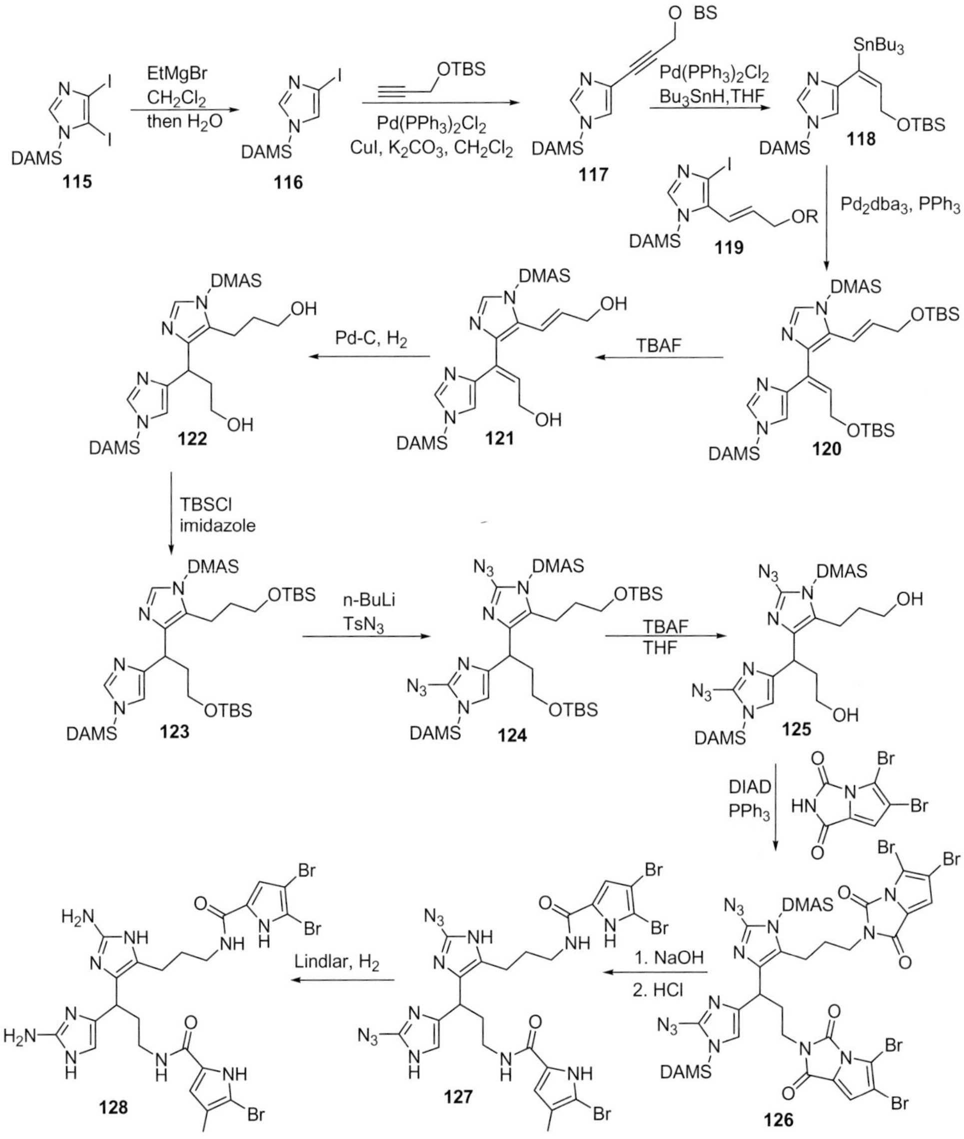

528 | 14 Pyrimidine and Imidazole

Scheme 14.15

Acknowledgment

The financial support from CSIR New Delhi under Emeritus Scientist Scheme (MPM) and DST New Delhi under Fast Track Young Scientist Scheme (VK) is gratefully acknowledged.

References

1. Haskell, T.H. (1958) Amicetin, bamicetin and plicacetin. Chemical studies. *J. Am. Chem. Soc.*, **80** (3), 747–751.
2. Gu, Z. and Lovett, P.S. (1995) A gratuitous inducer of cat-86, amicetin, inhibits bacterial peptidyl transferase. *J. Bacteriol.*, **177** (12), 3616–3618.
3. Alarcón, B., Lacal, J.C., Fernández-Souza, J.M. and Carrasco, L. (1984) Screening for new compounds with antiherpes activity. *Antiviral Res.*, **4** (5), 231–244.
4. Noecker, L.A., Martino, J.A., Foley, P.J., Rush, D.M., Giuliano, R.M. and Villani, F.J., Jr. (1998) Synthesis of amicetose by three enantioselective methods. *Tetrahedron Asymmetry*, **9** (2), 203–212;Stevens, C.L., Nemec, J. and Ransford, G.H. (1972) Total synthesis of the amino sugar nucleoside antibiotic, plicacetin. *J. Am. Chem. Soc.*, **94** (9), 3280–3281.
5. Tanaka, F., Takeuchi, S., Tanaka, N., Yonehara, H., Umezawa, H. and Sumiki, Y. (1961) Bacimethrin, a new antibiotic produced by *B. megatherium*. *J. Antibiot.*, **14**, 161.
6. Reddick, J.J., Saha, S., Lee, J.-M., Melnick, J.S., Perkins, J. and Begley, T.P. (2001) The mechanism of action of bacimethrin, a naturally occurring thiamin antimetabolite. *Bioorg. Med. Chem. Lett.*, **11** (17), 2245–2248.
7. Zilles, J.L., Croal, L.R. and Downs, D.M. (2000) Action of the thiamine antagonist bacimethrin on thiamine biosynthesis. *J. Bacteriol.*, **182** (19), 5606–5610.
8. Ulbricht, T.L.V. and Price, C.C. (1956) The synthesis of some pyrimidine metabolite analogs. *J. Org. Chem.*, **21** (5), 567–571.
9. Perandones, F. and Soto, J.L. (1998) Synthesis of pyrido[2,3-d]pyrimidines from aminopyrimidine carbaldehydes. *J. Heterocycl. Chem.*, **35** (2), 413–419.
10. Takeuchi, S., Hirayama, K., Ueda, K., Sakai, H. and Yonehara, H. (1958) Blasticidin S, a new antibiotic. *J. Antibiot.*, **11**, 1.
11. Kimura, M. and Yamaguchi, I. (1996) Recent development in the use of blasticidin S, a microbial fungicide, as a useful reagent in molecular biology. *Pestic. Biochem. Physiol.*, **56** (3), 243–248.
12. Ichikawa, Y., Hirata, K., Ohbayashi, M. and Isobe, M. (2004) Total synthesis of (+)-blasticidin S. *Chem. A Eur. J.*, **10** (13), 3241–3251.
13. Boger, D.L., Teramoto, S. and Cai, H. (1996) Synthesis and evaluation of deglycobleomycin A$_2$ analogues containing a tertiary N-methyl amide and simple ester replacement for the L-histidine secondary amide: direct functional characterization of the requirement for secondary amide metal complexation. *Bioorg. Med. Chem.*, **4** (2), 179–193.
14. Takimoto, C.H. and Calvo, E. (2008) Principles of oncologic pharmacotherapy, in *Cancer Management: A Multidisciplinary Approach*, 11 edn (eds R. Pazdur, L.D. Wagman, K.A. Camphausen and W.J. Hoskins), CMP United Business Media.
15. Katano, K., An, H., Aoyagi, Y., Overhand, M., Sucheck, S.J., Stevens, W.C., Jr., Hess, C.D., Zhou, X. and Hecht, S.M. (1998) Total synthesis of bleomycin group antibiotics. Total syntheses of bleomycin demethyl A$_2$, bleomycin A$_2$, and decarbamoyl bleomycin demethyl A$_2$. *J. Am. Chem. Soc.*, **120** (44), 11285–11296.
16. Gerlinck, R.G.S., Braekman, J.C., Daloze, D., Bruno, I., Riccio, R., Rogeau, D. and Amade, P. (1992) Crambines C1 and C2: two further ichthyotoxic guanidine alkaloids from the sponge *Crambe crambe*. *J. Nat. Prod.*, **55** (4), 528–532.
17. Chang, L.C., Whittaker, N.F. and Bewley, C.A. (2003) Crambescidin 826 and dehydrocrambine A: new polycyclic guanidine alkaloids from the marine sponge *Monanchora* sp. that inhibit HIV-1 fusion. *J. Nat. Prod.*, **66** (11), 1490–1494.
18. Aoki, S., Kong, D.X., Matsui, K. and Kobayashi, M. (2004) Erythroid differentiation in K562 chronic myelogenous cells induced by crambescidin 800, a pentacyclic guanidine alkaloid. *Anticancer Res.*, **24** (4), 2325–2330.

19 Palagiano, E., Demarino, S., Minale, L., Riccio, R., Zollo, F., Iorizzi, M., Carre, J.B., Debitus, C., Lucarain, L. and Provost, J. (1995) Ptilomycalin A, crambescidin 800 and related new highly cytotoxic guanidine alkaloids from the starfishes *Fromia monilis* and *Celerina heffernani*. Tetrahedron, 51 (12), 3675–3682.

20 Berlinck, R.G.S., Braekman, J.C., Daloze, D., Bruno, I., Riccio, R., Ferri, S., Spampinato, S. and Speroni, E. (1993) Polycyclic guanidine alkaloids from the marine sponge *Crambe crambe* and Ca^{++} channel blocker activity of crambescidin 816. J Nat Prod., 56 (7), 1007–1015.

21 Moore, C.G., Murphy, P.J., Williams, H.L., McGown, A.T. and Smith, N.K. (2007) Synthetic studies towards ptilomycalin A: total synthesis of crambescidin 359. Tetrahedron, 63 (47), 11771–11780.

22 Ohtani, I., Moore, R.E. and Runnegar, M.T.C. (1992) Cylindrospermopsin: a potent hepatotoxin from the blue-green alga *Cylindrospermopsis raciborskii*. J. Am. Chem. Soc., 114 (20), 7941–7942.

23 Banker, R., Teltsch, B., Sukenik, A. and Carmeli, S. (2000) 7-Epicylindrospermopsin, a toxic minor metabolite of the cyanobacterium *Aphanizomenon ovalisporum* from Lake Kinneret, Israel. J. Nat. Prod., 63 (3), 387–389.

24 Fastner, J., Heinze, R., Humpage, A.R., Mischke, U., Eaglesham, G.K. and Chorus, I. (2003) Cylindrospermopsin occurrence in two German lakes and preliminary assessment of toxicity and toxin production of *Cylindrospermopsis raciborskii* (Cyanobacteria) isolates. Toxicon, 42 (3), 313–321.

25 Keen, S.P. and Weinreb, S.M. (2000) Studies on total synthesis of cylindrospermopsin: new constructions of uracils from α,β-unsaturated esters. Tetrahedron Lett., 41 (22), 4307–4310.

26 Sakata, K., Sakurai, A. and Tamura, S. (1977) Isolation and antimicrobial activities of ezomycins B_1, B_2, C_1, C_2, D_1 and D_2. Agric. Biol. Chem., 41 (10), 2027–2032.

27 Gooday, G.W. (1990) *Biochemistry of Cell Walls and Membranes in Fungi* (eds P.J. Kuhn, A.P. Trinci, M.J. Jung, M.W. Goosey and L.G. Copping), Springer, Berlin, Germany;Ruiz-Herrera, J. and San-Blas, G. (2003) Chitin synthesis as a target for antifungal drugs. Curr. Drug Targets Infect. Disord., 3 (1), 77–91.

28 Knapp, S. and Gore, V.K. (2000) Synthesis of the ezomycin nucleoside disaccharide org. Lett., 2 (10), 1391–1393.

29 Koepfli, J.B., Mead, J.F. and Brockman, J.J.A. (1949) Alkaloids of *Dichroa febrifuga*. I. Isolation and degradative studies. J. Am. Chem. Soc., 71 (3), 1048–1054.

30 Jiang, S., Zeng, Q., Gettayacamin, M., Tungtaeng, A., Wannaying, S., Lim, A., Hansukjariya, P., Okunji, C.O., Zhu, S. and Fang, D. (2005) Antimalarial activities and therapeutic properties of febrifugine analogs. Antimicrob. Agents Chemother., 49 (3), 1169–1176.

31 Takeuchi, Y., Azuma, K., Takakura, K., Abe, H. and Harayama, T. (2000) Asymmetric synthesis of febrifugine and isofebrifugine using yeast reduction. Chem. Commun., (17), 1643–1644;Kobayashi, S., Ueno, M., Suzuki, R. and Ishitani, H. (1999) Catalytic asymmetric synthesis of febrifugine and isofebrifugine. Tetrahedron Lett., 40 (11), 2175–2178.

32 Cheng, C.C. (1976) Structural similarity between febrifugine and chloroquine. J. Theor. Biol., 59 (2), 497–501.

33 Harbour, G.C., Tymiak, A.A., Rinehart, K.L., Jr., Shaw, P.D., Hughes, R., Jr., Mizsak, S.A., Coats, J.H., Zurenko, G.E., Li, L.H. and Kuentzel, S.L. (1981) Ptilocaulin and isoptilocaulin, antimicrobial and cytotoxic cyclic guanidines from the Caribbean sponge *Ptilocaulis aff. P. spiculifer*. J. Am. Chem. Soc., 103 (18), 5604–5606.

34 Porse, B.T., Kirillov, S.V., Awayez, M.J., Ottenheijm, H.C.J. and Garrett, R.A. (1999) Direct crosslinking of the antitumor antibiotic sparsomycin, and its derivatives, to A2602 in the peptidyl transferase center of 23S-like rRNA within ribosome-t-RNA complexes. Proc. Natl. Acad. Sci. USA, 96 (16), 9003–9008.

35 Jahn, T., König, G.M., Wright, A.D., Wörheide, G. and Reitner, J. (1997) Manzacidin D: an unprecedented

secondary metabolite from the "living fossil" sponge *Astrosclera willeyana*. *Tetrahedron Lett.*, **38** (22), 3883–3884.

36 Dembitsky, V.M. (2002) Bromo- and iodo-containing alkaloids from marine microorganisms and sponges. *Russ. J. Bioorg. Chem.*, **28** (3), 170–182.

37 Wehn, P.M. and Bois, J.D. (2002) Enantioselective synthesis of the bromopyrrole alkaloids manzacidin A and C by stereospecific C–H bond oxidation. *J. Am. Chem. Soc.*, **124** (44), 12950–12951; Drouin, C., Woo, J.C.S., MacKay, D.B. and Lavigne, R.M.A. (2004) Total synthesis of (±)-manzacidin D. *Tetrahedron Lett.*, **45** (39), 7197–7199.

38 Franco, L.H., Joffé, E.B.K., Puricelli, L., Tatian, M., Seldes, A.M. and Palermo, J.A. (1998) Indole alkaloids from the tunicate *Aplidium meridianum*. *J. Nat. Prod.*, **61** (9), 1130–1132.

39 Gompel, M., Leost, M., Joffé, E.B.D.K., Puricelli, L., Franco, L.H., Palermo, J. and Meijer, L. (2004) Meridianins, a new family of protein kinase inhibitors isolated from the Ascidian *Aplidium meridianum*. *Bioorg. Med. Chem. Lett.*, **14** (7), 1703–1707.

40 Fresneda, P.M., Molina, P. and Bleda, J.A. (2001) Synthesis of the indole alkaloids meridianins from the tunicate *Aplidium meridianum*. *Tetrahedron*, **57** (12), 2355–2363.

41 Isono, K., Asahi, K. and Suzuki, S. (1969) Polyoxins, antifungal antibiotics. XIII. Structure of polyoxins. *J. Am. Chem. Soc.*, **91** (26), 7490–7505.

42 Krainer, E., Becker, J.M. and Naider, F. (1991) Synthesis and biological evaluation of dipeptidyl and tripeptidyl polyoxin and nikkomycin analogs as anticandidal prodrugs. *J. Med. Chem.*, **34** (1), 174–180.

43 Plant, A., Thompson, P. and Williams, D.M. (2009) Application of the ugi reaction for the one-pot synthesis of uracil polyoxin C analogues. *J. Org. Chem.*, **74** (13), 4870–4873.

44 Ubakata, M., Morita, T.-I., Uramoto, M. and Osada, H. (1996) Sparoxomycins Al and A2, new inducers of the flat reversion of NRK cells transformed by temperature sensitive rous sarcoma virus II. Isolation, physico-chemical properties and structure elucidation. *J. Antibiot.*, **49** (1), 65–70.

45 Zemlicka, J. and Bhuta, A. (1982) Sparsophenicol: a new synthetic hybrid antibiotic inhibiting ribosomal peptide synthesis. *J. Med. Chem.*, **25** (10), 1123–1125.

46 Nakajima, N., Enomoto, T., Matsuura, N. and Ubukata, M. (1998) Synthesis and morphological reversion activity on src^{ts}NRK cells of pyrimidinyl propanamide antibiotics, sparsomycin, sparoxomycin A_1, A_2, and their analogues. *Bioorg. Med. Chem. Lett.*, **8** (23), 3331–3334.

47 Baxter, R.L., Hanley, A., Bryan, C. and Henry, W.S. (1990) Thiamine biosynthesis in yeast – evaluation of 4-hydroxy-5-hydroxymethyl-2-methylpyrimidine as a precursor. *J. Chem. Soc. Perkin Trans. I*, **9**, 2963–2966.

48 Trimurtulu, G., Faulkner, D.J., Perry, N.B., Ettouati, L., Litaudon, M., Blunt, J.W., Munro, M.H.G. and Jameson, G.B. (1994) Alkaloids from the Antarctic sponge *Kirkpatrickia varialosa*. Part 2: variolin A and N(3′)-methyl tetrahydrovariolin B. *Tetrahedron*, **50** (13), 3993–4000.

49 Perry, N.B., Ettouati, L., Litaudon, M., Blunt, J.W., Munro, M.H.G., Parkin, S. and Hope, H. (1994) Alkaloids from the antarctic sponge *Kirkpatrickia varialosa*. Part 1: Variolin B, a new antitumour and antiviral compound. *Tetrahedron*, **50** (13), 3987–3992.

50 Anderson, R.J. and Morris, J.C. (2001) Total synthesis of variolin B. *Tetrahedron Lett.*, **42** (49), 8697–8699; Molina, P., Fresneda, P.M., Delgado, S. and Bleda, J.A. (2002) Synthesis of the potent antitumoral marine alkaloid variolin B. *Tetrahedron Lett.*, **43** (6), 1005–1007.

51 Kiefer, P.A., Schwartz, R.E., Koker, M.E.S., Hughes, R.G., Jr., Rittschof, D. and Rinehart, K.L. (1991) Bioactive bromopyrrole metabolites from the Caribbean sponge *Agelas conifera*. *J. Org. Chem.*, **56** (9), 2965–2975 and references cited.

52 Williams, D.H. and Faulkner, D.J. (1996) N-Methylated ageliferins from the sponge *Astrosclera willeyana* from Pohnpei. *Tetrahedron*, **52** (15), 5381–5390.

53 O'Malley, D.P., Li, K., Maue, M., Zografos, A.L. and Baran, P.S. (2007) Total synthesis of dimeric pyrrole–imidazole alkaloids: sceptrin, ageliferin, nagelamide E, oxysceptrin, nakamuric acid, and the axinellamine carbon skeleton. *J. Am. Chem. Soc.*, **129** (15), 4762–4775.

54 Tsuda, M., Yasuda, T., Fukushi, E., Kawabata, J., Sekiguchi, M., Fromont, J. and Kobayashi, J. (2006) Agesamides A and B, bromopyrrole alkaloids from sponge *Agelas* species: application of DOSY for chemical screening of new metabolites. *Org. Lett.*, **8** (19), 4235–4238.

55 Trost, B.M. and Dong, G. (2007) Asymmetric annulation toward pyrrolopiperazinones: concise enantioselective syntheses of pyrrole alkaloid natural products. *Org. Lett.*, **9** (12), 2357–2359.

56 Urban, S., Leone, P.A., Carroll, A.R., Fechner, G.A., Smith, J., Hooper, J.N.A. and Quinn, R.J. (1999) Axinellamines A–D, novel imidazo–azolo–imidazole alkaloids from the Australian marine sponge *Axinella* sp. *J. Org. Chem.*, **64** (3), 731–735.

57 Monks, N.R., Lerner, C., Henriques, A.T., Farias, F.M., Schapoval, E.E.S., Suyenaga, E.S., Rocha, A.B., Schwartsmann, G. and Mothes, B. (2002) Anticancer, antichemotactic and antimicrobial activities of marine sponges collected off the coast of Santa Catarina, southern Brazil. *J. Exp. Mar. Biol. Ecol.*, **281** (1), 1–12.

58 Sivappa, R., Hernandez, N.M., He, Y. and Lovely, C.J. (2007) Studies toward the total synthesis of axinellamine and massadine. *Org. Lett.*, **9** (20), 3861–3864.

59 Pettit, G.R., Collins, J.C., Knight, J.C., Herald, D.L., Nieman, R.A., Williams, M.D. and Pettit, R.K. (2003) Antineoplastic agents. 485. Isolation and structure of cribrostatin 6, a dark blue cancer cell growth inhibitor from the marine sponge *Cribrochalina* sp. *J. Nat. Prod.*, **66** (4), 544–547.

60 Pettit, R.K., Fakoury, B.R., Knight, J.C., Weber, C.A., Pettit, G.R., Cage, G.D. and Pon, S. (2004) Antibacterial activity of the marine sponge constituent cribrostatin 6. *J. Med. Microbiol.*, **53**, 61–65.

61 Markey, M.D. and Kelly, T.R. (2008) Synthesis of cribrostatin 6. *J. Org. Chem.*, **73** (19), 7441–7443.

62 Capon, R.J., Rooney, F., Murray, L.M., Collins, E., Sim, A.T.R., Rostas, J.A.P., Butler, M.S. and Carroll, A.R. (1998) Dragmacidins: new protein phosphatase inhibitors from a Southern Australian deep-water marine sponge, *Spongosorites* sp. *J. Nat. Prod.*, **61** (5), 660–662.

63 Cutignano, A., Bifulco, G., Bruno, I., Casapullo, A., Gomez-Paloma, L. and Riccio, R. (2000) Dragmacidin F: a new antiviral bromoindole alkaloid from the mediterranean sponge *Halicortex* sp. *Tetrahedron*, **56** (23), 3743–3748.

64 Garg, N.K., Caspi, D.D. and Stoltz, B.M. (2004) The total synthesis of (+)-dragmacidin F. *J. Am. Chem. Soc.*, **126** (31), 9552–9553.

65 Cafieri, F., Fattorusso, E., Mangoni, A., Sacfati, O.T. and Carnuccio, R. (1995) A novel bromopyrrole alkaloid from the sponge *Agelas longissima* with antiserotonergic activity. *Bioorg. Med. Chem. Lett.*, **5** (8), 799–804.

66 Daninos-Zeghal, S., Mourabit, A.A., Ahond, A., Poupat, C. and Potier, P. (1997) Synthèse de métabolites marins 2-aminoimidazoliques: hyménidine, oroïdine et kéramadine. *Tetrahedron*, **53** (22), 7605–7614.

67 Araki, A., Tsuda, M., Kubota, T., Mikami, Y., Fromont, J. and Kobayashi, J. (2007) Nagelamide J, a novel dimeric bromopyrrole alkaloid from a sponge *Agelas* species. *Org. Lett.*, **9** (12), 2369–2371.

68 Bhandari, M.R., Sivappa, R. and Lovely, C.J. (2009) Total synthesis of the putative structure of nagelamide D. *Org. Lett.*, **11** (7), 1535–1538.

69 Kinnel, R.B., Gehrken, H.P. and Scheuer, P.J. (1993) Palau'amine: a cytotoxic and immunosuppressive hexacyclic bisguanidine antibiotic from the sponge *Stylotella agminata*. *J. Am. Chem. Soc.*, **115** (8), 3376–3377.

70 Overman, L.E., Rogers, B.N., Tellew, J.E. and Trenkle, W.C. (1997) Stereocontrolled synthesis of the tetracyclic core of the bisguanidine alkaloids palau'amine and styloguanidine. *J. Am. Chem. Soc.*, **119** (30), 7159–7160.

71 Nakao, Y., Yoshida, S., Matsunaga, S. and Fusetani, N. (2003) (Z)-Sarcodictyin A, a new highly cytotoxic diterpenoid from the soft coral *Bellonella albiflora*. *J. Nat. Prod.*, **66** (4), 524–527.

72 Lindel, T., Jensen, P.R., Fenical, W., Long, B.R., Casazza, A.M., Carboni, J. and Fairchild, C.R. (1997) Eleutherobin, a new cytotoxin that mimics paclitaxel (taxol) by stabilizing microtubules. *J. Am. Chem. Soc.*, **119** (37), 8744–8745.

73 Nicolaou, K.C., Xu, J.Y., Kim, S., Pfefferkorn, J., Ohshima, T., Vourloumis, D. and Hosokawa, S. (1998) Total synthesis of sarcodictyins A and B. *J. Am. Chem. Soc.*, **120** (34), 8661–8673.

74 Rahbaek, L., Anthoni, U., Christophersen, C., Nielsen, P.H. and Petersen, B.O. (1996) Marine alkaloids. 18. Securamines and securines, halogenated indole-imidazole alkaloids from the marine bryozoan *Securiflustra securifrons*. *J. Org. Chem.*, **61** (3), 887–889.

75 Korakas, P., Chaffee, S., Shotwell, J.B., Duque, P. and Wood, J.L. (2004) Natural product synthesis special feature: Efficient construction of the securine A carbon skeleton. *Proc. Natl. Acad. Sci. USA*, **101** (33), 12054–12057.

76 Bao, B., Sun, Q., Yao, X., Hong, J., Lee, C., Cho, H.Y. and Jung, J.H. (2007) Bisindole alkaloids of the topsentin and hamacanthin classes from a marine sponge *Spongosorites* sp. *J. Nat. Prod.*, **70** (1), 2–8.

77 Murai, K., Morishita, M., Nakatani, R., Kubo, O., Fujioka, H. and Kita, Y. (2007) Concise total synthesis of (–)-spongotine A. *J. Org. Chem.*, **72** (23), 8947–8949.

Part Three
Natural Products Containing Medium and Large Ring-Sized Heterocyclic Systems

15
Oxepines and Azepines
Darren L. Riley and Willem A.L. van Otterlo

15.1
Introduction

Oxepines and azepines are ubiquitous motifs found in the structures of natural products. A structure-based search on Beilstein Commander results in excess of 6700 compounds, isolated from biological samples, using the simple oxepine structure **I** (X = O) as a core, and over 2500 compounds for the azepine structure **I** (X = N, Figure 15.1). A substructure search then reveals that researchers have associated these core structures with biological activity as many of these compounds, >2000 for oxepine and >369 for azepine, have associated pharmacological and bioactivity data[1].

It is therefore not surprising that oxepine- and azepine-containing natural products have attracted the focused attention of synthetic and medicinal chemists alike. This chapter aims to highlight the importance of the oxepine and azepine structures by identifying a number of natural products from each class that exhibit interesting biological activities, as well as the majority chosen having been the target of a "total synthesis" endeavor. The examples chosen in Table 15.1 (oxepines) and Table 15.2 (azepines) will attempt to span the skeletal possibilities, ranging from natural products containing one heteroatom (i.e., structure **I**, Figure 15.1), two heteroatoms (i.e., **II** and **III**) and their corresponding aromatic-fused counterparts, **IV–VI** and **VII–X**, respectively (Figure 15.1). In addition, a few examples of rarer compounds containing an oxazapine in their structure, that is, with both a nitrogen and oxygen atom in a seven-membered ring, will be described. Finally,

---

1) Search was performed on CrossFire Commander (Version 7.1 SR1) on 8 December 2009. Query structures entered as in I (Figure 15.1) with "all atoms" being marked as "Free Sites." Search refinement was performed with Field names = "INP" and ='Pharm.

Heterocycles in Natural Product Synthesis, First Edition. Edited by Krishna C. Majumdar and Shital K. Chattopadhyay.
© 2011 Wiley-VCH Verlag GmbH & Co. KGaA. Published 2011 by Wiley-VCH Verlag GmbH & Co. KGaA.

15 Oxepines and Azepines

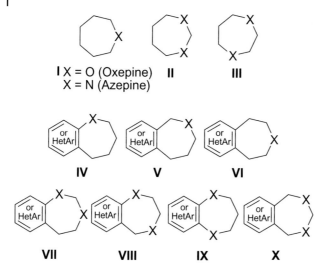

Figure 15.1 Core structures of oxepines and azepines described in this chapter (X = N or O).

the key synthetic strategies used in making the seven-membered heterocyclic rings in the *total synthesis* of a select number of the seven-membered oxepine, azepine or oxazapine natural products in the tables will then be briefly described. Where possible, emphasis will be placed on recent scientific contributions in the literature[2].

15.2
Synthesis of the Heterocyclic Core of Selected Natural Products Containing Oxepines

In Section 15.2 the key reactions affording the oxepine cores of some of the natural products listed in Table 15.1 will be described.

---

2) In compiling this book chapter it quickly became evident that it would be very difficult to do justice to all the work done in this area of research—we thus sincerely apologize to any researcher's contributions which were omitted due to space constraints. In addition, we have unfortunately been unable to add valuable contributions describing the synthesis of analogs, particularly for medicinal chemistry, or even descriptions of synthetic methodology *towards* the synthesis of the natural products chosen in the tables.

Table 15.1 Representative examples of oxepine-containing natural products.

Serial No.	Type (see Figure 1)	Trivial name	Structure	Source	Isolation [Ref]	Biological activity	Total synthesis [Ref]
1	I	Brevetoxin A **1**	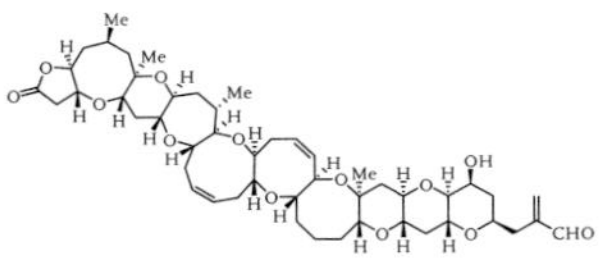	*Gymnodinium breve* Davis (marine dinoflagellate)	[3, 4]	Potent neurotoxin [3]	[4]
2	I	Crambescidin 359 **2**	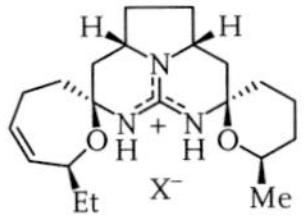	*Monanchora unguiculata* (marine sponges)	[5]	Moderate cytotoxicity (K562, A2780, H-460 and P388) [6]	[6–8]
3	II	Candicanoside A **3**	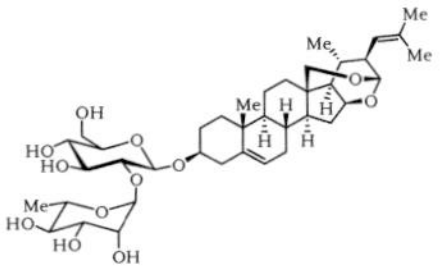	*Galtonia candicans* (bulbs)	[9]	Potent anticancer activity (HL-60 and others) [9]	[10, 11]
4	II	Formamicin **4**	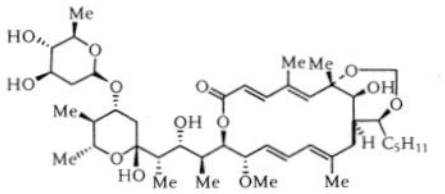	*Saccharothrix* sp. MK27-91F2 (fungal soil actinomycete)	[12, 13]	Antibacterial, antifungal and cytotoxicity against a variety of murine tumor cell lines [12]	[14]
5	III	(Z)-Isolaureatin **5**	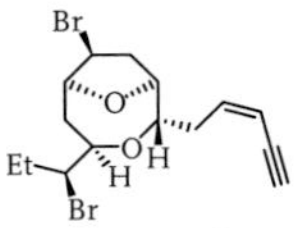	*Laurencia nipponica* (red algae)	[15]	Insecticidal activity against mosquito larvae [16]	[17]

(Continued)

Table 15.1 (Continued)

Serial No.	Type (see Figure 1)	Trivial name	Structure	Source	Isolation [Ref]	Biological activity [Ref]	Total synthesis [Ref]
6	III	Sorangicin A 6	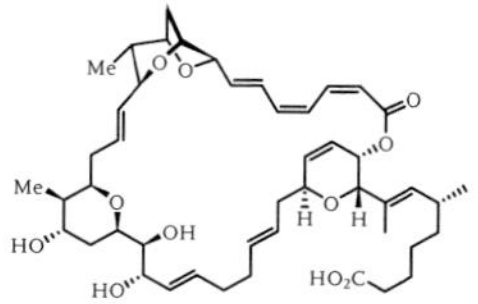	*Sorangium cellulosum* (fungal soil myxobacterium)	[18, 19]	Antibacterial [20]	[21]
7	IV	(E)-Pterulone 7	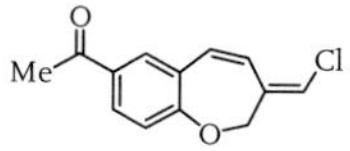	*Pterula* sp. 82168 (basidiomycete)	[22, 23]	Antifungal and inhibitor of eukaryotic respiration [22]	[24–27]
8	IV	Bauhinoxepin J 8	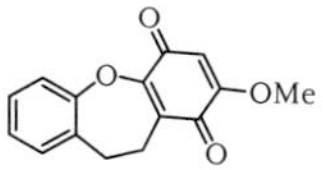	*Bauhinia purpurea* ("shrubby tree")	[28]	Cytotoxicity (KB and BC cell lines) and antimycobacterial activity [28]	[29]
9	IV	Aristoyagonine 9	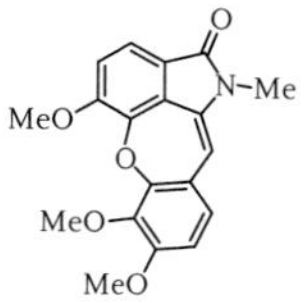	*Sarcocapnos enneaphylla* (plant)	[30]	Analogs cytotoxic against various cell lines [31]	[32]
10	V	Cladoacetal A 10	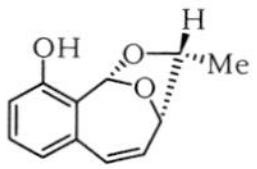	'A fungicolous hyphomycete resembling *Cladosporium*'	[33]	Inhibited *Staphylococcus aureus* [33]	

15.2 Synthesis of the Heterocyclic Core of Selected Natural Products Containing Oxepines

11	VI	Kosinostatin 11		Actinomycete *Micromonospora* sp. TP-A0468 (marine actinomycete)	[34, 35]	Antibacterial, cytotoxic against cancer cell lines and inhibited DNA topoisomerase IIα [34]
12	VII	Paeciloquinone E 12		*Paecilomyces carneus* P-177 (fungus)	[36, 37]	Kinase inhibitor (EGFR, c-Src) [36]
13	VIII/IX	Granulatin 13		*Pseudocyphellaria granulata* and *P. faveolata* (lichen)	[38]	HIV-1 integrase inhibitory activity [39] [40]
14	IX	Paraherquamide A 14		*Penicillium paraherquei* or *P. cluniae* Quintanilla (mold)	[41, 42]	Potent antihelmintic activity [42] [43]
15	X/VII	Cercosporin 15		*Cercospora kikuchii* (fungus)	[44]	Protein kinase C inhibitor [45] [46]

It should be noted that a recent review by Peczuh and co-workers [47] provides a comprehensive description of modern approaches to the synthesis of oxepines. In addition, other earlier reviews should also be consulted [48, 49].

The first example described is brevetoxin A **1** (Scheme 15.1), a member of a family of marine-derived toxins that have long fascinated scientists because of their complex structures and potent neurotoxicity [50]. Brevetoxin A **1** is of interest as it contains no fewer than five different ring sizes, including a seven-membered oxepine ring (D, in Scheme 15.1), the topic of this chapter. In a seminal paper describing the total synthesis of this compound, Nicolaou and co-workers synthesized the D-ring during a double Yamaguchi cyclization of substrate **16** to afford compound **17** [4]. It was important that rings B and D were the second set of cycles to be constructed in the strategy, after which the rest of the complex molecule was synthesized.

Ptilomycalin A **18** is from a family of guanidinium alkaloids that have been isolated from warm water marine sponges (Scheme 15.2) [51, 52]. This compound possessed potent cytoxicity against human cancer cell lines and was synthesized in 1995 by Overman and co-workers [53]. It is interesting that the core guanidinium portion of **18** has also been isolated from the marine sponge *Monanchora unguiculata* [5] as crambescidin 359 **2** (X = Cl) and found to be significantly less active against human cancer cell lines than **18** [6]. A number of groups [6–8] have synthesized crambescidin 359 **2** and Scheme 15.2 shows the work of Nagasawa and co-workers [8]. In their approach pyrrolidine **19** was reacted with bis-*N*-Boc-thiourea, followed by oxidation of the diols to afford diketone **20**. Subsequent removal of the protecting groups with concomitant acetalization afforded **2** as the camphorsulfonate anion which was exchanged with tetrafluoroborate. Overman's research group has published the successful syntheses of a number of natural products containing the crambescidin core. Scheme 15.2 also shows how Aron

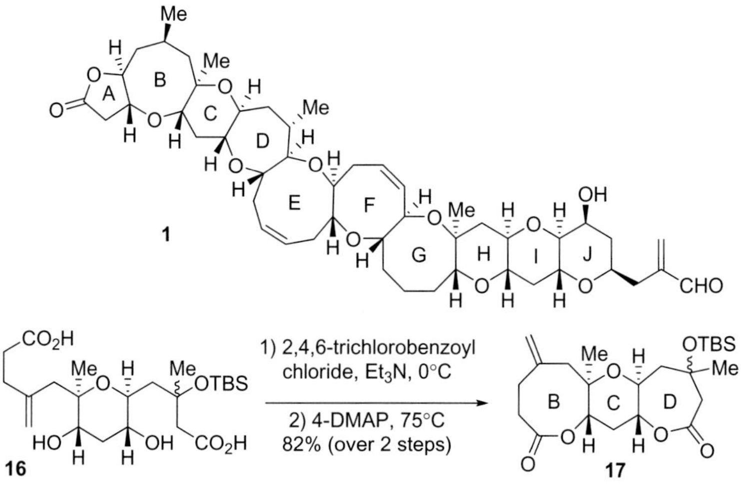

Scheme 15.1

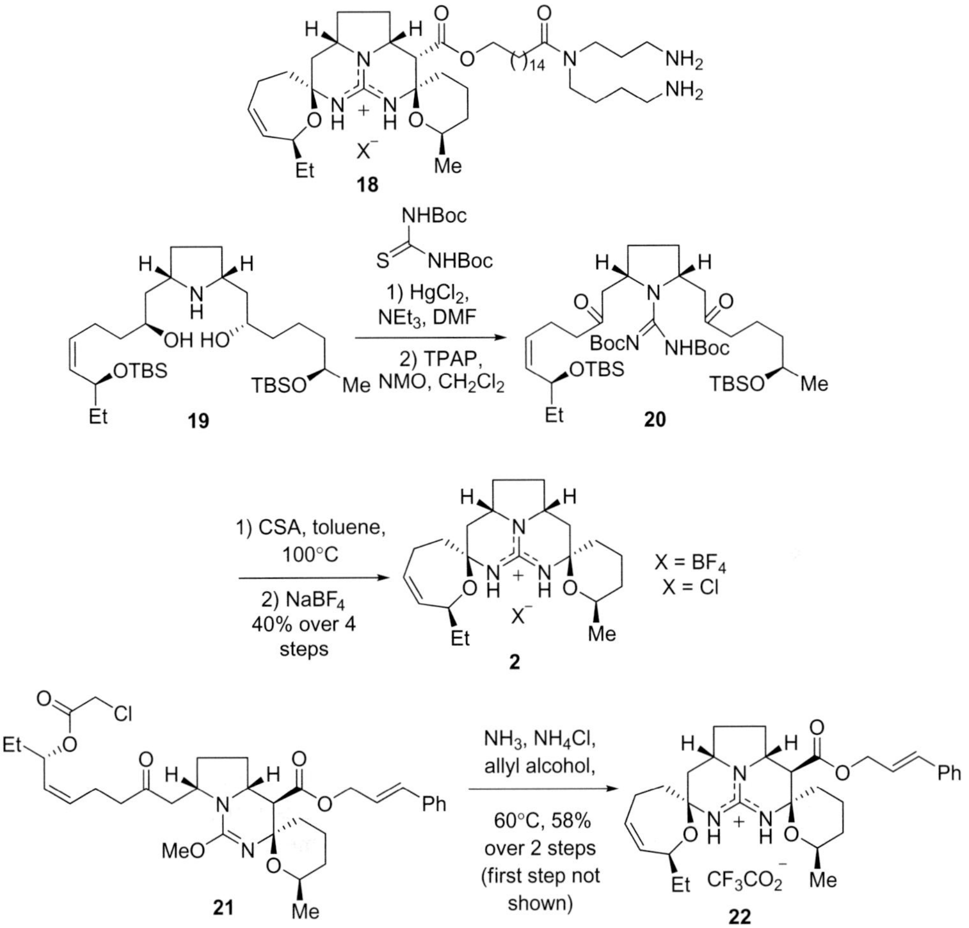

15.2 Synthesis of the Heterocyclic Core of Selected Natural Products Containing Oxepines

Scheme 15.2

and Overman constructed the oxepine core of crambescidin by treating precursor **21** with buffered ammonia [7]. Subsequent deprotection of the cinnamyl ester, followed by decarboxylation then afforded crambescidin 359 **2**.

The "rearranged cholestane disaccharide" candicanoside A **3** (Scheme 15.3) was discovered during a program to investigate the anticancer bioactivities of compounds contained in a Southern African genus, *Ornithogalum* [9]. As a result, candicanoside A **3** was isolated from the bulbs of *Galtonia candicans* and proved to be highly active against human leukemia cells (HL-60 and others). The structure of **3** contains a fused seven-membered ring acetal (characterized as type II in Table 15.1) and this portion of the molecule became one of the main synthetic challenges in the total synthesis of this compound. After a number of unsuccessful attempts using other synthetic approaches [11], Tang and Yu successfully

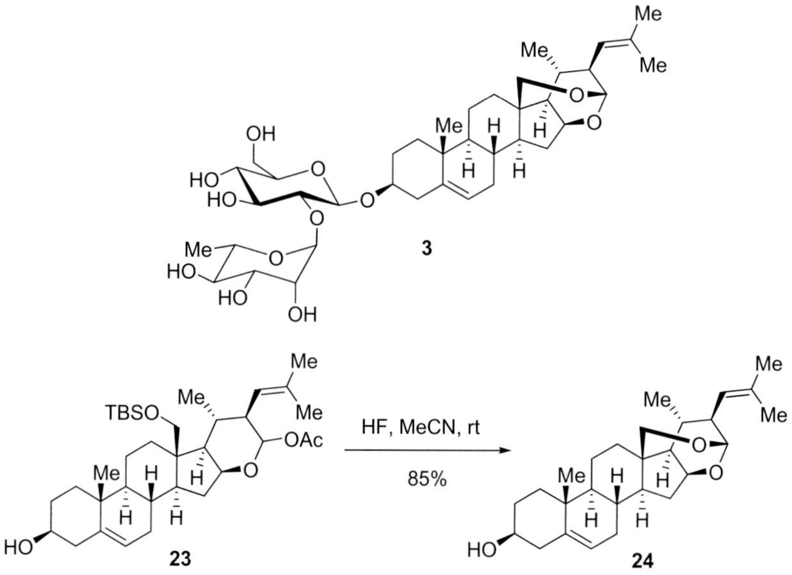

Scheme 15.3

synthesized the crucial acetal by removing the *t*-butyldimethylsilyl protecting group from precursor **23** with hydrogen fluoride which resulted in the spontaneous formation of the desired seven-membered ring acetal in compound **24** [10, 11]. The total synthesis of the anticancer saponin **3** was then achieved by coupling of **24** with a suitably protected glucosyl imidate, followed by further elaboration of the carbohydrate portion into the natural product **3**.

In 1997 formamicin **4**, a member of the plecomacrolide family [54], was isolated from the culture broth of *Saccharothrix* sp. MK27-91F2 by Igarashi and co-workers (Scheme 15.4) [12, 13]. This compound was found to display potent cytotoxicity against a variety of murine tumor cell lines and because of its structural complexity, which included a novel seven-membered methylene acetal (i.e., a dioxepine), was only recently synthesized by Roush and co-workers [14]. In this particular work, a number of cores containing the desired dioxepine structure were synthesized, of which the conversion of substrate **25** to **26** using triethylsilyl triflate (TESOTf) is an example (Scheme 15.4) [55]. This particular transformation involved the transketalization of the tertiary methoxymethyl ether and the *p*-methoxybenzyl ether to afford the seven-membered methylene acetal **26**. This compound was then converted into vinyl iodide **27**, a key intermediate in the total synthesis of formamicin **4**; in addition, a related strategy was used to synthesize the aglycon of **4**, formamicinone [56].

(*Z*)-Isolaureatin **5** (Scheme 15.5) is another natural product in which the oxepine ring (Type III, Figure 15.1) forms a core part of its structure [15]. In a recent total synthesis by Kim and co-workers, the key step involved the introduction of an oxolane ring, giving rise to the bridged oxepine ring in the process [17]. This

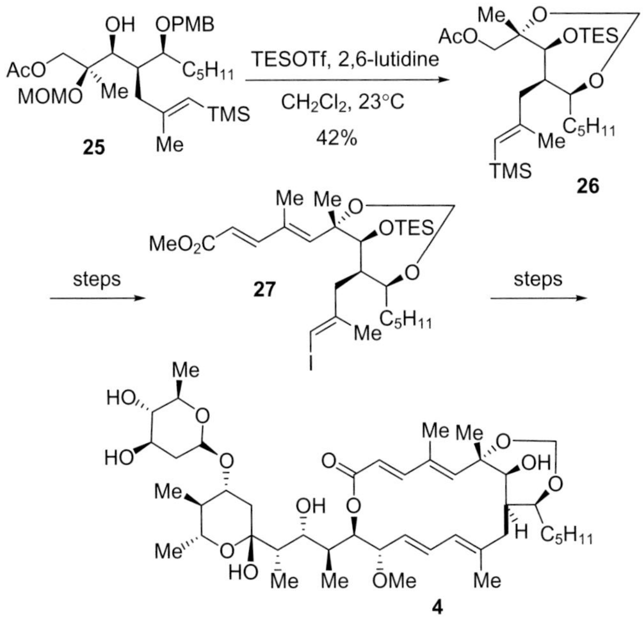

15.2 Synthesis of the Heterocyclic Core of Selected Natural Products Containing Oxepines

Scheme 15.4

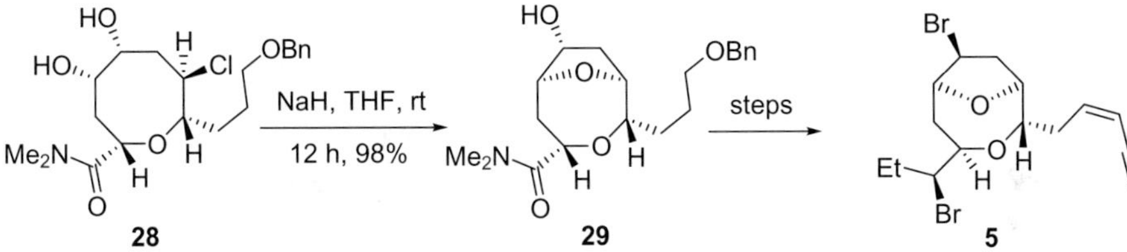

Scheme 15.5

occurred by nucleophilic displacement by the hydroxyl anion, regioselectively generated with sodium hydride, of the chloride in compound **28**, affording the desired structure **29** in excellent yield. Compound **29** then turned out to be a crucial intermediate in the synthesis of (Z)-isolaureatin **5**.

Sorangicin A **6** (Scheme 15.6) was isolated from myxobacteria *Sorangium cellulosum* by Jansen, Höfle and co-workers and found to have an intriguing dioxabicyclo[3.2.1]octane skeleton, which contains the 1,4-dioxepin core [18, 19]. In addition, the compound exhibited impressive broad spectrum antibiotic activity. Recently Smith and co-workers were able to complete the challenging total synthesis of **6** [21] and during this work they published two synthetic approaches to the dioxabicyclo[3.2.1]octane core of sorangicin A **6**. The first published route described specifically how the drop-wise addition of epoxide **30** to excess $BF_3 \cdot OEt_2$

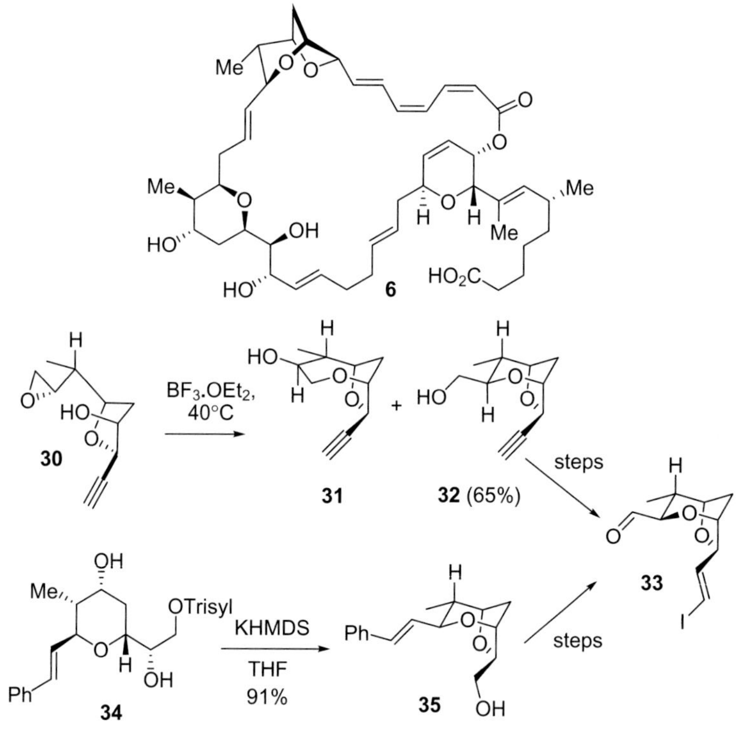

Scheme 15.6

resulted in a reasonable amount of the desired compound (−)-**32** (Scheme 15.6) [57]. The second route entailed a KHMDS-promoted epoxide ring formation-ring opening cascade to afford dioxabicyclo[3.2.1]octane **35** from **34** [58]. Compounds **32** and **35** were then readily converted into the important building block **33** which was used in the successful total synthesis of natural product **6**.

(*E*)-Pterulone **7**, (*Z*)-pterulinic acid **36** and (*E*)-pterulone B **37** are three related halogenated 2,3-dihydro-1-benzoxepines (Type IV, Figure 15.1) that have been isolated from fermentation broths (Scheme 15.7). A number of researchers have published total syntheses of these compounds using a variety of synthetic techniques, with the most popular target being pterulone **7**. For example, Balme and co-workers used a base-mediated cyclization on substrate **38** to afford the benzoxepine **39** in good yield [24]. Oxidative cleavage of the exocyclic double bond, followed by the use of base then afforded the core seven-membered enone **40**, which was used to synthesize pterulone **7** and related compounds. This family of natural products has also been synthesized by other means [26], including the application of ring-closing metathesis (RCM) [25] and a tandem S_N2/Wittig reaction [59].

Dibenz[*b,f*]oxepines are a class of compounds which are increasingly being identified as bioactive constituents in natural products and accordingly more syn-

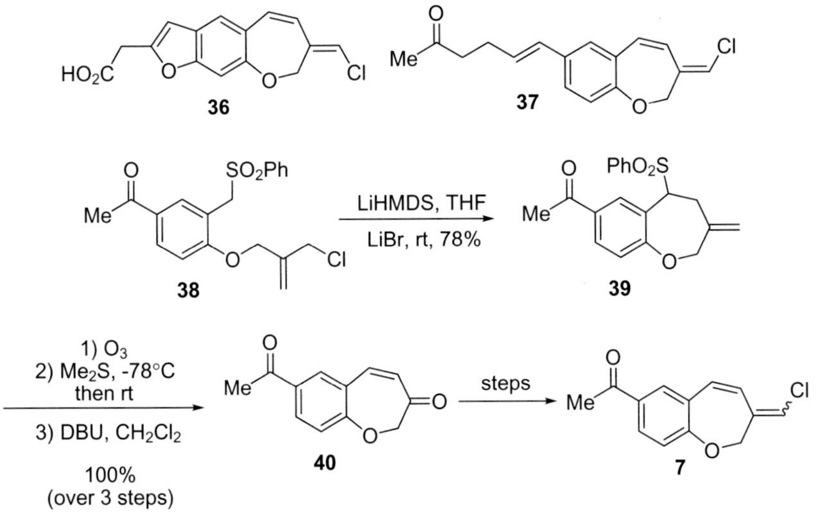

Scheme 15.7

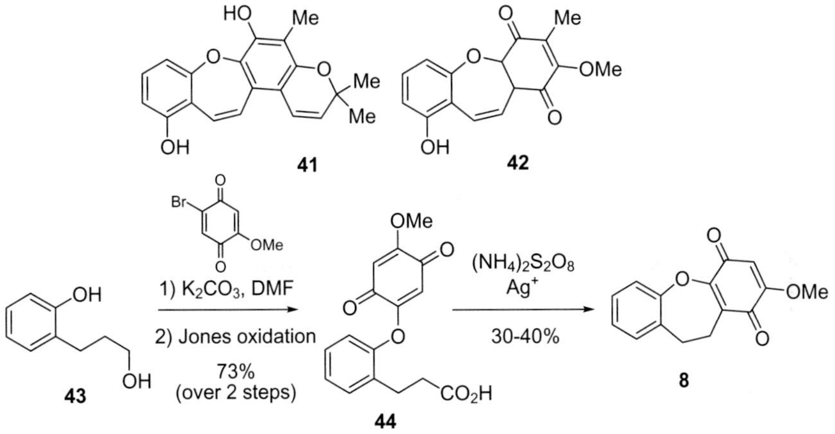

Scheme 15.8

thetic routes towards their syntheses have been developed [60]. Of interest to this chapter are compounds containing this structure such as bauhinoxepin A **41** [61] and bauhiniastatin 1 **42** [62], isolated from *Bauhinia saccocalyx* and *B. purpurea*, respectively (Scheme 15.8). In addition, both compounds possessed interesting anticancer activity. Recent syntheses of these types of compounds have included intramolecular Ullman cyclizations [63] and directed metallation/cyclizations [64]. Another compound belonging to this family, bauhinoxepin J **8** [28], was found to have a dihydrodibenz[b,f]oxepine skeleton. Recently, it succumbed to a short innovative synthesis by Kraus and co-workers which included a key intramolecular persulfate-mediated radical addition of quinone **44**, itself synthesized from the simple precursor **43**, to afford the natural product **8** in few steps [29].

Aristoyagonine **9** is a member of the diverse aristocularine alkaloid family (which in turn is part of the cularinoid family), the majority of these members having the characteristic benzoxepine nucleus in different oxidation states (see for example oxosarcophylline **45**). Aristoyagonine **9** is particular in that it contains a lactam ring as seen in Scheme 15.9. A number of synthetic strategies have been used to synthesize compound **9**, albeit with varying success. The methodology shown in Scheme 15.9, used by Couture and co-workers, applied a copper-mediated diaryl ether coupling to synthesize the benzazepine-containing **9** from **46** in good yield [32]. An advantage of this approach was that **46** could be readily obtained in only four synthetic steps with an overall yield of 26%.

A dioxepine ring system (Type IX, Figure 15.1) is found in many natural products as it biosynthetically related to a prenylated catechol system. In this section a commonly used synthetic strategy resulting in the formation of the dioxepine ring, as part of a more complex structure, will be highlighted. The paraherquamide, and the closely related marcfortine compounds belong to classes of natural products isolated from numerous *Penicillium* species, and a number of their members contain a benzodioxepine ring. Examples include paraherquamide A **14**, a compound with potent antiparasitic and antinematodal activity, and marcfortine B **47**. In Scheme 15.10 the key steps in the construction of the core dioxepine portion of paraherquamide A **14** [43] and marcfortine B **47** [65] are summarized. In the first sequence, Williams and co-workers epoxidized the prenylated oxindole **48** with *m*-CPBA, followed by a reductive cyclization with SnCl$_4$, which afforded the key intermediate **49**. This compound was then used in the total synthesis of paraherquamide A **14** [43]. Trost and co-workers used a similar strategy, albeit in the end-game of the total synthesis of marcfortine B **47** [65]. The final step involved the elimination of the secondary alcohol formed during the *endo*-cyclization of

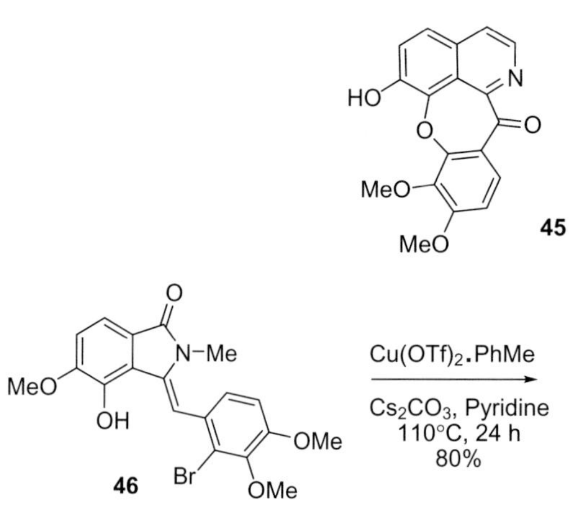

Scheme 15.9

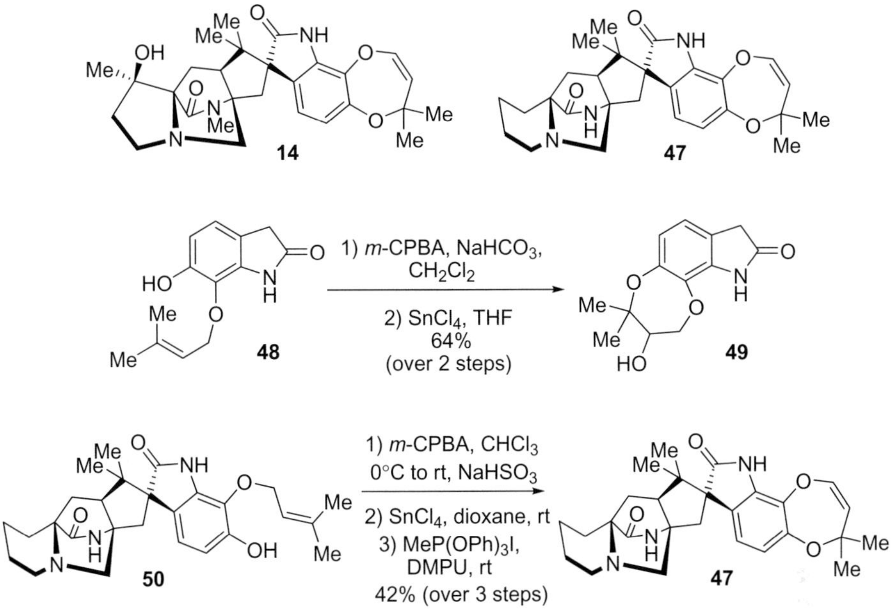

Scheme 15.10

compound **50** to afford the desired natural product **47**. This particular strategy for the synthesis of dioxepine rings was developed in the early 1990s by Williams and Cushing [66].

Finally, in this section describing the syntheses of oxepines, the recent synthesis of cercosporin **15**, a member of a series of natural perylenequinones with potent protein kinase C inhibitory activity, will be described. Other well known compounds from this family include (–)-phleichrome **51** and hypocrellin A **52**. In this interesting work, Kozlowski and co-workers synthesized the compound **54**, containing the desired dioxepine ring, by reacting dinaphthol **53** with the alkylating agent bromochloromethane (Scheme 15.11) [46]. Further steps, which included a ruthenium-mediated decarbonylation, then afforded the desired natural product **15**, and completed the first total synthesis of this compound.

15.3
Synthesis of the Heterocyclic Core of Selected Natural Products Containing Azepines

In Section 15.3 the emphasis will shift towards the synthesis of naturally-derived compounds with an azepine core as part of their structure. To this end a number of examples from Table 15.2 will be highlighted and the key reactions towards the synthesis of the nitrogen-containing heterocycle in their structure will be discussed.

Table 15.2 Representative examples of azepine-containing natural products.

Serial No.	Type (see Figure 1)	Trivial name	Structure	Source	Isolation [Ref]	Biological activity	Total synthesis [Ref]
16	I	Tuberostemonine 55	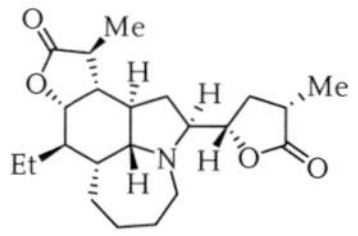	*Stemona tuberosa* and *S. sessifolia* (plant)	[67]	Insecticidal activity [67, 68].	[69, 70]
17	I	Fawcettimine 56	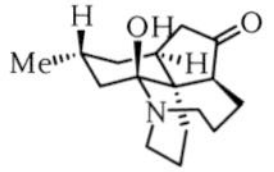	*Lycopodium fawcetti* (plant)	[71]	Acetylcholine esterase (AChE) inhibition [72]	[73]
18	II	Antipathine 57	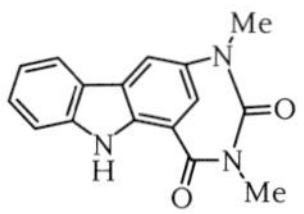	*Antipathes dichotoma* (zoanthid black coral)	[74]	Moderate to weak cytotoxicity (human stomach and liver carcinomas – SGC-7901 and Hep_G2) [74]	
19	II	Monanchorin 58	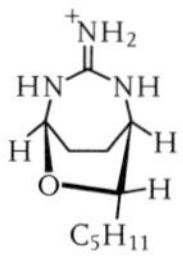	*Monanchora unguiculata* (marine sponge)	[75]	Weak cytotoxic activity (IC2 murine mast cell line) [75]	[76]
20	III	Liposidomycin A-(III) 59	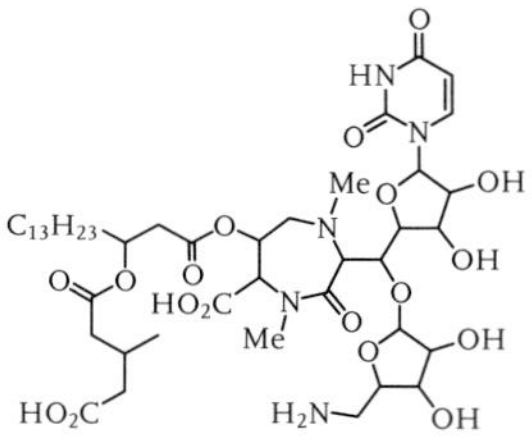	*Streptomyces sp.* SN-1061M (bacteria)	[77, 78]	Inhibitors of bacterial peptidoglycan synthesis [79]	

15.3 Synthesis of the Heterocyclic Core of Selected Natural Products Containing Azepines | 551

21	III	(−)-Aplaminal **60**	*Aplysia kurodai* (sea hare)	[80]	Cytotoxicity against HeLa S₃ cells [80]	[81]
22	IV/V	Silvaticamide **61**	*Aspergillus silvaticus*	[82]	Toxic to mice [82]	
23	V	Hymenialdisine **62**	*Axinella verrucosa, Acanthella aurantiaca, Hymeniacidon* sp. and *Stylissa messa* (marine sponges)	[83–87]	Kinase inhibitor (GSK-3β, CDK family, ErK1, ErK2, CK and MEK) [87, 88]	[89, 90]
24	VI	Lennoxamine **63**	*Berberis darwinii* (plant)	[91]		[92–94]
25	VI	Clavizepine **64**	*Corydalis claviculata* (L.) DC (plant)	[95]		[96–98]
26	VII	Pentostatin **65**	*Streptomyces antibioticus* (bacterium)	[99]	Potent adenosine deaminase inhibitory activity [99, 100]	[101]

(Continued)

Table 15.2 (Continued)

Serial No.	Type (see Figure 1)	Trivial name	Structure	Source	Isolation [Ref]	Biological activity	Total synthesis [Ref]
27	VIII	Sclerotigenin 66	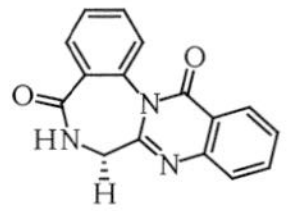	*Penicillium sclerotigenum* (fungus)	[102]	Insecticidal activity against crop pest *Helicoverpa zea* [102]	[103–105]
28	VIII	Cyclopenin 67	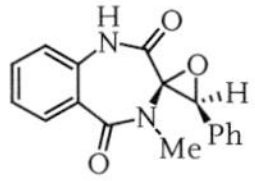	*Penicillium cyclopium* and *P. viridicatum* (fungus)	[106, 107]	Weakly nematicidal [108]	[109, 110]
29	IX	Asmarine A 68	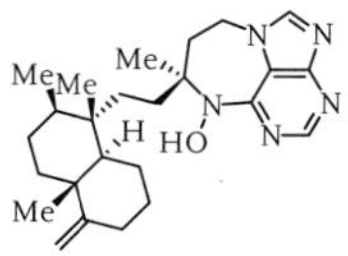	*Raspailia* sp. (marine sponge)	[111, 112]	Cytotoxic activity (P-388, A-549, HT-29 and MEL-28 cell lines) [112]	
30	IX	Neosurugatoxin A 69	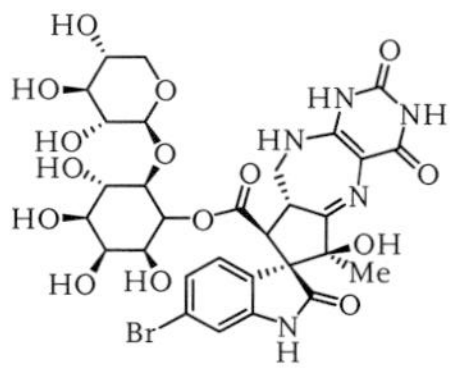	*Babylonia japonica* (Japanese ivory shell)	[113]	Possesses potent antinicotinic activity [113]. Causes mydriasis in mice [114]	[115]

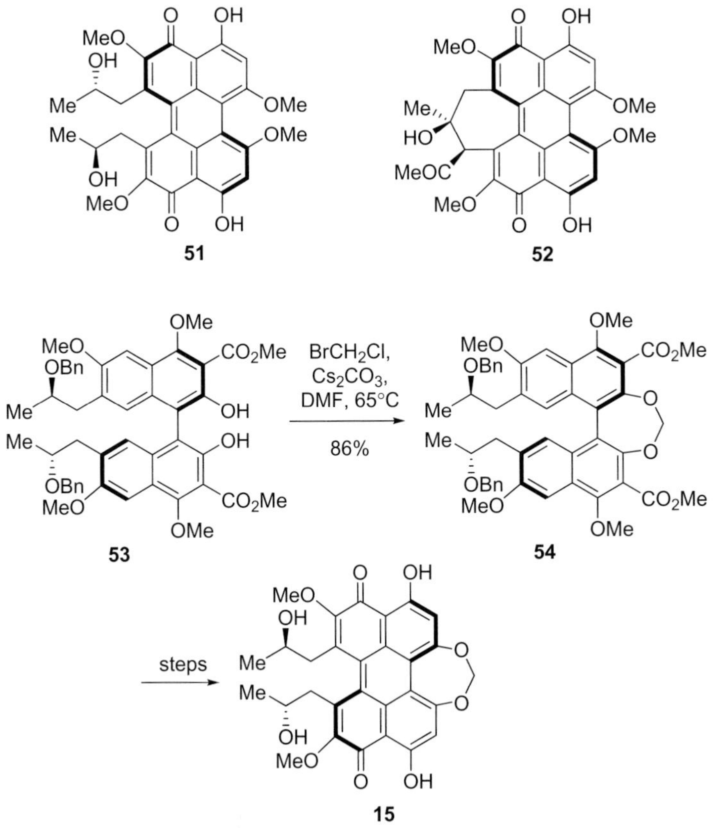

15.3 Synthesis of the Heterocyclic Core of Selected Natural Products Containing Azepines

Scheme 15.11

The first example is (−)-tuberostemonine **55**, a member of the *Stemona* alkaloid family comprising of compounds with a characteristic pyrrolo[1,2-a]azepine nucleus (Type I, Figure 15.1) [116]. These compounds have a wide range of interesting biological activities, and in particular their insecticidal properties have attracted attention from synthetic and medicinal chemists [117]. In terms of total syntheses of this compound, Wipf and co-workers have described an efficient approach to the seven-membered azepine ring system in **42**, in which they used a reliable RCM reaction with the Grubbs' second generation catalyst to give compound **71** from substrate **70** (Scheme 15.12) [69, 70].

Fawcettimine **56** and the related fawcettidine **72** are members of the *Lycopodium* family of alkaloids that contain a distinctive azepine ring system [72]. These compounds have been found to have a particular ability to inhibit the enzyme acetycholine esterase [72]. A recent total synthesis of (+)-fawcettidine **72** by Kozak and Dake used a Ramberg–Bäcklund reaction for a late-stage construction of the desired azepine ring, that is, **73**→**74** (Scheme 15.13) [118]. This compound was then readily converted into the natural product **72** in three steps. A more

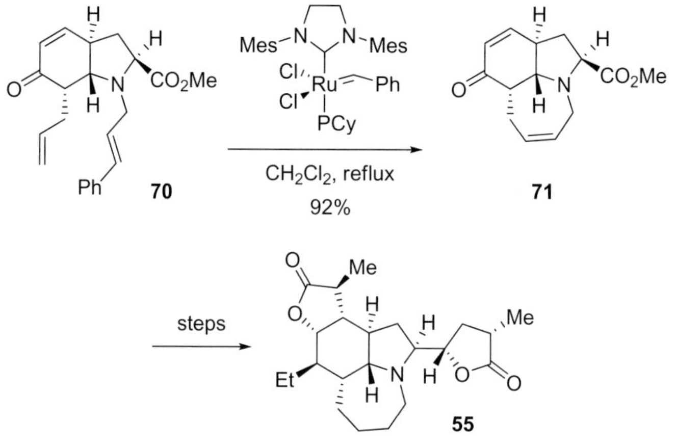

Scheme 15.12

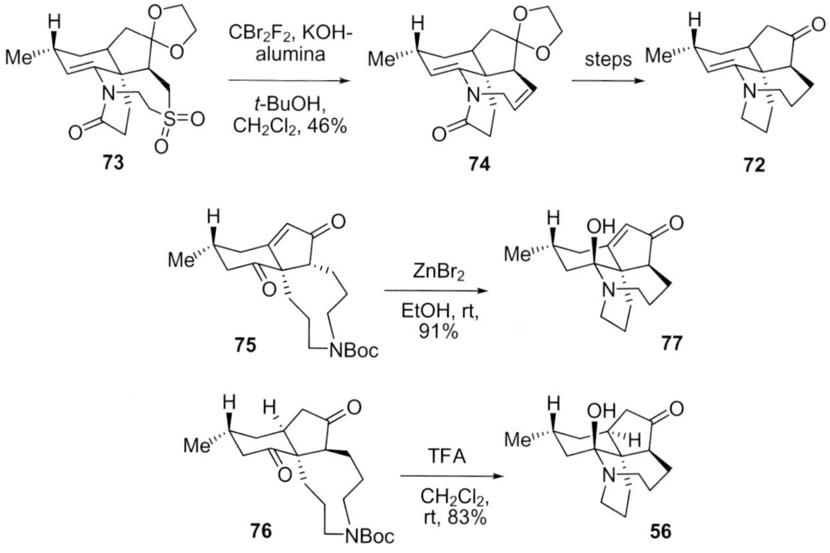

Scheme 15.13

traditional end-game for the construction of the fawcettimine-type skeleton was used by Takayama and co-workers [119]. In this particular synthesis, removal of the Boc group from intermediate **75**, with concomitant isomerization at C-4, resulted in the hemiaminal functionality and the desired construction of the azepine ring system of the fawcettimine-related *Lycopodium* alkaloid, lycoposerramine C **77**. Toste and co-workers used a similar final step to synthesize (+)-fawcettimine **56** from bis-ketone **76** in good yield [73].

A natural product named monanchorin **58** was isolated in 2004 by McKee and co-workers from the sponge *Monanchora unguiculata* (Scheme 15.14) [75]. This

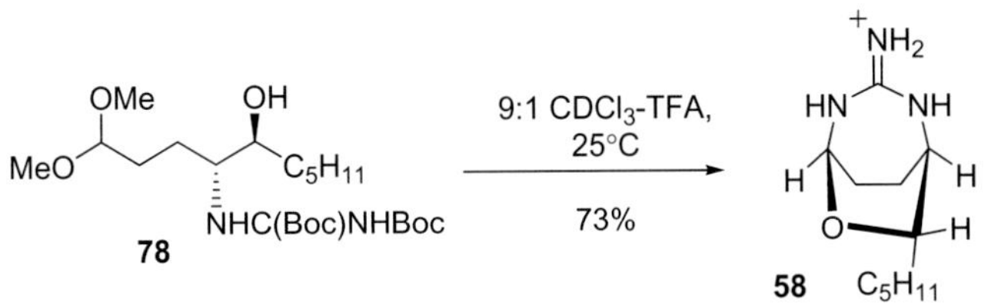

15.3 Synthesis of the Heterocyclic Core of Selected Natural Products Containing Azepines | 555

Scheme 15.14

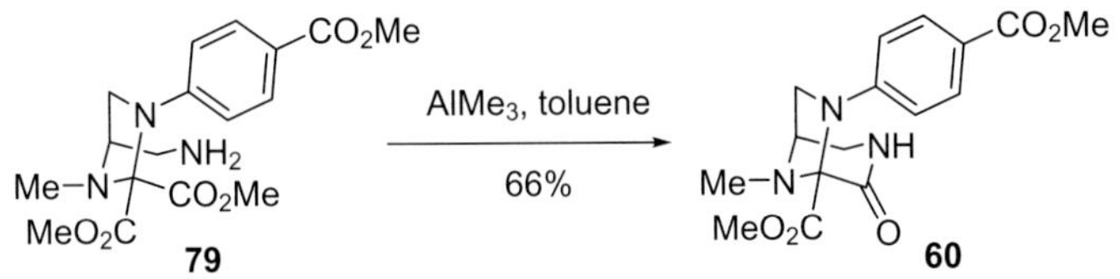

Scheme 15.15

compound contains a 6-oxa-2,4-diazabicyclo[3.2.2]nonane skeleton and was found to be mildly cytotoxic against IC2 murine mast cell lines. Recently this compound was synthesized by Yu and Snider. The key reaction forming the bridged skeleton, which contains the diazepine core, involved deprotection of the two Boc groups and cleavage of the methyl acetal in **78** under acidic conditions to form the aminal functionality in **58** (see Scheme 15.14) [76].

(−)-Aplaminal **60** is an interesting metabolite isolated in 2008 from a sea hare species called *Aplysia kurodai* (Scheme 15.15) [80]. This compound has an unprecedented 3,7,8-triazabicycle[3.2.1]octane structure, in which each bridge contains a nitrogen atom, and which contains the 1,4-diazepine motif of interest in this section (Type **III**, Figure 15.1). Due to its fascinating structure, and the fact that aplaminal **60** demonstrated modest anticancer activity, Smith III and Liu recently published the first total synthesis of this compound [81]. The final ring-closure in the synthesis afforded the completed 3,7,8-triazabicycle[3.2.1]octane structure and involved the treatment of diester **79** with AlMe₃ to afford the natural product **60** in 66% yield. Note that the total synthesis of **60** was achieved in nine linear steps from N-Boc-D-serine in an impressive overall yield of 19%.

The aldisine **80** alkaloids are a group of marine sponge-derived natural compounds containing an azepine structure fused to a pyrrole ring [120], 10Z-hymenialdisine **62** being another representative member of this family (Scheme 15.16). In terms of biological activity these compounds have attracted attention by their ability to significantly inhibit cytoplasmic kinases, including kinases such as MEK-1 and others known to be important in oncologic processes [87, 88]. A typical synthesis of a brominated aldisine core structure **81** (and the associated regioisomer **82** due to an acid-mediated 1,2-Br shift) was used by Annoura and Tatsuoka in their synthesis of hymenialdisine **62** [89]. Their approach used a PPA-mediated cyclization of the acid derivative of substrate **83** (Scheme 15.16), a methodology also used later by Papeo and co-workers for the synthesis of debromo derivatives

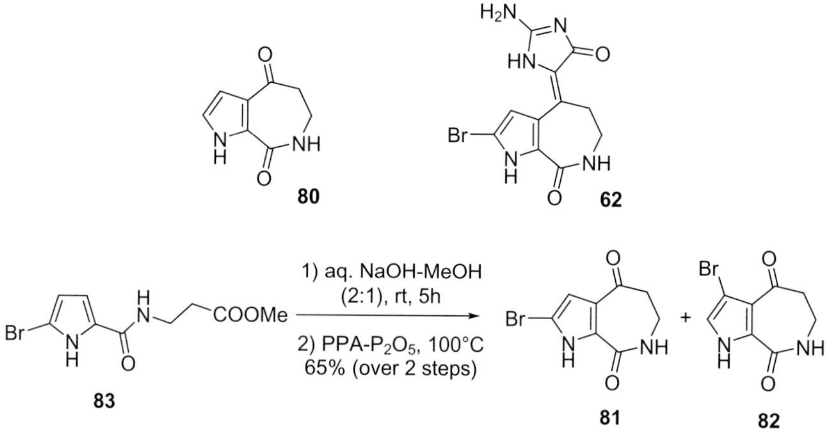

Scheme 15.16

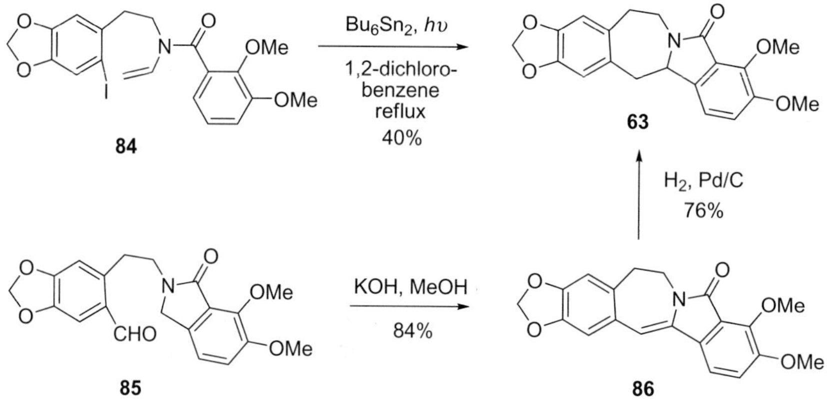

Scheme 15.17

[90]. The same group overcame the bromide migration problem observed by Annoura and Tatsuoka by using an aluminum-mediated Friedel–Crafts cyclization to form the azepine ring structure typical of the hymenialdisines.

Lennoxamine **63** is a natural product with a 1*H*-benzo[*d*]azepine core that has attracted considerable attention from synthetic chemists, despite not having any documented biological activity! Two examples of recent syntheses of this compound are summarized in Scheme 15.17. The first by Ishibaschi and co-workers uses an innovative radical cascade on substrate **84** and makes use of an aryl radical-induced 7-*endo* cyclization, followed by a homolytic aromatic substitution to afford lennoxamine **63** [92]. Secondly, the scheme shows how Sahakitpichan and Ruchirawat applied a base-mediated intramolecular condensation of aldehyde isoindolone **85** to afford dehydroisoindolobenzazepine **86**, which was readily hydrogenated to give the desired natural product **63** [93].

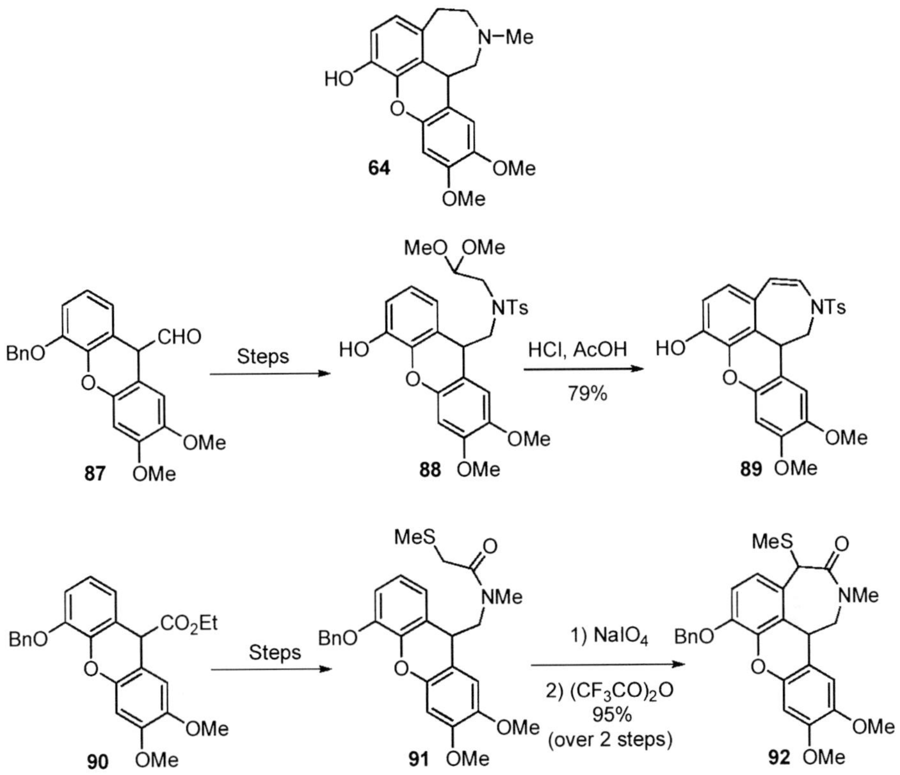

15.3 Synthesis of the Heterocyclic Core of Selected Natural Products Containing Azepines

Scheme 15.18

In 1986, Boente et al. isolated a novel dibenzopyranazepine alkaloid from *Corydalis claviculata* (L.) DC, a compound which they named clavizepine **64** (Scheme 15.18). Although there is no reported biological activity for this compound it has succumbed to a number of syntheses, all starting from appropriately substituted xanthenes. Domínguez and co-workers synthesized the substituted xanthene-9-carboxaldehyde **87** and readily converted it into dimethylacetal **88** [96]. An acid-mediated cyclization reaction then afforded the benzazepine core **89**. Ishibashi et al. used the substituted ethyl xanthene-9-carboxylate **90** in their construction of sulfide **91** [98]. Subsequent oxidation to the sulfoxide (intermediate not shown), followed by a trifluoroacetic anhydride-mediated Pummerer-type ring closure then afforded clavizepine intermediate **92** in excellent yield over the two steps. Both compounds **89** and **92** were then readily converted into the racemic natural product **64**.

Pentostatin **65** [101] and chloropentostatin **93** [121], compounds containing 6,7-dihydroimidazo[4,5-*d*][1,3]diazepin-8(3*H*)-one motifs (Type VII, Figure 15.1), have been shown to be a potent inhibitors of human erythrocytic adenosine deaminase, a potential drug target in both viral diseases and cancer [100]. To synthesize the required heterocycle, Baker and co-workers treated diamine **94** with an excess of triethyl orthoformate in dry dimethyl sulfoxide to afford compound **95** in good yield (Scheme 15.19) [101]. A subsequent glycosylation reaction using this heterocycle then successfully afforded the desired natural adenosine analogs.

In 1999 a benzodiazepine **66** with interesting insecticidal properties was isolated from the sclerotia of the fungi *Penicillium sclerotigenum*, and named sclerotigenin (Scheme 15.20) [102]. This compound had already been synthesized more than 20 years before its isolation by Harrison et al [104]. The related alkaloids circumdatin F **96** [122] and circumdatin C **97** [123] were also synthesized by Witt and Bergman, among others [103]; for instance, the benzoxazine core **96** was synthesized by the dehydration of compound **98** (Scheme 15.20) [124].

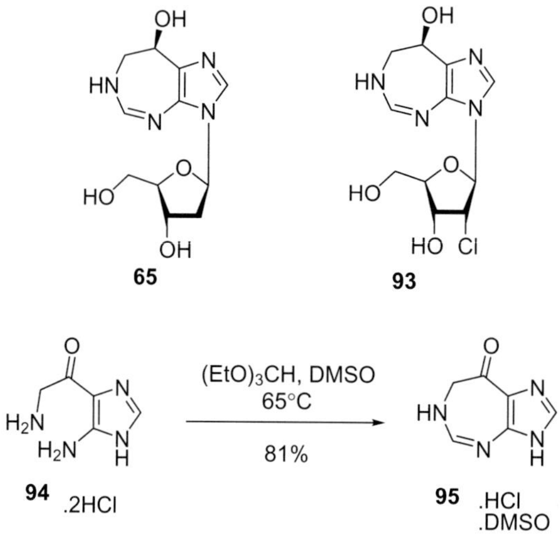

Scheme 15.19

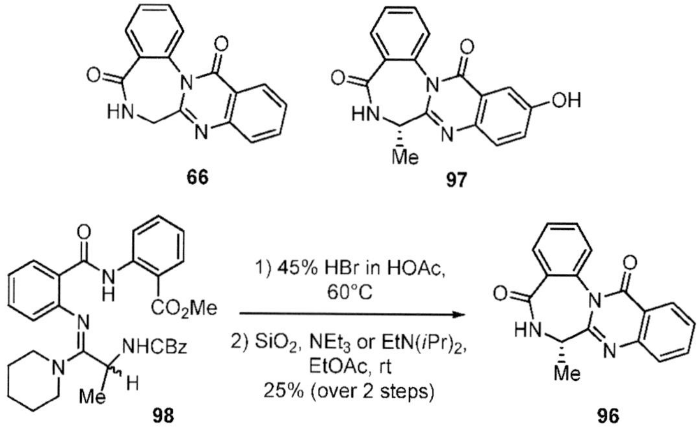

Scheme 15.20

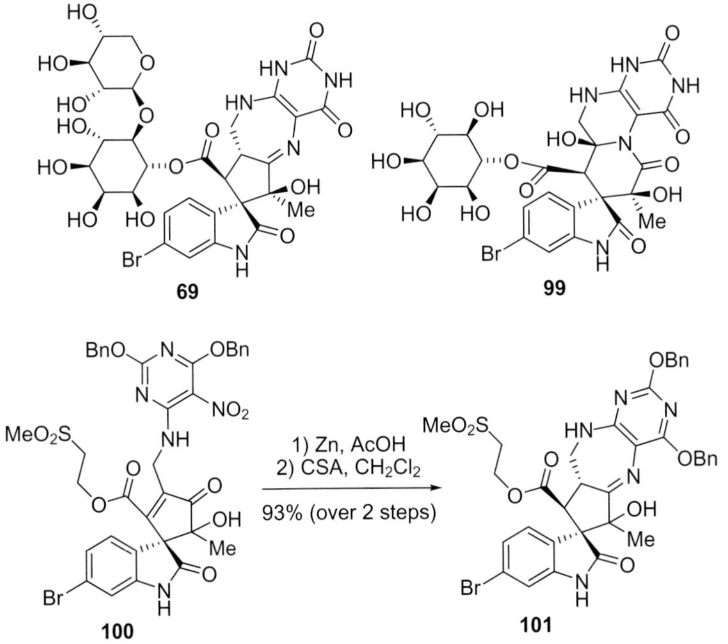

15.4 Synthesis of the Heterocyclic Core of Selected Natural Products Containing Oxazapines

Scheme 15.21

Finally in Section 15.3, the synthesis of a compound named neosurugatoxin **69**, containing a novel pyrimido[5,4-b][1,4]diazepine core, will be described. Together with surugatoxin **99** [125], this compound was isolated from the toxic Japanese ivory shellfish, *Babylonia japonica* which reportedly had caused food poisoning in humans (Scheme 15.21) [113]. Due to the small amounts of natural product available for testing, Inoue and co-workers completed the total synthesis of this compound [115]. To synthesize the key diazepine ring these researchers first reduced the nitro functional group of compound **100** with zinc metal in acetic acid and then treated the resulting amine with camphor sulfonic acid to afford **101** as a mixture of four interconvertible isomers in excellent yield over the two steps. This advanced intermediate was then converted into neosurugatoxin **69** over a number of steps.

15.4
Synthesis of the Heterocyclic Core of Selected Natural Products Containing Oxazapines

In Section 15.4, the synthesis of only one structure containing a seven-membered ring system with an oxygen and nitrogen atom will be discussed. The paucity of examples in this particular category exemplifies how scarce these particular structural motifs are in natural compounds, relative to the oxepine- and azapine-containing skeletons (Table 15.3).

Table 15.3 Representative examples of oxazapine-containing natural products.

Serial No.	Trivial name	Structure	Source	Isolation [Ref]	Biological activity	Total synthesis [Ref]
31	(±) Batrachotoxinin A **102**		*Phyllobates bicolor* and *P. aurotaenia* (poison arrow frogs)	[126, 127]	Potent neurotoxin [127]	[128]
32	Concavine **103**		*Clitocybe concave* (fungus)	[129]	Weak antibacterial activity (*Bacillus cereus* and *B. subtilis*) [129]	

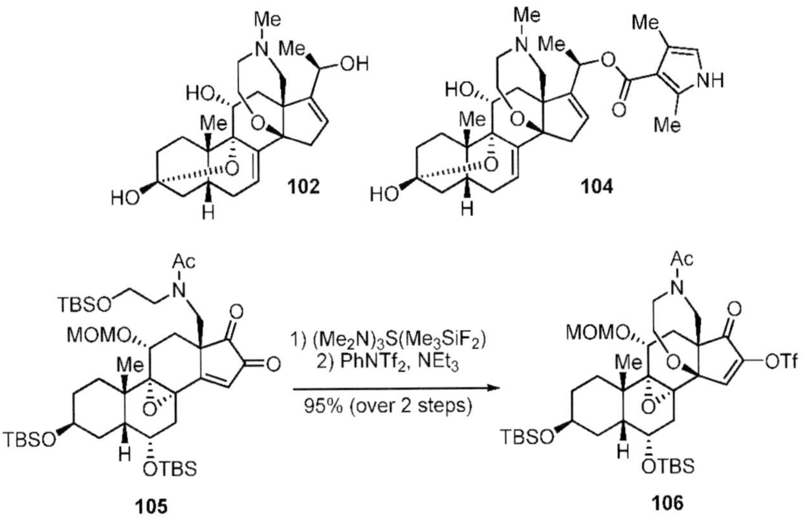

Scheme 15.22

(±) Batrachotoxinin A **102** and the related batrachotoxin **104** are members of a family of steroidal alkaloids which have been isolated from the skins of frogs originating from New Guinea. Closely related compounds have also been isolated from the feathers of birds [130]. As expected, with alkaloids derived from poison arrow frogs the compounds **102** and **104** displayed potent neurotoxicity. Of interest to this chapter is that these compounds contain a rare seven-membered oxazapine ring as part of their structure. The first total synthesis of **102**, and a formal synthesis of **104**, was accomplished by Kishi and co-workers in 1998 (Scheme 15.22) [128]. The key reactions in constructing the seven-membered N,O-ring involved the de-protection of the primary silyl group in intermediate **105**, followed by conversion into the corresponding triflate with PhNTf$_2$, to afford compound **106** in excellent yield. Further synthetic steps then afforded the desired (±) batrachotoxinin A **102**.

15.5
Conclusion

In this chapter we have attempted to describe the variety of natural products containing scaffolds with oxepine or azepine motifs. In addition, most of the examples chosen have been shown to possess a wide variety of bioactivities. Because of these two factors, and the interesting challenge of forming the oxepine and azepine structures, these compounds have attracted the attention of synthetic and medicinal chemists alike. It is our hope that this chapter will stimulate more scientific research into this particular area, resulting in the total synthesis of more natural products (and analogs) with oxygen- and nitrogen-containing seven-membered ring systems.

Acknowledgments

This material is based upon work supported financially by the National Research Foundation, Pretoria. Any opinion, findings and conclusions or recommendations expressed in this material are those of the author(s) and therefore the NRF does not accept any liability in regard thereto. This work was also supported by the University of the Witwatersrand (University and Science Faculty Research Councils). We gratefully thank Prof. Charles B. de Koning and Prof. Joseph P. Michael (School of Chemistry, University of the Witwatersrand) for critical evaluation of this book chapter.

References

1 Shimizu, Y., Chou, H.-N., Bando, H., Van Duyne, G. and Clardy, J.C. (1986) Structure of brevetoxin A (GB-1 toxin), the most potent toxin in the florida red tide organism *Gymnodinium breve* (*Ptychodiscus brevis*). *J. Am. Chem. Soc.*, **108**, 514–515.

2 Shimizu, Y., Bando, H., Chou, H.-N., Van Duyne, G. and Clardy, J.C. (1986) Absolute configuration of brevetoxins. *J. Chem. Soc. Chem. Commun.*, 1656–1658.

3 Yasumoto, T. and Murata, M. (1993) Marine toxins. *Chem. Rev.*, **93**, 1897–1909.

4 Nicolaou, K.C., Yang, Z., Shi, G.-Q., Gunzner, J.L., Agrios, K.A. and Gärtner, P. (1998) Total synthesis of brevetoxin A. *Nature*, **392**, 264–269.

5 Braekman, J.C., Daloze, D., Tavares, R., Hajdu, E. and Van Soest, R.W.M. (2000) Novel polycyclic guanidine alkaloids from two marine sponges of the genus *Monanchora*. *J. Nat. Prod.*, **63**, 193–196.

6 Moore, C.G., Murphy, P.J., Williams, H.L., McGown, A.T. and Smith, N.K. (2007) Synthetic studies towards ptilomycalin A: total synthesis of crambescidin 359. *Tetrahedron*, **63**, 11771–11780.

7 Aron, Z.D. and Overman, L.E. (2005) Total synthesis and properties of the crambescidin core zwitterionic acid and crambescidin 359. *J. Am. Chem. Soc.*, **127**, 3380–3390.

8 Nagasawa, K., Georgieva, A., Koshino, H., Nakata, T., Kita, T. and Hashimoto, Y. (2002) Total synthesis of crambescidin 359. *Org. Lett.*, **4**, 177–180.

9 Mimaki, Y., Kuroda, M., Sashida, Y., Yamori, T. and Tsuruo, T. (2000) Candicanoside A, a novel cytotoxic rearranged cholestane glycoside from *Galtonia candicans*. *Helv. Chim. Acta*, **83**, 2698–2704.

10 Tang, P. and Yu, B. (2009) Total synthesis of candicanoside A, a rearranged cholestane disaccharide, and its 4'-O-(p-methoxybenzoate) congener. *Eur. J. Org. Chem.*, 259–269.

11 Tang, P. and Yu, B. (2007) Total synthesis of candicanoside A, a potent antitumor saponin with a rearranged steroid side chain. *Angew. Chem. Int. Ed.*, **46**, 2527–2530.

12 Igarashi, M., Kinoshita, N., Ikeda, T., Nakagawa, E., Hamada, M. and Takeuchi, T. (1997) Formamicin, a novel antifungal antibiotic produced by a strain of *Saccharothrix* sp. I. Taxonomy, production, isolation and biological properties. *J. Antibiot.*, **50**, 926–931.

13 Igarashi, M., Nakamura, H., Naganawa, H. and Takeuchi, T. (1997) Formamicin, a novel antifungal antibiotic produced by a strain of *Saccharothrix* sp. II. Structure elucidation of formamicin. *J. Antibiot.*, **50**, 932–936.

14 Durham, T.B., Blanchard, N., Savall, B.M., Powell, N.A. and Roush, W.R. (2004) Total synthesis of formamicin. *J. Am. Chem. Soc.*, **126**, 9307–9317.

15 Irie, T., Izawa, M. and Kurosawa, E. (1970) Laureatin and isolaureatin, constituents of *Laurencia nipponica* Yamada. *Tetrahedron*, **26**, 851–870.

16 Watanabe, K., Umeda, K. and Miyakado, M. (1989) Isolation and identification of three insecticidal principles from the red alga *Laurencia nipponica* Yamada. *Agric. Biol. Chem.*, **53**, 2513–2515.

17 Kim, H., Lee, H., Lee, D., Kim, S. and Kim, D. (2007) Asymmetric total syntheses of (+)-3-(Z)-laureatin and (+)-3-(Z)-isolaureatin by "lone pair-lone pair interaction-controlled" isomerization. *J. Am. Chem. Soc.*, **129**, 2269–2274.

18 Jansen, R., Irschik, H., Reichenbach, H., Schomburg, D., Wray, V. and Höfle, G. (1989) Antibiotics from gliding bacteria. 37. Sorangicin A, a highly-active antibiotic with novel macrolide-polyether structure from *Sorangium cellulosum*, SO ce12: spectroscopic structure elucidation, crystal and solution structure. *Liebigs Ann. Chem.*, 111–119.

19 Jansen, R., Wray, V., Irschik, H., Reichenbach, H. and Höfle, G. (1985) Isolation and spectroscopic structure elucidation of sorangicin A, a new type of macrolide-polyether antibiotic from gliding bacteria–30. *Tetrahedron Lett.*, **26**, 6031–6034.

20 Irschik, H., Jansen, R., Gerth, K., Höfle, G. and Reichenbach, H. (1987) Antibiotics from gliding bacteria. 32. The sorangicins, novel and powerful inhibitors of eubacterial RNA polymerase isolated from myxobacteria. *J. Antibiot.*, **40**, 7–13.

21 Smith, A.B. III, Dong, S., Brenneman, J.B. and Fox, R.J. (2009) Total synthesis of (+)-sorangicin A. *J. Am. Chem. Soc.*, **131**, 12109–12111.

22 Engler, M., Anke, T., Sterner, O. and Brandt, U. (1997) Pterulinic acid and pterulone, two novel inhibitors of NADH:ubiquinone oxidoreductase (complex I) produced by a *Pterula* species. I. Production, isolation and biological activities. *J. Antibiot.*, **50**, 325–329.

23 Engler, M., Anke, T. and Sterner, O. (1997) Pterulinic acid and pterulone, two novel inhibitors of NADH:ubiquinone oxidoreductase (complex I) produced by a *Pterula* species. II. Physico-chemical properties and structure elucidation. *J. Antibiot.*, **50**, 330–333.

24 Lemaire, P., Balme, G., Desbordes, P. and Vors, J.-P. (2003) Efficient syntheses of pterulone, pterulone B and related analogs. *Org. Biomol. Chem.*, **1**, 4209–4219.

25 Kahnberg, P., Lee, C.W., Grubbs, R.H. and Sterner, O. (2002) Alternative routes to pterulone. *Tetrahedron*, **58**, 5203–5208.

26 Kahnberg, P. and Sterner, O. (2001) Synthesis of the antifungal 1-benzoxepin pterulone. *Tetrahedron*, **57**, 7181–7184.

27 Huang, S.-T., Kuo, H.-S. and Chen, C.-T. (2001) Total synthesis of NADH:ubiquinone oxidoreductase (complex I) antagonist pterulone and its analog. *Tetrahedron Lett.*, **42**, 7473–7475.

28 Boonphong, S., Puangsombat, P., Baramee, A., Mahidol, C., Ruchirawat, S. and Kittakoop, P. (2007) Bioactive compounds from *Bauhinia purpurea* possessing antimalarial, antimycobacterial, antifungal, anti-inflammatory, and cytotoxic activities. *J. Nat. Prod.*, **70**, 795–801.

29 Kraus, G.A., Thite, A. and Liu, F. (2009) Intramolecular radical cyclizations onto quinones. A direct synthesis of bauhinoxepin J. *Tetrahedron Lett.*, **50**, 5303–5304.

30 Campello, M.J., Castedo, L., Domínguez, D., de Lera, A.R., Saá, J.M., Suau, R., Tojo, E. and Vidal, M.C. (1984) New oxidized isocularine alkaloids from *Sarcocapnos* plants. *Tetrahedron Lett.*, **25**, 5933–5936.

31 Suau, R., Rico, R., López-Romero, J.M., Nájera, F., Ruiz, A. and Ortiz-Lopez, F.J. (2002) Synthesis of 3,4-dioxocularine and aristocularine alkaloids in a convergent route from aryloxy-phenyl acetamides involving oxalyl chloride-Lewis acid. *Arkivoc*, **v**, 62–72.

32 Moreau, A., Couture, A., Deniau, E. and Grandclaudon, P. (2004) A new route to aristocularine alkaloids: total synthesis of aristoyagonine. *J. Org. Chem.*, **69**, 4527–4530.

33 Höller, U., Gloer, J.B. and Wicklow, D.T. (2002) Biologically active polyketide metabolites from an undetermined fungicolous hyphomycete resembling *Cladosporium. J. Nat. Prod.*, **65**, 876–882.

34 Furumai, T., Igarashi, Y., Higuchi, H., Saito, N. and Oki, T. (2002) Kosinostatin, a quinocycline antibiotic with antitumor activity from *Micromonospora* sp. TP-A0468. *J. Antibiot.*, **55**, 128–133.

35 Igarashi, Y., Higuchi, H., Oki, T. and Furumai, T. (2002) NMR analysis of quinocycline antibiotics: structure determination of kosinostatin, an antitumor substance from *Micromonospora* sp. TP-A0468. *J. Antibiot.*, **55**, 134–140.

36 Petersen, F., Fredenhagen, A., Mett, H., Lydon, N.B., Delmendo, R., Jenny, H.-B. and Peter, H.H. (1995) Paeciloquinone A, B, C, D, E and F: new potent inhibitors of protein tyrosine kinases produced by *Paecilomyces carneus*. 1. Taxonomy, fermentation, isolation and biological-activity. *J. Antibiot.*, **48**, 191–198.

37 Fredenhagen, A., Hug, P., Sauter, H. and Peter, H.H. (1995) Paeciloquinone a, b, c, d, e and f: new potent inhibitors of protein tyrosine kinase produced by *Paecilomyces carneus*. 2. Characterization and structure determination. *J. Antibiot.*, **48**, 199–204.

38 Goh, E.M. and Wilkins, A.L. (1979) Structures of the lichen depsidones granulatin and chlorogranulatin. *J. Chem. Soc. Perkin Trans.*, **1**, 1656–1658.

39 Neamati, N., Hong, H., Mazumder, A., Wang, S., Sunder, S., Nicklaus, M.C., Milne, G.W.A., Proksa, B. and Pommier, Y. (1997) Depsides and depsidones as inhibitors of HIV-1 integrase: discovery of novel inhibitors through 3D database searching. *J. Med. Chem.*, **40**, 942–951.

40 Pulgarin, C. and Tabacchi, R. (1989) Synthesis of methyl virensate. *Helv. Chim. Acta*, **72**, 1061–1065.

41 Yamazaki, M., Okuyama, E., Kobayashi, M. and Inoue, H. (1981) The structure of paraherquamide, a toxic metabolite from *Penicillium paraherquei. Tetrahedron Lett.*, **22**, 135–136.

42 Lopez-Gresa, M.P., González, M.C., Ciavatta, L., Ayala, I., Moya, P. and Primo, J. (2006) Insecticidal activity of paraherquamides, including paraherquamide H and paraherquamide I, two new alkaloids isolated from *Penicillium cluniae. J. Agric. Food Chem.*, **54**, 2921–2925.

43 Williams, R.M., Cao, J., Tsujishima, H. and Cox, R.J. (2003) Asymmetric, stereocontrolled total synthesis of paraherquamide A. *J. Am. Chem. Soc.*, **125**, 12172–12178.

44 Kuyama, S. and Tamura, T. (1957) Cercosporin. A pigment of *Cercosporina kikuchii* Matsumoto et Tomoyasu. I. Cultivation of fungus, isolation and purification of pigment. *J. Am. Chem. Soc.*, **79**, 5725–5726.

45 Tamaoki, T., Takahashi, I., Kobayashi, E., Nakano, H., Akinaga, S. and Suzuki, K. (1990) Calphostin (UCN1028) and calphostin related-compounds, a new class of specific and potent inhibitors of protein kinase C. *Biol. Med. Signal Transduct.*, **24**, 497–501.

46 Morgan, B.J., Dey, S., Johnson, S.W. and Kozlowski, M.C. (2009) Design, synthesis, and investigation of protein kinase C inhibitors: total syntheses of (+)-calphostin D, (+)-phleichrome, cercosporin, and new photoactive perylenequinones. *J. Am. Chem. Soc.*, **131**, 9413–9425.

47 Snyder, N.L., Haines, H.M. and Peczuh, M.W. (2006) Recent developments in the synthesis of oxepines. *Tetrahedron*, **62**, 9301–9320.

48 Elliott, M.C. (2002) Saturated oxygen heterocycles. *J. Chem. Soc. Perkin Trans. 1*, 2301–2323.

49 Hoberg, J.O. (1998) Synthesis of seven-membered oxacycles. *Tetrahedron*, **54**, 12631–12670.

50 Nicolaou, K.C., Frederick, M.O. and Aversa, R.J. (2008) The continuing saga of the marine polyether biotoxins. *Angew. Chem. Int. Ed.*, **47**, 7182–7225.

51 Berlinck, R.G.S., Burtoloso, A.C.B. and Kossuga, M.H. (2008) The chemistry and biology of organic guanidine derivatives. *Nat. Prod. Rep.*, **25**, 919–954.

52 Berlinck, R.G.S. and Kossuga, M.H. (2005) Natural guanidine derivatives. *Nat. Prod. Rep.*, **22**, 516–550.

53 Overman, L.E., Rabinowitz, M.H. and Renhowe, P.A. (1995) Enantioselective total synthesis of (–)-ptilomycalin A. *J. Am. Chem. Soc.*, **117**, 2657–2658.

54 Dai, W.-M., Guan, Y. and Jin, J. (2005) Structures and total syntheses of the plecomacrolides. *Curr. Med. Chem.*, **12**, 1947–1993.

55 Powell, N.A. and Roush, W.R. (2001) Studies on the total synthesis of formamicin: synthesis of the C(1)-C(11) fragment. *Org. Lett.*, **3**, 453–456.

56 Savall, B.M., Blanchard, N. and Roush, W.R. (2003) Total synthesis of the formamicin aglycon, formamicinone. *Org. Lett.*, **5**, 377–379.

57 Smith, A.B. III and Fox, R.J. (2004) Construction of a C(30–38) dioxabicyclo[3.2.1]octane subtarget for (+)-sorangicin A, exploiting a regio and stereocontrolled acid-catalyzed epoxide ring opening. *Org. Lett.*, **6**, 1477–1480.

58 Smith, A.B. III and Dong, S. (2009) An efficient, second-generation synthesis of the signature dioxabicyclo[3.2.1]octane core of (+)-sorangicin A and elaboration of the (Z,Z,E)-triene acid system. *Org. Lett.*, **11**, 1099–1102.

59 Lin, Y.-L., Kuo, H.-S., Wang, Y.-W. and Huang, S.-T. (2003) Efficient entry to 1-benzoxepine ring skeleton via tandem S_N2/Wittig reaction. Total synthesis of NADH: ubiquinone oxidoreductase (complex I) antagonist pterulinic acid. *Tetrahedron*, **59**, 1277–1281.

60 SanMartin, R., Churruca, F. and Dominguez, E. (2004) Dibenzo[b,f] oxepines: syntheses and applications. A review. *Org. Prep. Proc. Int.*, **36**, 297–330.

61 Kittakoop, P., Nopichai, S., Thongon, N., Charoenchai, P. and Thebtaranonth, Y. (2004) Bauhinoxepins A and B: new antimycobacterial dibenzo[b,f]oxepins from *Bauhinia saccocalyx. Helv. Chim. Acta*, **87**, 175–179.

62 Pettit, G.R., Numata, A., Iwamoto, C., Usami, Y., Yamada, T., Ohishi, H. and Cragg, G.M. (2006) Antineoplastic agents 551. Isolation and structures of bauhiniastatins 1–4 from *Bauhinia purpurea. J. Nat. Prod.*, **69**, 323–327.

63 Lin, J., Zhang, W., Jiang, N., Niu, Z., Bao, K., Zhang, L., Liu, D., Pan, C. and Yao, X. (2008) Total synthesis of bulbophylol B. *J. Nat. Prod.*, **71**, 1938–1941.

64 MacNeil, S.L., Gray, M., Gusev, D.G., Briggs, L.E. and Snieckus, V. (2008) Carbanionic Friedel-Crafts equivalents. Regioselective directed *ortho* and remote metalation-C-N cross coupling routes to acridones and dibenzo[b,f] azepinones. *J. Org. Chem.*, **73**, 9710–9719.

65 Trost, B.M., Cramer, N. and Bernsmann, H. (2007) Concise total synthesis of (±)-marcfortine B. *J. Am. Chem. Soc.*, **129**, 3086–3087.

66 Williams, R.M. and Cushing, T.D. (1990) Synthetic studies on paraherquamide: synthesis of the 2H-1,5-benzodioxepin ring system. *Tetrahedron Lett.*, **31**, 6325–6328.

67 Greger, H. (2006) Structural relationships, distribution and biological activities of *Stemona* alkaloids. *Planta Med.*, **72**, 99–113.

68 Brem, B., Seger, C., Pacher, T., Hofer, O., Vajrodaya, S. and Greger, H. (2002) Feeding deterrence and contact toxicity of *Stemona* alkaloids – a source of potent natural insecticides. *J. Agric. Food Chem.*, **50**, 6383–6388.

69 Wipf, P., Rector, S.R. and Takahashi, H. (2002) Total synthesis of (–)-tuberostemonine. *J. Am. Chem. Soc.*, **124**, 14848–14849.

70 Wipf, P. and Spencer, S.R. (2005) Asymmetric total syntheses of tuberostemonine, didehydrotuberostemonine, and 13-epituberostemonine. *J. Am. Chem. Soc.*, **127**, 225–235.

71 Burnell, R.H. (1959) Lycopodium alkaloids. 1. Extraction of alkaloids from *Lycopodium fawcettii*, Lloyd and Underwood. *J. Chem. Soc.*, 3091–3093.

72 Hirasawa, Y., Kobayashi, J. and Morita, H. (2009) The lycopodium alkaloids. *Heterocycles*, **77**, 679–729.

73 Linghu, X., Kennedy-Smith, J.J. and Toste, F.D. (2007) Total synthesis of (+)-fawcettimine. *Angew. Chem. Int. Ed.*, **46**, 7671–7673.

74 Qi, S.-H., Su, G.-C., Wang, Y.-F., Liu, Q.-Y. and Gao, C.-H. (2009) Alkaloids from the South China Sea black coral *Antipathes dichotoma*. *Chem. Pharm. Bull.*, **57**, 87–88.

75 Meragelman, K.M., McKee, T.C. and McMahon, J.B. (2004) Monanchorin, a bicyclic alkaloid from the sponge *Monanchora unguiculata*. *J. Nat. Prod.*, **67**, 1165–1167.

76 Yu, M. and Snider, B.B. (2009) Synthesis of (+)- and (−)-monanchorin. *Org. Lett.*, **11**, 1031–1032.

77 Kimura, K.-I., Ikeda, Y., Kagami, S., Yoshihama, M., Ubukata, M., Esumi, Y., Osada, H. and Isono, K. (1998) New types of liposidomycins that inhibit bacterial peptidoglycan synthesis and are produced by *Streptomyces*. II. Isolation and structure elucidation. *J. Antibiot.*, **51**, 647–654.

78 Kimura, K.-I., Kagami, S., Ikeda, Y., Takahashi, H., Yoshihama, M., Kusakabe, H., Osada, H. and Isono, K. (1998) New types of liposidomycins that inhibit bacterial peptidoglycan synthesis and are produced by *Streptomyces*. I. Producing organism and medium components. *J. Antibiot.*, **51**, 640–646.

79 Kimura, K.-I., Ikeda, Y., Kagami, S., Yoshihama, M., Suzuki, K., Osada, H. and Isono, K. (1998) Selective inhibition of the bacterial peptidoglycan biosynthesis by the new types of liposidomycins. *J. Antibiot.*, **51**, 1099–1104.

80 Kuroda, T. and Kigoshi, H. (2008) Aplaminal: a novel cytotoxic aminal isolated from the sea hare *Aplysia kurodai*. *Org. Lett.*, **10**, 489–491.

81 Smith, A.B., III and Liu, Z. (2008) Total synthesis of (−)-aplaminal. *Org. Lett.*, **10**, 4363–4365.

82 Yamazaki, M., Fujimoto, H., Ohta, Y., Iitaka, Y. and Itai, A. (1981) A new toxic fungal metabolite, silvaticamide from *Aspergillus silvaticus*. *Heterocycles*, **15**, 889–893.

83 Cimino, G., De Rosa, S., De Stefano, S., Mazzarella, L., Puliti, R. and Sodano, G. (1982) Isolation and X-ray crystal structure of a novel bromo-compound from two marine sponges. *Tetrahedron Lett.*, **23**, 767–768.

84 Mattia, C.A., Mazzarella, L. and Puliti, R. (1982) 4-(2-Amino-4-oxo-2-imidazolin-5-ylidene)-2-bromo-4,5,6,7-tetrahydropyrro lo-[2,3-c]azepin-8-one methanol solvate–a new bromo compound from the sponge *Acanthella aurantiaca*. *Acta Crystallogr. Sect. B Struct. Commun.*, **38**, 2513–2515.

85 Kitagawa, I., Kobayashi, M., Kitanaka, K., Kido, M. and Kyogoku, Y. (1983) Marine natural products. 12. On the chemical-constituents of the Okinawan marine sponge *Hymeniacidon aldis*. *Chem. Pharm. Bull.*, **31**, 2321–2328.

86 Denanteuil, G., Ahond, A., Guilhem, J., Poupat, C., Dau, E.T.H., Potier, P., Pusset, M., Pusset, J. and Laboute, P. (1985) Marine-invertebrates from *neo*-Caledonian lagoons. 5. Isolation and identification of metabolites from a new species of sponge, *Pseudaxinyssa cantharella*. *Tetrahedron*, **41**, 6019–6033.

87 Tasdemir, D., Mallon, R., Greenstein, M., Feldberg, L.R., Kim, S.C., Collins, K., Wojciechowicz, D., Mangalindan, G.C., Concepción, G.P., Harper, M.K. and Ireland, C.M. (2002) Aldisine alkaloids from the Philippine sponge *Stylissa massa* are potent inhibitors of mitogen activated protein kinase kinase-1 (MEK-1). *J. Med. Chem.*, **45**, 529–532.

88 Nguyen, T.N.T. and Tepe, J.J. (2009) Preparation of hymenialdisine, analogs and their evaluation as kinase inhibitors. *Curr. Med. Chem.*, **16**, 3122–3143.

89 Annoura, H. and Tatsuoka, T. (1995) Total syntheses of hymenialdisine and debromohymenialdisine: stereospecific construction of the 2-amino-4-oxo-2-imidazolin-5(Z)-disubstituted ylidene ring system. *Tetrahedron Lett.*, **36**, 413–416.

90 Papeo, G., Posteri, H., Borghi, D. and Varasi, M. (2005) A new glycociamidine ring precursor: syntheses of (Z)-hymenialdisine, (Z)-2-debromohymenialdisine, and (±)-*endo*-2-debromohymenialdisine. *Org. Lett.*, **7**, 5641–5644.

91 Valencia, E., Freyer, A.J., Shamma, M. and Fajardo, V. (1984) ±-Nuevamine, an isoindoloisoquinoline alkaloid, and (±)-lennoxamine, an

isoindolobenzazepine. *Tetrahedron Lett.*, **25**, 599–602.

92 Taniguchi, T., Iwasaki, K., Uchiyama, M., Tamura, O. and Ishibashi, H. (2005) A short synthesis of lennoxamine using a radical cascade. *Org. Lett.*, **7**, 4389–4390.

93 Sahakitpichan, P. and Ruchirawat, S. (2004) A practical and highly efficient synthesis of lennoxamine and related isoindolobenzazepines. *Tetrahedron*, **60**, 4169–4172.

94 Comins, D.L., Schilling, S. and Zhang, Y.C. (2005) Asymmetric synthesis of 3-substituted isoindolinones: application to the total synthesis of (+)-lennoxamine. *Org. Lett.*, **7**, 95–98.

95 Boente, J.M., Castedo, L., Domínguez, D. and Ferro, M.C. (1986) Clavizepine, the 1st dibenzopyranazepine alkaloid. *Tetrahedron Lett.*, **27**, 4077–4078.

96 de la Fuente, M.C., Castedo, L. and Domínguez, D. (1996) A synthetic route to (+/−)-clavizepine through a dibenzoxepine intermediate. *J. Org. Chem.*, **61**, 5818–5822.

97 Ishibashi, H., Takagaki, K., Imada, N. and Ikeda, M. (1994) First total synthesis of the benzopyranobenzazepine alkaloid (±)-clavizepine. *Synlett*, 49–50.

98 Ishibashi, H., Takagaki, K., Imada, N. and Ikeda, M. (1994) Total synthesis of (±)-clavizepine. *Tetrahedron*, **50**, 10215–10224.

99 Woo, P.W.K., Dion, H.W., Lange, S.M., Dahl, L.F. and Durham, L.J. (1974) A novel adenosine and ara-A deaminase inhibitor, (R)-3-(2-deoxy-β-D-erythropentofuranosyl)-3,6,7,8-tetrahydroimidazo[4,5-d][1,3]diazepin-8-ol. *J. Heterocyclic Chem.*, **11**, 641–643.

100 Agarwal, R.P., Spector, T. and Parks, R.E. (1977) Tight-binding inhibitors. IV. Inhibition of adenosine deaminases by various inhibitors. *Biochem. Pharmacol.*, **26**, 359–367.

101 Chan, E., Putt, S.R., Showalter, H.D.H. and Baker, D.C. (1982) Total synthesis of (8R)-3-(2-deoxy-β-D-erythropentofuranosyl)-3,6,7,8-tetrahydroimidazo[4,5-d][1,3]diazepin-8-ol (pentostatin), the potent inhibitor of adenosine deaminase. *J. Org. Chem.*, **47**, 3457–3464.

102 Joshi, B.K., Gloer, J.B., Wicklow, D.T. and Dowd, P.F. (1999) Sclerotigenin: a new antiinsectan benzodiazepine from the sclerotia of *Penicillium sclerotigenum*. *J. Nat. Prod.*, **62**, 650–652.

103 Snider, B.B. and Busuyek, M.V. (2001) Synthesis of circumdatin F and sclerotigenin. Use of the 2-nitrobenzyl group for protection of a diketopiperazine amide; synthesis of *ent*-fumiquinazoline G. *Tetrahedron*, **57**, 3301–3307.

104 Harrison, D.R., Kennewell, P.D. and Taylor, J.B. (1977) Novel fused-ring derivatives of 2-methyl-3-o-tolyl-4(3H) quinazolone; quinazolino[3,2-a][1,4] benzodiazepines. *J. Heterocycl. Chem.*, **14**, 1191–1196.

105 Tseng, M.-C., Yang, H.-Y. and Chu, Y.-H. (2010) Total synthesis of asperlicin C, circumdatin F, demethylbenzomalvin A, demethoxycircumdatin H, sclerotigenin, and other fused quinazolinones. *Org. Biomol. Chem.*, **8**, 419–427.

106 Bracken, A., Pocker, A. and Raistrick, H. (1954) Studies in the biochemistry of micro-organisms. 93. Cyclopenin, a nitrogen-containing metabolic product of *Penicillium cyclopium* Westling. *Biochem. J.*, **57**, 587–595.

107 Birkinshaw, J.H., Luckner, M., Stickings, C.E., Mohammed, Y.S. and Mothes, K. (1963) Studies in biochemistry of micro-organisms. 114. Viridicatol and cyclopenol, metabolites of *Penicillium viridicatum* Westling and *Penicillium cyclopium* Westling. *Biochem. J.*, **89**, 196–202.

108 Kusano, M., Koshino, H., Uzawa, J., Fujioka, S., Kawano, T. and Kimura, Y. (2000) Nematicidal alkaloids and related compounds produced by the fungus *Penicillium* cf. *simplicissimum*. *Biosci. Biotechnol. Biochem.*, **64**, 2559–2568.

109 White, J.D., Haefliger, W.E. and Dimsdale, M.J. (1970) Stereospecific synthesis of *dl*-cyclopenin and *dl*-cyclopenol. *Tetrahedron*, **26**, 233–242.

110 Martin, P.K., Rapoport, H., Smith, H.W. and Wong, J.L. (1969) Synthesis of cyclopenin and isocyclopenin. *J. Org. Chem.*, **34**, 1359–1363.

111 Yosief, T., Rudi, A. and Kashman, Y. (2000) Asmarines A-F, novel cytotoxic compounds from the marine sponge *Raspailia* species. *J. Nat. Prod.*, **63**, 299–304.

112 Yosief, T., Rudi, A., Stein, Z., Goldberg, I., Gravalos, G.M.D., Schleyer, M. and Kashman, Y. (1998) Asmarines A-C; three novel cytotoxic metabolites from the marine sponge *Raspailia* sp. *Tetrahedron Lett.*, **39**, 3323–3326.

113 Kosuge, T., Tsuji, K., Hirai, K., Yamaguchi, K., Okamoto, T. and Iitaka, Y. (1981) Isolation and structure determination of a new marine toxin, neosurugatoxin, from the japanese ivory shell, *Babylonia japonica*. *Tetrahedron Lett.*, **22**, 3417–3420.

114 Hayashi, E., Isogai, M., Kagawa, Y., Takayanagi, N. and Yamada, S. (1984) Neosurugatoxin, a specific antagonist of nicotinic acetylcholine-receptors. *J. Neurochem.*, **42**, 1491–1494.

115 Inoue, S., Okada, K., Tanino, H. and Kakoi, H. (1994) Total synthesis of neosurugatoxin. *Tetrahedron*, **50**, 2753–2770.

116 Pilli, R.A. and de Oliveira, M. (2000) Recent progress in the chemistry of the *Stemona* alkaloids. *Nat. Prod. Rep.*, **17**, 117–127.

117 Alibes, R. and Figueredo, M. (2009) Strategies for the synthesis of *Stemona* alkaloids. *Eur. J. Org. Chem.*, 2421–2435.

118 Kozak, J.A. and Dake, G.R. (2008) Total synthesis of (+)-fawcettidine. *Angew. Chem. Int. Ed.*, **47**, 4221–4223.

119 Nakayama, A., Kogure, N., Kitajima, M. and Takayama, H. (2009) First asymmetric total syntheses of fawcettimine-type lycopodium alkaloids, lycoposerramine C and phlegmariurine A. *Org. Lett.*, **11**, 5554–5557.

120 Jin, Z. (2005) Muscarine, imidazole, oxazole and thiazole alkaloids. *Nat. Prod. Rep.*, **22**, 196–229.

121 Schaumberg, J.P., Hokanson, G.C., French, J.C., Smal, E. and Baker, D.C. (1985) 2'-Chloropentostatin, a new inhibitor of adenosine deaminase. *J. Org. Chem.*, **50**, 1651–1656.

122 Rahbaek, L. and Breinholt, J. (1999) Circumdatins D, E, and F: further fungal benzodiazepine analogs from *Aspergillus ochraceus*. *J. Nat. Prod.*, **62**, 904–905.

123 Rahbaek, L., Breinholt, J., Frisvad, J.C. and Christophersen, C. (1999) Circumdatin A, B, and C: three new benzodiazepine alkaloids isolated from a culture of the fungus *Aspergillus ochraceus*. *J. Org. Chem.*, **64**, 1689–1692.

124 Witt, A. and Bergman, J. (2001) Total syntheses of the benzodiazepine alkaloids circumdatin F and circumdatin C. *J. Org. Chem.*, **66**, 2784–2788.

125 Kosuge, T., Zenda, H., Ochiai, A., Masaki, N., Noguchi, M., Kimura, S. and Narita, H. (1972) Isolation and structure determination of a new marine toxin, surugatoxin from the Japanese ivory shell, *Babylonia japonica*. *Tetrahedron Lett.*, 2545–2548.

126 Daly, J.W., Witkop, B., Bommer, P. and Biemann, K. (1965) Batrachotoxin. The active principle of Colombian arrow poison frog, *Phyllobates bicolor*. *J. Am. Chem. Soc.*, **87**, 124–126.

127 Albuquerque, E.X., Daly, J.W. and Witkop, B. (1971) Batrachotoxin: chemistry and pharmacology. *Science*, **172**, 995–1002.

128 Kurosu, M., Marcin, L.R., Grinsteiner, T.J. and Kishi, Y. (1998) Total synthesis of (±)-batrachotoxinin A. *J. Am. Chem. Soc.*, **120**, 6627–6628.

129 Arnone, A., Bava, A., Fronza, G., Nasini, G. and Ragg, E. (2005) Concavine, an unusual diterpenic alkaloid produced by the fungus *Clitocybe concava*. *Tetrahedron Lett.*, **46**, 8037–8039.

130 Dumbacher, J.P., Beehler, B.M., Spande, T.F., Garraffo, H.M. and Daly, J.W. (1992) Homobatrachotoxin in the genus *Pitohui*: chemical defense in birds? *Science*, **258**, 799–801.

16
Bioactive Macrocyclic Natural Products
Siti Mariam Mohd Nor, Zhengshuang Xu and Tao Ye

16.1
General

Heterocyclic compounds have been characterized as substructures in many natural products originating from living organisms (terrestrial, marine and microorganisms). A variety of heterocycle-containing natural products have provided compounds that show useful biological properties and can serve as leading compounds for drug discovery. Over 20 new drugs derived from natural products have been introduced to the market, and 36 compounds derived from natural products are in Phase III clinical trials from 2003–2008 (Figure 16.1) [1a].

In light of their potential in medicine, it is not surprising that synthetic investigations and biological evaluations of these secondary metabolites have been intensive. By synthesizing natural compounds and closely-related synthetic analogs, scientists gain valuable knowledge on how the natural compounds work and how to produce even better compounds as drug candidates.

16.2
Natural Products Containing Azoles

The methods to prepare thiazoline and thiazole and their derived natural products are discussed in Chapter 13. Oxazoline and oxazole can be synthesized in similar approaches and are discussed in Chapter 12.

16.2.1
Apratoxin A

Apratoxins were isolated from the cyanobacterial *Lyngbya* sp. collected in Guam and Palau by Moore and Paul and their respective co-workers [2a]. Apratoxin A **1**, (Figure 16.2) is cytotoxic against the LoVo and KB cancer cell lines (IC$_{50}$ = 0.36–0.52 nM) and is one of the most cytotoxic cyclodepsipeptides among those discovered in marine cyanobacteria. Further studies revealed that apratoxin A **1** mediates

Heterocycles in Natural Product Synthesis, First Edition. Edited by Krishna C. Majumdar and Shital K. Chattopadhyay.
© 2011 Wiley-VCH Verlag GmbH & Co. KGaA. Published 2011 by Wiley-VCH Verlag GmbH & Co. KGaA.

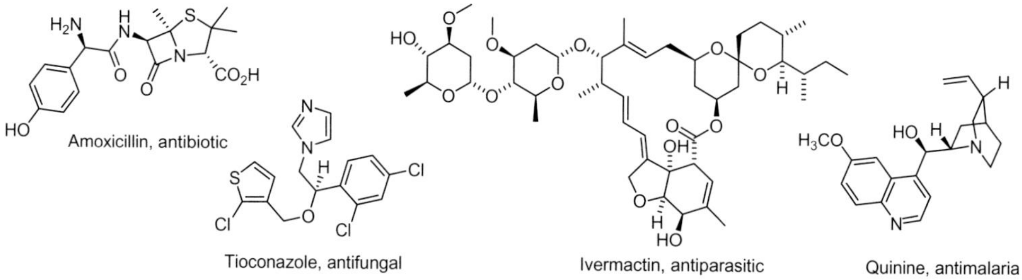

Figure 16.1 Drugs derived from natural products containing heterocycles.

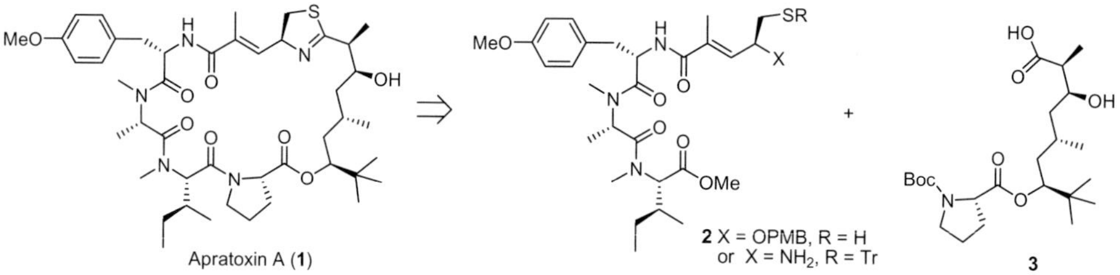

Figure 16.2 Structure of apratoxin A and its retrosynthetic analysis.

its antiproliferative activity through the induction of G1 cell cycle arrest and an apoptotic cascade [3]. The construction of the thiazoline moiety from the Cys residue proved to be a very challenge task during the total synthesis of apratoxin A **1**. Thiazoline-formation based on the cyclodehydration of β-hydroxythioamide was not applicable in this particular case. This was because the thioamide prepared by thionation or thioacylation readily underwent an intramolecular Michael addition to the α,β-unsaturated double bond before the cyclodehydration process. Cyclodehydration of the related thioesters also proved to be problematic, because epimerization of the stereogenic centers, β-eliminations of the hydroxyl group at the side-chain and decomposition of the starting material occurred during the process. An effective total synthesis was first achieved by Forsyth and co-workers and the thiazoline moiety was constructed by the use of a Staudinger–aza-Wittig reaction [4a].

Thus condensation of thiol **2** with acid **3** in the presence of DPPA produced the thioester **4** in 80% yield, oxidative removal of PMB ether followed by Mitsunobu reaction installed the azide functional group, after transformation of the TBS ether to the corresponding TES ether, key intermediate **5** was obtained in 78% yield. Treatment of the azido thioester **5** with Ph$_3$P facilitated the Staudinger–aza-Wittig reaction and produced thiazoline **6** in 63% yield. Macrolactonization and subsequently removal of silyl ether completed the total synthesis of apratoxin A **1**, see Scheme 16.1.

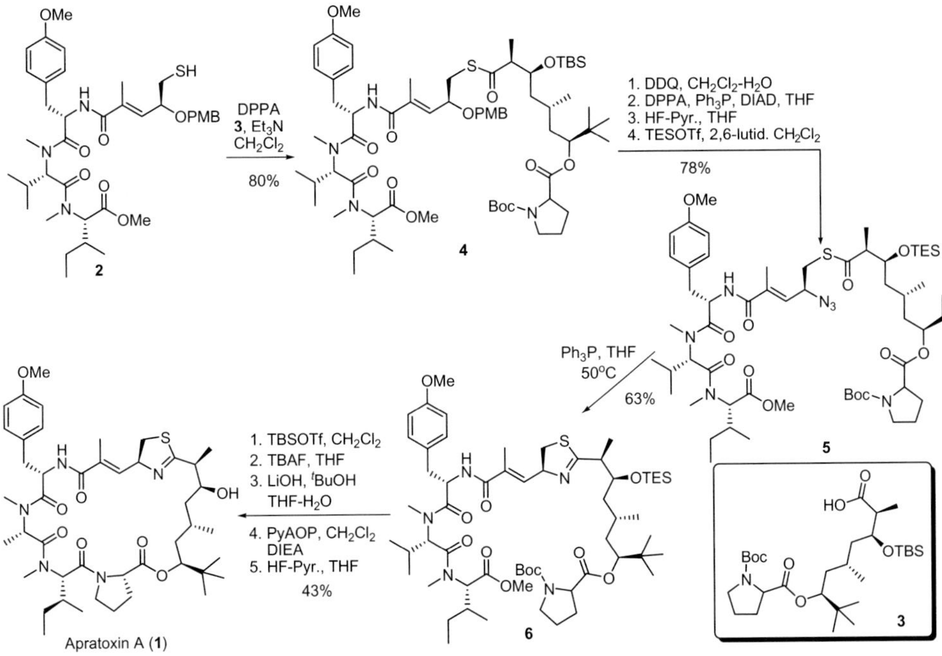

Scheme 16.1 Forsyth's total synthesis of apratoxin A.

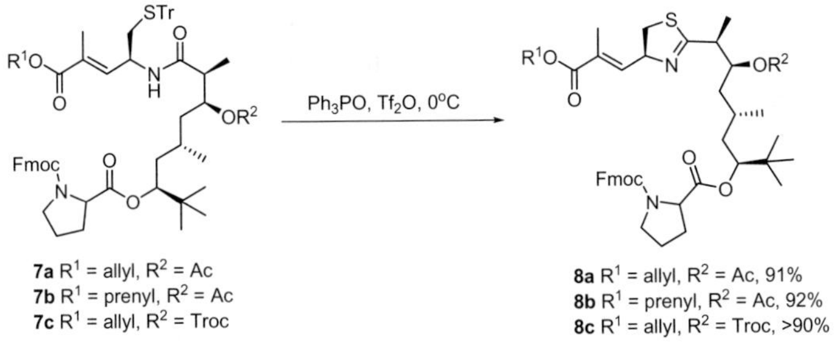

Scheme 16.2 Synthesis of thiazoline of apratoxin A using Kelly's procedure.

Ma, Takahashi and Doi also completed the total synthesis of apratoxin A **1** by the use of Kelly's biomimetic thiazoline formation method (Scheme 16.2) [5a]. Ma started from thioether **7a** and **7b**, after treatment with triphenylphosphine oxide and triflic anhydride at 0 °C, the desired thiazoline fragment **8a** and **8b** were obtained in excellent yields. Takahashi and Doi elected Troc as a protecting group for the β-hydroxy group **7c**, which produced the corresponding thiazoline **8c** in comparable isolation yield.

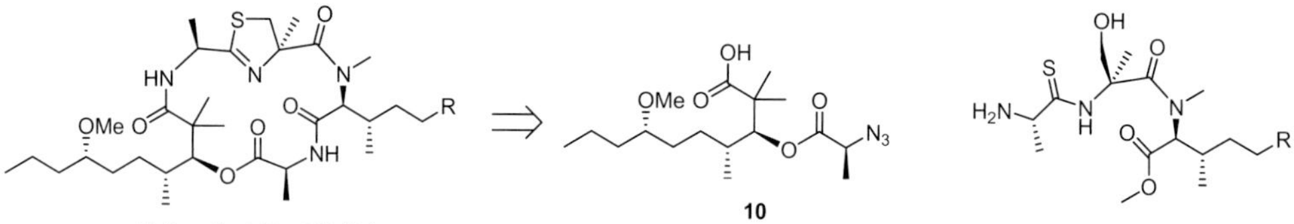

Halipeptin A R = OH (9a)
Halipeptin D R = H (9b)

11a R = OTBDPS
11b R = H

Figure 16.3 Halipeptins A and D and retrosynthetic analysis.

16.2.2
Halipeptins A and D

Halipeptins A and B were first isolated by the Gomez-Paloma group from the sponge *Haliclona* sp. in 2001, and it was proposed that the structure contains an unusual oxazetidine-type heterocycle [6]. Along with their report on halipeptin C [7], they subsequently revised the oxazetidine-type heterocycle moiety to a thiazoline ring, which was also confirmed by the isolation of halipeptin D from a different sponge *Leiosella* cf. *arenifibrosa* by another research group [8]. Among these natural products, halipeptins A **9a** and D **9b** are more active than halipeptins B and C, and halipeptin D **9b** exhibits potent cytotoxic properties against human colon cancer HCT-116 cell line and BMS ODCP (oncology diverse cell panel) of tumor cell lines at IC_{50} of 7 nM and 420 nM (average).

Synthetic efforts have mainly focused on halipeptins A and D. Successful total syntheses of halipeptins have been achieved by Nicolaou, Ma, Hamada and their respective co-workers. According to these synthetic studies, the stereogenic center on alanine attached to the thiazoline ring readily undergoes racemization regardless of the methods employed for thiazoline formation. Toward the total synthesis of halipeptin A, Nicolaou and co-workers employed thiolysis reaction to prepare the thioamide followed by DAST-mediated cyclodehydration protocol (Figure 16.3) [9a].

Thus aza-Wittig reaction of the azide in tripeptide **12** in the presence of trimethylphosphine produced oxazoline **13**. After reductive removal of Cbz group in **13**, straightforward thiolysis with H_2S produced the desired thioamide **11**. Coupling reaction with acid moiety **10** in the presence of PyAOP furnished the linear precursor **14**, which was first subjected to thiazoline formation using DAST as dehydration agent to give **15**. Further tactical and sapiential functional transformation completed the total synthesis of halipeptins A and D, although extensive epimerization at the C-2 exomethine position of the thiazoline and the α-position of amide for macrocyclization reaction had hampered the total synthesis (Scheme 16.3).

Ma and co-workers applied the thioacylation reaction to produce the thioamide, followed by the same cyclodehydration process for the construction of a thiazoline ring of halipeptins [10], while Hamada and co-workers took the advantage of Kelly's Ph_3PO/Tf_2O-mediated deprotection–cyclodehydration of the trityl-protected

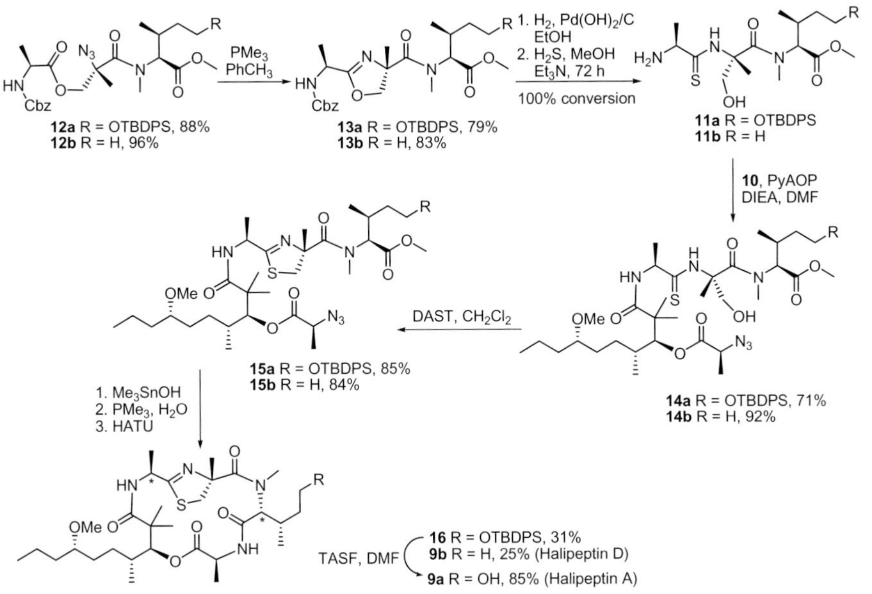

Scheme 16.3 Nicolaou's total synthesis of halipeptins A and D.

α-methylcysteine amide that had been introduced in the total synthesis of apratoxin A [11].

16.2.3
Largazole

Largazole **17**, which consists of a thiazoline, a thiazole, (S)-valine and acylated (S)-3-hydroxy-7-mercaptohept-4-enoic acid, was isolated from the cyanobacterium *Symploca* sp. in 2008. Largazole **17** exhibits remarkably selective biological activities against human or murine-derived cancer cells (melanoma cell lines, IC_{50} = 45–315 nM) [12a]. Largazole is a pro-drug that can be converted into its active metabolite, largazole thiol **18**; this has a 3-hydroxy-7-mercaptohept-4-enoic acid unit, which was also found in several cytotoxic natural products, and is known as a histone deacetylase inhibitor (HDACi) [13a]. Detailed studies demonstrated that largazole thiol **18** is an extraordinarily potent Class I HDACs (HDAC1 and HDAC2 with K_i = 70 pM, respectively) as compared with largazole **17** (Figure 16.4) [13d].

As a star molecule of 2008, the total synthesis of largazole **17** was a challenging task since all methods for macrolactonization and ring closure *via* a late stage thiazoline installation failed. Luesch *et al.* first completed the total synthesis of largazole **17** [12b]. Cyanide **23** was condensed with **24** to produce the thiazole-thiazoline motif **19** in 51% yield; N-deprotection of **19** followed by condensation with **20** gave

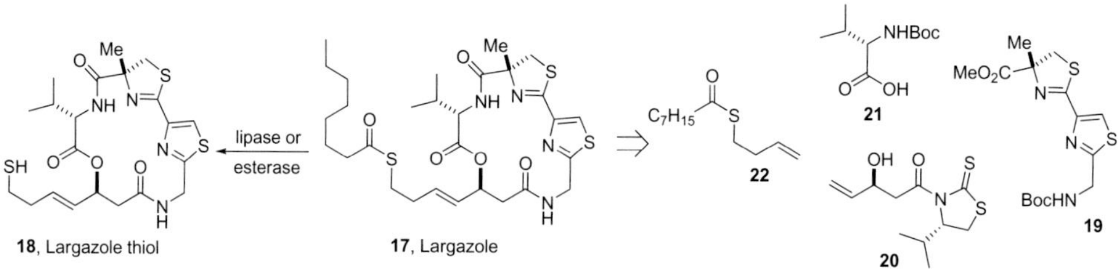

Figure 16.4 Largazole and its retrosynthetic analysis.

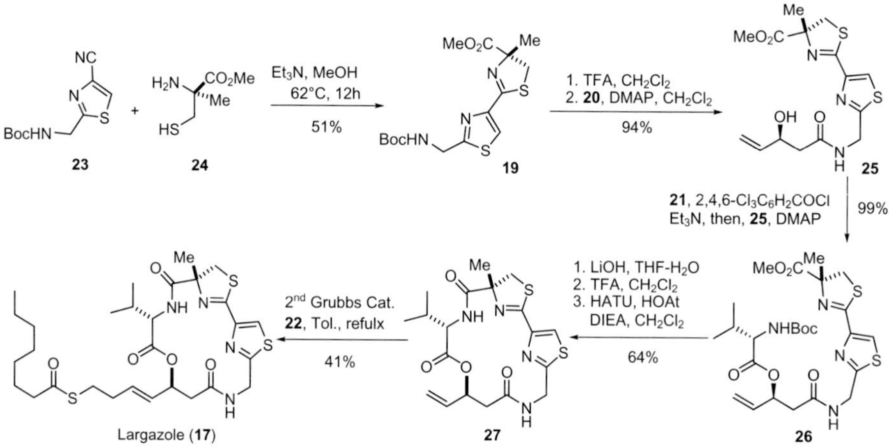

Scheme 16.4 Total synthesis of largazole.

25. Yamaguchi esterification was employed to produce the linear precursor **26** and HATU was proved to be suitable for the macrocyclization to provide macrocycle **27** in 64% yield. Alkene cross-metathesis was then employed to accomplish the total synthesis in 41% yield (Scheme 16.4).

Later, Cramer [14], Williams [13d], Ye [15], Doi [16], Phillips [17], Ghosh [18], Forsyth [19] and Jiang [20] all completed the total synthesis of largazole. To construct the thiazole-thiazoline moiety, Cramer, Williams, Ye, and Phillips adopted a similar protocol to Leusch, while Doi and Ghosh took advantage of Kelly's biomimetic process (Ph_3PO/Tf_2O). Jiang and co-workers in their latest attempt, used a solid phase based synthesis and employed the Heathcock's $TiCl_4$ mediated cyclization of bis-thiolamine to produce bis-thiazoline precursor, followed by activated-MnO_2 oxidation. Apart from this work on total synthesis, some SAR studies were also conducted, which paved the road for further medicinal investigation using largazole **17** as the lead compound [21a].

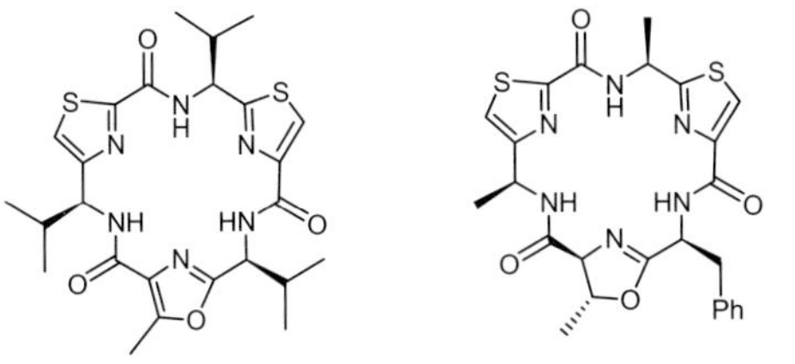

Figure 16.5 Bistratamide H and didmolamide A.

16.2.4
Bistratamide H and Didmolamide A

Bistratamide H **28** together with seven other bistratamides (C, D, E, F, G, I and J) were isolated by Faulkner et al. from ascidian *Lissoclinum bistratum*. The bistratamide H **28**, bistratamide J and bistratamide E showed cytotoxic activity against the human colon tumor (HCT-116) cell line with IC$_{50}$ value of 1.7 µg mL^{-1}, 1 µg mL^{-1} and 7.9 µg mL^{-1}, respectively [22]. Cytotoxic activity against T24 bladder carcinoma cells was shown in bistratamide A with IC$_{50}$ = 60 µg mL^{-1} [23].

Didmolamide A **29** and didmolamide B, (Figure 16.5) two cyclic hexa-peptides were isolated from *Didemnum molle* in 2003. These metabolites also showed cytotoxic effects against tumor cell lines (A549, HT29 and MEL28) with IC$_{50}$ values of 10–20 µg mL^{-1} [24]. Bistratamide H **28** consists of two thiazole rings and one oxazole ring, whereas in didmolamide A **29** the oxazole ring was replaced with an oxazoline ring.

Kelly and co-workers completed the total synthesis of bistratamides E–J. They prepared the thiazole moieties by a two-step sequence involving the use of bis(triphenyl)oxodiphosphonium trifluoromethanesulfonate-mediated thiazoline formation followed by oxidation with activated manganese dioxide [25a].

Recently, Pattenden et al. examined the mixed cyclooligomerization towards bistratamide H **28** and didmolamide A **29** instead of a straightforward strategy involving the coupling reaction to form a linear precursor followed by macrolactamization [26].

Thus, thiazoles **30**, **31** and **32** were prepared from the corresponding amino acid using a modified Hantzsch protocol (Scheme 16.5). A coupling reaction between thiazole acid **31** and thiazole amine **32** first produced bis-thiazole, which was subjected to global deprotection to give free bis-thiazoles **33a** and **33b** in good yield. The bistratamide H **28** and didmolamide A **29** were assembled from the corresponding oxazole **35** and oxazoline **38**, respectively. Oxazole **35** and oxazoline **38** were synthesized from oxazoline **34** and **36**, respectively. Saponification of methyl ester of **34** and removal of Boc-protecting group produced *trans*-oxazoline **35** in 85% yield. The oxazoline **36** was also further oxidized with BrCCl$_3$/DBU to give oxazole **37**. Subsequent deprotection of **37** using LiOH and 4 M HCl resulted in fully deprotected oxazole **38** (Scheme 16.6).

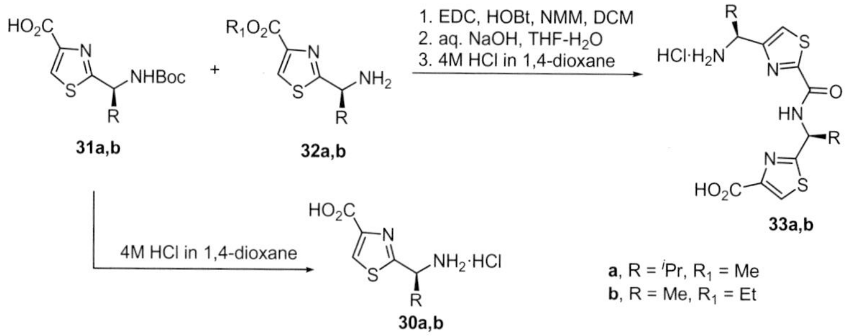

Scheme 16.5 Synthesis of thiazole-containing fragments.

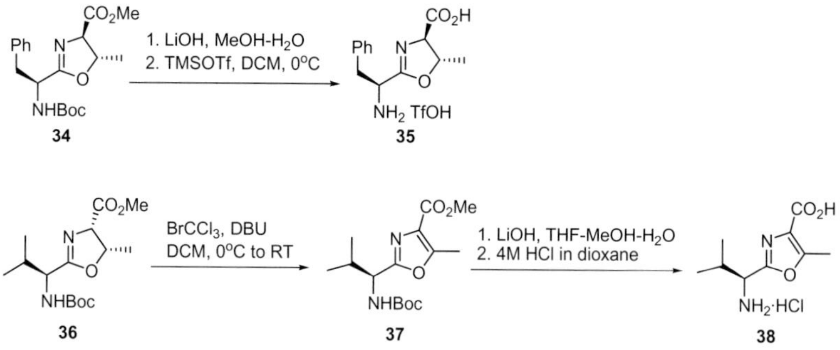

Scheme 16.6 Synthesis of oxazole-containing fragments.

A 1:1 mixture of the oxazole **38** and the bis-thiazole **33a** was reacted in the presence of FDPP/DIEA leading to the formation of bistratamide H **28** in 36% yield. In a similar manner, the coupling of oxazole **38** and thiazole **30a** produced the bistratamide H **28** in 25% yield. Treatment of bis-thiazole **33b** or thiazole **30b** with oxazoline **35** in the presence of FDPP or DPPA with NMM in DMF produced didmolamide A **29** in very low (2%) yield. The azole-based cyclic trimer and cyclic tetramer were also obtained from all reactions. The results showed that the cyclo-oligomerization involving an oxazoline provided a lower yield of the desired natural product (Scheme 16.7).

These studies involved the assembly of preformed azole amino acids, which served as useful synthetic strategy to establish small molecule libraries of novel natural and non-natural cyclic peptides [27a].

16.2.5
IB-01211

IB-01211 **39** was originally isolated from the marine microorganism *Thermoactinomyces* genus strain ES7-008 in 2005 [28]. A compound with the same

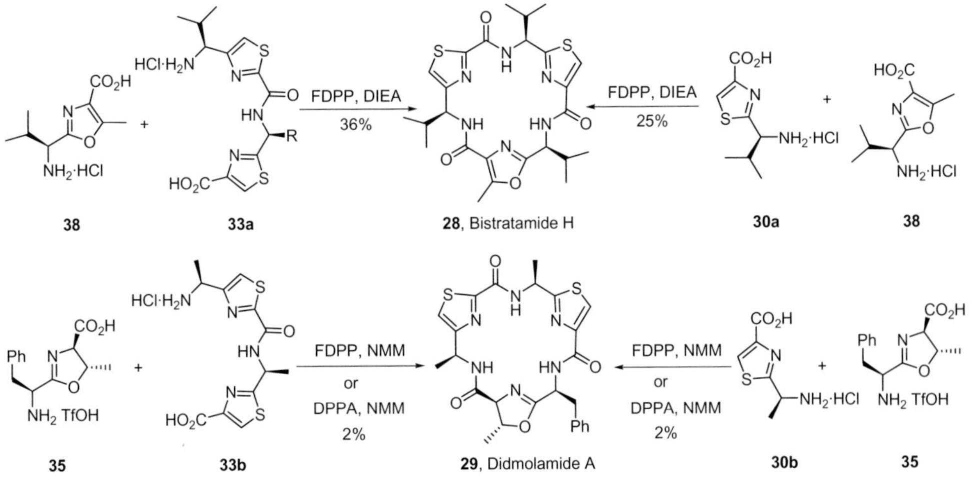

Scheme 16.7 Completion the total synthesis of bistratamide H and didmolamide A.

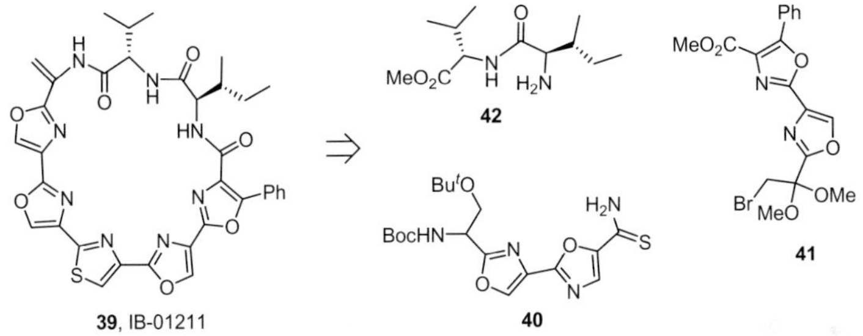

Figure 16.6 Structure of IB-01211 and its retrosynthetic analysis.

structure called mechercharmycin A was isolated from the bacterium *Thermoactinomyces* sp. YM3-251 collected at Mecherchar, Republic of Palau by Kanoh's group later in the same year [29]. The structure of IB-01211 **39** contains four consecutive oxazoles, one thiazole and tripeptide residue (Figure 16.6). IB-01211 was found to exhibit high cyctotoxicity against several tumor cell lines (IC$_{50}$: human lung cancer A549 cells = 4.0×10^{-8} M, human leukemia Jurkat cells = 4.6×10^{-8} M).

The first total synthesis of IB-01211 **39**, and the only one reported to date, was designed by Hernandez *et al.* in 2007 [30], which featured a Hantzsch reaction in the late stage of the total synthesis to build the thiazole hetero-ring and complete the macrocyclization simultaneously. The convergent synthesis of IB-01211 **39** started from peptides **42**, **43** and **46**, which were prepared under standard carbodiimide-coupling reactions. Cyclization–oxidation procedures converted

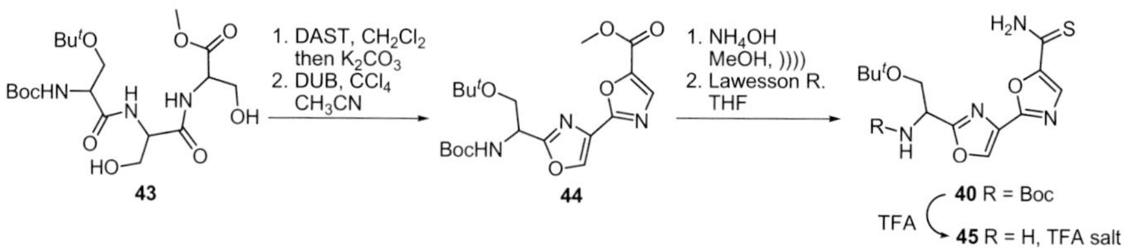

Scheme 16.8 Synthesis of fragment **45**.

tripeptide **43** into bis-oxazole **44**, which was subjected to aminolysis and thionylation to produce thioamide **40**. Treatment of **40** with TFA liberated the amine as its TFA salt **45** with the *t*-butyl ether untouched (Scheme 16.8).

Treatment of dipeptide **46** with DAST–pyridine followed by dehydrogenation afforded the desired oxazole. Deprotection with TFA followed by coupling with the bromopyruvic acid dimethyl acetal **47** afforded **48**. Again, DAST–DBU mediated oxazole ring formation provided bis-oxazole **41**. Hydrolysis of the methyl ester of **41** with LiOH and condensation with dipeptide **42** in the presence of EDCI produced compound **49**. Further saponification of **49** followed by coupling with fragment **45** produced linear precursor **50** in good yield. Formation of the macrocycle **39** was achieved in 11% yield in one pot reaction through double deprotection of **50** with formic acid, dehydration and dehydrogenation by using Hantzsch reaction conditions (Scheme 16.9).

16.2.6
(R)-Telomestatin

Telomestatin **51**, a novel macrocycle, was isolated from *Streptomyces anulatus* 3533-SV4 by Kazuo's group (Figure 16.7) [31]. This unique structure contains two 5-methyloxazoles, five oxazoles and one thiazoline ring. It has proven to be a very potent telomerase inhibitor that interacts specifically with the G-quadruplex and stabilizes it, without affecting DNA polymerases or reverse transcriptases. Telomestatin **51** has shown inhibition with $IC_{50} = 5.0$ nM compared with other G-quadruplex-interactive molecules such as anthraquinones, cationic porphyrin and fluoroquinophenox-azines. Telomestatin **51** showed its antiproliferative and proapoptotic effects in ARD, MM1S and ARP myeloma cells by inhibition of telomerase activity and reduction in telomerase length [32].

Three total syntheses of telomestatin **51** had been reported [33a]. Takahashi's group employed a convergent route to complete the total synthesis and confirmed that the cysteine residue was (*R*)-configuration [33c]. The peptide **52**, oxazole **53** and **54** were selected as starting materials for the total synthesis (Figure 16.7).

Oxidation of dipeptide **52** with $SO_3 \cdot Py$/DIEA, followed by cyclodehydration with $PPh_3\text{-}I_2$ provided the corresponding oxazoline, which was saponified with LiOH to

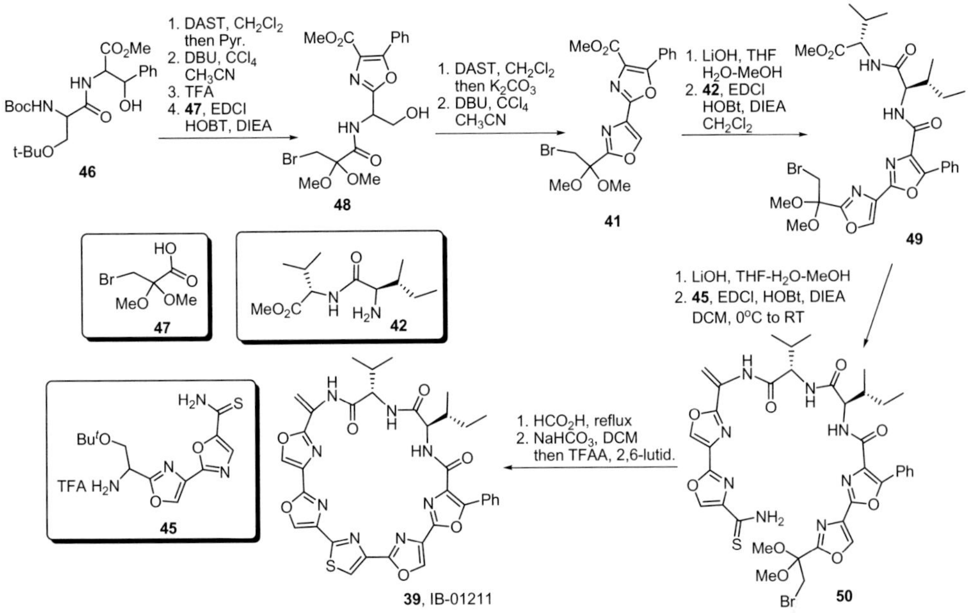

Scheme 16.9 Completion of the total synthesis of IB-01211.

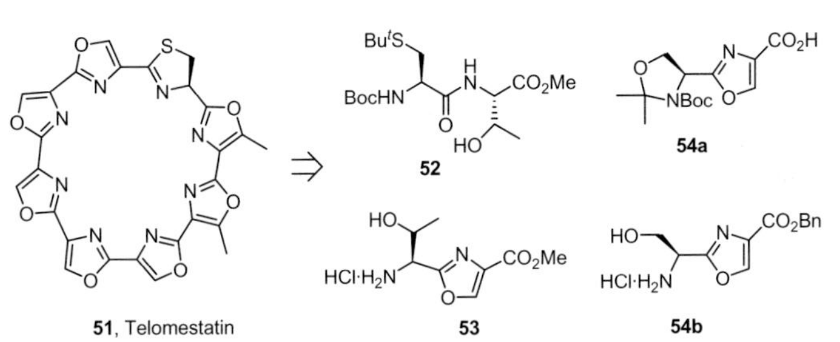

Figure 16.7 Structure of telomestatin and its retrosynthetic analysis.

produce the acid **55**. Condensation of **55** with amine **53** using PyBroP/DIEA afforded the bis-oxazole **56**. Treatment of **56** with Burgess reagent followed by reaction with BrCCl$_3$-DBU and Boc-cleavage provided tris-oxazole **57** in 75% yield. The synthesis of the second tris-oxazole unit **59** began with bis-oxazole **58** which was prepared by the coupling reaction between the free amine of **54b** and acid **54a**. Cyclodehydration of **58** was achieved with DAST to afford the corresponding oxazoline. BrCCl$_3$-DBU promoted oxazole formation and hydrogenolysis of the benzyl-protecting group produced tris-oxazole **59**. Coupling reaction of **57** with **59**

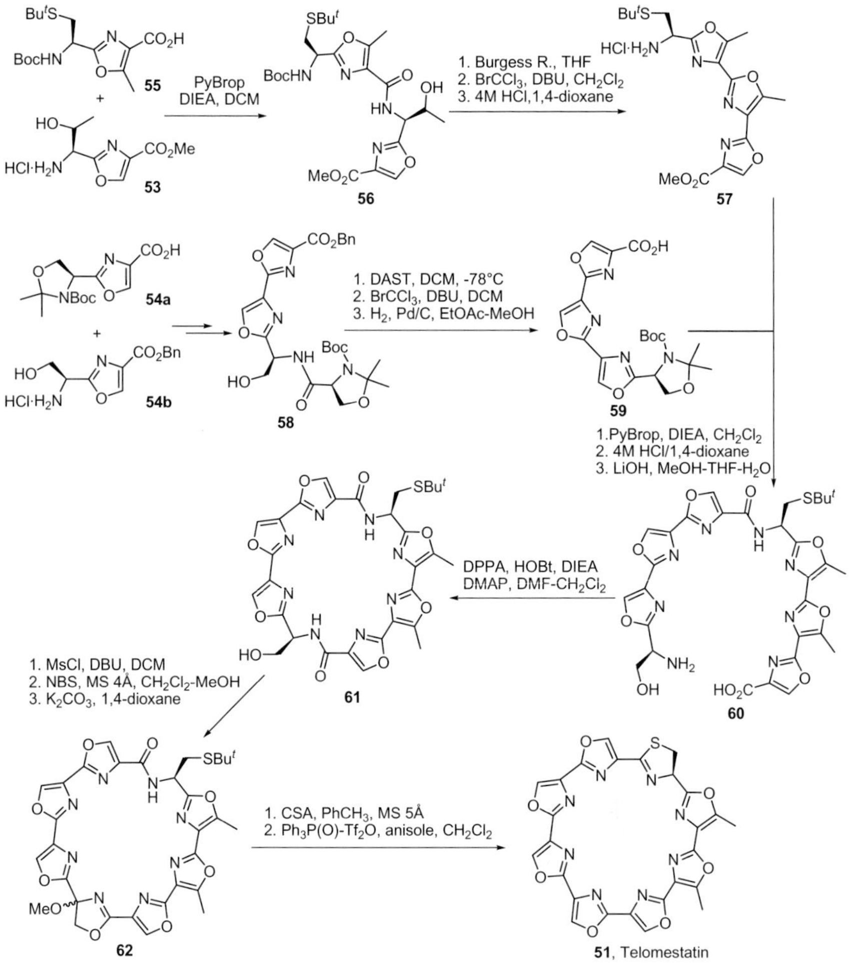

Scheme 16.10 Total synthesis of telomestatin.

resulted in the corresponding amide followed by removal of the Boc-group and saponification of methyl ester to afford hexa-oxazole **60**. Treatment of **60** with DPPA-HOBt in the presence of DIEA/DMAP then produced the macrocycle **61** in 48% overall yield (Scheme 16.10).

Cyclodehydration of **61** proved to be problematic either with DAST or Burgess reagent. Finally, mesylation of **61** with MsCl/DBU, followed by treatment with NBS in the presence of MS 4Å, and later the cyclization with K_2CO_3 led to the oxazoline **62** as a mixture of diastereoisomers (79%, 1:1) [34]. Elimination of MeOH in **62** with CSA then produced the corresponding oxazole. Subsequent deprotection of *t*-butyl and cyclodehydration was done using Kelly's method with modification in the presence of anisole, which was essential in the reaction to complete the total synthesis of telomestatin **51** [35].

16.3
Pyridine- and Piperidine-Containing Natural Products

Thiostreptons share some common structural features: a tri- or tetra-substituted pyridine core structure, consecutive heterocycles, including the thiazole/oxazole rings and dehydroamino acids. This kind of thiopeptide antibiotic has been known for decades. They usually inhibit bacterial protein synthesis in both gram-positive and gram-negative bacteria by binding to either the L11 binding domain or the Ef-Tu protein complex [36a]. The *in vitro* antibacterial activity of these thiopeptide antibiotics is comparable to that of penicillin, with little or no adverse toxicological effects to mammalian cells. In addition, some of them exhibit selective cytotoxicity to certain cancer cells and potent immunosuppressive properties [37a].

Representative thiostreptons are illustrated in Figure 16.8. Amythiamicin D **63** was isolated from a strain of *Amycolatopsis* sp. MI481-42F4 [38], while thiostrepton **64** was first isolated in 1954 from *Streptomyces azureus* ATCC 14921, and was subsequently isolated from fermentation extracts of *S. hawaiiensis* ATCC 12236 and *S. laurentii* ATCC 31255. Its total synthesis was completed by Nicolaou and co-workers in 2004 [39]. Siomycin A **65** was isolated from the culture broth of *S. sioyaensis*. Its total synthesis was achieved in 2007 by Hashimoto and Nakata and their co-workers [40]. Cyclothiazomycin **66**, originated from the fermentation broth of *Streptomyces* sp. NR0516, exhibits renin inhibitory activity with IC_{50} of 1.7 µM [41]. Promothiocins **67** were isolated from *Streptomyces* sp. SF2741 and Moody and co-workers pioneered the total synthesis work [42]. The total synthesis of micrococcin **68** and GE2270s **69**, two of these thiostreptons, were reported after 2006.

16.3.1
Micrococcin P1

Micrococcin P1 (MP1, **68**) was isolated from *Bacillus pumilus* collected from East African soil by Fuller in 1955. Micrococcin P1 displays potent antibiotic activity towards microorganisms; it was a potent growth inhibitor of the human malaria parasite *Plasmodium falciparum* (50% inhibitory concentration of 35 nM) [43]. It binds to ribosomes and disrupts protein synthesis [44]. Micrococcin P1 is very similar and probably identical to a compound known as micrococcin which was isolated from a strain of *Mocrococcus* sp. from sewage by Su in 1948 [45].

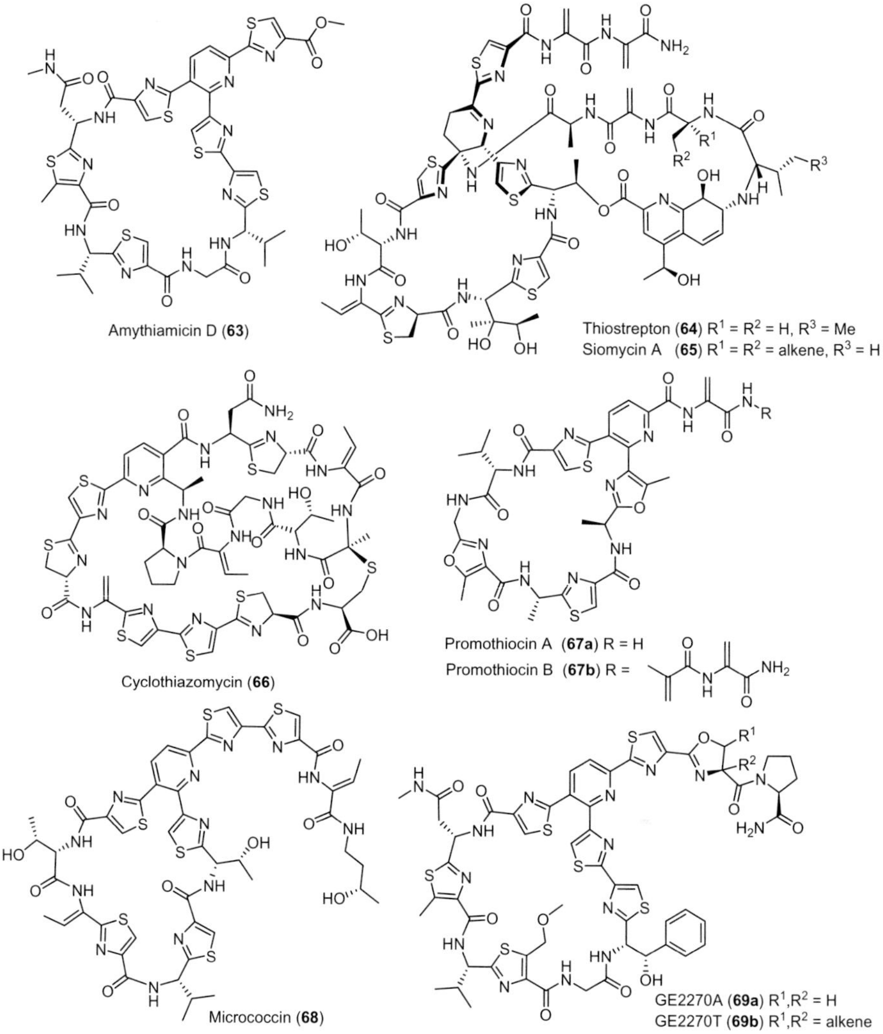

Figure 16.8 Thiostreptons-containing natural products.

Several synthetic studies on MP1 **68** have been reported [46a]. In 2009, Ciufolini *et al.* published the total synthesis of micrococcin P1 **68** and established the stereochemistry (Figure 16.9) [47]. A challenging part of the synthesis of **68** was the assembly of the central pyridine-thiazole moiety *via* the Hantzsch reaction. The building blocks, bis-thiazole **70** and **71**, were prepared according to the established method. Thus, Michael reaction of **71** with lithium enolate of **70** produced diketone **76**. Treatment of **76** with $NH_4OAc/EtOH$ and then DDQ in toluene produced

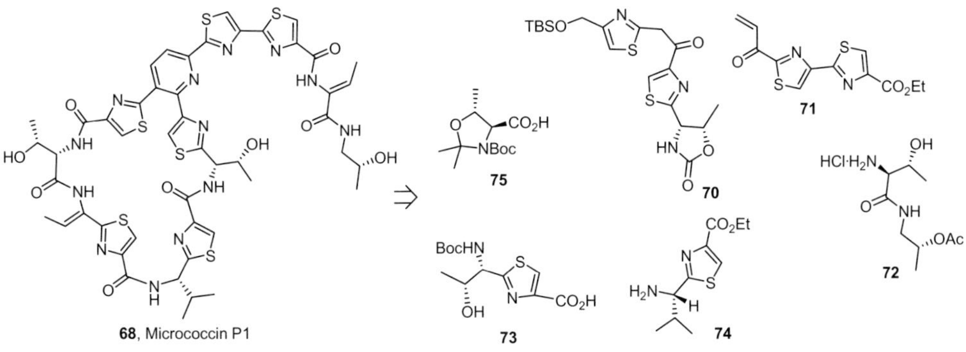

16.3 Pyridine- and Piperidine-Containing Natural Products

Figure 16.9 Retrosynthetic analysis of micrococcin P1

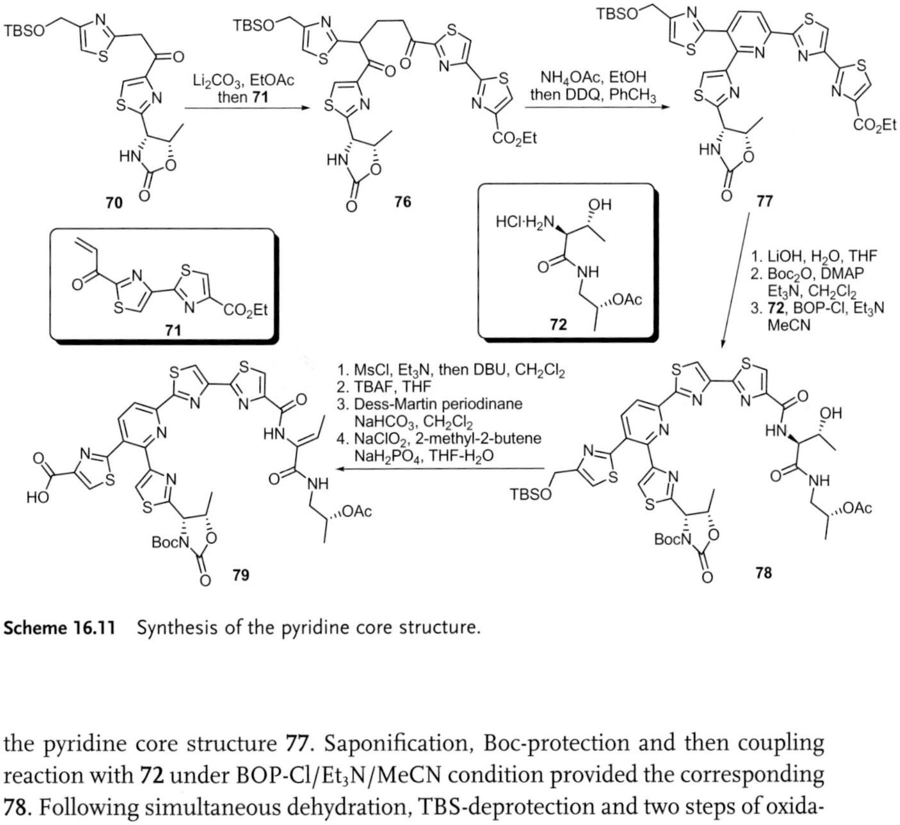

Scheme 16.11 Synthesis of the pyridine core structure.

the pyridine core structure **77**. Saponification, Boc-protection and then coupling reaction with **72** under BOP-Cl/Et$_3$N/MeCN condition provided the corresponding **78**. Following simultaneous dehydration, TBS-deprotection and two steps of oxidation then furnished the acid **79** (Scheme 16.11).

The bis-thiazole amine **80** was prepared starting with the DCC-mediated coupling reaction of thiazole **73** and thiazole **74**. Subsequent Boc-removal, coupling

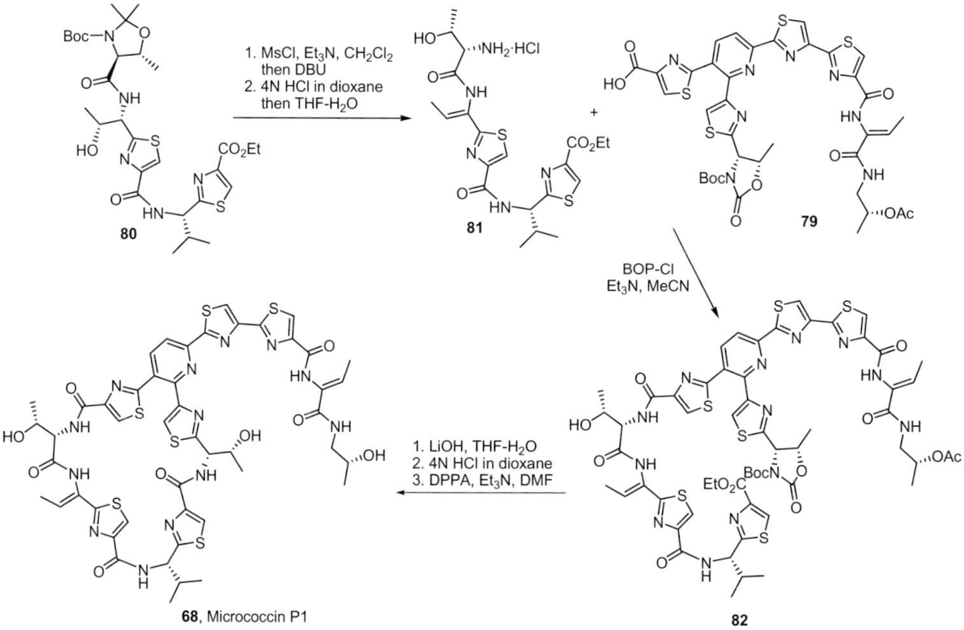

Scheme 16.12 Total synthesis of micrococcin P1.

with **75**, mesylation of the secondary alcohol with MsCl, then dehydration by DBU led to the corresponding alkene, which was then treated with 4 N HCl to afford amine **81**. The amine **81** was then coupled to the acid **79** with BOP-Cl in MeCN to give acyclic precursor **82** in 73% yield. Finally, deprotection of ethyl ester and Boc of **82** and macrolactamization with DPPA furnished micrococcin P1 **68** in 41% yield over three steps (Scheme 16.12).

16.3.2
GE2270s

GE2270s were first isolated from the fermentation broth of *Planobispora rosea* ATCC53773 [48a]. The stereochemistry of GE2270A was further elucidated in 1995 and 2005 by Tavecchia and Bach, respectively [49a], and in 2006 the co-crystallization of GE2270A and the bacterial elongation factor EF-Tu was disclosed by Parmeggiani [36c]. Like the amythiamicin, GE2270s have no dehydroamino acid moieties in their gross structure.

Nicolaou and co-workers completed the total synthesis of GE2270A **69a** and GE2270T **69b** in 2006 (Figure 16.10) [48d]. The key reaction for the construction of pyridine core was the aza-Diels–Alder dimerization of intermediate **92**, which was derived from thiazolindine **91** by treatment with silver carbonate and DBU in the presence of pyridine and benzylamine, and the isolation yield was as high as 60%. Treatment of racemic **93** with DBU in hot ethyl acetate produced the aro-

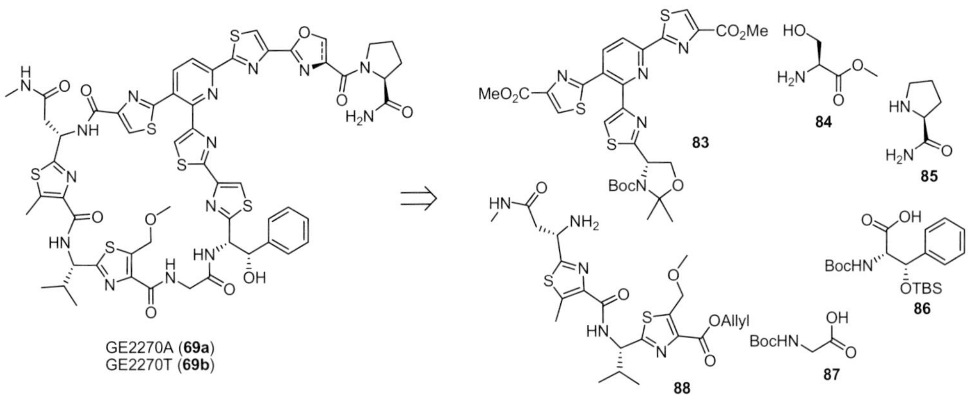

16.3 *Pyridine- and Piperidine-Containing Natural Products* | **585**

Figure 16.10 Retrosynthetic analysis of GE2270s.

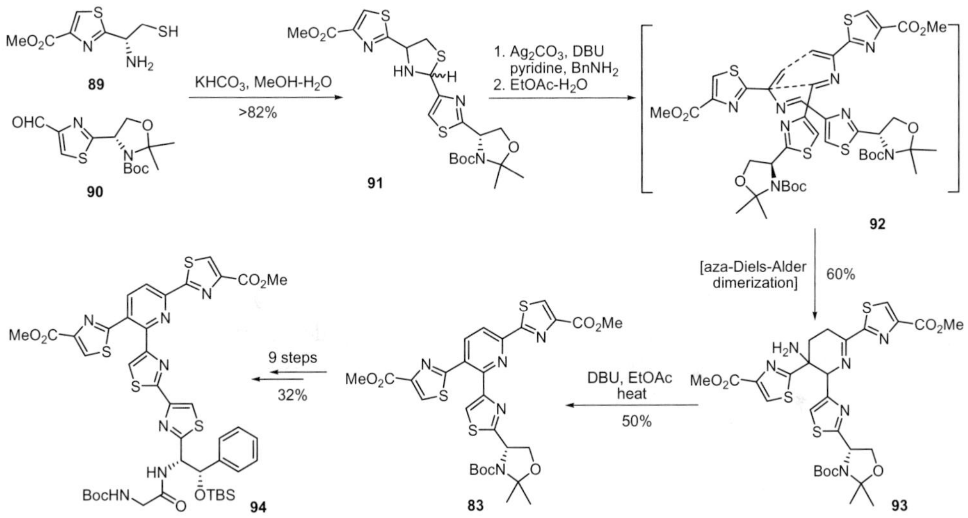

Scheme 16.13 Aza-Diels-Alder dimerization for the pyridine core structure.

matic key intermediate **83** in 50% yield. Further standard peptide coupling reactions elongated the side-chain from the lower serine terminal. The additional thiazole ring was constructed to produce **94** by a sequential reactions including thionation, DAST mediated cyclization and dehydrogenation of the corresponding thiazoline intermediate with Williams' protocol (Scheme 16.13).

With **94** in hand, trimethyltin hydroxide stimulated hydrolysis gave the monoacid with no regio-selectivity. However, after the two isomers' condensation with amine **88** under the same reaction conditions, only **96** gave the desired macrocycle **97** (Scheme 16.14).

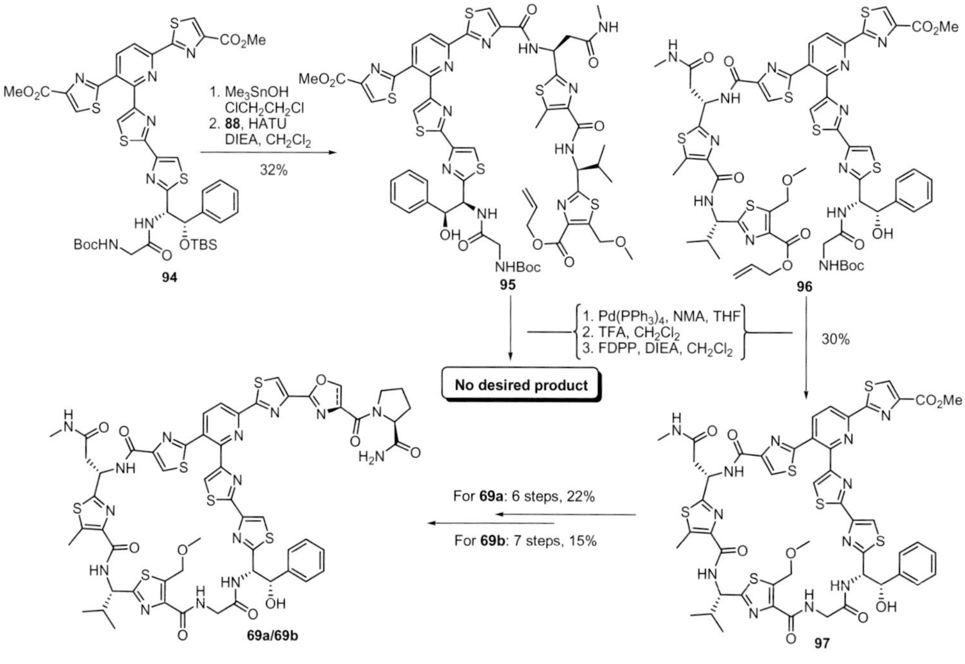

Scheme 16.14 Total synthesis of GE2270s by Nicolaou *et al.*

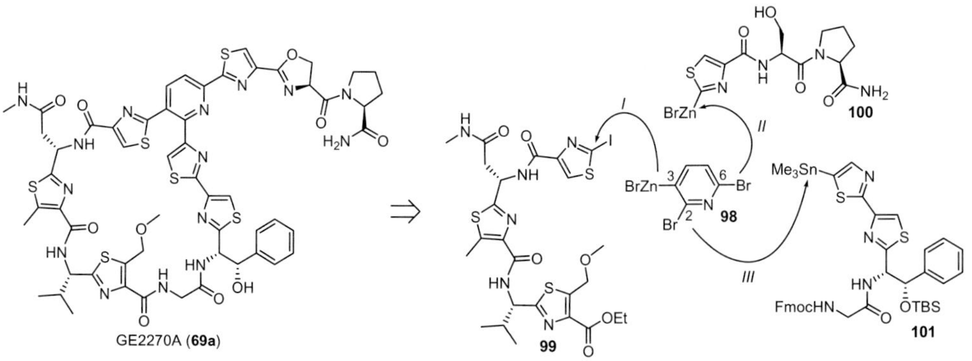

Figure 16.11 Bach's retrosynthetic analysis for GE2270A.

Bach and co-workers completed the total synthesis of GE2270A **69a** in 2007, the piperidine core structure was constructed *via* three steps of Stille cross-coupling reactions (*I–III*) by using **98–101** as advanced intermediates (Figure 16.11). The only oxazoline ring was formed at the end of the total synthesis in the presence of DAST [50].

16.4
Indole- and Imidazole-Containing Natural Products

16.4.1
Celogentin C

Celogentin C **102**, a bicyclic octa-peptide, was isolated from the seeds of *Celosia argentea* by Kobayashi *et al.* in 2001 [51]. It inhibits the growth of some cancer cell lines (SR leukemia cell lines = 35%, MDA-MB-435 melanoma cell lines = 23%, HS 578T breast cancer lines = 30% and MDA-MB-468 breast cancer lines = 34%). Celogentin C **102** showed antimitotic activity with $IC_{50} = 0.8\,\mu M$ as compared with the anticancer agent vinblastine ($IC_{50} = 3.0\,\mu M$). Several synthetic studies toward **102** had been reported [52a]. Castle *et al.* reported the total synthesis of celogentin C **102** which involved a left-to-right strategy in 2010 (Figure 16.12) [53].

The left ring of the macrocycle was accomplished by intermolecular Knoevenagel condensation, a radical conjugate addition and nitro reduction [54]. An indole-imidazole oxidative coupling reaction played a central role in the construction of the right ring. The tryptophan **103** was synthesized from the corresponding glycinate Schiff base and propargyl bromide by modified Larock-type annulation with iodoaniline [54h,i]. The methyl ester in **103** was then converted into aldehyde **107** in three steps under Braslau modification of McFayden–Steven reaction (Scheme 16.15) [55].

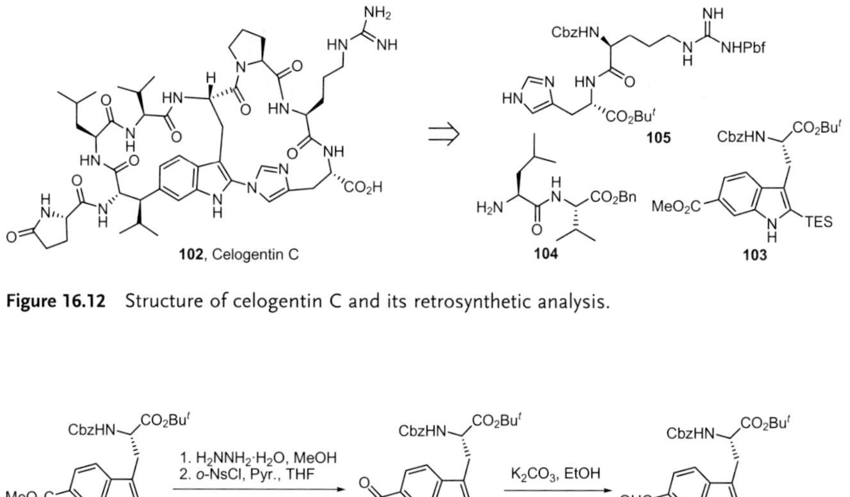

Figure 16.12 Structure of celogentin C and its retrosynthetic analysis.

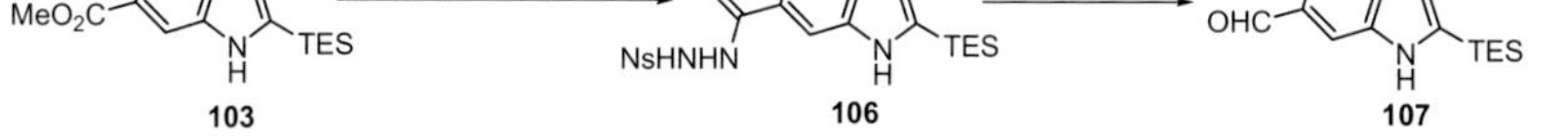

Scheme 16.15 Synthesis of the indole moiety of celogentin C.

The nitroacetamide **109** was prepared in two steps from the H-Leu-Val-OBn **104**. Treatment of **104** with Rajappa's dithioketene acetal [56a] in the presence of TsOH provided vinyl sulfide **108**. Reaction of **108** with $HgCl_2$ later produced nitro peptide **109** in 74% yield. Coupling of **109** and **107** with $TiCl_4$/NMM/THF-Et_2O (2:1) produced the alkene **110**. Subsequent nitro reduction and radical conjugate addition of **110** provided a mixture of amine **111**. Then coupling reaction of **111** with pyroglutamic acid followed by Bn-deprotection furnished the peptide **112**. Macrocyclization was achieved by HBTU/HOBt in DMF in 91% yield, followed by treatment with bromocatecholborane to produce the left ring **113** (Scheme 16.16).

Reaction of hexa-peptide **113** with Pro-OBn/EDCI/HOBt, oxidative coupling of the resulted product with Pbf-*N*-Arg-His dipeptide **105** followed by transfer

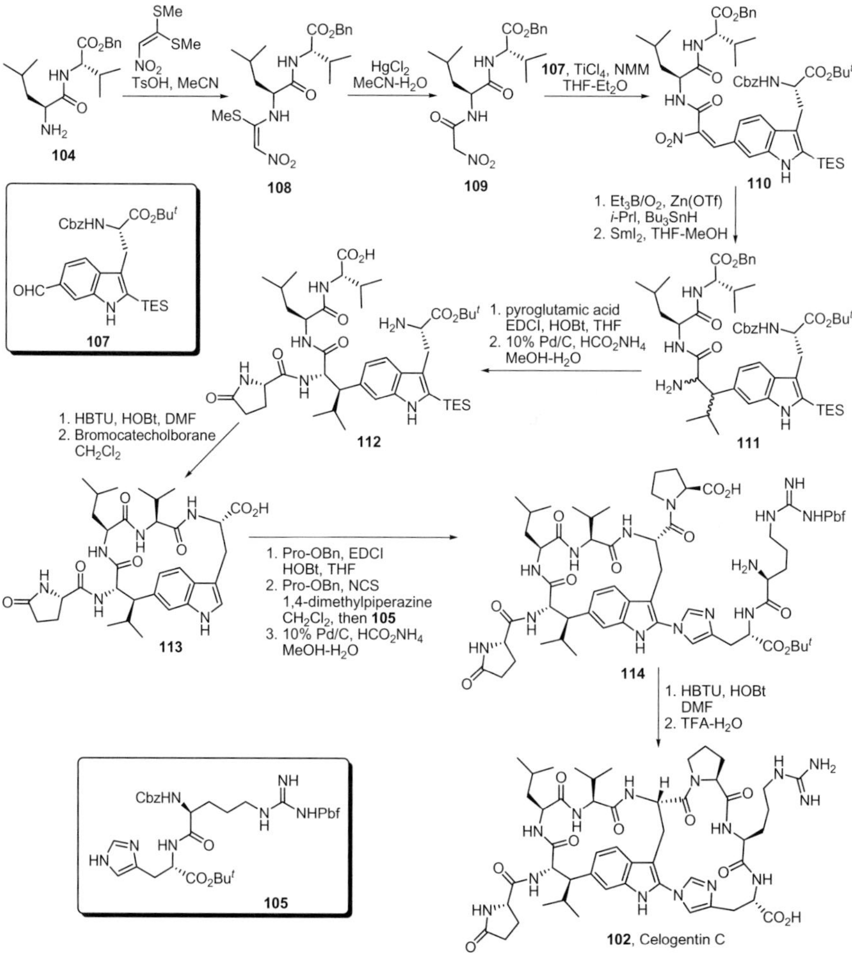

Scheme 16.16 Total synthesis of celogentin C.

hydrogenation produced octa-peptide **114**. Treatment of **114** with HBTU produced efficient macrolactamization in 83% yield, which then underwent deprotection of both *t*-butyl-group and Pbf-group, leading to the target celogentin C **102**.

16.4.2
Complestatin (Chloropeptin II)

Complestatin **115**, isolated from *Streptomyces lavendulae* SANK 60477 by Kaneko and co-workers, is a potent inhibitor of the alternate pathway of human complement (50% inhibition at $0.7\,\mu g\,mL^{-1}$) [57a]. Seto *et al.* reported that complestatin **115** showed antibacterial activity only to a few gram-positive bacteria at very high concentration (*ca.* $2000\,\mu g\,mL^{-1}$) [58]. Compound **115**, which is structurally close to glycopeptide antibiotics, also shown a potent inhibitor against HIV-1 infectivity *in vitro* (HIV-1 induced cytopathicity, HIV-1 antigen expression in MT-4, HIV-1 induced cell fusion cells and inhibited focus formation in HT4-6C with IC_{50} were $2.0\,\mu g\,mL^{-1}$, $1.5\,\mu g\,mL^{-1}$, $0.9\,\mu g\,mL^{-1}$ and $0.9\,\mu g\,mL^{-1}$, respectively) [59]. In 2009, the group of Boger reported the first total synthesis of complestatin **115** by Larock indole macrocyclization approach (Figure 16.13) [60].

Styren **119** was converted to diaryl **121** *via* a three-step sequence: conversion to aryl boronic acid, Suzuki coupling [61] with 2-bromo-5-iodoaniline **120**, and acetylation. Deprotection of **121** and EDCI mediated coupling reaction with **122**, followed by deprotection of Fmoc and again coupling reaction with silyl-alkyne **123** in the presence of EDCI, produced amide **124**. Larock cyclization of **124** with Pd(OAC)$_2$/DtBPF provided indole **125** in 71% yield [54h,i]. Hydrogenation and a two-step oxidation of **125**, followed by global demethylation with BBr$_3$ and Boc-protection of the amine, afforded desired acid **117**. The acid **117** and tripeptide **116** were coupled using EDCI and HOAt, the peptide was then cyclized to give bicycle compound **126** upon treatment with K$_2$CO$_3$ in THF. Boc-cleavage in **126** followed by coupling with 2-(3,5-dichloro-4-hydroxyphenyl)-2-oxoacetic acid **118**, two-step reaction removal of NO$_2$ group and saponification finally produced complestatin **115** in 60% yield. (Scheme 16.17).

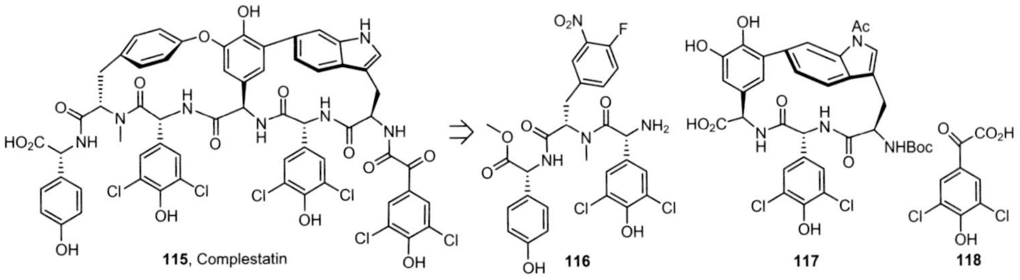

Figure 16.13 Structure of complestatin and its retrosynthetic analysis.

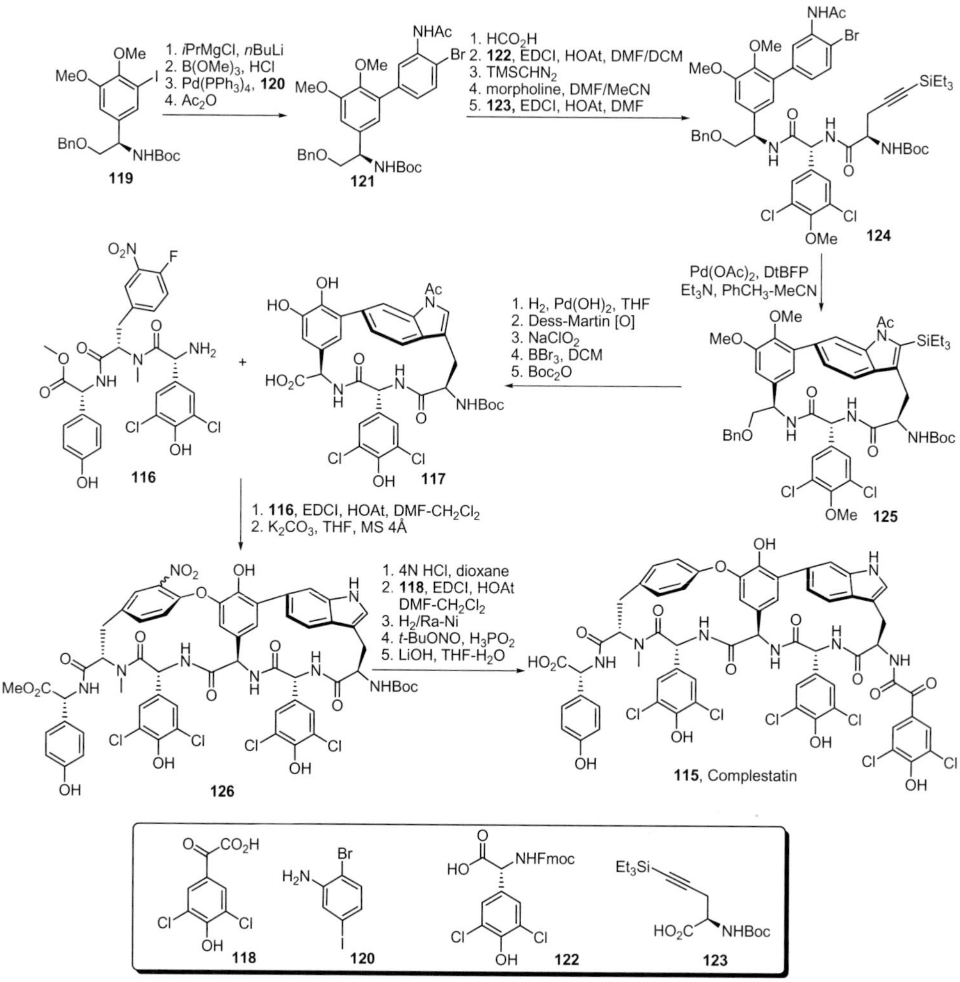

Scheme 16.17 Total synthesis of complestatin.

16.5
Pyran- and Furan-Containing Natural Products

16.5.1
Phorboxazole B

Phorboxazole B **127** along with phorboxazole A (epimer at C13), was first isolated from the marine sponge of *Phorbas* sp. and then from *Raspailia* sp [62a]. This macrolide **127** contains four substituted hydropyran rings, two oxazole rings, a 21-membered macrolactone skeleton and an unsaturated side-chain. Phorboxazole B shows significant antifungal and antibiotic activities against *Candida albicans*

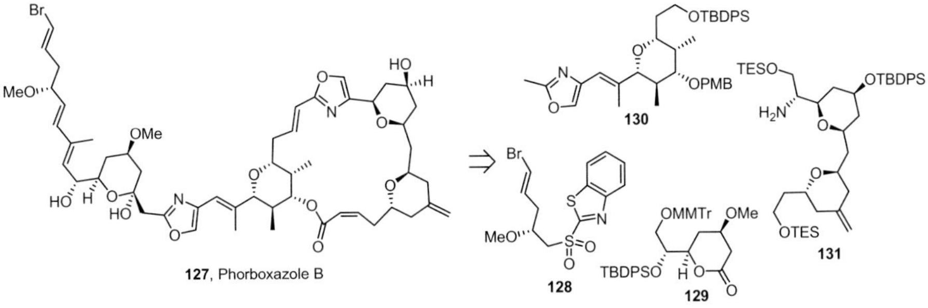

Figure 16.14 Structure of phorboxazole B and its retrosynthetic analysis.

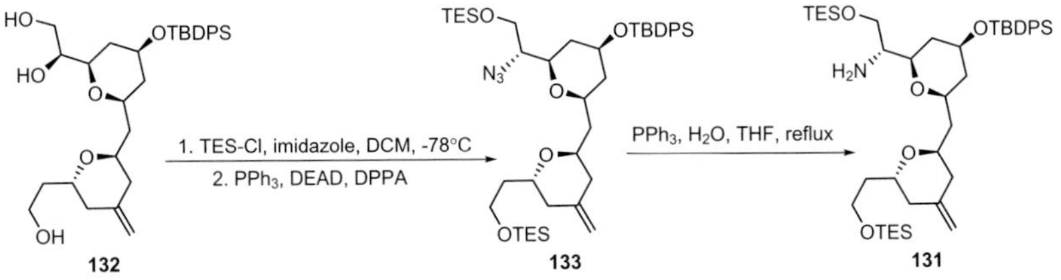

Scheme 16.18 Synthesis of amine **131**.

and *Saccharomyces carlsbergensis*. Phorboxazole B **127** is a potent inhibitor against cancer cell lines (colon HCT-116 and breast MCF7 with an observed GI_{50} value of 4.36×10^{-10} M and 5.62×10^{-10} M, respectively). It also exhibits against the NCI panel of 60 tumor cell lines with $GI_{50} = 1.58 \times 10^{-9}$ M [63]. To date, four total syntheses of phorboxazole B **127** and a number of synthetic studies have been reported [64a]. In 2007, Burke and co-workers completed the total synthesis of **127** *via* directional synthesis/desymmetrization strategy and a catalytic enantioselective hetero-Diels-Alder approach (Figure 16.14) [64a].

The synthesis began with the introduction of the TES-group into triol **132**. The secondary alcohol was then converted into azide **133** through Mitsunobu reaction followed by reduction of the azide-group to produce amine **131** (Scheme 16.18).

Alkene **134** was prepared from Diels–Alder reaction of the corresponding mannitol-derived aldehyde with Brassard diene in the presence of Eu^{3+} catalyst. Hydrogenation of **134** and selective protection of the corresponding alcohol with the MMTr-group and TBDPS-group afforded lactone **129** in 71% yield. Aldehyde **136** was produced from the coupling of thiazole **130** and lactone **129** in a sequence of five steps. Olefination reaction of sulfone **128** under Williams' condition [65], selective removal of TBDPS ether, oxidation to aldehyde, Wittig reaction and saponification produced acid **137** as a mixture of E/Z isomer. Coupling of **137** and bis-hydropyran **131** with HOBt/EDCI and removal of the TES-group provided an amide, which was oxidized and dehydrated to produce oxazole **138**. Further

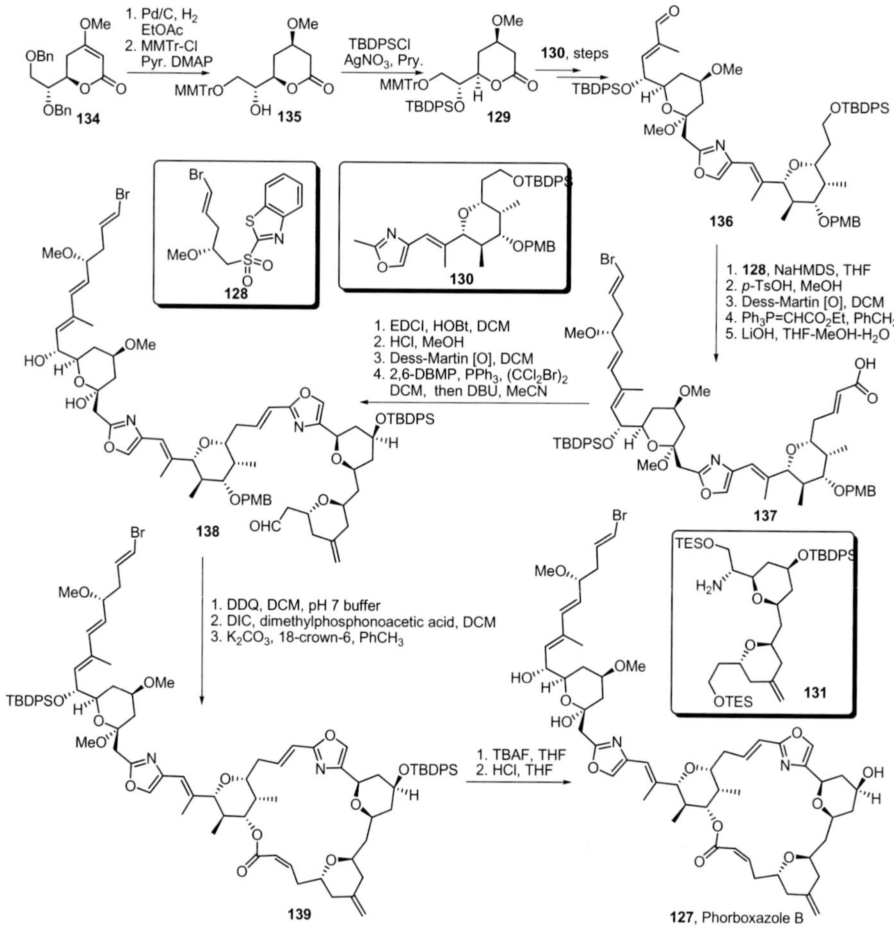

Scheme 16.19 Completion the total synthesis of phorboxazole B.

oxidative deprotection of the PMB ether, installation of the dimethylphosphonoacetate using DIC, followed by treatment with K_2CO_3 produced cyclic **139**. Finally, phorboxazole B **127** (83% yield) was achieved by subsequent silyl-group cleavage and hydrolysis of **139** (Scheme 16.19).

16.5.2
Sorangicin A

(+)-Sorangicin A **140** was isolated from the *Sorangium cellulosum* strain So ce 90 by Hofle [66]. By inhibition of the RNA polymerase, sorangicin A displays remarkable antibiotic activity against both gram-positive and negative bacteria at low concentration. Further studies demonstrated that sorangicin A is potent inhibitor to the rifampicin-resistant microbial mutants (Figure 16.15) [67a].

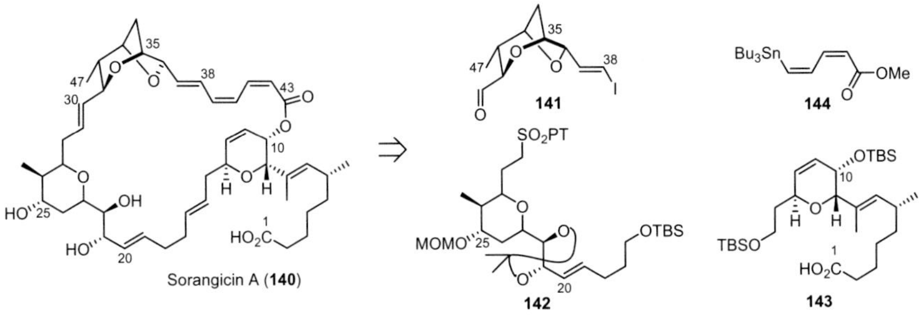

16.5 Pyran- and Furan-Containing Natural Products

Figure 16.15 Structure of sorangicin A and its retrosynthetic analysis.

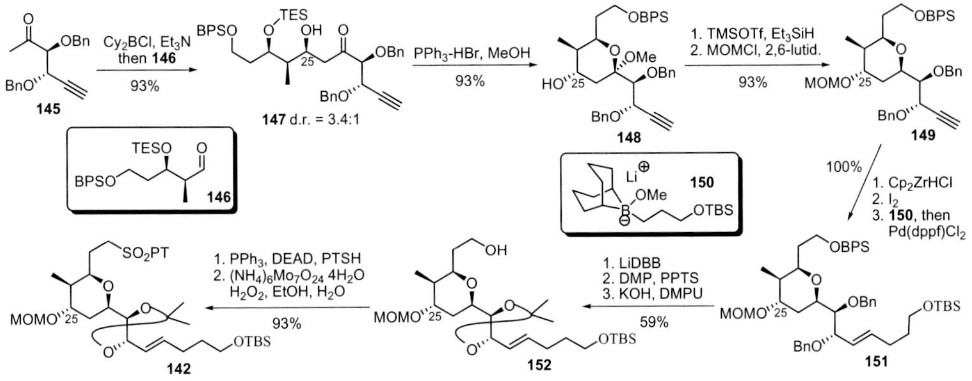

Scheme 16.20 Preparation of sulfone **142**.

The synthesis of the pyran-containing fragment **142** started with the 1,4-*anti*-aldol reaction between **145** and **146** that produced alcohol **147** in a 3.4:1 diastereoisomer ratio. The desired stereoisomer was treated with triphenylphosphine hydrobromide complex to stimulate deprotection of TES ether and ketal formation in excellent yield. The ketal **148** was reduced with triethyl silane and TMSOTf and the C-25 hydroxyl was protected with MOM to afford tetrahydropyran (+)-**149** in 93% yield as a single diastereomer [68]. Hydrozirconation/iodination of alkyne, followed by a Suzuki–Miyaura coupling furnished olefin *E*-**151** in 59% yield. The two benzyl groups were removed and the vicinal diol was protected with acetonide. Selective removal of the *t*-butyldiphenylsilyl (BPS) group followed by Mitsunobu reaction with 1-phenyltetrazole-5-thiol (PTSH) gave the corresponding thioether, which was oxidized to the sulfone **142**. Sulfone **142** was prepared in 17 steps with 17% overall yield (Scheme 16.20).

Michael addition of higher-order cuprate derived from bromide **153** to enone **154**, followed by protection of the enol led to enol ether **155** with high

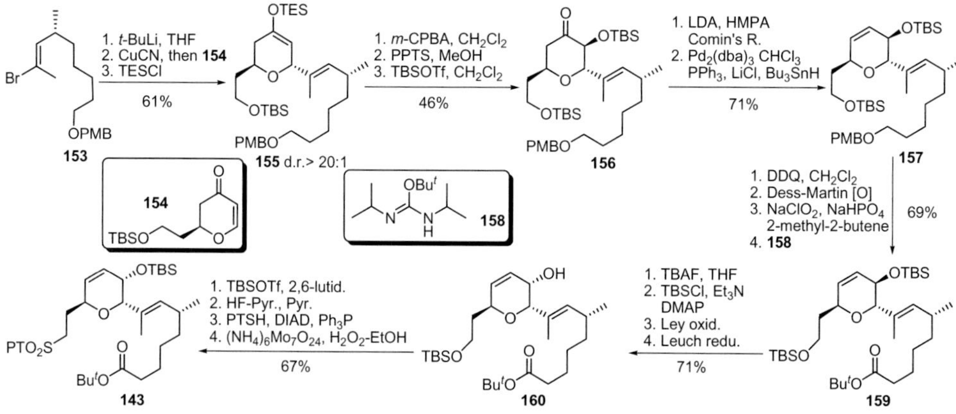

Scheme 16.21 Preparation of sulfone **143**.

diastereo-selectivity. Treatment of the enol ether with *m*-CPBA produced the hydroxy ketone protected as its TES ether, conversion of the TES ether to TBS ether furnished **156** in 46% yield. Kinetic deprotonation of ketone **156** and trap the enol with Comin's reagent [*N*-(5-chloro-2-pyridyl)triflimide] gave enol triflate, which after reductive elimination produced the dehydropyrane **157** in 71% yield. DDQ oxidative removal of PMB ether, followed by a two-step oxidation and esterification furnished ester **159**. Due to the absolute configuration at C-10 was revised to *S* configuration; this prepared fragment was transformed into the desired stereoisomer *via* an oxidation–reduction reaction sequence. After protection of the secondary alcohol as TBS ether, the primary hydroxy group was released and converted into sulfone **143** in 67% yield, and was ready for fragment assembly toward total synthesis (Scheme 16.21).

With the key intermediates in hand, Smith took advantage of Julia olefination reaction to couple fragments **141**, **142** and **143** to get compound **162**, then a Stille coupling reaction between fragments **162** and **144** gave linear precursor **163** in 88% yield. Macrocyclization of seco-acid derived from **163**, promoted with the modified Mukaiyama reagent to produce the corresponding macrolactone **164** in 73% overall yield. Sorangicin A **140** was obtained in 70% yield after removal of protective groups (Scheme 16.22).

16.5.3
Kendomycin

Kendomycin **164** was first isolated from *Streptomyces violaceoruber* by Funahashi and co-workers. Its absolute stereochemistry was established by X-ray analysis after the Zeeck group isolated it from various strains of *Actinomycetes*. Kendomycin exhibits antibacterial activities toward the drug resistant strains of *Staphylococcus aureus*, and also shows remarkable cytotoxicity against cancer cells including

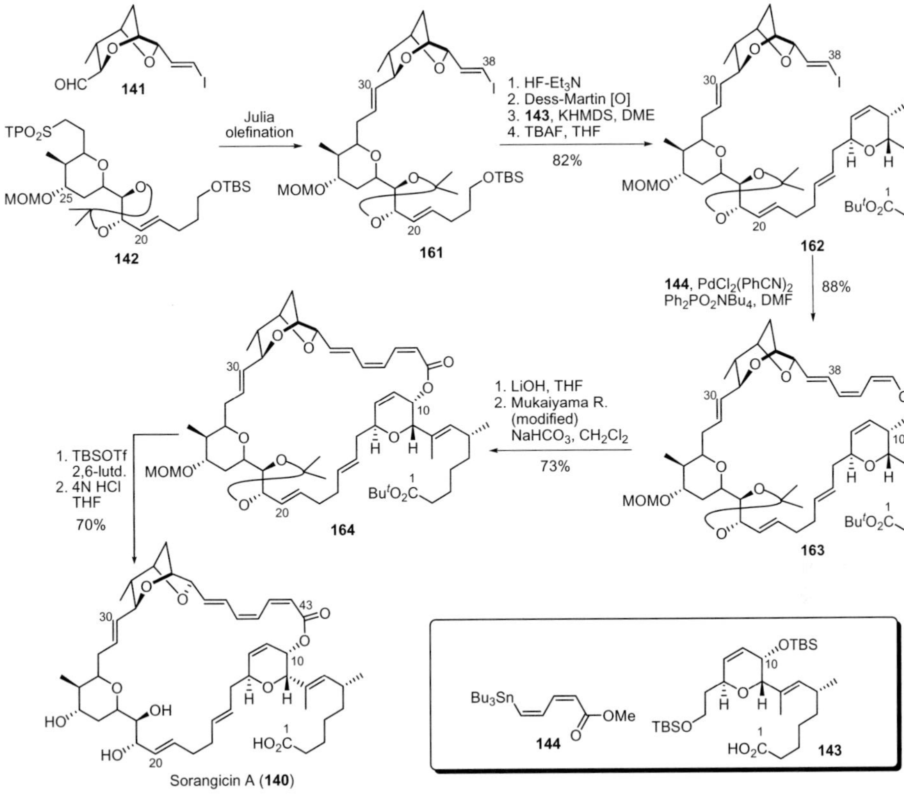

16.5 Pyran- and Furan-Containing Natural Products

Scheme 16.22 Completion of the total synthesis of sorangicin A.

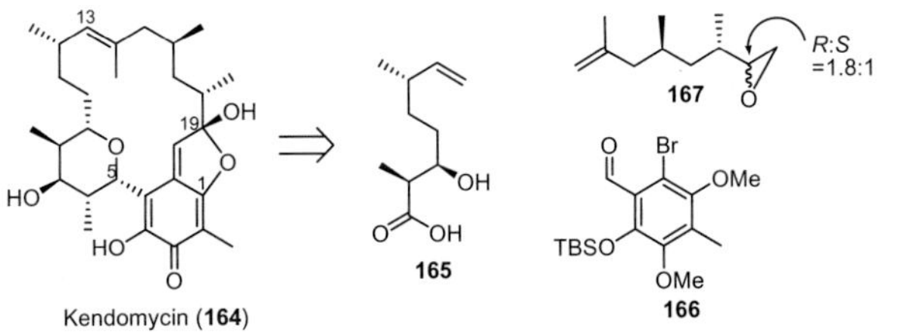

Figure 16.16 Structure of kendomycin and its retrosynthetic analysis.

HMO2, HEP G2, and MCF7 ($GI_{50} < 0.1\,\mu M$) [69a]. The molecule contains a pentasubstituted pyran ring, which along with its other structure features has resulted in significant challenges to its total synthesis. Smith and co-workers employed a Petasis–Ferrier rearrangement/ring-closing olefin metathesis strategy to achieve the total synthesis of kendomycin in 2006 (Figure 16.16) [70].

Condensation of β-hydroxy acid **165** and aldehyde **166** in the presence of *i*-PrOTMS and TMSOTf, followed by the treatment with Tebbe's reagent, produced enol-acetal **168** as a single isomer. The Petasis–Ferrier rearrangement reaction was performed efficiently to afford tetrahydropyran **169** in 85% yield. Substrate-controlled methylation, stereoselective reduction and protection of the secondary alcohol finished the construction of the penta-substituted pyran core structure **170** in 65% yield.

Lithiation of the aromatic bromide **170**, regioselective ring opening of epoxide **167** in the presence of $BF_3 \cdot Et_2O$ produced the desired diene. RCM reaction proved to be stereoselective, only the *S*-isomer at C-19 reacted smoothly to give the macrocycle with *Z*-configuration at the C-C double bond. After a four-step reaction sequence, the *Z*-double bond was isomerized to *E*-double bond with concomitant removal of the TBS group on the phenol oxygen atom, producing compound **172** in 44% yield. Selective removal of TES group, followed by Dess–Martin oxidation furnished the desired ketone, the aromatic ring was also oxidized to quinone, and simultaneous removal of the TBS and methyl ether completed the total synthesis of kendomycin **164** (Scheme 16.23).

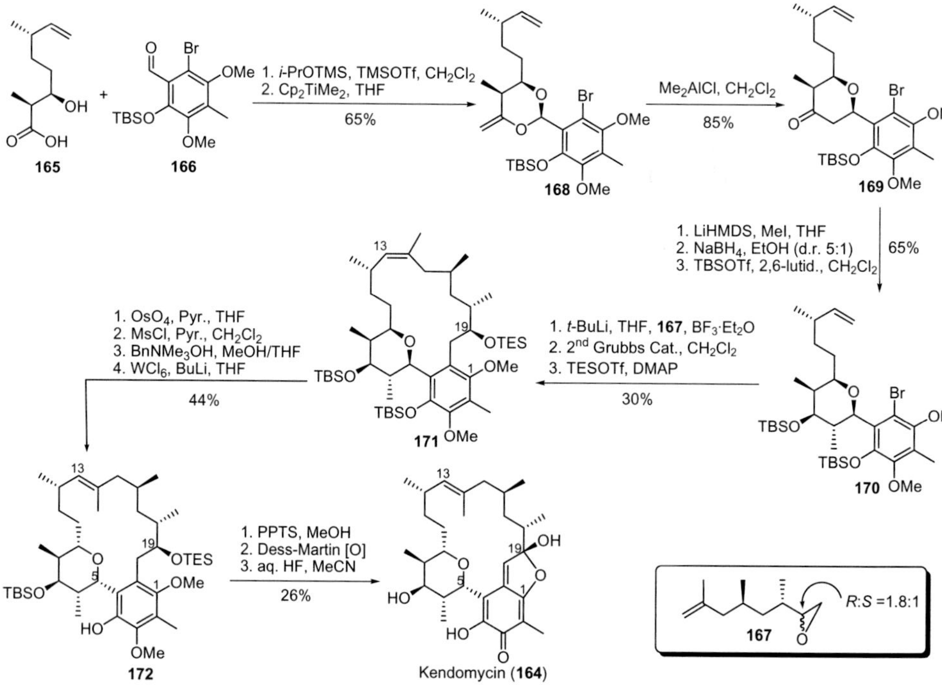

Scheme 16.23 Total synthesis of kendomycin.

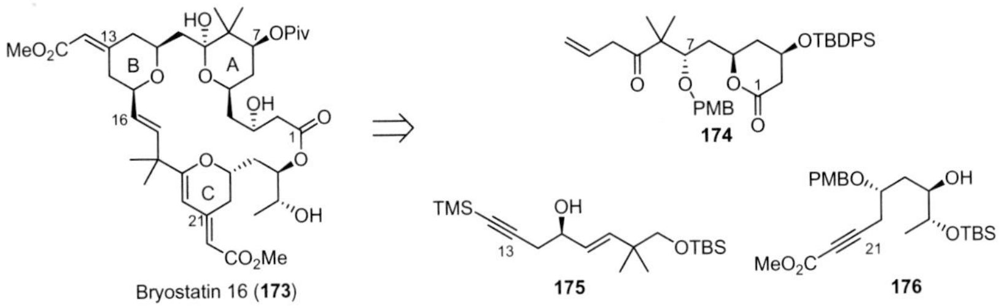

Figure 16.17 Structure of bryostatin 16 and its retrosynthetic analysis.

16.5.4
Bryostatin 16

Pettit and his co-workers have studied the marine bryozoan *Bugula neritina* L. (family: Bugulidae) for more than 40 years and found that this specific bryozoan is a rich source of new anticancer drugs of the bryostatin class [71]. Bryostatin 1 is in phase I and II clinical trials for anticancer and treatment of Alzheimer's disease [72]. Bryostatin 16 was isolated by the same research group in 1996. It shows significant growth inhibitory activity against murine P388 lymphocytic leukemia with ED_{50} at $9.3 \times 10^{-3}\,\mu g\,mL^{-1}$. [73] The sensitive structure of Bryostatin 16 **173** contains three highly substituted pyran rings (A, B and C), an atom-economical and chemoselective total synthesis approach was disclosed in 2009 by Trost and co-workers (Figure 16.17) [74].

Cationic ruthenium complex-catalyzed tandem alkene–alkyne coupling/Michael addition between intermediate **174** and **175** led to the generation of tetrahydropyran **177** with high chemoselectivity. Subsequently, the exo-cyclic vinyl silane was bromonated and the camphorsulfonic acid-catalyzed transesterification/methyl ketalization/desilylation was completed in one step, which produced **178** containing both A and B pyran rings in over 90% yield. The vinyl bromide was converted into methyl ester, while the primary alcohol was transformed to alknyl *via* a two-step procedure that led to compound **179**. The acid derived from **179** was condensed with alcohol **176** using Yamaguchi method to afford ester **180** in 53% yield. Oxidative removal two PMB ether followed by a macrocyclization at the alkyne and alkynoate in the presence of Pd(OAc)$_2$ and tris(2,6-dimethoxyphenyl)phosphine (TDMPP) afforded **181** in 56% yield. Next, the free adjacent hydroxy group underwent the 6-*endo-dig* cyclization in the presence of [Au(PPh$_3$)]SbF$_6$ and sodium bicarbonate, producing the bryostatin 16 skeleton, C-ring pyran **182** in 73% yield. Finally, pivalation under forced reaction conditions and subsequent removal of all silyl protection groups and the methyl ketal completed the total synthesis of bryostatin 16 **173** (Scheme 16.24).

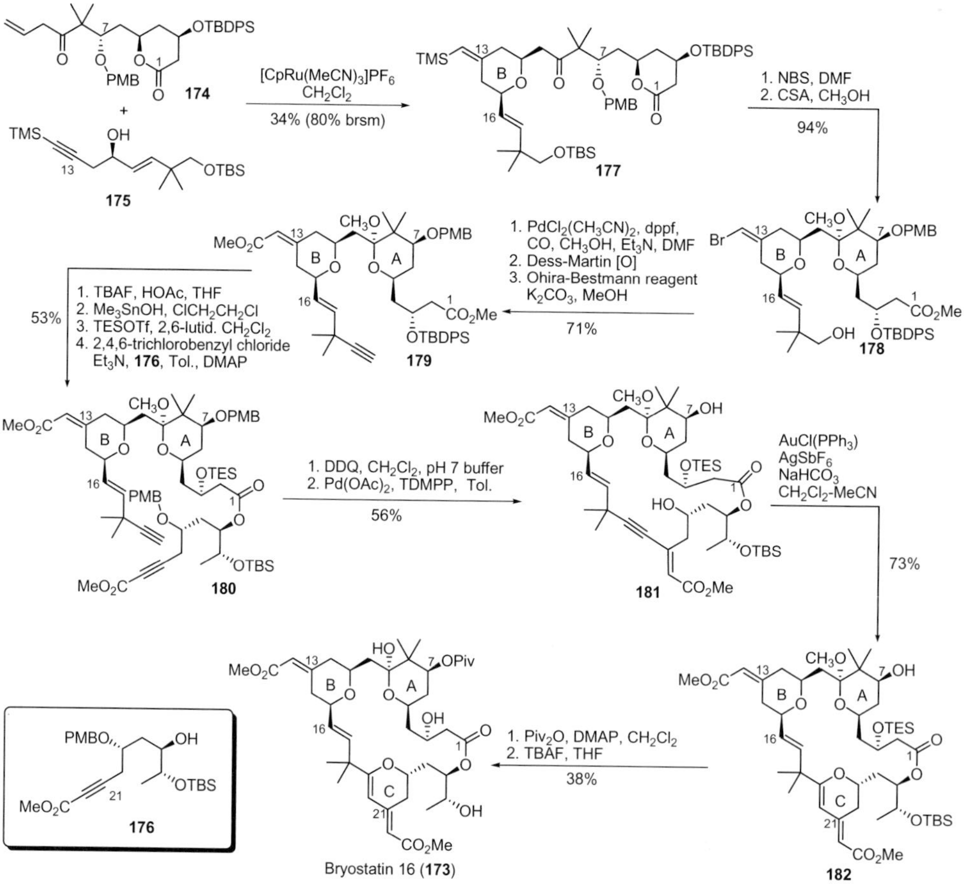

Scheme 16.24 Total synthesis of bryostatin 16.

16.5.5
IKD-8344

IKD-8344, **183** was first isolated from an unidentified alkalophilic Actinomycete (strain no. 8344), and later found in the *Streptomyces* sp. A6792 [75a]. As a novel antibiotic macrodilide, IKE-8344 has asymmetric structure with six tetrahydrofuran rings. It exhibits potent anthelmintic activity, antifungal activity and strong cytotoxicity against L5178Y mouse leukemia cell ($IC_{50} = 0.54\,ng\,mL^{-1}$) (Figure 16.18).

Mesylation of the hydroxy group of compound **187**, followed by acid mediated acetonide deprotection and treatment with NaHMDS produced the first *threo-trans* oxolane **188** in 66% yield [76]. After the primary alcohol was mesylated, the left-wing was subjected to two stereocontrolled allylations to produce compound **190** with the mesylate reduced to the methyl group on its right-wing. The terminal

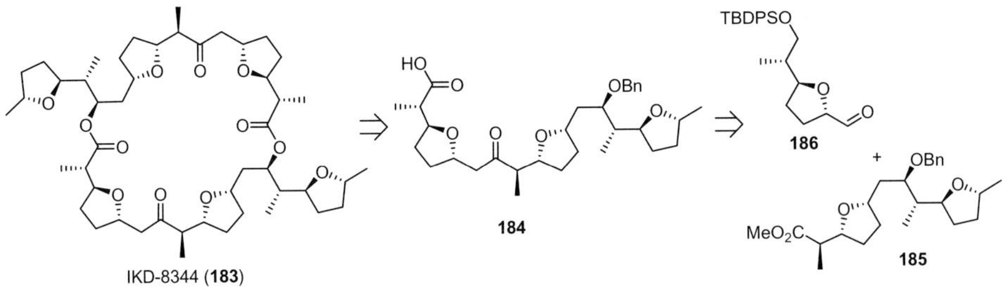

16.5 Pyran- and Furan-Containing Natural Products | 599

Figure 16.18 Structure of IKD-8344.

double bond was cleaved to aldehyde by ozonolysis, after reduction and selective tosylation of the resulted primary alcohol, the secondary alcohol was condensed with acetal **191** in the presence of PPTS. S_N2 substitution with sodium iodide, the tosylate was converted to iodide **192** in 72% yield over four steps from compound **190**. Radical cyclization reaction proceeded smoothly to produce the second tetrahydrofuran ring in 91% yield and 10:1 diastereoselectivity for the desired compound **185**. Ester of **185** was transformed to Wittig salt, which after condensation with aldehyde **186** produced the trioxolane linear precursor **193** in 86% yield. Intermediate **193** was converted into the corresponding acid **184** by a sequence of reactions including epoxidation of alkene with m-CPBA, regioselective epoxide-ring opening reaction with LiAlH$_4$ with concomitant removal of the TBDPS, diol oxidation with Dess–Martin periodinane, and further chlorite oxidation of the resultant aldehyde. Silylation of the carboxylic acid and catalytic hydrogenation removal of Bn group produced another coupling partner **194**. Condensation of **184** and **194** under modified Yamaguchi conditions afforded the linear precursor, which was converted into natural product IKD-8344 **183** by unmasking orthogonal protecting groups of the linear seco-acid and macrocyclization using modified Yamaguchi conditions (Scheme 16.25).

16.5.6
Deoxypukalide

Deoxypukalide **195** was isolated from *Leptogorgia* sp. in 2007 by Darias and co-workers [77]. Its total synthesis was accomplished by Donohoe and co-workers in 2008 [78]. The total synthesis featured two RCM reactions to construct the butenolide and furan rings (Figure 16.19).

The synthesis commenced with (*S*)-perillyl alcohol **199**. After the primary alcohol was protected as its TIPS ether, selective ozonolysis of the trisubstituted alkene was performed in the presence of a stoichiometric amount of pyridine and excess isoprene. The resulted aldehyde was reacted with the aluminum allenolate to give the allylic alcohol **200** in 48% yield. Acetal transformation followed by RCM reaction of triene **198** and acid-catalyzed aromatization afford furan derivative **201** in 85% yield. A further three-step functional group transformation produced the key

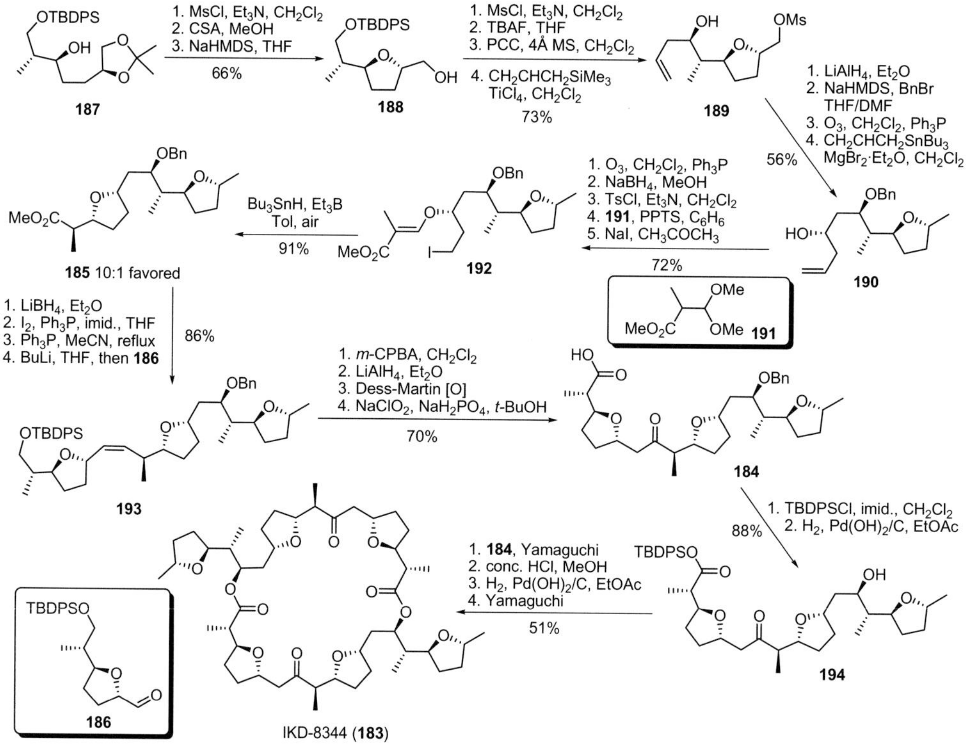

Scheme 16.25 Total synthesis of IKD-8344.

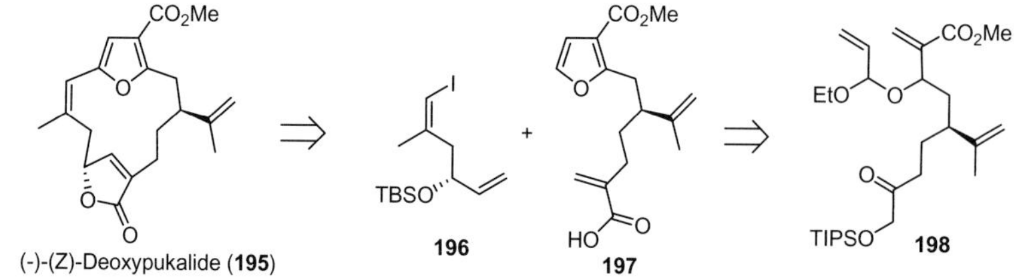

Figure 16.19 Structure of deoxypukalide and its retrosynthetic analysis.

intermediate **197**, and Negishi cross coupling reaction with vinyl iodide **196** followed by treatment with TBAF produced the linear precursor **202** ready for macrocyclization. Treatment of **202** with 2-methyl-6-nitrobenzoic anhydride (MNBA) in the presence of DMAP and triethylamine gave macrolactone **203** in 73% yield. Finally, Grubbs II catalyst was proved efficient to establish the butenolide ring and afford deoxypukalide **195** in 72% yield (Scheme 16.26).

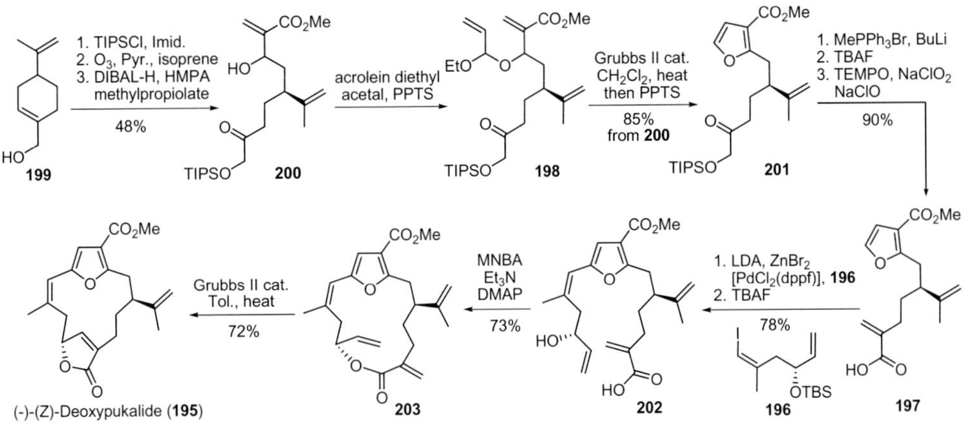

Scheme 16.26 Total synthesis of deoxypukalide.

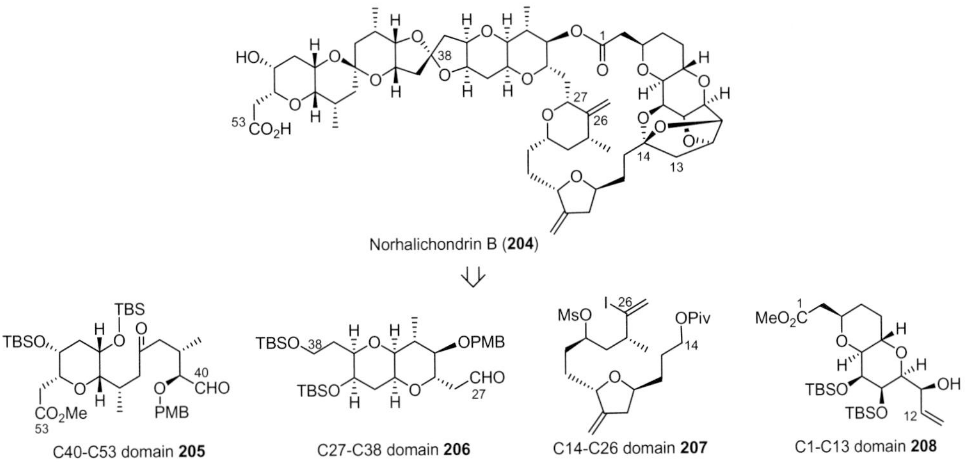

Figure 16.20 Structure of norhalichondrin B and its retrosynthetic analysis.

16.5.7
Norhalichondrin B

Halichondrins, including norhalichondrin B **204**, are a family of antitumor polyether macrolide originally isolated from *Hulichondria okadui* Kadota and later found in sponge *Lissodendoryx* sp [79a]. All these marine natural products are strongly inhibitory against the murine leukemia cell line P388, and one of the synthetic analogs, E7389, was under its way to become a new anticancer drug [80a]. Phillips and co-workers completed the total synthesis of norhalichondrin B **204** *via* a highly convergent strategy (Figure 16.20) [81].

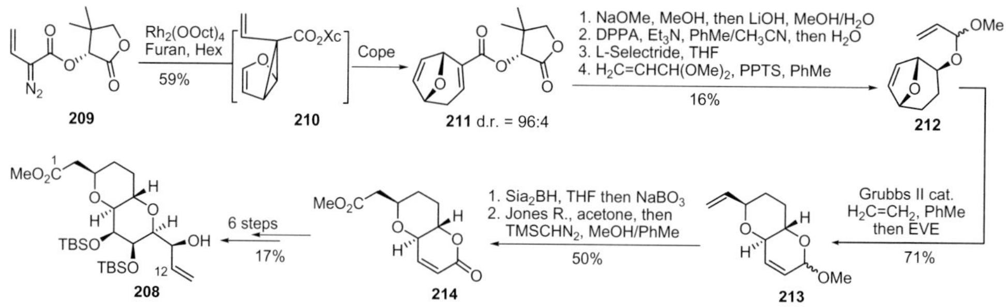

Scheme 16.27 Preparation of C1-C13 fragment **208**.

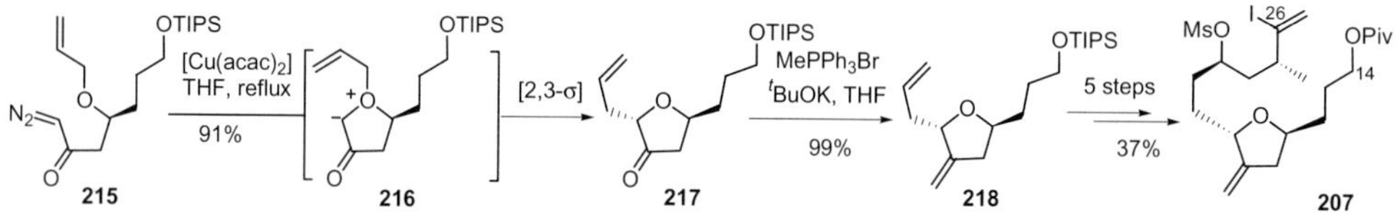

Scheme 16.28 Preparation of C14-C26 fragment **207**.

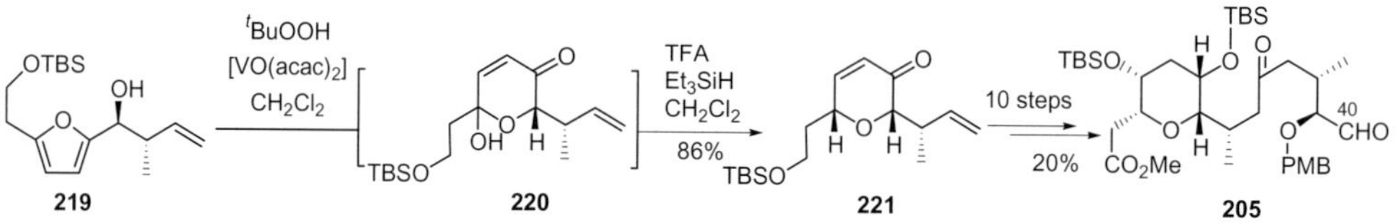

Scheme 16.29 Preparation of C40-C53 fragment **205**.

The rhodium-catalyzed cyclopropanation of **209** with furan followed by Cope rearrangement furnished the oxabicyclo[3.2.1]ocetene **211** in 59% yield. After transesterification, the carboxylic acid moiety was converted into acetal **212** over a four-step reaction with 16% yield. Grubbs catalyst mediated ROM-RCM protocol produced the fused pyranopyran **213** in 71% yield, which comprises the core structure of fragment **208** (Scheme 16.27).

Exposing diazoketone **215** to copper catalyst in refluxed THF, the [2,3]-sigmatropic rearrangement occurred smoothly to afford tetrahydrofuran **217** in 91% yield. After the *exo*-methylene was installed, diene **218** was delivered to vinyl iodide **207** in five steps and 37% yield (Scheme 16.28).

Ester **205** was synthesized from furan derivative **219**. Achmatowicz oxidation produced pyranone hemiacetal **220**, which underwent acid-mediated ionic hydrogenation to afford pyranone **221** in 86% yield. Chain elongation and functional group transformation within 10 steps provided the key intermediate **205** in 20% yield (Scheme 16.29).

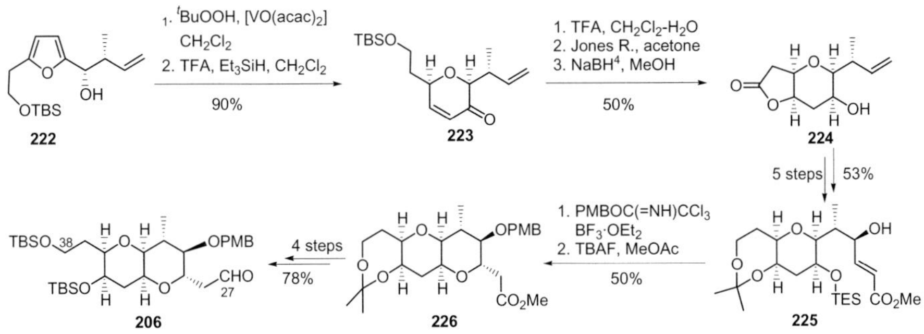

Scheme 16.30 Preparation of C27-C38 fragment **206**.

Another pyranopyran-containing fragment **206** was also synthesized using a diastereoisomer of **219** and similar strategy for **205**. Desilylation of pyranone **223** and Jones oxidation converted the primary alcohol into carboxylic acid, which underwent the Michael addition to enone, and after stereoselective reduction produced intermediate **224** in 50% yield. Further functional group transformation produced α,β-unsaturated ester **225**. Protection of the secondary hydroxy group with PMB ether and subsequent removal of the TES group facilitated the oxa-Michael adduct **226** in 50% yield. A further four steps of functional group manipulation afforded to the desired precursor **206** in 78% yield (Scheme 16.30).

The assembly of fragments commenced with condensation of intermediates **207** and **206**, via NHK reaction and the tandem S_N2 substitution, to afford intermediate **227** in 59% yield. A conventional chemical process ensured the installation of the enone **228** in 70% yield, and cross-metathesis with fragment **208** produced **229** in 62% yield. Global deprotection of silyl ethers with TBAF using acetic acid buffer and concomitant hetero-Michael addition provided first the tetrahydrofuran, and subsequently acid-catalyzed ketal formation afforded the desired 2,6,9-trioxa tricyclo[3.3.2.0^{3,7}]decane **230** in 64% yield. Protection of the free alcohols with TBS ether, and then removal of PMB ether and methyl ester produced the corresponding *seco*-acid, and the macrolactonization was proved to be effective to give the macrocycle **231** in 47% yield. Strategical functional groups transformation ensured the preparation of **232**, which underwent a Horner-Wadsworth-Emmons condensation with aldehyde **205** to produce the advanced intermediate **233** in 83% yield (Scheme 16.31).

Sequential removal of TBS ethers and PMB ether under precisely controlled reaction conditions afforded the methyl ester of norhalichondrin B, which after lithium hydroxide stimulated saponification completed the total synthesis of norhalichondrin B **204** in 39% yield.

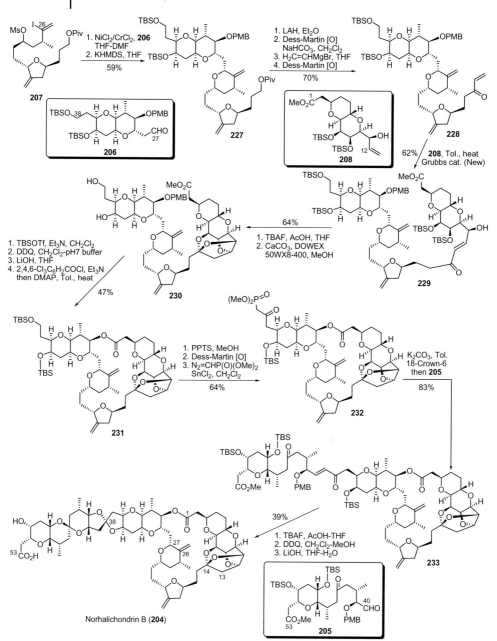

Scheme 16.31 Completion the total synthesis of norhalichondrin B.

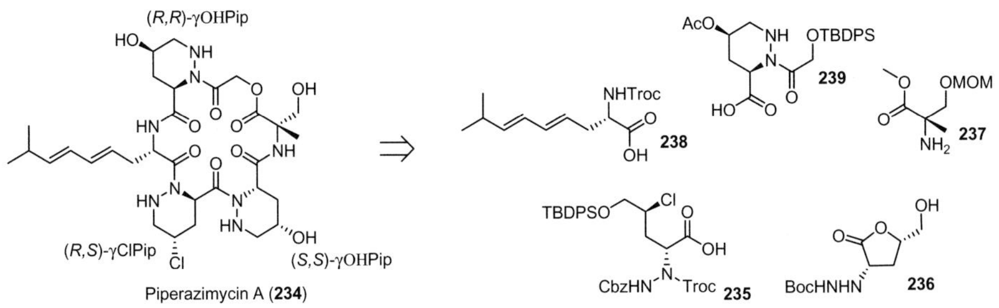

Figure 16.21 Structure of piperazimycin A and its retrosynthetic analysis.

16.6
Piperazic Acid-Containing Natural Products

16.6.1
Piperazimycin A

Piperazimycin A **234** was isolated by Fenical's group from cultivated marine sediments from the Island of Guam. Piperazimycin A exhibits *in vitro* cytotoxicity toward multiple tumor cell lines with a mean GI_{50} value of 100 nM [82]. Studies also proved that piperazimycin A is more potent against solid tumors than against leukemia cell lines. Piperazimycin A is an 18-membered cyclic hexadepsipeptide that contains two enantiomeric γ-hydroxypiperazic acids and one γ-chloropiperazic acid [83]. Challenges towards the total synthesis of piperazimycin A lie in the difficult coupling reaction between the piperazic acids, especially dipeptide unit derived from two piperazic acids. Ma and co-workers employed a highly practical strategy for the total synthesis of piperazimycin A (Figure 16.21) [83].

The hydroxy group of compound **240** was activated and transformed to hydrazide **241** in 81% yield. After protection of the α-nitrogen atom, the lactone was hydrolyzed, the acid was esterified and the secondary hydroxy group was converted into chloride **242**. This is the precursor of the γ-chloropiperazic acid moiety. Coupling reaction between carboxylic acid **235**, which was derived from **242**, and hydrazide **236** was accomplished using acyl chloride activation method and produced dipeptide **243** in 74% yield. After the primary alcohol was activated as a leaving group, the liberation of the hydrazine by the removal of Boc with TFA completed the first intramolecular cyclization to produce the right-bottom γ-OHpip. Saponification and allyl ester formation afforded compound **244**. After the secondary alcohol was masked as an acetate, the allyl ester was removed and the carboxylic acid was coupled with compound **237** in the presence of HATU to provide **245** in 87% yield. Selective removal of the Troc and Cbz groups from **245** and protection of the hydrazine with FmocOSu gave the desired Fmoc-hydrazide intermediate, which was coupled with *N*-Troc-(*S*)-2-amino-8-methyl-4,6-nonadienoic acid **238** using the acyl chloride protocol to produce the tetrapeptide **246** in 54% yield. TBAF-mediated desilylation of **246** afforded the corresponding alcohol, which was subjected to a one-pot cyclization process including the deprotection of the Fmoc group and cyclization under Mitsunobu condition to furnish compound **247** in 96% yield. The Troc protecting group was removed under reductive condition and the resulting free amine was subjected to coupling reaction with the dipeptide acid **239** using HATU as dehydration reagent to produce the linear precursor **248**. Cleavage of the TBDPS ether followed by chlorination with hexachloroacetone (HCA) provided the corresponding chloride **249**. Saponification of the methyl ester and the acetyl protecting groups, followed by treatment with base and sodium iodide in DMF afforded the macrocycle, which was subjected to de-protection of MOM ether and to produce piperazimycin A **234** in 59% yield (Scheme 16.32).

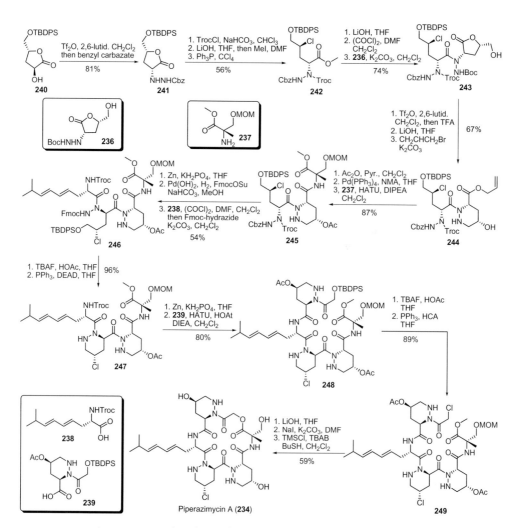

Scheme 16.32 Total synthesis of piperazimycin A.

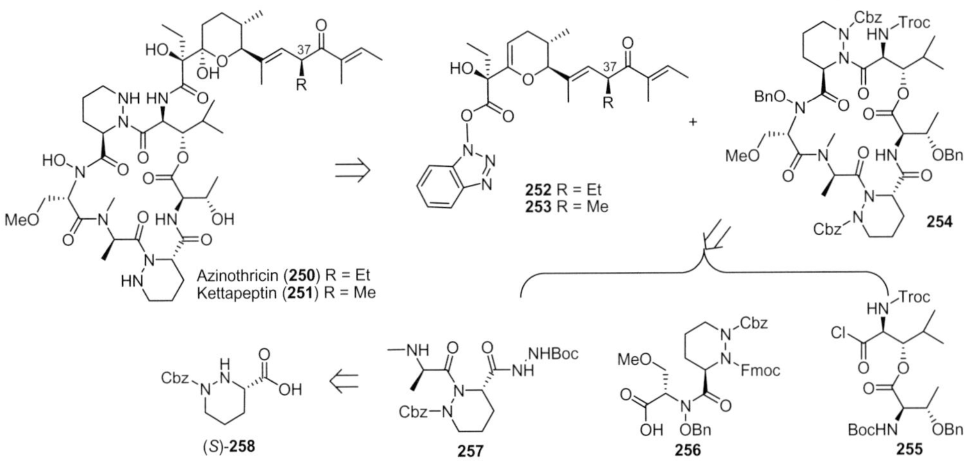

Figure 16.22 Structures of azinothricin and kettapeptin and their retrosynthetic analysis.

16.6.2
Azinothricin and Kettapeptin

Azinothricin **250** was isolated in 1986 from the culture filtrate of *Stretomyces* sp. X-14950. It is a 19-membered macrocycle composed by six non-natural amino acids and a novel pyran-containing side-chain. It is active against several gram-positive microorganisms, such as 31 strains of *Staphylococcus aureus* (MIC: <0.008–0.016 µg mL^{-1}), 16 strains of *Enterococcus faecalis* (MIC: 0.063 µg mL^{-1}), two strains of *Streptococcus pyogenes* (MIC: 0.016 µg mL^{-1}) and two strains of *Streptococcus pneumoniae* (MIC: <0.008 µg mL^{-1}) [84]. Kettapeptin **251** was isolated from the ethyl acetate extract of the *Streptomyces* sp. isolate GW99/1572. It exhibits significant biological activity against gram-positive bacteria (Figure 16.22) [85].

The total synthesis of azinothricin **250** and kettapeptin **251** accomplished by Hale and co-workers was summarized in Schemes 16.33 [86]. A silver cyanide assisted amidation successfully promoted the coupling reaction between alkoxyamine **259** and acid chloride **260** to give the corresponding dipeptide ester, which was treated with phenylsilane and catalytic amount of Pd(PPh$_3$)$_4$ to afford acid **256** in 77% yield. Fragment **257** was also obtained from a silver cyanide assisted amidation in excellent yield. Condensation of fragments **256** and **257** was effected by BOPCl to produce tetrapeptide **262** in 79% yield. After removal of Fmoc protective group in **262**, the amine was coupled with acylchloride **255** under silver cyanide-mediated condition to furnish the linear precursor **263** in 55% yield. Both Boc groups were removed with TFA, and the acylhydrazide was cleaved with NBS to afford the corresponding hexapeptide salt, which was subjected to a macrocyclization promoted with HATU in the presence of N-ethylmorpholine (NEM) to afford core structure **254** in 45% yield.

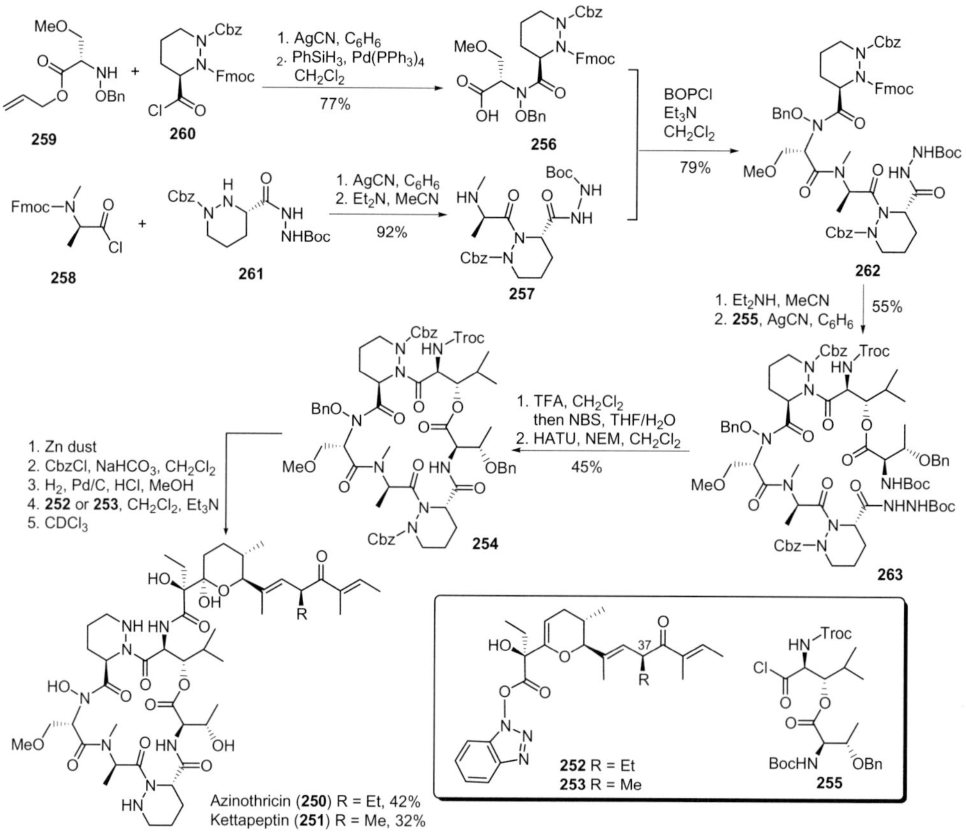

Scheme 16.33 Total synthesis of azinothricin and kettapeptin.

The activated esters **252** and **253** were prepared *via* an improved approach which was originally developed by Hale and co-workers for their synthesis of A83586C [87a]. Protecting group manipulation and global deprotection of the macrocycle **254** followed by regioselective acylation with **252/253** and treatment of the corresponding products with CDCl₃, produced azinothricin **250** and kettapeptin **251** in 42% and 32% yield, respectively. The above synthetic strategy was also further applied to the synthesis of analogs of A83586C by Hale and co-workers [88].

16.7
Mixed Heterocyclic Systems

16.7.1
(−)-Nakadomarin A

(−)-Nakadomarin A **264** was isolated from the sponge *Amphimedon* sp. off the coast of the Kerama Islands, Okinawa by Kobayashi and co-workers in 1997 [89a]. This

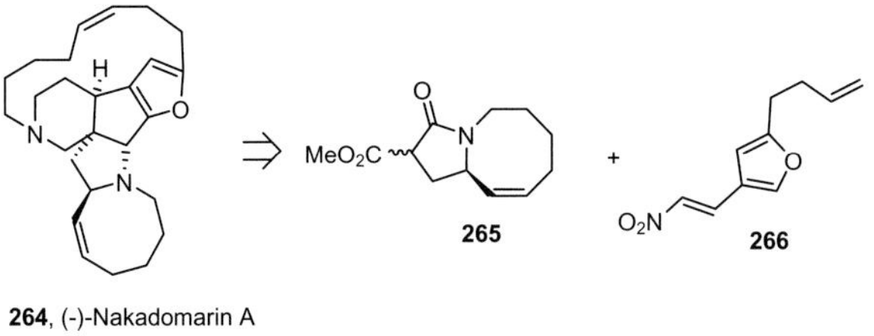

264, (−)-Nakadomarin A

Figure 16.23 Structure of nakadomarin A and its retrosynthetic analysis.

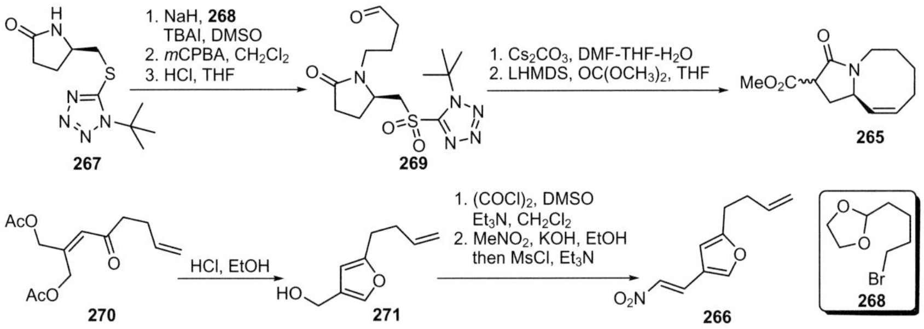

Scheme 16.34 Preparation of nitro olefin **266**.

hexa-cyclic alkaloid consists of six cyclic systems. (−)-Nakadomarin A shows cytotoxic activity against murine lymphoma L1210 cells and inhibition of cyclin dependent kinase 4 with IC_{50} 1.3 µg mL^{-1} and 9.9 µg mL^{-1}, respectively. It also exhibits antimicrobial activity against the fungus *Trichophyton mentagrophtes* and gram-positive bacterium *Corynebacterium xerosis* (MIC = 23 µg mL^{-1} and 11 µg mL^{-1}, respectively) (Figure 16.23).

To date, only two total syntheses of (−)-nakadomarin A **264** had been reported by Nishida *et al*. and Dixon *et al*. [90]. Dixon's strategy involved a multiple catalyst-controlled carbon–carbon bond forming and cascade sequences. The formation of six-membered ring took place in the late stage of the synthesis. Lactam **265** was constructed from the sulfide **267**. N-alkylation of **267** with bromide **268**, followed by oxidation of the sulfide to the corresponding sulfone and deprotection of acetal produced sulfone **269**. Intramolecular Julia–Kocienski olefination of **269** promoted by Cs_2CO_3 and further C-acylation with dimethylcarbonate led to bicyclic **265** in 82% yield. In parallel, an acid hydrolysis of enone **270** produced furanyl alcohol **271**, which was converted into nitro **266** by Swern oxidation and Henry-type condensation (Scheme 16.34).

With building blocks **265** and **266** in hand, bifunctional cinchona catalyst **272** [91a] was employed to facilitate a Michael addition reaction of **265** and **266** to afford **273** with high diastereoselectivity (d.r. = 91 : 9) in 57% yield. Subjection of **273** to

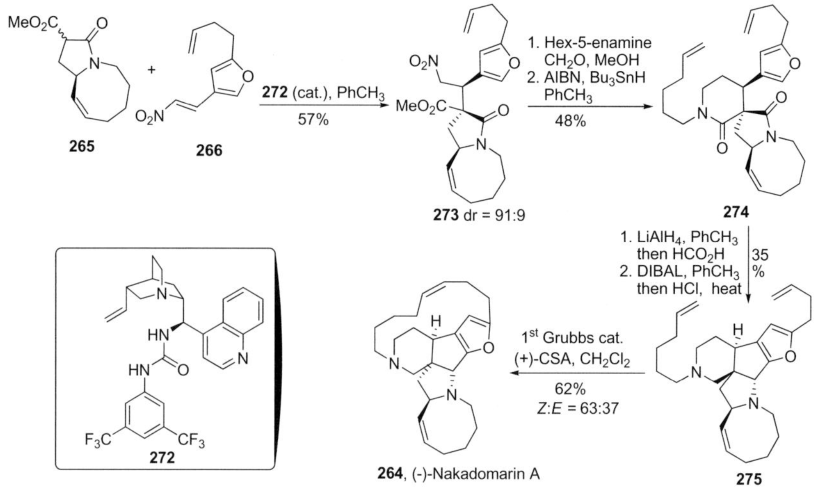

Scheme 16.35 Completion the total synthesis of nakadomarin A.

a three-component nitro-Mannich/lactamization cascade afforded the corresponding lactam adduct, which then underwent a traceless reduction of nitro group to produce **274** in 48% overall yield. Strictly controlled reaction conditions facilitated the selective LiAlH$_4$ and DIBAL-H reduction of lactam **274** then produced the pentacycle **275** in 35% yield. Finally, (−)-nakadomarin A **264** macrocycle was accomplished upon olefin metathesis by using First Grubbs' catalyst in the presence of CSA (Scheme 16.35).

16.8
Conclusions

Due to their widespread occurrence in nature, diverse biological activity and interesting chemical properties, heterocycle-containing macrocyclic natural products have become attractive targets for organic synthesis. As demonstrated in the foregoing illustrative examples, various strategies and methodologies have been applied to achieve the total synthesis of these macrocyclic natural products. Considerable advances have been made during the past decade. The chapter incorporates the relevant literature for the past five years enabling the reader to appreciate the need and scope for innovative research in this area. The successful total syntheses of apratoxin A, largazole, bistratamide H, didmolamide A, IB-01211, telomestatin, micrococcin P1, GE2270, celogentin C, complestatin, phorboxazole B, sorangicin A, kendomycin, bryostatin 16, IKD-8344, deoxypukalide, norhalichondrin B, piperazimycin A, azinothricin, kettapeptin, and nakadomarin A highlight the state of the art of contemporary organic synthesis, particularly in correctly installing the stereocenters by exploiting substrate and reagent control as well as handling a multitude of delicately balanced chemoselectivity issues in such complex systems. These completed syntheses also served to confirm or establish the absolute stereostructures of these different classes of macrocyclic natural products.

References

1 (a) Buss, A.D. and Butler, M.S. (2010) Natural product chemistry for drug discovery (RSC Biomolecular Sciences No. 18). *J. Med. Chem.*, **53**, 2329–2329; (b) Li, J.W.-H. and Vederas, J.C. (2009) Drug discovery and natural products: end of an era or an endless frontier? *Science*, **325**, 161–165; (c) Butler, M.S. (2008) Natural products to drugs: natural product-derived compounds in clinical trials. *Nat. Prod. Rep.*, **25**, 475–516; (d) Hughes, B. (2009) 2008 FDA drug approvals. *Nat. Rev. Drug Discov.*, **8**, 93–96; (e) Rosen, J., Gottfries, J., Muresan, S., Backlund, A. and Oprea, T.I. (2009) Novel chemical space exploration *via* natural products. *J. Med. Chem.*, **52**, 1953–1962; (f) Chin, Y.-W., Balunas, M.J., Chai, H.B. and Kinghorn, A.D. (2006) Drug discovery from natural sources. *AAPS J.*, **8**, 239–253; (g) Butler, M.S. (2004) The role of natural product chemistry in drug discovery. *J. Nat. Prod.*, **67**, 2141–2153; (h) Balunas, M.J. and Kinghorn, A.D. (2005) Drug discovery from medicinal plants. *Life Sci.*, **78**, 431–441.

2 (a) Luesch, H., Yoshida, W.Y., Moore, R.E., Paul, V.J. and Corbett, T.H. (2001) Total structure determination of apratoxin A, a potent novel cytotoxin from the marine cyanobacterium *Lyngbya majuscula*. *J. Am. Chem. Soc.*, **123**, 5418–5423; (b) Luesch, H., Yoshida, W.Y., Moore, R.E., Paul, V.J. and Corbett, T.H. (2002) New apratoxins of marine cyanobacterial origin from Guam and Palau. *Bioorg. Med. Chem.*, **10**, 1973–1978.

3 Luesch, H., Chanda, S.K., Raya, R.M., Dejesus, P.D., Orth, A.P., Walker, J.R., Belmonte, J.C.I. and Schultz, P.G. (2006) A functional genomics approach to the mode of action of the antiproliferative natural product apratoxin A. *Nat. Chem. Biol.*, **2**, 158–167.

4 (a) Chen, J. and Forsyth, C.J. (2003) Total synthesis of apratoxin A. *J. Am. Chem. Soc.*, **125**, 8734–8735; (b) Chen, J. and Forsyth, C.J. (2004) Total synthesis of the marine cyanobacterial cyclodepsipeptide apratoxin A. *Proc. Natl. Acad. Sci. U.S.A.*, **101**, 12067–12072.

5 (a) Ma, D., Zou, B., Cai, G., Hu, X. and Liu, J.O. (2006) Total synthesis of the cyclodepsipeptide apratoxin A and its analogs and assessment of their biological activities. *Chem. Eur. J.*, **12**, 7615–7626; (b) Doi, T., Numajiri, Y., Munakata, A. and Takahashi, T. (2006) Total synthesis of apratoxin A. *Org. Lett.*, **8**, 531–534; (c) Numajiri, Y., Takahashi, T. and Doi, T. (2009) Total synthesis of (−)-apratoxin A, 34-epimer, and its oxazoline analog. *Chem. Asian J.*, **4**, 111–125.

6 Randazzo, A., Bifulco, G., Giannini, C., Bucci, M., Debitus, C., Cirino, G. and Gomez-Paloma L. (2001) Halipeptins A and B: two novel potent anti-inflammatory cyclic depsipeptides from the Vanuatu marine sponge *Haliclona* species. *J. Am. Chem. Soc.*, **123**, 10870–10876.

7 Della Monica C., A. Randazzo, G. Bifulco, P. Cimino, M. Aquino, I. Izzo, F. De Riccardis F. and Gomez-Paloma L. (2002) Structural revision of halipeptins: synthesis of the thiazoline unit and isolation of halipeptin C. *Tetrahedron Lett.*, **43**, 5707–5710.

8 Nicolaou, K.C., Schlawe, D., Kim, D., Longbottom, D.A., de Noronha, R.G., Lizos, D.E., Rao Manam, R. and Faulkner, D.J. (2005) Total synthesis of halipeptins: isolation of halipeptin D and synthesis of oxazoline halipeptin analogs. *Chem. Eur. J.*, **11**, 6197–6211.

9 (a) Nicolaou, K.C., Kim, D.W., Schlawe, D., Lizos, D.E., de Noronha, R.G. and Longbottom, D.A. (2005) Total synthesis of halipeptins A and D and analogs. *Angew. Chem. Int. Ed.*, **44**, 4925–4929; (b) Nicolaou, K.C., Lizos, D.E., Kim, D.W., Schlawe, D., de Noronha, R.G., Longbottom, D.A., Rodriquez, M., Bucci, M. and Cirino, G. (2006) Total synthesis and biological evaluation of halipeptins A and D and analogs. *J. Am. Chem. Soc.*, **128**, 4460–4470.

10 Yu, S., Pan, X., Lin, X. and Ma, D. (2005) Total synthesis of halipeptin A, a potent anti-inflammatory cyclic depsipeptide. *Angew. Chem. Int. Ed.*, **44**, 135–138.

11 Hara, S., Makino, K. and Hamada, Y. (2006) Total synthesis of halipeptin A, a potent anti-inflammatory cyclodepsipeptide from a marine sponge. *Tetrahedron Lett.*, **47**, 1081–1085.

12 (a) Taori, K., Paul, V.J. and Luesch, H. (2008) Structure and activity of largazole, a potent antiproliferative agent from the floridian marine cyanobacterium *Symploca* sp. *J. Am. Chem. Soc.*, **130**, 1806–1807; (b) Ying, Y., Taori, K., Kim, H., Hong, J. and Luesch, H. (2008) Total synthesis and molecular target of largazole, a histone deacetylase inhibitor. *J. Am. Chem. Soc.*, **130**, 8455–8459.

13 (a) Masuoka, Y., Nagai, A., Shin-Ya, K., Furihata, K., Nagai, K., Suzuki, K., Hayakawa, Y. and Seto, H. (2001) Spiruchostatins A and B, novel gene expression-enhancing substances produced by *Pseudomonas* sp. *Tetrahedron Lett.*, **42**, 41–44; (b) Shigematsu, N., Ueda, H., Takase, S. and Taneka, H. (1994) FR901228, a novel antitumor bicyclic depsipeptide produced by chromobacterium *Violaceum* no. 968. II. Structure determination. *J. Antibiot.*, **47**, 311–314; (c) Yoshida, M., Kijama, M., Akita, M. and Beppu, T. (1990) Potent and specific inhibition of mammalian histone deacetylase both *in vivo* and *in vitro* by trichostatin A. *J. Biol. Chem.*, **265**, 17174–17179; (d) Bowers, A., West, N., Taunton, J., Schreiber, S.L., Bradner, J.E. and Williams, R.M. (2008) Total synthesis and biological mode of action of largazole: a potent class I histone deacetylase inhibitor. *J. Am. Chem. Soc.*, **130**, 11219–11222; (e) Yurek-George, A., Cecil, A.R.L., Mo, A.H.K., Wen, S., Rogers, H., Habens, F., Maeda, S., Yoshida, M., Packham, G. and Ganesan, A. (2007) The first biologically active synthetic analogs of FK228, the depsipeptide histone deacetylase inhibitor. *J. Med. Chem.*, **50**, 5720–5726; (f) Greshock, T.J., Johns, D.M., Noguchi, Y. and Williams, R.M. (2008) Improved total synthesis of the potent *HDAC* inhibitor FK228 (FR-901228). *Org. Lett.*, **10**, 613–616.

14 Seiser, T., Kamena, F. and Cramer, N. (2008) Synthesis and biological activity of largazole and derivatives. *Angew. Chem. Int. Ed.*, **47**, 6483–6485.

15 Ren, Q., Dai, L., Zhang, H., Tan, W., Xu, Z. and Ye, T. (2008) Total synthesis of largazole. *Synlett*, 2379–2383.

16 Numajiri, Y., Takahashi, T., Takagi, M., Shin-ya, K. and Doi, T. (2008) Total synthesis of largazole and its biological evaluation. *Synlett*, (16), 2483–2486.

17 Nasveschuk, C.G., Ungermannova, D., Liu, X. and Phillips, A.J. (2008) A concise total synthesis of largazole, solution structure, and some preliminary structure activity relationships. *Org. Lett.*, **10**, 3595–3598.

18 Ghosh, A.K. and Kulkarni, S. (2008) Enantioselective total synthesis of (+)-largazole, a potent inhibitor of histone deacetylase. *Org. Lett.*, **10**, 3907–3909.

19 Wang, B. and Forsyth, C.J. (2009) Total synthesis of largazole – devolution of a novel synthetic strategy. *Synthesis*, 2873–2880.

20 Zeng, X., Yin, B., Hu, Z., Liao, C., Liu, J., Li, S., Li, Z., Nicklaus, M.C., Zhou, G. and Jiang, S. (2010) Total synthesis and biological evaluation of largazole and derivatives with promising selectivity for cancers cells. *Org. Lett.*, **12**, 1368–1371.

21 (a) Bowers, A.A., West, N., Newkirk, T.L., Troutman-Youngman, A.E., Schreiber, S.L., Wiest, O., Bradner, J.E. and

Williams, R.M. (2009) Synthesis and histone deacetylase inhibitory activity of largazole analogs: alteration of the zinc-binding domain and macrocyclic scaffold. *Org. Lett.*, **11**, 1301–1304; (b) Bowers, A.A., Greshock, T.J., West, N., Estiu, G., Schreiber, S.L., Wiest, O., Williams, R.M. and Bradner, J.E. (2009) Synthesis and conformation–activity relationships of the peptide isosteres of FK228 and largazole. *J. Am. Chem. Soc.*, **131**, 2900–2905; (c) Ying, Y., Liu, Y., Byeon, S.R., Kim, H., Luesch, H. and Hong, J. (2008) Synthesis and activity of largazole analogs with linker and macrocycle modification. *Org. Lett.*, **10**, 4021–4024; (d) Chen, F., Gao, A.-H., Li, J. and Nan, F.-J. (2009) Synthesis and biological evaluation of C7-demethyl largazole analogs. *Chem. Med. Chem.*, **4**, 1269–1272.

22 Perez, L.J. and Faulkner, D.J. (2003) Bistratamides E-J, modified cyclic hexapeptides from the Philippines ascidian *Lissoclinum bistratum*. *J. Nat. Prod.*, **66**, 247–250.

23 Hawkins, C.J., Watters, D.J., Lavin, M.F., Parry, D.L. and McCaffrey, E.J. (1990) WO 9005731 A1.

24 Rudi, A., Chill, L., Aknin, M. and Kasman, Y. (2003) Didmolamide A and B, two new cyclic hexapeptides from the marine ascidian *Didemnum molle*. *J. Nat. Prod.*, **66**, 575–577.

25 (a) You, S.L. and Kelly, J.W. (2004) Highly efficient biomimetic total synthesis and structural verification of bistratamides E and J from *Lissoclinum bistratum*. *Chem. Eur. J.*, **10**, 71–75; (b) You, S.L. and Kelly, J.W. (2005) The total synthesis of bistratamides F-I. *Tetrahedron*, **61**, 241–249.

26 Bertram, A., Maulucci, N., New, O.M., Mohd Nor, S.M. and Pattenden, G. (2007) Synthesis of libraries of thiazole, oxazole and imidazole-based cyclic peptides from azole-based amino acids. A new synthetic approach to bistratamides and didmolamides. *Org. Biomol. Chem.*, **5**, 1541–1553.

27 (a) Bertram, A. and Pattenden, G. (2000) Self-assembly of amino acid-based thiazoles and oxazoles. Total synthesis of dendroamide A, a cyclic hexapeptide from the cyanobacterium *Stigonema dendroideum*. *Synlett*, 1519–1521; (b) Bertram, A. and Pattenden, G. (2002) Dendroamide A, nostocyclamide and related cyclopeptides from cyanobacteria. Total synthesis, together with organised and metal-templated assembly from oxazole and thiazole-based amino acids. *Heterocycles*, **58**, 521–561; (c) Bertram, A. and Pattenden, G. (2001) Synthesis and metal-templated assembly of oxazole and thiazole-based amino acids. Total synthesis of nostocyclamide and related cyclic peptides. *Synlett*, 1873–1874.

28 Romero, F., Malet, L., Cañedo, M.L., Cuevas, C. and Reyes, J. (2005) WO 2005000880 A2.

29 Kanoh, K., Matsuo, Y., Adachi, K., Imagawa, H., Nishizawa, M. and Shizuri, Y. (2005) Mechercharmycins A and B, cytotoxic substances from marine-derived *Thermoactinomyces* sp. YM3-251. *J. Antibiot.*, **58**, 289–292.

30 Hernández, D., Vilar, G., Riego, E., Cañedo, M.L., Cuevas, C., Albericio, F. and Alvarez, M. (2007) Synthesis of IB-01211, a cyclic peptide containing 2,4-concatenated thia- and oxazoles, via Hantzsch macrocyclization. *Org. Lett.*, **9**, 809–811.

31 Shin-Ya, K., Wierzba, K., Matsuo, K., Ohtani, T., Yamada, Y., Furihata, K., Hayakawa, Y. and Seto, H. (2001) Telomestatin, a novel telomerase inhibitor from *Streptomyces anulatus*. *J. Am. Chem. Soc.*, **123**, 1262–1263.

32 Shammas, M.A., Reis, R.J.S., Li C., Koley, H., Hurley, L.H., Anderson, K.C. and Munshi N.C. (2004) Telomerase inhibition and cell growth arrest after telomestatin treatment in multiple myeloma. *Clin. Cancer Res.*, **10**, 770–776.

33 (a) Yamada, S., Shigeno, K., Kitagawa, K., Okajima, S. and Asao, T. (2002) WO 200248153; (b) Deeley, J., Bertram, A. and Pattenden, G. (2008) Novel polyoxazole-based cyclopeptides from *Streptomyces* sp. Total synthesis of the cyclopeptide YM-216391 and synthetic studies towards telomestatin. *Org. Biomol. Chem.*, **6**, 1994–2010; (c) Doi, T., Yoshida, M., Shin-Ya, K. and Takahashi, T. (2006) Total synthesis of (R)-telomestatin. *Org. Lett.*, **8**, 4165–4167.

34 Endoh, N., Tsuboi, K., Kim, R., Yonezawa, Y. and Shin, C. (2003) Useful synthesis of the longer array oxazole rings for telomestatin. *Heterocycles*, **60**, 1567–1572.

35 You, S., Razavi, H. and Kelly, J.W. (2003) A biomimetic synthesis of thiazolines using hexaphenyl-oxodiphosphonium trifluoromethanesulfonate. *Angew. Chem. Int. Ed.*, **42**, 83–85.

36 (a) Anborgh, P.H. and Parmeggiani, A. (1993) Probing the reactivity of the GTP- and GDP-bound conformations of elongation factor Tu in complex with the antibiotic GE2270 A. *J. Biol. Chem.*, **268**, 24622–24628; (b) Heffron, S.E. and Jurnak, F. (2000) Structure of an EF-Tu complex with a thiazolyl peptide antibiotic determined at 2.35 Å resolution: atomic basis for GE2270A inhibition of EF-Tu. *Biochemistry*, **39**, 37–45; (c) Parmeggiani, A., Krab, I.M., Okamura, S., Nielsen, R.C., Nyborg, J. and Nissen, P. (2006) Structural basis of the action of pulvomycin and GE2270 A on elongation factor Tu. *Biochemistry*, **45**, 6846–6857.

37 (a) McConkey, G.A., Rogers, M.J. and McCutchan, T.F. (1997) Inhibition of *Plasmodium falciparum* protein synthesis targeting the plastid-like organelle with thiostrepton. *J. Biol. Chem.*, **272**, 2046–2049.; (b) Jonghee, K. (2002) PCT Int. Appl. WO 2002066046; (c) Ueno, M., Furukawa, S., Abe, F., Ushioda, M., Fujine, K., Johki, S., Hatori, H. and Ueda, H. (2004) Suppressive effect of antibiotic siomycin on antibody production. *J. Antibiot.*, **57**, 590–596.

38 Amythiamicin D: Isolation: (a) Shimanaka, K., Takahashi, Y., Iinuma, H., Naganawa, H. and Takeuchi, T. (1994) Novel antibiotics, amythiamicins. II. Structure elucidation of amythiamicin D. *J. Antibiot.*, **47**, 1145–1152; (b) Shimanaka, K., Takahashi, Y., Iinuma, H., Naganawa, H. and Takeuchi, T. (1994) Novel antibiotics, amythiamicins. III. Structure elucidations of amythiamicins A, B and C. *J. Antibiot.*, **47**, 1153–1159; (c) Shimanaka, K., Iinuma, H., Hamada, M., Ikeno, S., Tsuchiya, K.S., Arita, M. and Hori, M. (1995) Novel antibiotics, amythiamicins IV. A mutation in the elongation factor Tu gene in a resistant mutant of *B. subtilis*. *J. Antibiot.*, **48**, 182–184; Total synthesis: (d) Hughes, R.A., Thompson, S.P., Alcaraz, L. and Moody, C.J. (2005) Total synthesis of the thiopeptide antibiotic amythiamicin D. *J. Am. Chem. Soc.*, **127**, 15644–15651.

39 Thiostrepton: Isolation: (a) Anderson, B., Hodgkin, D.C. and Viswamitra, M.A. (1970) The structure of Thiostrepton. *Nature*, **225**, 233–235; Total synthesis: (b) Nicolaou, K.C., Safina, B.S., Zak, M., Estrada, A.A. and Lee, S.H. (2004) Total synthesis of thiostrepton, Part 1: construction of the dehydropiperidine/thiazoline-containing macrocycle. *Angew. Chem. Int. Ed.*, **43**, 5087–5092; (c) Nicolaou, K.C., Zak, M., Safina, B.S., Lee, S.H. and Estrada, A.A. (2004) Total synthesis of thiostrepton, Part 2: construction of the quinaldic acid macrocycle and final stages of the synthesis. *Angew. Chem., Int. Ed.*, **43**, 5092–5097.

40 Siomycin A: Isolation: (a) Nishimura, H., Okamoto, S., Mayama, M., Ohtsuka, H., Nakajima, K., Tawara, K., Shimohira, M. and Shimaoka, N. (1961) Siomycin, a new thiostrepton-like antibiotic. *J. Antibiot.*, **14**, 255–263; Total synthesis: (b) Mori, T., Higashibayashi, S., Goto, T., Kohno, M., Satouchi, Y., Shinko, K., Suzuki, K., Suzuki, S., Tohmiya, H., Hashimoto, K. and Nakata, M. (2007) Total synthesis of siomycin a. *Tetrahedron Lett.*, **48**, 1331–1335; (c) Higashibayashi, S., Hashimoto, K. and Nakata, M. (2002) Synthetic studies on the thiostrepton family of peptide antibiotics: synthesis of the tetrasubstituted dehydropiperidine and piperidine cores. *Tetrahedron Lett.*, **43**, 105–110.

41 Cyclothiazomycin: Aoki, M., Ohtsuka, T., Yamada, M., Ohba, Y., Yoshizaki, H., Yasuno, H., Sano, T., Watanabe, J., Yokose, K., Seto, H. (1991) Cyclothiazomycin, a novel polythiazole-containing peptide with renin inhibitory activity taxonomy, fermentation, isolation and physico-chemical characterization. *J. Antibiot.*, **44**, 582–588.

42 Promothiocins: Isolation: (a) Yun, B.S., Hidaka, T., Furihata, K. and Seto, H.

(1994) Promothiocins A and B. novel thiopeptides with a tip a promoter inducing activity produced by *Streptomyces* sp. SF2741. *J. Antibiot.*, **47**, 510–514; Total synthesis: (b) Bagley, M.C., Bashford, K.E., Hesketh, C.L. and Moody, C.J. (2000) Total synthesis of the thiopeptide promothiocin A. *J. Am. Chem. Soc.*, **122**, 3301–3313; (c) Moody, C.J. and Bagley, M.C. (1998) The first synthesis of promothiocin A. *Chem. Commun.*, 2049–2050.

43 Fuller, A.T. (1955) A new antibiotic of bacterial origin. *Nature*, **175**, 722–722.

44 (a) Harms, J.M., Wilson, D.N., Schluenzen, F., Connell, S.R., Stachelhaus, T., Zaborowska, Z., Spahn, C.M.T. and Fucini, P. (2008) Translational regulation via L11: molecular switches on the ribosome turned on and off by thiostrepton and micrococcin. *Mol. Cell*, **30**, 26–38; (b) Cundliffe, E. and Thompson, J. (1981) Concerning the mode of action of micrococcin upon bacterial protein synthesis. *Eur. J. Biochem.*, **118**, 47–52; (c) Rosendahl, G. and Douthwaite, S. (1994) The antibiotics micrococcin and thiostrepton interact directly with 23S rRNA nucleotides 1067A and 1095A. *Nucleic Acids Res.*, **22**, 357–363.

45 Su, T.L. (1948) Micrococcin, an antibacterial substance formed by a strain of *Micrococcis*. *Brit. J. Exp. Path.*, **29**, 473–481.

46 (a) Merritt, E.A. and Bagley, M.C. (2007) Convergent synthesis of the central heterocyclic domain of micrococcin P1. *Synlett*, 954–958; (b) Yonezawa, Y., Konn, A. and Shin, C.-G. (2004) Useful synthesis of 2,3,6-tri- and 2,3,5,6-tetrasubstituted pyridine derivatives from aspartic acid. *Heterocycles*, **63**, 2735–2746; (c) Okumura, K., Shigekuni, M., Nakamura, Y. and Shin, C.-G. (1996) Useful synthesis of 2,3,6-polythiazolesubstituted pyridine skeleton [fragment A–C] of peptide antibiotic, micrococcin P. *Chem. Lett.*, 1025–1026.

47 Lefranc, D. and Ciufolini, M.A. (2009) Total synthesis and stereochemical assignment of micrococcin P1. *Angew. Chem. Int. Ed.*, **48**, 4198–4201.

48 (a) Selva, E., Beretta, G., Montanini, N., Saddler, G.S., Gastaldo, L., Ferrari, P., Lorenzetti, R., Landini, P., Ripamonti, F., Goldstein, B.P., Berti, M., Montanaro, L. and Denaro, M. (1991) Antibiotic Ge2270 A: a novel inhibitor of bacterial protein synthesis. I. Isolation and characterization. *J. Antibiot.*, **44**, 693–701; (b) Kettenring, J., Colombo, L., Ferrari, P., Tavecchia, P., Nebuloni, M., K. Vekey, Gallo G.G. and Selva, E. (1991) Antibiotic Ge2270 A: a novel inhibitor of bacterial protein synthesis. II. Structure elucidation. *J. Antibiot.*, **44**, 702–715; (c) Selva, E., Ferrari, P., Kurz, M., Tavecchia, P., Colombo, L., Stella, S., Restelli, E., Goldstein, B.P., Ripamonti, F. and Denaro, M. (1995) Components of the GE2270 complex produced by *Planobispom rosea* ATCC 53773. *J. Antibiot.*, **48**, 1039–1042; Total synthesis: (d) Nicolaou, K.C., Zou, B., Dethe, D.H., Li, D.B., Chen, D. and Chen, Y.K. (2006) Total synthesis of antibiotics GE2270A and GE2270T. *Angew. Chem. Int. Ed.*, **45**, 7786–7792; (e) Delgado, O., Müller, H.M. and Bach, T. (2008) Concise total synthesis of the thiazolyl peptide antibiotic GE2270 A. *Chem. Eur. J.*, **14**, 2322–2339.

49 (a) Tavecchia, P., Gentili, P., Kurz, M., Sottani, C., Bonfichi, R., Selva, E., Lociuro, S., Restelli, E. and Ciabatti, R. (1995) Degradation studies of antibiotic MDL 62879 (GE2270A) and revision of the structure. *Tetrahedron*, **51**, 4867–4890; (b) Heckmann, G. and Bach, T. (2005) Synthesis of the heterocyclic core of the GE 2270 antibiotics and structure elucidation of a major degradation product. *Angew. Chem. Int. Ed.*, **44**, 1199–1201.

50 Muller, H.M., Delgado, O. and Bach, T. (2007) Total synthesis of the thiazolyl peptide GE2270 A. *Angew. Chem. Int. Ed.*, **46**, 4771–4774.

51 Kobayashi, J., Suzuki, H., Shimbo, K., Takeya, K. and Morita, H. (2001) Celogentins A–C, new antimitotic bicyclic peptides from the seeds of *Celosia argentea*. *J. Org. Chem.*, **66**, 6626–6633.

52 (a) Yuen, A.K.L., Jolliffe, K.A. and Hutton, C.A. (2006) Preparation of the central tryptophan moiety of the

celogentin/moroidin family of antimitotic cyclic peptides. *Aust. J. Chem.*, **59**, 819–826; (b) Michaux, J., Retailleau, P. and Campagne, J.-M. (2008) Synthesis of the central tryptophan-leucine residue of celogentin C. *Synlett*, 1532–1536.

53 Ma, B., Banerjee, B., Litvinor, D.N., He, L. and Castle, S.L. (2010) Total synthesis of the antimitotic bicyclic peptide celogentin C. *J. Am. Chem. Soc.*, **132**, 1159–1171.

54 For Knoevenagel condensation: (a) Zhang, Y., Wada, T. and Sasabe, H. (1996) Synthesis of a new carbazole cyclic dimer *via* knoevenagel condensation. *Tetrahedron Lett.*, **37**, 5909–5912; (b) Chrétien, F., Khaldi, M. and Chapleur, Y. (1997) A unified strategy for the synthesis of enantiomerically pure branched-chain cyclohexenones and cyclopentenones from a single progenitor. *Tetrahedron Lett.*, **38**, 5977–5980; (c) Srikanth, G.S.C. and Castle, S.L. (2004) Synthesis of β-substituted α-amino acids *via* Lewis acid promoted radical conjugate additions to α,β-unsaturated α-nitro esters and amides. *Org. Lett.*, **6**, 449–452; (d) Fornicola, R., Oblinger, E. and Montgomery, J. (1998) New synthesis of α-amino acid derivatives employing methyl nitroacetate as a versatile glycine template. *J. Org. Chem.*, **63**, 3528–3529; For radical conjugate addition see: (e) Srikanth, G.S.C. and Castle, S.L. (2005) Advances in radical conjugate additions. *Tetrahedron*, **61**, 10377–10441; For nitro reduction see: (f) Kende, A.S. and Mendoza, J.S. (1991) Controlled reduction of nitroalkanes to alkyl hydroxylamines or amines by samarium diiodide. *Tetrahedron Lett.*, **32**, 1699–1702; (g) Sturgess, M.A. and Yarberry, D.J. (1993) Rapid stereoselective reduction of thermally labile 2-aminonitroalkanes. *Tetrahedron Lett.*, **34**, 4743–4746; (h) Larock, R.C. and Yum, E.K. (1991) Synthesis of indoles *via* palladium-catalyzed heteroannulation of internal alkynes. *J. Am. Chem. Soc.*, **113**, 6689–6690; (i) Larock, R.C., Yum, E.K. and Refvik, M.D. (1998) Synthesis of 2,3-disubstituted indoles *via* palladium-catalyzed annulation of internal alkynes. *J. Org. Chem.*, **63**, 7652–7662.

55 Braslau, R., erson, M.O., Rivera, F., Jimerez, A., Haddad, T. and Axon, J.R. (2002) Acyl hydrazines as precursors to acyl radicals. *Tetrahedron*, **58**, 5513–5523.

56 (a) Manjunatha, S.G., Chittori, P. and Rajappa, S. (1991) Nitroacetyl group as a peptide synthon: synthesis of Dipeptides with an α,α-bisallylglycine residue at the N-terminus. *Helv. Chim. Acta*, **74**, 1071–1080; (b) Manjunatha, S.G., Reddy, K.V. and Rajappa, S. (1990) Nitroketene-S,N-acetals as precursors for nitroacetamides and the elusive nitrothioacetahides. *Tetrahedron Lett.*, **31**, 1327–1330.

57 (a) Kaneko, I., Kamoshida, K. and Takahashi, S. (1989) Complestatin, a potent anti-complement substance produced by *Streptomyces lavendulae*. I. Fermentation, isolation and biological characterization. *J. Antibiot.*, **42**, 236–241; (b) Kaneko, I., Fearon, D.T. and Ausfen, K.F. (1980) Inhibition of the alternative pathway of human complement *in vitro* by a natural microbial product, complestatin. *J. Immunol.*, **124**, 1194–1198.

58 Seto, H., Fujioka, T., Furihana, K., Kaneko, I. and Takahashi, S. (1989) Structure of complestatin, a very strong inhibitor of protease activity of complement in the human complement system. *Tetrahedron Lett.*, **30**, 4987–4990.

59 Momota, K., Kaneko, I., Kimura, S., Mitamura, K. and Shimada, K. (1991) Inhibition of human immunodeficiency virus type-1-induced syncytium formation and cytopathicity by complestatin. *Biochem. Biophys. Res. Commun.*, **179**, 243–250.

60 Garfunkle, J., Kimball, F.S., Trzupek, J.D., Takizawa, S., Shimamura, H., Tomishima, M. and Boger, D.L. (2009) Total synthesis of chloropeptin II (complestatin) and chloropeptin I. *J. Am. Chem. Soc.*, **131**, 16036–16038.

61 Miyaura, N. and Suzuki, A. (1995) Palladium-catalyzed cross-coupling reactions of organoboron compounds. *Chem. Rev.*, **95**, 2457–2483.

62 (a) Searle, P.A. and Molinski, T.F. (1995) Phorboxazoles A and B: potent cytostatic macrolides from marine sponge *Phorbas* species. *J. Am. Chem. Soc.*, **117**,

8126–8131; (b) Searle, P.A., Molinski, T.F., Brzezinski, L.J. and Leahy, J.W. (1996) Absolute configuration of phorboxazole A and B from the marine sponge *Phorbas* sp. 1. Macrolide and hemiketal rings. *J. Am. Chem. Soc.*, **118**, 9422–9423; (c) Molinski, T.F. (1996) Absolute configuration of phorboxazoles A and B from the marine sponge, *Phorbas* sp. 2. C43 and complete stereochemistry. *Tetrahedron Lett.*, **37**, 7879–7880; (d) Capon, R.J., Skene, C., Liu, E.H., Lacey, E., Gill, J.H., Heiland, K. and Friedel, T. (2004) Esmodil: an acetylcholine mimetic resurfaces in a southern australian marine sponge *Raspailia* (raspailia) sp. *Nat. Prod. Res.*, **18**, 305–309.

63 Smith, A.B., Razler, T.M., Ciavarri, J.P., Hirose, T., Ishikawa, T. and Meis, R.M. (2008) A second-generation total synthesis of (+)-phorboxazole A. *J. Org. Chem.*, **73**, 1192–1200.

64 (a) Lucas, B.S., Gopalsamuthiram, V. and Burke, S.D. (2007) Total synthesis of phorboxazole B. *Angew. Chem. Int. Ed.*, **46**, 769–772; (b) Li, D.-R., Zhang, D.-H., Sun, C.-Y., Zhang, J.-W., Yang, L., Chen, J., Liu, B., Su, C., Zhou, W.-S. and Lin, G.-Q. (2006) Total synthesis of phorboxazole B. *Chem. Eur. J.*, **12**, 1185–1204; (c) Evans, D.A., Fitch, D.M., Smith, T.E. and Cee, V.J. (2000) Application of complex aldol reactions to the total synthesis of phorboxazole B. *J. Am. Chem. Soc.*, **122**, 10033–10046; (d) Evans, D.A. and Fitch, D.M. (2000) Asymmetric synthesis of phorboxazole B, Part II: synthesis of the C1-C19 subunit and fragment assembly. *Angew. Chem. Int. Ed.*, **39**, 2536–2540; (e) Lucas, B.S., Luther, L.M. and Burke, S.D. (2004) Synthesis of the C1–C17 segment of phorboxazole B. *Org. Lett.*, **6**, 2965–2968; (f) Paterson, I., Steven, A. and Luckhurst, C.A. (2004) Phorboxazole B synthetic studies: construction of C(1–32) and C(33–46) subtargets. *Org. Biomol. Chem.*, **2**, 3026–3038; (g) Greer, P.B. and Donaldson, W.A. (2002) Synthetic studies directed toward the phorboxazoles: preparation of the C3–C15 bisoxane segment and two stereoisomers. *Tetrahedron*, **58**, 6009–6018.

65 Williams, D.R., Kiryanor, A.A., Emde, U., Clark, M.P., Berliner, M.A. and Reeves, J.T. (2004) Studies of stereocontrolled allylation reactions for the total synthesis of phorboxazole A. *Prod. Natl. Acad. Sci. U.S.A.*, **101**, 12058–12063.

66 Jansen, R., Wray, V., Irschik, H., Reichenbach, H. and Hofle, G. (1985) Isolation and spectroscopic structure elucidation of sorangicin a, a new type of macrolide-polyether antibiotic from gliding bacteria – XXX. *Tetrahedron Lett.*, **26**, 6031–6034.

67 (a) Irschik, H., Jansen, R., Gerth, K., Hofle, G. and Reichenbach, H. (1987) Sorangicins, novel and powerful inhibitors of eubacterial rna polymerase isolated from myxobacteria. *J. Antibiot.*, **40**, 7–13; (b) Campbell, E.A., Pavlova, O., Zenkin, N., Leon, F., Irschik, H., Jansen, R., Severinov, K. and Darst, S.A. (2005) Structural, functional, and genetic analysis of sorangicin inhibition of bacterial RNA polymerase. *EMBO J.*, **24**, 674–682. And references cited therein.

68 Smith, A.B., III, Dong, S., Brenneman, J.B. and Fox, R.J. (2009) Total synthesis of (+)-sorangicin A. *J. Am. Chem. Soc.*, **131**, 12109–12111.

69 (a) Bode, H.B. and Zeeck, A. (2000) Structure and biosynthesis of kendomycin, a carbocyclic *ansa*-compound from *Streptomyces*. *J. Chem. Soc., Perkin Trans. 1*, 323–328; (b) Bode, H.B. and Zeeck, A. (2000) Biosynthesis of kendomycin: origin of the oxygen atoms and further investigations. *J. Chem. Soc. Perkin Trans. 1*, 2665–2670.

70 Smith, A.B., III, Mesaros, E.F. and Meyer, E.A. (2006) Evolution of a total synthesis of (−)-kendomycin exploiting a petasis-ferrier rearrangement/ring-closing olefin metathesis strategy. *J. Am. Chem. Soc.*, **128**, 5292–5299.

71 Pettit, G.R. (1991) No. 57, The bryostatins, in *Progress in the Chemistry of Organic Natural Products* (eds W. Herz, G.W. Kirby, W. Steglich and C. Tamm), Springer-Verlag, New York, USA, pp. 153–195.

72 Blanchette Rockefeller NeuroSciences Institute (2008) Safety, efficacy, pharmacokinetics, and pharmacodynamics study of bryostatin 1

in patients with Alzheimer's disease. http://clinicaltrials.gov (accessed 11 January 2011).

73 Pettit, G.R., Gao, F., Blumberg, P.M., Herald, C.L., Coll, J.C., Kamano, Y., Lewin, N.E., Schmidt, J.M. and Chapuis, J.-C. (1996) Antineoplastic agents. 340. Isolation, structural elucidation of bryostatins 16–18. *J. Nat. Prod.*, **59**, 286–289.

74 Trost, B.M. and Dong, G. (2008) Total synthesis of bryostatin 16 using atom-economical, chemoselective approaches. *Nature*, **456/27**, 485–488.

75 (a) Minami, Y., Yoshida, K., Azuma, R., Nishii, M., Inagaki, J. and Nohara, F. (1992) Structure of a novel macrodiolide antibiotic IKD-8344. *Tetrahedron Lett.*, **33**, 7373–7376; (b) Ishida, T., In, Y., Nishii, M. and Minami, Y. (1994) Stereochemistry of a novel macrodiolide antibiotic IKD-8344. *Chem. Lett.*, 1321–1322; (c) Hwang, E.I., Yun, B.S., Yeo, W.H., Lee, S.H., Moon, J.S., Kim, Y.K., Lim, S.J. and Kim, S.U. (2005) Compound IKD-8344, a selective growth inhibitor against the mycelial form of *Candida albicans*, isolated from *Streptomyces* sp. A6792. *J. Microbiol. Biotechnol.*, **15**, 909–912.

76 Kim, W.H., Hong, S.K., Lim, S.M., Ju M.-A., Jung S.K., Kim, Y.W., Jung, J.H., Kwon, M.S. and Lee E. (2006) Total synthesis of IKD-8344. *Angew. Chem. Int. Ed.*, **45**, 7072–7075.

77 Dorta, E., Diaz-Marrero A.R., Brito I., Cueto M., D'Croz L. and Darias J. (2007) The oxidation profile at C-18 of furanocembranolides may provide a taxonomical marker for several genera of octocorals. *Tetrahedron*, **63**, 9057–9062.

78 Donohoe, T.J., Ironmonger, A. and Kershaw, N.M. (2008) Synthesis of (−)-(Z)-deoxypukalide. *Angew. Chem. Int. Ed.*, **47**, 7314–7316.

79 (a) Uemura, D., Takahashi, K., Yamamoto, T., Katayama, C., Tanaka, J., Okumura, Y. and Hirata, Y. (1985) Norhalichondrin A: an antitumor polyether macrolide from a marine sponge. *J. Am. Chem. Soc.*, **107**, 4796–4798; (b) Hirata, Y. and Uemura, D. (1986) Halichondrins-antitumor polyether macrolides from a marine sponge. *Pure Appl. Chem.*, **58**, 701–710; (c) Pettit, G.R., Tan, R., Gao, F., Williams, M.D., Doubek, D.L., Boyd, M.R., Schmidt, J.M., Chapuis, J.C. and Hamel, E. (1993) Isolation and structure of halistatin 1 from the Eastern Indian ocean marine sponge *Phakellia carteri*. *J. Org. Chem.*, **58**, 2538–2543; (d) Litaudon, M., Hickford, S.J.H., Lill, R.E., Lake, R.J., Blunt, J.W. and Munro, M.H.G. (1997) Antitumor polyether macrolides: new and hemisynthetic halichondrins from the New Zealand deep-water sponge *Lissodendoryx* sp. *J. Org. Chem.*, **62**, 1868–1871.

80 (a) Yu, M.J., Kishi, Y. and Littlefield, B.A. (2005) *Anticancer Agents from Natural Products* (eds G.M. Cragg, D.G.I. Kingston and D.J. Newman), CRC, Boca Raton, FL, USA; (b) Newman, S. (2007) Eribulin, a simplified ketone analog of the tubulin inhibitor halichondrin B, for the potential treatment of cancer. *Curr. Opin. Invest. Drugs*, **8**, 1057–1066.

81 Jackson, K.L., Henderson, J.A., Motoyoshi, H. and Phillips, A.J. (2009) A total synthesis of norhalichondrin B. *Angew. Chem. Int. Ed.*, **48**, 2346–2350.

82 Miller, E.D., Kauffman, C.A., Jensen, P.R. and Fenical, W. (2007) Piperazimycins: cytotoxic hexadepsipeptides from a marine-derived bacterium of the genus *Streptomyces*. *J. Org. Chem.*, **72**, 323–330.

83 Li, W., Gan, J. and Ma, D. (2009) Total synthesis of piperazimycin A: a cytotoxic cyclic hexadepsipeptide. *Angew. Chem. Int. Ed.*, **48**, 8891–8895.

84 Maehr, H., Liu, C., Palleroni, N.J., Smallheer, J., Todaro, L., Williams, T.H. and Blount, J.F. (1986) Microbial products VIII. Azinothricin, a novel hexadepsipeptide antibiotic. *J. Antibiot.*, **39**, 17–25.

85 Maskey, R.P., Fotso, S., Sevvana, M., Uson, I., Grun-Wollny, I. and Laatsch, H. (2006) Kettapeptin: isolation, structure elucidation and activity of a new hexadepsipeptide antibiotic from a terrestrial *Streptomyces* sp. *J. Antibiot.*, **59**, 309–314.

86 Hale, K.J., Manaviazar, S., George, J.H., Walters, M.A. and Dalby, S.M. (2009) Total synthesis of (+)-azinothricin and

(+)-kettapeptin. *Org. Lett.*, **11**, 733–736 and references cited therein.
87 (a) Hale, K.J. and Cai, J. (1997) Asymmetric total synthesis of antitumour antibiotic A83586C. *Chem. Commun.*, 2319–2320; (b) Hale, K.J. and Cai, J. (1996) Synthetic studies on the azinothricin family of antitumour antibiotics. 5. Asymmetric synthesis of two activated esters for the northern sector of A83586C. *Tetrahedron Lett.*, **37**, 4233–4236.
88 (a) Hale, K.J., Manaviazar, S., Lazarides, L., George, J., Walters, M.A., Cai, J., Delisser, V.M., Bhatia, G.S., Peak, S.A., Dalby, S.M., Lefranc, A., Chen, Y.-N.P., Wood, A.W., Crowe, P., Erwin, P. and El-Tanani, M. (2009) Synthesis of A83586C analogs with potent anticancer and β-catenin/tcf4/osteopontin inhibitory effects and insights into how A83586C modulates E2Fs and pRb. *Org. Lett.*, **11**, 737–740.
89 (a) Kobayashi, J., Watanabe, D., Kawasaki, N. and Tsuda, M. (1997) Nakadomarin A, a novel hexacyclic manzamine-related alkaloid from *Amphimedon* sponge. *J. Org. Chem.*, **62**, 9236–9239; (b) Kobayashi, J., Tsuda, M. and Ishibashi, M. (1999) Bioactive products from marine micro- and macro-organisms. *Pure Appl. Chem.*, **71**, 1123–1126.
90 (a) Ono, K., Nakagawa, M. and Nishida, A. (2004) Asymmetric total synthesis of (−)-nakadomarin A. *Angew. Chem. Int. Ed.*, **43**, 2020–2023; (b) Jakubec, P., Cockfield, D.M. and Dixon, D.J. (2009) Total synthesis of (−)-nakadomarin A. *J. Am. Chem. Soc.*, **131**, 16632–16633.
91 (a) Ye, J., Dixon, J.D. and Hyres, P.S. (2005) Highly enantioselective Michael addition of malonate esters to nitro olefins using a cinchonine-derived bifunctional organic catalyst. *Chem. Commun.*, 4481–4483; (b) McCooey, S.H. and Connon, S.J. (2005) Urea- and thiourea-substituted cinchona alkaloid derivatives as highly efficient bifunctional organocatalysts for the asymmetric addition of malonate to nitroalkenes: inversion of configuration at C9 dramatically improves catalyst performance. *Angew. Chem. Int. Ed.*, **44**, 6367–6370; (c) Okina, T., Hoashi, Y. and Takemoto, Y. (2003) Enantioselective michael reaction of nalonates to nitroolefins catalyzed by bifunctional organocatalysts. *J. Am. Chem. Soc.*, **125**, 12672–12673; (d) Jakubec, P., Halliwell, M. and Dixon, D.J. (2008) Cyclic imine nitro-mannich/lactamization cascades: a direct stereoselective synthesis of multicyclic piperidinone derivatives. *Org. Lett.*, **10**, 4267–4270.

Index

a

acanthamide A 188
acetylenic thiophene
– Rossi synthesis 386
Achmatowicz oxidation 172, 602
– furan ring 113ff.
acridine 358ff.
acridone 358ff.
acridone alkaloid 358
acromelic acid A 26
acronycine 359ff.
actinomycin D 20
actinophyllic acid 222ff.
– Overman's strategy 225f.
N-acyl aziridines 28
2-acyloxymethyl thiazole 483
aflatoxin B_1 136
aflatoxin B_2 127
– [3 + 2] cycloaddition with quinones 127
Agami–Couty N-debenzylation 283
agelastatin 200
– (–)-agelastatin 203
agelastatin A 202
– (–)-agelastatin A 18
– (+)-agelastatin A 19f.
– Trost synthesis 203
ageliferin 521
agesamide 18, 520f.
agesamide A 18
agesamide B 18
ajudazol A 408
ajudazol B 408
aldisine 555
alkaloid 221
– dipolar cycloaddition 126f.
– pyrrolidine-containing 13
alkaloid (–)-205B 14
alkylation
– thiazole 483

alkylideneglutamic acid 15
2-alkylpyridine 278
3-alkylpyridine 281
3-alkylpyridinium 281
3-alkyltetrahydropyridine compound 281
alliacol 119
allosecurinine 118ff.
π-allyl palladium complex 43
AM-2282 348
ambuic acid 65
– (+)-ambuic acid 70
– synthesis 70
amicetin 509
amide
– thionation 468
amide thiol
– hexaphenyloxodiphosphonium trifluoromethanesulfonate-mediated cyclization 466
– Mb(IV)-mediated cyclization 466
– Ti(IV)-mediated cyclization 464f.
– Ti(IV)-mediated cyclization (Heathcock) 464
– Ti(IV)-mediated cyclization (Kelly's modification) 465
aminacrine 359
9-aminoacridine 359
amino alcohol
– thiazoline from vicinal amino alcohol 468
– vicinal 468
amoxicillin 570
amphimedine 359ff.
amphimedoside A 269
amsacrine (m-AMSA) 359ff.
amythiamicin 484
amythiamicin D 476ff., 487, 581f.
AN-1609 155
anabaseine 269

Heterocycles in Natural Product Synthesis, First Edition. Edited by Krishna C. Majumdar and Shital K. Chattopadhyay.
© 2011 Wiley-VCH Verlag GmbH & Co. KGaA. Published 2011 by Wiley-VCH Verlag GmbH & Co. KGaA.

anabasine   269
anastrephin   139
– (–)-anastrephin   140
angelmarin   100
anhydrochantacin   131f.
anhydrolycorinone   129f.
14,15-anhydropristinamycin II$_B$ 418
(–)-anisomycin   28
annularin H   119
ansatrienol   124
anthrathiophene
– Kelly synthesis   386
anthrax tetrasaccharide   115
antiostatin   353
– retrosynthetic analysis   353
antiostatin A$_1$ 342
antiostatin A$_2$ 342
antiostatin A$_3$ 342
antiostatin A$_4$ 342
antiostatin B$_2$ 342ff.
antiostatin B$_3$ 342
antiostatin B$_4$ 342
antiostatin B$_5$ 342
antipathine   550
(–)-aplaminal   551ff.
aplidinone A   383
apratoxin   486
apratoxin A   467, 569ff.
– Forsyth's total synthesis   571
– Kelly's procedure   571
– retrosynthetic analysis   570
arborinine   359
archazolid   480ff.
archazolid A   491
archazolid B   491
archazolid C   491
archazolid D   491
arcyriaflavin A   342, 357
arcyriaflavin B   342
arcyriaflavin C   342
arcyriaflavin D   342
arcyriarubin A   357f.
arenastatin A   64
– synthesis   70
arenosclerin A   269
argemonine   301
argyrin B   481ff.
aristeromycin   117
aristocularine alkaloid family   548
aristoyagonine   540ff.
arylpyrrole   193
asiminocin   100
asmarine A   552
aspergillide B   154

aspergillide C   154
Aspidosperma alkaloid   126
aspirochlorine   383, 395
ataphyllidine   359
ataphylline   359
atebrine   360
axinellamine A   520f.
aza-Cope–Mannich rearrangement   226
aza-Diels-Alder dimerization   585
aza-Wittig reaction   474, 572
azepine   537, 549ff.
azetidin-2-one
– structural description   50
– synthetic methodologies for the formation   50
azetidine   41ff.
– derivative   41ff.
– γ-hydroxymethyl-substituted   43
– structural description   41
azetidine ring   43
L-azetidine-2-carboxylic acid   45
azimic acid   116
azinothricin   607
– retrosynthetic analysis   607
– total synthesis   608
aziridine
– [3 + 2] cycloaddition   26
– intramolecular [3 + 2] cycloaddition   25
– iodide-mediated rearrangement   27
– nucleophilic ring-opening for natural product synthesis   10
– [2,3]-Wittig rearrangement   27
aziridine-2,3-dicarboxylic acid   3ff.
– (2S,3S)-(+)-aziridine-2,3-dicarboxylic acid   75
azole   569
azomycin   519

b

bacimethrin   508f.
bacitracin A   486ff.
– *ent*-bacitracin A   491
Baeyer–Villiger reaction   197
(–)-balanol   19ff.
balsoxin   406
banminth   379
Baran's synthesis
– psychotrimine   247
Barrish–Singh procedure   412ff.
batrachotoxin   561
(±)-batrachotoxinin A   560f.
bauhiniastatin 1   547
bauhinoxepin A   547
bauhinoxepin J   540

beetle sex pheromone   132
bengazole A   437f.
benzo[c]phenanthridine   331
berberine   301
(–)-berkelic acid   168
berlambine   301
berninamycin A   425
bestatin   24
bhimamycin B   101
(+)-biotin   381ff.
– Baggiolini synthesis   393
– Goldberg and Sternbach synthesis   392
bipinnatin J   109
– (–)-bipinnatin J   135
bipyrrole
– Gribble synthesis   204
2,2′-bipyrrole   204
– Gribble synthesis   204
bipyrrole aldehyde   207
2,2′-bipyrrole aldehyde   206
Birch reduction
– pyrrole dicarboxylate   210
Birman approach   200
Bischler–Napieralsky condensation   326
Bischler–Napieralsky condensation/
    dehydrogenation procedure   315
Bischler–Napieralsky cyclization   323
bis(dithiepanethione)   239
[bis(2-methoxyethyl)amino]-sulfur trifluoride
    (Deoxo-Fluor™)   473f.
bis-oxazole   420, 439
– conjugated   429
bistramide A   158
bistratamide   428, 467ff.
bistratamide A   487
bistratamide B   487
bistratamide C   487
bistratamide D   481ff.
bistratamide E   487, 575
bistratamide F   487, 575
bistratamide G   487, 575
bistratamide H   487, 575ff.
bistratamide I   487, 575
bistratamide J   487, 575
blasticidin-S   509
bleomycin   509
bleomycin A_2 480ff., 492f.
(–)-blepharocalyxin D   171
Boger and Patel route
– bipyrrole aldehyde   207
Boger synthesis
– isochrysohermidine   213
– prodigiosin   206
Bohlmann–Rahtz pyridine synthesis   406

(+)-bourgeanic acid   74
(+)-bractazonine   111
branimycin   139
Brassard diene   591
brefeldin A   137f.
brevetoxin A   539ff.
breynin A   381
breynolide
– Smith synthesis   394
3-bromoindoline   238
2-(bromomethyl)-propenamide   51
(+)-bromoxone   24
bryoanthrathiophene   380
bryostatin 16   161, 597
– retrosynthetic analysis   597
– total synthesis   598
Bucherer–Bergs hydantoin formation   520
Buchwald–Hartwig amination   341ff.
Burgess cyclodehydration   419
Burgess dehydration-oxidation method   446
Burgess reagent   473
– polyethylene glycol-linked (PEG)   472ff.
(+)-burseran   112
5-(3-buten-1-ynyl)-2,2′-bithienyl   378ff.
N-t-butoxycarbonyl α-hydroxy
    pyrrolidine   48
butylcycloheptyl prodigiosin   207
– Reeves' synthesis   208

c
β-caboline   314
caerulomycin C   269ff.
calicheamicin   23
calothrixin A   343
calothrixin B   343
calyculin   410
calyculin A   410
calyculin B   410
calyculin C   410
calyculin D   410
calyculin E   410
calyculin F   410
calyculin G   410
calyculin H   410
camptothecine   301
canadine   301, 327
– (+)-canadine   328
candicanoside A   539ff.
cannabichromeorcinic acid   169
cannabinodiol   170
cannabinol   170
cantharidin   101
L-epi-capreomycidin   16
carazostatin   343

carbalexin C   343
carbazole   341ff.
carbazole alkaloid   341, 353
carbazole synthesis
– Fe-mediated   350ff.
– Pd-catalyzed   350
carbazomadurin A   343
carbazomadurin B   343
carbazomycin A   344
carbazomycin B   344
carbazomycin C   344
carbazomycin D   344
carbazomycin E   344
carbazomycin F   344
carbazomycin G   344
carbazomycin H   344
β-carboline   317
2-carboxamide-aziridine   23
2-carboxylate-N-nosyl aziridine   23
cardinalin 3   154
carquinostatin A   344ff.
(+)-carvone   167
(–)-castoramine   12
α-cedrene   30
celogentin C   587
– retrosynthetic analysis   587
– total synthesis   588
centrolobine   154
cephalosporin
– synthesis   54ff.
cephalosporin C   42, 54ff.
cephalostatin   101
ceratospongamide   476ff., 487
– cis,cis-ceratospongamide   476
cercosporin   541
(+)-chelidonine
– synthesis   76
Chida cyclization   202
(+)-chimonanthine   237
chivosazole   416
(–)-chloramphenicol   21
7-chloroarctinone-b   382
chlorodysinosin A   24f.
chlorohyellazole   344
chloropentostatin   557
chloropeptin II   589
cinchonine   301
circumdatin C   558
circumdatin F   558
cladoacetal A   540
clausenol   344
clausine I   345
clausine K   345
clausine L   345ff.

clausine Z   345
clauszoline-J   345
clavizepine   551ff.
(–)-cleistenolide
– synthesis   80
Clemmensen reduction   358
clotrimazole   519
codeine   302
(–)-codonopsinine
– synthesis   77
Cognex   360
Comin's reagent   594
complestatin (chloropeptin II)   589
– retrosynthetic analysis   589
– total synthesis   590
concavine   560
condensation
– vicinal amino thiol with a
  nitrile   462
– vicinal amino thiol with an iminium
  triflate   463
– vicinal amino thiol with an iminoether
  462f.
coniine   269, 285
– (–)-coniine   286f.
– (S)-coniine   13, 286
Corey-Bakshi-Sibata reagent   214
coriolin   65
cortamidine oxide   269
(+)-cortistatin A   165
CP-263,114   101
crambescidin   508
crambescidin 359   539ff.
crambescin   508ff.
cribochaline   281
cribochaline A   269
cribrostatin   520ff.
(±)-crinane   28f.
crisamicin A   170
(+)-croomine   31f.
cryptophycin 1   67
(–)-cubebol
– synthesis   79
cularinoid family   548
curacin A   470ff., 488
Curtius rearrangement   313
cyclic polyheterocyclic metabolites
  containing single oxazole
  residues   426
cyclization
– hexaphenyloxodiphosphonium
  trifluoromethanesulfonate-mediated
  cyclization of amide thiol   466
– intramolecular   43

- Mb(IV)-mediated cyclization of amide thiol 466
- thioester 467
- Ti(IV)-mediated cyclization of amide thiol 464f.
cycloaddition reaction 25
- dipolar 126f.
- furan 124
[2 + 1] cycloaddition
- synthesis of natural product 124f.
[2 + 2] cycloaddition 124
- enantioselective 43
- intermolecular 43
- metal-catalyzed 43
- synthesis of natural product 125
[3 + 2] cycloaddition 124
- aflatoxin B_2 127
- intramolecular 25
[4 + 2] cycloaddition 127ff.
[5 + 2] cycloaddition 134f., 168
[6 + 4] cycloaddition 134
- intramolecular 133
[8 + 2] cycloaddition 134
cyclodehydration 472
- β-hydroxy thioamide 473f.
cyclodidemnamide 476, 487
cyclohaliclonamine A 270
cyclooroidin 18
- (S)-(−)-cyclooriodin 200
cyclopenin 552
cyclopentadiene[c]pyrrole-1,3-diol 188
(±)-cyclophellitol 128
cyclostellettamine A 270
cyclothiazomycin 582
[2 + 2 + 2] cyclotrimerization 169
- Ru-catalyzed 169
cylindradine A 188
cylindrospermopsin 510
cystothiazole 479ff.
cystothiazole A 489
- (+)-cystothiazole A 464
cystothiazole B 489
cytisine 270
- (−)-cytisine 287
- (±)-cytisine 287f.

d

daminin 188
DAST 473
daurichromenic acid 169
dearomatization reaction 137
decarestricine J
- synthesis 80

dehydrogenation
- activated MnO_2 476
- chemical MnO_2 (CMD) 477
- Williams' method 477f.
N-demethyl mesembrine 26
dendroamide A 426, 467ff., 481ff.
deoxocassine 114ff.
Deoxo-Fluor™ 473f.
2′-deoxymugineic acid 42ff.
(−)-deoxynupharidine 12
deoxypukalide 599f.
- (−)-Z-deoxypukalide 109f.
- retrosynthetic analysis 600
- total synthesis 601
depentylperhydrohistrionicotoxin 31
depentylperhydrohistrionicotoxin derivative 31
(±)-desoxyeseroline 15
Dess–Martin oxidation 233, 422ff., 439, 596
Dess–Martin periodinane 599
2,4-diaminopteridine 388
diarylheptanoid 171
1,7-diazaanthraquinone framework 364
diazonamide A 433ff.
dibenz[b,f]oxepine 546
dibenzopyranazepine alkaloid 557
dibromoagelaspongine 200
dibromoisophakellin 200
dibromophakellin 200
2,4-dibromothiazole 485ff.
dictamnine 302
dictyoxetane
- dioxatricyclic segment 86
didehydromirabazole 480
didehydromirabazole A 494
11,11′-dideoxyverticillin 223
- Movassaghi 237
(+)-11,11′-dideoxyverticillin 237ff.
- Movassaghi's biomimetic synthesis 238
didmolamide 467ff.
didmolamide A 487, 575ff.
didmolamide B 487, 575
Diels–Alder reaction 127ff.
- inverse electron demand 196, 313
- synthesis of actinorhodin monomeric unit 128
Diels–Alder-retro Diels–Alder strategy 212
(6S)-5,6-dihydro-6-([2R]-2-hydroxy-6-phenylhexyl)-2H-pyran-2-one
- synthesis 74
dihydroclerodin 120
- (+)-dihydroclerodin 118

(−)-dihydrocodeinone 321
(S)-dihydrokavain
– synthesis 82
(−)-dihydropinidine 11f.
dihydrosesamin
– synthesis 77
trans-dihydroxygirinimbine 345
(±)-dihydroxyheliotridane 26
5,7-dihydroxy-1-methoxycarbonyl-6-oxo-6*H*-anthra[1,9-*bc*]thiophene 380
dimethyldioxirane (DMDO) 122
diospongin B 154
dipyrrolic natural product 203
discorhabdin A 381
– Kita synthesis 391
discorhabdin B 381
discorhabdin D 381
disorazole C_1 420f.
disparlure 65
dodoneine
– synthesis 71
dolastatin E 429, 487
dolastatin I 429
dragmacidin 521ff.
dragmacidin D 227
dragmacidin F 223ff.
– (+)-dragmacidin F 227ff., 524ff.
– Stoltz synthesis 226f.
(Z)-dysidazirine 4f.
– synthesis 5
dysiherbaine 155

e
ebrifugine 514
echinothiophene 380
echinoynethiophene A 380
electron-transfer-initiated cyclization (ETIC) method 163
electrophile
– furan derivative 118ff.
eleutherobin 101ff.
ellipticine 345ff.
– Ho's and Hsieh's synthesis 356
emetine 302, 320
– (−)-emetine 322ff.
enyne metathesis 194
(±)-1-epiaustraline 210f.
epibatidine 270
– Simpkins' total synthesis 211
epibreynin B 382
epicalyxin F 170f.
– (+)-epicalyxin F 171
epicoccin I 382
epidithiapiperazine-2,5-dione 395

(−)-12-*epi*-fisherindole I 235ff.
– Baran's synthesis 236
epi-indolizidine 223A 119
epi-oxetin
– synthesis 85
8a-epi-swainsonine 119
epocarbazolin A 67, 345
– total synthesis 71
epocarbazolin B 345
epolactaene 67
epothilone 480ff.
epoxide 63ff.
– natural product synthesis 63
epoxomicin 66
epoxyquinols 128
(+)-11,12-epoxysarcophytol A 67
– synthesis 68
erogorgiaene
– synthesis 88
erysotramidine 302
erythraline 302
D-*erythro*-sphingosine 19
Z-4-ethylene (2S)-glutamic acid 15
ethyloxirane 480
euchrestifoline 345ff.
eustifoline-A 346
eustifoline-B 346
eustifoline-C 346
eustifoline-D 346
ezomycin 510

f
fawcettidine 553
fawcettimine 550ff.
febrifugine 510ff., 514–517
feldamycin 16f.
Feldman annulations 203
Fischer–Borsche synthesis 341
foetithiophene A 379
formamicin 539ff.
formamicinone 544
2-formylpyrrole
– 4,5-disubstituted 207
FR-66979 4ff.
(−)-FR182877 173f.
FR-900482 4ff.
FR900490 16f.
FR901464 155
Friedel–Crafts acylation
– intramolecular 361
Friedlander cyclization 315
funebrine 188
furan 99ff., 590
– cycloaddition 124

furan derivative   99ff.
– chelator   140
– chiral auxiliary   140
– electrophiles   118ff.
– nucleophiles   118ff.
– reduction   112
furan ring
– Achmatowicz oxidation   113ff.
– natural product   100
– natural product synthesis   116f.
furoclausine A   346
furoscrobiculin   101
furostifoline   346
Fürstner pyrrole synthesis   195
Fürstner synthesis
– roseophilin   209
– streptorubin B   195
(R)-furyl carbinol   114
(S)-furyl carbinol   114
fuzanin C   270

g

Garner's acid   408
Garner's aldehyde   47
GE2270   479ff., 581ff.
– retrosynthetic analysis   585
– total synthesis   586
GE2270A   484, 582ff.
– Bach's retrosynthetic analysis   586
GE2270T   582ff.
gelsemine   166
germacradienol   109
– (–)-1-(10),5-germacradien-4-ol   111
gerrardine   383
GEX1A   155
ginkgolide B   101, 125f.
girinimbine   346ff.
glaciapyrrole A   188
glaucine   302
gliocladine A   385
(–)-gloeosporone
– synthesis   82
glycomaurin   346
glycomaurrol   346
glyfoline   360
goniofufurone   101
Graebe–Ullmann synthesis   341
grandisine A   175
(+)-grandisol   125
granulatin   541
grassypeptolide   487
Gribble synthesis
– bipyrrole   204
– 2,2′-bipyrrole   204

griseoviridin   418
Grubbs catalyst   602
guaiane   133
(–)-guanacastepene E   117f.
Gupton polycitone synthesis   198
– lamellarin   198

h

hachijodine   281
hachijodine F   270
haemanthamine   302
(±)-halenaquinol   131f.
halichondrin B   171
haliclamine A   270, 284f.
haliclonacyclamine A   270
halipeptin   486
halipeptin A   467ff., 572
– Nicolaou's total synthesis   573
– retrosynthetic analysis   572
halipeptin B   572
halipeptin D   572f.
– Nicolaou's total synthesis   573
– retrosynthetic analysis   572
haminol-1   271
Handy synthesis
– lamellarin G   198
– lamellarin G trimethyl ether   199
hanishin   18
Hantzsch method   479
– Holzapfel's modification   480f.
– Kelly-modified   480
– Meyers' modification   481f.
– Nicoloau-modified   482
– Panek-modified   421
– Schmidt's modification   480f.
– thiazole formation   480f.
(±)-hastanecine   26
hectochlorin   487
(±)-heliotridine   26
hemlock alkaloid   13, 285
(–)-hennoxazole A   429ff.
herboxidiene   155
hetero Diels–Alder reaction   279, 362ff.
heterocyclic system
– mixed   608
hongoquercin A   155
Horne annulation
– isophakellin   202
Horner-Wadsworth-Emmons condensation   603
Hunig base   197
hydrastine   303
hydroquinine   303

9-hydroxyacridine 358
erythro-β-hydroxy-α-amino acid 46
(6*R*)-6-hydroxy-3,4-dihydromilbemycin E 119
(±)-6β-hydroxyeuropsin 122
5′-hydroxymethyl-5-[butyl-3-en-l-yn]-2,2′-bithiophene isovaleroxy ester 383
hydroxymethyl-functionalized α-methylene azetidin-2-ones 51
γ-hydroxymethyl-substituted azetidine 43
3-hydroxy-11-norcytisine 271
3-hydroxy pipecolic acid 119
β-hydroxy thioamide 472
– cyclodehydration 473f.
hyellazole 347
hymenialdisine 188, 200, 551
hymenidine 521
hypocrellin A 549

i

IB-01211 445, 577ff.
– retrosynthetic analysis 577
– total synthesis 579
IKD-8344 598f.
– total synthesis 600
ikimine 281
ikimine A 271
imerubrine 133
imidazole 518ff., 587
iminium triflate 463
iminoether 463
iminosugar
– dipolar cycloaddition 126
indole 221ff., 587
– synthesis of natural product 222
indole alkaloid 247
indoline 237
indolizidine 221ff., 249
– synthesis of natural product 222
indolizidine alkaloid 14
(±)-indolizidine 195B 27
indolizidine 209B 27
indolizidine 209D 27
(−)-indolizidine 223AB 14
ineupatoriol 380
ingenol 133f.
inorganic Grignard reagent 278
intricarene 135
iodide-mediated rearrangement 27
1-iodomukonidine 351
ipomoeassin B 119
Ireland-Claisen rearrangement 138
iromycin 288

iromycin A 271, 288f.
isobatzelline E 189
5-isobutyl-1-nitrobiuret 354
isochrysohermidine 213
isofebrifugine 510ff., 516
(Z)-isolaureatin 539ff.
isophakellin
– Horne annulation 202
isoquinoline 299ff.
– biochemical precursor 300
– natural product 317
(±)-isoretronecanol 26
(+)-isoretronecanol
– Robins' synthesis 210f.
ivermactin 570
ixabepilone (Ixempra®) 490

j

jamtine 303
jerangolid D 156
jimenezin 156
(−)-jorumycin 12
Julia–Kocienski olefination 279
– intramolecular 609
Julia–Lythgoe olefination 410
Julia olefination 411, 594
jussiaeiine A 271

k

K-252c 349
(−)-kainic acid 26
kalkitoxin 470, 486ff.
kapakahine B 249
kapakahine F 249
kendomycin 162, 594f.
– retrosynthetic analysis 595
– total synthesis 596
3-ketoadociaquinone B 384
kettapeptin 607
– retrosynthetic analysis 607
– total synthesis 608
Knoevenagel condensation
– intermolecular 587
Knoevenagel-Stobbe condensation 316
Knoevenagel-type cyclization 387
Knölker synthesis
– pentabromopseudilin 194
Knorr cyclization 364
kosinostatin 541
kuraramine 271

l

(+)-L-733,060
– synthesis 76

La-[R]-BINOL 73
(+)-lactacystin 21
β-lactam antibiotics 52
lamellarin 195ff.
– Gupton synthesis 198
lamellarin A 189
lamellarin D 196
– Ishibashi synthesis 196
lamellarin G
– Handy synthesis 198
lamellarin G trimethyl ether 198
– Handy synthesis 199
– Steglich synthesis 197
lamellarin H 196
– Faulkner synthesis 197
– Ishibashi synthesis 196
largazole 467, 479ff., 573f.
– retrosynthetic analysis 574
– total synthesis 574
Larock-type annulation 587
lasonolide A 156
laulimalide 66
(+)-(Z)-laureatin 64
– total synthesis 87
lavanduquinocin 347
lavendamycin 303ff.
Lawesson's reagent 468f.
lennoxamine 551ff.
(±)-lepadiformine 13f.
leucamide A 424
leucascandrolide A 163, 413
leucascandrolide B 413
Lindel cyclization 201
Lindlar reduction 413ff.
liphagal 123
liposidomycin A-(III) 550
lissoclibadin 1 385
lissoclibadin 3 385
lissoclinamide 469ff., 487
lissoclintoxin A 385
litseaverticillols 114ff.
longamide 18
longamide B 18, 189
– Papeo synthesis 202
longamide B methyl ester
– Papeo synthesis 201
lophocereine 303
Lovely's conjugation 201
luciferin 460
lukianol 196
Lycopodium alkaloid 252f., 554
lycoposerramine C 554
lycorine 303
lyngbyabellin 476, 487

lyngbyabellin A 465
lyngbyabellin B 465ff.

m

macrocyclic natural product
– bioactive 569ff.
macrosphelide B 114ff.
(–)-madumycin II 416f.
maitotoxin 114
makaluvamine F 379ff.
– Kita synthesis 390
malloapeltine 271
manganese dioxide
– dehydrogenation with activated MnO_2 476
– dehydrogenation with chemical MnO_2 (CMD) 477
– oxidation of thiazolidine with CMD 476
– oxidation of thiazoline with activated MnO_2 476
6-C-β-manno pyranosylapigenin 100
manumycin A 64
manzacidin 511
manzacidin A 189
manzacidin C 512
manzacidin D 513
marcfortine B 548
marineosin 189
marinopyrrole 204
marinopyrrole B 189
maritidine 303
Martin's sulfurane 233
McFayden–Steven reaction
– Braslau modification 587
mechercharmycin A 445, 481ff.
medermycin 102
melanostatin 16f.
melithiazole 486ff.
melithiazole A 489
menisdaurilide 121
mepacrine 360
meridianin 511
meridianin A 516
meridianin E 516
merrilactone A 65
(–)-merrilactone A 86
(±)-mesembrine 28f.
metacycloprodigiosin
– Thomson synthesis 206f.
metallation 107
– reaction 108ff.
2-(1-methanesulfonyloxy) propenamide 51
12-methoxydihydrochelerythrine 119

β-methoxyl acrylate 489
β-methoxytyrosine 24
α-methylene azetidin-2-ones
– hydroxymethyl-functionalized 51
4-methylene (2S)-glutamic acid 15
5′-methyl-5-(4-[3-methyl-l-oxobutoxy]-1-butynyl)-2,2′-bithiophen 382
methyl sarcophytoate 156
2-[(methylthiomethyl)dithio]pyridine-N-oxide 271
metronidazole 519
Meyers' nickel peroxide method 442
micrococcin P1 480ff., 581f.
– retrosynthetic analysis 583
– total synthesis 584
micromeline 347
minfiensine 224, 240
– MacMillan's synthesis 243f.
– (+)-minfiensine 240ff.
– (±)-minfiensine 240
– Overman's synthesis 240f.
– Qin's synthesis 242f.
mirabazole 486ff.
mirabazole A 494
mirabazole B 494
mirabazole C 494
– (–)-mirabazole C 493
miridazole 460
misonidazole 519
mitomycin
– synthesis 5f.
mitomycin A 4ff.
mitomycin C 4ff.
mitomycin K 4ff.
(–)-mitorubrin 175
(–)-mitorubrinal 175
(–)-mitorubrinic acid 175
(–)-mitorubrinol 175
Mitsunobu N-alkylation 198
Mitsunobu reaction 472, 520
mollamide 486
monanchorin 550
monesin A 137f.
monocillin I 66
(±)-monomorine 27, 209f.
(+)-monomorine I 13
mono-oxazole 404, 436
monopyrrolic natural product 193
montipyridine 272ff.
moracin O 102
morphine 102, 304, 317
– (–)-morphine 319f.
(R,R)-Mosher ester 356
mugineic acid 42ff.

Mukaiyama-aldol condensation
– Carreira enantioselective vinylogous 417
Mukaiyama reagent 594
mukanadin A 189
mukonidine 347
multiplolide 66
multiplolide A
– synthesis 71
murrayacine 347
murrayanine 347
muscopyridine 272
muscoride A 432
mycalazal 190
mycothiazole 476ff.
myrmicarin 215A (M215A) 224, 249ff.
– Movassaghi's synthesis 249
myrmicarin 215B (M215B) 224, 249ff.
– Movassaghi's synthesis 249
myrmicarin 217 224, 249
– (–)-M217 249
– (+)-M217 249ff.
– Movassaghi's synthesis 249
myrmicarin 430A (M430A) 250
myrmicarin alkaloid 249ff.
myxothiazole 480ff.

n

nagelamide 522ff.
nagelamide A 190
nagelamide D 200
nagelamide J 527
nakadomarin A 123, 608ff.
– (–)-nakadomarin A 608
– retrosynthetic analysis 609
– total synthesis 610
nakinadine A 272
nandinine 304
nantenine 304
napyradiomycin A1 157
narcotine 304
natural polyether synthesis 68
natural product
– azepine 549
– azole 569
– furan 104
– furan-containing 590
– furan ring 100
– imidazole-based 518ff., 587
– indole-containing 587
– isoquinoline-containing 317
– oxazapine 559
– oxepine 538
– piperazic acid-containing 605
– piperidine-containing 581

- pyran-containing 590
- pyridine-containing 581
- pyrimidine-based 507ff.
- quinoline-containing 308
- thiazole 485
- thiazoline 485
- thiophene-containing 378
- thiostrepton-containing 582
natural product synthesis
- aziridine unit 3ff.
- [2 + 1] cycloaddition 124
- [4 + 3] cycloaddition 133
- [6 + 4] cycloaddition 134
- [8 + 2] cycloaddition 134
- dearomatization reaction 137
- epoxide moiety 68
- furan derivative as reagent 106
- furan derivative as nucleophile 122
- isoquinoline 300
- oxetane 85f.
- oxidation of furan rings 116ff.
- pyran moiety 158
- pyridine moiety 268
- quinoline 300
- radical reaction 136
- retro-Diels–Alder reaction 132
- sulfur heterocycle 393
- thiophene-based substrate 378
- thiophene nucleus 386
- transformation of an aziridine moiety 10ff.
- transformation of an aziridinium moiety 31f.
- transformation of the epoxide moiety 71
- transformation of the oxetane moiety 88
navenone B 272
Neber rearrangement 229, 524
Negishi reagent 406
nemertelline 268ff.
(+)-nemorensic acid 112
nemorensine 112
neocarazostatin B 347
- R-(–)-neocarazostatin B 355f.
neodysiherbaine A 155
neolamellarin A 190
neopeltolide 164, 413
neopyrrolomycin B 190
neosurugatoxin 559
neosurugatoxin A 552
niacin 268
nicotine 268ff.
nicotinic acid 268
ningalin 196
niphatesine 281

niphatesine A 272
niphatoxin A 273
niphatyne 281
niphatyne A 273
nitidine 304, 331f.
njaoaminium A 273
nojirimycin 126
(Z)-nonenolide 78
non-pyrrole natural product
- pyrrole derivative 209
noranabasamine 268ff.
- (S)-(–)-noranabasamine 276
norfluorocurarine 224, 245
- total synthesis 245f.
- Vanderwal's retrosynthesis 245
norhalichondrin 112f.
norhalichondrin B 102, 171ff., 601ff.
- retrosynthetic analysis 601
- total synthesis 604
N-norlaudanosine 304
(–)-norsecurinine 118ff.
noscapine (narcotine) 304
nostocyclamide 427
Noyori (S,S) catalyst 115
nucleophile
- carbon-centered 11
- furan derivative 118ff.
- halogen 24
- nitrogen-centered 15
- oxygen-centered 18
nucleophilic ring-opening
- aziridine for natural product synthesis 10
nuphar alkaloid 12
(–)-nupharolutine 12

o

oleandomycin 65
olefin cross-metathesis 175
olivacine 348
onionin A 382
oroidin 199f.
Ohira–Bestmann reagent 421
oxa [3+3] annulation
- intramolecular 167
oxa-Pictet–Spengler reaction 168
oxazapine 559
oxazole 403ff.
- derivative 403ff.
oxazole residue
- cyclic polyheterocyclic metabolite 426
oxazoline
- thiolysis 469f.
oxepine 537ff.
- synthesis 538

oxetane 63
– natural product synthesis 85
oxetanocin 67
oxetin 65
oxidation 111
– furan ring 116ff.
– thiazolidine 476
– thiazoline 475f.
oxidative cyclization
– Pd(II)-catalyzed 350
oximidine I 66
oxindole 222
oxirane 63

p

Paal–Knorr condensation 204
Paal–Knorr pyrrole synthesis 196
paecilomycine A 176
paeciloquinone E 541
palau'amine 520, 522
palladium complex
– π-allyl 43
pallescensin A 102
palmarumycin JC1 67
palmatine 305
pamamycin 102
panacene 102
papaverine 305
Papeo synthesis
– longamide B 202
paraherquamide A 541
paraherquamide A 548
(–)-paroxetine 32
pateamine 481ff.
patungensin 273
Pauson–Khand reaction
– intramolecular 176
pederin 158f.
pellotine 305
penaresidin
– synthesis 46f.
penaresidin A 42ff.
penaresidin B 42ff.
D-penicillamine 53
penicillin
– synthesis 52
penitrem D 223ff.
– (–)-penitrem D 230
– Smith's strategy 230
penmacric acid 212
pentabromopseudilin 194
pentachloropseudilin 194
– Knölker synthesis 194

2-(penta-1,3-diynyl)-5-(3,4-dihydroxybut-1-ynyl)thiophene 380
penta-oxazole 444
pentostatin 551ff.
pentothal 507f.
(±)-perhydrohistrionicotoxin 30
(–)-pericosin B
– synthesis 73
Petasis–Ferrier reaction 422
Petasis–Ferrier rearrangement 163
Petasis–Ferrier rearrangement/ring-closing olefin metathesis 595
Petasis–Tebbe methylidenation 163
Petasis–Tebbe rearrangement 422
(–)-peucedanol
– synthesis 73
pharmaceuticals
– thiazole-containing 460
2-phenyl-5-(3-buten-1-ynyl) thiophene 378
1,2-phenylenediamine-derived reagent 471
phenylpyridineylbutenol 273
phenylpyrrole 190
threo-phenylserine 28
(–)-phleichrome 549
phloeodictine A1 132
phomactin A 167
– (+)-phomactin A 168
phomoidride D 109f.
phorboxazole A 421
phorboxazole B 421, 590ff.
– retrosynthetic analysis 591
phormidinine A 273
(–)-phorocantholide-J 85
– synthesis 83
phosphorus pentasulfide 468f.
(±)-physostigmine 15
piericidin A1 273ff.
pimprinin 404
(9S,12R,13S)-pinellic acid 78
pinusolide 119
pipecoline 273, 285
– (–)-pipecoline 286f.
– (R)-pipecoline 286
piperazic acid-containing natural product 605
piperazimycin A 605
– retrosynthetic analysis 605
– total synthesis 606
piperidin-3-ol 515
piperidine 285, 581
piperidine alkaloid 285
(–)-pironetin
– synthesis 76

pityriazole 348ff.
– retrosynthetic analysis 351
platensimycin 166
– (–)-platensimycin 167
(±)-platynecine 26
plecomacrolide family 544
podophyllotoxin 102
polycitone 196
– Gupton synthesis 198
polycitone B 190
polycitrin
– Steglich synthesis 213
polycitrin A 212
(–)-polygalolide A 168
(–)-polygalolide B 168
poly-oxazole 440
polyoxin 103, 511
(–)-posticlure 64
– synthesis 68
Pratt synthesis 193
– pyrrolinitrin 194
(+)-preussin 13
Prins cyclization 163ff., 424
prodigiosin 187, 205ff.
– Boger synthesis 206
– Rapoport synthesis 206
prodigiosin R1 191
proflavine 360
promothiocin 581
promothiocin A 425, 582
promothiocin B 582
protoberberine 326
protoemetine 305
– (–)-protoemetine 324
protoemetinol 305
– (–)-protoemetinol 324
(–)-protopraesorediosic acid 124f.
proximicin C 103
(+)-PS-5 11, 29
(–)-pseudoconhydrin 29ff.
(S)-pseudoconhydrin 32
pseudodehydrothyrsiferol 172ff.
(+)-pseudodistomin D 29
(S,S)-(+)-pseudoephedrine 286
psychotrimine 224, 247f.
– Baran's synthesis 247
psychotrine 305
(+)-psymberin 159
(Z)-pterulinic acid 546
(E)-pterulone 540ff.
(E)-pterulone B 546
ptilomycalin A 542
pulo'upone 274
Pummerer-type ring closure 557

pyran 153ff., 590
– derivative 153ff.
pyran compound
– aliphatic ring 171
– aromatic ring 168
– 2,6-cyclic 161
– fused 168ff.
pyran natural product
– 2,6-disubstituted 158
pyran-γ-lactone
– fused 169f.
pyranicin 160
pyranigrin D 191
pyrano[2,3-c]acridone 361
pyranone 172
(S)-β-pyrazolylalanine 16f.
pyridine 267ff., 581
– Bohlmann–Rahtz synthesis 406
– derivative 267ff.
pyrido[4,3-b]carbazole alkaloid 356
pyrido[3′,2′:4,5]pyrrolo[1,2-c]pyrimidine 514
pyridone 287
pyridoxine (vitamin B_6) 268
pyrimethamine 507f.
pyrimidine 507ff.
pyrinadine A 274ff.
pyrinodemin A 274ff.
(S)-pyroglutamic acid 412
pyrrole 187ff.
– derivative 187ff.
– Fürstner synthesis 195
– natural product 193
– non-pyrrole natural product 209
– Paal–Knorr 206
pyrrole-containing small molecule 187
pyrrole dicarboxylate
– stereodivergent Birch reduction 210
pyrrole-2,5-dicarboxylate 209
pyrrolic aldehydes 208
pyrrolidinoindoline alkaloid 237
pyrrolinitrin 193f.
– Pratt synthesis 194
pyrrolizidine 211
pyrrolomycin B 1901
pyrrolo[2,1-c]oxazine-carboxaldeyde 191
pyrrolo[1,2-a]pyrrolone 191
meta-pyrrolophane 194f.
pyrrolostatin 191

q
quaterpyridine 268
quinacrine 360
(–)-quinic acid 227, 524
quinidine 306

quinine 306ff., 570
quinoline 299ff.
– biochemical precursor 300
– natural product 308
d-quinotoxine 306ff.
dl-quinotoxine 310
(S)-quiscalic acid 16

r
radical oxidation process
– Cu(I/II)-mediated 477f.
Ramberg–Bäcklund reaction 553
Rapoport synthesis
– prodigiosin 206
rearrangement 25
– iodide-mediated 27
rebeccamycin 348
reboxetine 32
reduction 25, 111
– furan derivative 112
Reeves' synthesis
– butylcycloheptyl prodigiosin 208
renieramycin 13
(–)-renieramycin G 12
(–)-renieramycin M 12
reserpine 221
resiniferatoxin 134f.
reticuline 306
retro-Diels–Alder reaction 132
retronecine 27
– (±)-retronecine 26
rhazinicine 191
rhizonin A 103
rhizoxin 409
(+)-rhopaloic acid B 160
ricciocarpin B 109f.
rigidin 192
riluzole 460
L-ristosamine 25
ritonavir 460
Robinson annulation 167
Robinson–Gabriel cyclization 413, 436
Robinson–Gabriel cyclodehydration 434ff.
Robinson–Gabriel method 409
Robinson–Gabriel oxidation-ring closure 422
roseophilin 208
– Fürstner synthesis 209
rubioncolin B 171
rubioncolin B methyl ether 171
rugulovasine 121f.
rumbrin 192
(–)-rutacridone 360

s
salinomycin 114ff., 137ff.
salvinorin A 103
sandramycin 306ff.
– (–)-sandramycin 311f.
sanguinarine 306
saponin 544
sarcodictyin 522
(±)-sarracenin
– synthesis 89
sceptrin 199
(+)-Schweinfurthin
– synthesis 79
(+)-Schweinfurthin E
– synthesis 79
scleritodermin A 470ff., 487
sclerotigenin 552
(–)-scyphostatine 64
– synthesis 68
securamine 522
securinine 103, 136
septa-oxazole-thiazoline 444
serratezomine A 224
– Johnston's synthesis 252f.
– (+)-serratezomine A 252f.
sesamin
– synthesis 77
shermilamine B 384
siamenol 348
silvaticamide 551
simplakidine A 274
simplexidine 274
siomycin A 582
siphonazole 439
skimmianine 307
solenopsin A 274
solsodomine A 192
Sonogashira coupling 196
Sonogashira–Hagihara coupling
– Pd(0)-catalyzed 353
sorangicin A 540ff., 592f.
– retrosynthetic analysis 593
– (+)-sorangicin A 592
– total synthesis 595
sparsomycin 511ff.
D-erythro-sphingosine 19ff., 32, 75
D-erythro-sphingosine triacetate 22
(L)-threo sphingosine 28
spiroketal enol ether
– thiophene substituted 387
spirolide 104
spirastrellolide A 103, 157
spirochlorine
– Williams synthesis 396

spiruchostatin A   383, 395ff.
spongidine   277f.
spongidine A   274
spongotine   522
spongotine A   520
Staudinger aza-Wittig reaction
– thiazoline   475
Staudinger reaction/intramolecular aza-Wittig (S-AW) process
– nondehydrative thiazoline formation   473
Staudinger reduction   474
(+)-staurosporine   348
staurosporinone   349ff.
Steglich synthesis
– polycitrin   213
Stemona alkaloid   553
(±)-stenine   129f.
stenusine   275ff.
stepholidine   307, 328
– (–)-(S)-stepholidine   330
Stille alkylation   489
Stille coupling   431
– intramolecular   418
Still–Gennari coupling   414
Still–Gennari homologation   413
streptogramin group A antibiotic   416
streptonigrin   275, 307
streptorubin B   194
– Fürstner synthesis   195
strychnine   130f., 221
Strychnos alkaloid   240ff.
styloguanidine   520
styren   589
sulfathiazole   460
sulfolane   382
sulfomycin fragment
– synthesis   406
sulfur heterocycle   377ff.
(±)-supinidine   26ff.
surugatoxin   559
Suzuki coupling reaction   198
Suzuki–Miyaura coupling   350, 893
– Pd(0)-catalyzed   268, 356
sventrin   192
synfurylcarbinol   112

t
(–)-tabersonine   31
tacrine (Cognex)   360
talipexole   460
tallysomycin   480ff.
tallysomycin A   492f.
tallysomycin B   492f.
tambjamine   204

tambjamine A   192
tandem intramolecular Diels–Alder (IMDA)-intramolecular hetero-Diels–Alder (IMHDA) reaction   173
tandemthiazoline   493
tantazole   486ff.
tantazole A   494
tantazole B   468, 494
tantazole F   494
tarchonanthuslactone   213f.
taurodispacamide   200
taxol   65
– synthesis   87
tecleanthine   360
telomestatin   444, 467, 487, 578ff.
– retrosynthetic analysis   579
– (R)-telomestatin   443, 578
– total synthesis   580
tembjjamine G
– Banwell synthesis   205
tenuecyclamide   467ff.
tenuecyclamide A   428, 487
tenuecyclamide B   428, 487
tenuecyclamide C   428, 487
tenuecyclamide D   428, 487
1,2,3,4-tetrahydroisoquinoline   308
tetrahydrolathyrine   16
tetrahydropalmatine   307, 326
– (–)-tetrahydropalmatine   327
tetrapyrrole   187
tetrathioaspirochlorine   385
tetronasin sodium salt   139f.
tetronomycin   118ff.
teuscorolide   119
texaline   406
texamine   405
– relatives   405
thalictricavine   307
– (+)-thalictricavine   328
theonelladine   281
theonelladine A   275
theopederin D   157
thiangazole   419, 486ff.
– Pattenden synthesis   462
– (–)-thiangazole   464
thiaplidiaquinone A   384
thiaplidiaquinone B   384
thiazole   459ff.
– alkylation   483
– carbon chain elongation of 2/4-halothiazole   483
– Cu(I/II)-mediated radical oxidation process   478
– derivative   459ff., 483

- π-electron density of thiazole ring 483
- Hantzsch formation and modifications 479ff.
- multicomponent reaction 484
- natural product 485
- peptide 491
- pharmaceuticals 460
- polyketide 485
- 5-substituted 484f.
- synthesis 460, 474ff.

thiazolidine
- dehydrogenation 474
- oxidation with chemical MnO_2 (CMD) 476

thiazoline 459ff.
- dehydrogenation 474ff.
- derivative 459ff.
- Kelly's procedure 571
- natural product 485
- oxidation 475f.
- peptide 491
- polyketide 485
- Staudinger aza-Wittig reaction 475
- synthesis 460ff.
- vicinal amino alcohol 468

thiazoline formation
- iminium triflate-mediated 464
- Lewis acid-catalyzed 463
- Staudinger reaction/intramolecular aza-Wittig (S-AW) process 473

thieno[2,3-b]pyrazine 388
thioacylating reagent 470f.
thioacyl-N-benzimidazolinone 470
thioacyl-N-phthalimide 471
thioamide 471ff.
thiobenzotrazole 471
thioester
- cyclization 467
thiolysis
- oxazoline 470
- oxazoline (Wipf's method) 469
thionation
- amide 468
thiophene 377ff.
thiophene substituted spiroketal enol ether
- Wu synthesis 387
thiostrepton 481ff., 582
Thomson synthesis
- metacycloprodigiosin 206f.
L-allo-threonine 21
tioconazole 570
tjipanazole B 349
tjipanazole D 349
tjipanazole E 349
tjipanazole F1 349

tjipanazole F2 349
tjipanazole I 349
tonghaosu 136f.
tonghaosu analog 379
N-tosyl aziridine 14
toxopyrimidine 511
(±)-trachelantamidine 26
(±)-trehazolamine 22
2,3,6-trideoxy-3-amino-hexopyranose 25
trimethoprim 507f.
tripyrrole 187
tripyrrolic natural product 205
tris-oxazole 420, 440ff.
trunkamide A 472, 486
Tsuji–Trost alkylation 202
tuberostemonine 550
- (−)-tuberostemonine 553
tubocurarin 307
tubulysin 470, 476ff.
tubulysin A 491f.
tubulysin B 484ff.
tubulysin C 492
tubulysin D 491f.
tubulysin U 484ff.
tubulysin V 484ff.
tubuphenylalanine (Tup) 491
tubutyrosine (Tut) 491
tubuvaline (Tuv) 491
(±)-turneforcidine 26

u

ulapualide A 440ff.
Ullmann–Jourdan coupling 362
untenine A 275
uprolide D 104
urothione 381ff.
- Taylor and Reiter synthesis 388
ustiloxin 23
ustiloxin D 23
ustiloxin F 23

v

validamine analogues 121
variolin 511
variolin B 514
(−)-velbanamine 32
verbalactone
- synthesis 74
verruculotoxin 17
(−)-vincadifformine 31
vincorine 243
vindorosine 126f.
vinyl aziridine 29
- [2,3]-Wittig rearrangement 25ff.

(E)-vinylstannane 489, 527
virantmycin 25
viscosaline 275
viscosamine 275
vitamin B_1 459
vitamin B_3 268
vitamin B_6 268

w

Wadsworth–Emmons homologation 408
Wadsworth–Horner–Emmons olefination 279f.
Weinreb approach 201
welwitindolinone A isonitrile 223ff.
– Baran's strategy 235ff.
– (+)-welwitindolinone A isonitrile 235
– Wood's strategy 232ff.
trans-Whiskey lactone
– synthesis 88
Wieland–Miescher ketone 231
Williams' method
– dehydrogenation 477
(+)-WIN 64821 237
Wipf cyclodehydration 409
Wipf–Williams method 414ff., 427, 446
Wittig olefination 524
[2,3]-Wittig rearrangement 27
– aziridine 27
– vinyl aziridine 25ff.

WS75624 480
WS75624 A 486ff.
WS75624 B 486ff.

x

xanthohumol 104
xanthopappin A 379
xanthopappin C 383
xanthoxoline 360
xestamine 281
xestamine A 275
xestamine C 281f.
xestamine D 282
xestamine E 281f.
xestamine H 281
XH-14 104

y

Yamada–Shioiri reagent 313
yangambin 104
YM-216391 445f., 476, 487

z

(–)-zamifenacin 32
zaragozic acid 104
zidovudine 507f.
Zincke aldehyde 245f.
zincophorin 157
zyzzyanones B 192